The Human Radiation Experiments

Final Report

OF THE

Advisory Committee

ON

Human Radiation Experiments

New York   Oxford
OXFORD UNIVERSITY PRESS
1996

OXFORD UNIVERSITY PRESS

Oxford   New York

Athens   Auckland   Bangkok   Bombay
Calcutta   Cape Town   Dar es Salaam   Delhi
Florence   Hong Kong   Istanbul   Karachi
Kuala Lumpur   Madras   Madrid   Melbourne
Mexico City   Nairobi   Paris   Singapore
Taipei   Tokyo   Toronto

and associated companies in
Berlin   Ibadan

Published by Oxford University Press, Inc.,
198 Madison Avenue, New York, New York 10016

Oxford is a registered trademark of Oxford University Press

ISBN   978-0-19-510792-0

The three supplemental volumes to the Final Report of the Advisory Committee on Human
Radiation Experiments may be purchased from the Superintendent of Documents, U.S.
Government Printing Office. These are Supplemental Volume 1, Ancillary Materials (stock
number 061-000-0850-1-9), 840 pages; Supplemental Volume 2, Sources and Documentation (stock number 061-000-00851-9), 163 pages; and Supplemental Volume 3, (stock
number 061-000-0852-7), 902 pages. All telephone orders should be directed to Superintendent of Documents, U.S. Government Printing Office, Washington, D.C. (202) 512-
1800; fax orders to (202) 512-2250; and mail orders to U.S. Government Printing Office,
P.O. Box 371954, Pittsburgh, PA 15250-7954.

ADVISORY COMMITTEE ON HUMAN RADIATION EXPERIMENTS

1726 M STREET, N.W., SUITE 600

WASHINGTON, D.C. 20036

October 1995

To the Members of the Human Radiation Interagency Working Group:

Secretary Hazel O'Leary, Department of Energy
Secretary William Perry, Department of Defense
Attorney General Janet Reno, Department of Justice
Secretary Donna Shalala, Department of Health and Human Services
Secretary Jesse Brown, Department of Veterans Affairs
Director Alice Rivlin, Office of Management and Budget
Director John Deutch, Central Intelligence Agency
Administrator Daniel Goldin, National Aeronautics and Space Administration

On behalf of the Advisory Committee on Human Radiation Experiments, it is my privilege to transmit to you our Final Report.

Since the Committee's first meeting in April 1994 we have been able to conduct an intensive inquiry into the history of government-sponsored human radiation experiments and intentional environmental releases of radiation that occurred between 1944 and 1974. We have studied the ethical standards of that time and of today and have developed a moral framework for evaluating these experiments. Finally, we have examined the extent to which current policies and practices appear to protect the rights and interests of today's human subjects. This report documents our findings and makes recommendations for your consideration.

The committee listened to the testimony of more than 200 public witnesses who appeared before us. We are deeply grateful to all these witnesses, who overcame the obstacles of geography and emotions to assist us.

Our work and this report would not have been possible without the extraordinary effort the President and you put forward to open the government's records to our inquiry and thus to the nation. We are especially pleased that, through our joint efforts, the American people now have access to the tens of thousands of documents that bear on this important history.

None of our conclusions came easily. We endeavored, both as individuals and as a committee, to live up to the responsibility with which we were entrusted. This report represents the consensus of fair-minded people who gave the best they had to offer to their fellow citizens.

We thank President Clinton for this opportunity and for his courage and leadership in appointing the Advisory Committee.

Ruth R. Faden
Chair, Advisory Committee
on Human Radiation Experiments

Contents

Advisory Committee on Human Radiation Experiments

Ruth R. Faden, Ph.D., M.P.H.–Chair
Philip Franklin Wagley Professor of Biomedical Ethics and Director
The Bioethics Institute
Johns Hopkins University
Baltimore, Maryland

Senior Research Scholar
Kennedy Institute of Ethics
Georgetown University
Washington, D.C.

Kenneth R. Feinberg, J.D.
Kenneth R. Feinberg & Associates
Washington, D.C.

Eli Glatstein, M.D.
Professor and Chair
Department of Radiation
    Oncology
The University of Texas
Southwestern Medical Center at
    Dallas
Dallas, Texas

Jay Katz, M.D.
Elizabeth K. Dollard Professor
    Emeritus of Law, Medicine
    and Psychiatry
Harvey L. Karp Professorial
    Lecturer in Law and Psycho-
    analysis
Yale Law School
New Haven, Connecticut

Patricia A. King, J.D.
Professor of Law
Georgetown University Law Center
Washington, D.C.

Susan E. Lederer, Ph.D.
Associate Professor
Department of Humanities
The Pennsylvania State University
    College of Medicine
Hershey, Pennsylvania

Ruth Macklin, Ph.D.
Professor of Bioethics
Department of Epidemiology &
    Social Medicine
Albert Einstein College of
    Medicine
Bronx, New York

Lois L. Norris
Second Vice President of Omaha
    National Bank and Omaha
    National Corporation (Retired)
Omaha, Nebraska

Nancy L. Oleinick, Ph.D.
Professor of Radiation
    Biochemistry
Division of Radiation Biology
Case Western Reserve University
    School of Medicine
Cleveland, Ohio

Henry D. Royal, M.D.
Professor of Radiology
Associate Director, Division of
    Nuclear Medicine
Mallinckrodt Institute of
    Radiology
Washington University Medical
    Center
St. Louis, Missouri

Philip K. Russell, M.D.
Professor, Department of
    International Health
Johns Hopkins University
School of Hygiene and Public
    Health
Baltimore, Maryland

Mary Ann Stevenson, M.D., Ph.D.
Assistant Professor of Radiation
    Oncology
Joint Center for Radiation Therapy
Harvard Medical School
Boston, Massachusetts

Deputy Chief
New England Deaconess Hospital
Department of Radiation
    Oncology
Boston, Massachusetts

Duncan C. Thomas, Ph.D.
Director, Biostatistics Division
Department of Preventive
    Medicine
University of Southern California
School of Medicine
Los Angeles, California

Reed V. Tuckson, M.D.
President
Charles Drew University of
    Medicine and Science
Los Angeles, California

Staff of Advisory Committee on Human Radiation Experiments

Dan Guttman
Executive Director

Jeffrey Kahn
Associate Director

Anna Mastroianni
Associate Director

Stephen Klaidman
Director of Communications
Counselor to the Committee

Sarah Flynn
Editor

Staff*

Senior Policy and Research Analysts

Barbara Berney James David John Harbert Gregg Herken Jonathan Moreno
Ronald Neumann Gary Stern Jeremy Sugarman Donald Weightman Gilbert Whittemore

Research Analysts

Jonathan Engel Patrick Fitzgerald Mark Goodman Deborah Holland Denise Holmes
Michael Jasny Gail Javitt Wilhelmine Miller Patricia Perentesis Kathy Taylor
Sandra Thomas Faith Weiss

Research Associates

Miriam Bowling Praveen Fernandes Sara Chandros Hull Valerie Hurt John Kruger
Ellen Lee Shobita Parthasarathy Nöel Theodosiou

Information Services

David Saumweber, Director Robin Cochran, Librarian Tom Wisner, Senior Technology Consultant

Communications and Outreach

Lanny Keller Kristin Crotty

Committee and Staff Affairs

Jerry Garcia Jeanne Kepper

Consultants

Jeffrey Botkin Allen Buchanan Gwen Davis Gail Geller Steve Goodman
Jon Harkness Rebecca Lowen Suzanne White Junod Nancy Kass Charles McCarthy
Monica Schoch-Spana Patricia Stewart-Henney John Till E. W. Webster

---

*includes both full-time and part-time staff

Acknowledgments

The Committee's work over the past year and a half would have been impossible without the assistance of an extraordinary number of individuals and groups from all corners of the United States, and beyond. We wish to express the depth of our gratitude to the many people who assisted, informed, and advised us.

Some of these people are identified by name elsewhere in this report and its supplemental volumes. An appendix in this volume lists the more than 200 witnesses who appeared before the Committee at our public meetings in Washington, D.C., Cincinnati, Knoxville, San Francisco, Santa Fe, and Spokane. The supplemental volumes identify the dozens of individuals who agreed to formal, taped interviews in connection with the Committee's oral history projects. We thank all these people and many more:

- The hundreds of people who contacted the Committee with information about their own experiences or the experiences of their family members. Many of these people shared not only their personal stories but also the information they had collected in the course of conducting their own research into government archives.
- The representatives of many groups whose interests coincided with the work of the Committee. These include organizations of former subjects of biomedical radiation experiments (and their families), downwinders, atomic veterans, uranium miners, and workers in and around atomic energy communities. These groups, as well, shared the accumulated information and perspective of years of experience and research.
- The numerous professionals in fields related to our research who gave of their time and expertise to provide information or comment

on the myriad factual, technical, and policy questions before the Committee. These experts provided help in understanding areas ranging from military and human rights law to the laws of the atom, from the history of the government's use of secrecy to the history of radiation science.

- The dozens of universities and independent hospitals, located in all regions of the country, that willingly provided us with the documents we needed to conduct our Research Proposal Review Project.
- The nearly 1,900 individuals who graciously participated in our Subject Interview Study, and the university hospitals, veterans hospitals and community hospitals that permitted us to conduct the study.
- The numerous chairs of institutional review boards and radiation safety committees who were kind enough to share with us their views about the current status of human subject protections.
- Archivists at public and private libraries, universities, and research institutions, who assisted the Committee in our search for information.
- The many journalists and scholars who have previously researched and written about the subjects covered in this report, for sharing the knowledge and wisdom embodied in their own many years of inquiry and reflection.
- A variety of state and local agencies for sharing with the Committee the results of their own reviews of activities that we explored.
- Members of Congress and congressional staff, including the staffs of the General Accounting Office and the Office of Technology Assessment, for sharing the product of their own prior inquiries into many of the areas discussed in this report.

- The members of the Human Radiation Interagency Working Group, who provided invaluable assistance. We are particularly grateful to the many employees at the Department of Energy, the Department of Defense, the Department of Health and Human Services, the Department of Veterans Affairs, the National Aeronautics and Space Administration, and the Central Intelligence Agency, who aided us in the search and retrieval of the many thousands of documents that provide the backbone for the Committee's review of human radiation experiments that took place between 1944 and 1974 and the history of government requirements for the conduct of that research. We are also grateful to the staffs of the Nuclear Regulatory Commission and the National Archives and Records Administration for their invaluable assistance. Many of the same people, as well as others, also provided advice and information as we undertook our evaluation of the conduct of research involving human subjects today.

We wish to thank both the professional and administrative members of our staff who worked so hard and showed such dedication to our task. Their talent, energy, and commitment provided the foundation for our work. It is impossible to overstate our gratitude and appreciation for their extraordinary efforts.

Finally, we wish to acknowledge our indebtedness to President Clinton for the honor he bestowed upon us when he selected us to serve on the Advisory Committee.

Documentary Note

In fulfilling its mandate, the Advisory Committee on Human Radiation Experiments (ACHRE) relied on several thousand separate sources: primary and secondary published monographs, journal articles, historical records and manuscripts, original correspondence and surveys, interviews, specially constructed databases, searches of public and commercial databases, and documentary films. Only a fraction of these, however, is represented in the final report. More extensive information may be found in the supplemental volume *Sources and Documentation,* which contains a full account of the ACHRE research program, a finding aid to the complete research document collection, a bibliography of published sources used, an index to significant documents and identified experiments, and other auxiliary materials. Further information both about the sources used by the Advisory Committee generally and about the particular sources cited in this volume should be sought there.

The unpublished documents referenced in this report are identified by their places in the ACHRE Research Document Collection. These identifiers, or *ACHRE document numbers,* have four parts: originating institution, date of receipt, order of receipt, and document number. For example, DOE-051094-A-123 is the 123d document described in the first ("A") Department of Energy ("DOE") shipment (or *accession)* received on May 10, 1994 ("051094"). One of the appendices, A *Citizen's Guide to the Nation's Archives,* provides instructions for using references to the ACHRE collection to find documents there and in the collections of the National Archives and at the agencies.

Remarks by President William J. Clinton in Acceptance of Human Radiation Final Report

OCTOBER 3, 1995
OLD EXECUTIVE
OFFICE BUILDING

Let me begin with a simple thank you to everyone who participated in this extraordinary project and to everyone who supported them.

I want to thank Secretary O'Leary for her extraordinary devotion to this cause. And you heard in her remarks basically the way that she views this. It's a part of her ongoing commitment to finish the end of the Cold War. And perhaps no Energy Secretary has ever done as much as she has to be an advocate, whether it is for continued reforms within the Energy Department or her outspoken endorsement of the strongest possible commitment on the part of the United States to a Comprehensive Test Ban Treaty, which I believe we will achieve next year, in no small measure thanks to the support of the Secretary of Energy.

And, of course, I want to thank Dr. Ruth Faden for her extraordinary commitment of about a year and a half of her life to this unusual but important task.

And all of you who served on the committee—I remember the first time we put this committee together. I said, that's a pretty distinguished outfit. I do thank you so much for the work you have done.

I saw this committee as an indispensable part of our effort to restore the confidence of the American people in the integrity of their government. All of these political reform issues to me are integrated. When I became the President, I realized we had great new economic challenges, we had profound social problems, that a lot of these things had to be done by an energized American citizenry, but that our national government had a role to play in moving our country through this period of transition. And in order to do it, we needed to increase the capacity of the government to do it through political reform, but we also needed, as much as anything else, to increase the confidence of the American people so that, at the very least, they could trust the United States government to tell the truth and to do the right things.

So you have to understand that, for me, one reason this is so important is that it is part of our ongoing effort to give this government back to the American people—Senator Glenn's long effort to get Congress to apply to itself the same laws it imposes on the private sector; the restrictions that I imposed on members of my administration in high positions for lobbying for foreign governments; and when the lobby bill failed in the Congress, I just imposed it by Executive Order on members of the Executive Branch. All these efforts at political reform, it seems to me, are important.

But none of these efforts can succeed unless people believe that they can rely on their government to tell them the truth and to do the right thing. We have declassified thousands of government documents, files from the second World War, the Cold War, President Kennedy's assassination. These actions are not only consistent with our national security, they are essential to advance our values.

So, to me, that's what this is all about. And to all those who represent the families who have been involved in these incidents, let me say to you, I hope you feel that your government has kept its commitment to the American people to tell the truth and to do the right thing.

We discovered soon after I entered office that with the specter of an atomic war looming like

Armageddon far nearer than it does today, the United States government actually did carry out on our citizens experiments involving radiation. That's when I ordered the creation of this committee. Dr. Faden and the others did a superb job. They enlisted many of our nation's most significant and important medical and scientific ethicists. They had to determine first whether experiments conducted or sponsored by our government between 1944 and 1974 met the ethical and scientific standards of that time and of our time. And then they had to see to it that our research today lives up to nothing less than our highest values and our most deeply held beliefs.

From the beginning, it was obvious to me that this energetic committee was prepared to do its part. We declassified thousands of pages of documents. We gave committee members the keys to the government's doors, file cabinets and safes. For the last year and a half, the only thing that stood between them and the truth were late nights and hard work.

This report I received today is a monumental document—in more ways than one. But it is a very, very important piece of America's history, and it will shape America's future in ways that will make us a more honorable, more successful and more ethical country.

What this committee learned I would like to review today because I think it must be engraved on our national memory. Thousands of government-sponsored experiments did take place at hospitals, universities and military bases around our nation. The goal was to understand the effects of radiation exposure on the human body.

While most of the tests were ethical by any standards, some were unethical, not only by today's standards, but by the standards of the time in which they were conducted. They failed both the test of our national values and the test of humanity.

In one experiment, scientists injected plutonium into 18 patients without their knowledge. In another, doctors exposed indigent cancer patients to excessive doses of radiation, a treatment from which it is virtually impossible that they could ever benefit.

The report also demonstrates that these and other experiments were carried out on precisely those citizens who count most on the government for its help—the destitute and the gravely ill. But the dispossessed were not alone. Members of the military—precisely those on whom we and our government count most—were also test subjects.

Informed consent means your doctor tells you the risk of the treatment you are about to undergo. In too many cases, informed consent was withheld. Americans were kept in the dark about the effects of what was being done to them. The deception extended beyond the test subjects themselves to encompass their families and the American people as a whole, for these experiments were kept secret. And they were shrouded not for a compelling reason of national security, but for the simple fear of embarrassment, and that was wrong.

Those who led the government when these decisions were made are no longer here to take responsibility for what they did. They are not here to apologize to the survivors, the family members or the communities whose lives were darkened by the shadow of the atom and these choices.

So today, on behalf of another generation of American leaders and another generation of American citizens, the United States of America offers a sincere apology to those of our citizens who were subjected to these experiments, to their families, and to their communities.

When the government does wrong, we have a moral responsibility to admit it. The duty we owe to one another to tell the truth and to protect our fellow citizens from excesses like these is one we can never walk away from. Our government failed in that duty, and it offers an apology to the survivors and their families and to all the American people who must be able to rely upon the United States to keep its word, to tell the truth, and to do the right thing.

We know there are moments when words alone are not enough. That's why I am instructing my Cabinet to use and build on these recommendations, to devise promptly a system of relief, including compensation, that meets the standards of justice and conscience.

When called for, we will work with Congress to serve the best needs of those who were harmed. Make no mistake, as the committee report says, there are circumstances where compensation is appropriate as a matter of ethics and principle. I am committed to seeing to it that the United States of America lives up to its responsibility.

Our greatness is measured not only in how we so frequently do right, but also how we act when we know we've done the wrong thing; how we confront our mistakes, make our apologies, and take action.

That's why this morning, I signed an Executive Order instructing every arm and agency of our government that conducts, supports or regulates research involving human beings to review immediately their procedures, in light of the recommendations of this report, and the best knowledge and standards available today, and to report back to me by Christmas.

I have also created a Bioethics Advisory Commission to supervise the process, to watch over all such research, and to see to it that never again do we stray from the basic values of protecting our people and being straight with them.

The report I received today will not be left on a shelf to gather dust. Every one of its pages offers a lesson, and every lesson will be learned from these good people who put a year and a half of their lives into the effort to set America straight.

Medical and scientific progress depends upon learning about people's responses to new medicines, to new cutting-edge treatments. Without this kind of research, our children would still be dying from polio and other killers. Without responsible radiation research, we wouldn't be making the progress we are in the war on cancer. We have to continue to research, but there is a right way and a wrong way to do it.

There are local citizens' review boards, there are regulations that establish proper informed consent and ensure that experiments are conducted ethically. But in overseeing this necessary research, we must never relax our vigilance.

The breathtaking advances in science and technology demand that we always keep our ethical watchlight burning. No matter how rapid the pace of change, it can never outrun our core convictions that have stood us so well as a nation for more than 200 years now, through many different scientific revolutions.

I believe we will meet the test of our times—that as science and technology evolve, our ethical conscience will grow, not shrink. Informed consent, community right-to-know, our entire battery of essential human protections—all these grew up in response to the health and humanitarian crises of this 20th century. They are proof that we are equal to our challenges.

Science is not ever simply objective. It emerges from the crucible of historical circumstances and personal experience. Times of crisis and fear can call forth bad science, even science we know in retrospect to be unethical. Let us remember the difficult years chronicled in this report, and think about how good people could have done things that we know were wrong.

Let these pages serve as an internal reminder to hold humility and moral accountability in higher esteem than we do the latest development in technology. Let us remember, too, that cynicism about government has roots in historical circumstances. Because of stonewallings and evasions in the past, times when a family member or a neighbor suffered an injustice and had nowhere to turn and couldn't even get the facts, some Americans lost faith in the promise of our democracy. Government was very powerful, but very far away and not trusted to be ethical.

So today, by making ourselves accountable for the sins of the past, I hope more than anything else, we are laying the foundation stone for a new era. Good people, like these Members of Congress who have labored on this issue for a long time, and have devoted their careers to trying to do the right thing, and having people justifiably feel confidence in the work of their representatives, they will continue to work to see that we implement these recommendations.

Under our watch, we will no longer hide the truth from our citizens. We will act as if all that we do will see the light of day. Nothing that happens in Washington will ever be more important in anyone's life affected by these experiments, perhaps, than this report we issue today. But all of us as Americans will be better off because of the larger lesson we learned in this exercise and because of our continuing effort to demonstrate to our people that we can be faithful to their values.

Thank you very much.

Preface

On January 15, 1994, President Clinton created the Advisory Committee on Human Radiation Experiments in response to his concern about the growing number of reports describing possibly unethical conduct of the U.S. government, and institutions funded by the government, in the use of, or exposure to, ionizing radiation in human beings at the height of the Cold War. He directed us to uncover the history of human radiation experiments and intentional environmental releases of radiation; to identify the ethical and scientific standards for evaluating these events; and to make recommendations to ensure that whatever wrongdoing may have occurred in the past cannot be repeated.

The Advisory Committee is composed of fourteen members: a citizen representative and thirteen experts in bioethics, radiation oncology and biology, epidemiology and statistics, public health, history of science and medicine, nuclear medicine, and law. We report to a Cabinet-level group convened by the President (the Human Radiation Interagency Working Group), whose members are the secretaries of defense, energy, health and human services, and veterans affairs; the attorney general; the administrator of the National Aeronautics and Space Administration; the director of the Central Intelligence Agency; and the director of the Office of Management and Budget.

On April 21, 1994, at the end of the first day of our opening meeting, President Clinton invited us to the White House to personally communicate his commitment to the process we were about to undertake. He urged us to be fair, thorough, and unafraid to shine the light of truth on this hidden and poorly understood aspect of our nation's past. Our most important task, he said, was to tell the full story to the American public. At the same time, we were also to examine the present, to determine how the conduct of human radiation research today compares with that of the past and to assess whether, in the light of this inquiry, changes need to be made in the policies of the federal government to better protect the American people. This report and the accompanying supplemental volumes constitute the Committee's attempt to tell the story of the past and to report on our inquiry into the present.

WHY THE COMMITTEE WAS CREATED

Past research with human subjects, including human radiation research, has been a source of life-saving knowledge. Research involving human subjects continues to be essential to the progress of medical science, since most advances in medicine must at some point in their development be tested in human subjects. Every one of us who has been either a patient or a loved one of a patient has benefited from knowledge gained through research with human subjects. But medical science, like all science, does not proceed or progress without the taking of risks. In medical research, these risks often fall on the human subject, who sometimes does not stand to benefit personally from the knowledge gained. This is the source of the moral tension at the core of the enterprise of research involving human subjects. In order to secure important collective goods—scientific knowledge and advances in medicine—individuals are put in harm's way. The moral challenge is how to protect the rights and interests of these individuals while enabling and encouraging the advancement of science.

The Committee had its origins when public controversy developed surrounding human radiation experiments that were conducted half a century ago. In November 1993, the *Albuquer-*

que Tribune published a series of articles that, for the first time, publicly revealed the names of Americans who had been injected with plutonium, the man-made material that was a key ingredient of the atom bomb. Reporter Eileen Welsome put a human face to what had previously been anonymous data published in official reports and technical journals. As World War II was ending, she wrote, doctors in the United States injected a number of hospitalized patients with plutonium, very likely without their knowledge or consent. The injections were part of a group of experiments to determine how plutonium courses through the human body. The experiments, and the very existence of plutonium, were shrouded in secrecy. They were conducted at the direction of the U.S. government, with the assistance of university researchers in Berkeley, Chicago, and Rochester (New York), with the expectation that the information gained could be used to limit the hazards to the thousands of workers laboring to build the bomb.

On reading the articles, Secretary of Energy Hazel O'Leary expressed shock, first to her staff, and then in response to a question posed at a press conference. She was particularly concerned because the Department of Energy had its earliest origins in the agencies responsible for building the atomic bomb and sponsoring the plutonium experiments. During the Cold War, these agencies had continued to do much of their work in the twilight zone between openness and secrecy. Now, the Cold War was over. The time had come, Secretary O'Leary determined, to make public anything that remained to be told about the plutonium experiments.

Subsequent press reports soon noted that the plutonium injections were not the only human radiation experiments that had been conducted during the war and the decades that followed. In Massachusetts, the press reported that members of the "science club" at the Fernald School for the Retarded had been fed oatmeal containing minute amounts of radioactive material. In Ohio, news articles revived an old controversy about University of Cincinnati researchers who had been funded by the Defense Department to gather data on the effects of "total-body irradiation" on cancer patients. In the Northwest, the

papers retold the story of Atomic Energy Commission funding of researchers to irradiate the testicles of inmates in Oregon and Washington prisons in order to gain knowledge for use in government programs. The virtually forgotten 1986 report prepared by a subcommittee headed by U.S. Representative Edward Markey, "American Nuclear Guinea Pigs: Three Decades of Radiation Experiments on U.S. Citizens," was also recalled to public attention.[1]

Coincidentally, the fact that the environment had also been used as a secret laboratory became a subject of controversy. A November 1993 congressional report uncovered thirteen cases in which government agencies had intentionally released radiation into the environment without notifying the affected populations.[2] At various times, tests were conducted in Tennessee, Utah, New Mexico, and Washington state. This report had been prepared at the request of Senator John Glenn in his capacity as chair of a committee that had undertaken a comprehensive oversight investigation of the nuclear weapons complex. As a young marine in 1945, the senator was in a squadron being trained for possible deployment to Japan when the atomic bomb ended the war; as an astronaut, he had been the subject of constant testing and medical monitoring by space administration flight surgeons; as a senator he was at the center of the country's efforts to understand and control nuclear weapons. Senator Glenn understood the importance of national security, but he found it "inconceivable . . . that, even at the height of the communist threat, some of our scientists and doctors and military and perhaps political leaders approved some of these experiments to be conducted on an unknowing and unwitting public."[3]

In the immediate aftermath of Secretary O'Leary's press conference and the further press reports, thousands of callers flooded the Department of Energy's phone lines to recount their own experiences and those of friends and family members.

Underlying the outrage and concern expressed by government officials and members of the public were many unanswered questions. How many human radiation experiments were conducted? No one knew if the number was closer to 100 or 1,000. Were all the human radiation experiments done in secret, and were any of them still secret? Are any secret or controversial studies still ongoing? Scientists and science journalists pointed out that some of the highly publicized experiments had long ago been the subject of technical journal articles, even press accounts, and were old news; other commentators countered that, for most of the public, articles in technical journals might as well be secret.

How, why, and from what population groups were subjects selected for experiments? Some suspected that subjects were disproportionately chosen from the most vulnerable populations—children, hospitalized patients, the retarded, the poor—those too powerless to resist the government and its researchers.

Did the experiments benefit the American people through the advancement of science and the enhancement of the ability to treat disease?

How many intentional releases took place, and how many people were unknowingly put at risk? The answer here was sketchy; the releases identified in the November 1993 Glenn report had all been performed in secret, and much information about them was still secret.

How great were the risks to which people were exposed? Many pointed out that radiation is not only present in our natural environment, but that, as a result of biomedical research, most people routinely rely on radiation as a means of diagnosing and treating disease. Others noted that while this is so, radiation can be abused, and the potential dangers of low-level exposure are still not well understood.

What did our government and the medical researchers it sponsored do to ensure that the subjects were informed of what would be done to them and that they were given meaningful opportunities to consent? Today, federal government rules require the prior review of proposed experiments, to ensure that the risks and potential benefits have been considered and that subjects will be adequately informed and given the opportunity to consent. But the standards of today, many historians and scholars of medical ethics noted, are not those of yesterday. Others, however, declared that it was self-evident that no one should be experimented upon without his or her voluntary consent. Indeed, it was pointed

out that this very principle was proclaimed aloud to the world in 1947, as the plutonium experiments were coming to a close. It was the American judges at the international war crimes trials in Nuremberg, Germany, who invoked the principle in finding doctors guilty of war crimes for their vile experiments on inmates of Nazi concentration camps. How could yesterday's standard have been less strict than that of today? How, moreover, could the standard not have been known by the government that sponsored the experiments and the researchers who conducted them?

Finally, there were questions about how human experiments are conducted today. Insofar as wrong things happened in the past, how confident should we be that they could not happen again? Have practices changed? Do we have the right rules, and are they implemented and enforced?

THE PRESIDENT'S CHARGE

The Advisory Committee was created under the Federal Advisory Committee Act of 1972, which provides that committee meetings and basic decision making be conducted in the open. The Committee's charter[4] defined human radiation experiments to include

(1) experiments on individuals involving intentional exposure to ionizing radiation. This category does not include common and routine clinical practices. . . .
(2) experiments involving intentional environmental releases of radiation that (A) were designed to test human health effects of ionizing radiation; or (B) were designed to test the extent of human exposure to ionizing radiation.

The Committee was mandated to review experiments conducted between 1944 and 1974, the latter being the year that the U.S. Department of Health, Education, and Welfare issued rules for the protection of human subjects of federally sponsored research. The Committee was asked to determine the ethical and scientific standards by which to evaluate the pre-1974 experiments and the extent to which these experiments were consistent with such standards. We were also to "consider whether (A) there was a clear medical or scientific purpose for the experiments; (B) appropriate medical follow-up was conducted; and (C) the experiments' design and administration adequately met the ethical and scientific criteria, including standards of informed consent, that prevailed at the time of the experiments and that exists today." The charter also directed that, upon completing our review, the Committee may recommend that subjects (or families) be notified of potential health risks and the need for medical follow-up and also that we "may recommend further policies, as needed, to ensure compliance with recommended ethical and scientific standards for human radiation experiments."

In order to inform the public about the conduct of research involving human subjects taking place today, we were authorized to sample and consider examples of research with human subjects currently under way.

In essence, we were to answer several fundamental questions: (1) What was the federal government's role in human radiation experiments conducted from 1944 to 1974? (2) By what standards should the ethics of these experiments be evaluated? and (3) What lessons learned from studying past and present research standards and practices should be applied to the future?

In addition, while the Committee was not expressly charged with considering issues relating to remedies, including financial compensation, we have felt obliged to address the type of remedies that we believe the government, as an ethical matter, should provide to subjects of experiments where the circumstances warranted such a response.

THE COMMITTEE'S APPROACH

When those of us selected by President Clinton to serve on the Committee read about human radiation experiments in our hometown newspapers during the 1993 holiday season, none of us imagined that within months we would be embarking on such an intense and challenging investigation of an important aspect of our nation's past and present, requiring new insights

and difficult judgments about enduring ethical questions.

On April 21 and 22, 1994, the Committee held its first meeting, and most of us met each other for the first time. As we listened to opening statements by Cabinet members and members of Congress, as well as the first witness from the general public, it became clear how daunting a task we were undertaking. We realized that our ability to reconstruct the story of past radiation experiments required both the capacity to join with the agencies in the search through thousands of boxes for documents and the intuition to recognize which documents were important. We knew that the ability to tell that story depended on our ability to understand the full range of technically complex, often emotionally charged issues related to human radiation experiments. We could not understand, much less tell, the story until we sought out all who could enhance our understanding, a difficult job because the voices to which we had to listen spoke in the varied languages of medicine, a multiplicity of scientific disciplines, the military, policymakers, philosophers, patients, healthy subjects, family members of former subjects, and individuals in a variety of other roles.

Finally, we were also convinced that an important determinant of our success in keeping faith with the American people would be to understand not only how human subject research was conducted in the past but also how it is being conducted in the present.

Reaching In and Reaching Out

As we began our work, Committee members first sought to educate one another. Early meetings included basic presentations on such topics as research ethics, radiation, the history of human experimentation, the law of remedies, and the debate over the effects of low levels of radiation.

Then we determined to search broadly for those who could contribute to our understanding. We hired a staff with the expertise and experience need for the Committee's myriad tasks. Finally, we sought to make ourselves available to those who wanted to speak to us directly, especially people who felt they or their loved ones were harmed, or might have been harmed, by human radiation-related research or exposure. Each of the Committee's meetings reserved a period for public comment. Since April 1994, the full Committee held sixteen public meetings, each of two to three days' duration. Fifteen of those meetings were held in Washington, D.C., and one was in San Francisco. In addition, subsets of Committee members presided over public forums in Cincinnati, Knoxville, Santa Fe, and Spokane. We traveled to these different cities in order to hear from people who could not come to Washington, D.C., and lived in communities where, or near where, experiments or intentional releases of interest to the Committee had taken place. We further sought to reach out to those who could not attend our meetings. By phone, mail, and personal visit, we and our staff communicated with members of the public, researchers, attorneys, investigative reporters, authors, and representatives of dozens of groups of interested people who shared some aspect of the Committee's concern.

The Records of Our Past: The Search for Documents

One of the most difficult tasks before the Committee was determining how many federally sponsored human radiation experiments occurred between 1944 and 1974 and who conducted them. When President Clinton established the Committee, he also directed the Human Radiation Interagency Working Group to provide us with all relevant documentary information in each of the agencies' files. Teams were formed to identify the hundreds of government sites where relevant documents might be located. We discovered there was no easy way to identify how many experiments had been conducted, where they took place, and which government agencies had sponsored them. The location and retrieval of documents thus required an extraordinary effort, and we appreciate the assistance of all our collaborators.

We began with documents that were assembled during the 1980s and that provided the basis for the Markey report. But review of those materials confirmed that, even for this relatively well-known group of experiments, basic infor-

mation was lacking. We found that the Department of Health and Human Services (DHHS), which is the primary government sponsor of research involving human subjects, reported that, as permitted by federal records laws, it had long since discarded files on experiments performed decades ago. Furthermore, the capsule descriptions of research that remained sometimes did not make clear whether the subjects of research had been humans or animals. To complicate matters further, the DHHS also pointed out that much research documentation had originated and been retained only in the files of nonfederal grantee institutions and investigators. Other agencies did provide some lists of experiments; in many cases, however, there was no information on basic questions of concern (for example, who the subjects were and what, if anything, they were told).

What rules or policies, if any, existed to govern federally sponsored experiments in the pre-1974 period? The prevailing assumption was that, with a few notable exceptions, it was not until the mid-1960s that federal agencies began to develop such policies in any significant way. Most scholarship focused on divisions of the (then) Department of Health, Education, and Welfare. Little was known about approaches to human experimentation at the Atomic Energy Commission and the Department of Defense. Yet it was clear from the outset of our inquiry that these agencies, as well as the DHEW, were central to the story of human radiation experiments and that many of the experiments of interest predated by decades the mid-1960s' interest in human subject protections.

As we began our search into the past, we found that it was necessary to reconstruct a vanished world. The Committee and the agencies had to collect information scattered in warehouses throughout the country. At the same time, we had to create and test the framework needed to ensure that there would be a "big picture" into which all the pieces of the puzzle would fit.

After a few months, the outlines of a world that had been almost lost began to reemerge. Working with the Defense Department, we discovered that long-forgotten government entities had played central roles in the planning of midcentury atomic warfare-related medical research and experimentation. These groups, the piecing together of long-lost or forgotten records would show, debated the ethics of human experimentation and discussed possible human radiation experimentation: Similarly, working with the Department of Energy, we pieced together the minutes, and even many transcripts, of the key medical advisory committee to the Atomic Energy Commission. We sought to mine agency histories, when they existed: for example, at the Committee's request the Defense Nuclear Agency (the heir to the part of the Manhattan Project that was transferred to the Defense Department) made public portions of the more than 500 internal histories that chronicle its story, most of which had previously been available only to those with security clearances.

Despite these successes, it became evident that the records of much of our nation's recent history had been irretrievably lost or simply could not be located. The Department of Energy told the Committee that all the records of the Intelligence Division of its predecessor, the Atomic Energy Commission, had been destroyed—mainly during the 1970s, but in some cases as late as 1989. The CIA explained, as had been previously reported, that records of the program known as MKULTRA, in which unwitting subjects were experimented upon with a variety of substances, had been destroyed during the 1970s, when the program became a widely publicized scandal. Though documents related to the program referred to radiation, the CIA concluded that human experiments using ionizing radiation never took place under that program, based on currently available evidence.

We also turned to nongovernmental archives throughout the country. Cryptic notes and fragments of correspondence located in private and university archives were fitted into our growing outline. For example, a copy of an important 1954 Army surgeon general research policy statement, referenced in Defense Department documents, was found at Yale University among the papers of a Nobel laureate.

By the end of our term, the Committee had received, organized, and reviewed hundreds of thousands of pages of documents from public and private archives. This collection will be available to individuals and scholars who wish to pursue the great many stories that remain to be told, and we view this as one of our most significant contributions.[5]

The Records of Our Past: The Memories of the People

The Committee listened to the testimony of more than 200 public witnesses who appeared before us. We heard from people or their family members who had been subjects in controversial radiation experiments, including the plutonium injections, total-body irradiation experiments, and experiments involving the use of radioactive tracers with institutionalized children. We heard from "atomic veterans": soldiers who had been marched to ground zero at atomic bomb tests, sailors who had walked the decks of ships contaminated by radioactive mist, and pilots who had flown through radioactive mushroom clouds. We also heard from their widows. We heard from people who lived "downwind" from nuclear weapons tests in Nevada and intentional releases of radioactive material in Washington state. We heard from Navajo miners who had served the country in uranium mines filled with radioactive dust, from native Alaskans who had been experimented upon by a military cold weather research lab, and from Marshall Islanders, whose Pacific homeland had been contaminated by fallout after a 1954 hydrogen bomb test.

We heard from officials and researchers responsible for human research today and from those who were present at or near the dawn of the Cold War. We heard from individuals who, on their own time, had long been seeking to piece together the story of human radiation experiments and offered to share their findings. We heard from scholars, from members of Congress, and from people who wanted to bear witness for those who could no longer speak. We heard from a woman who, as a high-school student intern decades ago, attended at the bedside while a terminally ill patient was injected with uranium and from a powerfully spoken veteran of the nuclear weapons work force who told of the "body snatching" of dead friends in the name of science.

Most important, we heard from many people who believed that something involving the government and radiation happened to them or their loved ones decades ago; most had been unable to find out exactly what had happened, or why, and now they wanted to know the truth. These witnesses spoke eloquently of their pain, their frustration, and the reasons they do not trust the government. Their very appearance before the Committee testified to a commitment to the country and to the value of the nation's effort to understand its past. We are deeply grateful to all of these witnesses, who overcame the obstacles of geography and emotions to participate in this work.

We combined our public meetings with additional efforts to interview, and record for the nation's archives, those who could shed light on Cold War human radiation experiments and on the ethics of biomedical experimentation. Dozens of interviews were conducted with former government officials responsible for programs that included radiation research, as well as with radiation researchers.

In Mississippi we talked with a retired general who served as a military assistant to secretaries of defense in the 1940s and 1950s; in Berkeley, we talked with the chemist who was one of the discoverers of plutonium; in Rhode Island we talked with the physicist who served as the link between the civilian health and safety agencies and the Cold War military research efforts; in Florida we talked with a pioneer in health physics, a discipline created to provide for the safety of nuclear weapons workers; in San Francisco and Washington, D.C., we talked to the lawyers who advised the Atomic Energy Commission at its postwar creation; in New York we talked with the Navy radiation researcher who was rousted from his Maryland laboratory to respond to the emergency created by the exposure of the Marshall Islanders; in San Diego we talked with a researcher whose own career and massive history of radiation research had covered much of the Committee's territory.

We also launched a special effort, called the Ethics Oral History Project, to learn from eminent physicians who were beginning their careers in academic medicine in the 1940s and 1950s about how research with human subjects was then conducted. The Ethics Oral History Project also included interviews with two people who had been administrators of the National Insti-

tutes of Health during the 1950s, since they were intimately involved with ethical and legal aspects of research involving human subjects at the time.

We listened to all these people and more, and through their testimony, this report is informed.

Bounds of Our Inquiry

In the course of listening to public testimony, it became clear to us that confusion exists about what an experiment is and whether it can be distinguished from other activities in which people are put at risk and information is gathered about them. The biomedical community, for example, struggles with the distinction between scientific research and related activities. In a medical setting, it is sometimes hard to distinguish a formal experiment designed to test the effectiveness of a treatment from ordinary medical care in which the same treatment is being administered outside of a research project. The patient receiving the treatment may discern no difference between the two, but the distinction is relevant to questions of ethics. The physician-investigator may face conflicts between the obligation to do what is best for each individual patient and the requirements of scientific research, whereas the physician involved only in clinical care has a responsibility solely to the patient.

Similarly, in an occupational setting in which employees are put at risk, it is often difficult to distinguish formal scientific efforts to study effects on the health of employees from routine monitoring of employees' exposure to hazards in the work place for purposes of ensuring worker safety. In the first case, the rules of research ethics apply; in the second they do not. And yet, here too, the worker may discern no difference between the two activities. A further complication for the Committee to consider was the fact that research in occupational settings rarely takes the form of a classic experiment, in which the investigator controls the variable under study and then randomly assigns subjects to be in the "treatment" or "control" group. Instead, most occupational research employs observational and statistical methods, drawing most heavily from the field of epidemiology. These distinctions were unimportant, however, to the representatives of atomic veterans, uranium miners, and residents of the Marshall Islands, who told us of their belief that they, or those they spoke for, were subjects of research.

The Committee struggled with how strictly to define human radiation experiments for purposes of our inquiry. There is no single, clear definition of an experiment that is widely subscribed to by every member of the biomedical community. Even our description above of a classic experiment is open to contest. Today, as well as in the past, the scientific community has rarely employed the term *experiment* in discussions of biomedical research; other terms, not necessarily synonymous—such as *clinical study, clinical investigation, quasi experiment,* and *case control study*—are all used. We concluded that it was not possible to interpret our charge by stipulating an artificial definition of *human radiation experiment.* Instead, in keeping with the realities of biomedical research, we decided to interpret our charge broadly, as including both research involving human subjects in which the research design called for exposing subjects to ionizing radiation *and* research designed to study the effects of radiation exposure resulting from nonexperimental activities.

This latter category includes the research involving uranium miners and Marshall Islanders. In these cases we quickly determined that it was in some respects impossible to isolate the ethical questions raised by the research from the ethics of the context in which the research was conducted. A central issue was the exposure of people to risk, regardless of whether they were clearly understood to be subjects of research. This characterization is true, as well, of the experience of atomic veterans. As a consequence, we considered events that might be said to be on the boundary between research and some other activity. Our inquiry underscored the importance for social policy of the need to keep focused on questions of risk and well-being regardless of what side of that boundary the activity producing the risk falls.

Human Experimentation Today

In tandem with the reconstruction of the past, we undertook three projects to examine the current state of human radiation experiments.

First, we studied how each agency of the federal government that currently conducts or funds research involving human subjects regulates this activity and oversees it. We surveyed what the operative rules are, how they are implemented, and how they are enforced.

Second, from among the very large number of research projects involving human subjects currently supported by the federal government, we randomly selected 125 research projects for scrutiny by the Committee. For each of these projects, we reviewed all available relevant documentation to assess how well it appeared the rights and interests of the subjects participating in these projects were being protected. The success of this review required the cooperation of private research institutions all over the country, on whom we were dependent for access to important documents. We had expected that perhaps no more than half of those asked to cooperate would agree to do so, but with little hesitation, all of the research centers that we approached agreed to cooperate.

Third, to learn from the subjects themselves, the Committee interviewed almost 1,900 patients receiving medical care in outpatient facilities of private and federal hospitals throughout the country. We asked patients about their attitudes toward medical research with human subjects and about the meaning they attach to the different terms used to explain medical research to potential subjects. We ascertained, and attempted to verify, how many of these patients were currently or ever had been subjects of research. Patient-subjects were asked about their reasons for agreeing to join research projects; patients who reported having refused offers to enter research projects were asked why they had decided against participating.

In all three of these projects, we focused not only on human radiation experiments but on human research generally. In critical (but not all) respects, the government regulations that apply to human radiation research do not differ from those that govern other kinds of research involving human subjects. Moreover, the underlying ethical principles that should guide the conduct of research are identical, whether one is considering human radiation research or all research with human subjects. Finally, the Committee hoped to learn whether, in practice, there are any differences between the conduct of radiation and nonradiation experiments.

LESSONS FROM HISTORY: LOOKING TO THE FUTURE

What we have found is a story about the government's attempt to serve two critical purposes: safeguarding national security and advancing medical knowledge. One-half century ago, the U.S. government and its experts in the fields of radiation and medicine were seeking to learn more about radiation in order to protect workers, service personnel, and the general public against potential atomic war and individuals against the menace of disease.

Toward these laudable ends, the government used patients, workers, soldiers, and others as experimental subjects. It acted through the experts to whom we regularly entrust the well-being of our country and our selves: elected officials, civil servants, generals, physicians, and medical researchers.

Moreover, the government acted with full knowledge that the use of individuals to serve the ends of government raises basic ethical questions. If, as we look back, there could be doubt about the importance of the matter to the leaders of the time, we need only look to the appearance before the U.S. Senate of David Lilienthal, who had been nominated to serve as the first chairman of the Atomic Energy Commission, the civilian successor to the Manhattan Project and the predecessor to today's Department of Energy. In his testimony, Lilienthal forcefully stated:

... all Government and private institutions must be designed to promote and protect and defend the integrity and the dignity of the individual.... Any forms of government ... which make men means rather than ends in themselves ... are contrary to this conception; and therefore I am deeply opposed to them.... The fundamental tenet of communism is that the state is an end in itself, and that therefore the powers which the state exercises over the individual are without any ethical standards to limit them. This I deeply disbelieve.[6]

What did happen when individuals were sometimes used as means to achieve national goals? How well were the national goals of preserving the peace and advancing medical science reconciled with the equally important end of respect for individual dignity and health? What rules were followed to protect people and how well did they work? Was the public let in on the balancing of collective and individual interest? In what sense did the public, in general, and individuals, in particular, know what was happening and have the opportunity to provide their meaningful consent?

In this report we try to convey our understanding of how, when only good was sought, when its pursuit was entrusted to the experts on whom we most relied, and when missions were substantially accomplished, distrust, as well as accomplishment, remains.

We focus on the ways in which the government and its experts recognized the interest of individual dignity and sought to strike a balance with the national interests being pursued. We focus equally on the extent to which the public was privy to this balancing. In particular, we try to show how individuals' understanding and participation were limited by the conjunction of government secrecy and expert knowledge.

All Americans should experience immense satisfaction in the strides that have been made toward accomplishing both our national security and our medical research goals. However, as attested to by the many thousands of letters and calls that led to the Committee's creation, and the eloquent statements of the witnesses who appeared before us, this pride is diluted by a bitter aftertaste—distrust by many Americans of the federal government and those who served it.

The government has the power to create and keep secrets of immense importance to us all. Secret keeping is a part of life. Secret keeping by the government may be in the national interest. However, if government is to be trusted, it is important to know, at the very least, the basic rules of secrecy and to know that they are reasonable and that they are being followed.

Similarly, experts, by training and experience, have knowledge that individual people must, as a practical matter, rely on. However, legitimate questions arise when experts wear multiple hats or when they are relied on in areas beyond their expertise.

Where official secrecy is coupled with expert authority, and both are focused on a public that is not privy to secrets and does not speak the languages of experts, the potential for distrust is substantial.

In telling the story, and asking the questions, we have kept our eyes open for ways in which lost trust can be restored. It might be presumed that the past we report on here is so different from the present that it will be of little use in understanding research involving human subjects today. In fact, as we shall see, basic questions posed by the story of human radiation experiments conducted during the 1944-1974 period are no less relevant today. Then, as now, there were standards; the question is how they worked to protect individuals and the public. Then, as now, the ethical impulse was complexly alloyed with concerns for legal liability and public image. Then, as now, the most difficult questions often concerned the scope and practical meaning of ethical rules, rather than their necessity. The country has come to recognize, from its experience of the past half century, that tinkering with the regulations that govern publicly supported institutions, imposing ethical codes on experts, and altering the balance between secrecy and openness are important but not always sufficient means of reform. The most important element is a citizenry that understands the limits of these activities. That is why the purpose of this story is not simply to learn which changes to make in rules or policies that apply to government or professionals, but to begin to learn something more about how the Cold War world worked, as the most important means to making the world of tomorrow work better.

HOW THIS REPORT IS ORGANIZED

Though this report is addressed largely to those who can affect future policy in light of the information the Advisory Committee has gathered, specifically the Human Radiation Interagency Working Group, it has been written in such a way that it should be accessible to a wide range of interested readers.

We begin with an introduction, titled "The Atomic Century," which describes the intersection of several developments: the birth and remarkable growth of radiation science; the parallel changes in medicine and medical research; and the intersection of these changes with government programs that called on medical researchers to play important new roles beyond that involved in the traditional doctor-patient relationship. The introduction concludes with a section titled "The Basics of Radiation Science" for the lay reader.

The remainder of the text is divided into four parts. Each part is preceded by an overview.

Part I, "Ethics of Human Subjects Research: A Historical Perspective," which contains four chapters, explores how both federal government agencies and the medical profession approached human experimentation in the period 1944 through 1974. We begin with the story of the principles stated at midcentury at the highest levels of the Cold War medical research bureaucracies and what we have ascertained about whether these principles were translated into federal rules or requirements. We then turn to the norms and practices engaged in at the time by medical researchers themselves. It is in this chapter that we report the results of our Ethics Oral History Project. In chapter 3, we review the development of formal and public regulations concerning research involving human subjects in the 1960s and 1970s. In the last chapter in part I we present our framework for evaluating the ethics of human radiation experiments, grounded in both history and philosophical analysis.

Part II, "Case Studies," approaches particular experiments from several angles, each of which raises overlapping ethical questions. The chapters on the plutonium injections and total-body irradiation consider the use of sick patients to provide data needed to protect the health of workers engaged in the production of nuclear weapons; the chapter on prisoners considers the use of healthy subjects for this purpose; the chapter on children considers experimentation with particularly vulnerable people; and the chapter on the AEC program of radioisotope distribution considers the institutional safeguards that underlay the conduct of thousands of human radiation experiments. The chapters on intentional releases, atomic veterans, and observational studies consider, in common, situations in which entire groups of people were exposed to risk as a consequence of government-sponsored Cold War programs. The section concludes with a review of the degree to which secrecy impaired, and may still impair, our ability to understand human radiation experiments and intentional releases conducted in the 1944-1974 period.

Part III, "Contemporary Projects," reports the findings of our three inquiries into the present. We begin by describing what we have learned about how the different federal agencies that sponsor human research regulate and oversee this activity. Next, we report the results of our Research Proposal Review Project, followed by the results of our Subject Interview Study. Part III concludes with the Committee's synthesis of the implications of the results of all three of these projects for the current state of human subject research.

Part IV, "Coming to Terms with the Past, Looking Ahead to the Future," reports the Committee's findings and recommendations.

A FINAL NOTE

The Committee's findings and recommendations represent our best efforts to distill almost eighteen months of inquiry into, debate about, and analysis of human radiation experiments. But what they cannot fully express is the appreciation we developed for how much damage was done to individuals and to the American people during the period we investigated and how this damage endures today. The damage we speak of here is not physical injury, although this too did occur in some cases. Rather, the damage is measured in the pain felt by people who believe that they or their loved ones were treated with disrespect for their dignity and disregard for their interests by a government and a profession in which they had placed their trust. It is measured in a too-often cynical citizenry, some of whom have lost faith in their government to be honest brokers of information about risks to the public and the purposes of government actions. And it is measured in the confusion among patients that remains today about the differences between medical research and medical care—differences that can impede the ability of patients to determine what is in their own best interest.

In the period that we examined, extraordinary advances in biomedicine were achieved and a foundation was laid for fifty years without a world war. At the same time, however, it was a time of arrogance and paternalism on the part of government officials and the biomedical community that we would not under any circumstances wish to see repeated.

As we listened to the heart-rending testimony of many public witnesses, we came to feel great sorrow about the suffering they described. Our most difficult task was determining what to recommend as the appropriate national response to these emotions and the events that stimulated them. What can best precipitate the healing of wounds and the restoration of trust? Appropriate remedies for those who were wronged or harmed were of critical importance, but remedies alone speak only to the past, not the future. It is equally important that, the historical record having been spelled out and appropriate remedies identified, we as a nation move forward and take action to prevent similar occurrences from happening in the future. In the end, if trust in government is to be restored, those in power must always act in good faith in their dealings with the citizenry. At the same time, however, we must recognize that unless we have expectations of honesty and fairness from our government and unless we are vigilant in holding the government to those expectations, trust will never be restored.

Finally, we hope that this report conveys the sense of gratitude and honor that we experienced as citizens serving on the Advisory Committee. We were provided by the President with extraordinary access to the records of our past and given complete liberty to deliberate on what we found. Although some of what we report is a matter for national regret, our freedom of inquiry and the cooperation we received from officials and fellow citizens of all perspectives, confirms that our nation's highest traditions are not things of the past but live very much in the present.

NOTES

1. U.S. House of Representatives, Committee on Energy and Commerce. Subcommittee on Energy Conservation and Power, November 1986, "American Nuclear Guinea Pigs: Three Decades of Radiation Experiments on U.S. Citizens" (ACHRE No. CON-050594-A-1).

2. U.S. Senate, Committee on Governmental Affairs. 11 November 1993, "Nuclear Health and Safety: Examples of Post World War II Radiation Releases at U.S. Nuclear Sites," GAO/RCED-94-51-FS (ACHRE No. CON-042894-A-4).

3. Advisory Committee on Human Radiation Experiments, proceedings of 21 April 1994, transcript, 112-113.

4. The full text of the Committee's charter appears at the end of this report. See table of contents for page number.

5. For further information on access to this collection, see "A Citizen's Guide to the Nation's Archives" at the end of this report.

6. David E. Lilienthal, *The Journals of David E. Lilienthal: 1945-1950*, 2 vols. (New York: Harper and Row, 1961), as quoted in David McCullough, *Truman* (New York: Simon and Schuster, 1992), 537-538.

The Human Radiation Experiments

Introduction:
The Atomic Century

ONE hundred years ago, a half century before the atomic bombing of Hiroshima and Nagasaki, the discovery of x rays spotlighted the extraordinary promise, and peril, of the atom. From that time until 1942, atomic research was in private hands. The Second World War and the Manhattan Project, which planned and built the first atomic bombs, transformed a cottage industry of researchers into the largest and one of the most secretive research projects ever undertaken. Scientists who had once raced to publish their results learned to speak in codes accessible only to those with a "need to know." Indeed, during the war the very existence of the manmade element plutonium was a national secret.

After the war's end, the network of radiation researchers, government and military officials, and physicians mobilized for the Manhattan Project did not disband. Rather, they began working on government programs to promote both peaceful uses of atomic energy and nuclear weapons development.

Having harnessed the atom in secret for war, the federal government turned enthusiastically to providing governmental and nongovernmental researchers, corporations, and farmers with new tools for peace—radioisotopes—mass-produced with the same machinery that produced essential materials for the nation's nuclear weapons. Radioisotopes, the newly established

Atomic Energy Commission (AEC) promised, would create new businesses, improve agricultural production, and through "human uses" in medical research, save lives.

From its 1947 creation to the 1974 reorganization of atomic energy activities, the AEC produced radioisotopes that were used in thousands of human radiation experiments conducted at universities, hospitals, and government facilities.[1] This research brought major advances in the understanding of the workings of the human body and the ability of doctors to diagnose, prevent, and treat disease.

The growth of radiation research with humans after World War II was part of the enormous expansion of the entire biomedical research enterprise following the war. Although human experiments had long been part of medicine, there had been relatively few subjects, the research had not been as systematic, and there were far fewer promising interventions than there were in the late 1940s.

With so many more human beings as research subjects, and with potentially dangerous new substances involved, certain moral questions in the relationship between the physician-researcher and the human subject—questions that were raised in the nineteenth century—assumed more prominence than ever: What was there to protect people if a researcher's zeal for

1

data gathering conflicted with his or her commitment to the subjects' well-being? Was the age-old ethical tradition of the doctor-patient relationship, in which the patient was to defer to the doctor's expertise and wisdom, adequate when the doctor was also a researcher and the procedures were experimental?

While these questions about the role of medical researchers were fresh in the air, the Manhattan Project, and then the Cold War, presented new ethical questions of a different order.

In March 1946, former British Prime Minister Winston Churchill told an audience in Fulton, Missouri, that an "iron curtain" had descended between Eastern and Western Europe —giving a name to the hostile division of the continent that had existed since the end of World War II. By the following year, *Cold War* was the term used to describe this state of affairs between the United States and its allies on the one hand and the Soviet bloc on the other. A quick succession of events underscored the scope of this conflict, as well as the stakes involved: In 1948 a Soviet blockade precipitated a crisis over Berlin; in 1949, the American nuclear monopoly ended when the Soviet Union exploded its first atomic bomb; in 1950, the Korean War began.

The seeming likelihood that atomic bombs would be used again in war, and that American civilians as well as soldiers would be targets, meant that the country had to know as much as it could, as quickly as it could, about the effects of radiation and the treatment of radiation injury.

This need for knowledge put radiation researchers, including physicians, in the middle of new questions of risk and benefit, disclosure and consent. The focus of these questions was, directly and indirectly, an unprecedented public health hazard: nuclear war. In addressing these questions, medical researchers had to define the new roles that they would play.

As advisers to the government, radiation researchers were asked to assist military commanders, who called for human experimentation to determine the effects of atomic weapons on their troops. But these researchers also knew that human experimentation might not readily provide the answers the military needed.

As physicians, they had a commitment to prevent disease and heal. At the same time, as government advisers, they were called upon to participate in making decisions to proceed with weapons development and testing programs that they knew could put citizens, soldiers, and workers at risk. As experts they were asked to ensure that the risks would not be excessive. And as researchers they saw these programs as an opportunity for gathering data.

As researchers, they were often among the first to volunteer to take the risks that were unavoidable in such research. But the risks could not always be disclosed to members of the public who were also exposed. In keeping with the tradition of scientific inquiry, these researchers understood that their work should be the subject of vigorous discussion, at least among other scientists in their field. But, as government officials and advisers, they understood that their public statements had to be constrained by Cold War national security requirements, and they shared in official concern that public misunderstanding could compromise government programs and their own research.

Medical researchers, especially those expert in radiation, were not oblivious to the importance of the special roles they were being asked to play. "Never before in history," began the 1949 medical text *Atomic Medicine*, "have the interests of the weaponeers and those who practice the healing arts been so closely related."[2] This volume, edited by Captain C. F. Behrens, the head of the Navy's new atomic medicine division, was evidently the first treatise on the topic. It concluded with a chapter by Dr. Shields Warren, the first chief of the AEC's Division of Biology and Medicine, who would become a major figure in setting policy for postwar biomedical radiation research. While the atomic bomb was not "of medicine's contriving," the book began, it was to physicians "more than to any other profession" that atomic energy had brought a "bewildering array of new problems, brilliant prospects, and inescapable responsibilities." The text, a prefatory chapter explained, treats "not of high policy, of ethics, of strategy or of international control [of nuclear materials], as physicians these matters are not for us."[3] Yet what many readers of *Atomic Medicine* could not know in 1949 was that Behrens, along with Warren and other biomedical experts, was already engaged in vigorous but secret discussions of the ethics underlying human radiation experiments. At the heart of these discussions lay difficult choices at the intersection of geopolitics, science, and medicine that would have a fundamental impact on the federal government's relationship with the American people.

This chapter provides a brief survey of the development of radiation research and the changing roles of the biomedical researcher, from the discovery of x rays by a single individual to the complex world of government-sponsored human radiation experimentation. Finally, at the end of this chapter, an aid to the reader titled "The Basics of Radiation Science" provides information needed to understand technical concepts in this report.

BEFORE THE ATOMIC AGE: "SHADOW PICTURES," RADIOISOTOPES, AND THE BEGINNINGS OF HUMAN RADIATION EXPERIMENTATION

Radiation has existed in nature from the origins of the universe, but was unknown to man until a century ago. Its discovery came by accident. On a Friday evening, November 8, 1895, the German physicist Wilhelm Roentgen was studying the nature of electrical currents by using a cathode ray tube, a common piece of scientific equipment. When he turned the tube on, he noticed to his surprise that a glowing spot appeared on a black paper screen coated with fluorescent material that was across the room. Intrigued, he soon determined that invisible but highly penetrating rays were being produced at one end of the cathode ray tube. The rays could expose photographic plates, leaving shadows of dense objects, such as bone.

After about six weeks of experimenting with his discovery, which he called x rays, Roentgen sent a summary and several "shadow pictures" to a local scientific society. The society published the report in its regular journal and wisely printed extra copies. News spread rapidly; Roentgen sent copies to physicists throughout Europe. One Berlin physicist "could not help thinking that I was reading a fairy tale . . . only the actual photograph proved to everyone that this was a fact."[4]

Physicians immediately recognized these rays as a new tool for diagnosis, a window into the interior of the body. The useless left arm of German Emperor Wilhelm II was x-rayed to reveal the cause of his disability, while Queen Amelia of Portugal used x rays of several of her court ladies to vividly display the dangers of "tight-lacing."[5] Physicians began to use x rays routinely for examining fractures and locating foreign objects, such as needles swallowed by children or bullets shot into adults.[6] During World War I, more than 1.1 million wounded soldiers were treated with the help of diagnostic x rays.[7]

In 1896, Roentgen's insight led to the discovery of natural radioactivity. Henri Becquerel, who had been studying phosphorescence, discovered that shadow pictures were also created when wrapped photographic plates were exposed to crystals partly composed of uranium. Could this radioactive property be concentrated further by extracting and purifying some as-yet-unknown component of the uranium crystals? Marie and Pierre Curie began laborious chemical analyses that led to the isolation of the element polonium, named after Marie's native Poland.[8] Continuing their work, they isolated the element radium. To describe these elements' emission of energy, they coined the word *radioactivity*.[9]

As with x rays, popular hopes and fears for natural radioactivity far exceeded the actual applications. One 1905 headline captures it all: "Radium, as a Substitute for Gas, Electricity, and as a Positive Cure for Every Disease."[10] Following initial enthusiasm that radiation could, by destroying tumors, provide a miracle cure for cancer, the reappearance of irradiated tumors led to discouragement. Despite distressing setbacks, research into the medical uses of radiation persisted. In the 1920s French researchers, performing experiments on animals, discovered that radiation treatments administered in a series of fractionated doses, instead of a single massive dose, could eliminate tumors without causing permanent damage. With the new method of treatment, doctors began to report impressive survival rates for patients with a variety of cancers. Fractionation became, and remains, an accepted approach to cancer treatment.[11]

Along with better understanding of radiation's benefits came a better practical appreciation of its dangers. Radiation burns were quickly apparent, but the greater danger took longer to manifest itself. Doctors and researchers were frequently among the victims. Radiation researchers were also slow to take steps to protect themselves from the hidden danger. One journal opened its April 1914 issue by noting that "[w]e have to deplore once more the sacrifice of a radiologist, the victim of his art."[12]

Clear and early evidence of tragic results sharpened both expert and public concern. By 1924, a New Jersey dentist noticed an unusual rate of deterioration of the jawbone among local women. On further investigation he learned that all at one time had jobs painting a radium solution onto watch dials. Further studies revealed that as they painted, they licked their brushes to maintain a sharp point. Doing so, they absorbed radium into their bodies. The radium gradually revealed its presence in jaw deterioration, blood disease, and eventually, a painful, disfiguring deterioration of the jaw.[13] There was no question that radium was the culprit. The immediate outcome was a highly publicized crusade, investigation, lawsuits, and payments to the victims. Despite the publicity surrounding the dial painters, response to the danger remained agonizingly slow. Patent medicines containing radium and radium therapies continued.[14]

The tragedy of the radium dial painters and similar cases of patients who took radium nostrums have provided basic data for protection standards for radioactive substances taken into the body. One prominent researcher in the new area of radiation safety was Robley Evans. Evans was drawn into the field by the highly publicized death in 1932 of Eben Byers, following routine consumption of the nostrum Radiothor. Byers's death spurred Evans, then a California Institute of Technology physics graduate student, to undertake research that led to a study of the effects on the body of ingesting radium; this study would continue for more than half a century.[15]

Evans's study and subsequent studies of the effects of radium treatments provided the anchor in human data for our understanding of the effects of radiation within the human body. As the dangers of the imprudent use of x rays and internal radiation became clear, private scientific advisory committees sprang up to develop voluntary guidelines to promote safety among those working with radiation. When the government did enter the atomic age, it often referred to the guidelines of these private committees as it developed radiation protection standards.[16]

The Miracle of Tracers

In 1913, the Hungarian chemist Georg von Hevesy began to experiment with the use of radioactive forms of elements (radioisotopes) to trace the behavior of the normal, nonradioactive forms of a variety of elements. Ten years later Hevesy extended his chemical experiments to biology, using a radioisotope of lead to trace the movement of lead from soil into bean plants. In 1943, Hevesy won the Nobel Prize for his work on the use of radioisotopes as tracers.

Previously, those seeking to understand life processes of an organism had to extract molecules and structures from dead cells or organisms, and then study those molecules by arduous chemical procedures, or use traceable chemicals that were foreign to the organism being studied but that mimicked normal body chemicals in some important way. Foreign chemicals could alter the very processes being measured and, in any case, were often as difficult to measure precisely as were normal body constituents. The radioactive tracer—as *Our Friend the Atom*, a book written by Dr. Heinz Haber for Walt Disney productions, explained in 1956 to readers of all ages—was an elegant alternative: "Making a sample of material mildly radioactive is like putting a bell on a sheep. The shepherd traces the whole flock around by the sound of the bell. In the same way it is possible to keep tabs on tracer-atoms with a Geiger counter or any other radiation detector."[17]

By the late 1920s the tracer technique was being applied to humans in Boston by researchers using an injection of dissolved radon to measure the rate of blood circulation, an early example of using radioactivity to observe life processes.[18] However, research opportunities were limited by the fact that some of the elements that are most important in living creatures do not possess naturally occurring radioactive isotopes.

The answer to this problem came simultaneously at faculty clubs and seminars in Berkeley and Boston in the early 1930s. Medical researchers realized that the famed "atom smasher," the cyclotron invented by University of California physicist Ernest Lawrence, could be used as a factory to create radioisotopes for medical research and treatment. "Take an ordinary needle," *Our Friend the Atom* explained, "put it into an atomic reactor for a short while. Some of the ions contained in the steel will capture a neutron and be transformed into a radioisotope of iron. . . . Now that needle could be found in the proverbial haystack without any trouble."[19]

In 1936, two of Lawrence's Berkeley colleagues, Drs. Joseph Hamilton and Robert Stone, administered radiosodium to treat several leukemia patients. In 1937, Ernest Lawrence's brother, physician John Lawrence, became the first to use radiophosphorus for the treatment of leukemia. This application was extended the following year to the treatment of polycythemia vera, a blood disease. This method soon became a standard treatment for that disease. In 1938, Hamilton and Stone also began pioneering work in the use of cyclotron-produced neutrons for the treatment of cancer. The following year, not long before the war in Europe began, Ernest Lawrence unveiled a larger atom smasher, to be used to create additional radioisotopes and hence dubbed the "medical cyclotron."[20] The discovery that some radioisotopes deposited selectively in different parts of the body—the thyroid, for example—inspired a spirited search for a radioactive "magic bullet" that might treat, or even cure, cancer and other diseases.

In Cambridge, the age of "nuclear medicine" is said to have begun in November 1936 with a lunchtime seminar at Harvard, at which MIT President Karl Compton talked on "What Physics Can Do for Biology and Medicine." Robley Evans, by that time at MIT, is reported to have helped prepare the portion of the talk from which medical researchers at the Massachusetts General Hospital's thyroid clinic came to realize that MIT's atom smasher could produce a great research tool for their work—radioisotopes. Soon, doctors at the thyroid clinic began a series of experiments, including some involving humans, that would lead to the development of radioiodine as a standard tool for diagnosing and treating thyroid disease.[21]

In late 1938, the discovery of atomic fission in Germany prompted concern among physicists in England and the United States that Nazi Germany might be the first to harness the power of the atom—as a propulsion method for submarines, as radioactive poison, or most worrisome of all, as a bomb capable of unimagined destruction. In the United States, a world-famous physicist, Albert Einstein, and a recent émigré from Hungary, Leo Szilard, alerted President Franklin D. Roosevelt to the military implications of the German discovery in an August 1939 letter.

Assigning his own science adviser, Vannevar Bush, to the task of determining the feasibility of an atomic bomb, Roosevelt's simple "O.K.," scrawled on a piece of paper, set in motion the chain of events that would lead to the largest and most expensive engineering project in history. Soon, Ernest Lawrence's Radiation Laboratory and its medical cyclotron were mobilized to aid in the nationwide effort to build the world's first atomic bomb. In a related effort, Drs. Stone and Hamilton, and others, would turn their talents to the medical research needed to ensure the safety of those working on the bomb.

THE MANHATTAN PROJECT: A NEW AND SECRET WORLD OF HUMAN EXPERIMENTATION

In August 1942, the Manhattan Engineer District was created by the government to meet the goal of producing an atomic weapon under the pressure of ongoing global war. Its central mission became known as the Manhattan Project. Under the direction of Brigadier General Leslie Groves of the Army Corps of Engineers, who recently had supervised the construction of the Pentagon, secret atomic energy communities were created almost overnight in Oak Ridge, Tennessee, at Los Alamos, New Mexico, and in Hanford, Washington, to house the workers and gigantic new machinery needed to produce the bomb. The weapon itself would be built at the Los Alamos laboratory, under the direction of physicist J. Robert Oppenheimer.

Plucked from campuses around the country, medical researchers came face to face with the need to understand and control the effect upon the thousands of people, doctors included, of radioactive materials being produced in previously unimaginable quantities.

In November 1942 General Groves, through the intermediation of an Eastman Kodak official, paid a call on University of Rochester radiologist Stafford Warren. Rochester, like MIT and Berkeley, was another locale where radiation research had brought together physicists and physicians. "They wanted to know what I was doing in radiation. So I discussed the cancer work and some of the other things," Warren told an interviewer in the 1960s. Then "[w]e got upstairs and they looked in the closet and they closed the transom and they looked out the window. . . . Then they closed and locked the door and said, 'Sit down.'"[22]

Soon thereafter, Dr. Warren was made a colonel in the U.S. Army and the medical director of the Manhattan Project. As his deputy, Warren called on Dr. Hymer Friedell, a radiologist who had worked with Dr. Stone in California. Dr. Stone himself had meanwhile moved to the University of Chicago, where he would play a key role in Manhattan Project-related medical research.

Initially, researchers knew little or nothing about the health effects of the basic bomb components, uranium, plutonium, and polonium.[23] But, as a secret history written in 1946 stated, they knew the tale of the radium dial painters:

The memory of this tragedy was very vivid in the minds of people, and the thoughts of potential dangers of working in areas where radiation hazards existed were intensified because the deleterious effects of radiation could not be seen or felt and the results of over-exposure might not become apparent for long periods after such exposure.[24]

The need for secrecy, Stafford Warren later recalled, compounded the urgency of understanding and controlling risk. Word of death or toxic hazard could leak out to the surrounding community and blow the project's cover.[25]

The need to protect the Manhattan Project workers soon gave rise to a new discipline, called health physics, which sought to understand radiation effects and monitor and protect nuclear worker health and safety. The Project was soon inundated with data from radiation-detection instruments, blood and urine samples, and physical exams. The "clinical study of the personnel," Robert Stone wrote in 1943, "is one vast experiment. Never before has so large a collection of individuals been exposed to so much radiation."[26] Along with these data-gathering efforts came ethical issues.

Would disclosure of potential or actual harm to the workers, much less the public, impair the program? For example, a July 1945 Manhattan Project memo discussed whether to inform a worker that her case of nephritis (a kidney disease) may have been due to her work on the Project. The issue was of special import because, the memo indicated, the illness might well be a precursor of more cases. The worker, the memo explained, "is unaware of her condition which now shows up on routine physical check and urinalysis."[27]

As this memo showed, there was an urgent need for decisions on how to protect the workers, while at the same time safeguard the security of the project: "The employees must necessarily be rotated out, and not permitted to resume further exposure. In frequent instances no other type of employment is available. Claims and litigation will necessarily flow from the circumstances outlined." There were also, the memo concluded, "Ethical considerations":

The feelings of the medical officers are keenly appreciated. Are they in accordance with their canons of ethics to be permitted to advise the patient of his true condition, its cause, effect, and probable prognosis? If not on ethical grounds, are they to be permitted to fulfill their moral obligations to the individual employees in so advising him? If not on moral grounds, are those civilian medical doctors employed here bound to make full disclosure to patients under penalty of liability for malpractice or proceeding for revocation of license for their failure to do so?[28]

It is not clear what was decided in this case. However, the potential conflict between the government doctors' duty to those working on government projects and the same doctors' obligations to the government would not disappear. Following the war, as we see in chapter 12, this conflict would be sharply posed as medical researchers studied miners at work producing uranium for the nation's nuclear weapons.

Another basic question was the extent to which human beings could or should be studied to obtain the data needed to protect them. The radium dial painter data served as a baseline to determine how the effects of exposures in the body could be measured. But this left the question of whether plutonium, uranium, and polonium behaved more or less like radium. Research was needed to understand how these elements worked in the body and to establish safety levels. A large number of animal studies were conducted at laboratories in Chicago, Berkeley, Rochester, and elsewhere; but the relevance of the data to humans remained in doubt.

The Manhattan Project contracted with the University of Rochester to receive the data on physical exams and other tests from Project sites and to prepare statistical analyses. While boxes of these raw data have been retrieved, it is not clear what use was made of them.[29] Accidents, while remarkably few and far between, became a key source of the data used in constructing an understanding of radiation risk. But accidents were not predictable, and their occurrence only enhanced the immediacy of the need to gain better data.

In 1944, the Manhattan Project medical team, under Stafford Warren and with the evident concurrence of Robert Oppenheimer, made plans to inject polonium, plutonium, uranium, and possibly other radioactive elements into human beings. As discussed in chapter 5, the researchers turned to patients, not workers, as the source of experimental data needed to protect workers. By the time the program was abandoned by the government, experimentation with plutonium had taken place in hospitals at the Universities of California, Chicago, and Rochester, and at the Army hospital in Oak Ridge, and further experimentation with polonium and uranium had taken place at Rochester.

The surviving documentation provides little indication that the medical officials and researchers who planned this program considered the ethical implications of using patients for a purpose that no one claimed would benefit them, under circumstances where the existence of the substances injected was a wartime secret. Following the war, however, the ethical questions raised by these experiments would be revisited in debates that themselves were long kept secret.

In addition to experimentation with internally administered radioisotopes, external radiation was administered in human experiments directed by Dr. Stone at Chicago and San Francisco and by others at Memorial Hospital in New York City. Once again, the primary subjects were patients, although some healthy subjects were also involved. In these cases, the researchers may have felt that the treatment was of therapeutic value to the patients. But, in addition to the question of whether the patients were informed of the government's interest, this research raised the question of whether the government's interest affected the patients' treatment. As discussed in chapter 8, these questions would recur when, beginning in 1951, and for two decades thereafter, the Defense Department would fund the collection of data from irradiated patients.

Ensuring safety required more, however, than simply studying how radioactive substances moved through and affected the human body. It also involved studying how these substances moved through the environment. While undetectable to the human senses, radiation in the environment is easily measurable by instruments. When General Groves chose Hanford, on the Columbia River in Washington state, as a site for the plutonium production facility, a secret research program was mounted to understand the fate of radioactive pollution in the water, the air, and wildlife.[30]

Outdoor research was at times improvisational. Years after the fact, Stafford Warren would recall how Manhattan Project researchers had deliberately "contaminated the alfalfa field" next to the University of Rochester medical school with radiosodium, to determine the shielding requirements for radiation-measuring equipment. Warren's associate Dr. Harold Hodge recalled that a shipment of radiosodium was received by plane from Robley Evans at MIT, mixed with water in a barrel, and poured into garden sprinklers:

We walked along and sprinkled the driveway. This was after dark. . . . The next thing, we went out and sprayed a considerable part of the field. . . . It was sprayed and then after a while sprayed again, so there was a second and third application. We were all in rubber, so we didn't get wet with the stuff . . . then Staff [Warren] said that one of the things we needed was to see what would be the effect on the inside of a wooden building. So we took the end of the parking garage, and we sprinkled that up about as high as our shoulders, and somebody went inside and made measurements, and we sprinkled it again. Then we wanted to know about the inside of a brick building, and so we sprinkled the side of the animal house. . . . I had no idea what the readings were. . . I hadn't the foggiest idea of what we were doing, except that obviously it was something radioactive.[31]

Outdoor releases would put at risk unsuspecting citizens, even communities, as well as workers. There were no clear policies and no history of practice to guide how these releases should be conducted. As we explore in chapter 11, this would be worked out by experts and officials in secret, on behalf of the workers and citizens who might be affected.

THE ATOMIC ENERGY COMMISSION AND POSTWAR BIOMEDICAL RADIATION RESEARCH

On August 6, 1945, when the atomic bomb was dropped on Hiroshima, the most sensitive of secrets became a symbol for the ages. A week later, the bomb was the subject of a government report that revealed to the public the uses of plutonium and uranium.[32] Immediately, debate began over the future of atomic energy. Could it be controlled at the international level? Should it remain entirely under control of the military? What role would industry have in developing its potential? Although American policymakers failed to establish international control of the bomb, they succeeded in creating a national agency with responsibility for the domestic control of atomic energy.

The most divisive question in the creation of the new agency that would hold sway over the atom was the role of the military. Following congressional hearings, the Atomic Energy Commission was established by the 1946 McMahon Act, to be headed by five civilian commissioners. President Truman appointed David Lilienthal, former head of the Tennessee Valley Authority, as the first chairman of the AEC, which took over responsibilities of the Manhattan Engineer District in January 1947.

Also in 1947, under the National Security Act, the armed services were put under the authority of the newly created National Military Establishment (NME), to be headed by the secretary of defense. In 1949 the National Security Act was amended, and the NME was transformed into an executive department—the Department of Defense.[33] The Armed Forces Special Weapons Project, which would coordinate the Defense Department's responsibilities in the area of nuclear weapons, became the military heir to the Manhattan Engineer District. The Military Liaison Committee was also established as an intermediary between the Atomic Energy Commission and the Defense Department; it was also to help set military requirements for the number and type of nuclear weapons needed by the armed services.

Even before the AEC officially assumed responsibility for the bomb from the Manhattan Project, the Interim Medical Advisory Committee, chaired by former Manhattan Project medical director Stafford Warren, began meeting to map out an ambitious postwar biomedical research program. Former Manhattan Project contractors proposed to resume the research that had been interrupted by the war and to continue wartime radiation effects studies upon human subjects.[34]

In May 1947, Lilienthal commissioned a blue-ribbon panel, the Medical Board of Review, that reported the following month on the agency's biomedical program. In strongly recommending a broad research and training program, the board found the need for research "both urgent and extensive." The need was "urgent because of the extraordinary danger of exposing living creatures to radioactivity. It is urgent because effective defensive measures (in the military sense) against radiant energy are not yet known." The board, pointing to the AEC's "absolute monopoly of new and important tools for research and important knowledge," noted the commensurate responsibilities—both to employees and others who could suffer from "its negligence or ignorance" and to the scientific world, with which it was obliged to "share its acquisitions . . . whenever security considerations permit."[35] In the fall of 1947, as recommended by the Medical Board of Review, the AEC created a Division of Biology and Medicine (DBM) to coordinate biomedical research involving atomic energy and an Advisory Committee for Biology and Medicine (ACBM), which reported directly to the AEC's chairman.[36]

Not surprisingly, the DBM and ACBM became gathering places for the luminaries of radiation science. The ACBM was headed by a Rockefeller Foundation official, Dr. Alan Gregg. It settled on Dr. Shields Warren, a Harvard-trained pathologist, to serve as the first chief of the DBM. Warren, as we shall see, would play a central role in developments related to radiation research and human experimentation. In the 1930s, focusing on cancer research, and influenced by the work of Hevesy and the pioneering radioisotope work being done in Berkeley and Boston, Warren turned to the question of the effects of radiation on animals and the treatment of acute leukemia, the "most hopeless . . . of tumors at that time." As the war neared, Warren enlisted in the Naval Reserve. He continued medical work for the Navy, turning down an invitation to join Stafford Warren (no relation) on "a project . . . that he couldn't tell me anything about [the Manhattan Project]."[37]

While most of the AEC's budget would be devoted to highly secret weapons development and related activities, the biomedical research program represented the commission's proud public face. Even before the AEC opened its doors, Manhattan Project officials and experts had laid the groundwork for a bold program to encourage the use of radioisotopes for scientific research, especially in medicine. This program was first presented to the broad public in a September 1946 article in the *New York Times Magazine*. The article began dramatically by describing the use of "radioactive salt" to measure circulation in a crushed leg, so that a decision on whether to amputate below or above the knee could be made.[38]

By November 1946, the isotope distribution program was well under way, with more than 200 requests approved, about half of which were designated for "human uses." From the beginning, the AEC's Isotope Division at Oak Ridge had in its program director, Paul Aebersold, a veritable Johnny Appleseed for radioelements.[39] In presentations before the public and to researchers, Aebersold, dubbed "Mr. Isotope," touted the simplicity and low cost with which scientists would be provided with radioisotopes: "The materials and services are made available . . . with a minimum of red tape and under conditions which encourage their use."[40] At an international cancer conference in St. Louis in 1947, the AEC announced that it would make radioisotopes available without cost for cancer research and experimental cancer treatment. This, Shields Warren later recalled, had a "tremendous effect" and "led to a revolution in the type of work done in this field."[41]

To AEC administrators, Aebersold emphasized the benefits to the AEC's public image: "Much of the Commission's success is judged by the public and scientists . . . on its willingness to carry out a wide and liberal policy on the distribution of materials, information, and services," he wrote in a memo to the AEC's general manager.[42]

The AEC biomedical program as a whole also provided for funding of cancer research centers, research equipment, and numerous other research projects. Here, too, were advances that would save many lives. Before the war, radiotherapy had reached a plateau, limited by the cost of radium and the inability of the machines of the time to focus radiation precisely on tumors to the exclusion of surrounding healthy tissue. AEC facilities inherited from the Manhattan Project could produce radioactive cobalt, a cheaper substitute for radium. As well, the AEC's "teletherapy" program funded the development of new equipment capable of producing precisely focused high-energy beams.[43]

The AEC's highly publicized peacetime medical program was not immune to the pressures of the Cold War political climate. Even the lives of young researchers in the AEC Fellowship Program conducting nonclassified research were subject to Federal Bureau of Investigation review despite protests from commission members. Congressionally mandated Cold War requirements such as loyalty oaths and noncommunist affidavits, Chairman Lilienthal declared, would have a chilling effect on scientific discussion and could damage the AEC's ability to recruit a new generation of scientists.[44] The reach of the law, the Advisory Committee for Biology and Medicine agreed, was like a "blighting hand; for thoughtful men now know how political domination can distort free inquiry into a malignant servant of expediency and authoritarian abstraction."[45] Nonetheless, the AEC accepted the congressional conditions for its fellowship program and determined to seek the program's expansion.[46]

The AEC's direct promotional efforts were multiplied by the success of Aebersold and his colleagues in carrying the message to other government agencies, as well as to industry and private researchers. This success led, in turn, to new programs.

In August 1947, General Groves urged Major General Paul Hawley, the director of the medical programs of the Veterans Administration, to address medical problems related to the military's use of atomic energy. Soon thereafter, Hawley appointed an advisory committee, manned by Stafford Warren and other medical researchers. The advisers recommended that the VA create both a "publicized" program to promote the use of radioisotopes in research and a "confidential" program to deal with potential liability claims from veterans exposed to radiation hazards.[47] The "publicized" program soon mushroomed, with Stafford Warren, Shields Warren, and Hymer Friedell among the key advisers. By 1974, according to VA reports, more than 2,000 human radiation experiments would be performed at VA facilities,[48] many of which would work in tandem with neighboring medical schools, such as the relationship between the UCLA medical school, where Stafford Warren was now dean, and the Wadsworth (West Los Angeles) VA Hospital.

While the AEC's weapons-related work would continue to be cloaked in secrecy, the isotope program was used by researchers in all corners of the land to achieve new scientific understanding and help create new diagnostic and therapeutic tools. It was, however, only a small part of an enormous institution. By 1951 the AEC would employ 60,000 people, all but 5,000 through contractors. Its land would encompass 2,800 square miles, an area equal to Rhode Island and Delaware combined. In addition to research centers throughout the United States, its operations "extend[ed] from the ore fields of the Belgian Congo and the Arctic region of Canada to the weapons proving ground at Enewetak Atoll in the Pacific and the medical projects studying the after-effects of atomic bombing in . . . Japan."[49] The Isotope Division, however, would employ only about fifty people and, when reactor production time was accounted for, occupy only a fraction of its budget and resources.[50]

THE TRANSFORMATION IN GOVERNMENT-SPONSORED RESEARCH

The AEC's decision to proceed with a biomedical research program was part of an even greater transformation, in which government continued and expanded wartime support for research in industry and at universities. Before World War II, biomedical research was a small enterprise in which the federal government played a minor role. During the war, however, large numbers of American biomedical researchers were mobilized by the armed forces. These researchers played an important role in advancing military medicine in a wide range of areas, including blood substitutes, antimalarial drugs and, as noted above, in nurturing the infant science of nuclear medicine.

As the war was drawing to a close, President Roosevelt asked for advice from his Office of Scientific Research and Development (OSRD) on how to convert the nation's military research effort to a peacetime footing, and whether the government should take an activist role in promoting research. The OSRD, under Vannevar Bush, responded in July 1945, after Roosevelt's death, with a report called "Science, the Endless Frontier." Bush and his colleagues recommended among other things the establishment of a National Science Foundation (NSF) to support basic research in all areas including the biomedical sciences. While the principle that the federal government should fund medical research came to seem self-evident, this was hardly the case at the time. In a personal reminiscence published in 1970, Bush wrote:

To persuade the Congress of these pragmatically inclined United States to establish a strong organization to support fundamental research would seem to be one of the minor miracles. We in this country have supported well those pioneers who have created new gadgetry for our use or our amusement. But we have not had during our formative years the respect for scientific endeavors, for scholarship generally, to the extent it had been present in Europe."[51]

Congress worked Bush's small miracle and passed relevant legislation, but President Harry Truman vetoed the bill. When the bill passed again, however, Bush persuaded Truman to sign it.[52]

At the new AEC, and elsewhere, a key element of the support for science was the determination to fund extramural research, that is, research outside the agency. Prior to the war, federal support for private researchers was limited. The Manhattan Project was only one of several wartime efforts that drew private researchers into government service and that provided federal funds for those who remained in private research centers. Following the war, as researchers returned to universities, laboratories, and hospitals, the continued federal support of their efforts transformed the relationship between government and science and the dimensions of the scientific effort.[53]

During the war, the Committee on Medical Research (CMR) of the OSRD operated entirely by funding external research. In 1944, Congress empowered the surgeon general of the Public Health Service to make grants to universities, hospitals, laboratories, and individuals, which provided the legislative basis for the postwar National Institute of Health (NIH) extramural program.[54] In 1948, Congress authorized the National Heart Institute to join the decade-old National Cancer Institute, and NIH became the National Institutes of Health.

By the late 1960s, the annual appropriations of NIH exceeded $1 billion.[55] Research involving medical uses of radioisotopes and external radiation was among the newer fields benefiting from the increased funding. As discussed in more detail in chapter 6, government-supported radioisotope research has proved profoundly important in the development of techniques for medical diagnosis and treatment.

Federal research funding has also continued to be essential to the development of the use of external sources of radiation. For example, the crude images made possible by Roentgen's discovery of x rays have been replaced by higher resolution, three-dimensional pictures, such as those produced by computerized tomographic (CT) scanning and magnetic resonance imaging (MRI).

Today, the benefits of federally sponsored medical research are often taken for granted. To many of those in the midst of the postwar planning and advocacy, however, the result was not foreordained. "Fortunately," Shields Warren recalled years later, postwar "momentum" kept AEC research budgets on track until, in 1957, the Soviet launch of Sputnik (the first space satellite) jolted the American people into a renewed commitment to the support of scientific research.[56]

THE AFTERMATH OF HIROSHIMA AND NAGASAKI: THE EMERGENCE OF THE COLD WAR RADIATION RESEARCH BUREAUCRACY

While promoting the beneficial uses of radiation, the government also wished to continue and expand research on its harmful effects. Three days after the destruction of Hiroshima, Robert Stone wrote two letters to Stafford Warren's deputy, and Stone's former student, Hymer Friedell. The first expressed hope that the contribution of medical researchers could now be made public, so that people would know what they had done during the war.[57] The second letter described Stone's "mixed feelings" at the success that had been achieved and his fear that the lingering effects of radiation from the bomb had been underestimated: "I could hardly believe my eyes," Stone wrote, "when I saw a series of news releases said to be quoting Oppenheimer, and giving the impression that there is no radioactive hazard. Apparently all things are relative."[58]

Friedell and other researchers, including Stafford Warren and Shields Warren, soon traveled to Hiroshima and Nagasaki to begin what became an extensive research program on survivors. The data from that project quickly became and still remain the essential source of information on the long-term effects of radiation on populations of human beings. It was not long, however, before there were additional real-life data on the bomb, from postwar atomic tests. In 1946, the United States undertook the first peacetime nuclear weapons tests at Bikini Atoll in the Marshall Islands. Operation Crossroads, conducted before journalists and VIPs from around the world, was intended to test the ability of a flotilla of unmanned ships to withstand the blast. Since most of the ships remained afloat, the Navy declared Crossroads a triumph.[59]

Behind the scenes, however, Crossroads medical director Stafford Warren expressed horror at the level of contamination on the ships due to the underwater atomic blast.[60] When the ships returned to the West Coast from the Pacific, they were extensively studied to assess the damage and contamination from the atomic bombs. The government created the Naval Radiological Defense Laboratory (NRDL) to study the effects of atomic bombs on ships and to design ways to protect them. "Crossroads," according to an NRDL history, "left no doubt that man was faced with the necessity for coping with strange and unprecedented problems for which no solutions were available."[61]

Hiroshima and Nagasaki, it now seemed, were only the beginning, not the end, of human exposure to bomb-produced radiation. As Crossroads confirmed with the lingering problem of contaminated ships, what the bomb did not obliterate it might still damage by radiation over the course of days or years. It was no longer enough to know about the effects of radioactive materials on American nuclear weapons workers; now there was the urgent need to understand the effects on American soldiers, sailors, and even citizens as well.

Largely invisible to the public, an ad hoc bureaucracy sprang up to address the medical and radiation research problems of atomic warfare. This bureaucracy brought together former wartime radiation researchers, who were joined by junior colleagues, to advise, and participate in, the government's growing radiation research program. Other, already established groups—

such as the AEC's Division of Biology and Medicine and its advisory committee—also had important places in the new network.

Beyond considering fallout from the testing of atomic bombs, these groups also looked at how radiation itself might be used as a weapon. During the war, scientists like J. Robert Oppenheimer had speculated on the possibility that fission products (radioactive materials produced by the bomb or by reactors) could be dispersed in the air and on the ground to kill or incapacitate the enemy. In 1946, the widespread contamination of ships at Crossroads by radioactive mist gave dramatic evidence of the potential of so-called radiological warfare, or RW. In 1947, the military created a committee of experts to study the problem. The following year, a blue-ribbon panel of physicians and physicists looked at the prospects, both offensive and defensive, of what the Pentagon termed "Rad War." The work of these panels would lead to dozens of intentional releases of radiation into the environment at the Army's Dugway, Utah, testing grounds from the late 1940s to the early 1950s. The very fact that the government was engaged in RW tests was a secret. Indeed, the records of the RW program —including, as we shall see in chapter 11, the debate on what the public should be told about the program—would remain largely secret for almost fifty years.

In 1949, a military program to build a nuclear-powered airplane led to a set of proposed human radiation experiments. The NEPA (Nuclear Energy for the Propulsion of Aircraft) program had its origins in 1946 as a venture that included the Manhattan Project's Oak Ridge site, the military, and private aircraft manufacturers. Robert Stone, as we shall see in chapter 8, was a leading proponent of experiments involving healthy volunteers, as a key to answering questions about the radiation hazard faced by the crew of the proposed airplane.

The NEPA and RW groups considered important, but still discrete, projects. Where did the "big picture" discussions take place? The Advisory Committee has pieced together the records of the Armed Forces Medical Policy Council, the Committee on Medical Sciences, and the Joint Panel on the Medical Aspects of

Atomic Warfare.[62] These three Defense Department groups, all chaired by civilian doctors, guided the government on both the broad subject of military-related biomedical research and the new and special problems posed by atomic warfare.

If the surviving records are an indication, from its creation in 1949 to its evident demise with the reorganization of the Defense Department in 1953, the Joint Panel quickly became the hub of atomic warfare-related biomedical research. The Joint Panel gathered information about relevant research from all corners of the government, provided guidance for Defense Department programs, and reviewed and coordinated policy in the matter of human experimentation using atomic energy.

By charter, the group was to be headed by a civilian. Harvard's Dr. Joseph Aub, a long-standing member of the Boston-based medical research community who had worked with Robley Evans on the study of the radium dial painters and had also studied lead toxicity, served as chair. Those who served with Aub included Evans, Hymer Friedell, and Louis Hempelmann, Oppenheimer's Manhattan Project medical aide. Other government participants came from the AEC, the Public Health Service, the National Institutes of Health, the Veterans Administration, and the CIA. (The charter provided that the Joint Panel should collect information on relevant research conducted abroad, which the CIA evidently provided.)[63]

This bureaucracy provided the venue for secret discussions that linked the arts of healing and war in ways that had little precedent. At one and the same time, for example, doctors counseled the military about the radiation risk to troops at the site of atomic bomb tests, advised on the need for research on the "psychology of panic" at such bomb tests, and debated the need for rules to govern atomic warfare-related experimentation. (See chapter 10.)

The records of the Joint Panel show that, during the height of the Cold War, the resources of civilian agencies were part of the mobilization of resources to serve national security interests. For example, Dr. Howard Andrews, trained as a physicist, was the National Institutes of Health's representative to the Joint Panel, and

in the 1950s he worked with the DOD and the AEC in monitoring safety measures and measuring fallout from nuclear tests.[64]

In 1950 President Truman ordered federal agencies, including the Public Health Service and NIH, to focus their resources on activities that would benefit national security needs. On paper, at least, PHS and NIH policymakers sought to direct resources to questions of radiation injury, civil defense, and worker health and safety.[65] For example, a 1952 internal planning memo explained that NIH "will not wait for formal requests by the armed forces . . . to undertake research which NIH staff knows to be of urgent military and civilian defense significance. Limited selective conversion of research to work directly related to biological warfare, shock, radiation injury and thermal burns will begin immediately. . . ."[66] The fragmentary surviving documentation, however, does not show the extent to which PHS- and NIH-funded researchers actually redirected their investigations or merely recast the purpose of ongoing work.

NEW ETHICAL QUESTIONS FOR MEDICAL RESEARCHERS

As medical researchers became fixtures in the Cold War research bureaucracy, they assumed roles that, if not entirely new, raised ethical questions with which they had rarely dealt before. The surviving records of the period reveal that frank and remarkable discussions took place among military and civilian officials and researchers, all of whom had to balance the benefits of gaining knowledge needed to fight and survive an atomic war with the risks that had to be taken to gain this knowledge. They had to consider, and even debated, whether human radiation experimentation was justified, what kinds of risks entire populations could be exposed to, and what the public could and should be told.

Whether to Experiment with Humans: The Debate Is Joined

Spurred by proposals for human radiation experiments connected with the nuclear-powered airplane (NEPA) project, AEC and DOD medical experts in 1949 and 1950 engaged in debate on the need for human experimentation. The transcript of a 1950 meeting among AEC biomedical officials and advisers and military representatives provides unique insight into the mix of moral principles and practical concerns.[67]

The participants in the debate included many of the key medical figures in the Manhattan Project and the postwar radiation research bureaucracy. For the Navy, for example, Captain Behrens, the editor of *Atomic Medicine*, made the point that an atomic bomb might contaminate, but not sink, ships. The Navy would need to know the risk of sending rescue or salvage parties into the contaminated area. There were questions of "calculated risk which all of the services are interested in, and not only the services but probably the civilians as well."[68] Brigadier General William H. Powell, Jr., of the Office of the Air Force Surgeon General, added further questions: How does radiation injure tissue? Can equipment protect against the bomb's effects? Is there a way to treat radiation injury? How should mass casualties be handled?[69]

These questions were hardly abstract. Operation Crossroads had demonstrated that postblast contamination of Navy ships was a serious hazard. The use of the atomic bomb as a tactical weapon, declared Brigadier General James Cooney of the AEC's Division of Military Applications, "has now gone beyond the realm of possibility and into the realm of probability."[70] This meant that "we have a responsibility that is tremendous," Cooney added. "If this weapon is used tactically on a corps or division, and we have, say, 5,000 troops who have received 100 R[oentgens] radiation, the Commander is going to want from me, 'Is it all right for me to reassemble these men and take them into combat?' I don't know the answer to that question."[71] Commanders needed to know "How much radiation can a man take?"[72]

Cooney argued that human experimentation was necessary. He invoked the military's tradition of experimentation with healthy volunteers, dating back to Walter Reed's famous work on yellow fever at the turn of the century. Cooney urged that the military seek volunteers within its ranks—"both officer and enlisted"—to be exposed to as much as 150 R of whole-body radiation.[73]

The AEC's Shields Warren took the other side in this debate. Warren raised two basic points in response to Cooney. First, human experimentation was not essential because animal research would be adequate to find the answers. Second, data from human experimentation would likely be scientifically useless. "We have," Warren declared, "learned enough from animals and from humans at Hiroshima and Nagasaki to be quite certain that there are extraordinary variables in this picture. There are species variables, genetics variables within species, variations in condition of the individual within that species." The danger of failing to provide data had to be weighed against the danger of providing misleading data: "It might be almost more dangerous or misleading to give an artificial accuracy to an answer that is of necessity an answer that spreads over a broad range in light of these variables."[74]

There were, moreover, political obstacles to the program Cooney had proposed. Satisfactory answers, Warren concluded, would require "going to tens of thousands of individuals." But America was not the Soviet Union: "If we were considering things in the Kremlin, undoubtedly it would be practicable. I doubt that it is practicable here."[75]

At the heart of Warren's objections to Cooney's proposal was a concern about employing "human experimentation when it isn't for the good of the individual concerned and when there is no way of solving the problem."[76] To Cooney's invocation of Walter Reed, Warren responded that, in the case of yellow fever, humans were needed as subjects because there was no nonhuman host to the disease.

Cooney did not disagree with Warren "that statistically we will prove nothing." But, he pointed out, "[G]enerals are hard people to deal with. . . . If we had 200 cases whereby we could say that these men did or did not get sick up to 150 R, it would certainly be a great help to us."[77]

Even then, Warren rejoined, the data might not be of great use: "I can think in terms of times when even if everybody on a ship was sea-sick, you would still have to keep the ship operating."[78]

The 1950 debate over NEPA provides clear evidence that midcentury medical experts gave thought before engaging in human experimentation that involved significant risk and was not intended to benefit the subject. On paper, the debate was decided in Shields Warren's favor. Following Warren's and DBM's opposition, Cooney and the military agreed that "human experimentation" on healthy volunteers would not be approved. However, even as this policy was declared, the Defense Department, with Warren's apparent acquiescence, proceeded to contract with private hospitals to gather data on sick patients who were being treated with radiation. The government's use of sick patients for research, as we shall see in chapter 8, raised difficult ethical questions of its own.

Whether to Put Populations at Risk: The Debate Continues

As the medical experts debated the issue of whether to put individual human subjects at risk in radiation experiments on behalf of NEPA, they were also engaged in secret discussions about whether to proceed with the testing of nuclear weapons, which might put whole populations at risk.

It was also in 1950 that the decision was made to carry out atomic bomb testing at a site in the continental United States. President Truman chose the Nevada desert as the location for the test site. Shields Warren's Division of Biology and Medicine was assigned the job of considering the safety of early tests. Like the earlier transcript, an account of a May 1951 meeting at Los Alamos, convened by Warren, provides a window onto the balancing of risks and benefits by medical researchers.

The meeting focused on the radiological hazards to populations downwind from underground testing planned at the Nevada Test Site. Those in attendance realized that the testing could be risky. "I would almost say from the discussion this far," Warren summarized, "that in light of the size and activity of some of these particles, their unpredictability of fallout, the possibility of external beta burns is quite real."[79] Committee members considered the testing a "calculated risk" for populations downwind, but they thought that the information they could gain made the risk worthwhile. According to the record of the meeting, Warren summarized the view of Dr. Gioacchino Failla, a Columbia University radiological physicist: "[T]he time has come when we should take some risk and get

some information . . . we are faced with a war in which atomic weapons will undoubtedly be used, and we have to have some information about these things . . . if we look for perfect safety we will never make these tests."[80] Worried about the potential consequences of miscalculation, the AEC's Carrol Tyler observed, "We have lost a continental site no matter where we put it." Still, Tyler argued, "If we are going to gamble it might as well be done where it is operationally convenient."[81] A proposed deep underground test did not take place, and a test evidently considered less risky was substituted. Ultimately, in a summary prepared at the end of the 1951 test series, the Health Division leader of the AEC's Los Alamos Laboratory recorded that perhaps only good fortune had averted significant contamination: "Thanks to the kindness of the winds, no significant activity was deposited in any populated localities. It was certainly shown however," he wrote, "that significant exposures at considerable distances could be acquired by individuals who actually were in the fallout while it was in progress."[82]

The NEPA debate and the advent of nuclear testing confronted biomedical experts with a set of conflicting, and even contradictory, objectives. First, they were called upon to offer advice on decisions that might inevitably put people at some risk. The risk had to be balanced against the benefit, which in most instances was defined as connected with the nation's security. In many cases, the experts agreed, it was better to bear the lesser risk now, in order to avoid a greater risk later. Second, these experts were also called upon, as in the 1951 Nevada test, to provide advice on minimizing risk. Third, as in the Nevada test, these same experts saw the tests as opportunities to gather data that might ultimately be used to reduce risk for all.

Whether and What the Public Should Be Told About Government-Created Radiation Risk

Scientific research had a long and celebrated tradition of open publication in the scientific literature. But several factors caused Cold War researchers to limit their public disclosures. These included, preeminently, concern with national security, which necessarily required secrecy. But they also included the concern that the release of research information would undermine needed programs because the public could not understand radiation or because the information would embarrass the government.

The tension between the publicizing of information and the limits on disclosure was a constant theme in Cold War research. When, in June 1947, the Medical Board of Review appointed by David Lilienthal reported on the AEC's biomedical program, it declared that secrecy in scientific research is "distasteful and in the long run contrary to the best interests of scientific progress."[83] As shown by its organization of the medical isotope program, the AEC acted quickly to make sure that the great preponderance of biomedical research done under its auspices would be published in the open literature.

However, recently retrieved documents show that the need for secrecy was also invoked where national security was not endangered. At the same time that biomedical officials, such as those on the Medical Board of Review, spoke openly of the need to limit national security restrictions, internally they sometimes sided with those who would restrict information from the public even where release admittedly would not directly endanger national security. Thus, as we shall see in chapter 13, Shields Warren and other AEC medical officials agreed to withhold data on human experiments from the public on the grounds that disclosure would embarrass the government or could be a source of legal liability.

A further important qualification to what the public could know related to research connected with the atomic bomb—including the creation of a worldwide network to gather data on the effects of fallout from nuclear tests. In 1949, the AEC undertook Project Gabriel, a secret effort to study the question of whether the tests could threaten the viability of life on earth. In 1953, Gabriel led to Project Sunshine, a loose confederation of fallout research projects whose human data-gathering efforts, as we see in chapter 13, operated in the twilight between openness and secrecy.

Finally, while documents show that medical experts and officials shared an acute awareness of the importance of public support to the success of Cold War programs, this awareness was coupled with concern about the American public's ability to understand the risks that had to be borne to win the Cold War. The concern that citizens could not understand radiation risk is illustrated by a recently recovered NEPA transcript. In July 1949, the nuclear airplane project gathered radiation experts and psychologists to consider psychological problems connected to radiation hazard. To the assembled experts the greatest unknown was not radiation itself, but the basis for public fear and misunderstanding of radiation.

"I believe," General Cooney proposed, "that the general public is under the opinion that we don't know very much about this condition [radiation]. . . . We know," he ventured, "just about as much about it as we do about many other diseases that people take for granted . . . even tuberculosis."[84]

Yet, said the Navy's Captain Behrens, "there are some peculiar ideas relative to radiation that are related to primitive concepts of hysteria and things in that category. . . . There is such a unique element in it; for some it begins to border on the mystical."[85] A good deal of the public's fear of radiation, declared Berkeley's Dr. Karl M. Bowman, a NEPA medical adviser, "is essentially the fear of the unknown. The dangers have been enormously magnified." As Dr. Bowman and others noted, the public's perception was not without reason, for "we have emphasized for purposes of getting funds for research how little we know."[86]

The perspective expressed in the NEPA transcript would lead, as shown in chapter 10, to the use of atomic bomb tests to perform human research on the psychology of panic and, as shown in other case studies, to decisions to hold information closely out of concern that its release could create public misunderstanding that would imperil important government programs.

CONCLUSION

In the atomic age, Captain Behrens's *Atomic Medicine* pointed out, radiation research was both the agent and the beneficiary of dramatic developments at the intersection of government and medicine. When ethical questions were raised by these developments, radiation researchers would be on the front line in having to deal with them. The burgeoning government-funded biomedical research, including human radiation research, required a reexamination of the traditional doctor-patient relationship. At the same time, the evolving role of medical researchers as government officials and advisers also posed questions about the place of doctors, and more generally of scientists, in service to government.

THE BASICS OF RADIATION SCIENCE

The ethical and historical issues of human radiation experiments cannot be understood without a basic grasp of the underlying science. This requires more than a glossary defining technical terms. At least an intuitive understanding of the natural laws and scientific techniques of radiation science is necessary. Obviously, acquiring a professional level of knowledge would require far more time than most readers can afford; indeed, entire careers are devoted to studying just one aspect of the field. To serve the interests of democracy in a technological world, however, we must provide sufficient technical background for all citizens to become active participants in considering the ethical and political dimensions of scientific research.

What follows is an attempt to provide such a background for the events and issues discussed in this report, directed toward those readers less familiar with "the basics" of radiation science. This task was deemed important enough to deserve a distinct section of this Introduction.

What Is Ionizing Radiation?

What is radiation?

Radiation is a very general term, used to describe any process that transmits energy through space or a material away from a source. Light, sound, and radio waves are all examples of radiation. When most people think of radiation, however, they are thinking of *ionizing radiation*—radiation that can disrupt the atoms and molecules within the body. While scientists think of these emissions in highly mathematical terms, they can be visualized either as subatomic particles or as rays. Radiation's effects on humans can best be understood by first examining the effect of radiation on *atoms*, the basic building blocks of matter.

What is ionization?

Atoms consist of comparatively large particles (protons and neutrons) sitting in a central nucleus, orbited by smaller particles (electrons): a miniature solar system. Normally, the number of protons in the center of the atom equals the number of electrons in orbit. An *ion* is any atom or molecule that does not have the normal number of electrons. *Ionizing radiation* is any form of radiation that has enough energy to knock electrons out of atoms or molecules, creating ions.

How is ionizing radiation measured?

Measurement lies at the heart of modern science, but a number by itself conveys no information. Useful measurement requires both an instrument for measurement (such as a stick to mark off length) and an agreement on the *units* to be used (such as inches, meters, or miles). The units chosen will vary with the *purpose* of the measurement. For example, a cook will measure butter in terms of tablespoons to ensure the meal tastes good, while a nutritionist may be more concerned with measuring calories, to determine the effect on the diner's health.

The variety of units used to measure radiation and radioactivity at times confuses even scientists, if they do not use them every day. It may be helpful to keep in mind the *purpose* of various units. There are two basic reasons to measure radiation: the study of physics and the study of the biological effects of radiation. What creates the complexity is that our instruments measure *physical* effects, while what is of interest to some are *biological* effects. A further complication is that units, as with words in any language, may fade from use and be replaced by new units.

Radiation is not a series of distinct events, like radioactive decays, which can be counted individually. Measuring radiation in bulk is like measuring the movement of sand in an hour-glass; it is more useful to think of it as a continuous flow, rather than a series of separate events. The *intensity* of a beam of ionizing radiation is measured by counting up how many ions (how much electrical charge) it creates in air. The *roentgen* (named after Wilhelm Roentgen, the discoverer of x rays) is the unit that measures the ability of x rays to ionize air; it is a unit of exposure that can be measured directly. Shortly after World War II, a common unit of measurement was the *roentgen equivalent physical (rep)*, which denoted an ability of other forms of radiation to create as many ions in air as a roentgen of x rays. It is no longer used, but appears in many of the documents examined by the Advisory Committee.

What are the basic types of ionizing radiation?

There are many types of ionizing radiation, but the most familiar are *alpha, beta,* and *gamma/x-ray* radiation. *Neutrons,* when expelled from atomic nuclei and traveling as a form of radiation, can also be a significant health concern.

Alpha particles are clusters of two neutrons and two protons each. They are identical to the nuclei of atoms of helium, the second lightest and second most common element in the universe, after hydrogen. Compared with other forms of radiation, though, these are very heavy particles—about 7,300 times the mass of an electron. As they travel along, these large and heavy particles frequently interact with the electrons of atoms, rapidly losing their energy. They cannot even penetrate a piece of paper or the layer of dead cells at the surface of our skin. But if released within the body from a radioactive atom inside or near a cell, alpha particles can do great damage as they ionize atoms, disrupting living cells. Radium and plutonium are two examples of alpha emitters.

Beta particles are electrons traveling at very high energies. If alpha particles can be thought of as large and slow bowling balls, beta particles can be visualized as golf balls on the driving range. They travel farther than alpha particles and, depending on their energy, may do as much

damage. For example, beta particles in fallout can cause severe burns to the skin, known as beta burns. Radiosotopes that emit beta particles are present in fission products produced in nuclear reactors and nuclear explosions. Some beta-emitting radioisotopes, such as iodine 131, are administered internally to patients to diagnose and treat disease.

Gamma and *x-ray* radiation consists of packets of energy known as *photons*. Photons have no mass or charge, and they travel in straight lines. The visible light seen by our eyes is also made up of photons, but at lower energies. The energy of a gamma ray is typically greater than 100 kilo-electron volts (keV—"k" is the abbreviation for *kilo*, a prefix that multiplies a basic unit by 1,000) per photon, more than 200,000 times the energy of visible light (0.5 eV). If alpha particles are visualized as bowling balls and beta particles as golf balls, photons of gamma and x-radiation are like weightless bullets moving at the speed of light. Photons are classified according to their origin. Gamma rays originate from events within an atomic nucleus; their energy and rate of production depend on the radioactive decay process of the radionuclide that is their source. X rays are photons that usually originate from energy transitions of the electrons of an atom. These can be artificially generated by bombarding appropriate atoms with high-energy electrons, as in the classic x-ray tube. Because x rays are produced artificially by a stream of electrons, their rate of output and energy can be controlled by adjusting the energy and amount of the electrons themselves. Both x rays and gamma rays can penetrate deeply into the human body. How deeply they penetrate depends on their energy; higher energy results in deeper penetration into the body. A 1 MeV ("M" is the abbreviation for *mega*, a prefix that multiplies a basic unit by 1,000,000) gamma ray, with an energy 2,000,000 times that of visible light, can pass completely through the body, creating tens of thousands of ions as it does.

A final form of radiation of concern is *neutron* radiation. Neutrons, along with protons, are one of the components of the atomic nucleus. Like protons, they have a large mass; unlike protons, they have no electric charge, allowing them to slip more easily between atoms. Like a Stealth fighter, high-energy neutrons can travel farther into the body, past the protective outer layer of the skin, before delivering their energy and causing ionization.

Several other types of high-energy particles are also ionizing radiation. Cosmic radiation that penetrates the Earth's atmosphere from space consists mainly of protons, alpha particles, and heavier atomic nuclei. Positrons, mesons, pions, and other exotic particles can also be ionizing radiation.

What Is Radioactivity?

What causes radioactivity?

As its name implies, *radioactivity* is the act of emitting radiation spontaneously. This is done by an atomic nucleus that, for some reason, is unstable; it "wants" to give up some energy in order to shift to a more stable configuration. During the first half of the twentieth century, much of modern physics was devoted to exploring why this happens, with the result that nuclear decay was fairly well understood by 1960. Too many neutrons in a nucleus lead it to emit a negative *beta* particle, which changes one of the neutrons into a proton. Too many protons in a nucleus lead it to emit a positron (positively charged electron), changing a proton into a neutron. Too much energy leads a nucleus to emit a *gamma* ray, which discards great energy without changing any of the particles in the nucleus. Too much mass leads a nucleus to emit an *alpha* particle, discarding four heavy particles (two protons and two neutrons).

How is radioactivity measured?

Radioactivity is a physical, not a biological, phenomenon. Simply stated, the radioactivity of a sample can be measured by counting how many atoms are spontaneously decaying each second. This can be done with instruments designed to detect the particular type of *radiation* emitted with each "decay" or disintegration. The actual number of disintegrations per second may be quite large. Scientists have agreed upon common

units to use as a form of shorthand. Thus, a curie (abbreviated "Ci" and named after Pierre and Marie Curie, the discoverers of radium[87]) is simply a shorthand way of writing "37,000,000,000 disintegrations per second," the rate of disintegration occurring in 1 gram of radium. The more modern International System of Measurements (SI) unit for the same type of measurement is the *becquerel* (abbreviated "Bq" and named after Henri Becquerel, the discoverer of radioactivity), which is simply a shorthand for "1 disintegration per second."

What is radioactive half-life?

Being unstable does not lead an atomic nucleus to emit radiation immediately. Instead, the probability of an atom disintegrating is constant, as if unstable nuclei continuously participate in a sort of lottery, with random drawings to decide which atom will next emit radiation and disintegrate to a more stable state. The time it takes for half of the atoms in a given mass to "win the lottery"—that is, emit radiation and change to a more stable state—is called the *half-life*. Half-lives vary greatly among types of atoms, from less than a second to billions of years. For example, it will take about 4.5 billion years for half of the atoms in a mass of uranium 238 to spontaneously disintegrate, but only 24,000 years for half of the atoms in a mass of plutonium 239 to spontaneously disintegrate. Iodine 131, commonly used in medicine, has a half-life of only eight days.

What is a radioactive decay chain?

Stability may be achieved in a single decay, or a nucleus may decay through a series of states before it reaches a truly stable configuration, a bit like a Slinky toy stepping down a set of stairs. Each state or step will have its own unique characteristics of half-life and type of radiation to be emitted as the move is made to the next state. Much scientific effort has been devoted to unraveling these decay chains, not only to achieve a basic understanding of nature, but also to design nuclear weapons and nuclear reactors. The unusually complicated decay of uranium 238, for example—the primary source of natural radioactivity on earth—proceeds as follows:[88]

U-238 emits an alpha
↓
Thorium 234 emits a beta
↓
Protactinium 234 emits a beta
↓
Uranium 234 emits an alpha
↓
Thorium 230 emits an alpha
↓
Radium 226 emits an alpha
↓
Radon 222 emits an alpha
↓
Polonium 218 emits an alpha
↓
Lead 214 emits a beta
↓
Bismuth 214 emits a beta
↓
Polonium 214 emits an alpha
↓
Lead 210 emits a beta
↓
Bismuth 210 emits a beta
↓
Polonium 210 emits an alpha
↓
Lead 206, which is stable

How can radioactivity be caused artificially?

Radioactivity can occur both naturally and through human intervention. An example of artificially induced radioactivity is *neutron activation*. A neutron fired into a nucleus can cause nuclear *fission* (the splitting of atoms). This is the basic concept behind the atomic bomb. Neutron activation is also the underlying principle of boron-neutron capture therapy for certain brain cancers. A solution containing boron is injected into a patient and is absorbed more by the cancer than by other cells. Neutrons fired at the area of the brain cancer are readily absorbed (captured) by the boron nuclei. These nuclei then

become unstable and emit radiation that attacks the cancer cells. Simple in its basic physics, the treatment has been complex and controversial in practice and after half a century is still regarded as highly experimental.

What Are Atomic Number and Atomic Weight?

What is an element?

Chemical behavior is what originally led scientists to classify matter into various *elements*. Chemical behavior is the ability of an atom to combine with other atoms. In more technical terms, chemical behavior depends upon the *type* and *number* of the *chemical bonds* an atom can form with other atoms. In classroom kits for building models of molecules, *atoms* are usually represented by colored spheres with small holes for pegs and the *bonds* are represented by the small pegs that can connect the spheres. The number of peg holes signifies the maximum number of bonds an atom can form; different types of bonds may be represented by different types of pegs. Atoms that have the same number of peg holes may have similar chemical behavior. Thus, atoms that have identical chemical behavior are regarded as atoms of the same element. For example, an atom is labeled a "carbon atom" if it can form the same number, types, and configurations of bonds as other carbon atoms. Although the basics are simple to explain, how atoms bind to each other becomes very complex when studied in detail; new discoveries are still being made as new types of materials are formed.

What is atomic number?

An atom may be visualized as a miniature solar system, with a large central nucleus orbited by small electrons. The bonding capacity of an atom is determined by the *electrons*. For example, atoms that in their normal state have one electron are hydrogen atoms and will readily (and sometimes violently) bond with oxygen. This bonding capacity of hydrogen was the cause of the explosion of the airship Hindenburg in

1937. Atoms that in their normal state have two electrons are helium atoms, which will not bond with oxygen and would have been a better choice for filling the Hindenburg.

We can pursue the question back one step further: What determines the number of electrons? The *number of protons* in the nucleus of the atom. Here, the analogy between an atom and the solar system breaks down. The force that holds the planets in their orbits is the gravitational attraction between the planets and the sun. However, in an atom what holds the electrons in their orbit is the *electrical attraction* between the electrons and the protons in the nucleus. The basic rule is that *like charges repel and opposite charges attract.* Although a proton has more mass than an electron, they both have the same amount of electrical charge, but opposite in kind. Scientists have designated *electrons* as having a *negative* charge and *protons* as having a *positive charge.* One positive proton can hold one negative electron in orbit. Thus, an atom with one proton in its nucleus normally will have one electron in orbit (and be labeled a hydrogen atom); an atom with ninety-four protons in its nucleus will normally have ninety-four electrons orbiting it (and be labeled a plutonium atom).

The number of protons in a nucleus is called the *atomic number* and always equals the number of electrons in orbit about that nucleus (in a nonionized atom). Thus, all atoms that have the same number of protons—the atomic number—are atoms of the same element.

What is atomic weight?

The nuclei of atoms also contain *neutrons,* which help hold the nucleus together. A neutron has *no electrical charge* and is slightly more massive than a proton. Because a neutron can decay into a proton plus an electron (the essence of beta decay), it is sometimes helpful to think of a neutron as an electron and a proton blended together, although this is at best an oversimplification. Because a neutron has no charge, *a neutron has no effect on the number of electrons orbiting the nucleus.* However, because it is even more massive than a proton, a neutron can add significantly to the weight of an atom. The total weight of an atom is called the *atomic weight.* It is approximately equal to the number of protons and neutrons, with a little extra added by the electrons. The stability of the nucleus, and hence the atom's *radioactivity,* is heavily dependent upon the number of neutrons it contains.

What notations are used to represent atomic number and weight?

Each atom, therefore, can be assigned both an atomic number (the number of protons equals the number of electrons) and an atomic weight (approximately equaling the number of protons plus the number of neutrons). A normal helium atom, for example, has two protons and two neutrons in its nucleus, with two electrons in orbit. Its *chemical behavior* is determined by the atomic number 2 (the number of protons), which equals the normal number of electrons; the stability of its nucleus (that is, its radioactivity) varies with its atomic weight (approximately equal to the number of protons and neutrons). The most well-known form of plutonium, for example, has an atomic number of 94, since it has 94 protons, and with the 145 neutrons in its nucleus, an atomic weight of 239 (94 protons plus 145 neutrons). In World War II, its very existence was highly classified. A code number was developed: the last digit of the atomic number (94) and the last digit of the atomic weight (239). Thus, in some of the early documents examined by the Advisory Committee, the term *49* refers to plutonium.

Styles of notation vary, but usually isotopes are written as:

atomic number Chemical abbreviation atomic weight

or as

atomic weight Chemical abbreviation

Thus, the isotope of plutonium just discussed would be written as:

$$_{94}\text{Pu}^{239} \qquad \text{or as} \qquad ^{239}\text{Pu}$$

Since the atomic weight is what is often the only item of interest, it might also be written simply as Pu-239, plutonium 239, or Pu^{239}.

Radioisotopes: What Are They and How Are They Made?

What are isotopes?

The isotopes of an element are all the atoms that have in their nucleus the number of protons (atomic number) corresponding to the chemical behavior of that element. However, the isotopes of a single element vary in the number of neutrons in their nuclei. Since they still have the same number of protons, all these isotopes of an element have *identical chemical behavior*. But since they have different numbers of neutrons, these isotopes of the same element may have *different radioactivity*. An isotope that is radioactive is called a *radioisotope* or *radionuclide*. Two examples may help clarify this.

The most stable isotope of uranium, U-238, has an atomic number of 92 (protons) and an atomic weight of 238 (92 protons plus 146 neutrons). The isotope of uranium of greatest importance in atomic bombs, U-235, though, has three fewer neutrons. Thus, it also has an atomic number of 92 (since the number of protons has not changed) but an atomic weight of 235 (92 protons plus only 143 neutrons). The *chemical behavior* of U-235 is identical to all other forms of uranium, but its nucleus is less stable, giving it *higher radioactivity* and greater susceptibility to the chain reactions that power both atomic bombs and nuclear fission reactors.

Another example is iodine, an element essential for health; insufficient iodine in one's diet can lead to a goiter. Iodine also is one of the earliest elements whose radioisotopes were used in what is now called nuclear medicine. The most common, stable form of iodine has an *atomic number* of 53 (protons) and an *atomic weight* of 127 (53 protons plus 74 neutrons). Because its nucleus has the "correct" number of neutrons, it is stable and is not radioactive. A less stable form of iodine also has 53 protons (this is what makes it behave chemically as iodine) but four extra neutrons, for a total atomic weight of 131 (53 protons and 78 neutrons). With "too many" neutrons in its nucleus, it is unstable and radioactive, with a half-life of eight days. Because it behaves *chemically* as iodine, it travels throughout the body and localizes in the thyroid gland just like the stable form of iodine. But, because it is radioactive, its presence can be detected. Iodine 131 thus became one of the earliest *radioactive tracers*.

How can different isotopes of an element be produced?

How can isotopes be produced—especially radioisotopes, which can serve many useful purposes? There are two basic methods: separation and synthesis.

Some isotopes occur in nature. If radioactive, these usually are radioisotopes with very long half-lives. Uranium 235, for example, makes up about 0.7 percent of the naturally occurring uranium on the earth.[89] The challenge is to *separate* this very small amount from the much larger bulk of other forms of uranium. The difficulty is that all these forms of uranium, because they all have the same number of electrons, will have identical chemical behavior: they will bind in identical fashion to other atoms. Chemical separation, developing a chemical reaction that will bind only uranium atoms, will separate out uranium atoms, but not distinguish among different isotopes of uranium. The only difference among the uranium isotopes is their *atomic weight*. A method had to be developed that would sort atoms according to weight.

One initial proposal was to use a *centrifuge*. The basic idea is simple: spin the uranium atoms as if they were on a very fast merry-go-round. The heavier ones will drift toward the outside faster and can be drawn off. In practice the technique was an enormous challenge: the goal was to draw off that very small portion of uranium atoms that were lighter than their brethren. The difficulties were so enormous the plan was abandoned in 1942.[90] Instead, the technique of *gaseous diffusion* was developed. Again, the basic idea was very simple: the rate at which gas passed (*diffused*) through a filter depended on the weight of the gas molecules: lighter molecules diffused more quickly. Gas molecules that contained U-235 would diffuse slightly faster than gas molecules containing the more common but also heavier U-238. This method also presented formidable technical challenges, but was eventually implemented in the gigantic gas diffusion plant at Oak Ridge, Tennessee. In this process, the uranium was chemically combined with fluorine to form a hexafluoride gas prior to separation by diffusion. This is not a practical method for extracting radioisotopes for scientific and medical use. It was extremely expensive and could only supply naturally occurring isotopes.

A more efficient approach is to artificially manufacture radioisotopes. This can be done by firing high-speed particles into the nucleus of an atom. When struck, the nucleus may absorb the particle or become unstable and emit a particle. In either case, the number of particles in the nucleus would be altered, creating an isotope. One source of high-speed particles could be a cyclotron. A cyclotron accelerates particles around a circular race track with periodic pushes of an electric field. The particles gather speed with each push, just as a child swings higher with each push on a swing. When traveling fast enough, the particles are directed off the race track and into the target.

A cyclotron works only with charged particles, however. Another source of bullets are the neutrons already shooting about inside a nuclear reactor. The neutrons normally strike the nuclei of the fuel, making them unstable and causing the nuclei to split (fission) into two large fragments and two to three "free" neutrons. These free neutrons in turn make additional nuclei unstable, causing further fission. The result is a chain reaction. Too many neutrons can lead to an uncontrolled chain reaction, releasing too much heat and perhaps causing a "meltdown." Therefore, "surplus" neutrons are usually absorbed by "control rods." However, these surplus neutrons can also be absorbed by targets of carefully selected material placed in the reactor. In this way the surplus neutrons are used to create radioactive isotopes of the materials placed in the targets.

With practice, scientists using both cyclotrons and reactors have learned the proper mix of target atoms and shooting particles to "cook up" a wide variety of useful radioisotopes.

How Does Radiation Affect Humans?

Radiation may come from either an external source, such as an x-ray machine, or an internal source, such as an injected radioisotope. The impact of radiation on living tissue is complicated by the type of radiation and the variety of tissues. In addition, the effects of radiation are not always easy to separate from other factors, making it a challenge at times for scientists to isolate them. An overview may help explain not only the effects of radiation but also the motivation for studying them, which led to much of the research examined by the Advisory Committee.

What effect can ionizing radiation have on chemical bonds?

The functions of living tissue are carried out by molecules, that is, combinations of different types of atoms united by *chemical bonds*. Some of these molecules can be quite large. The proper functioning of these molecules depends upon their *composition* and also their *structure* (shape). Altering chemical bonds may change composition or structure. Ionizing radiation is powerful enough to do this. For example, a typical ionization releases six to seven times the energy needed to break the chemical bond between two carbon atoms.[91] This ability to disrupt chemical bonds means that ionizing radiation focuses its impact in a very small but crucial area, a bit like a karate master focusing energy to break a brick. The same amount of raw energy, distributed more broadly in nonionizing form, would have much less effect. For example, the amount of energy in a lethal dose of ionizing radiation is roughly equal to the amount of thermal energy in a single sip of hot coffee.[92] The crucial difference is that the coffee's energy is broadly distributed in the form of nonionizing heat, while the radiation's energy is concentrated in a form that can ionize.

What is DNA?

Of all the molecules in the body, the most crucial is *DNA* (deoxyribose nucleic acid), the fundamental blueprint for all of the body's structures. The DNA blueprint is encoded in each cell as a long *sequence* of small molecules, linked together into a chain, much like the letters in a telegram. DNA molecules are enormously long chains of atoms wound around proteins and packed into structures called *chromosomes* within the cell nucleus. When unwound, the DNA in a single human cell would be more than 2 meters long. It normally exists as twenty-three pairs of chromosomes packed within the cell nucleus, which itself has a diameter of only 10 micrometers (0.00001 meter).[93] Only a small part of this DNA needs to be read at any one time to build a specific molecule. Each cell is continually reading various parts of its own DNA as it constructs fresh molecules to perform a variety of tasks. It is worth remembering that the structure of DNA was not solved until 1953, nine years after the beginning of the period studied by the Advisory Committee. We now have a much clearer picture of what happens within a cell than did the scientists of 1944.

What effect can ionizing radiation have on DNA?

Ionizing radiation, by definition, "ionizes," that is, it pushes an electron out of its orbit around an atomic nucleus, causing the formation of electrical charges on atoms or molecules. If this electron comes from the DNA itself or from a neighboring molecule and directly strikes and disrupts the DNA molecule, the effect is called *direct action*. This initial ionization takes place very quickly, in about 0.000000000000001 of a second. However, today it is estimated that about two-thirds of the damage caused by x rays is due to *indirect action*. This occurs when the liberated electron does not directly strike the DNA, but instead strikes an ordinary water molecule. This ionizes the water molecule, eventually producing what is known as a *free radical*. A free radical reacts very strongly with other molecules as it seeks to restore a stable configuration of electrons. A free radical may drift about up to 10,000,000,000 times longer than the time needed for the initial ionization (this is still a very short time, about 0.00001 of a second), increasing the chance of it disrupting the crucial DNA molecule. This also increases the possibility that other substances could be introduced that would neutralize free radicals before they do damage.[94]

Neutrons act quite differently. A fast neutron will bypass orbiting electrons and occasionally crash directly into an atomic nucleus, knocking out large particles such as alpha particles, pro-

tons, or larger fragments of the nucleus. The most common collisions are with carbon or oxygen nuclei. The particles created will themselves then set about ionizing nearby electrons. A slow neutron will not have the energy to knock out large particles when it strikes a nucleus. Instead, the neutron and the nucleus will bounce off each other, like billiard balls. In so doing, the neutron will slow down, and the nucleus will gain speed. The most common collision is with a hydrogen nucleus, a proton that can excite or ionize electrons in nearby atoms.[95]

What immediate effects can ionizing radiation have on living cells?

All of these collisions and ionizations take place very quickly, in less than a second. It takes much longer for the biological effects to become apparent. If the damage is sufficient to kill the cell, the effect may become noticeable in hours or days. Cell "death" can be of two types. First, the cell may no longer perform its function due to internal ionization; this requires a dose to the cell of about 100 gray (10,000 rad). (For a definition of gray and rad, see the section below titled "How Do We Measure the Biological Effects of Radiation?") Second, "reproductive death" (mitotic inhibition) may occur when a cell can no longer reproduce, but still performs its other functions. This requires a dose of 2 gray (200 rad), which will cause reproductive death in half the cells irradiated (hence such a quantity is called a "mean lethal dose.")[96] Today we still lack enough information to choose among the various models proposed to explain cell death in terms of what happens at the level of atoms and molecules inside a cell.[97] If enough crucial cells within the body totally cease to function, the effect is fatal. Death may also result if cell reproduction ceases in parts of the body where cells are continuously being replaced at a high rate (such as the blood cell-forming tissues and the lining of the intestinal tract). A very high dose of 100 gray (10,000 rad) to the entire body causes·death within twenty-four to forty-eight hours; a whole-body dose of 2.5 to 5 gray (250 to 500 rad) may produce death within several weeks.[98] At lower or more localized doses, the effect will not be death, but specific symptoms due to the loss of a large number of cells. These effects were once called nonstochastic; they are now called *deterministic*.[99] A beta burn is an example of a deterministic effect.

What long-term effects can radiation have?

The effect of the radiation may not be to kill the cell, but to alter its DNA code in a way that leaves the cell alive but with an error in the DNA blueprint. The effect of this *mutation* will depend on the nature of the error and when it is read. Since this is a random process, such effects are now called *stochastic*.[100] Two important stochastic effects of radiation are cancer, which results from mutations in nongerm cells (termed *somatic cells*), and heritable changes, which result from mutations in germ cells (eggs and sperm).

How can ionizing radiation cause cancer?

Cancer is produced if radiation does not kill the cell but creates an error in the DNA blueprint that contributes to eventual loss of control of cell division, and the cell begins dividing uncontrollably. This effect might not appear for many years. Cancers induced by radiation do not differ from cancers due to other causes, so there is no simple way to measure the rate of cancer due to radiation. During the period studied by the Advisory Committee, great effort was devoted to studies of irradiated animals and exposed groups of people to develop better estimates of the risk of cancer due to radiation. This type of research is complicated by the variety of cancers, which vary in radiosensitivity. For example, bone marrow is more sensitive than skin cells to radiation-induced cancer.[101]

Large doses of radiation to large numbers of people are needed in order to cause measurable increases in the number of cancers and thus determine the differences in the sensitivity of different organs to radiation. Because the cancers can occur anytime in the exposed person's lifetime, these studies can take seventy years or more to complete. For example, the largest and scientifically most valuable epidemiologic study of radiation effects has been the ongoing study of the Japanese atomic bomb survivors. Other important studies include studies of large groups exposed to radiation as a consequence of their occupation (such as uranium miners) or as a consequence of medical treatment. These types of studies are discussed in greater detail in the section titled "How Do Scientists Determine the Long-Term Risks from Radiation?"

How can ionizing radiation produce genetic mutations?

Radiation may alter the DNA within any cell. Cell damage and death that result from mutations in somatic cells occur only in the organism in which the mutation occurred and are therefore termed *somatic* or *nonheritable* effects. Cancer is the most notable long-term somatic effect. In contrast, mutations that occur in germ cells (sperm and ova) can be transmitted to future generations and are therefore called *genetic* or *heritable* effects. Genetic effects may not appear until many generations later. The genetic effects of radiation were first demonstrated in fruit flies in the 1920s. Genetic mutation due to radiation does not produce the visible monstrosities of science fiction; it simply produces a greater frequency of the same mutations that occur continuously and spontaneously in nature.

Like cancers, the genetic effects of radiation are impossible to distinguish from mutations due to other causes. Today at least 1,300 diseases are known to be caused by a mutation.[102] Some mutations may be beneficial; random mutation is the driving force in evolution. During the period studied by the Advisory Committee, there was considerable debate among the scientific community over both the extent and the consequences of radiation-induced mutations. In contrast to estimates of cancer risk, which are based in part on studies of human populations, estimates of heritable risk are based for the most part upon animal studies plus studies of Japanese survivors of the atomic bombs.

The risk of genetic mutation is expressed in terms of the *doubling dose:* the amount of radiation that would cause additional mutations equal in number to those that already occur naturally from all causes, thereby doubling the naturally occurring rate of mutation.

It is generally believed that mutation rates depend linearly on dose and that there is no threshold below which mutation rates would not be increased. Spontaneous mutation (unrelated to radiation) occurs naturally at a rate of approximately 1/10,000 to 1/1,000,000 cell divisions per gene, with wide variation from one gene to another.

Attempts have been made to estimate the contribution of ionizing radiation to human mutation rates by studying offspring of both exposed and nonexposed Japanese atomic bomb survivors. These estimates are based on comparisons of the rate of various congenital defects and cancer between exposed and nonexposed survivors, as well as on direct counting of mutations at a small number of genes. For all these endpoints, no excess has been observed among descendants of the exposed survivors.

Given this lack of direct evidence of any increase in human heritable (genetic) effects resulting from radiation exposure, the estimates of genetic risks in humans have been compared with experimental data obtained with laboratory animals. However, estimates of human genetic risks vary greatly from animal data. For example, fruit flies have very large chromosomes that appear to be uniquely susceptible to radiation. Humans may be less vulnerable than previously thought. Statistical lower limits on the doubling dose have been calculated that are compatible with the observed human data. Based on our inability to demonstrate an effect in humans, the lower limit for the genetic doubling dose is thought to be less than 100 rem.[103]

How Do We Measure the Biological Effects of External Radiation?

The methods of measuring radiation and radioactivity, purely physical events, were discussed earlier. In studying the effect of radiation on living organisms, a biological event, the crucial data are the *amount of energy absorbed by a specific amount and type of tissue.* This requires first measuring the amount of energy left behind by the radiation in the tissue and, second, the amount and type of tissue.

What is an absorbed dose *of radiation?*

The risk posed to a human being by any radiation exposure depends partly upon the *absorbed dose,* the amount of energy absorbed per gram of tissue. Absorbed dose is expressed in *rad.* A *rad* is equal to 100 ergs of energy absorbed by 1 gram of tissue. The more modern, internationally adopted unit is the *gray* (named for the English medical physicist L. H. Gray); one gray equals 100 rad. Almost all the documents from the time period studied by the Advisory Committee use the term *rad* rather than *gray.* It is important to realize that absorbed dose refers to energy *per gram of absorbing tissue,* not total energy. Someone absorbing 1 gray (100 rad) in a small amount of tissue, such as a thyroid gland, will absorb much less total energy than someone absorbing 1 gray (100 rad) throughout his or her entire body. Thus, when speaking of absorbed dose, it is crucial to know the amount of tissue being exposed, not simply the number of gray or rad.

What is an equivalent dose *of radiation?*

Even the *rad* or *gray,* though, are still units that measure a purely physical event: the amount of energy left behind in a gram of tissue. It does not directly measure the *biological effect* of that radiation. The biological effect of the same amount of absorbed energy may vary according to the type of radiation involved. This biological effect can be computed by multiplying the absorbed dose (in rad or gray) by a number indicating the *quality factor* of the particular type of radiation. For photons and electrons the quality factor is defined to be 1; for neutrons it ranges from 5 to 20 depending on the energy of the neutron; for alpha particles it is 20.[104] Thus, 1 gray (100 rad) of alpha particles is currently judged to have an effect on living tissue that is twenty times more than 1 gray (100 rad) of x rays. Multiplying the absorbed dose (in rad or gray) by the quality factor (also known as the radiation weighting factor) produces what is called the *equivalent dose.* For the period studied by the Advisory Committee, this was expressed in terms of a unit called the *rem,* an acronym for *roent-*

EXPERIMENTAL AND NONEXPERIMENTAL DOSES *

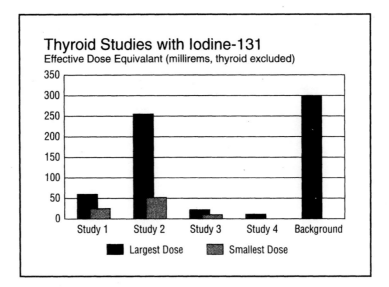

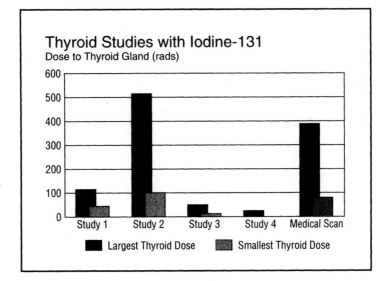

*The experiments themselves are discussed in chapter 7. These graphs are reproduced with permission from Task Force on Human Subject Research, Commonwealth of Massachusetts Department of Mental Retardation, April 1994, "A Report on the Use of Radioactive Materials in Human Subject Research that Involved Residents of State-Oriented Facilities within the Commonwealth of Massachusetts from 1942–1973" (ACHRE No. MASS-072194-A), 17, and the Working Group on Human Subject Research, Commonwealth of Massachusetts Department of Mental Retardation, June 1994, "The Thyroid Studies: A Follow-up Report on the Use of Radioactive Materials in Human Subject Research that Involved Residents of State-Operated Facilities within the Commonwealth of Massachusetts from 1942-1973" (ACHRE No. MASS-072194-A), 14.

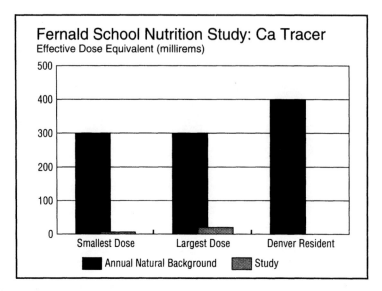

Fernald School Nutrition Study: Ca Tracer
Effective Dose Equivalent (millirems)

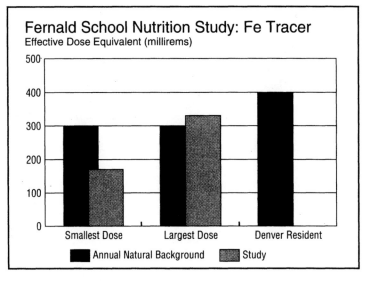

Fernald School Nutrition Study: Fe Tracer
Effective Dose Equivalent (millirems)

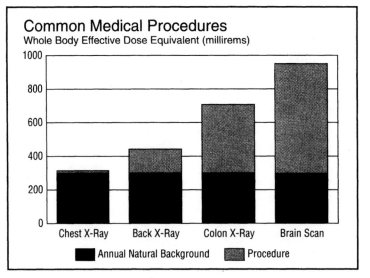

Common Medical Procedures
Whole Body Effective Dose Equivalent (millirems)

gen equivalent man.[105] (The term *equivalent* simply meant that an absorbed dose expressed in rem would have equivalent biological effects, regardless of the type of radiation. Thus, 10 rem of x rays should have the same biological effect as 10 rem of neutrons absorbed by the same part of the body.) The modern unit is the *sievert* (abbreviated Sv and named for the prominent Swedish radiologist, Rolf Sievert), which is equal to 100 rem. Thus, an equivalent dose of 200 rem would today be expressed as 2 sievert.

What is an effective dose of radiation?

Finally, the biological effect of radiation depends on the *type of tissue* being irradiated. As with different types of radiation, a weighting or quality factor is introduced depending on the type of tissue. The more sensitive the tissue is to radiation, the higher the factor. The *effective dose* is the sum of the *equivalent doses* of the various types of irradiated tissue, each properly weighted for its sensitivity to radiation. Tissue weighting factors are determined from the relative incidence of cancers in different tissues in the Japanese survivors of the atomic bombs.

Calculating the *effective dose* makes it possible to readily compare different exposures, as illustrated by the accompanying graphs.

How Do We Measure the Biological Effects of Internal Emitters?

The general principles just described require further refinement when applied to doses from internal emitters.

What information is needed to calculate absorbed dose of a radionuclide inside the body?

Calculating the absorbed dose from a radionuclide inside the body is complex since it involves both the physics of radioactive decay and the biology of the body's metabolism. Six important factors that must be considered are these:

1. The amount of the radionuclide administered.
2. The type of radiation emitted during the decay process.
3. The physical half-life of the radionuclide.
4. The chemical form of the radionuclide.
5. The fraction of the radionuclide that accumulates in each organ.
6. The length of time that the radionuclide remains in the organ (the biological half-life).

How varied are the types of radiation that different radionuclides emit?

Radionuclides can emit several types of radiation (e.g., gamma rays, beta or alpha particles). Each radionuclide emits its own unique mixture of radiations; indeed, scientists identify radioactive materials by using these unique mixtures as if they were fingerprints. The mix of radiations for a specific radionuclide is always the same, regardless of whether the radionuclide is located on a bench in a physicist's laboratory or inside the human body. This means that the type of radiation of each radionuclide can be measured outside the body with great precision by laboratory instruments. A *quality factor*, discussed earlier, is used to adjust for the difference in the biological effects of different types of radiation.

What determines how long a radionuclide will irradiate the body?

The combination of the physical and biological half-life (the effective half-life) determines how long a radionuclide will continue to pump out energy into surrounding tissue. If the physical and biological half-lives of a particular chemical form of a radionuclide are very long, the radionuclide will continue to expose an individual to radiation over his or her lifetime. The total lifetime radiation exposure, expressed in rem, is called the *committed dose equivalent*.

The *physical half-life* is the length of time it will take for half of the atoms in a sample to decay to a more stable form. The physical half-life of each radionuclide can be measured precisely in the laboratory. A shorter half-life means that the miniature power source will "run down" sooner. Sometimes, however, a radionuclide will not decay immediately to a stable form, but to a second, still unstable, form. A full calculation, therefore, must also include the types of radiation and physical half-lives of any decay products.

The biological half-life does not depend on the radionuclide but rather on the *chemical form* of the radionuclide. One chemical form of the radionuclide might be rapidly eliminated from the body whereas other chemical forms may be slowly eliminated.

To measure the biological half-life of a particular chemical form of a radionuclide, that chemical form needs to be studied in animals. Since the biological processes of different animals vary considerably, an accurate determination of the biological half-life requires that each chemical form of the radionuclide be studied in each animal of interest. Prior to studying a chemical form of a radionuclide in a human being, animal studies are performed to get some idea of what to expect.

Once the results of animal studies are available, scientists are able to predict what amount of that chemical form of the radionuclide can be safely injected into humans. An accurate determination of what fraction of each chemical form of the radionuclide accumulates in each organ and how long it stays in each organ in humans can only be determined by studying humans. These type of studies are called biodistribution studies.

What is the tissue weighting factor?

Some chemical forms of radionuclides are highly concentrated in one small organ (e.g., iodine in the thyroid gland). When this happens, that organ will absorb most of the radiated energy, and little energy will be deposited in the remainder of the body. Thus, for each chemical form of a radionuclide, there is an organ that will receive the highest dose from that radionuclide. Since organs also vary greatly in their sensitivity to radiation, the biological consequences of the radiation dose differ depending on the organ. This difference in sensitivity to radiation is represented by what is called a tissue weighting factor.

What is the difference between committed equivalent dose *and* committed effective dose?

An estimate of the risk posed by a radionuclide in the body depends on its chemical form, its biodistribution, its physical properties (how it decays), and the sensitivity of the organs exposed. When all these factors are considered in the calculation of risk for a single radionuclide, the total lifetime exposure is called the *committed equivalent dose*. If more than one radioisotope is present, the sum of all the committed equivalent doses is called the *committed effective dose*. Both are expressed in *rem* or the more modern units *sieverts*.[106] These calculations provide a basis for comparing the risk posed by different isotopes.

How do radiation risks compare with chemical risks?

It should be noted that radiation is not the only possible hazard resulting from the medical use of radionuclides. Few radioisotopes, whether intentionally or accidentally introduced into the body, enter in a chemically pure form. The radioactive atoms are usually part of a larger chemical compound. The chemical form of the radioisotope may pose its own hazards of *chemical toxicity*. Chemical toxicity depends upon the chemical effect of the compound on the body, quite independent of any effects of radiation. Determining chemical toxicity is an entire field of science on its own.

How Do Scientists Determine the Long-Term Risks from Radiation?

Where did the risk estimates in this report come from?

Throughout this report, the reader will find numerous statements estimating the risks of cancer and other outcomes to individuals exposed to various types of radiation. These estimates were obtained from various scientific advisory committees that have considered these questions in depth.[107] Their estimates in turn are based on syntheses of the scientific data on observed effects in humans and animals.

How are risk estimates expressed?

Epidemiologists usually express the risk of disease in terms of the number of new cases (*incidence rate*) or deaths (*mortality rate*) in a population in some period of time. For example, an incidence rate might be 100 new cases per 100,000 people per year; a mortality rate might be 15 deaths per 100,000 people per year. These rates vary widely by age, conditions of exposure, and various other factors. To summarize this complex set of rates, government regulatory bodies often consider the lifetime risk of a particular outcome like cancer. When relating a disease, such as cancer, to one of its several causes, a more useful concept is the *excess lifetime risk* expected from one particular pattern of exposure, such as continuous exposure to 1 rad per year.

It is well established that cancer rates begin to rise above the normal background rate only some time after exposure, the *latent period*, which varies with the type of cancer and other factors such as age. Even after the latent period has passed and radiation effects begin to appear, not all effects are due to radiation. The excess rate may still vary by age, latency, or other factors, but for many cancers it tends to be roughly proportional to the rate in the general population. This is known as the constant relative risk model, and the ratio of rates at any given age between exposed and unexposed groups is called the *relative risk*. Many advisory committees have based their risk estimates on models for the relative risk as a function of dose and perhaps other factors. Other committees, however, have based their estimates on the difference in rates between exposed and unexposed groups, a quantity known as the *absolute risk*. This quantity also varies with dose and other factors, but when this variation is appropriately accounted for, either approach can be used to estimate lifetime risk.

What are the types of data on which such estimates are based?

Human data are one important source, discussed below. Two other important sources of scientific data are experiments on animals and on cell cultures. Because both types of research are done in laboratories, scientists can carefully control the conditions and many of the variables. For the same reason, the experiment can be repeated to confirm the results. Such research has contributed in important ways to our understanding of basic radiobiological principles. It also has provided quantitative estimates of such parameters as the relative effectiveness of different types of radiation and the effects of dose and dose rate. In some circumstances, where human data are limited or nonexistent, such laboratory studies may provide the only basis on which risks can be estimated.

Why are human data preferable to data on animals or tissue cultures for most purposes?

Most scientists prefer to base risk estimates for humans on human data wherever possible. This is because in order to apply animal or tissue culture data to humans, scientists must *extrapolate* from one species to another or from simple cellular systems to the complexities of human physiology. This requires adjusting the data for differences among species in life span, body size, metabolic rates, and other characteristics. Without actual human data, extrapolation provides no guarantee that there are no unknown factors also at work. It is not surprising that there is no clear consensus as to how to extrapolate risk estimates from one species to another. This problem is not unique to radiation effects; there are countless examples of chemicals having very different effects in different species, and humans can differ quite significantly from animals in their reaction to toxic agents.

How have human data been obtained?

There are serious ethical issues with conducting experiments on humans, as discussed elsewhere in the report. However, most of the human data that are used to estimate risks, not just risk from radiation, come from *epidemiologic studies* on populations that already have been exposed in various ways. For radiation effects, the most important human data come from studies of the Japanese atomic bomb survivors carried out by the Radiation Effects Research Foundation (formerly the Atomic Bomb Casualty Commission) in Hiroshima. Other valuable sources of data include various groups of medically exposed patients (such as radiotherapy patients) and occupationally exposed workers (such as the uranium miners, discussed in chapter 12).[108]

Why is it necessary to compare exposed populations with unexposed populations?

Unlike a disease caused by identifiable bacteria, no "signature" has yet been found in cancerous tissue that would link it definitively to prior radiation exposure. Radiogenic cancers are identical in properties, such as appearance under a microscope, growth rate, and potential to metastasize, to cancers occurring in the general population. Finding cancers in an exposed population is not enough to prove they are due to radiation; the same number of cancers might have occurred due to the natural frequency of the disease. The challenge is to separate out the effects of radiation from what would otherwise have occurred. A major step in this direction is to develop follow-up (or *cohort*) studies, in which an exposed group is followed over time to observe their disease rates, and these rates are then compared with the rates for the general population or an unexposed control group.[109]

Why is the analysis of epidemiologic data so complicated?

Simply collecting data on disease rates in exposed and control populations is not enough; indeed, casual analysis may lead to serious errors in understanding. Sophisticated data-collection techniques and mathematical models are needed to develop useful risk estimates for several reasons:

1. Random variation due to sample size.
2. Multiple variables.
3. Limited time span of most studies.
4. Problems of extrapolation.

In addition, individual studies may also be *biased* in their design or implementation.

What is random variation?

The observed proportion of subjects developing disease in any randomly selected subgroup (sample) of individuals with similar exposures is subject to the vagaries of *random variation.*

A simple-minded example of this is the classic puzzle of determining, in a drawer of 100 socks, how many are white and how many are black, by pulling out one sock at a time. Obviously, if we pull out all the socks, we know for certain. In most areas of study, though, "pulling out all the socks" is far too expensive and time-consuming. But if we pull only 10, with what degree of confidence can we predict the color of the others? If we pull 20, we will have more confidence. In other words, *the larger the sample, the greater our confidence.* Using statistical techniques, our degree of confidence can be calculated from the size of the entire population (in this case 100 socks) and the size of the actual sample. The result is popularly called the *margin of error.*

The most common examples of this in everyday life are the public opinion polls continually quoted in the news media. As can be seen in the simple example of the drawer of socks, the highest degree of confidence can be achieved simply by pulling all the socks out of the drawer. For public opinion polls, this would be far too expensive; instead, a small sample is selected at random from the population. Nowadays it is common to report not only the actual results, but also the sample size and the *margin of error.* The margin of error depends not only on the sample size, but also on how high a *degree of confidence* we desire. The degree of confidence is the probability that our sample has provided a true picture of the entire population. For example, the margin of error will be smaller for 80 percent degree of confidence than for 95 percent. Even where a study covers an entire exposed population, such as the atomic bomb survivors, the issue of random variation remains when we wish to generalize the findings to other populations.

What are multiple variables?

The effects of radiation will depend upon, or *vary,* with the dose of radiation received. However, these effects also may vary with other factors—other *variables*—that are not dependent upon the radiation dose itself. Examples of such variables are age, gender, latency (time since exposure), and smoking. Data on these other variables must be collected as well as data on the basic elements of radiation dose and disease. The challenge is to then distinguish between disease rates due to radiation and those due to other factors. For example, if the population studied were all heavy smokers, this might explain in part a higher rate of lung cancer. Much of the science of epidemiology is devoted to choosing what factors to collect data on and then developing the *multivariate mathematical models* needed to separate out the effect of each variable. Radiation effects vary considerably across subgroups and over time or age. Because of this, direct estimates of risk for particular subgroups would be very unstable. Mathematical models must be used. These models allow all the data to be used to develop risk estimates that, while based on sufficiently large estimates to be stable, will be applicable to particular subgroups.

A more subtle problem is *misspecification* of the model finally chosen to calculate risks. The model may weigh selected factors in a manner that best fits the data from a statistical viewpoint. This model, while fitting the data, may not actually be a "correct" view of nature; another model that does not fit the data quite as well may actually better describe the as-yet-unknown underlying mechanisms.

Why does a limited time span reduce the value of a study?

The most pronounced effects of large exposures to radiation manifest themselves quickly in symptoms loosely termed *radiation sickness*.

However, another concern is understanding the effects of much lower levels of radiation. Unlike the more acute effects of large exposures, these may not appear for some time. Some cancers, for example, do not appear until many years after the initial exposure. These *latent effects* may continue to appear in a population throughout their entire lifetimes. Calculating the *lifetime risk* of an exposure requires following the entire sample until all its members have died. Thus far, none of the exposed populations have yet been followed to the ends of their lives, although the radium dial painter study for the group painting before 1930 essentially has been completed, and the follow-up has been closed out.[110]

Why does extrapolation among human populations pose problems?

As discussed earlier, extrapolating results from one species to another is problematic due to differences in how species respond to radiation.

Even though humans are all members of the same species, there are similar problems when extrapolating results from one group of humans to another group. Within the human species, different groups can have different rates of disease. For example, stomach cancer is much more common and breast cancer much rarer among Japanese than among U.S. residents.

How then should estimates of the radiation-induced excess of cancer among the atomic bomb survivors be applied to the U.S. population? Assumptions are needed to "transport" risk estimates from one human population to another human population that may have very different "normal" risks.

Why does extrapolation from high to low doses pose problems?

Acquiring high-quality human data on low-dose exposure is difficult. Past studies indicate that the effects of low doses are small enough to be lost in the "noise" of random variation. In other words, the *random variation* due to sample size may be greater than the effects of the radiation. Thus, to estimate the risks of low doses, it is necessary to *extrapolate* from the effects of high doses down to the lower range of interest. As with extrapolation among species or among human populations, assumptions must be made.

The basic assumption concerns the *dose effect*. Is the effect of a dose *linear*? This would mean that half the dose would produce half the effect; one-tenth of the dose would produce one-tenth of the effect, and so forth. Nature is not always so reasonable, however. There are many instances in nature of *nonlinear* relationships. A nonlinear dose effect, for example, could mean that half the dose would produce 75 percent of the effects. Or, going in the other direction, a nonlinear dose effect could mean that half the dose would produce only 10 percent of the effect. Reliable data are too sparse to settle the issue empirically. Much of the ongoing controversy over low-dose effects concerns which *dose effect* relationship to assume. Concerning dose response, most radiation advisory committees assume that radiation risks are linear in doses at low levels, although these risks may involve non-linear terms at higher doses.

Another assumption concerns the effect of *dose rate*. It is generally agreed that the effect of high-dose x rays is reduced if the radiation is received over a period of time instead of all at once. (This reduction in acute effects, due to the cell's ability to repair itself in between exposures, is one of the reasons that modern protocols for radiotherapy use several fractionated doses.) The degree to which this also happens at low doses is less clear. There are few human data on the effect of dose rate on cancer induction. Most estimates of the effect come from animal or cell culture experiments. There is also evidence of quite different dose-rate effects for alpha radiation and neutrons.

How can a specific study be biased?

When applied to an epidemiologic study, the term *bias* does not refer to the personal beliefs of the investigators, but to aspects of the study design and implementation. There are several possible sources of bias in any study.

What is called a *confounding bias* may result if factors other than radiation have affected disease rates. Such factors, as mentioned earlier, might be a rate of smoking higher than the general population.

A *selection bias* may result if the sample was not truly a random selection from the population under study. For example, the results of a study that includes only employed subjects might not be applicable to the general population, since employed people as a group are healthier than the entire population.

An *information bias* may result from unreliability in a source of basic data. For example, basing the amount of exposure on the memory of the subjects may bias the study, since sick people may recall differently than healthy people. Dose, in particular, can be difficult to determine when studies are conducted on populations exposed prior to the study, since there usually was no accurate measurement at the time of exposure. Sometimes when dose measurements were taken, as in the case of the atomic veterans, the data are not adequate by today's standards.[111]

Finally, any study is subject to the *random variation* discussed earlier, which depends on how large the sample is. This is more important for low-dose than for high-dose studies, since the low-dose effects themselves are small enough to be lost amid random variations if the sample is too small.

To summarize, multiple studies may produce somewhat different results because there is an actual difference in the response between populations or because studies contain spurious results due to their own inadequacies. In addition, it must be recognized that the entire body of scientific literature is itself subject to a form of bias known as *publication bias*, meaning an overreporting of findings of excess risk. This is because studies that demonstrate an excess risk may be more likely to be published than those that do not.

In view of all these uncertainties, what risk estimates did the committee choose?

Despite all these uncertainties, it must be pointed out that more is known about the effects of ionizing radiation than any other carcinogen.

The BEIR V Committee of the National Academy of Sciences estimated in 1990 that the lifetime risk from a single exposure to 10 rem of whole-body external radiation was about 8 excess cancers (of any type) per 1,000 people. (This number is actually an average over all possible ages at which an individual might be exposed, weighted by population and age distribution.) For continuous exposure to 0.1 rem per year throughout a lifetime, the corresponding estimate was 5.6 excess cancers (that is, over and above the rate expected in a similar, but nonexposed population) per 1,000 people. It is widely agreed that for x rays and gamma rays, this latter figure should be reduced by some factor to allow for a cell's ability to repair DNA, but there is considerable uncertainty as to what figure to use; a figure of about 2 or 3 is often suggested.[112]

The estimates of lifetime risk from the BEIR V report have a range of uncertainty due to *ran-*

dom variation of about 1.4-fold. The additional uncertainties, due to the factors discussed earlier, are likely to be larger than the random variation.

In comparison, for most chemical carcinogens, the uncertainties are often a factor of 10 or more. This agreement among studies of radiation effects is quite remarkable and reflects the enormous amount of scientific research that has been devoted to the subject, as well as the large number of people who have been exposed to doses large enough to show effects.

NOTES

1. In 1974 the AEC's regulatory activities for civilian nuclear power and the use (including medical research) of radioisotopes produced in nuclear reactors were transferred to the Nuclear Regulatory Commission and its research and weapons-development activities to the Energy Research and Development Administration (ERDA). In 1977 ERDA was incorporated into the new Department of Energy.

2. Captain C. F. Behrens, ed., *Atomic Medicine* (New York: Thomas Nelson and Sons, 1949), 3.

3. Ibid., 7.

4. Otto Glasser, *William Conrad Roentgen and the Early History of the Roentgen Rays* (Springfield, Ill., and Baltimore: Charles C. Thomas, 1934), 29; Glasser is quoting O. Lummer of Berlin.

5. Ibid., 271.

6. Ibid., 244–282.

7. Robert Reid, *Marie Curie* (New York: E. P. Dutton, 1974), 241.

8. Ibid., 86–87.

9. P. Curie and M. S. Curie, "Radium: A New Body, Strongly Radio-Active, Contained in Pitchblende," *Scientific American* (28 January 1899): 60. The term *hyperphosphorescence* was suggested by Silvanus Thompson. Reid, *Marie Curie*, 87. See also Susan Quinn, *Marie Curie: A Life* (New York: Simon and Schuster, 1995).

10. *New York Journal*, 21 June 1905, reproduced in David J. DiSantis, M.D., and Denise M. DiSantis, "Radiologic History Exhibit: Wrong Turns on Radiology's Road of Progress," *Radiographics* (1991): 1121–1138, figure 17.

11. Henry S. Kaplan, "Historic Milestones in Radiobiology and Radiation Therapy," *Seminars in Oncology* 6, no. 4 (December 1979): 480.

12. "Autopsy of a Radiologist," *Archives of the Roentgen Ray* 18 (April 1914): 393.

13. Reid, *Marie Curie*, 274; Barton C. Hacker, *The Dragon's Tail: Radiation Safety in the Manhattan Project, 1942-1946* (Berkeley, Calif.: University of California Press, 1987), 22–23.

14. The marketing of one nostrum containing radium, Radiothor, was not officially shut down by the Federal Trade Commission until 1932. "With the institution of regulations, the radioactive patent medicine industry collapsed overnight." Roger M. Macklis, "The Great Radium Scandal," *Scientific American* 269 (March 1993): 94–99. In the 1920s, the use of capsules containing radium inserted into the nose was introduced as a means of shrinking lymphoid tissue in children to treat middle ear obstructions and infections. During World War II this procedure was used on submariners and Air Force personnel as treatment and, in the case of several hundred submariners, on an experimental basis to test the effectiveness of nasopharyngeal irradiation in shrinking lymphoid tissue and equalizing external and middle ear pressure. In the late 1940s, the observation that no controlled study had ever been conducted to test the treatment's effectiveness in preventing deafness in children led Johns Hopkins researchers to begin the experimental treatment of several hundred children. As the Advisory Committee began its work in 1994, controversy over the long-term effects of this treatment still swirled. Samuel Crane, "Irradiation of Nasopharynx," *Annals of Otology, Rhinology, and Laryngology* 55 (1946): 779–788; H. L. Holmes and J. D. Harris, "Aerotitis Media in Submariners," *Annals of Otology, Rhinology, and Laryngology* 55 (1946): 347–371. See chapter 7 and ACHRE Briefing Book, vol. 13, tab E, April 1995, for fuller discussion.

15. Macklis, "The Great Radium Scandal," 94–99.

16. The National Council on Radiation Protection began as the American Committee on X Ray and Radium Protection in 1928, under the aegis of the International Congress of Radiology. A private organization, its members were physicians, physicists, and representatives of the equipment manufacturers. Prior to World War II its main function was to issue recommendations on radiological safety, which were published by the National Bureau of Standards (a federal agency). At times this arrangement created confusion, leading people to believe the publications were official recommendations. After the war, the private group was revived as the National Committee on Radiation Protection. In 1956 it was renamed the National Committee on Radiation Protection and Measurements. In the early 1960s, it received a congressional charter and was renamed the National Council on Radiation Protection and Measurements. Throughout its history it has coordinated its activities with other groups, such as the International Commission on Radiological Protection and committees of the National Academy of Sciences (known as the BEAR and BEIR Committees). The most complete record of the NCRP's activities is Lauriston S. Taylor, *Organization for Radiation Protection: The Operations of the ICRP and NCRP, 1928–1974* (Washing-

ton, D.C.: Office of Technical Information, U.S. Department of Energy.) Lauriston Taylor, a physicist at the National Bureau of Standards, served as the executive director of the organization from its founding in 1928 to 1974. For further background on the history of radiation protection, see Daniel P. Serwer, *The Rise of Radiation Protection: Science, Medicine and Technology in Society, 1896-1935* (Ph.D. diss. in the History of Science, Princeton University, 1976) (Ann Arbor: University Microfilms 77-14242, 1977); Gilbert F. Whittemore, *The National Committee on Radiation Protection, 1928–1960: From Professional Guidelines to Government Regulation* (Ph.D. diss. in the History of Science, Harvard University, 1986) (Ann Arbor: University Microfilms 87-04465, 1987); J. Samuel Walker, "The Controversy Over Radiation Safety: A Historical Overview," *Journal of the American Medical Association* 262 (1989): 664-668; D. C. Kocher, "Perspective on the Historical Development of Radiation Standards," *Health Physics* 61, no. 4 (October 1994).

17. Heinz Haber, *The Walt Disney Story of Our Friend the Atom* (New York: Simon and Schuster, 1956), 152. The German-born Dr. Haber had come to the United States in 1947 to work for the Air Force School of Aviation Medicine and was a cofounder of the field of space medicine. In the early 1950s he joined the faculty of the University of California at Los Angeles. As Spencer Weart, a historian of the images of the atomic age has recorded, the accompanying Walt Disney movie *Our Friend the Atom*, which was shown on television and in schools beginning in 1957, was probably the most effective of educational films on the perils and potential of atomic energy. "The great storyteller introduced the subject as something 'like a fairy tale,' indeed the tale of a genie released from a bottle. The cartoon genie began as a menacing giant. . . . But scientists turned the golem into an obedient servant, who wielded the 'magic power' of radioactivity. . . ." Spencer R. Weart, *Nuclear Fear: A History of Images* (Cambridge, Mass.: Harvard University Press, 1988), 169.

18. Marshall Brucer, *Chronology of Nuclear Medicine* (St. Louis: Heritage Publications, 1990), 199–200. Radon is a gas at room temperature. Doctors developed an innovative system for capturing radon from used cancer therapy vials and dissolving it in a saline solution, which was then injected.

19. Haber, *Our Friend the Atom*, 152.

20. J. L. Heilbron and Robert W. Seidel, *Lawrence and His Laboratory: A History of the Lawrence Berkeley Laboratory*, vol. 1 (Los Angeles: University of California Press, 1989). The birth and development of nuclear medicine at the University of California's Berkeley and San Francisco branches is the subject of a case study in a supplemental volume to this report.

21. John Stanbury, *A Constant Ferment* (Ipswich, N.Y.: Ipswich Press, 1991), 57-67.

22. Stafford Warren, interview by Adelaide Tusler (Los Angeles: University of California), 23 June 1966 in *An Exceptional Man for Exceptional Challenges, Vol. 2* (Los Angeles: University of California, 1983) (ACHRE No. UCLA-101794-A-1), 421–422.

23. Manhattan Project researchers focused on polonium in the development of the initiator for the bomb. See Richard Rhodes, *The Making of the Atomic Bomb* (New York: Simon and Schuster, 1986), 578–580.

24. Manhattan District Program, 31 December 1946 (book 1, "General," volume 7, "Medical Program") (ACHRE No. NARA-052495-A-1), 2.2.

25. Stafford Warren in *Radiology in World War II*, ed. Arnold Lorentz Ahnfeldt (Washington, D.C.: GPO, 1966), 847.

26. Robert Stone, 10 May 1943 ("Health Radiation and Protection") (ACHRE No. DOE-011195-B-1).

27. Philip J. Close, Second Lieutenant, JAGD, to Major C. A. Taney, Jr., 26 July 1945 ("Determination of Policy on Cases of Exposure to Occupational Disease") (ACHRE No. DOE-120894-E-96), 1.

28. Ibid., 3.

29. Response to ACHRE Request No. 012795-B, Oak Ridge Associated Universities, D. M. Robie to A. ("Tony") P. Polendak, 15 June 1979 ("Storage of records—Shipment 1161").

30. The story of this early Hanford research is told in Neal D. Hines, *Proving Ground: An Account of the Radiobiological Studies in the Pacific 1946–61* (Seattle: University of Washington Press, 1962). As Hines explains, the initial study of the effect of radioactivity on aquatic organisms was undertaken by a University of Washington researcher. The program could not be identified with the Columbia River, and the research was to be conducted in a normal campus setting. The project's name ("Applied Fisheries Laboratory") was selected to disguise its work. The primary researcher initially did not know the true purpose, and the university accepted the work for undisclosed purpose on the assurance that national security required it.

31. Harold Hodge, interview by J. Newell Stannard, transcript of audio recording, 22 October 1980 (ACHRE No. DOE-061794-A-4), 21–22. Stafford Warren, interview by J. Newell Stannard, transcript of audio recording, 7 February 1979 (ACHRE No. DOE-061794-A), 3.

32. Henry DeWolf Smyth, *Atomic Energy for Military Purposes: The Official Report on the Development of the Atomic Bomb under the Auspices of the United States Government, 1940-45* (Princeton, N.J.: Princeton University Press, 1945).

33. The organizational history of the Department of Defense is chronicled in *The Department of Defense: Documents on Establishment and Organization 1944-1978*, eds. Alice C. Cole, Alfred Goldberg, Samuel A. Tucker, Rudolf A. Winnacker (Washington, D.C.: Office of the Secretary of Defense, Historical Office, 1978).

34. The program expanded from the base of Manhattan Project research sites such as Oak Ridge, Hanford, Chicago, and the Universities of California, Chicago, and Rochester to take in a growing portion of the university research establishment. The minutes of the January 1947 meeting record an ambitious program to focus on the physical measurement of radiation, the biological effects of radiation, methods for the detection of radiation damage, methods for the prevention of radiation injury, and protective measures. There followed an itemized list of the work to be done at Argonne National Laboratory, Los Alamos, Monsanto, Columbia University, and the Universities of Michigan, Rochester, Tennessee, California, and Virginia.

The University of Rochester was to be the largest university contractor, receiving more than $1 million, followed by the University of California (about one-half million for UCLA, where Stafford Warren was dean of the new medical school, and Berkeley, to which Stone had returned to join Hamilton), Western Reserve (to which Warren's deputy Hymer Friedell was headed), and Columbia (more than $100,000). Argonne received an amount comparable to Rochester; other labs, including Los Alamos National Laboratory and Clinton Laboratories (now Oak Ridge National Laboratory), were scheduled for $200,000 or less. Stafford Warren, Interim Medical Committee, proceedings of 23-24 January 1947 (ACHRE No. UCLA-111094-A-26). See also ACHRE Briefing Book, vol. 3, tab F, document H.

35. "Report of the Board of Review," 20 June 1947, attached to letter from David Lilienthal, Chairman, AEC, to Dr. Robert F. Loeb, Chairman, AEC Medical Board of Review, 27 June 1947 ("At the conclusion of the deliberations . . .") (ACHRE No. DOE-051094-A-191), 3-4.

36. The Advisory Committee has assembled the minutes of the meetings, and such transcripts as have been retrieved.

37. Shields Warren, interview by Dr. Peter Olch, National Library of Medicine, transcript of audio recording, 10-11 October 1972, 59.

38. Harry H. Davis, "The Atom Goes to Work for Medicine," *New York Times Magazine*, 26 September 1946 (ACHRE No. DOE-051094-A-408).

39. Marshall Brucer, M.D., Chairman, Medical Division, Oak Ridge Institute for Nuclear Studies, wrote:

Paul Aebersold's isotopes division was the only safely nonsecret part of AEC. Aebersold had unlimited funds, unlimited radioisotopes, and seemingly unlimited energy to promote the unlimited cures that had been held back from the American public for too long. The liberal establishment was in the depths of shame for having ended the war by killing people. Radioisotopes didn't kill people; they cured cancer.

Aebersold spoke at every meeting of one person or more that had one minute or more available on its program.

No matter what the meeting's subject, Aebersold's topic was always the same. He sold isotopes.

Marshall Brucer, "Nuclear Medicine Begins with a Boa Constrictor," *Journal of Nuclear Medicine* 19, no. 6 (1978): 595.

40. Isotopes Division, prepared for discussion with general manager, "Present and Future Scope of Isotope Distribution," 4 March 1949 (ACHRE No. DOE-011895-B-1).

41. Interview with Shields Warren, 10-11 October 1972, 76.

42. Isotopes Division, 4 March 1949, 2.

43. See Kaplan, "Historic Milestones," 480.

44. "Summary of Congressional Hearings on Fellowship Issue," 16 May 1949 (ACHRE No. DOE-061395-D-1).

45. Advisory Committee for Biology and Medicine, proceedings of 10 September 1949 (ACHRE No. DOE-072694-A), 18.

46. Ibid, 19.

47. For a further discussion of the contemplated secret record keeping by the VA, see chapter 10. As noted there, a VA investigation concluded that the "confidential" division was never activated.

48. The VA provided the Advisory Committee with capsule descriptions of experiments, which appeared in periodic VA reports of the time. In fact, the number of descriptions exceeded 3,000 for the portion of the 1944–74 period the reports covered. However, further information on the vast majority of the experiments was typically unavailable, and the VA noted that some of the descriptions may be redundant (or reflect refunding of a single experiment), and some may not have involved humans. Therefore, the "more than 2,000" reflects a very rough estimate adjusted for these considerations.

49. Paul C. Aebersold, address before Rocky Mountain Radiological Society, 9 August 1951 ("The United States Atomic Energy Program: Part I—Overall Progress") (ACHRE No. TEX-101294-A-1), 6.

50. By 1955 the program was receiving 8,000 applications a year, including hundreds from abroad. A July 1955 Aebersold summary of accomplishments pronounced that, as a result of the program, there were now 100 companies in the radiation instrument business, two dozen suppliers of commercially labeled compounds, pharmaceutical companies, hundreds of isotope specialists, a half-dozen waste disposal firms, and ten safety monitoring companies. Also, 2,693 U.S. institutions had received isotope authorization, including 1,126 industrial firms, 1,019 hospitals and private physicians, 220 colleges and universities, 244 federal and state laboratories, and 47 foundations. "Capsule Summary of Isotopes Distribution Program," July 1955 (ACHRE No. TEX-101294-A-2).

51. Vannevar Bush, *Pieces of the Action* (New York: William Morrow and Company, 1970), 65.

52. Ibid.

53. In addition to direct grants to private institutions the AEC pioneered the creation of research con-

sortia. In 1946, for example, the University of Tennessee and a consortium of southeastern universities urged the Manhattan Project to establish the Oak Ridge Institute of Nuclear Studies (ORINS). Following the creation of the AEC, ORINS operated under AEC contract to train researchers and to operate a clinical research facility focused on cancer. In 1966 ORINS became known by the name of its operating contractor, the Oak Ridge Associated Universities, and the research facility is now known as the Oak Ridge Institute for Science and Education (ORISE).

54. Donald C. Swain, "The Rise of a Research Empire: NIH, 1930 to 1950," *Science* 138, no. 3546 (14 December 1962): 1235. The National Institutes of Health began as the Laboratory of Hygiene in 1887. It was renamed the National Institutes of Health in 1948.

55. Assistant Director, Office of Extramural Research, National Institutes of Health, to Anna Mastroianni, Advisory Committee, 16 July 1995 ("Comments on Draft Chapters of ACHRE Final Report").

56. Interview with Shields Warren, 10-11 October 1972, 78.

57. Robert S. Stone, M.D., to Lieutenant Colonel H. L. Friedell, U.S. Engineer Corps, Manhattan District, 9 August 1945 ("In reading through the releases . . .") (ACHRE No. DOE-121494-D-2).

58. Robert S. Stone, M.D., to Lieutenant Colonel H. L. Friedell, U.S. Engineer Corps, Manhattan District, 9 August 1945 ("As you and many others are aware, a great many of the people . . .") (ACHRE No. DOE-121494-D-1).

59. Jonathan M. Weisgall, *Operation Crossroads: The Atomic Tests at Bikini Atoll* (Annapolis, Md.: Naval Institute Press, 1994). For a contemporary account by a doctor who served as a radiation monitor, see David Bradley, *No Place to Hide* (Boston: Little, Brown and Co., 1948).

60. Weisgall, *Operation Crossroads*, 266-270.

61. "History of the U.S. Naval Radiological Defense Laboratory, 1946-58" (ACHRE No. DOD-071494-A-1), 1.

62. The Joint Panel was the child of the Committee on Medical Science and Committee on Atomic Energy (hence the term *Joint*), both of which, in turn, were committees of the Defense Department's Research and Development Board. That board served as the secretary of defense-level coordinator of departmentwide R&D.

63. The Committee has assembled the charter, agenda, reports, and available minutes of the Joint Panel. ACHRE Research Collection Series, Library File, Compilation of the Minutes of the Joint Panel on Medical Aspects of Atomic Warfare, 1948-1953 (1994).

64. Howard Andrews, interview by Gilbert Whittemore (ACHRE staff), transcript of audio recording, 3 December 1994 (ACHRE Reseach Project, Interview Series, Targeted Interview Project).

65. In a February 1950 paper, the Public Health Service explained its role in national defense:

During and since WW II, science and technology have introduced new weapons and whole new industries whose effects on human health have not been precisely determined and effective methods against these hazards have not yet been developed.

If, for example, an atomic bomb were to burst over a large city in this country, tens of thousands of burned and injured people could not be given effective treatment because science has not yet found the practical means. . . . The operation of atomic piles and related facilities also presents a variety of problems as to human tolerance of radiation and the disposition of radioactive substances.

"The U.S. Public Health Service and National Defense," February 1950 (ACHRE No. HHS-071394-A-2), 1.

66. National Institutes of Health, 2 August 1952 ("Assumptions Underlying NIH Defense Planning") (ACHRE No. HHS-071394-A-1).

67. Advisory Committee for Biology and Medicine, transcript (partial) of proceedings of 10 November 1950 (ACHRE No. DOE-012795-C-1). While the document is undated, discussion of the meeting appears in the November 1950 ACBM minutes (12); a letter from Alan Gregg, Chairman, ACBM, to Gordon Dean, Chairman, AEC, 30 November 1950 ("The Advisory Committee for Biology and Medicine held their twenty-fourth . . .") (ACHRE No. DOE-072694-A); and a letter from Marion W. Boyer, AEC General Manager, to Honorable Robert LeBaron, Chairman, Military Liaison Committee, 10 January 1951 ("As you know, one of the important problems . . .") (ACHRE No. DOE-040395-A-1).

68. Behrens, transcript, proceedings of 10 November 1950, 2.

69. Powell, transcript, proceedings of 10 November 1950, 8-10.

70. Cooney, transcript, proceedings of 10 November 1950, 6.

71. Ibid., 7.

72. Ibid., 6.

73. Ibid., 7-8.

74. Warren, transcript, proceedings of 10 November 1950, 13.

75. Ibid., 14.

76. Ibid., 15.

77. Cooney, transcript, proceedings of 10 November 1950, 15.

78. Ibid., 16.

79. "Notes on the Meeting of a Committee to Consider the Feasibility and Conditions for a Preliminary Radiological Safety Shot for Operation 'Windsquall,'" 21–22 May 1951 (ACHRE No. DOE-030195-A-1), 41.

80. Ibid., 40.

81. Ibid., 19.

82. T. L. Shipman, Health Division Leader, to Alvin Graves, J-Division Leader, 27 December 1951 ("Summary Report Rad Safe and Health Activities at Buster-Jangle") (ACHRE No. DOE-033195-B-1).

83. [AEC] Board of Review to the Atomic Energy Commission, 20 June 1947 ("Report of the Board of Review") (ACHRE No. DOE-071494-A-4), 10.

84. NEPA Medical Advisory Panel, Subcommittee No. IX, "An Evaluation of Psychological Problem of Crew Selection Relative to the Special Hazards of Irradiation Exposure," 22 July 1949 (ACHRE No. DOD-121494-A-2), 20.

85. Ibid., 27.

86. Ibid., 22.

87. Definition of "curie," *The Compact Edition of the Oxford English Dictionary* (Oxford, England: Oxford University Press, 1971), 3937.

88. J. Newell Stannard, *Radioactivity and Health: A History* (Oak Ridge, Tenn.: Office of Scientific and Technical Information, 1988), 9.

89. Hanson Blatz, ed., *Radiation Hygiene Handbook* (New York: McGraw-Hill Book Co., 1959), 6–185.

90. Richard G. Hewlett and Oscar E. Anderson, *The New World: A History of the Atomic Energy Commission, Vol. I: 1939–1946* (Berkeley: University of California Press, 1990), reprint of 1962 edition, 107–108.

91. Eric Hall, *Radiobiology for the Radiologist*, 4th ed. (Philadelphia: J. B. Lippincott, 1994), 3.

92. Ibid., 5.

93. The DNA strand would be about 5 centimeters (cm) long; the average cell diameter is about 20 microns (0.002 cm). Bruce Alberts et al., eds., *Molecular Biology of the Cell* (New York: Garland, 1983), 385-388.

94. Hall, *Radiobiology for the Radiologist*, 4th ed., 9–10.

95. Ibid.

96. Ibid., 30.

97. Ibid., 32–33.

98. Ibid., 312.

99. Ibid., 324.

100. Ibid.

101. International Commission on Radiological Protection, *Recommendations: ICRP Publication No. 60* (New York: Pergamon Press, 1991), cited in Hall, *Radiobiology for the Radiologist*, 4th ed., 456.

102. Hall, *Radiobiology for the Radiobiologist*, 4th ed., 355.

103. Committee on the Biological Effects of Ionizing Radiation, National Research Council, *Health Effects of Exposure to Low Levels of Ionizing Radiation: BEIR V* (Washington, D.C.: National Academy Press, 1990), 2–4.

104. International Commission on Radiological Protection, *Recommendations: ICRP Publication No. 60*, quoted in Hall, *Radiobiology for the Radiologist*, 4th ed., 455.

105. ". . . *roentgen equivalent man, or mammal (rem)*. The dose of any ionizing radiation that will produce the same biological effect as that produced by one roentgen of high-voltage x-radiation." Blatz, ed., *Radiation Hygiene Handbook*, 2–19.

106. Hall, *Radiobiology for the Radiobiologist*, 4th ed., 458.

107. These include the National Council on Radiation Protection and Measurement (NCRP), the International Commission on Radiation Protection (ICRP), the United Nations Scientific Committee on the Effects of Atomic Radiation (UNSCEAR), the Committee on the Biological Effects of Ionizing Radiation (BEIR) of the National Research Council, and the Environmental Protection Agency (EPA).

108. In addition, there have been a number of studies of people exposed to low levels of radiation, including military personnel and residents subject to fallout from nuclear weapons testing, workers at and residents near nuclear facilities, patients given diagnostic x rays, and regions with high natural background radiation. Most of these either have not produced convincing positive results or are unsuitable for risk assessment because of the statistical instability of their estimates.

109. Some indirect estimates have been based on "case control" studies, in which diseased cases are compared with unaffected controls to look for differences in their past exposures that could account for their different outcomes.

General reference works include D. G. Kleinbaum, W. Kupper, and H. Morgenstern, *Epidemiologic Research: Principles and Quantitative Methods* (Belmont, Calif.: Lifetime Learning Publications, 1982), and J. D. Boice, Jr., and J. E. Fraumeni, Jr., *Radiation Carcinogenesis: Epidemiology and Biological Significance* (New York: Raven Press, 1984).

110. Dr. Shirley Fry to Bill LeFurgy, 31 August 1995 ("HRE Draft Final Report"), 8, contained in Ellyn Weiss, Special Counsel and Director, Office of Human Radiation Experiments, DOE, to Anna Mastroianni, ACHRE, 11 September 1995.

111. "[T]he NTPR dose data are not suitable for dose-response analysis." Institute of Medicine, "A Review of the Dosimetry Data Available in the Nuclear Test Personnel Review (NTPR) Program, An Interim Letter Report of the Committee to Study the Mortality of Military Personnel Present at Atmospheric Tests of Nuclear Weapons to the Defense Nuclear Agency" (Washington, D.C.: Institute of Medicine, May 15, 1995), 2.

112. Committee on the Biological Effects of Ionizing Radiation, National Research Council, *Health Effects of Exposure to Low Levels of Ionizing Radiation: BEIR V* (Washington, D.C.: National Academy Press, 1990), 22, 162.

Part I

Ethics of Human Subjects Research: A Historical Perspective

PART I
OVERVIEW

WHEN the Advisory Committee began work in April 1994 we were charged with determining whether "the [radiation] experiments' design and administration adequately met the ethical and scientific standards, including standards of informed consent, that prevailed at the time of the experiments and that exist today" and also to "determine the ethical and scientific standards and criteria by which it shall evaluate human radiation experiments."

Although this charge seems straightforward, it is in fact difficult to determine what the appropriate standards should be for evaluating the conduct and policies of thirty or fifty years ago. First, we needed to determine the extent to which the standards of that time are similar to the standards of today. To the extent that there were differences we needed to determine the relative roles of each in making moral evaluations.

In chapter 1 we report what we have been able to reconstruct about government rules and policies in the 1940s and 1950s regarding human experiments. We focus primarily on the Atomic Energy Commission and the Department of Defense, because their history with respect to human subjects research policy is less well known than that of the Department of Health, Education, and Welfare (now the Department of Health and Human Services). Drawing on records that were previously obscure, or only recently declassified, we reveal the perhaps surprising finding that officials and experts in the highest reaches of the AEC and DOD discussed requirements for human experiments in the first years of the Cold War. We also briefly discuss the research policies of DHEW and the Veterans Administration during these years.

In chapter 2 we turn from a consideration of government standards to an exploration of the norms and practices of physicians and medical scientists who conducted research with human subjects during this period. We include here an analysis of the significance of the Nuremberg Code, which arose out of the international war crimes trial of German physicians in 1947. Using the results of our Ethics Oral History Project, and other sources, we also examine how scientists of the time viewed their moral responsibilities to human subjects as well as how this translated into the manner in which they conducted their research. Of particular interest are the differences in professional norms and practices between research in which patients are used as subjects and research involving so-called healthy volunteers.

In chapter 3 we return to the question of government standards, focusing now on the 1960s and 1970s. In the first part of this chapter, we review the well-documented developments that influenced and led up to two landmark events in the history of government policy on research involving human subjects: the promulgation by DHEW of comprehensive regulations for oversight of human subjects research and passage by Congress of the National Research Act. In the latter part of the chapter we review developments and policies governing human research in agencies other than DHEW, a history that has received comparatively little scholarly attention. We also discuss scandals in human research conducted by the DOD and the CIA that came to light in the 1970s and that influenced subsequent agency policies.

With the historical context established in chapters 1 through 3, we turn in chapter 4 to the core of our charge. Here we put forward and defend three kinds of ethical standards for evaluating human radiation experiments conducted from 1944 to 1974. We embed these standards in a moral framework intended to clarify and facilitate the difficult task of making judgments about the past.

1

Government Standards for Human Experiments: The 1940s and 1950s

WHEN the Advisory Committee began its work, a central task was the reconstruction of the federal government's rules and policies on human experiments from 1944 through 1974. The history of research rules at the Department of Health, Education, and Welfare (DHEW) was well known, at least from 1953 on, when DHEW's National Institutes of Health (NIH) adopted a policy on human subjects research for its newly opened research hospital, the Clinical Center. In the 1960s, the DHEW and some other executive branch agencies undertook regulation of research involving human subjects. These were early steps of a process that culminated, in 1991, in the comprehensive federal policy known as the "Common Rule."[1] The historical background of this process, including a well-publicized series of incidents and scandals that motivated it, was also widely known and much discussed (see chapter 3).[2]

By contrast to DHEW, much less was known about the history of research rules for other agencies also involved in research with human subjects during this period, including the Department of Defense (DOD), the Atomic Energy Commission (AEC), and the Veterans Administration (VA). From the perspective of the charge to the Advisory Committee, these agencies were at least as important as DHEW. It was known that in 1953 the secretary of defense is-

sued, in Top Secret, a memorandum on human subjects based on the Nuremberg Code.[3] In 1947 an international tribunal had declared the Nuremberg Code the standard by which a group of doctors in Nazi Germany should be judged for their horrific wartime experiments on concentration camp inmates. However, the actual impact of the Nuremberg Code on the biomedical community in the United States, both inside and outside of government, is a matter of some disagreement (see chapter 2). The general view was that, despite some developments in the 1940s and 1950s, there was little activity within the federal government on issues of human subjects research before the 1960s.

But while scholars have known of the 1953 secretary of defense memorandum, which was declassified in 1975, other relevant Department of Defense documents remained classified or had lain buried in archives. Moreover, relevant records of the Atomic Energy Commission were largely unexplored and in some cases still classified. These records are important because, from its creation in 1947, the AEC distributed radioisotopes that would be used in thousands of human radiation experiments, and it was a funding source for many other experiments (see Introduction). Along with the DOD, also created in 1947, the AEC was searching for biomedical information needed to understand the effects of radiation as it prepared for the possibility of atomic warfare. Although the AEC was thus the catalyst for a considerable amount of human experimentation after World War II, there has been literally no scholarship on the AEC's position on the use of human beings in radiation-related research.

Now that previously obscure, even classified, records are being made public, it appears that in the first years of the Cold War, officials and experts in the AEC and DOD did discuss the requirements for human experiments. In this chapter we tell what we have learned about those discussions.

We begin by telling the story of the AEC general manager's early declarations on human research, which included a requirement that consent be obtained from patient-subjects. This story requires a careful look at a series of letters and memorandums exchanged in the late 1940s. Together these documents paint a clearly important but nonetheless confusing picture of a new agency's attempts to come to grips with the complexities of human experimentation. We consider not only what these documents say, but what we can piece together about what they meant in the context of the times. Central questions include the precise scope of the activities covered by the requirements and whether and how these 1947 statements were communicated and put into effect in the AEC's burgeoning contract research and radioisotope distribution programs.

We turn next to the Department of Defense, where we trace the history of rules on the use of healthy "normal volunteer" subjects in military research from the time of Walter Reed through the secretary of defense's 1953 memorandum, and beyond. This memorandum is the earliest known instance in which a federal agency that sponsored human experiments adopted the Nuremberg Code. What is known about how the memorandum was interpreted and implemented by the military establishment takes up much of the rest of this chapter. Here, as in the case of the AEC, key questions concern the scope of the activities covered by requirements and the extent to which they were put into effect.

Finally, we briefly discuss how research involving human subjects was addressed at the National Institutes of Health and the Veterans Administration in the 1950s. The evolution of policies governing human research at DHEW has been well documented and is only summarized here.[4] We now know that NIH's 1953 policy was not the earliest federal requirement that consent be obtained from patients as well as healthy subjects. However, in contrast with the 1940s declarations by the AEC, it was a far more visible statement issued by an agency that was emerging as the leading sponsor of human subjects research. In contrast with what is known about NIH, the extent to which there were research rules at the VA in the 1940s and 1950s remains unclear.

A recurring theme in this chapter is the uncertainty about the significance within government agencies of many of the official statements that are discussed. While these statements emanated from high and responsible officials and committees, often they cannot be linked to fuller expressions of commitment by the agencies. Some of these statements were not widely disseminated, and there were no implementing guidelines or regulations and no sanctions for failures to abide by them. Thus, it is sometimes unclear what formal, legal significance these statements had. We are no less interested, however, in what these statements can tell us about how government officials and advisers saw human research at the time and how they understood the obligations surrounding it.

THE ATOMIC ENERGY COMMISSION: A REQUIREMENT FOR "CONSENT" IS DECLARED AT THE CREATION

Even before the AEC came into existence on January 1, 1947, Manhattan Project researchers and officials had begun to lay the groundwork for the expansion of the government's support of biomedical radiation research conducted under federal contract. By the time the AEC began operations, the parallel program to distribute federally produced radioisotopes to research institutions throughout the country was already well under way.

The planning for these undertakings required both reflection on high-level matters of policy and attention to matters of small but critical legal and bureaucratic detail. Both legal rules and administrative processes were uncharted. For example, who would be responsible if things went awry and subjects were injured? When could the government tell private doctors or researchers how to conduct treatment or research? The need for rules seemed obvious, but the particular rules that would be arrived at were not.

In April 1947 and again in November, Carroll Wilson, the general manager of the new agency, wrote letters first to Stafford Warren and then to Robert Stone, both of whom played prominent roles in Manhattan Project medical research, Warren as medical director, and Stone as a key member of the Chicago branch of the project. In these letters, Wilson maintained that "clinical testing" with patients could go forward only where there was a prospect that the patient could benefit medically and only after that patient had been informed about the testing and there was documentation that the patient had consented. What was the origin of this position, and what was its reach? It appears that these letters were the products of an agency that was not only seeking to devise rules for new programs but also was trying to glean lessons from the experience with the secret research that had been conducted during the Manhattan Project. In the course of setting rules for the future, the AEC and its research community had to confront whether and how to proceed with human experimentation in the face of human experiments, including plutonium injections, conducted under the auspices of the Manhattan Project, experiments that were conducted in secret and that had the potential for both negative public reaction and litigation.

The First Wilson Letter

General Manager Wilson's first 1947 letter on human research, dated April 30, was, at least in part, a straightforward effort to define the rules according to which the AEC would provide contractors with research funding. The need for such rules had been discussed by the AEC's Interim Medical Advisory Committee, chaired by Stafford Warren, in January 1947 when it met to consider whether "clinical testing" should be

part of the AEC contract research program. The report of the meeting records projects involving human subjects at the University of Rochester and the University of California at Berkeley, and perhaps others.[5] In a January 30 letter to General Manager Wilson, Stafford Warren reported the committee's conclusion that in the study of health hazards and the use of fissionable and radioactive materials, "final investigations by clinical testing of these materials" would be needed. Warren therefore requested that the AEC legal department determine the "financial and legal responsibility" of the AEC when such "clinical investigations" are carried out under AEC-approved and -financed programs.[6] (The term *experiment* was not used, and the precise meaning of *clinical testing* is not clear.)

A month later, in early March, Warren met with Major Birchard M. Brundage, chief of the AEC's Medical Division, and two AEC lawyers to consider the terms for the resumption of "clinical testing." In a memorandum for the record, the lawyers summarized the meeting. In the case of "clinical testing" the lawyers

expressed the view that it was most important that it be susceptible of proof that any individual patient, prior to treatment, was in an understanding state of mind and that the nature of the treatment and possible risk involved be explained very clearly and that the patient express his willingness to receive the treatment.[7]

Initially, the lawyers had proposed that researchers obtain a "written release" from patients. However, "on Dr. Warren's recommendation," the lawyers agreed that it would be sufficient if "at least two doctors certify in writing to the patient's state of mind to the explanation furnished him and to the acceptance of the treatment."[8]

In his April 30 letter to Stafford Warren, Wilson announced that the AEC had approved Warren's committee's recommendations for a "program for obtaining medical data of interest to the Commission in the course of treatment of patients, which may involve clinical testing."[9] Wilson's letter spelled out ground rules that were agreed upon. The commission understood that "treatment (which may involve clinical testing) will be administered to a patient only when there is expectation that it may have therapeutic effect."

In addition, the commission adopted the requirement for documentation of consent agreed upon in Warren's meeting with the lawyers:

[I]t should be susceptible of proof from official records that, prior to treatment, each individual patient, being in an understanding state of mind, was clearly informed of the nature of the treatment and its possible effects, and expressed his willingness to receive the treatment.[10]

The commission deferred to Warren's request that written releases from the patient not be required. However,

it does request that in every case at least two doctors should certify in writing (made part of an official record) to the patient's understanding state of mind, to the explanation furnished him, and to his willingness to accept the treatment.[11]

Carroll Wilson's April letter was sent to Stafford Warren as head of the Interim Medical Advisory Committee, which was responsible for advising the AEC on its contract research program, and forwarded to Major Brundage at the Oak Ridge office.[12] Stafford Warren was at this point dean of the medical school at the University of California at Los Angeles, one of the dozen research institutions involved in the AEC contract research program. With one exception the Advisory Committee on Human Radiation Experiments did not locate documentation that the letter or its contents were communicated to any other research institutions involved with the AEC's contract research program. The exception is the University of California at San Francisco, where there is indirect evidence that someone at that institution had been apprised of Wilson's April letter. Of the eighteen plutonium injections, only the last one, that involving Elmer Allen, or "CAL-3," took place after the April letter. In Mr. Allen's medical chart, there is a notation signed by two physicians indicating that the "experimental nature" of the procedure was explained and that the patient "agreed."[13] Although the note in Mr. Allen's chart suggests an effort on the part of the researchers to comply with Wilson's April letter, the researchers did not comply with the other provision of the Wilson letter, that "treatment (which may involve clinical testing) will be administered to a patient only when there is

expectation that it may have therapeutic effect."[14] As is discussed in more detail in chapter 5, there was no expectation at the time that Mr. Allen would benefit medically from an injection of plutonium.[15]

The Second Wilson Letter

The context of the second Wilson letter, as well as its precise terms, further indicates that the April 1947 letter was given little distribution and effect. In the fall of 1947, the AEC laboratory at Oak Ridge requested advice from Carroll Wilson's office on the rules for experiments involving human subjects. Just as the AEC's Washington headquarters had embarked on the funding of a new research program, Oak Ridge was also in the midst of considering the rules governing the expansion of its own medical research program and the distribution of isotopes, which was then headquartered at Oak Ridge. In September 1947, the manager of Oak Ridge Operations wrote to Wilson, asking, "What responsibilities does the AEC bear for human administration of isotopes (a) by private physicians and medical institutions outside the Project, and (b) by physicians within the project. . . . What are the criteria for future human use?"[16]

Two weeks later, Oak Ridge sent a memorandum to the Advisory Committee for Biology and Medicine (ACBM). The ACBM had succeeded both Stafford Warren's Interim Medical Advisory Committee and the Medical Board of Review, a group appointed by AEC Chairman David Lilienthal to review the AEC's medical program. The memorandum emphasized the need for "medico-legal criteria" for "future human tracer research" because some of that research would be "of no immediate therapeutic value to the patient." The memorandum outlined the pros and cons of "tracer studies":

Pro-
1. Tracer research is fundamental to toxicity studies.
2. The adequacy of the health protection which we afford our present employees may in a large measure depend upon information obtained using tracer techniques.

3. New and improved medical applications can only be developed through careful experimentation and clinical trial.

4. Tracer techniques are inherent in the radioisotope distribution program.

Con-

1. Moral, ethical and medico-legal objections to the administration of radioactive material without the patient's knowledge or consent.

2. There is perhaps a greater responsibility if a federal agency condones human guinea pig experimentation.

3. Publication of such researches in some instances will compromise the best interests of the Atomic Energy Commission.

4. Publication of experiments done by Atomic Energy Commission contractor's personnel may frequently be the source of litigation and be prejudicial to the proper functioning of the Atomic Energy Commission Insurance Branch.[17]

The questions raised by Oak Ridge were discussed by the ACBM at its October 11, 1947, meeting, which decided to give the "matter more study."[18] The minutes of the October 11 meeting record that "human experimentation" was then discussed in the context of a request by Dr. Robert Stone to release "classified papers containing certain information on human experimentation with radioisotopes conducted within the AEC research program."[19] The request was part of a continuing effort by Stone and other scientists to obtain permission to publish the research, including the plutonium experiments, that they had conducted in secret during the Manhattan Project. Earlier in 1947, the AEC had reversed a decision to declassify a report on the plutonium injections, citing the potential for public embarrassment and legal liability (see chapter 5). The question of what to do with these requests continued to fester.

The minutes explain that the "problem" raised by Stone had been dealt with by Chairman Lilienthal's Medical Board of Review in June. In a cryptic statement, the minutes record the ACBM's agreement that papers on human experiments "should remain classified unless the

stipulated conditions laid down by the Board of Review were complied with."[20]

The "stipulated conditions" referred to are contained in General Manager Wilson's November 5, 1947, letter to Stone. According to Wilson's letter, at a June meeting the Medical Board of Review concluded that "the matter of human experimentation" would remain classified where certain "conditions" were not satisfied. Wilson then quoted from the "preliminary unpublished and restricted draft of the [Medical Board] report read to the Commissioners" as follows:

The atmosphere of secrecy and suppression makes one aspect of the medical work of the Commission especially vulnerable to criticism. We therefore wish to record our approval of the position taken by the medical staff of the AEC in point of their studies of the substances dangerous to human life. We [the Medical Board of Review] believe that no substances known to be, or suspected of being, poisonous or harmful should be given to human beings unless all of the following conditions are fully met: (a) that a reasonable hope exists that the administration of such a substance will improve the condition of the patient, (b) *that the patient give his complete and informed consent in writing*, and (c) that the responsible next of kin give in writing a similarly complete and informed consent, revocable at any time during the course of such treatment [emphasis added].[21]

In other words, the opinion of the Medical Board of Review was presented by Wilson in his November letter as both a prescription for the future conduct of human experiments and a presentation of the criteria that must be met for the declassification of past research. Wilson again referenced these conditions in a letter to ACBM Chairman Alan Gregg, also on November 5. "I am sure," Wilson wrote Gregg, "that this information will assist Dr. Stone in evaluating the present problem and inform him as to the conditions that must be met in future experiments."[22] Thus, as discussed in more detail in chapters 5 and 13, the requirement that research proceed only with consent appears to have been coupled with the decision to withhold from the public information about experiments that failed to meet that standard.

Two points should be made about the term *informed consent*, which appears in the November letter from Wilson to Stone. First, it is not clear what meaning Wilson and the members of the Medical Board of Review attributed to the term. No further explanation was given. Second, it is nevertheless a matter of some historical interest that this term is used at all. Previous scholarship had attributed its first official usage to a landmark legal opinion in a medical malpractice case that was issued a decade later.[23]

The April and November 1947 Wilson letters have some common elements, in spite of their differences in detail. They both provided that research with humans proceed (1) only where there is reasonable hope of therapeutic effect; and (2) with documentary proof that the patient-subject was informed of the treatment and its possible effects and had consented to its administration.

But there are many remaining mysteries about the AEC's 1947 statements. In interviews with Advisory Committee staff, Joseph Volpe, who served as an AEC attorney in its early days and became general counsel in 1949, explained that a letter authored by General Manager Wilson could state AEC policy and confidently recollected that informed consent from research subjects would have been required by the first AEC general counsel. This requirement, Volpe maintained, should be reflected in the commission's minutes.[24] However, Committee and DOE review of the commission's minutes did not reveal evidence that the "consent" policy was expressly addressed.

Even more troubling is that both Wilson letters precluded research that did not offer patient-subjects a prospect of direct medical benefit. In the context of the concern about the plutonium injections and the other "nontherapeutic" research conducted during the Manhattan Project experiments, this provision readily makes sense. Yet, as Oak Ridge's inquiry to Washington noted, nontherapeutic research in the form of tracer studies had been, and would continue to be, a mainstay of AEC-sponsored isotope research. How could it be that the Wilson letters were intended to ban exactly the kind of research that at the same time the AEC was so actively promoting? It is conceivable that the requirement of the isotope distribution program for risk review prior to the human use of radioisotopes

was a means of addressing this notion. However, if the equation between that risk review procedure and the provision in the November Wilson letter seems implicit, the documentary evidence does not provide an express link between the requirement stated in the Wilson letter and the rules of the isotope distribution program.

From Statements to Policy: A Failure of Translation

Despite the fact that they were developed in response to a need for clarity in the way that human research should be conducted, we have found little evidence of efforts to communicate or implement the rules stated by Wilson in coordination with the AEC's biomedical advisory groups and other AEC officials. In some cases the evidence described in the following paragraphs suggests that policies for consent from subjects were established and implemented, while in other cases it suggests that, if there were any such policies, they were unknown or lost. Taken together, however, this evidence further supports the view that the *ideas* present in General Manager Wilson's 1947 statements were available to those working in the field during this time, albeit perhaps in a primitive form.

Consider, for example, a 1951 exchange between the AEC's Division of Biology and Medicine (DBM), which directed the AEC's medical research program, and the commission's Los Alamos Laboratory, which was in routine contact with Washington. An information officer at Los Alamos, Leslie Redman, who was charged to review papers that involved human experimentation, asked the DBM for a "definite AEC policy" on "human experimentation." In the course of his work, Redman wrote, he had been advised by "various persons" at Los Alamos that "regulations or policies of the AEC" on human experimentation were available, but he had been unable to locate more than general information about these regulations. According to his letter, his understanding was that

these regulations are comparable to those of the American Medical Association: that an experiment be performed under the supervision of an M.D., with the permission of the patient, and for the purpose of seeking a cure.[25]

Redman's characterization of the American Medical Association's guidelines, as we shall see in chapter 2, is partly incorrect. The requirement of a therapeutic intent is absent from the AMA guidelines. The possibility of direct therapeutic benefit for the patient was, however, a condition of research according to both of General Manager Wilson's 1947 letters.

Shields Warren, the DBM chief, responded to Redman by citing Wilson's November 5, 1947, letter to Stone and by excerpting the conditions quoted above.[26] But Warren did not term these conditions "standards" or "requirements." Rather, Warren's response to Los Alamos "urges" compliance with these "guiding principles."[27]

Though Los Alamos was provided with the criteria stated by Wilson in November 1947, General Manager Wilson's statements were not routinely communicated in response to requests for guidance from non-AEC researchers. In an April 1948 letter to the DBM, a university researcher explained that the Isotopes Division had approved his request to use phosphorus 32 for "experimental procedures in the human . . . simply for investigational purposes and not for treatment of disease." What, the researcher wanted to know, should be done about "medical-legal aspects" and "permission forms"?[28] The request could have been answered by referring to Wilson's 1947 statements about consent. Instead, the DBM simply referred the researcher to the Isotopes Division at Oak Ridge.[29] In its response, the Isotopes Division did not indicate that consent should be solicited, as Wilson had stipulated. The Isotopes Division, stating it could be "of little assistance," declined to provide "legal advice," save to note that "we understand that most hospitals do require patients to sign general releases before entering into treatment."[30]

From 1947 onward, the AEC had ample opportunity to disseminate a research policy. The AEC routinely provided educational and administrative materials to applicants for AEC funding and to the far greater number of applicants for AEC-produced radioisotopes. The isotopes distribution program, in particular, included a sophisticated structure of regulation, replete with review committees, training courses, and informational brochures (see chapter 6). At the federal level, this included the Subcommittee on Human Applications of the Committee on Isotope Distribution, whose very purpose was "to review all initial requests for radioisotopes to be used *experimentally* or otherwise in human beings [emphasis added]."[31] The AEC Subcommittee on Human Applications was supplemented by similar committees at the research institutions where the work was conducted.

In principle, there does not seem to be any reason these local committees could not have been instructed by the Isotopes Division on consent requirements.[32] Some evidence suggests that in March 1948 the Subcommittee on Human Applications discussed consent requirements for healthy subjects and patient-subjects. In a document dated March 29, 1948, the Subcommittee on Human Applications appeared to resolve that

1. Radioactive materials should be used in experiments involving human subjects when information obtained will have diagnostic value, therapeutic significance, or will contribute to knowledge on radiation protection.

2. Radioactive materials may be used in normal human subjects provided
   a. The subject has full knowledge of the act and has given his consent to the procedure.
   b. Animal studies have established the assimilation, distribution, selective localization and excretion of the radioisotope or derivative in question.

3. Radioactive materials may be used in patients suffering from diseased conditions of such nature that there is no reasonable probability of the radioactivity employed producing manifest injury provided:
   a. Animal studies have established the assimilation, distribution, selective localization and excretion of the radioisotope or derivative in question.
   b. The subject is of sound mind, has full knowledge of the act and has given his consent to the procedure. . . .

4. Investigations are approved (1) by medical director or his equivalent at the installation responsible for the investigation, (2) by the Director, Division of Biology and Medicine, and (3) full written descriptions of experimen-

tal procedures and calculated estimates of radiation to be received by body structure and organs must be submitted.[33]

We were unable to locate any further references to this document and do not know whether it represented a policy that was adopted. Perhaps it represents the consensus of the Subcommittee on Human Applications, as it had met shortly before that, or perhaps it is simply a draft document prepared by staff.

Whatever the ultimate disposition of this document, it provides some idea of the problems that were under consideration at the time and indicates that views on human use were unsettled. The first numbered item, for example, appears to recommend human radiation experiments when they will offer diagnostic value and therapeutic significance *or* knowledge about radiation protection. If the document had in fact been adopted, the recognition that isotope experimentation could be undertaken to "contribute to knowledge" (item 1) would appear to revise the Wilson letters' prohibition of nontherapeutic experimentation. The third item also addresses consent and risk of injury to patient-subjects without indicating that there should be any potential benefit. Another peculiarity is found in the second item, which refers to consent from normal human subjects but does not rule out experiments that present risk to the subject.

In any event, at a 1948 meeting the Subcommittee on Human Applications articulated a consent requirement as part of a decision to permit patients suffering from serious diseases to receive "larger doses for investigative purposes."[34] This requirement was disseminated to all radioisotope purchasers in 1949.[35] The subcommittee allowed investigators to administer "larger doses" to seriously ill patients but only with the patient's consent. While it is possible that the basis for permitting larger doses was an assumption that smaller ones would be of no potential benefit to subjects, item 3 of the just-quoted March 1948 document suggests the assumption was rather that in seriously ill patients other disease processes would be more likely to take their course before radiation injury was manifested.

There is evidence that at least one AEC-funded entity did routinely provide some form of disclosure and consent in the early 1950s. From its opening in 1950 the AEC-sponsored Oak Ridge Institute for Nuclear Studies (ORINS), a research hospital, advised incoming patients that procedures were experimental. Additionally, patients were given written information that advised them that "probable benefit, if any, cannot always be predicted in advance."[36] Patients were also asked to sign a form that indicated that they were "fully advised" about the "character and kind of treatment and care," which would be "for the most part experiments with no definite promise of improvement in my physical condition."[37] Thus, at least in the case of ORINS, and perhaps at other AEC facilities, a local process was instituted apart from any known communication of the statements by AEC officials.

Nonetheless, there is other evidence that the AEC did not communicate the requirements detailed in General Manager Wilson's 1947 letters to its own contract research organizations, which, as in the cases of Argonne, Los Alamos, Brookhaven, and Oak Ridge, had significant biomedical programs and were engaged in human research. When the Division of Biological and Medical Research at Argonne National Laboratory met in January 1951 to discuss beginning a program of human experimentation in cancer research, one of its members asserted that the ACBM had not established a "general policy concerning human experimentation." The minutes of the meeting at Argonne record that the ACBM "has been approached several times in the past for a general policy and has refused to formulate one."[38]

In 1956, Los Alamos asked the DBM to "restate its position on the experimental use of human volunteer subjects" for tracer experiments.[39] The DBM responded by stating that tracer doses might be administered under certain conditions, which included the provision that subjects be volunteers who were fully informed. The focus of this position seems to have been research with healthy people and not patients, and no reference was made to the provisions of the Wilson letters.[40] The DBM's 1956 formulation was given "staff distribution" by Los Alamos and restated in 1962.[41]

Also in 1956, the Isotopes Division did state a requirement for healthy subjects. All subjects were to be informed volunteers. As part of its "Recommendations and Requirements" guidebook for the medical uses of radioisotopes, which was distributed to all medical users of radioisotopes, the Isotopes Division stated:

Uses of radioisotopes in normal subjects for experimental purposes shall be limited to:

a. Tracer doses which do not exceed the permissible total body burden for the radioisotope in question. In all instances the dose should be kept as low as possible.
b. Volunteers to whom the intent of the study and the effects of radiation have been outlined.
c. Volunteers who are unlikely to be exposed to significant additional amounts of radiation.[42]

These requirements apparently applied to all uses of AEC radioisotopes, whether government or private researchers were involved. The "experimental or nonroutine" use of radioisotopes in any human subjects was limited to institutional programs where local review committees existed to oversee the risk to which subjects were exposed. In stating these requirements, the AEC reiterated that "patients" in whom "there is no reasonable probability of producing manifest injury" may be used in some experiments not normally permitted, but did not reiterate the requirement that consent should be obtained from these patients, as was stated in 1948.

What, then, can be said about the rules and policies of the AEC in the 1940s and 1950s? General Manager Wilson's 1947 letters clearly stipulate a requirement of "informed consent" from patient-subjects, at least where potentially "poisonous or harmful" substances are involved. But with the exception of ORINS there is little indication that this requirement was imposed as binding policy on any AEC facility, contractor, or recipient of radioisotopes. By contrast, later requirements that healthy subjects be informed volunteers and that seriously ill patients be permitted to receive higher doses only with their consent appear to have been more broadly communicated and enforced. The only evidence of general attention to matters of consent from patient-subjects comes from ORINS, whose policies and practices show a striking similarity to those that, as we shall see, were being contemporaneously employed at another facility essentially devoted to experimental work, the NIH's Clinical Center. At the same time, there is evidence of considerable attention in both policy and practice to issues of safety and acceptable risk (see chapter 6). Questions of subject selection, as in the case of seriously ill patients, emerge only in this context of safety; there is no evidence that issues of fairness or concerns about exploitation in the selection of subjects figured in AEC policies or rules of the period.

THE DEPARTMENT OF DEFENSE: CONSENT IS FORMALIZED

The story of research involving human subjects in the U.S. military began at least a century ago. Well before 1944, the beginning of the period of special interest to the Advisory Committee, the military needed healthy subjects to test means to prevent and treat infectious diseases to which military personnel might be exposed. The notion that consent should be obtained from human subjects was clearly part of this tradition; less clear is how consistently this was applied and what consent actually meant to those in authority.

The most famous example of the early use of subject consent in the military took place at the turn of the century. Walter Reed's successful research on yellow fever, the mosquito-borne disease that bedeviled Panama Canal construction efforts, employed healthy subjects who signed forms indicating their agreement. Whether the practice was required by the Army or self-imposed by Reed is unknown. In 1925 an Army regulation to promote infectious disease research noted that "volunteers" should be used in "experimental" research.[43]

The Navy also provided early requirements for human subject research. In 1932, the secretary of the Navy granted permission for the conduct of an experiment involving divers on condition that the subjects were "informed volunteers."[44] In 1943 the secretary of the Navy also required that all investigators seeking to conduct research with service personnel obtain prior approval from the secretary.[45]

As we have noted in the Introduction, during World War II, federally funded biomedical research related to the war effort (outside the Manhattan Project) was coordinated by the Committee on Medical Research (CMR) of the Office of Scientific Research and Development, which was part of the Executive Office of the President. The CMR supported a program of human research, during which the question of the rules for the conduct of human research was addressed. In 1942 a University of Rochester researcher, seeking to "work out a human experiment on the chemical prophylaxis of gonorrhea," asked the CMR for "an opinion that such human experimentation is desirable."[46] In an October 9, 1942, response, the CMR's chairman offered the following general statement, which was endorsed by the full committee:

[H]uman experimentation is not only desirable, but necessary in the study of many of the problems of war medicine which confront us. When any risks are involved, volunteers only should be utilized as subjects, and these only after the risks have been fully explained and after signed statements have been obtained which shall prove that the volunteer offered his services with full knowledge and that claims for damage will be waived. An accurate record should be kept of the terms in which the risks involved were described.[47]

In spite of the CMR's statement in response to this researcher's query, it supported other experiments that involved subjects whose capacity to give valid consent to participation was doubtful, including institutionalized people with cognitive disabilities.[48]

During the war, the Navy used consent forms in wartime experiments using prisoners and conscientious objectors, as a proposal for research on an influenza vaccine with prisoners at San Quentin in 1943 shows.[49] The form used in this case indicates that the subject is "acting freely and voluntarily without any coercion on the part of any person whomever."[50] To be sure, the forms located by the Advisory Committee were called "waiver" or "release" rather than "consent"

forms. Thus, the attestation to voluntary participation was punctuated by the release of experimenters from liability. However, at a time when free young men were routinely conscripted into the military, the requirement that subjects, including prisoners and conscientious objectors, must be volunteers seems remarkable.

In sharp contrast with these procedures, the Navy, too, sometimes functioned in a manner inconsistent with a voluntary consent policy for healthy subjects. Surviving subjects have reported that harmful mustard gas experiments on naval personnel at the Naval Research Laboratory in Washington, D.C., during World War II failed to adequately inform subjects and seem to have involved manipulation or coercion of "volunteers."[51] The lack of medical follow-up on the subjects of these experiments was sharply criticized in a 1993 report by the Institute of Medicine of the National Academy of Sciences.[52]

The NEPA Debate on the Ethics of Prisoner Experiments

Many of the researchers and officials who had been involved in Manhattan Project human experiments during the war and then in the 1947 AEC deliberations about human research policy also were engaged in 1949 and 1950 in discussions of the ground rules for research with human subjects in the development of new military technology. This time the forum was the joint AEC-DOD project on Nuclear Energy for the Propulsion of Aircraft (NEPA). The DOD convened an advisory panel of private and public officials to determine how to obtain data needed to answer questions such as whether the air crew would be put at undue risk by the nuclear-powered engine. The participants in the discussion included university researchers Hymer Friedell, Stafford Warren, Robert Stone, and Joseph Hamilton, and AEC officials Shields Warren and Alan Gregg. Shields Warren argued that human experimentation was not appropriate because the research could be done on animals and human data was not likely to produce scientifically valid results (see Introduction).

Robert Stone, the recipient of the November 1947 letter in which AEC General Manager Wilson called for "informed consent," emerged as the primary proponent of human experiments. In a January 1950 discussion paper, he focused on the "ethics of human experimentation."[53] After a recitation of a tradition that included Walter Reed's experience and the historical use of prisoners and medical students as research subjects, Stone cited requirements that had been publicized by the American Medical Association in 1946. These rules provided that subjects must give voluntary consent, that animal experimentation must precede human experimentation, and that human experiments should be "performed under proper medical protection and management."[54] (See chapter 2.) Stone argued that it would be possible to conduct NEPA-related experiments with prisoners in compliance with all three of these requirements.

Stone's proposal generated considerable discussion among DOD and AEC experts and officials. In April 1950, the DOD's Joint Panel on the Medical Aspects of Atomic Warfare endorsed the use of prisoners of "true volunteer status" as meeting "the requirements of accepted American standards for the use of human subjects for research purposes."[55]

However, AEC officials were less than enthusiastic. "Doesn't the prisoner proposal," ACBM Chairman Alan Gregg asked a military official in the course of one discussion, "fall in the category of cruel and unusual punishment?"[56] "Not," the official replied, "if they would carry out the work as they proposed. . . . It would be on an absolutely voluntary basis, and under every safety precaution that could be built up around it . . . it didn't strike me as being cruel and unusual." To which Shields Warren retorted: "It's not very long since we got through trying Germans for doing exactly the same thing."[57]

In December 1950 the AEC convened a panel to discuss what was known about potential radiation effects on service personnel and whether human research was needed. Joseph Hamilton, Robert Stone's colleague at the University of California, was unable to attend the meeting, and in his regrets he offered his thoughts on the matter. In a letter to Shields Warren, he noted that the proposal to use prisoner volunteers "would have a little of the Buchenwald touch" and reported that he had no "very constructive ideas as to where one would turn for such vol-

unteers should this plan be put into effect."[58] He suggested using large primates, even though, from a purely scientific viewpoint, the data collected would not be as useful as data from humans.[59]

Apparently Stone lost the debate. A decision was made not to conduct experiments with prisoners or other healthy subjects in connection with the NEPA project. However, as will be discussed in more detail in chapter 8, the military contracted with a private hospital to study patients who were being irradiated for cancer treatment, in the hopes of answering the same kinds of questions that would have been addressed if NEPA research with prisoners had gone forward.

Congress Provides for DOD Contractor Indemnification in the Case of Injury

In the aftermath of World War II, the military continued its long-standing program of infectious disease research using human subjects. During the late 1940s and early 1950s the Army Epidemiological Board (AEB) and its 1949 successor, the Armed Forces Epidemiological Board (AFEB), which was established to advise on medical research funded by the DOD and to direct some research undertaken with Army funds, sponsored studies with healthy subjects that focused on hepatitis, dengue fever, and other infectious diseases. Consistent with military tradition, at least some AEB-sponsored researchers were using written permission forms. The forms, frequently referred to as an "Agreement with Volunteer," or a "release," outlined the study and the risks to the subject and protected the DOD from liability.[60]

In the late 1940s, some university researchers expressed concern that they were not adequately protected from liability in the case of injury or death of their prisoner-subjects. The ensuing dialogue provides a window on the role of the written releases and the understanding of the rules governing human subject research. In response to a researcher's request to be reimbursed by the Army for a disability policy for the subjects, the Army lawyers replied that the Army could not provide indemnification in the absence of clear congressional authority. Army legal advisers recommended that the researcher "protect himself, the State of New Jersey [the research locale], and the Government by means of the usual waiver."[61]

In a February 1948 letter, the AEB director, John R. Paul, explained that the "world situation" had placed the rules for human experimentation up for grabs.[62]

At this stage in the world situation one should proceed cautiously, until standards are set by what ever body is in 'authority.' I am not sure just what the rules are but I understand that . . . some type of vigilance committee has laid down certain principles about volunteers in order to protect this country from the criticisms brought up in Germany during the Nuremberg trials. . . . During the war we more or less made our own policies on this, but I am not sure that this is possible today. . . .[63]

The allusion to a "vigilance committee" is unclear. It may be a reference to a committee established by the governor of Illinois to examine the use of prisoners as research subjects in that state and chaired by Andrew Ivy, the principal expert witness for the prosecution at the Nuremberg Medical Trial (see chapter 2). Given the date of the letter, February 18, 1948, it seems likely that Paul had just skimmed through his new copy of the *Journal of the American Medical Association*—the report of Ivy's committee was published in the February 14, 1948, issue.[64]

In April 1948, an AEB official made it plain to the researchers that the fact that state authorities or the prison warden gave permission for the experiment should be of little comfort to them. In case of a lawsuit, responsibility "would devolve entirely upon the individual experimenter."[65] Only Congress could provide a solution, but it would be a "dangerous course" to raise the matter publicly. "I have," the AEB official wrote,

given considerable thought to the matter of whether it would be advisable to approach individuals or groups in Congress with the idea of having laws passed relating to payment of compensation for disability or release of the experimenter from liability. I am afraid that this would be a dangerous course, and that it might in fact injure clinical investigations generally. There is a very real possibility that unfavorable publicity would quickly result.[66]

It appears that the relief sought by researchers was provided by Congress in 1952, however, under the umbrella of a law that provided indemnification for DOD research and development activities as a whole. In October 1952, following the death of a prisoner-subject in an AFEB-sponsored hepatitis study[67] and questions raised by the Army Chemical Corps about release forms for "human 'guinea pigs,'"[68] the AFEB administrator queried the DOD Legal Office about a recently passed federal law. The law provided authority for the military to indemnify contractors for risks undertaken in "research and development situations." Did the new law "afford relief to the immediate dependents of . . . prison volunteers when as [a] result of these experiments they should die[?]"[69] The answer was yes, but only by providing relief to the researchers first. "From the wording of the law, and from . . . the legislative history," the Legal Office replied, "it is a direct indemnification to the contractor and not to the individual human guinea pig."[70]

Thus, what appears to have been the first Cold War congressional enactment to deal with human subjects of research addressed the government's obligation to its contractors, not the government's and its researchers' obligations to the subjects. Moreover, the record indicates that a more direct approach was not sought by the DOD because of concerns about public relations. At the same time Congress was acting, however, the DOD itself was secretly debating a new policy for human experiments.

The Secretary of Defense Issues the Nuremberg Code in Top Secret

As the Korean War began in mid-1950, the military's interest in human experimentation—in connection with chemical and biological as well as atomic and radiation warfare—intensified. The need for a DOD-wide policy on the use of human subjects in research was noted by Colonel George Underwood, the director of the Office of the Secretary of Defense, in a February 1953 memorandum to the incoming administration of Dwight D. Eisenhower: "There is no DOD policy on the books which permits this type of research [human experiments in the field of atomic, biological, and chemical warfare]."[71]

From 1950 to 1953 discussions about human research and human research policy were held in several high-level DOD panels, including the Armed Forces Medical Policy Council (AFMPC), the Committee on Medical Sciences (CMS), and the Joint Panel on the Medical Aspects of Atomic Warfare. These groups were headed by civilian researchers, and, in at least the latter two cases, included representatives of the AEC, CIA, NIH, VA, and Public Health Service.

At its September 8, 1952, meeting, the AFMPC heard a presentation from the chief of preventive medicine of the Army Surgeon General's Office on the topic of biological warfare research:

It was pointed out that the research had reached a point beyond which essential data could not be obtained unless human volunteers were utilized for such experimentation. . . . Following detailed discussion, it was unanimously agreed that the use of human volunteers in this type of research be approved.[72]

At its October 13, 1952, meeting the AFMPC again took up the question of human experimentation. "It was resolved," the chairman wrote to the secretary of defense, "that the ten rules promulgated at the Nuremberg trials be adopted as the guiding principles to be followed. An eleventh rule [barring experiments with prisoners of war] was added by the legal advisor to the Council, Mr. Stephen S. Jackson."[73]

DOD attorney Jackson evidently was responsible for the inclusion of the Nuremberg Code in the AFMPC's proposed policy. In an October 13, 1952, memo to the chairman of the AFMPC, Jackson

recommended: that the attached principles and conditions for human experimentation, which were laid down by the Tribunal in the Nuremberg Trials, be adopted instead of those previously submitted by me.[74]

As an addendum to the Nuremberg Code, Jackson proposed a requirement that "consent be expressed in writing before at least one witness." This recommendation followed from the suggestion of Anna Rosenberg, assistant secretary of defense for manpower and personnel, who was an expert on labor relations.[75]

A letter written by the administrator of the Armed Forces Epidemiological Board documents Mr. Jackson's role and motivation:

It was on Mr. Jackson's insistence that the 'Nuremberg Principles' were used in toto in the document, since he stated, these already had international judicial sanction, and to modify them would open us to severe criticism along the line—"see they use only that which suits them."[76]

Thus, the DOD's counsel cited the 1947 Nuremberg military tribunal ruling as establishing an international legal precedent to which American researchers should be held.

It appears that in succeeding months the AFMPC proposal was received unenthusiastically by other DOD committees that reviewed it. In a November 12, 1952, memorandum, the executive director of the Committee on Medical Sciences pointed out that "human experimentation has been carried on for many years." He contended that

to issue a policy statement on human experimentation at this time would probably do the cause more harm than good; for such a statement would have to be "watered down" to suit the capabilities of the average investigator.[77]

"Human experimentation," the CMS executive director asserted, "has, in years past, and is at present governed by an unwritten code of ethics," which is "administered informally by fellow workers in the field [and] is considered to be satisfactory. . . . To commit to writing a policy on human experimentation would focus unnecessary attention on the legal aspects of the subject."[78]

Notwithstanding the reservations of the CMS and others,[79] the Nuremberg Code proposal had the support of President Truman's secretary of defense, Robert A. Lovett.[80] However, the secretary's aide, George V. Underwood, wrote in January 1953, "Since consequences of this policy will fall upon Mr. Wilson [President Eisenhower's nominee for secretary of defense, Charles Wilson], it might be wise to pass to him as a unanimous recommendation from the 'alumni.'"[81]

THE NUREMBERG CODE

1. The voluntary consent of the human subject is absolutely essential. This means that the person involved should have legal capacity to give consent; should be so situated as to be able to exercise free power of choice, without the intervention of any element of force, fraud, deceit, duress, overreaching, or other ulterior form of constraint or coercion; and should have sufficient knowledge and comprehension of the elements of the subject matter involved as to enable him to make an understanding and enlightened decision. The latter element requires that before the acceptance of an affirmative decision by the experimental subject there should be made known to him the nature, duration, and purpose of the experiment; the method and means by which it is to be conducted; all inconveniences and hazards reasonably to be expected; and the effects upon his health or person which may possibly come from his participation in the experiment. The duty and responsibility for ascertaining the quality of the consent rest upon each individual who initiates, directs or engages in the experiment. It is a personal duty and responsibility which may not be delegated to another with impunity.

2. The experiment should be such as to yield fruitful results for the good of society, unprocurable by other methods or means of study, and not random and unnecessary in nature.

3. The experiment should be so designed and based on the results of animal experimentation and a knowledge of the natural history of the disease or other problem under study that the anticipated results will justify the performance of the experiment.

4. The experiment should be so conducted as to avoid all unnecessary physical and mental suffering and injury.

5. No experiment should be conducted where there is an *a priori* reason to believe that death or disabling injury will occur; except, perhaps, in those experiments where the experimental physicians also serve as subjects.

6. The degree of risk to be taken should never exceed that determined by the humanitarian importance of the problem to be solved by the experiment.

7. Proper preparations should be made and adequate facilities provided to protect the experimental subject against even remote possibilities of injury, disability, or death.

8. The experiment should be conducted only by scientifically qualified persons. The highest degree of skill and care should be required through all stages of the experiment of those who conduct or engage in the experiment.

9. During the course of the experiment the human subject should be at liberty to bring the experiment to an end if he has reached the physical or mental state where continuation of the experiment seems to him to be impossible.

10. During the course of the experiment the scientist in charge must be prepared to terminate the experiment at any stage, if he has probable cause to believe, in the exercise of the good faith, superior skill, and careful judgment required of him, that a continuation of the experiment is likely to result in injury, disability, or death to the experimental subject.

In a January 13, 1953, memorandum for the new secretary, the AFMPC "strongly recommended that a policy be established for the use of human volunteers (military and civilian employees) in experimental research at Armed Forces facilities." The policy would render the research "subject to the principles and conditions laid down as a result of the Nuremberg trials."[82]

On February 26, 1953, Secretary of Defense Wilson signed off on the AFMPC policy. It was

THE WILSON MEMORANDUM

26 Feb 1953

Memorandum for the
    Secretary of the Army
    Secretary of the Navy
    Secretary of the Air Force
Subject: Use of Human Volunteers in Experimental Research

1. Based upon a recommendation of the Armed Forces Medical Policy Council, that human subjects be employed, under recognized safeguards, as the only feasible means for realistic evaluation and/or development of effective preventive measures of defense against atomic, biological or chemical agents, the policy set forth below will govern the use of human volunteers by the Department of Defense in experimental research in the fields of atomic, biological and/or chemical warfare.

2. By reason of the basic medical responsibility in connection with the development of defense of all types against atomic, biological and/or chemical warfare agents, Armed Services personnel and/or civilians on duty at installations engaged in such research shall be permitted to actively participate in all phases of the program, such participation shall be subject to the following conditions:

a. The voluntary consent of the human subject is absolutely essential.

(1) This means that the person involved should have legal capacity to give consent; should be so situated as to be able to exercise free power of choice, without the intervention of any element of force, fraud, deceit, duress, over-reaching, or other ulterior form of constraint or coercion; and should have sufficient knowledge and comprehension of the elements of the subject matter involved as to enable him to make an understanding and enlightened decision. This latter element requires that before the acceptance of an affirmative decision by the experimental subject there

should be made known to him the nature, duration, and purpose of the experiment; the method and means by which it is to be conducted; all inconveniences and hazards reasonably to be expected; and the effects upon his health or person which may possibly come from his participation in the experiment.

(2) The concept [*sic*] of the human subject shall be in writing; his signature shall be affixed to a written instrument setting forth substantially the aforementioned requirements and shall be signed in the presence of at least one witness who shall attest to such signature in writing.

(a) In experiments where personnel from more than one Service are involved the Secretary of the Service which is exercising primary responsibility for conducting the experiment is designated to prepare such an instrument and coordinate it for use by all the Services having human volunteers involved in the experiment.

(3) The duty and responsibility for ascertaining the quality of the consent rests upon each individual who initiates, directs or engages in the experiment. It is a personal duty and responsibility which may not be delegated to another with impunity.

b. The experiment should be such as to yield fruitful results for the good of society, unprocurable by other methods or means of study, and not random and unnecessary in nature.

c. The number of volunteers used shall be kept at a minimum consistent with item b., above.

d. The experiment should be so designed and based on the results of animal experimentation and a knowledge of the natural history of the disease or other problem under study that the anticipated results will justify the performance of the experiment.

e. The experiment should be so conducted

as to avoid all unnecessary physical and mental suffering and injury.

f. No experiment should be conducted where there is an a priori reason to believe that death or disabling injury will occur.

g. The degree of risk to be taken should never exceed that determined by the humanitarian importance of the problem to be solved by the experiment.

h. Proper preparation should be made and adequate facilities provided to protect the experimental subject against even remote possibilities of injury, disability, or death.

i. The experiment should be conducted only by scientifically qualified persons. The highest degree of skill and care should be required through all stages of the experiment of those who conduct or engage in the experiment.

j. During the course of the experiment the human subject should be at liberty to bring the experiment to an end if he has reached the physical or mental state where continuation of the experiment seems to him to be impossible.

k. During the course of the experiment the scientist in charge must be prepared to terminate the experiment at any stage, if he has probable cause to believe, in the exercise of the good faith, superior skill and careful judgment required of him that a continuation of the experiment is likely to result in injury, disability, or death to the experimental subject.

l. The established policy, which prohibits the use of prisoners of war in human experimentation, is continued and they will not be used under any circumstances.

3. The Secretaries of the Army, Navy and Air Force are authorized to conduct experiments in connection with the development of defenses of all types against atomic, biological and/or chemical warfare agents involving the use of human subjects within the limits prescribed above.

4. In each instance in which an experiment is proposed pursuant to this memorandum, the nature and purpose of the proposed experiment and the name of the person who will be in charge of such experiment shall be submitted for approval to the Secretary of the military department in which the proposed experiment is to be conducted. No such experiment shall be undertaken until such Secretary has approved in writing the experiment proposed, the person who will be in charge of conducting it, as well as informing the Secretary of Defense.

5. The addresses will be responsible for insuring compliance with the provisions of this memorandum within their respective Services.

/signed/
C. E. Wilson

copies furnished:

Joint Chiefs of Staff
Research and Development Board

Downgraded to
UNCLASSIFIED
22 Aug 75

TOP SECRET

---

issued in a Top Secret memorandum to the secretaries of the Army, Navy, and Air Force. The Wilson memorandum reiterates the principles of the Nuremberg Code, requires written and witnessed informed consent of research subjects, and prohibits the use of prisoners of war. The policy was to "govern the use of human volunteers by the Department of Defense in experimental research in the fields of atomic, biologi-

cal, and/or chemical warfare for defensive purposes."[83]

The basis for the classification of the 1953 memorandum is not clear. Since the memorandum dealt with atomic and other unconventional forms of warfare, its classification may have been routine. There is evidence that the DOD had a general desire to keep hidden from public view any indication that it was involved in biological and chemical warfare-related research; the Wilson memorandum, of course, was just such an indication. In September 1952, the Joint Chiefs of Staff advised the services to "[e]nsure, insofar as practicable, that all published articles stemming from BW [biological warfare] and CW [chemical warfare] research and development programs are disassociated from anything which might connect them with U.S. military endeavor."[84]

In one sense the memorandum is a landmark in its official recognition of the Nuremberg Code, but in another sense it also generates important questions. Having determined to recognize international principles of human rights, why, or how, could the secretary have limited their application to some, but not all, human experiments? Why was the policy directed exclusively to experiments related to "atomic, biological, and chemical warfare"? Moreover, was the policy intended to govern such research wherever it was conducted; for example, when it was performed by private contractors, as well as by intramural researchers? How was a directive issued in secret implemented?

Communicating the 1953 Wilson Memorandum

That there were problems in the dissemination of Secretary Wilson's Top Secret memorandum is evidenced in a memorandum containing queries by officials of the Armed Forces Special Weapons Project (AFSWP), within a year of the Wilson memorandum's issuance. The AFSWP, now the Defense Nuclear Agency (DNA), was at the hub of DOD nuclear weapons research. In the course of a routine review of research reports, an AFSWP official learned that "volunteers were injured as a consequence of taking part in [a] field experiment" of flashblindness conducted at an atomic bomb test before the Wilson memorandum was issued (see chapter 10). The AFSWP reviewer immediately concluded that a "definite need exists for guidance in the use of human volunteers as experimental subjects."[85]

On further inquiry, the AFSWP reviewer found that a policy already existed, but had not been disseminated to investigators. A follow-up

memorandum, evidently written in early 1954, records:

In November 53 it was learned that there existed a T/S [Top Secret] document signed by the Secretary of Defense which listed various requirements and criteria which had to be met by individuals contemplating the use of human volunteers in Bio-medical or other types of experimentation.... It was learned that although this document details very definite and specific steps which must be taken before volunteers may be used in experimentation, no serious attempt has been made to disseminate the information to those experimenters who have a definite need-to-know.[86]

"The lowest level at which it had been circulated," the AFSWP reviewer learned, "was that of the three Secretaries of the Services." Efforts by an assistant secretary to "downgrade" the document had "not been able to obtain concurrence." The reviewer hoped that "this letter shall point up the need for some relaxation of the grip in which this document is now held, at least on a definite need-to-know basis."[87] (The application of the Wilson memorandum to further experiments conducted at atomic bomb tests is discussed in chapter 10.)

Implementation in the Army

The Army did take substantial steps to put into effect the Wilson memorandum. In June 1953 the Army chief of staff, John C. Oakes, issued a memorandum implementing the secretary of defense's policy in toto. Referred to in the Army as CS:385, this memorandum was initially classified Top Secret, but was declassified the following year. In addition to the provisions of the Wilson memorandum, the Army document required the prior review and approval of both the surgeon general and the secretary of the Army. The Army's memorandum also contained legal analysis that explained the source of the Army's authority to perform human experiments in the first place and the limits that this authority put on the selection of subjects.[88] Even in the midst of the Korean War, the Army did not view it as self-evident that the DOD could engage in human experiments or choose any subjects it wished. The memorandum explained that the authority to experiment on humans came from congressional enactments, including provisions for research and development.[89]

Interestingly, choice of subjects was to be governed by the Army's ability to ensure compensation in the case of death or disability.[90] This could be provided, the lawyers declared, only upon express congressional action. In the case of military personnel and contractor employees there was such provision. But there was no such authority in the case of private citizens who offered their services. The Army lawyers recommended, and the CS:385 policy provided, that private citizens not employed by Army contractors could not serve as research subjects.[91]

On March 12, 1954, the Army Office of the Surgeon General (OSG) issued an unclassified statement entitled "Use of Human Volunteers in Medical Research: Principles, Policies, and Rules."[92] This document too restated the Nuremberg principles. In contrast with the Wilson and Oakes memorandums, it was *not* restricted to research related to atomic, biological, or chemical warfare. Instead, the OSG statement was directed to "medical research" with human volunteers generally.[93]

Moreover, while CS:385 did not state directly whether it applied to contract researchers, the 1954 OSG statement was transmitted to at least some university researchers with the prefatory note, "To be used as far as applicable as a non-mandatory guide for planning and conducting contract research."[94] There is evidence that the OSG's requirements were sometimes more than "non-mandatory guides." For example, in a June 27, 1956, letter to the the Armed Forces Epidemiological Board, a Tulane University public health researcher agreed that his vaccine experiments with prisoner subjects would be conducted only after written consent was obtained from the subjects.[95] The Tulane researcher indicated that, with respect to his application for funding, "I have held it up since Dr. Dingle indicated I be familiar with the statement of the Office of the Surgeon General re the use of human volunteers. . . . I have read it and believe that our past and future work have [*sic*] and will comply with the rules stipulated."[96] Moreover, this researcher provided a written statement to supplement his original proposal that explained how the OSG requirements would be met. In

another case, a proposal involving measles and normal children, an AFEB official advised the researcher to "take [the OSG policy] into consideration in writing the proposal."[97]

As discussed earlier, in 1952 the Army obtained congressional authority to indemnify contract researchers in the event that an experiment caused injury or death. There is evidence that the Army sought to link the grant of an indemnification clause (ASPR 7.203.22, "Insurance—Liability to Third Persons") to contractor acceptance of the principles stated by the Army surgeon general. In a March 1957 letter to the University of Pittsburgh, which was proposing to use medical student-volunteers in a (nonradiation) experiment, the Army told Pittsburgh that the provision of the clause was "contingent upon your adhering to the following [March 1954 Office of the Surgeon General] principles, policies, and rules for the use of human volunteers in performing subject medical research contracts."[98]

While the evidence clearly shows that Army officials sought to apply the Nuremberg Code policy to contractors, it did not meet with complete success, and the full extent of its efforts remains unclear. As we see in chapter 2, in the early 1960s Harvard successfully resisted the inclusion of the Nuremberg Code language in its medical research contracts with the Army. As we see in chapter 8, which discusses DOD funding of research on the effects of total-body irradiation, the indemnification language was included in at least some contracts in which the surgeon general's policy was not mentioned. By 1969, however, the policy may have become standard in Army contracts under the authority of the Medical Research and Development Command.[99]

There are several possible explanations for the seeming absence of widespread inclusion of the surgeon general's memo as a contractual requirement, at least where indemnification was provided for. First, as discussed below, it is possible that the 1954 policy was meant to apply to research with healthy subjects, and not sick patients. (However, even if that were generally the case, the provision of indemnification might be expected to have triggered reflection on this limitation.) Second, as a related matter, the evidence we are reviewing shows a tension between the government's declaration of a principle and its readiness to actively insist that the principle be honored within the privacy of the doctor-patient relationship.

Finally, Army imposition of the surgeon general's principles may also have depended on the nature of its interest in the research being done. An April 3, 1957, memo distinguished cases where the institution "because of its primary interest, would conduct the research even without support of the OSG," from cases where "the study is conducted at the insistence of OSG." In the former case the strategy would be to seek cost-sharing contracts, in which the institution would "assume all responsibility for any possible effects resulting from the experimentation." In the latter case, the indemnification clause would be provided, but the March 1954 policy would also be required and included in the contract directly or by reference.[100]

It is not clear that the 1954 OSG policy on human volunteers was intended to apply to research with patients. The term *volunteer* is ambiguous but at the time was commonly used to refer to healthy subjects. Nonetheless, a 1962 Army memorandum that declared that since World War II "by and large research has been conducted in strict accordance with the Nuremberg Code" mentions patients.[101] The memo reported that a recent survey of contract research found that the volunteers treated in accord with the Nuremberg Code included "3,000 students, 250 patients, and 300 prisoners." It is not known what kind of research these 250 patients were involved in, nor is it known what proportion of the patients who had been subjects of research supported or conducted by the Army since World War II were represented by these 250.

Unfortunately, the 1962 review's confident declaration that Army research complied with the Nuremberg Code was too sanguine. In 1975, following public revelations that the Army and the CIA had conducted LSD experiments on unwitting subjects, the Army inspector general reviewed the application of the June 1953 policy to drug testing. The inspector general's review led to the declassification of the 1953 Wilson memorandum. The inspector general found that

the Army had, with one or two exceptions, used only "volunteers" for its drug-testing program. However, the "volunteers were not fully informed, as required, prior to their participation, and the methods of procuring their services in many cases appeared not to have been in accord with the intent of Department of the Army policies governing use of volunteers in research."[102]

Additional DOD Research Requirements

While the Navy is not known to have taken specific action in response to the 1953 Wilson memorandum, we have already noted that the Navy had long since provided for prior review and voluntary participation in some cases. The 1951 Navy "Manual of the Medical Department" required secretarial approval of human experimentation and the use of volunteers. These requirements applied to "experimental studies of a medical nature" involving "personnel of the Naval Establishment (military and civilian)."[103] Participation was to be "on a voluntary basis only."[104] The manual also mandated prior review for research with patient-subjects. "Clinical research," including "research projects and therapeutic trials," was to be "authorized by" the Bureau of Medicine and Surgery.[105]

At least for research with radioisotopes, the requirement for voluntary participation may have applied to patient-subjects as well as healthy subjects. In 1951 the Navy debated adoption of a permission form for the use of radioisotopes for patients at naval hospitals.[106] This form, to be signed by either the patient or the responsible next of kin, authorized the use of "tracer-therapeutic" doses "obtained from the Atomic Energy Commission for research purposes."[107]

Although it is not clear that the Army rules implementing the 1953 Wilson memorandum applied to patient-subjects, there is some evidence that consent forms that were usually used for surgical procedures were used in patient-related experimental settings involving radioisotopes. In 1955 an official from the Letterman Army Hospital in San Francisco asked the Walter Reed Hospital about the need for written "permission" forms for "test doses" of radioisotopes.[108] In response, the Army indicated that a standard form used for operations and anesthesia should also be employed, at the physician's discretion, when "authorization for administration of radioisotope therapy is desired."[109]

In the Air Force, a 1952 regulation on clinical research mandated safety and administrative procedures for the use of humans in experiments at Air Force medical facilities.[110] This regulation required prior group review but did not mention consent provisions or refer to the subjects as volunteers. In 1958 a letter from the Air Force's Air Research and Development Command describes the policy for the use of humans in "hazardous research and development tests." This policy reiterated the requirement for prior review discussed in the 1952 regulation. In this context, however, subjects were to be "volunteer[s]" who "underst[ood] the degree of risk involved in the experiment."[111]

What, then, were the operative rules in the Department of Defense for research involving human subjects in the 1940s and 1950s? By the mid-1950s, for the entire DOD for research related to atomic, biological, and chemical warfare, and for all research involving "human volunteers" in the Army, the formal rules were the ten principles of the Nuremberg Code and the additions included in the secretary of defense's 1953 policy. According to the 1975 testimony of the surgeon general of the Army before the U.S. Senate and the internal review conducted by the Army inspector general, these principles were Army "policy."[112] At the same time, as the inspector general reported in 1975 and as we discuss further in chapter 10, these requirements were not always known or followed. While there were attempts to implement the Army surgeon general's 1954 policy, it is not known how the policy's provisions, including the requirement to obtain voluntary consent, were interpreted. The Navy's 1951 requirements for prior review and voluntariness applied to all research involving Navy personnel.

The extent to which research rules applied to patient-subjects in the clinical setting is less clear. There is some indication that in some cases standard consent forms, akin to the surgical permits in use at the time, were employed with patients at military hospitals who were administered "test doses" of radioisotopes.

THE NATIONAL INSTITUTES OF HEALTH AND THE VETERANS ADMINISTRATION

During the late 1940s and 1950s, the AEC and DOD were by no means the only agencies sponsoring research involving human subjects. The Department of Health, Education, and Welfare (DHEW), through two of its components, the Public Health Service and the NIH, was emerging during this period as the dominant government agency sponsoring human biomedical research. The Veterans Administration (VA) as well conducted a large medical research program that involved the use of radioisotopes in numerous human experiments.

In the early 1950s NIH participated in some of the discussions preceding the issuance of the 1953 secretary of defense memorandum. At the request of a DOD official for information on NIH's approach to the use of human subjects, NIH responded with an April 1952 letter that included a draft statement on the "Ethical Principles Underlying Investigations Involving Human Beings." Among its other provisions, the April 28, 1952, draft states that

[t]he person who is competent to give consent to an investigative procedure must do so. He must have legal capacity to give consent and be able to exercise free choice, without the intervention of any element of force, fraud, deceit, duress, constraint or coercion. He must have sufficient knowledge and comprehension of the nature of the investigation to enable him to make an understanding and enlightened decision. He must therefore be told the nature, duration, and purpose of the experiment; the method and means by which it is to be conducted; the inconveniences and hazards reasonably to be expected; and the effects upon his health or person which can reasonably be expected to come from his participation in the investigation. He should understand, furthermore, that by his participation he becomes a co-investigator with the physician.[113]

Although it is not known what became of this draft statement, around this time NIH had good reason to develop a policy on the use of human subjects. In 1953 NIH opened the Clinical Center, a state-of-the-science research hospital. The center adopted a policy requiring "voluntary agreement based on informed understanding" from all research subjects and written consent from some patient-subjects involved in research that the physician believed to be unusually hazardous.[114] Written consent was required from all healthy, "normal" subjects of research beginning in 1954.[115] Additionally, NIH began a system of group review of proposed research that became a model for today's institutional review boards (IRBs).[116] Thus, the NIH policy appears to be the first instance of a single policy that expressly provides for consent from all subjects, be they healthy or sick. Even so, the policy was still limited to research at the Clinical Center and did not apply to the considerable amount of NIH-funded research being undertaken by grantees (extramural research).

The question of whether "patients," as well as healthy, "normal" volunteers, should give written consent arose in the development of the NIH policy. Legal counsel at NIH advised that, "from a legal point of view," there should be a "written statement . . . indicating the patient's awareness of the nature of the particular investigation in which he was to participate and acceptance of any particular inconvenience or risk inherent in his participation."[117] A signed form offered the best proof that a "policy" of "informed consent" was followed for all subjects enrolled in studies at the center.

The NIH attorney wrote that while the Clinical Center's Medical Advisory Board did not disagree with the principle, it did disagree with the need for a written statement:

[O]f the members that expressed their views, and most did so, all rejected such a proposal. The rejection was due, as I understand it, not to any particular detail but rather a more basic objection to written, as opposed to oral, statements. There was apparently, therefore, no objection to providing the patient with enough information to permit him to exercise an informed choice of participation or refusal as long as not reduced to writing for his signature.[118]

Nonetheless, the principle that all research subjects, including healthy subjects in the "normal volunteer" program and patient-subjects, should

make an informed choice seems to be acknowledged in the Medical Advisory Board's position.

The NIH Clinical Center approach adopted by the mid-1950s—written consent from healthy subjects and from only certain patient-subjects—persisted through the early 1960s and was paralleled in policies of the DOD and the AEC. The view that written consent from patients might unnecessarily interfere with doctor-patient relationships prevailed.

Within the NIH, dialogue continued throughout the 1950s, setting the stage for the leading role DHEW was to take in formulating human research regulations in the 1960s (see chapter 3).[119]

Although the NIH was by far the dominant agency in research involving human subjects, a significant amount of radioisotope research occurred at the VA. The VA research program employing radioisotopes at VA medical centers began in 1948.[120] This program was limited to VA hospitals affiliated with medical schools. From its inception, this program involved a system of prior group review by local radioisotope committees, normally composed of non-VA-affiliated teaching staff of the affiliated medical school.[121] These committees reviewed all research proposals and approved all research conducted at VA radioisotope units.

In its formative years, the advisers to the new VA program included Stafford Warren, Shields Warren, and others who were likely to be familiar with the consent principles articulated by the AEC. Nonetheless, the earliest evidence of a consent policy at the VA comes in the form of a 1958 general counsel's opinion on whether the VA could participate in certain research. The general counsel asserted that

persons who participate [in human subject research] must voluntarily consent to the experiment on themselves. Such consent must rest upon an understanding of the hazards involved. The volunteer may withdraw from the experiment at any time. Moreover, before the experiment, steps to reduce the hazard, as for example, indicated research on animals, must be made.[122]

This opinion was written in response to two proposed research projects, and it is not known if it was implemented in the projects or applied to others.

CONCLUSION

Records now available show that at the highest reaches of Cold War bureaucracies officials discussed conditions under which human experimentation could take place. These discussions took place earlier and in greater, although by today's standards uncritical and less searching, detail than might have been assumed. Nonetheless, the stated positions that resulted were often developed in isolation from one another, were neither uniform nor comprehensive in their coverage, and were often limited in their effectuation. Several interrelated factors seem to have been prominent in causing these discussions to take place and in determining the scope of the requirements that were declared and the efforts that were undertaken to implement them. We summarize these key factors below.

Administrative and Legal Circumstance

The creation of new programs, or the qualitative expansion of old ones, impelled officials, lawyers, and researchers to reflect on the rules to govern them. While these rules were sometimes cast as "legal" or "financial" requirements, they often included provisions, such as a requirement for written consent, that appear similar to statements in requirements that govern the conduct of research today. The language used to describe these rules was often that of law or administration, such as "waiver" or "release" forms, or it may have had particular meaning to researchers at the time, such as "clinical testing." As a result, it is often hard to compare these rules to current requirements, which have benefited from intervening decades of linguistic and conceptual refinement.

Professional Cultures

Differing professions brought their own tools and perspectives to discussions of conditions under which human subjects research could pro-

ceed. For example, lawyers were likely to insist on obtaining documented evidence of patient consent, while medical professionals emphasized the importance of the trust that underlay the relationship between doctor and patient; they sometimes objected to the use and implications of written consent forms.

If consent procedures were a source of disagreement, the need to minimize risk to subjects was not. In creating and administering the AEC's radioisotope distribution program, physician investigators and other researchers placed a premium on controlling and minimizing risk in the "human use" of radioisotopes. This emphasis on the establishment of administrative and educational procedures to control risk, the details of which are discussed in chapter 6, embodied an essential principle of ethical research.

The requirement for prior review included in the isotope distribution program was, as we have seen, also present elsewhere. Even before 1944, approval of the secretary of the Navy was required for research with human subjects. The secretary of the Army required prior approval of research related to atomic, biological, and chemical warfare in 1953. In the Air Force, secretarial approval of human experiments was codified in 1952. At NIH, prior group review was employed as a policy from 1953 on. The VA, whose program developed under the eye of AEC experts and advisers, relied on local isotope committees.

The Nature of the Subjects

While voluntary consent was acknowledged as a condition of human research by some government agencies well before 1944, it was not as broadly applied as it is today. Requirements of voluntary consent were asserted most clearly and consistently where the subjects were healthy. As a practical matter, healthy subjects are not likely to participate in experiments without specific request, and as a legal matter the invasion of a person's body in the absence of a prior relationship that might justify it has long been unacceptable. Still more important, the arbitrary use of people in experiments is incompatible with respect for human dignity.

The use of *patients* in medical research appeared in a different historical context from that of *healthy subjects,* and the agencies appear to have responded accordingly. From the perspective of the medical profession, the age-old tradition of the doctor-patient relationship, as we shall see in the next chapter, provided a justification for research with the potential to benefit patients, but not, of course, for healthy subjects who were not under medical care. There is little evidence that the agencies questioned whether research with patients that *did not* offer a prospect of benefit warranted a different response. An exception is the position articulated by the AEC's general manager in 1947, which made the possibility of benefit to the patient-subject a condition of permissible research, at least where the research involved "poisonous or harmful" substances. However, there is little indication that this provision was ever implemented.

The period we reviewed in this chapter led to considerable public disquiet about the use of healthy subjects and about the use of ill and institutionalized people in research from which they could not possibly benefit. It was this disquiet, in the wake of several well-publicized incidents, that formed the basis of the mid-1960s reforms of federal policy governing research with human subjects (see chapter 3). The focus on the way that patient-subjects were used in clinical research that offered some prospect of benefit, and particularly on consent issues, came much later. The latter discussion is one that continues today, as is evident from the Advisory Committee's work on current research regulation that is described in part III.

The Degree of Risk

To the extent that there was discussion in the 1940s and the 1950s of consent for patient-subjects, it seemed to arise mainly in circumstances in which those who were ill would be put at unusual risk from the research.

As we have seen, the AEC's radioisotope distribution division concluded that consent was required where patients were being subjected to "larger doses for investigative purposes" that apparently posed unusually hazardous or unknown risks. Similarly, from its establishment at mid-century, the AEC's hospital at Oak Ridge, which focused on new and potentially risky experimen-

tal cancer treatment, did have routine requirements for consent. Likewise, from its 1953 birth, the NIH's Clinical Center established a policy that recognized that patient choice was important for all kinds of research with patients, and written consent was required when an experiment involved an unusual hazard.

Formal Policies and Public Morality

It is important not to get lost in the details of the various documents we have cited in this chapter. What is most significant about the discussions that took place in federal agencies from the mid-1940s through the 1950s is the fact that so many of the ideas and values with which we are familiar were apparent then. That does not mean that the same words were used or that when they were used they had the same meaning as they do for us today. But it does mean that there were certainly more or less rough ideas about voluntary consent and minimization of risk. As we have seen in this chapter, these ideas were very much in play in the culture of the time.

NOTES

1. The "Common Rule" applies requirements for voluntary consent, prior review, and risk analysis to all federally sponsored research. This rule is discussed in chapter 14.
2. David Rothman, *Strangers at the Bedside: A History of How Law and Bioethics Transformed Medical Decision Making* (New York: Basic Books, 1991), and Ruth Faden and Tom Beauchamp, *A History and Theory of Informed Consent* (New York: Oxford University Press, 1986).
3. George J. Annas and Michael A. Grodin, eds., *The Nazi Doctors and the Nuremberg Code; Human Rights in Human Experimentation* (New York: Oxford University Press, 1992), 343–345.
4. See Faden and Beauchamp, *A History and Theory of Informed Consent,* and Mark S. Frankel, "Public Policymaking for Biomedical Research: The Case of Human Experimentation" (Ph.D. diss., George Washington University, 9 May 1976).
5. Stafford L. Warren, Chairman, Interim Medical Advisory Board ("Report of the 23–24 January 1947 Meeting of the Interim Medical Committee of the United States Atomic Energy Commission") (ACHRE No. UCLA-111094–A-26). The report summarized "specific projects" at twelve institutions. The projects at the University of Rochester included

"Study of the Metabolism of Plutonium, polonium, radium, etc. in human subjects" (p. 8). In the case of Berkeley, the projects identified to Dr. Stone were

1. Studies in whole-body radiation of human subjects by external and internal radiation.
2. Studies on the metabolism of radioactive iodine in animals and man.
3. Joint studies with Dr. Joseph G. Hamilton to evaluate the therapeutic applications of the fission products and the fissionable elements.
4. Exploration and therapeutic application of other radioactive elements and compounds (p. 11).

A 14 March 1947 memorandum from Austin Brues, director of the Biology Division of the Argonne National Laboratory, records that "clinical testing programs" had only been authorized, at least for the time being, at Berkeley and Rochester. However, Brues urged that Argonne also be included. On behalf of this request he cited the University of Chicago's "work using human subjects" with specific reference to a report on plutonium injections. He further noted that human subject work also included the Argonne project list provided at the January meeting. A. M. Brues, Director, Biology Division, to N. Hilberry, Associate Laboratory Director, 14 March 1947 ("Clinical Testing") (ACHRE No. DOE-050195–B).

6. Stafford Warren, Chairman, Interim Medical Advisory Committee, to Carroll Wilson, General Manager, AEC, 30 January 1947 ("The opinion on Clinical Testing . . .") (ACHRE No. DOE-051094–A-439), 1.

7. John L. Burling, Deputy General Counsel's Office, AEC, to Edwin Huddleson, Jr., Deputy General Counsel, AEC, 7 March 1947 ("Clinical Testing") (ACHRE No. DOE-051094–A-468), 2–3.

8. Ibid., 3.

9. Carroll L. Wilson, General Manager of the AEC, to Stafford Warren, the University of California at Los Angeles, 30 April 1947 ("This is to inform you that the Commission is going ahead with its plans . . .") (ACHRE No. DOE-051094–A-439), 2.

10. Ibid.

11. Ibid.

12. Robert J. Buettner, Assistant to Chairman, Interim Medical Advisory Committee, AEC, to B. M. Brundage, Chief, Medical Division, AEC, 12 May 1947 ("Transmitted herewith for your information . . .") (ACHRE No. DOE-051094–A-439), 1.

13. Note in medical chart of Cal-3, dated 18 July 1947 ("Elmer Allen chart") (ACHRE No. DOE-051094–A-615). For more information on this case, see chapter 5.

14. Wilson to Warren, 30 April 1947.

15. University of California at San Francisco, February 1995 ("Report of the UCSF Ad Hoc Fact Finding Committee ") (ACHRE No. UCSF-022495–A-6), 27.

16. J. C. Franklin, Manager, Oak Ridge Operations, to Carroll Wilson, General Manager, AEC, 26 September 1947 ("Medical Policy") (ACHRE No. DOE-113094–B-3), 2. Although the motivation for Oak Ridge's inquiry is not entirely clear, it seems to have come in part from concerns of Albert Holland, M.D., who became the acting medical adviser at Oak Ridge after Major Brundage retired. Holland served on the committee that oversaw the use of radioisotopes in human research, discussed in chapter 6. In November 1947 Holland wrote, in regard to the isotopes distribution program: "How far does the AEC's moral responsibility extend in this program?" Albert Holland, Jr., Medical Adviser, Oak Ridge, to J. C. Franklin, Manager of Oak Ridge Operations, 7 November 1947 ("Medical and Operational Decisions") (ACHRE No. DOE-113095–B-10), 2.

17. Unknown author to the Advisory Committee for Biology and Medicine, 8 October 1947 ("It is the desire of the Medical Advisor's Office . . .") (ACHRE No. DOE-051094–A-502).

18. Atomic Energy Commission, Advisory Committee for Biology and Medicine, minutes of 11 October 1947 (ACHRE No. DOE-072694–A-1), 10.

19. Ibid.

20. Ibid.

21. Carroll Wilson, General Manager, AEC, to Robert Stone, University of California, 5 November 1947 ("Your letter of September 18 regarding the declassification of biological and medical papers was read at the October 11 meeting of the Advisory Committee for Biology and Medicine.") (ACHRE No. DOE-052295–A-1).

22. Carroll Wilson, General Manager, AEC, to Alan Gregg, Chairman of the AEC Advisory Committee for Biology and Medicine, 5 November 1947 ("I want to thank you for your letter of October 14 concerning the questions raised by Dr. Stone in his letter to me of September 18 regarding declassification of biological and medical papers containing information on the experimental use of radioisotopes in human beings conducted under AEC sponsorship.") (ACHRE No. DOE-052295–A-1).

23. *Salgo v. Leland Stanford Jr. University Board of Trustees*, 317 P.2d 170 (1957).

24. Joseph Volpe, interview by Gregg Herken, Dan Guttman, and Debra Holland (ACHRE), transcript of audio recording, 6 October 1994 (ACHRE Research Project Series, Interview Program Files, Targeted Interview Project), 24–42.

In a May 1995 interview, Volpe agreed that a letter written by the general manager constituted a "policy." The transcript of the interview records:

> Interviewer: . . . today there are regular procedures for getting something recognized as a policy, including publication and so forth. In 1947, when the general manager writes a letter, is that a policy?
>
> Mr. Volpe: Yes, Yes.

Mr. Volpe noted that while the question of the precise authority of the general manager was not without controversy, Chairman Lilienthal "believed in delegation of authority and so always took measures to strengthen the general manager's hand on these things." Joseph Volpe, interview by Barbara Berney, Steve Klaidman, Dan Guttman, Lanny Keller, Jonathan Moreno, Patrick Fitzgerald, and Gilbert Whittemore (ACHRE), transcript of audio recording, 18 May 1995 (ACHRE Research Project Series, Interview Program Files, Targeted Interview Project), 37–38.

25. Leslie M. Redman, Los Alamos Laboratory, to Dr. Alberto F. Thompson, Chief, Technical Information Service, DBM, 22 January 1951 ("I find myself concerned in the course of duty with the review of papers relating to human experimentation.") (ACHRE No. DOE-051094-A-609).

26. Warren did not cite the context for Wilson's discussion of these conditions, that is, the need for criteria for declassification.

27. Shields Warren, Director, DBM, to Leslie Redman, "D" Division, Los Alamos National Laboratory, 5 March 1951 (". . . to reply to your letter of January 22, 1951, concerning policies on human experimentation.") (ACHRE No. DOE-051094-A-603).

28. Everett Idris Evans, M.D., Medical College of Virginia, to John Z. Bowers, M.D., Assistant to the Director, DBM, AEC, 8 April 1948 ("We have recently obtained approval from the Isotopes Division for human use of P32. . .") (ACHRE No. DOE-051094-A-64).

29. John Z. Bowers, Assistant to Director, DBM, AEC, to Everett Idris Evans, M.D., Medical College of Virginia, 27 April 1948 ("Thank you for recent letter requesting information regarding isotopes.") (ACHRE No. DOE-050194-A-480).

30. Nathan H. Woodruff, Chief Technical Division, Isotopes Division, to Everett I. Evans, M.D., Medical College of Virginia, 14 May 1948 ("Your letter of April 8 to Dr. Bowers has been referred to me for answer.") (ACHRE No. NARA-082294-A-10).

31. U.S. Atomic Energy Commission, Advisory Committee for Biology and Medicine, agenda of 14 February 1948 (ACHRE No. DOE-072694-A), 2.

32. In addition to the document discussed above, there is some indication that the AEC Isotopes Division was charged with ensuring that consent was obtained. In the early 1970s, when the AEC conducted an investigation into the plutonium experiments, Shields Warren told the investigators that his recollection was that ethical issues were addressed at the time by the issuance of prospective policies. Warren stated:

I think the way it [concern about the plutonium injections] was handled was that Alan Gregg and I agreed the best way to do [it] was to see that the rules were properly drawn up by the . . . Human Applications Isotope Committee, which had then come into being, so that

use without full safeguards could not occur, and that we saw no point in bringing this up after the fact as long as we were sure that nothing of this sort could happen in the future.

Shields Warren, interview by L. A. Miazga, Sidney Marks, Walter Weyzen (AEC), transcript of audio recording, 9 April 1974, 10–11 (ACHRE No. DOE-121294-D-14).

33. Unknown author, unpublished draft, 29 March 1948 ("The Experimental Use of Radioactive Materials in Human Subjects at AEC Establishments") (ACHRE No. DOE-050194-A-267).

34. Subcommittee on Human Applications, minutes of 22–23 March 1948, as discussed in the minutes of the 13 March 1949 meeting. S. Allan Lough, Chief, Radioisotopes Branch, to H. L. Friedell, G. Failla, J. G. Hamilton, and A. H. Holland, 19 July 1949 ("Revised Tentative Minutes of March 13, 1949 Meeting of the Subcommittee on Human Applications of Committee of U.S. Atomic Energy Commission, AEC Building, Washington, DC") (ACHRE No. DOE-101194-A-13), 5.

35. The subcommittee was not definitive about when larger doses were permitted, however. The policy was to apply in "instances in which the disease from which a patient is suffering permits the administration of larger doses for investigative purposes." U.S. Atomic Energy Commission, Isotopes Division, September 1949 ("Supplement No. 1 to Catalogue and Price List No. 3, July 1949") (ACHRE No. DOD-122794-A-1), 3–4.

36. While these statements were perhaps more than was told to patient-subjects in other institutions, they did not necessarily provide details about the research. In the application for admission, the applicant agreed to "such operations and biopsies as are deemed necessary and advisable by the hospital." Oak Ridge Institute of Nuclear Studies, 1950 ("Application for Admission to the Medical Division Hospital") (ACHRE No. DOE-121494-C-1), 1.

Upon admission, the applicant was required to sign a "Waiver and Release" that did not describe the treatment, but included a lengthy release from the patient, the patient's "heirs, executors, administrators, and assigns," for any "causes of action, claims, demands, damages, loss, costs, and expenses, whether direct or consequential," associated with or resulting from the care of the hospital. This form notes that the hospital has described the "character and kind of treatment." Oak Ridge Institute of Nuclear Studies, 1950 ("Waiver and Release") (ACHRE No. DOE-121494-C-3), 1.

37. Oak Ridge Institute for Nuclear Studies, 1950 ("Waiver and Release") (ACHRE No. DOE-121494-C-3).

38. Program Committee of the Division of Biological and Medical Research of the Argonne National Laboratory, minutes of 22 January 1951 (ACHRE No. DOE-051095-B), 3.

39. Thomas Shipman, M.D., Health Division Leader, Los Alamos Laboratory, AEC, to Dr. Charles Dunham, Director, DBM, AEC, 18 June 1956 ("Two questions have recently arisen—one of them specific, the other general—wherein we need an opinion from you.") (ACHRE No. DOE-091994–B-1).

40. Charles Dunham, Director, DBM, AEC, to Thomas Shipman, Health Division Leader, Los Alamos Laboratory, 5 July 1956 ("This is in response to your letter of June 18.") (ACHRE No. DOE-091994–B-2). In addition to consent, Dunham indicated that the research should proceed so long as (a) the doses were small, "true tracer doses"; (b) the proposal was approved by a senior medical officer; and (c) the work was supervised by a licensed physician.

41. T. L. Shipman, Health Division Leader, Los Alamos Laboratory, to Staff Distribution, 12 July 1956 ("Administration of Tracer Doses to Humans") (ACHRE No. DOE-091994–B-3), 1. Also, T. L. Shipman, Health Division Leader, Los Alamos Laboratory, to "Distribution," 3 September 1963 ("Administration of Tracer Doses to Humans For Experimental Purposes") (ACHRE No. DOE-091994–B-4), 1.

42. Isotopes Extension, Division of Civilian Application, U.S. AEC, "The Medical Uses of Radioisotopes, Recommendations and Requirements of the Atomic Energy Commission" (Oak Ridge, Tenn.: AEC, February 1956), 15.

43. U.S. Department of the Army, AR 40–210, *The Prevention of Communicable Diseases of Man—General* (21 April 1925).

44. Charles W. Shilling, Medical Corps, USN, Retired, undated paper ("History of the Research Division, Bureau of Medicine and Surgery, USN") (ACHRE No. DOD-080295–A), 74.

45. The Secretary of the Navy to All Ships and Stations, 7 April 1943 ("Unauthorized Medical Experimentation on Service Personnel") (ACHRE No. DOD-091494–A-2).

46. J. E. Moore, M.D., to Dr. A. N. Richards, excerpt of letter dated 6 October 1942 ("I have recently received an inquiry from Dr. Charles M. Carpenter of the University of Rochester School of Medicine who believes that he may be able to work out a human experiment on the chemical prophylaxis of gonorrhea.") (ACHRE No. NARA-060794–A-1).

47. A. N. Richards to J. E. Moore, 31 October 1942 (" Revision of Dr. Richards' letter of October 9, 1942") (ACHRE No. NARA-060794–A-1). Stafford Warren, the Manhattan Project medical director, also came from the University of Rochester. It is not clear how, if at all, the CMR's views on human experiments were accounted for in Manhattan Project research.

48. Rothman, *Strangers at the Bedside,* 30–50.

49. The Chief of the Bureau of Medicine and Surgery to the Officer-in-Charge, Naval Laboratory Research Unit No. 1, University of California, Berkeley, California, 6 March 1943 ("Proposed Clinical Evaluation of Influenza Antiserum, and Messages concerning Influenza Virus Specimens") (ACHRE No. DOD-062194–C-1).

50. Ibid., 2.

51. Institute of Medicine, National Academy of Sciences, *Veterans at Risk: The Health Effects of Mustard Gas and Lewisite* (Washington, D.C.: National Academy Press, 1993), 66–69.

52. Ibid., 214.

53. Robert S. Stone, unpublished paper, "Irradiation of Human Subjects as a Medical Experiment," 31 January 1950 (ACHRE No. NARA-070794–A).

54. American Medical Association, Judicial Council, "Supplementary Report of the Judicial Council," *Journal of the American Medical Association* 132 (1946): 1090.

55. The Under Secretary of the Navy to the Secretary of Defense, 24 April 1950 ("Recommendation that the Armed Service conduct experiments on human subjects to determine effects of radiation exposure") (ACHRE No. NARA-070794–A).

56. Atomic Energy Commission, Advisory Committee for Biology and Medicine, transcript (partial) of meeting, 10 November 1950 (ACHRE No. DOE-012795–C-1), 28.

57. Ibid., 28–29.

58. J. G. Hamilton, University of California, to Shields Warren, DBM, AEC, 28 November 1950 ("Unfortunately, it will not be possible for me to be at the meeting on December 8 . . .") (ACHRE No. DOE-072694–B-45), 1.

59. Ibid.

60. Adam J. Rapalski, Administrator, the Armed Forces Epidemiological Board, DOD, to Chief, Legal Office, 5 January 1952 ("Draft of 'Agreement with Volunteer'") (ACHRE No. DOD-040895–A).

61. Lieutenant Colonel Robert J. O'Connor, Chief, Legal Officer, JAGD, to Colonel Frank L. Baier, Army Medical Research and Development, 23 October 1947 ("Protection of Research Project Volunteers") (ACHRE No. NARA-012395–A-4).

62. John R. Paul, Director, AEB, DOD, to Dr. Joseph Stokes, Jr., Children's Hospital, Philadelphia, Pennsylvania, 18 February 1948 ("This is in reply to your hand written request for a comment [from] me re your letter to Dr. Macleod dated 11 February on the subject of funds for the reimbursement of volunteer prisoners . . .") (ACHRE No. NARA-012395–A-1).

63. Ibid.

64. Committee Appointed by Governor Dwight H. Green of Illinois, "Ethics Governing the Service of Prisoners As Subjects In Medical Experiments," *Journal of the American Medical Association* 136, no. 7 (1948): 457–458.

65. C. J. Watson, M.D., Commission on Liver Disease, Army Epidemiological Board, to Colin MacLeod, President of the Board, AEB, 5 April 1948 ("I have given considerations in the past few weeks to the matter of using volunteers in penal institutions

for experimentation . . .") (ACHRE No. NARA-012395–A-2).

66. Ibid.

67. "Prisoner Dies After Injection in Disease Study," *Washington Post*, 6 May 1952, 3.

68. L. M. Harff, Contract Insurance Branch, to File, 25 April 1952 ("Research and Development Contracts—Medical Investigations) (ACHRE No. DOD-012295–A).

69. Adam J. Rapalski, Administrator, AEB, to Chief Legal Office, 14 October 1952 ("Applicability of Section 5, Public Law 557–82d Congress") (ACHRE No. NARA-012395–A).

70. Adam J. Rapalski, Administrator, AEB, to Members of the AEB, undated memorandum ("Applicability of Section 5, Public Law 557–82nd Congress") (ACHRE No. NARA-012395–A). In congressional hearings, the activities used to illustrate the purpose of the indemnification provision included test piloting, damage that might be caused by cloud modification research, and cataracts caused by the operation of a cyclotron. In addition, however, biomedical human experimentation was specifically addressed in the following exchange between Representative Edward Hebert and Colonel W. S. Triplet, from the Army Research and Development Division:

Mr. Hebert. Colonel, would you expand on the proposal to make the Government liable for losses and damages?

. . .

Colonel Triplet. There have been some experiments or types of research in the past which would have come under section 5 [the indemnification provision]. There are more coming up in the future. One of the early cases, long before the time of the bill, I would cite as an example is Dr. Reed in Cuba in 1900 utilized the services of 21 volunteers to study yellow fever, an extremely dangerous experiment. Two of these volunteers died. Eighteen of the others became seriously ill. As a result a special medal was awarded these people by Congress. That is an example of the type of experiment that at the present time is going on in the medical service.

Subcommittee Hearings on H. R. 1180 to Facilitate the Performance of Research and Development Work by and on Behalf of the Departments of the Army, the Navy, and the Air Force, and for Other Purposes; House of Representatives, Committee on Armed Services, Subcommittee no. 3, 6 June 1952, 621 (ACHRE No. NARA-10495–D).

71. Colonel George V. Underwood, Director, Executive Office, Office of the Secretary of Defense, to Mr. Kyes, Deputy Secretary of Defense, 5 February 1953 ("Use of Human Volunteers in Experimental Research") (ACHRE No. DOD-062194–A).

72. Melvin Casberg, Chairman, AFMPC, to the Secretary of Defense, 24 December 1952 ("Human Volunteers in Experimental Research") (ACHRE No. NARA-101294–A-3).

73. Ibid.

74. Jackson recommended changes to the Nuremberg Code: the elimination of the Nuremberg Code exception for self-experimentation by physicians and the express provision that prisoners, but not prisoners of war, could be used. We do not know what Jackson had "previously submitted." See Stephen Jackson, Assistant General Counsel in the Office of the Secretary of Defense and Counsel for the AFMPC, to Melvin Casberg, undated memorandum ("The standards and requirements to be followed in human experimentation") (ACHRE No. NARA-101294–A-3).

75. Ms. Rosenberg, a high-ranking official in the DOD, was an expert in labor relations and a New Dealer. Her role was recorded in Stephen Jackson to Melvin Casberg, Chairman, AFMPC, 22 October 1952 ("I discussed the attached with Mrs. Rosenberg . . .") (ACHRE No. NARA-101294–A-3).

76. Colonel Adam J. Rapalski, Administrator, Armed Forces Epidemiological Board, DOD, to Colin MacCleod, President, Armed Forces Epidemiological Board, DOD, 2 March 1953 ("The attached copy of letter I believe is self-explanatory.") (ACHRE No. NARA-012395–A-5).

77. F. Lloyd Mussells, Executive Director, Committee on Medical Sciences, RDB, DOD, to Floyd L. Miller, Vice Chairman, Research and Development Board, DOD, 12 November 1952 ("Human Experimentation") (ACHRE No. NARA-071194–A-2).

78. Ibid.

79. In a 10 November 1952 meeting the Committee on Chemical Warfare was read a draft of the AFMPC policy. One member remarked to general laughter: "If they can get any volunteers after that I'm all in favor of it." Committee on Chemical Warfare, RDB, DOD, transcript of the meeting of 10 November 1952 (ACHRE No. NARA-102594–A), 128. H. N. Worthley, Executive Director, Committee on Chemical Warfare, RDB, DOD, to the Director of Administration, Office of the Secretary of Defense, 9 December 1952 ("Use of Volunteers in Experimental Research") (ACHRE No. NARA-101294–A), 1.

80. This, at least, was the 1994 recollection of Lovett's military assistant, General Carey Randall, who served in the same role for Lovett's predecessor and successor. General Carey Randall, interview by Lanny Keller (ACHRE), transcript of audio recording, 20 September 1994 (ACHRE Research Project Series, Interview Program File, Targeted Interview Project), 17.

81. George V. Underwood, Director of the Executive Office of the Secretary of Defense, to Deputy Secretary of Defense Foster, 4 January 1953 ("I believe that Mr. Lovett has a considerable awareness of this proposed policy.") (ACHRE No. NARA-101294–A-1), 1.

82. Melvin A. Casberg, Chairman, Armed Forces Medical Policy Council, DOD, to the Secretary of Defense, 13 January 1953 ("Digest 'Use of Human Volunteers in Experimental Research'") (ACHRE No. DOD-042595–A), 1.

83. Secretary of Defense to the Secretary of the Army, Secretary of the Navy, Secretary of the Air Force, 26 February 1953 ("Use of Human Volunteers in Experimental Research") (ACHRE No. DOD-082394–A). The second paragraph of the memorandum stipulates its application to "Armed Services personnel and/or civilians on duty at installations engaged in such research. . . ." The Advisory Committee takes this stipulation to be in recognition of the separate authority of the medical services, as distinct from research and development commands.

84. W. G. Lalor, Secretary, Joint Chiefs of Staff, to Chief of Staff, U.S. Army, Chief of Naval Operations, Chief of Staff, U.S. Air Force, 3 September 1952 ("Security Measures on Chemical Warfare and Biological Warfare") (ACHRE No. NARA-012495–A-1).

85. Irving L. Branch, Colonel, USAF, Acting Chief of Staff, to the Assistant Secretary of Defense (Health and Medicine), 3 March 1954 ("Status of Human Volunteers in Bio-medical Experimentation") (ACHRE No. DOD-090994–C), 2.

86. Ibid., 3.

87. Ibid.

88. Brigadier General John C. Oakes, GS, Secretary of the General Staff, Department of the Army, to the Chief Chemical Officer and the Surgeon General, 30 June 1953 ("CS:385—Use of Volunteers in Research") (ACHRE No. DOD-022295–B-1) (CS385). This document was originally classified as Top Secret then downgraded to Confidential and declassified in June 1954. "Research Report Concerning the Use of Volunteers in Chemical Agent Research." Inspector General and Auditor General, 1975 (Army IG report), 77.

89. Oakes, sec. 3(a).

90. A series of memorandums from the Office of the Judge Advocate General preceded and shed light on the 30 June 1953 memorandum:

Colonel Robert H. McCaw, JAGC, Chief, Military Affairs Division, to the Chief, Research and Development, Office of the Chief of Staff, 6 April 1953 ("Volunteers for Biological Warfare Research") (ACHRE No. DOD-082294–B).

Colonel Robert H. McCaw, JAGC, Chief, Military Affairs Division, to the Chief, Research and Development, Office of the Chief of Staff, 10 April 1953 ("Volunteers for Biological Warfare Research") (ACHRE No. DOD-082294–B).

Colonel A. W. Betts, GS, Executive for the Chief of Research and Development, to Mr. J. N. Davis, Office of the Under Secretary of the Army, 15 April 1953 ("Use of Volunteers in Experimental Research") (ACHRE No. DOD-082294–B).

91. CS:385, sec. 3(d).

92. Army Office of the Surgeon General, 12 March 1954 ("Use of Volunteers in Medical Research, Principles, Policies, and Rules of the Office of the Surgeon General") (ACHRE No. DOD-120694–A-4).

93. Ibid., 1. A copy of this document was found in the files of John Enders, Ph.D., Nobel Laureate in Medicine and Physiology, 1954, Yale University.

94. Ibid.

95. John Fox, M.D., Professor of Epidemiology, Tulane University School of Medicine, to Captain R. W. Babione, Executive Secretary, AFEB, 27 June 1956 ("Finally I am able to complete and send to you the application for a research contract to study . . . ") (ACHRE No. NARA-012395–A).

96. Ibid.

97. W. McD. Hammon, M.D., Director, Commission on Viral Infections, AFEB, to John Enders, Children's Medical Center, 20 November 1958 ("This is to confirm our telephone call this morning, November 20th, regarding approval of the AFEB for the protocol of the experiment which you propose to carry out . . .") (ACHRE No. NARA-032495–B), 1.

98. Max H. Brown, Contracting Officer, to Vice Chancellor, Schools of the Health Professions, University of Pittsburgh, 12 March 1957 ("This is in reply to letter . . .") (ACHRE No. DOD NARA-012395–A-6) The DOD has not located the Pittsburgh contract itself, which may have been long since routinely destroyed; therefore, it cannot be said for certain that the 1954 surgeon general provisions were made a contract requirement.

99. Herbert L. Ley to Colonel Howie, 8 January 1969 ("Review of Department of the Army Policy on Use of Human Subjects in Research") (ACHRE No. DOD-063094–A).

100. Max H. Brown to Contracting Officer, OTSG, 5 August 1957 ("The Use of Human Test Subjects in Medical Research Supported by the Office of the Surgeon General") (ACHRE No. NARA-012395–A).

101. Donald L. Howie, Assistant Chief, Medical Research, 10 July 1962 ("Memorandum for the Record, Use of Volunteers for Army Medical Research") (ACHRE No. DOD-120694–A-3). It is worth noting that prior to this memorandum, in March 1962, the Army promulgated its first regulation specifically directed to the conduct of clinical research. This regulation (AR 70–25, 26 March 1962) specifically exempted "clinical research," which apparently included research conducted on patients. See chapter 3.

102. Army IG report, 1975.

103. Department of the Navy, Bureau of Medicine and Surgery, "Manual of the Medical Department," sec. IV, research article 1–17 (26 September 1951).

104. On the question of written documentation, interestingly, the manual stipulated: "[V]olunteers" should not "execute a release for future liability for negligence attributable to the Navy," but the manual required that a statement be "entered into the Individual's Health Record" indicating the project number and the physical and psychological effects, or lack of same, resulting from the investigation.

"Manual of the Medical Department," sec. IV, art. 1–17.

105. Ibid.

106. Loren B. Poush, Code 11, USN, to Code 74, USN (Bureau of Medicine and Surgery), 18 October 1951 ("Legal comments relative to proposed means of proper authorization and safeguard in use of radioisotopes") (ACHRE No. NARA-070794-A-4).

107. Code 74, USN, to Code 11, USN, 18 September 1951 ("Proposed Means of Proper Authorization and Use of Radioisotopes") (ACHRE No. NARA-070794-A-4), 2.

108. Paul O. Wells, Chief, Radiological Service, Letterman Army Hospital, to Elmer A. Lodmell, Chief, Radiological Service, Walter Reed Army Hospital, 14 January 1955 ("I am writing this letter at the suggestion of General Gillespie after having discussed with him the matter of requiring patients to sign a permit for radioisotope therapy.") (ACHRE No. DOD-012295-A).

109. Standard Form 522 (SF-522), "Clinical Record—Authorization for Administration of Anesthesia and Performance of Operations and Other Procedures," was proposed for use "in those instances when authorization for administration of radioisotope therapy is desired." Eugene L. Hamilton, Chief, Medical Statistics Division, to the Chiefs of the Medical Plans and Operations Division and the Legal Office, 3 August 1955 ("Permit for Radioisotope Therapy") (ACHRE No. DOD-012295-A).

In response to an inquiry from Walter Reed Army Hospital concerning the use of consent forms for patients, the Medical Statistics Division, recommending the use of SF-522, indicated that consent should be obtained when a procedure "carries an unusual risk." Additionally, the Medical Statistics Division recommended that patients should be "counselled as to the nature, expected results of, and risks involved in procedures." Eugene L. Hamilton, Chief, Medical Statistics Division, to the Chiefs of the Professional Division, Medical Plans and Operations Division, and the Legal Office, undated memorandum (probably November 1956) ("Forms for Authorization of Radiation Therapy") (ACHRE No. DOD-012295-A).

110. U.S. Air Force, Research and Development, "Clinical Research," AFR 80–22 (11 July 1952).

111. The Deputy Commander for Research and Development of the Air Force R&D Command to RADC, WADC, APGC, AFCRC, AFSWC, AFMTC, AFMDC, AFFTC, AFBMD (ARDC), AFOSR, 12 September 1958 ("Conduct of Hazardous Human Experiments") (ACHRE No. HHS-090794-A).

112. Richard R. Taylor, Surgeon General of the Department of the Army, testimony before the Subcommittee on Administrative Practice and Procedure of the Judiciary Committee and the Subcommittee on Health of the Labor and Public Welfare Committee, U.S. Senate, 94th Cong., 1st Sess., 10 September 1975 (ACHRE No. DOD-063094-A), 1.

See also, U.S. Army Inspector General, *Use of Volunteers in Chemical Agent Research* (Washington D.C.: GPO, 1975), 77.

113. Charles V. Kidd, Director, Research and Planning Division, NIH, to Rear Admiral Winfred Dana, Medical Corps, USN, 30 April 1952 ("In accordance with our telephone conversation of this afternoon I am enclosing a copy of draft statement which we have developed.") (ACHRE No. DOD-111594-A), 2–3. The context of this statement is not known. Perhaps it was formulated in response to an inquiry from the DOD about the NIH's research requirements during the discussions that led to the drafting of the Wilson memorandum.

114. National Institutes of Health, 17 November 1953 ("Group Consideration of Clinical Research Procedures Deviating from Accepted Medical Practice or Involving Unusual Hazard") (ACHRE No. HHS-090794-A), 4.

115. Director, NIH, to Institute Directors, 15 November 1954 ("Participation by NIH Employees as Normal Controls in Clinical Research Projects") (ACHRE No. HHS-090794-A), 1. Although this memorandum referred only to NIH employees, Advisory Committee staff and NIH staff have concluded it applied to all healthy volunteer subjects.

116. National Institutes of Health, policy statement of 17 November 1953 ("Group Consideration of Clinical Research Procedures Deviating From Accepted Medical Practice Or Involving Unusual Hazard") (ACHRE No. HHS-090794-A).

117. Edward J. Rourke, Legal Adviser, NIH, to Mr. John A. Trautman, Director, Clinical Center, 5 December 1952 ("At your invitation, I presented to the Medical Board of the Clinical Center on December 2 a proposal that, in view of several factors in some degree peculiar to the Clinical Center, it would be advisable from the legal point of view among others to accept certain procedures relating to patient admission that are more formal than might otherwise be considered necessary") (ACHRE No. DOD-111594-A), 1.

118. Ibid.

119. For a more detailed review of this history see Faden and Beauchamp, *A History and Theory of Informed Consent,* and Frankel, "Public Policymaking for Biomedical Research: The Case of Human Experimentation."

120. George M. Lyon, M.D., Assistant Chief Medical Director for Research and Education, presentation to the Committee on Veterans Medical Problems, National Research Council, 8 December 1952 ("Appendix II, Medical Research Programs of the Veterans Administration") (ACHRE No. VA-052595-A).

121. Ibid., 558.

122. Guy H. Birdsall, General Counsel, Veterans Administration, to Chief Medical Director, 25 June 1958, ("Op. G.C. 28–58, Legal Aspects of Medical Research") (ACHRE No. VA-052595-A).

2

Postwar Professional Standards
and Practices for Human Experiments

IN chapter 1, we explored government discussions of research involving human subjects in the 1940s and 1950s. We found that, at several junctures, government officials exhibited an awareness of the Nuremberg Code, the product of an international war crimes tribunal in 1947. If a requirement of voluntary consent of the subject was endorsed by the Nuremberg judges and was recognized at the highest reaches of the new Cold War bureaucracy, then how, a citizen might now ask, could there be any question about the use of this standard to judge experiments conducted during this time in the United States? And yet precisely this question has been raised in connection with human radiation experiments. Did American medical scientists routinely obtain consent from their subjects in the 1940s and 1950s, including those who were patients, and if not, how did these scientists square their conduct with the demands of the Nuremberg Code?

This chapter describes the Advisory Committee's efforts to answer these questions and what we learned. We begin with an examination of what, in fact, was argued at Nuremberg. We focus particularly on the testimony of Andrew Ivy, the American Medical Association's (AMA) official consultant to the Nuremberg prosecutors, and on the AMA's response to the report Dr. Ivy prepared about the trial for the organization.

We turn next to an analysis of the actual practices of American medical scientists during this period. In addition to reviewing contemporary documentation and present-day scholarship, the Advisory Committee conducted interviews with leading medical scientists and physicians who were engaged in research with human subjects in the 1940s and 1950s. These sources suggest a different, more nuanced picture of the principles and practices of human research than that presented at Nuremberg.

Of particular importance in this picture are the practical and moral distinctions that many researchers made between investigations with healthy subjects and those with sick patients. Those working with healthy subjects could cite a tradition of consent that dated, at least, to Walter Reed's turn-of-the-century experiments; those working with sick patients were in a clinical context that was conditioned by a tradition of faith in the wisdom and beneficence of physicians, a tradition that was dominant until at least the 1970s. Closely related to these distinctions was the tension between being a scientist and being a physician. This tension confronted members of a new, and rapidly growing, breed of medical professionals in the United States working to make careers in clinical research. The chapter goes on to explore whether these distinctions and tensions were reflected in the Nuremberg Code and why the trial may not have had much impact on the treatment of patient-subjects.

The rest of the chapter explores the emerging awareness of the moral complexities of research at the bedside and the limitations of the Nuremberg Code to address them. We close with a brief discussion of the Declaration of Helsinki, the international medical community's attempt to produce a code of conduct compatible with the realities of medical research.

THE AMERICAN EXPERT, THE AMERICAN MEDICAL ASSOCIATION, AND THE NUREMBERG MEDICAL TRIAL

In the fall of 1943, the United States, Great Britain, and the Soviet Union agreed that, once victorious, they would prosecute individuals among the enemy who might have violated international law during the war. On August 8, 1945—exactly three months after V.E. Day and two days after the bombing of Hiroshima— representatives of the American, British, French, and Soviet governments officially established the International Military Tribunal in Nuremberg, Germany. An assemblage of Allied prosecutors presented cases against twenty-four high-ranking German government and military officials, including Hermann Goering and Rudolph Hess, before this international panel of judges. Quite early in the course of these initial Nuremberg trials, which ran from October 1945 to October 1946, "it became apparent," according to the recent recollections of American prosecutor Telford Taylor, "that the evidence had disclosed numerous important Nazis, military leaders, and others" who should also be tried.[1] In January 1946, President Harry Truman approved a supplementary series of war crimes trials. These trials were to take place in the same Nuremberg courtroom, and international law would continue to be the standard by which guilt or innocence would be determined. America's wartime allies would not, however, participate; responsibility for prosecuting and judging defendants in the second set of Nuremberg trials was left exclusively to the United States.

The first of twelve cases that would eventually make up this second series of trials in Nuremberg is technically called *United States v. Karl Brandt et al.* More popularly, this trial is known by a variety of other names such as "The Doctors' Trial" and "The Medical Case." For the sake of convenience and consistency we will refer to the trial by another common name: the Nuremberg Medical Trial. This case began on December 9, 1946, when U.S. Chief of Counsel for War Crimes Telford Taylor delivered the prosecution's opening statement against the twenty-three defendants (twenty of whom were physicians). To one degree or another, Taylor charged the defendants with "murders, tortures, and other atrocities committed in the name of medical science." The trial ended in late August 1947 when the judges handed down a ruling that included the so-called Nuremberg Code and seven death sentences.[2]

In the spring of 1946, the American prosecution team preparing for the Medical Trial, which was made up of lawyers commissioned in the Army, cabled Secretary of War Robert P. Patterson with a request for a medical expert. Patterson consulted with Army Surgeon General Norman T. Kirk, who suggested turning to the American Medical Association. Kirk contacted the AMA, and, after some internal consultation, the association's Board of Trustees voted in May 1946 to appoint Dr. Andrew C. Ivy as the AMA's official consultant to the Nuremberg prosecutors.[3] Dr. Ivy was one of America's leading medical researchers at the time. Early in the war, Ivy was the civilian scientific director of the Naval Medical Research Institute in Bethesda, Maryland.[4] During the summer of 1946, he was in the process of moving from a position as head of the Division of Physiology and Pharmacology at Northwestern University Medical School to the University of Illinois, where he would serve as a vice president with responsibility for the university's professional schools in Chicago.

The precise rationale behind Ivy's selection as the AMA's adviser to the Nuremberg prosecutors remains unclear, but it is likely that the AMA turned to Ivy for at least two reasons. First, his wartime research interests corresponded in topic, though not in style, to some of the most shocking experiments that had taken place in the

Nazi concentration camps. Ivy supervised and carried out experiments in seawater desalination, sometimes using human subjects, with the intent of developing techniques to aid Allied pilots and sailors lost at sea. He also conducted some pioneering human experiments in aviation medicine dealing with the physiological challenges posed by high altitudes. These are two of the areas in which Nazi researchers had conducted especially gruesome human experiments. Second, Ivy was well known for his energetic defense of animal experimentation against American antivivisectionists. For example, he served for eight years as the founding secretary-treasurer of the National Society for Medical Research, an organization formed by scientists in 1946 to ward off challenges to medical research posed by antivivisectionists. It seems likely that the AMA Board of Trustees would have recognized Ivy as someone who possessed an unusual combination of familiarity with the scientific aspects of experiments carried out in the concentration camps and broad understanding of the moral issues at stake in medical research, whether the experimental subjects were animals or humans. Also, Ivy was almost certainly perceived as someone who could be trusted to look out for the interests of the American medical research community during the Nuremberg Medical Trial. The AMA Board of Trustees probably realized that the entire enterprise of medical research would, to some degree, be on trial in Germany.

In July or early August of 1946, Ivy went to Germany to meet with the Nuremberg prosecution team. Ivy offered technical assistance to the lawyers struggling with the scientific details of the experiments, but he also recognized, as he put it, that the prosecutors "appeared somewhat confused regarding the ethical and legal aspects" of human experimentation.[5]

After returning from his initial trip to Europe in aid of the Nuremberg prosecutors, Ivy offered a preliminary oral report to the Board of Trustees of the American Medical Association at the board's August 1946 meeting. After his presentation, the trustees asked Ivy to provide a written summary of his findings, so that the AMA's Judicial Council (a committee of five whose duties included deliberating on matters of medical ethics) could "make a report as to the manner in which these [Nazi] experiments [were] infringements of medical ethics."[6]

In mid-September, Ivy submitted a written report to the AMA as he had been directed.[7] At roughly the same time, he also turned over a copy of the twenty-two-page typescript to the Nuremberg prosecution team. In this piece, Ivy laid out "the rules" of human experimentation. He stated without equivocation that these standards had been "well established by custom, social usage and the ethics of medical conduct." Ivy's rules read as follows:

1. Consent of the human subject must be obtained. All subjects must have been volunteers in the absence of coercion in any form. Before volunteering the subjects have been informed of the hazards, if any. (In the U.S.A. during War, accident insurance against the remote chance of injury, disability and death was provided. [This was not true in all cases.])

2. The experiment to be performed must be so designed and based on the results of animal experimentation and a knowledge of the natural history of the disease under study that the anticipated results will justify the performance of the experiment. That is, the experiment must be such as to yield results for the good of society unprocurable by other methods of study and must not be random and unnecessary in nature.

3. The experiment must be conducted
   a. only by scientifically qualified persons, and
   b. so as to avoid all unnecessary physical and mental suffering and injury, and
   c. so, that, on the basis of the results of previous adequate animal experimentation, there is no *a priori* reason to believe that death or disabling injury will occur, except in such experiments as those on Yellow Fever where the experimenters serve as subjects along with non-scientific personnel.[8]

A comparison of these rules with the Nuremberg Code, which the Nuremberg Tribunal issued as part of its judgment on August 19, 1947, reveals that the three judges extracted important elements of clause 1 from Ivy's first rule and clauses 2, 3, 4, 5, and 8 almost verbatim from the rest of Ivy's formulation. Significantly, the judges also reiterated Ivy's assertion that these rules were *already* widely understood and followed by medical researchers.[9]

It is possible that the Nuremberg judges never read Ivy's report directly. During his testimony at the trial, Ivy essentially read his set of rules into the court record.[10] Also, the judges could have gained exposure to Ivy's thinking through two additional indirect sources. First, another medical expert who aided the prosecution, an American Army psychiatrist named Leo Alexander, submitted on April 15, 1947, a memorandum to the prosecutors entitled "Ethical and Non-Ethical Experimentation on Human Beings." In this memorandum, which would have been passed to the judges, Alexander repeated in very similar language significant.portions of Ivy's rules as outlined in the September 1946 report.[11] Second, American prosecutor James McHaney closely followed the text of Ivy's rules when setting before the judges the "prerequisites to a permissible medical experiment on human beings" during the prosecution's closing statement on July 14, 1947.[12]

But Ivy's standards for human experimentation served as even more than the primary textual foundation for the Nuremberg Code; his set of rules also undergirded the AMA's first formal statement on human experimentation. As the Board of Trustees had directed when asking Ivy to prepare his written report, the finished document was immediately forwarded to the AMA Judicial Council. The board gave the Judicial Council three months to prepare a presentation for the House of Delegates, the large policy-making body of the AMA that was scheduled to hold an annual meeting in early December 1946.[13] Unfortunately, records of the Judicial Council's consideration of Ivy's report have not survived, but published proceedings of the House of Delegates meeting reveal the results of the council's deliberations.[14] Dr. E. R. Cunniffe, chair of the Judicial Council, summarized his panel's response to Ivy's report at an executive session of the House of Delegates on December 10, 1946 (the day immediately following the prosecution's opening statement in the Nuremberg Medical Trial). Cunniffe condemned the Nazi experi-

ments described in Ivy's report as gross violations of standards that were *already inherent* in the existing "Principles of Medical Ethics of the American Medical Association," which had undergone only minor revision since the AMA adopted them in 1847, the first year of the association's existence. But in recognition of the fact that guidelines for human experimentation were not explicitly laid out in these "Principles," the Judicial Council offered the following distillation of Ivy's rules:

In order to conform to the ethics of the American Medical Association, three requirements must be satisfied: (1) the voluntary consent of the person on whom the experiment is to be performed [must be obtained]; (2) the danger of each experiment must be previously investigated by animal experimentation, and (3) the experiment must be performed under proper medical protection and management.[15]

These three rules became the official policy of the AMA when the House of Delegates voted its approval "section by section and as a whole" on the morning of December 11, 1946. The AMA's official governing body also added a general admonition: "This House of Delegates condemns any other manner of experimentation on human beings than that mentioned herein."[16] It is worth noting that in 1946 roughly 70 percent of American physicians belonged to the AMA. In absolute terms, 126,835 physicians belonged to the association, but it must be acknowledged that membership in the national association came automatically with membership in county and state medical societies, which was often necessary for professional privileges at local hospitals.[17] Each member of the AMA would have received a regular subscription to the *Journal of the American Medical Association,* and all of these subscribers would have had an opportunity to read the three rules for human experimentation approved by the House of Delegates. At the same time, however, these rules were not published prominently; they were set in small type along with a variety of other miscellaneous business items in the lengthy published minutes of the meeting. Only an exceptionally diligent member, or one with a special interest in medical ethics, is likely to have located this item.

In mid-June 1947, Ivy took the stand late in the Nuremberg Medical Trial as a rebuttal witness for the prosecution to counter the claims of the defense that standards for proper conduct in human experimentation had not been clearly established before the initiation of the trial. The contents of Ivy's September 1946 report, and the AMA standards that arose from it, played a major role during his three days of testimony. At one point, prosecution associate counsel Alexander G. Hardy carefully walked Ivy through a verbatim oral recitation of the rules for human experimentation contained in Ivy's report and the condensed version of his rules as approved by the AMA. After a reading of the AMA principles, Hardy and Ivy had the following exchange:

Q. . . . Now, [do these rules] purport to be the principles upon which all physicians and scientists guide themselves before they resort to medical experimentation on human beings in the United States?

A. Yes, they represent the basic principles approved by the American Medical Association for the use of human beings as subjects in medical experiments.[18]

Hearing this specific, and obviously important, claim about research with human subjects in the United States, Judge Harold E. Sebring interjected with a broad question about the international significance of Ivy's assertion: "How do the principles which you have just enunciated comport with the principles of the medical profession over the civilized world generally?" Ivy responded: "They are identical, according to my information."[19]

Later in his testimony, Ivy faced cross-examination by Fritz Sauter, counsel for two of the German medical defendants. Sauter pushed Ivy to acknowledge that the AMA guidelines had come into formal existence only as the Nuremberg Medical Trial was getting under way. In response to this attempt to diminish the legal force of the AMA standards with the obvious suggestion that the rules had been made up too recently to be of relevance, Ivy made an explicit claim in court that the ideas inherent in the AMA standards significantly predated their official formulation:

Q. You told us that . . . an association had made a compilation regarding the ethics of medical experi-

ments on human beings. . . . Can you recall what I am referring to?

A. Yes.

Q. That was in December 1946, I believe.

A. Yes, I remember. . . .

Q. Did that take place in consideration of this trial?

A. Well, that took place as a result of my relations to the trial, yes.

Q. Before December of 1946 were such instructions in printed form in existence in America?

A. No. They were understood only as a matter of common practice.[20]

Thus, if Ivy is to be taken literally, the standards he forcefully articulated during the Nuremberg Medical Trial, which were affirmed by the AMA House of Delegates as the trial was just beginning and codified by three American judges as the trial came to an end, were the standards of practice at the time.

THE "REAL WORLD" OF HUMAN EXPERIMENTATION

It would be historically irresponsible, however, to rely solely on records related directly to the Nuremberg Medical Trial in evaluating the postwar scene in American medical research. The panorama of American thought and practice in human experimentation was considerably more complex than Ivy acknowledged on the witness stand in Nuremberg. In general, it does seem that most American medical scientists probably sought to approximate the practices suggested in the Nuremberg Code and the AMA principles when working with "healthy volunteers." Indeed, a subtle, yet pervasive, indication of the recognition during this period that consent should be obtained from healthy subjects was the widespread use of the term *volunteer* to describe such research participants. Yet, as Advisory Committee member Susan Lederer has recently pointed out, the use of the word *volunteer* cannot always be taken as an indication that researchers intended to use subjects who had knowingly and freely agreed to participate in an experiment; it seems that researchers sometimes used *volunteer* as a synonym for *research subject*, with no special meaning intended regarding the decision of the participants to join in an experiment.[21]

Even with this ambiguity it is, however, quite clear that a strong tradition of consent has existed in research with healthy subjects, research that generally offered no prospect of medical benefit to the participant. In the United States much of this tradition has rested on the well-known example of Walter Reed's turn-of-the-century experiments, when he employed informed volunteers to establish the mosquito as the vector of transmission for yellow fever.[22] Indeed, it seems that a tradition of research with consenting subjects has been particularly strong among Reed's military descendants in the field of infectious disease research (which has frequently required the use of healthy subjects). For example, Dr. Theodore Woodward, a physician-researcher commissioned in the Army, conducted vaccine research during the 1950s with healthy subjects under the auspices of the Armed Forces Epidemiological Board. In a recent interview conducted by the Advisory Committee, Woodward recalled that the risks of exposure to diseases such as typhus were always fully disclosed to potential healthy subjects and that their consent was obtained. Since some of these studies were conducted in other countries with non-English-speakers, the disclosure was given in the volunteer's language.[23] Of his own values during this time, Woodward stated: "If I gave someone something that could make them sick or kill them and hadn't told them, I'm a murderer."[24] Similarly, Dr. John Arnold, a physician who conducted Army-sponsored malaria research on prisoners from the late 1940s through the mid-1950s, recalled that he always obtained written permission from his subjects.[25]

Not all the evidence on consent and healthy subjects comes from the military tradition. A particularly compelling general characterization of research with "normal volunteers" during this period comes from the "Analytic Summary" of a conference on the "Concept of Consent in Clinical Research," which the Law-Medicine Research Institute (LMRI) of Boston University convened on April 29, 1961. At this conference, twenty-one researchers from universities, hospitals, and pharmaceutical companies across the country were brought together "to explore prob-

lems arising from the legal and ethical requirements of informed consent of research subjects."[26] The LMRI project was what one might now call a fact-finding mission; the LMRI staff was attempting "to define and to analyze *the actual patterns of administrative practice* governing the conduct of clinical research in the United States" during the early 1960s.[27] Anne S. Harris, an LMRI staff member and author of the conference's final report, offered a simple but significant assessment of the handling of healthy participants in nontherapeutic research as expressed by the researchers at the meeting, whose careers included the decade and a half since the end of World War II: "The conferees indicated that normal subjects are usually fully informed."[28]

Even so, researchers who almost certainly knew better sometimes employed unconsenting healthy subjects in research that offered them no medical benefits. For example, Dr. Louis Lasagna, who has since become a respected authority on bioethics, stated in an interview conducted by the Advisory Committee that between 1952 and 1954, when he was a research fellow at Harvard Medical School, he helped carry out secret, Army-sponsored experiments in which hallucinogens were administered to healthy subjects without their full knowledge or consent:

The idea was that we were supposed to give hallucinogens or possible hallucinogens to healthy volunteers and see if we could worm out of them secret information. And it went like this: a volunteer would be told, 'Now we're going to ask you a lot of questions, but under no circumstances tell us your mother's maiden name or your social security number,' I forget what. I refused to participate in this because it was so mindless that a psychologist did the interviewing and then we'd give them a drug and ask them a number of questions and sure enough, one of the questions was 'What is your mother's maiden name?' Well, it was laughable in retrospect . . . *[The subjects] weren't informed about anything* [emphasis added].[29]

Lasagna, reflecting "not with pride" on the episode, offered the following explanation: "It wasn't that we were Nazis and said, 'If we ask for consent we lose our subjects,' it was just that we were so ethically insensitive that it never occurred to us that you ought to level with people that they were in an experiment."[30] This might have been true for Lasagna the young research fellow, but the explanation is harder to understand for the director of the research project, Henry Beecher. Beecher was a Harvard anesthesiologist who, as we will see later in this chapter and in chapter 3, would emerge as an important figure in biomedical research and ethics during the mid-1960s.[31]

If American researchers experimenting on healthy subjects *sometimes* did not strive to follow the standards enunciated at Nuremberg, research practices with sick patients seem even more problematic in retrospect. Advisory Committee member Jay Katz has recently argued that this type of research *still* gives rise to ethical dificulties for physicians engaged in research with patients, and he has offered an explanation: "In conflating clinical trials and therapy, as well as patients and subjects, as if both were one and the same, physician-investigators unwittingly become double agents with conflicting loyalties."[32]

It is likely that such confusion and conflict would have been as troublesome several decades ago, if not more troublesome, than it is today. The immediate postwar period was a time of vast expansion and change in American medical science (see Introduction). Clinical research was emerging as a new and prestigious career possibility for a growing number of medical school graduates. Most of these young clinical researchers almost certainly would have absorbed in their early training a paternalistic approach to medical practice that was not seriously challenged until the 1970s. This approach encouraged physicians to take the responsibility for determining what was in the best interest of their patients and to act accordingly. The general public allowed physicians to act with great authority in assuming this responsibility because of an implicit trust that doctors were guided in their actions by a desire to help their patients.

This paternalistic approach to medical practice can be traced to the Hippocratic admonition: "to help, or at least do no harm."[33] Another long-standing medical tradition that can be found in Hippocratic medicine is the belief that each patient poses a unique medical problem calling for creative solution. Creativity in the treatment of individuals, which was not commonly thought of as requiring consent, could be—and often was—called experimentation.

This tradition of medical tinkering without explicit and informed consent from a patient was intended to achieve proper treatment for an individual's ailments; but it seems also to have served (often unconsciously) as a justification for some researchers who engaged in large-scale clinical research projects without particular concern for consent from patients.

Members of the medical profession and the American public have today come to better understand the intellectual and institutional distinctions between organized medical research and standard medical practice. There were significant differences between research and practice in the 1950s, but these differences were harder to recognize because they were relatively new. For example, randomized, controlled, double-blind trials of drugs, which have brought so much benefit to medical practice by greatly decreasing bias in the testing of new medicines, were introduced in the 1950s. The postwar period also brought an unprecedented expansion of universities and research institutes. Many more physicians than ever before were no longer solely concerned, or even primarily concerned, with aiding *individual* patients. These medical scientists instead set their sights on goals they deemed more important: expanding basic knowledge of the natural world, curing a dread disease (for the benefit of many, not one), and in some cases, helping to defend the nation against foreign aggressors. At the same time, this new breed of clinical researchers was motivated by more pragmatic concerns, such as getting published and moving up the academic career ladder. But these differences between medical practice and medical science, which seem relatively clear in retrospect, were not necessarily easy to recognize at the time. And coming to terms with these differences was not especially convenient for researchers; using readily available patients as "clinical material" was an expedient solution to a need for human subjects.

As difficult and inconvenient as it might have been for researchers in the boom years of American medical science following World War II to confront the fundamental differences between therapeutic and nontherapeutic relationships with other human beings, it was not impossible. Otto E. Guttentag, a physician at the University of California School of Medicine in San Francisco, directly addressed these issues in a 1953 *Science* magazine article. Guttentag's article, and three others that appeared with it, originated as presentations in a symposium held in 1951 on "The Problem of Experimentation on Human Beings" at Guttentag's home institution. Guttentag constructed his paper around a comparison between the traditional role of the physician as healer and the relatively new role of physician as medical researcher. Guttentag referred to the former as "physician-friend" and the latter as "physician-experimenter." He explicitly laid out the manner in which medical research could conflict with the traditional doctor-patient relationship:

Historically, . . . one human being is in distress, in need, crying for help; and another fellow human being is concerned and wants to help and the desire for it precipitates the relationship. Here *both* the healthy and the sick persons are . . . fellow-companions, partners to conquer a common enemy who has overwhelmed one of them. . . . Objective experimentation to confirm or disprove some doubtful or suggested biological generalization is foreign to this relationship . . . for it would involve taking advantage of the patient's cry for help, and of his insecurity.[34]

Guttentag worried that a "physician-experimenter" could not resist the temptation to "tak[e] advantage of the patient's cry for help."[35] To prevent the experimental exploitation of the sick that he envisioned (or knew about), Guttentag suggested the following arrangement:

Research and care would not be pursued by the same doctor for the same person, but would be kept distinct. The physician-friend and the physician-experimenter would be two different persons as far as a single patient is concerned. . . . The responsibility for the patient as patient would rest, during the experimental period, with the physician-friend, unless the patient decided differently.

Retaining his original physician as personal adviser, the patient would at least be under less conflict than he is at present when the question of experimentation arises.[36]

Among physicians, Guttentag was nearly unique in medicine in those days in raising such problems in print. Another example of concern about the moral issues raised by research at the bedside comes from what might be an unex-

pected source: a Catholic theologian writing in 1945. In the course of a general review of issues in moral theology, John C. Ford, a prominent Jesuit scholar, devoted several pages to the matter of experimentation with human subjects. Ford was not a physician, but his thoughts on this topic—published a year before the beginning of the Nuremberg Medical Trial—suggest that a thoughtful observer could recognize, even decades ago, serious problems with conducting medical research on unconsenting hospital patients:

The point of getting the patient's consent [before conducting an experiment] is increasingly important, I believe, because of reports which occasionally reach me of grave abuses in this matter. In some cases, especially charity cases, patients are not provided with a sure, well-tried, and effective remedy that is at hand, but instead are subjected to other treatment. The purpose of delaying the well-tried remedy is, not to cure *this* patient, but to discover experimentally what the effects of the new treatment will be, in the hope, of course, that a new discovery will benefit later generations, and that the delay in administering the well-tried remedy will not harm the patient too much. . . . This sort of thing is not only immoral, but unethical from the physician's own standpoint, and is illegal as well.[37]

The transcripts and reports produced in the Law-Medicine Research Institute's effort during the early 1960s to gather information on ethical and administrative practices in research in medical settings suggest that by this time more researchers had come to recognize the troubling issues associated with using sick patients as subjects in research that could not benefit them. The body of evidence from the LMRI project also suggests that problems with this type of human experimentation had been widespread before the early 1960s and remained common at that time. The transcript of a May 1, 1961, closed-door meeting of medical researchers organized by LMRI to explore issues in pediatric research shows a medical scientist from the University of Iowa offering a revealing generalization from which none of his colleagues dissented. In order to understand this transcript excerpt one must know that item "A1" on the meeting agenda related to research "primarily directed toward the advancement of medical science" and item

"A2" referred to "clinical investigation . . . primarily directed toward diagnostic, therapeutic and/or prophylactic benefit to patients."

We have done a thousand things with an implied feeling [of consent]. . . . We wear two hats. Item A2 allows us to do A1 *but we feel uncomfortable about it.* The responsibility of the physician includes responsibility to advance in knowledge. Things are different now and this problem of a secondary role [i.e., to advance knowledge] is increasingly in front stage [emphasis added].[38]

This researcher acknowledged that many physicians during the period let themselves slide into nontherapeutic research with patients. He provided the additional, and significant, assessment that he and his colleagues felt guilty about this behavior, even though it was quite common.

An even more probing analysis of these issues had taken place two days earlier at the April 29, 1961, LMRI conference on "The Concept of Consent," referred to above in our discussion of research with healthy subjects. The participants at this meeting recognized that research with sick patients could be both therapeutic and nontherapeutic. Interestingly, they suggested that patients employed for research in which "there was the possibility of therapeutic benefit with minimal or moderate risk" *were* "usually informed" of the proposed study. The author of the conference report offered the plausible explanation that informing subjects in potentially beneficial research "is psychologically more comfortable for investigators [because] the [therapeutic] expectations of potential subjects coincide with the purpose and expected results of the experiment."[39] The conferees identified research in which "patients are used for studies unrelated to their own disease, or in studies in which therapeutic benefits are unlikely" as the most problematic. Those at the meeting "indicated that it is most often subjects in this category to whom disclosure is not made."[40] The conference report outlined an approach employed by many researchers (including some at the meeting), in which, rather than seeking consent from patients for research that offers them no benefit,

[t]he therapeutic illusion is maintained, and the patient is often not even told he is participating in research. Instead, he is told he is "just going to have a test." If the experimental procedure involves minimal risk, but some discomfort, such as hourly urine collection, "All you do is tell the patient: 'We want you to urinate every hour.' We merely let them assume that it is part of the hospital work that is being done."[41]

Again, it is important to note that the conference participants displayed some moral discomfort with this pattern of behavior, as can be seen from the following exchange:

Dr. X: There is a matter here of whether the patient is not informed because the risk is too trivial, or because it's too serious.
Dr. Y: I think you're getting right at it. There's a great difference in not telling the patient because you're afraid he won't participate and not telling him because you don't think there is a conceivable risk, and it's so trivial you don't bother to inform him.
Dr. Z: On the question of whether it's [acceptable] not to tell, we would say that it is not permissible on the grounds of refusal potential.[42]

It is also important to draw out of this transcript excerpt the general point that most researchers in this period appear not to have had great ethical qualms about enrolling an uninformed patient in a research project if the risk was deemed low or nonexistent. Of course, the varying definitions of "low risk" could lead to problems with this approach. Indeed, the participants at the "Concept of Consent" conference grappled at length with this very issue without ever reaching consensus. A minority steadfastly asserted that participants in an experiment should be asked for consent even if the risk would be extremely low, such as in only taking a small clipping of hair.

The Advisory Committee's Ethics Oral History Project[43] has provided extensive additional evidence that medical researchers sometimes (perhaps even often) took liberties with sick patients during the decades immediately following World War II. The element of opportunism was recounted in several interviews. Dr. Lasagna, who was involved in pain-management studies in postoperative patients at Harvard in the 1950s, explained rather bluntly:

[M]ostly, I'm ashamed to say, it was as if, and I'm putting this very crudely purposely, as if you'd ordered a bunch of rats from a laboratory and you had experimental subjects available to you. They were never asked by anybody. They might have guessed they were involved in something because a young woman would come around every hour and ask them how they were and quantified their pain. We never made any efforts to find out if they guessed that they were part of it.[44]

Other researchers told similar tales, with a similar mixture of matter-of-fact reporting and regretful recollection. Dr. Paul Beeson remembered a study he conducted in the 1940s, while a professor at Emory University, on patients with bacterial endocarditis, an invariably fatal disease at the time. He recalled that he thought it would be interesting to use the new technique of cardiac catheterization to compare the number of bacteria in the blood at different points in circulation:

[This is] something I wouldn't dare do now. It would do no good for the patient. They had to come to the lab and lie on a fluoroscopic table for a couple of hours, a catheter was put into the heart, a femoral needle was put in so we could get femoral arterial blood and so on. . . . All I could say at the end was that these poor people were lying there and we had nothing to offer them and it might have given them some comfort that a lot of people were paying attention to them for this one study. I don't remember ever asking their permission to do it. I did go around and see them, of course, and said, "We want to do a study on you in the X-ray department, we'll do it tomorrow morning," and they said yes. There was never any question. Such a thing as informed consent, that term didn't even exist at that time. . . . [I]f I were ever on a hospital ethics committee today, I wouldn't ever pass on that particular study.[45]

Radiologist Leonard Sagan recalled an experiment in which he assisted during his training on a metabolic unit at Moffett Hospital in San Francisco in 1956–1957.

At the time, the adrenal gland was hot stuff. ACTH [adreno-corticotropic hormone] had just become available and it was an important tool for exploring the function of the adrenal gland. . . . This was the project I was involved in during that year, the study of adrenal function in patients with thyroid disease, both hypo- and hyperthyroid disease. So what did we do? I'd find some patients in the hospital and I'd add a little ACTH to their infusion and collect urines and measure output of urinary corticoids. . . . I didn't consider it dangerous. But I didn't consider it necessary to inform them either. So far as they were concerned, this was part of their treatment. They didn't

know, and no one had asked me to tell them. As far as I know, informed consent was not practiced anyplace in that hospital at the time.[46]

Sagan viewed the above experiment as conforming not only with the practices of the particular hospital but also in accord with the high degree of professional autonomy and respect that was granted to physicians in this era:

In 1945, '50, the doctor . . . was king or queen. It never occurred to a doctor to ask for consent for anything. . . . People say, oh, injection with plutonium, why didn't the doctor tell the patient? Doctors weren't in the habit of telling the patients anything. They were in charge and nobody questioned their authority. Now that seems egregious. But at the time, that's the way the world was.[47]

Another investigator, Dr. Stuart Finch, who was a professor of medicine at Yale during the 1950s and 1960s, recalled instances when oncologists there were overly aggressive in pursuing experimental therapies with terminal patients.

[I]t's very easy to talk a terminal patient into taking that medication or to try that compound or whatever the substance is. . . . Sometimes the oncologists [got] way overenthused using it. It's very easy when you have a dying patient to say, "Look, you're going to die. Why don't you let me try this substance on you?" I don't think if they have informed consent or not it makes much difference at that point.[48]

Economically disadvantaged patients seem to have been perceived by some physicians as particularly appropriate subjects for medical experimentation. Dr. Beeson offered a frank description of a quid pro quo rationale that was probably quite common in justifying the use of poor patients in medical research: "We were taking care of them, and felt we had a right to get some return from them, since it wouldn't be in professional fees and since our taxes were paying their hospital bills."[49]

Another investigator, Dr. Thomas Chalmers, who began his career in medical research during the 1940s, identified sick patients as the most vulnerable type of experimental subjects—more vulnerable even than prisoners:

One of the real ludicrous aspects of talking about a prisoner being a captive, and therefore needing more protection than others, is, there's nobody more captive than a sick patient. You've got pain. You feel awful. You've got this one person who's going to help you. You do anything he says. You're a captive. You can't, especially if you're sick and dying, discharge the doctor and get another one without a great deal of trauma and possible loss of lifesaving measures.[50]

Thus, as compared with prisoners, who are now generally viewed to be vulnerable to coercion, those who are sick may be even more compromised in their ability to withstand subtle pressure to be research subjects. Appropriate protection for the sick who might be candidates for medical research has proved to be an especially troublesome issue in the era following Nuremberg.

NUREMBERG AND RESEARCH WITH PATIENTS

The record of conducting nontherapeutic research with unconsenting sick patients during the postwar period discussed above seems to stand in particularly sharp contrast with the claims about the conduct of research involving human subjects in the United States that Andrew Ivy made during his testimony in Nuremberg. We have seen how some observers, even before Nuremberg, recognized that employing uninformed, vulnerable sick patients solely as a means to a scientific end was simply wrong. We must, however, also acknowledge that the particulars of the Nuremberg Medical Trial did not call for careful attention to the issues surrounding research with sick patients. None of the German physicians at Nuremberg stood accused of exploiting *patients* for experimental purposes.

Nonetheless, it is likely that Andrew Ivy would have argued that consent was appropriate in virtually all instances of medical research. Dr. Herman Wigodsky, who worked closely under Ivy at Northwestern in the late 1930s and early 1940s, explicitly commented during an Ethics Oral History Project interview that he did not believe that his mentor drew any sort of ethical line between various types of clinical research: "I don't think he made any distinction [between research with sick patients and research with healthy subjects]. Research was research. It didn't make any difference."[51]

Additional evidence that Ivy would have supported standards of consent for research with ill as well as with healthy subjects comes from his response to a set of rules for human experimentation put forth by the German Ministry of Interior in 1931, presented to him after he had prepared his written report for the AMA in the fall of 1946. These rules appear to be considerably more comprehensive and sophisticated than the Nuremberg Code itself.[52] Most significantly, the 1931 German standards cover both therapeutic and nontherapeutic research, calling for consent in both types of medical investigation. For reasons that are not clear, the prosecution team at Nuremberg did not choose to place much emphasis on these German standards in constructing the case. Ivy did, however, attempt (without much help from the prosecution) to initiate a discussion of the 1931 standards during his testimony. It is clear from the trial transcript that *Ivy* saw a rough equivalence between the more detailed and extensive German rules and those formulated by the AMA, with his assistance. Shortly after discussing the AMA principles on the witness stand, Ivy had the following exchange with prosecutor Alexander G. Hardy:

Q. Do you have any further statements to make concerning rules of medical ethics concerning experimentation in human beings?

A. Well, I find that since making [my] report to the American Medical Association that a decree of the Minister of Public Welfare [Ivy should have said "the Minister of the Interior"] of Germany in 1931 on the subject of "Regulations for Modern Therapy for the Performance of Scientific Experiments on Human Beings" contains all the [AMA] principles which I have read.[53]

Hardy did not take what now seems an obvious opportunity to allow Ivy to expand further on these rules. However, a few minutes later, Ivy brought up the German standards again on his own (and again Hardy did not pursue the topic further). At this point, Ivy stated his general agreement with the German standards of 1931 even more firmly:

I cited the principles . . . from the Reich Minister of the Interior dated February 28, 1931 to indicate that the ethical principles for the use of human beings as subjects in medical experiments in Germany in 1931 were similar to these which I have enunciated and which have been approved by the House of Delegates of the American Medical Association.[54]

Ivy's assertion of "similarity" between the AMA principles and those in the 1931 German document may not meet with agreement among those who compare the two. Though they may be viewed as similar in philosophy and intent, the German interior ministry document is far more detailed and comprehensive than that of the AMA.

Contrary to Ivy's claims at Nuremberg, and the positioning of Ivy by the prosecution, he cannot in any full sense be taken as the embodiment of the entire American medical profession in the years immediately following World War II. Again, Dr. Wigodsky spoke to this point in his recent interview:

Well, I've always felt that that stuff that Ivy wrote up during the time of the trials was pretty much an expression of his *personal philosophy* about research. And . . . it was the kind of understanding that we had in working with him about how he felt. Voluntariness being number one—you had to volunteer and had to be in a situation where you could volunteer. And consent in the sense that you didn't do anything to anybody that they didn't know what you were doing. That you explained to people what it was you were going to do and why you were going to do it and that sort of thing [emphasis added].[55]

Even if it is true that Andrew Ivy would have wholeheartedly endorsed the notion of obtaining consent from any research subject—whether an experiment held the possibility of personal benefit or not; whether the subjects were sick or healthy—it seems likely that the AMA House of Delegates would have been hesitant to endorse a condensation of Ivy's principles of research ethics if they had been explicitly extended to cover all categories of clinical investigation. Obtaining consent from patients within the normal clinical relationship was not a common practice in late 1946. At that time, and for many years to come, patient trust and medical beneficence were viewed as the unshakable moral foundations on which meaningful interactions between professional healers and the sick should be built. In fact, it was not until 1981 that the

AMA's Judicial Council specifically endorsed "informed consent" as an appropriate part of the therapeutic doctor-patient relationship.[56]

But, in the end, it must be acknowledged that the facts of the Nuremberg Medical Trial did not force Andrew Ivy, the AMA House of Delegates, the Nuremberg prosecutors, or the judges to grapple with the distinctions between research with sick patients and research with healthy subjects, or therapeutic and nontherapeutic research. The Nuremberg defendants stood accused of ghastly experimental acts that were absolutely without therapeutic intent, and their unfortunate subjects were never under any illusion that they were receiving medical treatment. To rebut the claims of some of the medical defendants that obtaining consent from research subjects was not a clearly established principle, Ivy could, and did, offer a variety of examples on the witness stand from a long tradition of human experimentation on consenting healthy subjects.[57] Ivy and the members of the prosecution team were not faced with what might have been a more troubling process: finding examples of well-organized nontherapeutic experiments on sick patients in which the subjects had clearly offered consent. Simply put, the Nuremberg Medical Trial did not demand it.

AMERICAN MEDICAL RESEARCHERS' REACTIONS TO NEWS OF THE NUREMBERG MEDICAL TRIAL

It is important to have some understanding of the extent to which American medical scientists paid attention to the events of the Nuremberg Medical Trial and made connnections with the messages that emanated from the courtroom in Germany. The Nuremberg Medical Trial received coverage in the American popular press, but it would almost certainly be an exaggeration to refer to this attention as exhaustive. Historian David Rothman has provided the following summary of the trial's coverage in the *New York Times*:

Over 1945 and 1946 fewer than a dozen articles appeared in the *New York Times* on the Nazi [medical] research; the indictment of forty-two doctors in the fall of 1946 was a page-five story and the opening of the trial, a page-nine story. (The announcement of the guilty verdict in August 1947 was a front-page story, but the execution of seven of the defendants a year later was again relegated to the back pages.)[58]

The Advisory Committee's Ethics Oral History Project suggests that American medical researchers, perhaps like the American public generally, were not carefully following the daily developments in Nuremberg. For example, Dr. John Arnold, a researcher who, during the Medical Trial, was involved in malaria experiments on prisoners at Stateville Prison in Illinois, offered a particularly vivid (if somewhat anachronistic) recollection of the scant attention paid to the Nuremberg Medical Trial among American medical scientists: "We were dimly aware of it. And as you ask me now, I'm astonished that we [were not] hanging on the TV at the time, watching for each twist and turn of the argument to develop. But we weren't."[59] It might have been expected that the researchers at Stateville would have been particularly concerned with the events at Nuremberg because some of the medical defendants claimed during the trial that the wartime malaria experiments at the Illinois prison were analogous to the experiments carried out in the Nazi concentration camps.

The strongest statement of awareness came from Dr. Herbert Abrams, a radiologist who was in his residency at Montefiore Hospital in the Bronx throughout most of the trial:

[The Nuremberg Medical Trial] was part of the history of the day. And there was extensive reportage . . . so that the manner of human experimentation as it had been done by the Nazis was very much in the news. We were all aware of it. I think that people experienced this kind of revulsion about it that you might anticipate. . . . It was surely something, at least in the environment I was in, we were aware of and that affected the thinking of everyone who was involved in clinical investigation.[60]

It seems likely, however, that the "environment" this young physician was in would have caused a heightened awareness of a trial dealing with Nazi medical professionals. Montefiore is a traditionally Jewish hospital that was home to many Jewish refugee physicians who had fled the terror and oppression of the Nazi regime.[61] A trial of German physicians almost certainly would have been of particular interest in this setting.

Even among American medical researchers who might have been aware of events at Nuremberg, it seems that many did not perceive specific personal implications in the Medical Trial. Rothman has enunciated this historical view most fully. He asserts that "the prevailing view was that [the Nuremberg medical defendants] were Nazis first and last; by definition nothing they did, and no code drawn up in response to them, was relevant to the United States."[62] Jay Katz has offered a similar summation of the immediate response of the medical community to the Nuremberg Code: "It was a good code for barbarians but an unnecessary code for ordinary physicians."[63]

Several participants in the Ethics Oral History Project affirmed the interpretations of Rothman and Katz, using similar language. Said one physician: "There was a disconnect [between the Nuremberg Code and its application to American researchers]. . . . The interpretation of these codes [by American physicians] was that they were necessary for barbarians, but [not for] fine upstanding people."[64] This same physician later acknowledged that, in a sense, some American researchers did not pay attention to the lessons of the Nuremberg Medical Trial because it was not convenient to do so:

The connection between those horrendous acts [carried out by German medical scientists in the concentration camps] and our everyday investigations was not made [by American medical researchers] for reasons of self-interest, to be perfectly frank. As I see it now, I'm saddened that we didn't see the connection, but that's what was done. . . . It's hard to tell you now . . . how we rationalized, but the fact is we did.[65]

The popular press mirrored the view that human experimentation as practiced in the United States was not a morally troubling enterprise—it was as American as apple pie. Between 1948 and 1960 magazines such as the *Saturday Evening Post, Reader's Digest,* and the *American Mercury* ran "human interest" stories on "human guinea pigs." These stories generally focused on specific groups of healthy subjects—prisoners, conscientious objectors, medical students, soldiers—and described them as "volunteers." The articles explained the ordeals to which the volunteers had submitted themselves. "Among these men and women," the *New York Times* informed its Sunday readership in 1958, "you will find those who will take shots of the new vaccines, who will swallow radioactive drugs, who will fly higher than anyone else, who will watch malaria infected mosquitos feed on their bare arms."[66] The articles assured the public that the volunteers had plausible, often noble, reasons for volunteering for such seemingly gruesome treatment. The explanations included social redemption (especially in the case of prisoners), religious or other beliefs (particularly for conscientious objectors), the advancement of science, service to society, and thrill-seeking.[67] In sum, most articles in the popular press were uncritical toward experimentation on humans and assumed that those involved had freely volunteered to participate.

However, a smaller number of press reports in the late 1940s and 1950s did suggest some tension between the words at Nuremberg and the practices in America. As early as 1948, for example, *Science News* reported the Soviet claim that Americans were using "Nazi methods" in the conduct of prisoner experiments.[68] Concern also began to be voiced about the dangers to volunteer "guinea pigs." In October 1954, for another example, the magazine *Christian Century* called on the Army to halt, at the first sign of danger, experiments at the Fitzsimmons Hospital in Denver, where soldiers were called upon to eat foods exposed to cobalt radiation.[69]

It is also possible that press accounts of experiments with *patients* rather than healthy subjects were more inclined to be critical, even in the late 1940s. A *Saturday Evening Post* article from the January 15, 1949, issue describes how a VA physician kept quiet about streptomycin trials involving the medical departments of the Army, Navy, and VA

because of the risk of congressional chastisement from publicity-conscious members of the House and Senate who might have screamed: 'You can't experiment on our heroes,' if it had been known that Army and Navy veterans of former wars were being used in the medical investigation. This was a real worry of the doctors who formulated the clinical program.[70]

Evidence suggests that some American researchers were genuinely and deeply concerned with the issues surrounding human experimentation during the years immediately following World War II. One source of insight into the thinking of American physicians engaged in clinical research during the 1950s is found in the ground-breaking work of medical sociologist Renee C. Fox. For two five-month periods between September 1951 and January 1953, Fox spent long days "in continuous, direct, and intimate contact with the physicians and patients" in a metabolic research ward that she pseudonymously called "Ward F-Second." In 1959 Fox reported with remarkable sensitivity and eloquence on the ethical dilemmas faced by the physicians conducting research on this ward. She did not suggest that the scientists under her observation were unaware of the Nuremberg Code; instead she offered a point-by-point paraphrasing of the Code, which she identified as "the basic principles governing research on human subjects which the physicians of the Metabolic Group [her collective term for the researchers whom she studied] were required to observe." Rather than being unconscious or contemptuous of a set of principles intended for barbarians, Fox reported that the researchers on "Ward F-Second" were sometimes troubled by their inability to apply the high, but essentially unquestioned, standards enunciated at the Nuremberg Medical Trial:

The physicians of the Metabolic Group were deeply committed to these principles and conscientiously tried to live up to them in the research they carried out on patients. However like most norms, the "basic principles of human experimentation" are formulated on such an abstract level that they only provide general guides to actual behavior. Partly as a consequence, the physicians of the Metabolic Group often found it difficult to judge whether or not a particular experiment in which they were engaged "kept within the bounds" delineated by these principles.[71]

Sometimes private discussions among researchers about the ethical aspects of human experimentation led to public events. A good example from the early 1950s is the symposium held on October 10, 1951, at the University of California School of Medicine in San Francisco at which Otto Guttentag made the presentation discussed earlier. One of Guttentag's colleagues, Dr. Michael B. Shimkin, organized the symposium in response to some confidential criticism that he had received for research carried out under his direction with patients at the University of California's Laboratory of Experimental Oncology. The exact nature of this criticism is unclear from the records that remain of the episode, but Shimkin reported in a memoir that "remedial steps" were taken, including "written protocols for all new departures in clinical research, which we asked the cancer board of the medical school to review."[72] In his memoirs Shimkin also recalls that patients were screened carefully before they were admitted to the Laboratory of Experimental Oncology:

They had to understand the experimental nature of our work, and every procedure was again explained to them; the initial release form even included agreement to an autopsy. The understanding did not absolve us of negligence, nor deprive patients of recourse to legal actions, but did set the tone and nature of our relationships. In all our 5 years of operations, not a single threat or implied threat of action against us was voiced. Two patients did instruct us to terminate our attempts at therapy.[73]

The criticism Shimkin experienced also demonstrated to him that a more open discussion of clinical research might be of benefit to his colleagues. According to his recollection, "There was an almost visible thawing of attitude by the airing of the problem" at the symposium.[74]

Less than a year after Shimkin's 1951 San Francisco symposium, the organizers of the "First International Congress of the Histopathology of the Nervous System," which was held in Rome, were sufficiently concerned with ethical issues that they invited Pope Pius XII to address "The Moral Limits of Medical Methods of Research and Treatment." In a speech before 427 medical researchers from around the world (including 62 Americans), the pope firmly endorsed the principle of obtaining consent from research subjects—whether sick or healthy. He also pointed his audience to the relatively recent lessons of the Nuremberg Medical Trial, which he summed up as teaching that "man should not exist for the use of society; on the contrary, the community exists for the good of man."[75] In an interview in 1961, Dr. Thomas Rivers, a prominent American virus researcher,

recalled that the pope's words had been influential among medical scientists working during the 1950s:

[I]n September 1952, Pope Pius XII had given a speech at the First International Congress on the Histopathology of the Nervous System in which he outlined the Roman Catholic Church's position on the moral limits of human experimentation for purposes of medical research. That speech had a very broad impact on medical scientists both here and abroad.[76]

The growing influence of the Nuremberg Medical Trial can be seen by looking at two editions of the best-known textbook of American medical jurisprudence in the midtwentieth century. In the 1949 edition of *Doctor and Patient and the Law*, Louis J. Regan, a physician and lawyer, offered very little under the heading "Experimentation," and what he did offer made no reference to Nuremberg:

The physician must keep abreast of medical progress, but he is responsible if he goes beyond usual and standard procedures to the point of experimentation. If such treatment is considered indicated, it should not be undertaken until consultation has been had and until the patient has signed a paper acknowledging and assuming the risk.[77]

However, in Regan's next edition of the same text, published in 1956, his few lines on human experimentation had been expanded to three pages. He presented a lengthy paraphrasing of the Nuremberg Code, and he repeated verbatim (without quotation marks) the judges' preamble to the Code, stating that "all agree" about these principles. Regan characterized the standards enunciated by the judges at Nuremberg as "the most carefully developed set of precepts specifically drawn to meet the problem of human experimentation." Immediately following his discussion of Nuremberg, Regan laid out the 1946 standards of the American Medical Association, which, as he put it, researchers needed to meet "in order to conform with the ethics of the American Medical Association."[78]

NEW TIMES, NEW CODES

In the spring of 1959 the National Society for Medical Research (NSMR), an organization that Andrew Ivy had helped to found in 1946, sponsored a "National Conference on the Legal Environment of Medicine" at the University of Chicago. Human experimentation was one of the major topics presented for discussion by the 148 conference participants, primarily medical researchers, from around the country. The published report of this conference reveals that the many researchers who gathered in Chicago understood the Nuremberg Code well enough to use it as a point of departure for discussion. As a group, the conferees acknowledged that "[t]he ten principles [of the Nuremberg Code] have become the principal guideposts to the ethics of clinical research in the western world." Not all those in attendance, however, seemed to have been entirely pleased with this state of affairs. A "Committee on the Re-Evaluation of the Nuremberg Experimental Principles" reported general agreement with "the spirit of these precautions" but discomfort with a number of "particulars." For example, they suggested that the absolute requirement for consent in the Code's first principle might be softened by inserting "either explicit or reasonably presumed" before the word "consent." They also added a clause that would allow for third-party permission for "those not capable of personal consent."[79]

The 1959 NSMR conference strongly suggests that by the late 1950s many and perhaps even most American medical researchers had come to recognize the Nuremberg Code as the most authoritative single answer to an important question: What are the rules for human experimentation? The same conference also provides compelling evidence that many researchers who were giving the ethical issues surrounding human experimentation serious attention at this time were not entirely happy with the prospect of living by the letter of the Code. The sources of discomfort with the Nuremberg Code can be grouped, retrospectively, into three broad categories. First, some recognized the discrepancies between what they had come to know as *real* practices in research on patient-subjects and what they read in the lofty, idealized language of the Code. Others simply disagreed with some elements of the Code. Still others disliked the very idea of a single, concrete set of standards to guide behavior in such a complex matter as human experimentation.

Henry Beecher, the Harvard-based medical researcher who was Louis Lasagna's mentor in the early 1950s, published a paper, "Experimentation in Man," in the *Journal of the American Medical Association* only a few months before the NSMR conference in Chicago. In this lengthy piece, Beecher addressed a mixture of all three sources of discomfort with the Nuremberg Code. Beecher offered the assertion that "it is unethical and immoral to carry out potentially dangerous experiments without the subject's knowledge and consent" as the "central conclusion" of his paper.[80] But, even with this strong statement, he was not entirely happy with the first clause of the Code; he viewed the Nuremberg consent clause as too extreme and not squaring with the realities of clinical research:

> It is easy enough to say, as point one [of the Nuremberg Code] does, that the subject "should have sufficient knowledge and comprehension of the elements of the subject matter involved as to enable him to make an understanding and enlightened decision." *Practically,* this is often quite impossible . . . for the complexities of essential medical research have reached the point where the full implications and possible hazards cannot always be known to anyone and are often communicable only to a few informed investigators and sometimes not even to them. Certainly the full implications of work to be done are often not really communicable to lay subjects. . . . [P]oint one states a requirement very often impossible of fulfillment [emphasis added].[81]

Beecher's second form of difficulty with the Code can be found in his opinion of another Nuremberg clause, which states, in part, that a human experiment should not be "random and unnecessary in nature." Beecher cited "anesthesia, x-rays, radium, and penicillin" as important medical breakthroughs that had resulted from "random" experimentation. He further stated that he "would not know how to define experiments 'unnecessary in nature.'"[82] Finally, Beecher expressed skepticism in general that any code could provide effective moral guidance for researchers working with human subjects. Near the beginning of his paper he wrote that "the problems of human experimentation do not lend themselves to a series of rigid rules."[83] Later in the piece, he expanded on this thought:

[I]t is not my view that many rules can be laid down to govern experimentation in man. In most cases, these are more likely to do harm than good. Rules are not going to curb the unscrupulous. Such abuses as have occurred are usually due to ignorance and inexperience. The most effective protection for all concerned depends upon a recognition and an understanding of the various aspects of the problem.[84]

Another episode involving Henry Beecher further clarifies the medical profession's dissatisfaction with the construction of the Nuremberg Code. In the fall of 1961, Beecher and other members of the Harvard Medical School's Administrative Board, the school's governing body, were presented with a set of "rigid rules" that had begun to appear in Army medical research contracts. The members of the board quickly recognized the "Principles, Policies and Rules of the Surgeon General, Department of the Army, Relating to the Use of Human Volunteers in Medical Research" awarded by the Army as little more than a restatement of the Nuremberg Code. The Army Office of the Surgeon General's provisions, as we discussed in chapter 1, originally appeared in 1954. Given what we have just read of Beecher, it is not surprising that he was uncomfortable with the prospect of working in strict accordance with the Nuremberg Code if he were to receive funding from the Army, nor, as we see from the minutes of the Administrative Board meetings in which this matter came up for discussion, was Beecher alone in his opposition. At the October 6, 1961, meeting of the board, when the Army contract insertion was first mentioned, "some members . . . felt that with the minor changes the regulations were acceptable, while others described the regulations as vague, ambiguous and, in many instances, impossible to fulfill."[85]

One of Beecher's fellow board members, Assistant Medical School Dean Joseph W. Gardella, M.D., produced a thoroughgoing written critique of the "Principles, Policies, and Rules of the Surgeon General" (and, thus, of the Nuremberg Code) following the October 1961 meeting for the consideration of the other board members. Gardella opened his analysis with some general comments on the intended meaning of the Nuremberg Code:

The Nuremberg Code was conceived in reference to Nazi atrocities and was written for the specific purpose of preventing brutal excesses from being committed or excused in the name of science. The code, however admirable in its intent, and however suitable for the purpose for which it was conceived, is in our opinion not necessarily pertinent to or adequate for the conduct of medical research in the United States.[86]

After questioning the pertinence of the Nuremberg Medical Trial to American medical science, Gardella went on to raise a general question about the scope of the Nuremberg Code; he strongly suggested that the code was not meant to cover what he perceived as the morally distinct enterprise of conducting potentially therapeutic research with sick patients:

Does it refer only to healthy volunteers who have nothing to gain in terms of their health by participating as research subjects? Or does it include the sick, whose physicians foresee for them the possibility of personal benefit through their participation? The distinction is important in that we believe that it would be difficult and might prove to be impossible to devise one set of guiding principles that would apply satisfactorily to both of these two different categories.[87]

Gardella offered a variety of specific objections to the Army surgeon general's "Principles," but several of these points related directly to the general questions raised above. The first rule of the Army "Principles" stated (in a clear example of borrowing from the Nuremberg Code) that "the voluntary consent of the human subject is absolutely essential." Gardella, like Beecher, did not question the general spirit of this stricture; he worried about the practical application of this seemingly simple idea. Some of Gardella's worries arose specifically in the context of research with sick patients:

The concept of "voluntary consent" is of central importance in any code relating to experimentation on humans. . . . And yet the concept of "consent" is not satisfactorily defined [in the Army "Principles"]. . . . The quality of the subject's consent depends . . . upon an interpretation . . . of a factual situation which will frequently be complex. Could the subject comprehend what he was told? Did he in fact comprehend? How far was his consent influenced by his condition or by his trust in his physician? These questions may be easily answered in the case of the [healthy] volunteer. *They may be more difficult for the sick* [emphasis added].[88]

Perhaps the most significant addition to the Nuremberg Code found in the Army "Principles" was the requirement for *written* consent from research subjects. Gardella objected to this requirement in research on patients in a firm, and revealing, fashion:

This condition is . . . inappropriate except in connection with healthy normal volunteers. The legal overtones and implications attendant to such a requirement have no place in [a] patient-physician relationship based on trust. Here such faith and trust serve as the primary basis of the subject's consent. Moreover being asked to sign a somewhat formal paper is likely to provoke anxiety in the subject [i.e., patient] who can but wonder at the need for so much protocol.[89]

Dr. Gardella presented his analysis of the Army "Principles" to the other members of the Harvard Medical School Administrative Board on March 23, 1962. The minutes of that meeting document that Gardella's views were not extreme or exceptional among leading medical scientists in the early 1960s, at least at Harvard University: "The members of the Board were in general agreement with the objections and criticisms expressed in [Gardella's] critique."[90] At this same meeting, Henry Beecher "agreed, in an expansive moment, to attempt to capture in a paragraph or so the broad philosophical and moral principles that underlie the conduct of research on human beings at the Harvard Medical School."[91] The members of the board hoped that such a statement might satisfy the Army and that it would allow Harvard, as Gardella put it, "to avert the catastrophic impact of the Surgeon General's regulation."[92]

A few months later, Beecher had completed a two-and-a-half-page "Statement Outlining the Philosophy and Ethical Principles Governing the Conduct of Research on Human Beings at Harvard Medical School." At the June 8, 1962, board meeting, Beecher's colleagues "commended" and "reaffirmed" the views expressed in Beecher's document.[93] In this statement, as in his 1959 published paper, Beecher emphasized the significance of consent, but he also asserted that "it is folly to overlook the fact that valid, informed consent may be difficult to the point of impossible to obtain in some cases." More than consent, Beecher believed in the significance of "a special relationship of trust between subject or patient and the investigator." In the end, Beecher concluded that the only reliable foundation for this relationship was a virtuous medical researcher, with virtuous peers:

It is this writer's point of view that the best approach [to research with human subjects] concerns the character, wisdom, experience, honesty, imaginativeness and sense of responsibility of the investigator who in all cases of doubt or where serious consequences might remotely occur, will call in his peers and get the benefit of their counsel. Rigid rules will jeopardize the research establishments of this country where experimentation in man is essential.[94]

Available evidence suggests that, by offering Henry Beecher's replacement for the Nuremberg Code, representatives of Harvard Medical School were able to extract a clarification during a meeting with Army Surgeon General Leonard D. Heaton, on July 12, 1962, that the "Principles" being inserted into Harvard's research contracts with the Army were "guidelines" rather than "rigid rules."[95]

While the Harvard Medical School discussion of the Army's "Principles" took place behind closed doors and involved a policy of limited applicability, the leaders of the international medical community were simultaneously engaged in a far more visible and global attempt to bring the standards enunciated in the Nuremberg Code into line with the realities of medical research. The 1964 statement by the World Medical Association (WMA), commonly known as the Declaration of Helsinki, created two separate categories in laying out rules for human experimentation: "Clinical Research Combined with Professional Care" and "Non-therapeutic Clinical Research."[96] In the former category, physicians were required to obtain consent from patient-subjects only when "consistent with patient psychology." In the latter type of research, the consent requirements were more absolute: "Clinical research on a human being cannot be undertaken without his free consent, after he has been fully informed." Another noteworthy deviation from the Nuremberg Code is Helsinki's allowance (in both therapeutic and nontherapeutic research) for third-party permission from a legal guardian.[97]

As one might predict from the similarity between the changes introduced by the Declaration of Helsinki and the changes to the

Nuremberg Code suggested by the American participants at the NSMR conference in 1959, the WMA document met with widespread approval among researchers in this country. Organizations including the American Society for Clinical Investigation, the American Federation for Clinical Research, and the American Medical Association offered their quick and enthusiastic endorsements.[98] Compared with the lofty, idealized language of the Nuremberg Code, the Helsinki Declaration may have seemed more sensible to many researchers in the early 1960s because it offered rules that more closely resembled research practice in the clinical setting.

CONCLUSION

In the late 1940s American medical researchers seldom recognized that research with patient-subjects ought to follow the same principles as those applied to healthy subjects. Yet, as we have seen in this chapter, some of those few who asked themselves hard questions about their research work with patients concluded that people who are ill are entitled to the same consideration as those who are not. That some did in fact reach this conclusion is evidence that it was not beyond the horizon of moral insight at that time. Nevertheless, they were a minority of the community of physician researchers, and the organized medical profession did not exhibit a willingness to reconsider its responsibilities to patients in the burgeoning world of postwar clinical research.

While a slowly increasing number of investigators reflected on the ethical treatment of human subjects during the 1950s, it was not until the 1960s and a series of highly publicized events with names like "Thalidomide," "Willowbrook," and "Tuskegee" that it became apparent that a professional code, whether it originated in Nuremberg or Helsinki, did not provide sufficient protection against exploitation and abuse of human subjects of research. In the next chapter we examine how the federal government became intimately, extensively, and visibly involved in the regulation of research with human subjects.

NOTES

1. A detailed recounting of the first series of Nuremberg Trials can be found in Telford Taylor, *The Anatomy of the Nuremberg Trials: A Personal Memoir* (New York: Alfred A. Knopf, 1992). Taylor describes the motivation for the second series of Nuremberg Trials in the introduction to this book (p. xii). He also mentions that he "hope[s] later to write a description of these subsequent trials" (p. xii). Taylor served as an assistant to chief American prosecutor Robert H. Jackson at the first series of trials; he was the chief prosecutor for the second series, which eventually included twelve separate trials.

2. *United States v. Karl Brandt et al.*, "The Medical Case, Trials of War Criminals before the Nuremberg Military Tribunals under Control Council Law No. 10" (Washington, D.C.: U.S. Government Printing Office, 1949). This two-volume set contains an abridged set of transcripts from the Nuremberg Medical Trial. A general timeline for the trial can be found on p. 3 of volume 1; the quotation of Taylor's charges can be found in the reproduction of his opening statement in volume 1, p. 27. The published trial transcripts provide extensive detail on the experiments carried out by German medical scientists on inmates at Nazi concentration camps. These experiments included a long list of brutalities carried out in the name of medical science. Some of these were specifically related to the Nazi war effort. German investigators conducted high-altitude tolerance tests for the *Luftwaffe* using a low-pressure chamber. Scientists forced prisoners to enter the chamber and subjected them to extreme pressure changes that resulted in excruciating pain and, sometimes, death. Among these experiments were human twin studies related to genetics and germ warfare. For example, a series of experiments involved injecting one twin with a potential germ warfare agent to test the effects of that agent. If the twin injected with the germ died, the other twin was immediately killed to compare the the organs between the healthy and the sick twin. Another series of experiments related to downed airman and shipwrecked sailors who were faced with deprivation of potable water. In these tests, prisoners were divided into four groups: the first received no water; a second set was forced to drink ordinary seawater; the third would drink seawater processed to remove the salty taste (but not the actual salt); and fourth group could drink desalinated seawater. Many of the subjects in the first three groups died. German researchers also compelled prisoners to engage in a variety of other cruel experiments, many of which were concerned with infectious diseases such as malaria, epidemic jaundice, and typhus. More information can be found on the Nazi prison camp experiments in several sources including Robert Jay Lifton, *The Nazi Doctors: Medical Killing and the Psychology of Genocide* (New York: Basic Books, 1986); Robert

N. Proctor, *Racial Hygiene: Medicine under the Nazis* (Cambridge, Mass.: Harvard University Press, 1988); and George J. Annas and Michael A. Grodin, eds., *The Nazi Doctors and the Nuremberg Code: Human Rights in Human Experimentation* (New York: Oxford University Press, 1992).

Japanese medical scientists, especially those associated with a biological warfare (BW) research corps known as Unit 731, also conducted many cruel medical experiments during the war. Until recently, these experiments were virtually unknown because American military and medical officials struck a postwar deal with leading Japanese scientists associated with Unit 731: immunity from war crimes prosecution in exchange for exclusive American access to the results of the Japanese BW experiments. The Japanese experiments and the American cover-up have recently received coverage in Sheldon Harris's *Factories of Death: Japanese Biological Warfare, 1932–1945, and the American Cover Up* (London/New York: Routledge, 1994). See also Peter Williams and David Wallace, *Unit 731: The Japanese Army's Secret of Secrets* (London: Hodder and Stoughton, 1989); and John W. Powell, Jr., "Japan's Biological Weapons, 1930–1945," *Bulletin of the Atomic Scientists* 37 (October 1981): 44–53.

3. American Medical Association, Board of Trustees, minutes of the May 1946 meeting, AMA Archive, Chicago, Illinois (ACHRE No. IND-072595–A), 156–157.

4. A full-blown biography of Ivy remains to be written, but some biographical information can be found in the following brief notices: Carl A. Dragstedt, "Andrew Conway Ivy," *Quarterly Bulletin of the Northwestern University Medical School* 18 (Summer 1944): 139–140; Morton I. Grossman, "Andrew Conway Ivy (1893–1978)," *Physiologist* 21 (April 1978): 11–12; D. B. Bill, "A. C. Ivy—Reminiscences," *Physiologist* 22 (October 1979): 21–22.

5. The quotation is taken from Andrew C. Ivy, "Nazi War Crimes of a Medical Nature," *Federation Bulletin* 33 (May 1947): 133. Ivy first publicly offered this view of the Nuremberg prosecutors' confusion about the ethics and legality of human experimentation when he presented this paper at an annual meeting of the Federation of State Medical Boards of the United States on 10 February 1947—just a few months after the start of the Medical Trial. In this presentation Ivy said that he traveled to Germany in August 1946. In a similar description of his experiences with the Nuremberg prosecution team published a few years later Ivy reiterates a similar story except that the date of his initial travel is given as July 1946: A. C. Ivy, "Nazi War Crimes of a Medical Nature," *Journal of the American Medical Association* 139 (15 January 1949): 131. An editorial in *JAMA* confirms some of the essential elements of Ivy's early work with the Nuremberg prosecutors (his selection by the AMA Board of Trustees at the request of the federal government and his travel to Germany "a few

months" before November 1946): "The Brutalities of Nazi Physicians," *JAMA* 132 (23 November 1946): 714. The basic narrative of Ivy's selection by the Board of Trustees and his travel to Europe can also be found in R. L. Sensenich, "Supplementary Report of the Board of Trustees," *JAMA* 132 (21 December 1946): 1006.

6. American Medical Association, Board of Trustees, minutes of the 16 August 1946 meeting, AMA Archive, Chicago, Illinois (ACHRE No. IND-072595–A), 8–9.

7. American Medical Association, Board of Trustees, minutes of the 19 September 1946 meeting, AMA Archive, Chicago, Illinois (ACHRE No. IND-072595–A), 51–52.

8. A. C. Ivy, "Report on War Crimes of a Medical Nature Committed in Germany and Elsewhere on German Nationals and the Nationals of Occupied Countries by the Nazi Regime during World War II," 1946. This report was not published, but it is available at the National Library of Medicine. A copy also exists in the AMA Archive (ACHRE No. DOD-063094–A).

9. *United States v. Karl Brandt et al.*, "The Medical Case, Trials of War Criminals before the Nuremberg Military Tribunals under Control Council Law No. 10" (Washington, D.C.: U.S. Government Printing Office, 1949), 2: 181–182. The judges' preamble to the Code states that "[a]ll agree . . . that certain basic principles must be observed in order to satisfy the moral, ethical and legal" aspects of human experimentation.

10. Ivy's recitation of his own set of rules does not appear in the published abridged transcripts of the trial. See the complete transcripts of the trial, which are available on microfilm at the National Archives (National Archives Microfilm, M887, reel 9, 13 June 1947, pp. 9141–9142). Throughout this chapter, we cite the abridged transcripts wherever possible and the full transcripts only if necessary.

11. Leo Alexander later reproduced his 15 April 1947 memo in two publications: "Limitations in Experimental Research on Human Beings," *Lex et Scientia* 3 (January-March 1966): 20–22, and "Ethics of Human Experimentation," *Psychiatric Journal of the University of Ottawa* 1 (1976): 42–44. In the 1976 article, Alexander made a seemingly exaggerated claim to be "the original author of the Nuremberg Code" (p. 40). Side-by-side comparison of Ivy's rules, Alexander's memo, and the Nuremberg Code does, however, suggest that the judges drew two original contributions from Alexander's memo: clauses 6 and 7 of the Nuremberg Code are embedded in the 15 April memo (they do not appear in Ivy's rules).

12. McHaney's closing statement can be found in the complete microfilm transcripts of the trial available through the National Archives. McHaney's closing statement and Alexander's memorandum (and Alexander's claim to authorship of the Code) are also reproduced in Michael A. Grodin's "Historical Ori-

gins of the Nuremberg Code," in *The Nazi Doctors and the Nuremberg Code: Human Rights in Human Experimentation*, 134–137.

13. American Medical Association, Board of Trustees, minutes of the 19 September 1946 meeting, AMA Archive, Chicago, Illinois (ACHRE No. IND-072595–A).

14. The AMA reports that the records of the Judicial Council for all of the 1940s have been lost. Personal communication between Marilyn Douros, of the AMA Archives, and Jon M. Harkness (ACHRE), 19 January 1995.

15. "Supplementary Report of the Judicial Council," proceedings of the House of Delegates Annual Meeting, 9–11 December 1946, *JAMA* 132 (28 December 1946): 1090. The bracketed addition to rule 1 was added in the final version of statement, which was approved by the House of Delegates on 11 December 1946.

16. William A. Coventry, "Report of the Reference Committee on Miscellaneous Business," proceedings of the House of Delegates Meeting, 9–11 December 1946, *JAMA* 133 (4 January 1947): 35.

17. Robert Williamson, an AMA archivist, reports that in 1942, 65 percent of American physicians were members of the AMA, and in 1949, 75 percent of American physicians were members; he did not have percentage figures available for 1946. Williamson also provided the absolute number of members for 1946. Personal communication between Jon M. Harkness (ACHRE) and Robert Williamson, 4 January 1995.

18. *United States v. Karl Brandt et al.*, "The Medical Case, Trials of War Criminals before the Nuremberg Military Tribunal under Control Council Law No. 10," vol. 2, 83.

19. Ibid.

20. Complete transcripts of the Nuremberg Medical Trial, National Archives Microfilm, M887, reel 9, 13 June 1947, pp. 9168–9170.

21. Susan E. Lederer, *Subjected to Science: Experimentation in America before the Second World War* (Baltimore: Johns Hopkins University Press, 1995), 105.

22. Lederer recounts the historical details of the yellow fever experiment (pp. 19–23) and explores Reed's powerful legacy (pp. 132–134) in *Subjected to Science*.

23. Theodore Woodward, interview by Gail Javitt and Suzanne White-Junod (ACHRE), transcript of audio recording, 14 December 1994 (ACHRE Research Project Series, Interview Program File, Ethics Oral History Project), 6.

24. Interview with Woodward, 14 December 1994, 10.

25. John D. Arnold, interview by Jon M. Harkness (ACHRE), transcript of audio recording, 6 December 1994 (ACHRE Research Project Series, Interview Program File, Ethics Oral History Project), 18.

26. The list of participants exists in the extant records of the LMRI project available at the Center for Law and Health Sciences, School of Law, Boston University. The quotation explaining the goal of the meeting is taken from the first page of a summary of the conference prepared for the project's final report, which was not published: Anne S. Harris, "The Concept of Consent in Clinical Research: Analytic Summary of a Conference," chapter 6 in *A Study of the Legal, Ethical, and Administrative Aspects of Clinical Research Involving Human Subjects: Final Report of Administrative Practices in Clinical Research, [NIH] Research Grant No. 7039* Law-Medicine Research Institute, Boston University, 1963 (ACHRE No. BU-053194–A).

27. The National Institutes of Health awarded LMRI almost $100,000 on 1 January 1960 to begin this project, which concluded 31 March 1963. The general statement of the project's purpose appears in LMRI final report, chapter 1 ("Focus of the Inquiry"), 1.

28. LMRI final report, chapter 6, 48.

29. Louis Lasagna, interview by Jon M. Harkness and Suzanne White-Junod (ACHRE), transcript of audio recording, 13 December 1994 (ACHRE Research Project Series, Interview Program File, Ethics Oral History Project), 5.

30. Ibid., 11.

31. Extensive newspaper clippings related to the Nuremberg Medical Trial exist in Beecher's personal papers in the Special Collections Department, Countway Library, Harvard University. Beecher's first major publication on research ethics appeared in early 1959: Henry K. Beecher, "Experimentation in Man," *JAMA* 169 (31 January 1959): 461–478. Of course, he is best known for a 1966 article: Henry K. Beecher, "Ethics and Clinical Research," *New England Journal of Medicine* 274 (16 June 1966): 1354–1360. Significantly, Beecher acknowledged in a manuscript copy of the original version of the *NEJM* paper, which he presented at a conference for science journalists on 22 March 1965, that "in years gone by work in my laboratory could have been criticized." Beecher, "Ethics and the Explosion of Human Experimentation," 2a, Beecher Papers, Countway Library (ACHRE No. IND-072595–A).

32. Jay Katz, "Human Experimentation and Human Rights," *St. Louis University Law Journal* 38 (1993): 28.

33. Stanley Joel Reiser, Arthur J. Dyck, and William J. Curran, eds., *Ethics in Medicine: Historical Perspectives and Contemporary Concerns* (Cambridge, Mass.: The MIT Press, 1977), 7.

34. Otto E. Guttentag, "The Physician's Point of View," *Science* 117 (1953): 207–210; the quotation is from 208. Guttentag's article appeared in *Science* with three others that had been presented at the 1951 symposium: Michael B. Shimkin, "The Researcher Worker's Point of View," 205–207; Alexander M. Kidd, "Limits of the Right of a Person to Consent to Experimentation on Himself," 211–212; and W. H. Johnson, "Civil Rights of Military Personnel Regard-

ing Medical Care and Experimental Procedures," 212–215.

35. Guttentag, "The Physician's Point of View," 208.

36. Ibid., 210.

37. John C. Ford, "Notes on Moral Theology," *Theological Studies* 6 (December 1945): 534–535. Ford's discussion of human experimentation arose in a lengthy and discursive review of issues and ideas in moral theology. For several years, he contributed a similar review to each volume of *Theological Studies*.

38. Transcripts of "Social Responsibility in Pediatric Research" conference, 1 May 1961, 7. LMRI records, Center for Law and Health Sciences, School of Law, Boston University (ACHRE No. BU-053194–A).

39. "LMRI Final Report," chapter 6, 43.

40. Ibid., 43–44.

41. Ibid., 44.

42. Ibid., 46–47.

43. Committee member and historian Susan Lederer took principal responsibility for organizing the Ethics Oral History Project, with assistance from several members of the staff including two historians experienced in the techniques of oral history. The Committee also drew on advice from several outside experts, including historians and ethicists, to create a list of potential interviewees and to refine the list of questions that we wanted to explore during interviews. In total, the Committee conducted twenty-two interviews in the Ethics Oral History Project. Most of the subjects were medical researchers whose careers began in the late 1940s or early 1950s, but we also spoke with some research administrators. In general, we chose to interview researchers who had exhibited some particular interest in research ethics during their careers. But this does not mean that we held interviews only with researchers who viewed recent developments in research ethics in a positive fashion. The interviews were all recorded on audio tape and professionally transcribed. Interview subjects had an opportunity to review the transcripts. Complete sets of all transcripts can be found in the archival records of the Advisory Committee.

44. Interview with Lasagna, 13 December 1994, 13.

45. Paul Beeson, interview by Susan E. Lederer (ACHRE), transcript of audio recording, 20 November 1994 (ACHRE Research Project Series, Interview Program File, Ethics Oral History Project), 16–17.

46. Leonard Sagan, interview by Gail Javitt, Suzanne White-Junod, Sandra Thomas, and John Kruger (ACHRE), transcript of audio recording, 17 November 1994 (ACHRE Research Project Series, Interview Program File, Ethics Oral History Project), 13–14.

47. Ibid., 19–20.

48. Stuart Finch, interview by Gail Javitt, Suzanne White-Junod, and Valerie Hurt (ACHRE), transcript of audio recording, 6 December 1994 (ACHRE Re-search Project Series, Interview Program File, Ethics Oral History Project), 52.

49. Interview with Paul Beeson, 20 November 1994, 39.

50. Thomas Chalmers, interview by Jon Harkness (ACHRE), transcript of audio recording, 9 December 1994 (ACHRE Research Project Series, Interview Program File, Ethics Oral History Project), 75.

51. Herman Wigodsky, interview by Gail Javitt and Suzanne White-Junod (ACHRE), transcript of audio recording, 17 January 1995 (ACHRE Research Project Series, Interview Program File, Ethics Oral History Project), 14.

52. For an analysis and translation of the 1931 German rules see Hans-Martin Sass, "Reichsrundschreiben 1931: Pre-Nuremberg German Regulations Concerning New Therapy and Human Experimentation," *Journal of Medicine and Philosophy* 8 (1983): 99–111. A similar analysis and translation of the same set of rules appears in Grodin, "Historical Origins of the Nuremberg Code," 129–132.

53. Full trial transcripts, 9142.

54. Abridged trial transcripts, 83.

55. Interview with Dr. Herman Wigodsky, 17 January 1995, 16–17.

56. Ruth Faden and Tom Beauchamp, *A History and Theory of Informed Consent* (New York: Oxford University Press, 1986), 96.

57. Ivy's several examples ranged from Walter Reed's turn-of-the-century experiments with yellow fever to wartime malaria experiments in American state and federal prisons. See page 9119 of the full trial transcripts for Ivy's discussion of the Reed experiments and pages 9125–9129 for his description of the malaria experiments that had taken place in the United States during the war.

58. David J. Rothman, *Strangers at the Bedside: A History of How Law and Bioethics Transformed Medical Decision Making* (New York: Basic Books, 1994), 62.

59. Interview with John Arnold, 6 December 1994, 9–10.

60. Herbert Abrams, interview by Jon Harkness (ACHRE), transcript of audio recording, 12 January 1995 (ACHRE Research Project Series, Interview Program File, Ethics Oral History Project), 25.

61. Dorothy Levenson, *Montefiore: The Hospital as Social Instrument, 1884–1984* (New York: Farrar, Straus & Giroux, 1984). For information on the presence of Jewish refugee physicians at Montefiore, see pages 154–155.

62. Rothman, *Strangers at the Bedside*, 62–63.

63. Katz, "The Consent Principle of the Nuremberg Code," 228.

64. William Silverman, interview by Gail Javitt (ACHRE), transcript of audio recording 14 February 1995 (ACHRE Research Project Series, Interview Program File, Ethics Oral History Project), 61–62.

65. Ibid., 87–88.

66. "Why Human 'Guinea Pigs' Volunteer," *New York Times Magazine*, 13 April 1958, 62.

67. See, for example, John L. O'Hara, "The Most Unforgettable Character I've Met," *Reader's Digest*, May 1948, 30–35; Thomas Koritz, "I Was a Human Guinea Pig," *Saturday Evening Post*, 25 July 1953, 27, 79–80, 82; Don Wharton, "'A Treasure in the Heart of Every Man,'" *Reader's Digest*, December 1954, 49–53 (condensed from "Prisoners Who Volunteer, Blood, Flesh—and Their Lives," *American Mercury*, December 1954, 51–55); Howard Simons, "They Volunteer to Suffer," *Saturday Evening Post*, 26 March 1960, 33, 87–88.

68. "Experiments on Prisoners," *Science Newsletter* (also *Science News*), 21 February 1948, 53, 117.

69. "C.O.'s Offer Selves for Atomic Experiments," *Christian Century*, 20 October 1954, 1260.

70. Robert D. Potter, "Are We Winning the War Against TB?" *Saturday Evening Post*, 15 January 1949. Cited in Marcel C. LaFollette, *Making Science Our Own: Public Images of Science 1910–1955* (Chicago: University of Chicago Press, 1990), 138–140.

71. Renee C. Fox, *Experiment Perilous: Physicians and Patients Facing the Unknown* (Philadelphia: University of Pennsylvania Press, 1974, first published 1959). Fox describes her long days of observation on page 15; she discusses the Nuremberg Code at 46–47.

72. Michael B. Shimkin, *As Memory Serves: Six Essays on a Personal Involvement with the National Cancer Institute, 1938–1978* (Bethesda, Md.: U.S. Department of Health and Human Services, 1983), 127.

73. Ibid., 128.

74. Ibid., 127.

75. The quotation is a translation from the French in which Pius XII delivered the address: "Il faut remarquer que l'homme dans son être personnel n'est pas ordonné en fin de compte à l'utilité de la société, mais au contraire, la communauté est là pour l'homme." The French text can be found in the *Atti del Primo Congresso Internazionale di Istopatologia del Sistema Nervosa/Proceedings of the First International Congress of Neuropathology*, Rome, 8–13 September 1952. English translations of the pope's address appear in a variety of publications including *The Linacre Quarterly: Official Journal of the Federation of Catholic Physicians' Guilds* 19 (November 1952): 98–107 and *The Irish Ecclesiastical Record* 86 (1954): 222–230.

76. Saul Benison, *Tom Rivers: Reflections on a Life in Medicine and Science* (Cambridge, Mass.: MIT Press, 1967), 498.

77. Louis J. Regan, *Doctor and Patient and the Law*, 2d ed. (St. Louis: C. V. Mosby, 1949), 398.

78. Louis J. Regan, *Doctor and Patient and the Law*, 3d ed. (St. Louis: C. V. Mosby, 1956), 370–372.

79. *Report on the National Conference on the Legal Environment of Medicine, 27–28 May 1959* (Chicago: National Society for Medical Research, 1959); the quotations are from pages 91 and 88, respectively.

80. Henry K. Beecher, "Experimentation in Man," *Journal of the American Medical Association* 169 (1959): 118/470.

81. Ibid., 121/473.

82. Ibid., 122/474.

83. Ibid., 109/461.

84. Ibid., 119/471.

85. Harvard Medical School, Harvard Medical School Administrative Board, proceedings of the 6 October 1961 meeting (ACHRE No. HAR-062394–A-3).

86. Memorandum to "GPB" [Harvard Medical School Dean Berry] from "JWG" [Assistant Dean Gardella] ("Criticisms of 'Principles, Policies and Rules of the Surgeon General, Department of the Army, relating to the use of Human Volunteers in Medical Research Contracts awarded by the Army'") (ACHRE No. IND-072595–A), 1.

87. Ibid., 2.

88. Ibid.

89. Ibid., 3.

90. Harvard Medical School, Harvard Medical School Administrative Board, proceedings of 23 March 1962 (ACHRE No. HAR-062394–A-3).

91. Joseph W. Gardella, Assistant Dean, Harvard Medical School to Henry K. Beecher, Massachusetts General Hospital, 27 March 1962 ("I write to confirm my impression . . .") (ACHRE No. HAR-062394–A-4).

92. Ibid.

93. Harvard Medical School, Harvard Medical School Administrative Board, proceedings of 8 June 1962 (ACHRE No. HAR-062394–A-3).

94. Henry Beecher, undated ("Statement Outlining the Philosophy and Ethical Principles Governing the Conduct of Research on Human Beings at the Harvard Medical School") (ACHRE No. IND-072595–A).

95. Henry K. Beecher to Lieutenant General Leonard D. Heaton, 12 July 1962 ("I have just returned to Boston . . .") (ACHRE No. HAR-062394–A-2).

96. World Medical Association, "Declaration of Helsinki: Recommendations Guiding Medical Doctors in Biomedical Research Involving Human Subjects," adopted by the Eighteenth World Medical Assembly, Helsinki, Finland, 1964.

97. "Draft Code of Ethics on Human Experimentation," *British Medical Journal* 2 (1962): 1119; "Human Experimentation: Code of Ethics of the World Medical Association," *British Medical Journal* 2 (1964): 177.

98. Faden and Beauchamp, *A History and Theory of Informed Consent*, 156–157, and Paul M. McNeill, *The Ethics and Politics of Human Experimentation* (Cambridge, U.K.: Press Syndicate of the University of Cambridge, 1993), 44–47. For a more detailed comparison between the Nuremberg Code and the Declaration of Helsinki, see Jay Katz, "The Consent Principle of the Nuremberg Code," 231–234.

3

Government Standards for Human Experiments: The 1960s and 1970s

THE year 1974 marks the upper bound for the period of the Advisory Committee's historical investigation. That year two landmark events in the history of government policy on research involving human subjects took place: the promulgation by the Department of Health, Education, and Welfare (DHEW) of comprehensive regulations for oversight of human subject research and passage by Congress of the National Research Act. The DHEW regulations set rules for oversight of human subject research supported by the single largest funding source for such research, and the National Research Act authorized the establishment of the National Commission for the Protection of Human Subjects of Biomedical and Behavioral Research (also known as the National Commission), which was charged with examining the conduct of research involving human subjects. In the years following 1974, many of the rules promulgated by DHEW were subsequently adopted by various other government agencies, culminating in governmentwide regulations under the Common Rule in 1991.[1]

In the first part of this chapter, we trace the developments in the 1960s and early 1970s that influenced and led up to the DHEW regulations and the National Research Act. These developments included congressional hearings on the practices of the drug industry and the thalidomide tragedy, critical scholarly writings, interim policies at DHEW, public outcry over controversial cases of medical research, and the congressional hearings these cases occasioned. People were surprised and shocked to learn about practices and behaviors they knew to be wrong. While the ethical principles such practices violated may not have been well articulated specific to the enterprise of human research, they were part of individuals' moral consciousness. The history of these events has been well told before, and we only summarize it here, drawing heavily on the previous work of other authors.[2]

The 1974 regulations were promulgated by DHEW and applied only to that agency. Likewise, the National Research Act authorized the establishment of the National Commission and directed it to make recommendations to the secretary of DHEW. In the latter part of this chapter, we review developments in policies governing human research during this period in agencies other than DHEW. This is a history that has received comparatively little scholarly attention.

In the 1970s, just as DHEW was moving ahead with broad new regulations, scandal rocked the Department of Defense and the CIA. It was revealed that, with cooperation from university researchers, these agencies had engaged in secret experimentation on military and civilian subjects without their knowledge, sometimes with tragic results.[3] The discovery of the existence of these secret programs led to further congressional investigations and to a 1975 Department of the Army review of the effectiveness of the 1953 Secretary of Defense Wilson memorandum adopting the Nuremberg Code. This Army review led to the eventual declassification of the Wilson memorandum, which had been Top Secret upon its issuance and remained classified until 1975. It also led, much later, to litigation in which justices of the U.S. Supreme Court for the first time commented on the applicability of the Nuremberg Code to actions undertaken by the U.S. government.[4] The chapter concludes with a discussion of these important events.

THE DEVELOPMENT OF HUMAN SUBJECT RESEARCH POLICY AT DHEW

As the largest funding source in the federal government for human subject research, DHEW led the way in developing regulations aimed at protecting the rights and welfare of subjects. The evolution of the regulations, which would eventually be adopted on a government wide basis, was influenced by revelations of unethical research, congressional reaction to the revelations, and concern over public perception of such research. That regulations were eventually adopted at all by DHEW was influenced by the political realities of the time and the lack of congressional support for a standing regulatory body to oversee human subject research, as had been recommended by an influential federally appointed panel, the Tuskegee Syphilis Study Ad Hoc Panel. In a trade-off that would have major influence on the future of human subject research oversight, the proposed bill creating the standing regulatory body was withdrawn in exchange for the National Research Act, establishing the National Commission, and an understanding that DHEW would promulgate the aforementioned regulations. This historical backdrop is outlined in the remainder of this chapter.

The Thalidomide Tragedy and the Congressional Requirement for Patient Consent

In 1959 a Senate subcommittee chaired by Senator Estes Kefauver of Tennessee began hearings into the conduct of pharmaceutical companies. Testimony revealed that it was common practice for drug companies to provide samples of experimental drugs, whose safety and efficacy had not been established, to physicians, who were then paid to collect data on their patients taking these drugs. Physicians throughout the country prescribed these drugs to patients without their knowledge or consent as part of this loosely controlled research. These practices and others prompted calls by Kefauver and other senators for an amendment to the Food, Drug, and Cosmetic Act of 1938 to address the injuriousness and ineffectiveness of certain drugs. In 1961 the dangers of new drug uses were vividly exemplified by the thalidomide disaster in Europe, Canada, and to a lesser degree, the United States.[5] Starting in late 1957, the sedative thalidomide was given to countless pregnant women and caused thousands of birth defects in newborn infants (most commonly, missing or deformed limbs). The thalidomide disaster was widely covered by the television networks, and the visual impact of these babies stunned viewers and caused Americans to question the protections afforded those receiving investigational agents.

It is in large measure because of the thalidomide episode that the 1962 Kefauver-Harris amendments to the Food, Drug, and Cosmetic Act were passed,[6] requiring that informed consent be obtained in the testing of investigational drugs.[7] While such testing occurred mainly with patients, Congress carefully avoided interfering in the doctor-patient relationship and in the process severely reduced the effectiveness of the requirement. Consent was not required when it was "not feasible" or was deemed not to be in the best interests of the patient—both judgments made "according to the best judgment of the doctors involved."[8] Despite their being limited in scope, the Kefauver-Harris amendments were influential in advancing considerations of protections of research subjects first within the

DHEW and later throughout the rest of the government.

NIH and PHS Develop a Uniform Policy to Protect Human Subjects

In late 1963, concerns were raised within NIH by Director James Shannon after disturbing revelations about two research projects funded in part by the Public Health Service and NIH. One was the unsuccessful transplantation of a chimpanzee kidney into a human being at Tulane University, a procedure that promised neither benefit to the recipient nor new scientific information. The transplant was reportedly done with the consent of the patient, but without consultation or review by anyone other than the medical team involved.[9]

The second was research undertaken in mid-1963 at the Brooklyn Jewish Chronic Disease Hospital. There, investigators (the chief investigator, Dr. Chester M. Southam was a physician at the Sloan-Kettering Cancer Research Institute, and he received permission to proceed with the work from the hospital's medical director, Dr. Emmanuel E. Mandel) had undertaken a research project in which they injected live cancer cells into indigent elderly patients without their consent. The research went forward without review by the hospital's research committee and over the objections of three physicians consulted, who argued that the proposed subjects were incapable of giving adequate consent to participate.[10] The disclosure of the experiment served to make both PHS officials like Shannon and the Board of Regents of the University of the State of New York, which had jurisdiction over licensure of physicians, aware of the shortcomings of procedures in place to protect human subjects. They were further concerned over the public's reaction to disclosure of the research and the impact it would have on research generally and the institutions in particular. After a review, the Board of Regents censured the researchers. They suspended the licenses of Drs. Mandel and Southam, but subsequently stayed the suspension and placed the physicians on probation for one year.[11] There were no immediate repercussions for the hospital, Sloan-Kettering, the university, or PHS, but the case

nonetheless profoundly affected the subsequent development of federal guidelines to protect research subjects.

To add to the ferment, NIH officials had closely followed the work of the Law-Medicine Research Institute at Boston University, which issued survey findings in 1962 showing that few institutions had procedural guidelines covering clinical research.[12] And in the year after both the above-mentioned cases came to light, the World Medical Association issued its Declaration of Helsinki, which set standards for clinical research and required that subjects give informed consent prior to enrolling in an experiment.[13] Thus national and world opinion on matters related to the ethics of human subject research created a climate ripe for changes in policies and approaches toward research ethics.

Concern over disturbing cases and the growing attention paid to research ethics prompted NIH director James Shannon to create a committee in late 1963 under the direction of the NIH associate chief for program development, Robert B. Livingston, whose office supported centers at which NIH-funded research took place. The internal committee was charged with studying problems of inadequate consent and the standards of self-scrutiny involving research protocols and procedures. The committee was also to recommend a suitable set of controls for the protection of human subjects in NIH-sponsored research. The Livingston Committee recognized that ethically questionable research— exemplified by the research at the Jewish Chronic Disease Hospital—could wreak havoc on public perception, increase the likelihood of liability, and inhibit research.[14] These problems made it worthwhile to reconsider central oversight—or lack thereof—for research contracted out. However, the committee expressed concern over NIH taking too authoritarian a posture toward research oversight and so argued that it would be difficult for the agency to assume responsibility for ethics and research practices. When it issued its report in late 1964, the committee did not recommend any changes in the current NIH policies and, moreover, cautioned that "whatever NIH might do by way of designating a code or stipulating standards for acceptable clinical research would be likely to inhibit, delay, or distort the carrying out of clinical research. . . ."[15] In deference to physician autonomy and traditional regard for the sanctity of the doctor-patient relationship, the report concluded that NIH was "not in a position to shape the educational foundations of medical ethics. . . ."[16]

Director Shannon did not think the conclusions of the Livingston Committee went far enough, feeling as he did that NIH should take a position of increased responsibility for research ethics.[17] Especially in light of the Jewish Chronic Disease Hospital case and its implications for the NIH, both internally and in terms of public perception, he felt that a stronger reaction was needed. Thus, despite the committee's limited conclusions, Shannon and Surgeon General Luther Terry together decided in 1965 to propose to the National Advisory Health Council (NAHC), an advisory committee to the surgeon general of the Public Health Service,[18] that in light of recent problems, the NIH should assume responsibility for formal controls on individual investigators.[19] At the NAHC meeting, Shannon argued for impartial prior peer review of the risks research posed to subjects and questioned the adequacy of the protections of the rights of subjects.[20]

The council's members mostly agreed with Shannon's concerns and three months later issued a "resolution concerning research on humans" following Shannon's broad recommendations and endorsing the importance of obtaining informed consent from subjects:

Be it resolved that the National Advisory Health Council believes that Public Health Service support of clinical research and investigation involving human beings should be provided only if the judgment of the investigator is subject to prior review by his institutional associates to assure an independent determination of the protection of the rights and welfare of the individual or individuals involved, of the appropriateness of the methods used to secure informed consent, and of the risks and potential medical benefits of the investigation.[21]

What this statement did not do, however, was explain what would count as informed consent. The NAHC recommendations were accepted by the new surgeon general, William H. Stewart, and in February 1966 he issued a policy statement requiring PHS grantee institutions to address three topics by committee prior review for all proposed research involving human subjects:

This review should assure an independent determination (1) of the rights and welfare of the individual or individuals involved, (2) of the appropriateness of the methods used to secure informed consent, and (3) of the risks and potential medical benefits of the investigation.[22]

The 1966 PHS policy required that institutions give the funding agency a written "assurance" of compliance, but like the NAHC recommendations, the policy spoke strictly to the procedural aspects of informed consent and not to its meaning and criteria. Substantive informed consent criteria were established for research at the NIH Clinical Center shortly after the PHS policy was issued, but this new policy applied only to intramural research, that is, to research undertaken at the Clinical Center. The Clinical Center policy was important as the first federal research policy with a specific definition of what constituted informed consent requirements in the research context. The inclusion of specific consent requirements in policies applying to extramural research would not occur, however, until the mid-1970s.

The 1966 PHS policy is significant both for its recognition that patient-subjects, like healthy subjects, should be included in the consent provisions for federally sponsored human experimentation and for its attempt to strike a balance between federal regulation and local control, which continues to this day. Such a balancing continued the work begun by the AEC, in its provision for local human use committees as a condition for the use of AEC-supplied isotopes, and the DOD, in the provision for high-level review of proposed experimentation. Although a landmark in the government regulation of biomedical research, the 1966 policy was to be revised and changed throughout the decade as biomedical research drew greater attention and informed consent grew in importance.

While, from the outset, the PHS policy was revised periodically,[23] site visits by PHS employees to randomly selected institutions revealed a wide range of compliance.[24] These site visits found widespread confusion about how to assess risks and benefits, refusal by some researchers to cooperate with the policy, and in many cases, indifference by those charged with administering research and its rules at local institutions. Complaints of overworked review committees and requests for clarification and guidance came from research institutions all over the country.[25]

In response to continued questions about the scope and meaning of the policy, DHEW in 1971 produced *The Institutional Guide to DHEW Policy on Protection of Human Subjects.*[26] Better known as the "Yellow Book" because of its cover's color, this substantial guide contained both the requirements and commentary on how the requirements were to be understood and implemented. The guide provided that informed consent was to be obtained from anyone who "may be at risk as a consequence of participation" in research—including both patients and healthy volunteers.[27]

As the 1960s progressed, increased discussion of research practices appeared in both professional literature and the popular press. One person who advanced the debate in both arenas was Henry Beecher of Harvard Medical School.

Henry Beecher: The Medical Insider Speaks Out

Henry Beecher, as noted in chapter 2, was an active participant in professional discussions of ethics in research during the late 1950s and early 1960s. In March 1965, Beecher focused attention on the issues at a conference for science journalists sponsored by the Upjohn pharmaceutical company. There Beecher presented a paper discussing twenty-two examples of potentially serious ethical violations in experiments that he had found in recent issues of medical journals.[28] (Among them was the Brooklyn Jewish Chronic Disease Hospital study.) He explained this research had not taken place "in a remote corner, but [in] . . . leading medical schools, university hospitals, top governmental military departments, governmental institutes and industry."[29] He also acknowledged that his own conscience was not entirely clear: "Lest I seem to stand aside from these matters I am obliged to say that in years gone by work in my laboratory could have been criticized."[30] Beecher also explained the

consciousness-raising purpose of these revelations with stark clarity: "It is hoped that blunt presentation of these examples will attract the attention of the uninformed or the thoughtless and careless, the great majority of offenders."[31]

In making this presentation to a group of journalists, Beecher was clearly breaking with a professional expectation that such matters should be addressed within the biomedical community. After some reservations on the part of medical journals, the March 1965 paper having been rejected by at least the *Journal of the American Medical Association* (*JAMA*), Beecher published a revised version in the *New England Journal of Medicine* in June 1966.[32] That article, like his presentation at the conference, indicted the entire biomedical research community and the journals that published biomedical research results.

Beecher's efforts to focus professional, press, and therefore public awareness on the conduct of research involving human subjects met with some success. A July 1965 article in the *New York Times Magazine* was headlined "Doctors Must Experiment on Humans—But What Are the Patient's Rights?"[33] In February 1966, as the PHS issued its first uniform policy for biomedical research, more headlines, this time in the *Saturday Review*, asked, "Do We Need New Rules for Experimentation on People?"[34] In July 1966, following Beecher's article in the *New England Journal of Medicine* and an editorial in *JAMA*,[35] another article declared "Experiments on People—The Growing Debate."[36] Thus, by the mid- to late 1960s, professional, governmental, and public attention was all being drawn to issues of research on human subjects. Revelations of purportedly unethical treatment of research subjects would not be over by this time, but changes in policy largely driven by attention from so many corners were beginning to move toward a more comprehensive approach to research oversight.

Public Attention Is Galvanized: Willowbrook and Tuskegee

From 1956 to 1972 Dr. Saul Krugman of New York University led a study team at the Willowbrook State School for the Retarded, on Staten

Island, New York. The study was not secret or hidden. (It was one of the twenty-two projects Beecher discussed as ethically troublesome in his 1966 article.) The Willowbrook study was discovered by the media beginning in the late 1960s[37] and was the subject of further discussion of the case in separate places by Beecher,[38] theologian Paul Ramsey,[39] and physician Stephen Goldby.[40] Noting the high incidence of hepatitis among the residents of the school, nearly all of whom were profoundly mentally impaired children and adolescents, Krugman and his colleagues injected some of them with a mild form of hepatitis serum. The researchers justified their work on the grounds that the subjects probably would have become infected anyway, and they hoped to find a prophylaxis for the virus by studying it from the earliest stages of infection. Before beginning the work, Krugman discussed it with many physician colleagues and sought approval from the Armed Forces Epidemiological Board, which approved and funded the research,[41] and the executive faculty of the New York University School of Medicine, who approved the research. A review committee for human experimentation did not exist in 1955,[42] but later, when such a committee was formed, it too approved the research.

According to Krugman, the parents of each subject signed a consent form after receiving a detailed explanation of the research, without any pressure to enroll their child.[43] Some critics argued that the content of the consent form was itself deceiving, since it seemed to say that children were to receive a vaccine against the virus. Moreover, charges of coercion arose. It is alleged that parents who enrolled their children in the study were initially offered more rapid admission to the school through the hepatitis unit and later found, due to overcrowding, that the only route for admission of new patients was through the hepatitis unit.[44] Commentators further argued that the fault in the doctors' study lay in their deliberate attempt to infect the children, with or without parental consent, as opposed to studying the course of disease in children who naturally became sick.

Soon after Willowbrook, another research project, the Tuskegee syphilis study, provoked widespread public outcry when it was revealed the study had exposed people to unnecessary and serious harm with no prospect of direct benefit to them. Beginning in 1932, PHS physicians sought to trace the natural history of syphilis by observing some 400 African-American men affected by the disease and another group of approximately 200 African-American men without syphilis serving as controls. All the subjects lived in or around Tuskegee, Alabama. Originally designed to be a short-term study in the range of six to eight months, some investigators successfully argued that the potential scientific value of longer-term study was so great that the research ought to go on indefinitely. The subjects were enticed into the study with offers of free medical examinations. Many of those who came from around the area to be tested by "government doctors" had never had a blood test before and had no idea what one was.[45] Once selected to be subjects in the study, the men were not informed as to the nature of their disease or of the fact that the research held no therapeutic benefit for them. Subjects were asked to appear for "special free treatments," which included purely diagnostic procedures such as lumbar punctures.[46]

By the mid-1940s it was becoming clear that the death rate for the infected men in the study was twice as high as for those in the control group. This was the period in which penicillin was discovered and soon after began to be used to treat syphilis, at least in its primary stage. The study was reviewed by PHS officials and medical societies and reported by a number of journals from the early 1930s to 1970. In the 1960s a growing number of criticisms began to appear, although the study was not stopped until 1973.

Thus, men with a confirmed disease were not told of their diagnosis and were deceived into participating in the study under the guise of its being therapeutic for unspecified maladies. In addition to exposing the subjects to the additional harms of participation in the study, the false belief that treatment was being administered prevented subjects from otherwise seeking medical care for their disease. As at Willowbrook, a justification given after the fact for the research was that the disease had appeared in a way that was natural and inevitable and that the study would be of immense benefit to future patients.[47] Over this forty-year history, at least 28 participants died and approximately 100 more suffered blindness and insanity from untreated syphilis before the study was stopped.

In 1972, an account of the study was published on the front page of the *New York Times.*[48] In response, DHEW appointed the Tuskegee Syphilis Study Ad Hoc Panel to review the Tuskegee study as well as the department's policies and procedures for the protection of human subjects. The work of the ad hoc panel—which consisted of physicians, a university president, a theologian, an attorney, and a labor representative—contributed in large measure to the passage of the first comprehensive regulations for federally sponsored human subjects research. One member of the ad hoc panel who is also a member of the Advisory Committee, Jay Katz, expressed his dismay over the unwillingness or incapacity of society to mobilize the necessary resources for "treatment" at the beginning of the study and the deliberate efforts of the investigators to "obstruct the opportunity for treatment."[49]

Despite the fact that the PHS Policy for the Protection of Human Subjects had been in place for six years by the time the Tuskegee study was revealed, it was exposed by a journalist rather than by a review committee. Although an institutional committee had allegedly reviewed the Tuskegee study, the study was not discontinued until after the recommendation of the ad hoc panel.[50] The human rights abuses of the Tuskegee study demonstrated the need for both prior and ongoing review, in that the study was undertaken before prior review requirements were in place, and the prevailing review policies during the period of the study were so flawed that the study was allowed to continue.

As a result of their deliberations, the ad hoc panel found that neither DHEW nor any other agency in the government had adequate policies for oversight of human subjects research. The panel recommended that the Tuskegee study be stopped immediately and that remaining subjects be given necessary medical care resulting from their participation.[51] The panel also recommended that Congress establish "a permanent body with the authority to regulate at least all federally supported research involving human

subjects."[52] In summary, the panel concluded that despite the lessons of Nuremberg, the Jewish Chronic Disease Hospital case, and the Declaration of Helsinki, human subject research oversight and mechanisms to ensure informed consent were still inadequate and new approaches were needed to adequately protect the rights and welfare of human subjects.

Congressional Response to Abuses of Human Subjects: The National Research Act

Public attention to abuses such as those inflicted on the subjects of the Tuskegee study increased during the late 1960s and early 1970s. Following the initial revelations about the Tuskegee syphilis study, several bills were introduced in Congress to regulate the conduct of human experimentation. In February 1973 Senator Edward Kennedy held hearings on these bills;[53] the Tuskegee study; experimentation with prisoners, children, and poor women; and a variety of other issues related to biomedical research and the need for a national body to consider the ethics of research and advancing medical technology.[54] After the hearings, Senator Kennedy introduced an unsuccessful bill to create a National Human Experimentation Board, as recommended by the Tuskegee Syphilis Study Ad Hoc Panel. When it became clear, however, that the bill would not be successful, Senator Kennedy introduced the bill that would become the National Research Act, endorsing the regulations about to be promulgated by DHEW and establishing the National Commission for the Protection of Human Subjects of Biomedical and Behavioral Research, in return for DHEW's issuance of human subject research regulations.[55] The trade-off was clear: no national regulatory body in return for regulations applying to the research funded or performed by the government agency responsible for the greatest proportion of human subject research. This meant that the goal of oversight *of all* federally funded research would not be achieved and that whatever oversight did exist was left to the funding agencies rather than an independent body.

On May 30, 1974, DHEW published regulations for the use of human subjects in the *Fed-*

eral Register.[56] These regulations required that each grantee institution form a committee (what became known as an institutional review board, or IRB) to approve all research proposals before they were passed to DHEW for funding consideration. These committees were charged with reviewing the safety of the proposals brought to them as well as the adequacy of the informed consent obtained from each subject prior to participation in the research. Additionally, the regulations defined not only the procedure for obtaining informed consent but substantive criteria for it as well. Shortly after the announcement of the DHEW regulations, in July 1974, the National Research Act was passed, and with it came the establishment of the National Commission.[57]

The National Commission—charged with advising the secretary of DHEW (though the National Research Act did not require the secretary to follow the commission's recommendations)—existed over the next four years and published seventeen reports and appendix volumes. During its tenure, the commission did pioneering work as it addressed issues of autonomy, informed consent, and third-party permission, particularly in relation to research involving vulnerable subjects such as prisoners, children, and people with cognitive disabilities. It was also charged with examining the IRB system and procedures for informed consent, as background for proposing guidelines that would ensure that basic ethical principles were instituted in the research oversight system and in research involving vulnerable populations.

In the course of its deliberations, the commission identified three general moral principles—respect for persons, beneficence, and justice—as the appropriate framework for guiding the ethics of research involving human subjects. These three are known as the Belmont principles because they appeared in *The Belmont Report,* one of the commission's major publications.[58]

The National Commission was required to examine the "nature and definition" of informed consent as well as the "adequacy" of current practices. In its reports, the commission decisively argued that the basic justification for obligations to obtain informed consent is the moral prin-

ciple of respect for persons. This emphasis on respect for persons meant a great premium was put on autonomous decision making by the research subject, an emphasis that continues to the current day.

While it may not have been the intent of those who sponsored it, the National Research Act—because it was limited to DHEW-funded research—did not ensure that all federally sponsored research would be subject to requirements for informed consent and prior review. Nonetheless, by this time, as described below, published policies within the DOD, the AEC, the VA, and NASA did meet these requirements.

The passage of the National Research Act and the promulgation of DHEW's regulations were important milestones in the development of federal standards for the protection of human subjects of research. They represented the first national recognition of the need to protect human subjects. Moreover, they attempted to provide for that protection through the IRB requirement and establishment of the National Commission. The Advisory Committee's charter requires that it examine the standards for research between 1944 and 1974. These two landmark events in 1974 ushered in a new era in which the conduct and oversight of biomedical experimentation with humans remained a topic of national scrutiny and debate. Eventually, the approaches required by the 1974 DHEW regulations would be applied to nearly all federally sponsored human research, as described in chapter 14.

THE DEVELOPMENT OF REQUIREMENTS FOR HUMAN SUBJECT RESEARCH IN OTHER FEDERAL AGENCIES

The history and evolution of human subject research policy in the federal government is well documented for DHEW. However, many other agencies, most notably the military services, have important but less well-documented and less well-studied histories. Some of this history is described in chapter 1 of this report. Here we continue with a brief treatment of that history in the context of the evolution of human subject research policy.

Army Policy

In 1962 the Army, for the first time, issued as a formal regulation, Army Regulation (AR) 70–25, the 1953 policy embodied in the Wilson memorandum. The regulation made explicit, as the 1953 DOD and Army policies had only left implicit, basic issues about the scope of the DOD's rules. Unlike the Wilson memorandum, the new regulation applied to all types of research, not simply that related to atomic, biological, and chemical warfare. However, the regulation specifically excluded clinical research, that is, the research likely to be performed with patients at the Army's many hospitals. In 1963, an ad hoc committee of Army and civilian personnel concluded that the rule applied where research was done by contractors; however, tracer research (which arguably posed minimal risk) was excluded.[59] Despite the committee's recommendations, no immediate changes were made to the regulation. In 1963, however, the Army issued a regulation for radioisotope use that required local institutions to convene review committees and obtain approval from the secretary of the Army pursuant to AR 70–25 when radioisotopes were to be used with "volunteer" experimental subjects.[60]

The regulatory void apparently persisted until 1973, when another rule (AR 40–38, "Medical Services—Clinical Investigation Program") closed the gap. That rule clearly applied to "any person who may be at risk because of participation . . . [in] clinical investigation," including "patients" and "normal individuals."[61] It required that subjects of research be given an explanation of the proposal in understandable language and sign a "volunteer agreement."[62] Moreover, clinical research with patients, as well as healthy people, was to be reviewed by a "Human Use Committee."[63]

Navy Policy

As we saw in chapter 1, the Navy had required oral consent from research volunteers since at least 1951. Some evidence suggests that written consent was required in the mid-1960s; in a 1964 proposal to study the effects of hypoxia on service personnel it is indicated that a "signed

Consent to Voluntarily Participate in Research Experiment (NMRI Form 3)" would be used.[64] In 1967 a clear requirement for written consent appeared in the Navy's Medical Department manual.[65] It is unclear whether the policy drew a distinction between research on patients and research on healthy subjects. In 1969, in any event, the secretary of the Navy issued a comprehensive policy requiring written informed consent of research subjects, which appeared to cover both groups.[66]

Air Force Policy

In 1965 the Air Force promulgated AFR 169–8, "Medical Education and Research—Use of Volunteers in Aerospace Research," which required voluntary and written informed consent from all subjects in any "research, development, test, and evaluation" that may involve "distress, pain, damage to health, physical injury, or death."[67] As such, it seems inclusive of both healthy and patient-subjects.[68] Updating the language of the Nuremberg Code's first principle, the policy was based on the idea that the "voluntary informed consent of the human subject is absolutely essential."[69] Additionally, the regulation provided for the appointment of a committee to review all human research proposals at each originating facility.

NASA Policy

The National Aeronautics and Space Administration (NASA), created in 1958, inherited staff and research expertise from the DOD and other federal agencies. Before 1968, local centers at which research using radioisotopes was conducted—notably the Ames Research Center and the Manned Spacecraft Center (MSC)—were essentially autonomous. Each center established medical use subcommittees, as required by AEC rules.[70] Reorganization within NASA in 1968 combined the medical operations functions and the medical research functions at MSC into one medical research and operations directorate headed by Dr. Charles A. Berry.

By 1968, Ames had a policy requiring informed consent.[71] By definition, of course, the work of astronauts is frequently risky and experimental. The question of the proper boundary between experimental and occupational activities was one that could not be drawn easily. Consequently, the policy authorized the director of Ames to waive the consent requirement in several instances, including when obtaining consent would seriously hamper the research or when test pilots or astronauts were involved.[72]

Between 1968 and 1970, prior review for risk and subject consent was adopted at Ames in the form of the Human Research Experiments Review Board and indirectly at the MSC in accordance with the AEC requirements for a medical use committee.[73] In 1972 the prior review provisions and consent requirements of Ames and the MSC were reformulated in a NASA-wide policy.[74] This policy required voluntary and written informed consent from subjects prior to participation. The policy continued to provide waivers for "exceptional cases," as in the Ames policy, and did not apply to research conducted by NASA contractors or grantees.

The development of NASA's polices, like those at the PHS, NIH, and the DOD, appeared at a time when the public was becoming increasingly interested in biomedical research. In contrast with the 1940s and 1950s, bureaucratic developments during the 1960s and 1970s were mirrored by growing public debate about the adequacy of protections for human subjects.

SUPREME COURT DISSENTS INVOKE THE NUREMBERG CODE: CIA AND DOD HUMAN SUBJECTS RESEARCH SCANDALS

As we have seen, the development of federal legislation for government-sponsored research with human subjects arose in part because of institutional and governmental concern and public reaction to perceived abuses and failures by the government. Around the same time that the 1974 National Research Act was enacted, a scandal arose surrounding the discovery of secret Cold War chemical experiments conducted by the CIA and DOD. The review of these experiments led to the rediscovery of the previously secret 1953 Wilson memorandum and later to the first Supreme Court decision in which com-

ment was made, in dissent, on the application of the Nuremberg Code to the conduct of the U.S. government.

In December 1974, the *New York Times* reported that the CIA had conducted illegal domestic activities, including experiments on U.S. citizens, during the 1960s. That report prompted investigations by both Congress (in the form of the Church Committee) and a presidential commission (known as the Rockefeller Commission) into the domestic activities of the CIA, the FBI, and intelligence-related agencies of the military. In the summer of 1975, congressional hearings and the Rockefeller Commission report revealed to the public for the first time that the CIA and the DOD had conducted experiments on both cognizant and unwitting human subjects as part of an extensive program to influence and control human behavior through the use of psychoactive drugs (such as LSD and mescaline) and other chemical, biological, and psychological means. They also revealed that at least one subject had died after administration of LSD. Frank Olson, an Army scientist, was given LSD without his knowledge or consent in 1953 as part of a CIA experiment and apparently committed suicide a week later.[75] Subsequent reports would show that another person, Harold Blauer, a professional tennis player in New York City, died as a result of a secret Army experiment involving mescaline.[76]

The CIA program, known principally by the codename MKULTRA, began in 1950 and was motivated largely in response to alleged Soviet, Chinese, and North Korean uses of mind-control techniques on U.S. prisoners of war in Korea. Because most of the MKULTRA records were deliberately destroyed in 1973 by order of then-Director of Central Intelligence Richard Helms, it is impossible to have a complete understanding of the more than 150 individually funded research projects sponsored by MKULTRA and the related CIA programs.[77] Central Intelligence Agency documents suggest that radiation was part of the MKULTRA program and that the agency considered and explored uses of radiation for these purposes.[78] However, the documents that remain from MKULTRA, at least as currently brought to light, do not show that the CIA itself carried out any of these proposals on human subjects.

The congressional committee investigating the CIA research, chaired by Senator Frank Church, concluded that "[p]rior consent was obviously not obtained from any of the subjects."[79] The committee noted that the "experiments sponsored by these researchers . . . call into question the decision by the agencies not to fix guidelines for experiments."[80] (Documents show that the CIA participated in at least two of the DOD committees whose discussions, in 1952, led up to the issuance of the Wilson memorandum.) Following the recommendations of the Church Committee, President Gerald Ford in 1976 issued the first Executive Order on Intelligence Activities, which, among other things, prohibited "experimentation with drugs on human subjects, except with the informed consent, in writing and witnessed by a disinterested party, of each such human subject" and in accordance with the guidelines issued by the National Commission.[81] Subsequent orders by Presidents Carter and Reagan expanded the directive to apply to any human experimentation.[82]

Following on the heels of the revelations about CIA experiments were similar stories about the Army. In response, in 1975 the secretary of the Army instructed the Army inspector general to conduct an investigation.[83] Among the findings of the inspector general was the existence of the then-still-classified 1953 Secretary of Defense Wilson memorandum. In response to the inspector general's investigation, the Wilson memorandum was declassified in August 1975. The inspector general also found that the requirements of the 1953 memorandum had, at least in regard to Army drug testing, been essentially followed as written. The Army used only "volunteers" for its drug-testing program, with one or two exceptions.[84] However, the inspector general concluded that the "volunteers were not fully informed, as required, prior to their participation; and the methods of procuring their services, in many cases, appeared not to have been in accord with the intent of Department of the Army policies governing use of volunteers in research."[85] The inspector general also noted that "the evidence clearly reflected that every possible medical consideration was ob-

served by the professional investigators at the Medical Research Laboratories."[86] This conclusion, if accurate, is in striking contrast to what took place at the CIA.

The revelations about the CIA and the Army prompted a number of subjects or their survivors to file lawsuits against the federal government for conducting illegal experiments. Although the government aggressively, and sometimes successfully, sought to avoid legal liability, several plaintiffs did receive compensation through court order, out-of-court settlement, or acts of Congress. Previously, the CIA and the Army had actively, and successfully, sought to withhold incriminating information, even as they secretly provided compensation to the families.[87] One subject of Army drug experimentation, James Stanley, an Army sergeant, brought an important, albeit unsuccessful, suit. The government argued that Stanley was barred from suing it under a legal doctrine—known as the Feres doctrine, after a 1950 Supreme Court case, *Feres v. United States*—that prohibits members of the Armed Forces from suing the government for any harms that were inflicted "incident to service."[88]

In 1987, the Supreme Court affirmed this defense in a 5–4 decision that dismissed Stanley's case.[89] The majority argued that "a test for liability that depends on the extent to which particular suits would call into question military discipline and decision making would itself require judicial inquiry into, and hence intrusion upon, military matters."[90] In dissent, Justice William Brennan argued that the need to preserve military discipline should not protect the government from liability and punishment for serious violations of constitutional rights:

The medical trials at Nuremberg in 1947 deeply impressed upon the world that experimentation with unknowing human subjects is morally and legally unacceptable. The United States Military Tribunal established the Nuremberg Code as a standard against which to judge German scientists who experimented with human subjects. . . . [I]n defiance of this principle, military intelligence officials . . . began surreptitiously testing chemical and biological materials, including LSD.[91]

Justice Sandra Day O'Connor, writing a separate dissent, stated:

No judicially crafted rule should insulate from liability the involuntary and unknowing human experimentation alleged to have occurred in this case. Indeed, as Justice Brennan observes, the United States played an instrumental role in the criminal prosecution of Nazi officials who experimented with human subjects during the Second World War, and the standards that the Nuremberg Military Tribunals developed to judge the behavior of the defendants stated that the 'voluntary consent of the human subject is absolutely essential . . . to satisfy moral, ethical, and legal concepts.' If this principle is violated, the very least that society can do is to see that the victims are compensated, as best they can be, by the perpetrators.[92]

This is the only Supreme Court case to address the application of the Nuremberg Code to experimentation sponsored by the U.S. government.[93] And while the suit was unsuccessful, dissenting opinions put the Army—and by association the entire government—on notice that use of individuals without their consent is unacceptable. The limited application of the Nuremberg Code in U.S. courts does not detract from the power of the principles it espouses, especially in light of stories of failure to follow these principles that appeared in the media and professional literature during the 1960s and 1970s and the policies eventually adopted in the mid-1970s.

CONCLUSION

The 1960s and early 1970s witnessed an extraordinary growth in government, institutional, and public awareness of issues in the use of human subjects, fueled by scandals and an increasing emphasis on individual expression. The branches of the military had articulated policies during this period, in spite of numerous problems in implementation. By 1974 the DHEW had established a set of regulations and a system of local review, and Congress had established a commission to issue recommendations for further change to the DHEW. Together, these advances created a model and laid the groundwork for human subjects protections for all federal agencies.

Many conditions coalesced into the framework for the regulation of the use of human subjects in federally funded research that is the basis for today's system. Described further in chapter 14, this framework is undergirded by the three Belmont principles that were identified by the National Commission as governing the ethics of research with human subjects: *respect for persons, beneficence,* and *justice.* The federal regulations and the conceptual framework built on the Belmont principles became so widely adopted and cited that it might be argued that their establishment marked the end of serious shortcomings in federal research ethics policies. Whether this position is well supported is evaluated in light of the Advisory Committee's contemporary studies in part III.

By 1974, DHEW had extensive policies to protect human subjects within its purview. Policies were more variable among other government agencies. By 1975, the branches of the military set about developing their own more comprehensive policies for human subject research, and the CIA was required by executive order to comply with consent requirements in human subject research in light of scandalous practices in the past. In order to evaluate the adequacy of the efforts taken to protect people before these policies were established, we must take into account both the government's policies and rules and the norms and practices of medicine reviewed in chapters 1 through 3. The Advisory Committee's framework for the consideration of these factors is presented in the next chapter.

NOTES

1. For a discussion of the development of the Common Rule, see chapter 14.

2. We relied particularly on Ruth R. Faden and Tom L. Beauchamp, *A History and Theory of Informed Consent* (New York: Oxford University Press, 1986). Other excellent sources include Jay Katz, *Experimentation with Human Beings* (New York: Russell Sage Foundation, 1972), and Robert Levine, *Ethics and Regulation of Clinical Research* (Baltimore: Urban and Schwarzenberg, 1981).

3. U.S. Congress, The Select Committee to Study Governmental Operations with Respect to Intelligence Activities, *Foreign and Military Intelligence* [Church Committee report], report no. 94–755, 94th Cong., 2d Sess. (Washington, D.C.: GPO, 1976). Also, U.S. Army Inspector General, *Use of Volunteers*

in Chemical Agent Research [Army IG report] (Washington, D.C.: 1975).

4. In dissenting opinions, four justices of the U.S. Supreme Court (Brennan, Marshall, Stevens, and O'Connor) cited the Nuremberg Code. *United States et al. v. Stanley*, 483 U.S. 669, 687, 710 (1987).

5. Thalidomide was only available in clinical trials in the United States at that time, but was approved for use in a number of other countries.

6. Louis Lasagna, interview by Susan White-Junod and Jon Harkness (ACHRE), transcript of audio recording, 13 December 1994 (ACHRE Research Project Series, Interview Program Files, Ethics Oral History Project), 37–38. See also, Louis Lasagna, "1938–1968: The FDA, the Drug Industry, the Medical Profession, and the Public," in *Safeguarding the Public: Historical Aspects of Medicinal Drug Control*, ed. John B. Blake (Baltimore: The Johns Hopkins Press, 1970), 173.

7. *Food, Drug, and Cosmetic Act amendments*, 21 U.S.C. § 355 (1962).

8. *Congressional Record*, 87th Cong, 2d Sess., 22042, as cited in an attached memorandum, C. Joseph Stetler, Pharmaceutical Manufacturers Association, to James L. Goddard, M.D., Commissioner of Food and Drugs, DHEW, 11 October 1966 ("Regarding Statement Appearing in August 30, 1966 Federal Register Concerning Clinical Investigation of Drugs") (ACHRE No. HHS-090794–A).

9. Keith Reemtsma et al., "Reversal of Early Graft Rejection after Renal Heterotransplantation in Man," *Journal of the American Medical Association* 187 (1964): 691–696.

10. This research, conducted by Dr. Chester Southam of Sloan-Kettering Institute and Dr. Emmanuel Mandel of the Jewish Chronic Disease Hospital in 1963 and funded by the U.S. Public Health Service and the American Cancer Society, raised concern within PHS and brought about an investigation by the hospital. Drs. Mandel and Southam were subject to a disciplinary hearing before the Board of Regents of the University of the State of New York. The hospital's internal review and a suit against the hospital prompted concern and debate at the NIH. Edward J. Rourke, Assistant General Counsel, NIH, to Dr. Luther L. Terry, Surgeon General, 16 September 1965 ("Research Grants—Clinical— PHS responsibility—*Fink v. Jewish Chronic Disease Hospital* [New York Supreme Court, Kings County]") (ACHRE No. HHS-090794–A).

For a more thorough discussion of this case, see Katz, *Experimentation with Human Beings*, 9–65.

11. In 1967 Dr. Southam was elected vice president of the American Association for Cancer Research and became president the following year. Katz, *Experimentation with Human Beings*, 63 and 65.

12. For a fuller discussion of the Law-Medicine Research Institute, see chapter 2.

13. The development of the Declaration of Helsinki is discussed briefly in chapter 2.

14. Robert B. Livingston, Associate Chief for Program Development, Memorandum to the Director, NIH, 4 November 1964 ("Progress Report on Survey of Moral and Ethical Aspects of Clinical Investigation" [the Livingston report]) (ACHRE No. HHS-090795–A), 3.

15. Ibid., 7.

16. Ibid.

17. Mark S. Frankel, "Public Policymaking for Biomedical Research: The Case of Human Experimentation" (Ph.D. diss., George Washington University, 9 May 1976), 50–51.

18. The NAHC discussed the "general question of the ethical, moral, and legal aspects of clinical investigation" at its meetings of September and December 1965. Terry's interest in this was motivated in part by the concern of Senator Jacob K. Javits that the informed consent provisions of the 1962 Drug Amendments were not applicable to nondrug-related research. See (a) draft letter to Senator Javits from the Surgeon General, 15 October 1965; (b) Senator Javits to Luther L. Terry, Surgeon General, 15 June 1965; and (c) Edward J. Rourke, Assistant General Counsel, to William H. Stewart, Surgeon General, 26 October 1965. All in ACHRE No. HHS-090794–A.

19. Transcript of the NAHC meeting, Washington, D.C., 28 September 1965. See Faden and Beauchamp, *A History and Theory of Informed Consent*, 208.

20. Ibid.

21. Dr. S. John Reisman, the Executive Secretary, NAHC, to Dr. James A. Shannon, 6 December 1965 ("Resolution of Council") (ACHRE No. HHS-090794–A).

22. Surgeon General, Public Health Service to the Heads of the Institutions Conducting Research with Public Health Service Grants, 8 February 1966 ("Clinical research and investigation involving human beings") (ACHRE No. HHS-090794–A). This policy was distributed through Bureau of Medical Services Circular no. 38, 23 June 1966 ("Clinical Investigations Using Human Beings As Subjects") (ACHRE No. HHS-090794–A).

23. In December 1966 the policy was expanded to include behavioral as well as medical research. William H. Stewart, Surgeon General, Public Health Service, to Heads of Institutions Receiving Public Health Service Grants, 12 December 1966 ("Clarification of procedure on clinical research and investigation involving human subjects") (ACHRE No. HHS-072894–B), 2.

In 1967 the Public Health Service required that intramural research, including that conducted at NIH, abide by similar requirements. William H. Stewart, Surgeon General of the Public Health Service, to List, 30 October 1967 ("PHS policy for intramural programs and for contracts when investigations involving human subjects are included") (ACHRE No. HHS-072894–B), 2.

24. Frankel, "Public Policymaking for Biomedi-

cal Research: The Case of Human Experimentation," 161.

25. Ibid., 161–162.

26. U.S. Department of Health, Education, and Welfare, *The Institutional Guide to DHEW Policy on Protection of Human Subjects* (Washington, D.C.: GPO, 1971) (ACHRE No. HHS-090794–A).

27. Ibid., 1–2.

28. Beecher's criticism involved many aspects of the research, including the risk assessment, usefulness of the research, and the question of informed consent. On this last point, Beecher argued that while consent was important, he disputed the belief that it was easily obtainable. In his talk at Brook Lodge, Beecher questioned the "naive assumption implicit in the Nuremberg Code," that consent was readily obtainable. Beecher indicated the difficulty of obtaining truly informed consent may have led many researchers to treat the provision cavalierly and often to ignore it. Henry K. Beecher, "Ethics and the Explosion of Human Experimentation," unpublished manuscript of paper presented 22 March 1965, "a," Beecher Papers, Countway Library (ACHRE No. IND-072595–A).

29. Ibid.,"a" and "b."

30. Ibid., 2a.

31. Ibid., 2.

32. H. K. Beecher, "Ethics and Clinical Research," *New England Journal of Medicine* 274 (1966): 1354–1360.

33. W. Goodman, "Doctors Must Experiment on Humans—But What are Patients Rights?" *New York Times Magazine*, 2 July 1965, 12–13, 29–33, as cited in Faden and Beauchamp, *A History and Theory of Informed Consent*, 188.

34. J. Lear, "Do We Need New Rules for Experimentation on People?" *Saturday Review*, 5 February 1966, 61–70.

35. Henry K. Beecher, "Consent in Clinical Experimentation: Myth and Reality," *Journal of the American Medical Association* 195 (1966): 34–35.

36. J. Lear, "Experiments on People—The Growing Debate," *Saturday Review*, 2 July 1966, 41–43.

37. Both the *New York Times* and the *Wall Street Journal* ran stories on 24 March 1971. See *Medical World News*, 15 October 1971, "Was Dr. Krugman Justified in Giving Children Hepatitis?"

38. Beecher, *Research and the Individual: Human Studies* (Boston: Little, Brown, and Company, 1970), 122–127.

39. Paul Ramsey, *The Patient as Person: Explorations in Medical Ethics* (New Haven: Yale University Press, 1970), 51–55.

40. In a letter to the *Lancet*, Dr. Stephen Goldby called the work "unjustifiable" and asked, "Is it right to perform an experiment on a normal or mentally retarded child when no benefit can result to the individual?" (S. Goldby, "Letters to the Editor," *Lancet* 7702 [1971]: 749). The *Lancet* editors agreed with Goldby. On this side of the Atlantic, however, the edi-

tors of *NEJM* and *JAMA*, among others, defended Krugman's work.

41. Armed Forces Epidemiological Board, minutes of 24 May 1957 (ACHRE No. NARA-032495–B).

42. S. Krugman, "Ethical Practices in Human Experimentation," text of lecture presented at the Fifth Annual Midwest Student Medical Research Forum, 1 March 1974 (ACHRE No. IND-072895–A).

43. Ibid., 3–4.

44. Louis Goldman, "The Willowbrook Debate," *World Medicine* (September 1971 and November 1971): 23, 25.

45. James H. Jones, *Bad Blood* (New York: Free Press, 1993 edition), 114.

46. Jones, *Bad Blood* (1981), 69–71; Levine, *Ethics and Regulation of Clinical Research*, 70.

47. Charles J. McDonald, "The Contribution of the Tuskegee Study to Medical Knowledge," *Journal of the National Medical Association* (January 1974): 1–11, as cited in Faden and Beauchamp, *A History and Theory of Informed Consent*, 194–195.

48. Jean Heller, "Syphilis Victims in U.S. Study Went Untreated for 40 Years," *New York Times* (26 July 1972) 1, 8, as cited in Faden and Beauchamp, *A History and Theory of Informed Consent*, 195.

49. U.S. Department of Health, Education, and Welfare, *Final Report of the Tuskegee Syphilis Study Ad Hoc Panel* (Washington, D.C.: GPO, 1973), Jay Katz Concurring Opinion, 14.

50. Ibid.

51. Ibid., 21–32.

52. Ibid., 23.

53. Senator Jacob Javits introduced legislation that would have made the DHEW policy a regulation backed by federal law. S. 878 and S. 974, 93d Cong., 1st Sess. (1973).

Senator Hubert Humphrey introduced a bill to create a National Human Experimentation Standards Board—a separate federal agency with authority over research similar to the Securities and Exchange Commission's authority over securities transactions. S. 934, 93d Cong., 1st Sess. (1973).

Also, Senator Walter Mondale introduced a resolution to provide for a "study and evaluation of the ethical, social, and legal" aspects of biomedical research. S.J. Res. 71, 93d Cong., 1st Sess. (1973).

54. It is worth noting here that Senator Kennedy had convened similar hearings two years previously, in 1971, to consider the establishment of a national commission to examine "ethical, social, and legal implications of advances in biomedical research." Among the topics mentioned in this hearing was the total-body irradiation research sponsored by the Department of Defense at the University of Cincinnati, which we discuss in chapter 8.

55. Jay Katz, "Human Experimentation: A Personal Odyssey," *IRB* 9, no. 1 (January/February 1987): 1–6.

56. *Protection of Human Subjects*, 39 Fed. Reg.

105, 18914–18920 (1974) (to be codified at 45 C.F.R. § 46).

57. *National Research Act of 1974.* P.L. 348, 93d Cong., 2d Sess. (12 July 1974).

58. U.S. Department of Health, Education, and Welfare, Office for Protection from Research Risks, 18 April 1979, OPPR Reports [*The Belmont Report*] (ACHRE No. HHS-011795-A-2), 4–20.

59. Interestingly, this committee included Henry Beecher, who, as was discussed in part I, chapter 3, had objected to the imposition of these requirements to contract research in 1961. Beecher's presence on the committee testifies to the common relationship between military and private research during this time. Like many of the AFEB members and commissioners, many of the members of the ad hoc panel were nonmilitary consultants to the DOD.

60. Department of the Army, Army Regulation 40–37, 12 August 1963 ("Radioisotope License Program [Human Use]").

61. Department of the Army, AR 40–38, 23 February 1973 ("Medical Services—Clinical Investigation Program").

62. Ibid.

63. Ibid.

64. Commanding Officer, Naval Medical Research Institute, National Naval Medical Center, to Secretary of the Navy, 30 November 1964 ("Authorization to use human volunteers as subjects for study of effects of hypoxia on the visual field; request for") (ACHRE No. DOD-091494-A), 2.

65. Department of the Navy, "Manual of the Medical Department," 20–8, Change 36, 7 March 1967 ("Use of Volunteers in Medical or Other Hazardous Experiments") (ACHRE No. DOD-091494-A).

66. Department of the Navy, SecNav Instruction 3900.39, 28 April 1969 ("Use of volunteers as subjects of research, development, tests, and evaluation").

67. Department of the Air Force, AFR 169–8, 8 October 1965 ("Medical Education and Research—Use of Volunteers in Aerospace Research").

68. Ibid.

69. Ibid.

70. National Aeronautics and Space Administration, Manned Spacecraft Center, MSCI 1860.2, 12 May 1966 ("Establishment of MSC Radiological Control Manual and Radiological Control Committee") (ACHRE No. NASA-022895-A), 3.

National Aeronautics and Space Administration, "Ames Management Manual 7170–1," 15 January 1968 ("Human Research Planning and Approval") (ACHRE No. NASA-120894-A), 3.

71. Ames required the voluntary, written informed consent of the subject and stipulated that consent be informed by an

explanation to the subject in language understandable to him . . . [including] the nature, duration, and purpose of the human research; the manner in which it will be conducted; and all foreseeable risks, inconveniences and discomforts.

"Ames Management Manual 7170–1," 15 January 1968, 3.

72. The Ames director was authorized to waive the consent requirements (a) when the requirements would "not be necessary to protect the subject"; (b) when the research uses "classes of trained persons who knowingly follow a specialized calling or occupation which is generally recognized as hazardous," including "test pilots and astronauts"; and (c) when the research "would be seriously hampered" by compliance. "Ames Management Manual 7170–1," 15 January 1968, 3.

73. For example, one review from this group recommended changes in a consent form to include

[T]he part of the procedure you are consenting to which principally benefits the research program and is not part of your treatment is known as arterial puncture. . . . These risks will be explained to you in detail if you so desire. The entire procedure, including the diagnostic radioscan, takes about an hour.

Although this proposed consent form does not delineate the medical risks posed by the procedure, its statement that the patient's participation is incidental to treatment may provide a greater opportunity for the patient to make an informed decision about participation. George A. Rathert, Jr., Chairman, Human Research Experiments Review Board, ARC, to Director, 20 January 1969 ("Proposed Investigation entitled 'Measurement of Cerebral Blood Flow in Man by an Isotopic Technique Employing External Counting,' by Dr. Leo Sapierstein, Stanford University") (ACHRE No. NASA-022895-A), 4.

At MSC, the instruction establishing the Medical Uses Subcommittee was rescinded in 1968. In 1969, formal combination of the medical operations and medical research functions at MSC led to the reestablishment of the instruction as the Medical Isotopes Subcommittee at MSC. No evidence suggests what factors, other than risk, were considered in this form of prior review is available currently. National Aeronautics and Space Administration, Manned Spacecraft Center, MSCI 1860.2, 12 May 1966 ("Establishment of MSC Radiological Control Manual and Radiological Control Committee"); and National Aeronautics and Space Administration, NMI 1156.19, 28 August 1969 ("Medical Isotopes Subcommittee of the MSC Radiation Safety Committee") (ACHRE No. NASA-022895-A).

74. National Aeronautics and Space Administration, NMI 71008.9, 2 February 1972 ("Human Research Policy and Procedures") (ACHRE No. NASA-022895-A). See also, National Aeronautics and Space Administration, NMI 7100.9 ("Power and Authority—To Authorize Human Research and to Grant Certain Related Exceptions and Waivers") (ACHRE No. NASA-022895-A).

75. Commission on CIA Activities within the United States, *Report to the President,* (Washington, D.C.: GPO, 1975).

76. U.S. Congress, The Select Committee to Study Governmental Operations with Respect to Intelligence Activities, *Foreign and Military Intelligence* [Church Committee report], report no. 94–755, 94th Cong., 2d Sess. (Washington, D.C.: GPO, 1976), 394.

77. For general information on the CIA program, see the Church Committee report, 385–422, and J. Marks, *The Search for the "Manchurian Candidate": The CIA and Mind Control* (New York: Times Books, 1978).

78. Church Committee report, book 1, 389.

79. Church Committee report, book 1, 400, 402. In 1963 the CIA inspector general (IG) recommended that unwitting testing be terminated, but Deputy Director for Plans Richard Helms (who later became director of Central Intelligence) continued to advocate covert testing on the ground that "positive operational capability to use drugs is diminishing, owing to a lack of realistic testing. With increasing knowledge of the state of the art, we are less capable of staying up with the Soviet advances in this field." The Church Committee noted that "Helms attributed the cessation of the unwitting testing to the high risk of embarrassment to the Agency as well as the 'moral problem.' He noted that no better covert situation had been devised than that which had been used, and that 'we have no answer to the moral issue.'"

80. Ibid., 402.

81. Executive Order 11905 (19 February 1976).

82. Executive Order 12036, section 2–301 (26 January 1978) and Executive Order 12333, section 2.10 (4 December 1981).

83. U.S. Army Inspector General, *Use of Volunteers in Chemical Agent Research* [Army IG report] (Washington, D.C.: GPO, 1975), 2.

84. One noted exception involved using LSD as an interrogation devise on ten foreign intelligence agents, and one U.S. citizen suspected of stealing classified documents. Army IG report, 143.

85. Army IG report, 87.

86. Ibid.

87. The CIA paid death benefits to the Olson family after Frank Olson's death, and the Army secretly paid half of an $18,000 settlement that the Blauer family negotiated with the state of New York in 1955. The state ran the psychiatric institute that administered the drugs, but which never disclosed the Army's involvement. Both agencies feared that the resulting embarrassment and adverse publicity might undermine their ability to continue their secret research programs. *Barrett v. United States,* 660 F. Supp. 1291 (E. D. N.Y., 1987).

88. *Feres v. United States,* 340 U.S. 146 (1950).

89. *United States v. Stanley,* 483 U.S. 669 (1987).

90. 483 U.S. 669, 682.

91. 483 U.S. 669, 687–88.

92. 483 U.S. 669, 709–10.

93. George Annas, a scholar of human experimentation and biomedical ethics, has traced the history of the Nuremberg Code in the U.S. courts. The first express reference in a majority opinion, Annas found, was in a 1973 decision in the Circuit Court in Wayne County, Michigan. The decisions in which the Code has since been cited, Annas concluded, reflect the proposition that the Nuremberg Code is a "document fundamentally about nontherapeutic experimentation." Thus, the "types of experiments that U.S. judges have found the Nuremberg Code useful for setting standards have involved nontherapeutic experiments often conducted without consent. . . . Many of these experiments were justified by national security considerations and the cold war." George J. Annas, "The Nuremberg Code in U.S. Courts: Ethics versus Expediency," in George J. Annas and Michael A. Grodin, eds., *The Nazi Doctors and the Nuremberg Code: Human Rights in Human Experimentation* (New York: Oxford University Press, 1992), 218.

4

Ethics Standards
in Retrospect

ACCORDING to the mission set out in our charter, the Advisory Committee is in essence a national ethics commission. In this capacity we were obliged to develop an ethical framework for judging the human radiation experiments. This proved to be one of our most difficult tasks, for we were not only dealing with complex events that occurred decades ago, but also with some of the most controversial issues in moral philosophy. This chapter sets out the standards that we believe are appropriate for evaluating human radiation experiments and offers reasons for relying on them. It then applies these standards to the results of the historical research we have conducted and draws ethical conclusions.*

Fulfilling our charge to "determine the ethical and scientific standards and criteria" to evaluate human radiation experiments that took place between 1944 and 1974 requires consideration of a complex question: Is it correct to evaluate the events, policies, and practices of the past, and the agents responsible for them, against ethical standards and values that we accept as valid today but that may not have been widely accepted then? Or must we limit our ethical evaluation of the past to those standards and values that were widely accepted at the time? This is the problem of *retrospective moral judgment.*

Quite apart from the issue of the validity of projecting current standards onto the past, there is another question that this chapter must address: In a pluralistic society such as ours, is there *at present* a sufficiently broad consensus on ethical standards to make possible a public evaluation that is not simply the arbitrary imposition of one particular moral point of view among several or even many? This is the problem of *value pluralism.* The ethical framework the Advisory Committee employs takes both these issues into account.

This chapter is divided into two parts. In the first part we present and defend the ethical framework adopted by the Committee for the evaluation of human radiation experiments conducted from 1944 to 1974 and the agents responsible for them. We begin by identifying the types of moral judgments with which the Committee is concerned and the different kinds of ethical standards against which these judgments

---

*Some of the features of the moral framework presented in this chapter pertain to biomedical experiments only and not to intentional releases. A moral analysis of intentional releases involves somewhat different elements than a moral analysis of biomedical experiments, because they engage different ethical issues. For example, a requirement of individual informed consent is not applicable to the intentional releases, and the concepts of risk and benefit and national security have different implications for them. Ethical and policy issues specific to intentional releases are discussed in chapter 11.

can be made. We next address two challenges to the position that the Advisory Committee can use these, or any other, standards to make valid ethical judgments. These challenges are (1) that the diversity of views about ethics in American society invalidates any effort by a public body such as the Advisory Committee to make moral judgments and (2) that the diversity of views about ethics across time similarly invalidates our making defensible moral judgments about the past. Although the Committee does not accept these challenges as definitive, we discuss these as well as other factors that influence or limit ethical evaluation. We include here a discussion of an issue of particular relevance to our charge: what role, if any, considerations of national security should play in the Committee's ethical framework. We also consider factors that can mitigate the blame we would otherwise place on agents (whether individuals or collective entities) for having conducted morally wrong actions.

In the second part of the chapter, we explore how the Committee's ethical framework can be used to evaluate both experiments conducted in the past and the people and institutions that sponsored and conducted them. Drawing on the history presented in chapters 1 through 3, we illustrate how, when applied, the framework is specified by context and detail. This specification of the framework continues in part II of the report, when the framework is used to evaluate specific cases.

AN ETHICAL FRAMEWORK

Two Types of Moral Judgment

For purposes of the Committee's charge, there are two main types of moral judgment: judgments about the moral quality of actions, policies, practices, institutions, and organizations; and judgments about the praiseworthiness or blameworthiness of individual agents and in some cases entities such as professions and governments (insofar as these can be viewed as collective agents with powers and responsibilities). The first type contains several kinds of judgments. Actions may be judged to be obligatory, wrong, or permissible. Institutions, policies, and practices can be characterized as just or unjust,

equitable or inequitable, humane or inhumane. Organizations can be said to be responsible or negligent, fair-dealing or exploitative.

The second type of judgment about the praiseworthiness or blameworthiness of agents also contains a diversity of determinations. Agents, whether individual or collective, can be judged to be culpable or praiseworthy for this or that action or policy, to be generous or mean-spirited, responsible or negligent, to respect the moral equality of people or to discriminate against certain individuals or groups, and so on.

Three Kinds of Ethical Standards

A recognized way to make moral judgments is to evaluate the facts of a case in the context of ethical standards. The Committee identified three kinds of ethical standards as relevant to the evaluation of the human radiation experiments:[1]

1. Basic ethical principles that are widely accepted and generally regarded as so fundamental as to be applicable to the past as well as the present;
2. The policies of government departments and agencies at the time; and
3. Rules of professional ethics that were widely accepted at the time.

Basic Ethical Principles

Basic ethical principles are general standards or rules that all morally serious individuals accept. The Advisory Committee has identified six basic ethical principles as particularly relevant to our work: "One ought not to treat people as mere means to the ends of others"; "One ought not to deceive others"; "One ought not to inflict harm or risk of harm"; "One ought to promote welfare and prevent harm"; "One ought to treat people fairly and with equal respect"; and "One ought to respect the self-determination of others." These principles state moral requirements; they are principles of *obligation* telling us what we ought to do.[2]

Every principle on this list has exceptions, because all moral principles can justifiably be overridden by other basic principles in circumstances when they conflict. To give priority to one principle over another is not a moral mistake; it is a reality of moral judgment. The justifiability of such judgments depends on many factors *in the circumstance;* it is not possible to assign priorities to these principles in the abstract.

Far more social consensus exists about the acceptability of these basic principles than exists about any philosophical, religious, or political theory of ethics. This is not surprising, given the central social importance of morality and the fact that its precepts are embraced in some form by virtually all major ethical theories and traditions. These principles are at the deepest level of any person's commitment to a moral way of life.

It is important to emphasize that the validity of these basic principles is not typically thought of as limited by time: we commonly judge agents in the past by these standards. For example, the passing of fifty years in no way changes the fact that Hitler's extermination of millions of people was wrong, nor does it erase or even diminish his culpability. Nor would the passing of a hundred years or a thousand do so.

This is not to deny that it might be inappropriate to apply to the distant past some ethical principles to which we now subscribe. It is only to note that there are some principles *so basic* that we ordinarily assume, with good reason, that they are applicable to the past as well as the present (and will be applicable in the future as well). We regard these principles as basic because any minimally acceptable ethical standpoint must include them.

Policies of Government Departments and Agencies

The policies of departments and agencies of the government can be understood as statements of commitment on the part of those governmental organizations, and hence of individuals in them, to conduct their affairs according to the rules and procedures that constitute those policies. In this sense, policies create ethical obligations. When a department or agency adopts a particular policy, it in effect promises to make reasonable efforts to abide by it.[3]

At least where participation in the organization is voluntary, and where the organization's defining purpose is morally legitimate (it is not, for example, a criminal organization), to assume a role in the organization is to assume the obligations that attach to that role. Depending upon their roles in the organization, particular individuals may have a greater or lesser responsibility for helping to ensure that the policy commitments of the organization are honored. For example, high-level managers who formulate organizational policies have an obligation to take reasonable steps to ensure that these policies are effectively implemented. If they fail to discharge these obligations, they have done wrong and are blameworthy, unless some extenuating circumstance absolves them of responsibility. One sort of extenuating circumstance is that the policy in question is unethical. In that case, we would hold an individual blameless for not attempting to implement it (at least if the individual did so because of a recognition that the policy was unethical). Moreover, we might praise the individual for attempting an institutional reform at some professional or personal risk.

Different types of organizations have different defining purposes, and these differences determine the character of the department's or agency's role-derived obligations. All government organizations have special responsibilities to act impartially and to fairly protect all citizens, including the most vulnerable ones. These special obligations constitute a standard for evaluating the conduct of government officials.

Rules of Professional Ethics

Professions traditionally assume responsibilities for self-regulation, including the promulgation of certain standards to which all members are supposed to adhere. These standards are of two kinds: technical standards that establish the minimum conditions for competent practice, and ethical principles that are intended to govern the conduct of members in their practice. In exchange for exercising this responsibility, society implicitly grants professions a degree of autonomy. The privilege of this autonomy in turn creates certain special obligations for the profession's members.

These obligations function as constraints on professionals to reduce the risk that they will use their special power and knowledge to the detriment of those whom they are supposed to serve. Thus, physicians, whose special knowledge gives them opportunities for exploiting patients or breaching confidentiality, are obligated to act in the patient's best interest in general and to follow various prescriptions for minimizing conflicts of interest.

Unlike basic ethical principles that speak to the whole of moral life, rules of professional ethics are particularized to the practices, social functions, and relationships that characterize a profession. Rules of professional ethics are often justified by appeal to basic ethical principles. For example, as we discuss later in this chapter, the obligation to obtain informed consent, which is a rule of research and medical ethics, is grounded in principles of respect for self-determination, the promotion of others' welfare, and the non-infliction of harm.

In one respect, rules of professional ethics are like the policies of institutions and organizations: they express commitments to which their members may be rightly held by others. That is, rules of professional ethics express the obligations that collective entities impose on their members and constitute a commitment to the public that the members will abide by them. Absent some special justification, failure to honor the commitment to fulfill these obligations constitutes a wrong. To the extent that the profession as a collective entity has obligations of self-regulation, failure to fulfill these obligations can lead to judgments of collective blame.

Ethical Pluralism and the Convergence of Moral Positions

Although we have argued that there is broad agreement about and acceptance of basic ethical principles in the United States, such as principles that enjoin us to promote the welfare of others and to respect self-determination, people nevertheless disagree about the relative priority or importance of these principles in the moral life. For example, although any minimally acceptable ethical standpoint must include both these principles, some approaches to morality emphasize the importance of respecting self-de-

termination while others place a higher priority on duties to promote welfare. These differences in approaches to morality pose a problem for public moral discourse. How can a public body, such as the Advisory Committee, purport to speak on behalf of society as a whole and at the same time respect this diversity of views about ethics? The key to understanding how this is possible is to appreciate that different ethical approaches can and often do converge on the same ethical conclusions. People can agree about what ought to be done without necessarily appealing to the same moral arguments to defend their common position.

This phenomenon of convergence has been observed in the work of other public bodies whose charge was to make ethical evaluations on research involving human subjects, including the National Commission for the Protection of Human Subjects of Biomedical and Behavioral Research and the President's Commission for the Study of Ethical Problems in Medicine and Biomedical and Behavioral Research.[4] For example, both those who take the viewpoint that emphasizes obligations to promote welfare and to refrain from inflicting harm and those who accord priority to self-determination can agree that law and medical and research practice should recognize a right to informed consent for competent individuals. The argument for a requirement of informed consent based on promoting welfare and refraining from inflicting harm assumes that individuals are generally most interested in and knowledgeable about their own well-being. Individuals are thus in the best position to discern what will promote their welfare when deciding about participation in research or medical care. Allowing physicians or others to decide for them runs too great a risk of harm or loss of benefits. By contrast, an approach based on self-determination assumes that, at least for competent individuals, being able to make important decisions concerning one's own life and health is intrinsically valuable, independent of its contribution to promoting one's well-being. The most compelling case for recognizing a right of informed consent for competent subjects and patients draws upon both lines of justification, emphasizing that this requirement is necessary from the perspective of self-determination considered as valuable in itself *and* from the standpoint of promoting welfare and refraining from doing harm.

Therefore, although people may have different approaches to the moral life, which reflect different priorities among basic moral principles, these differences need not result in a lack of consensus on social policy or even on particular moral rules such as the rule that competent individuals ought to be allowed to accept or refuse participation in experiments. On the contrary, the fact that the same moral rules or social policies can be grounded in different basic moral principles and points of view greatly strengthens the case for their public endorsement by official bodies charged to speak for society as a whole.

The three kinds of ethical standards upon which the Committee relies for our ethical evaluations—the basic moral principles, government policies, and rules of professional ethics—also enjoy a broad consensus. They are not idiosyncratic to a particular ethical value system. Thus it would be a mistake to think that in order to fulfill our charge of ethical evaluation, the Advisory Committee must assume that there is only one uniquely correct ethical standpoint. A broad range of views can acknowledge that the medical profession should be held accountable for moral rules it publicly professes and that individual physicians can be held responsible for abiding by these rules of professional ethics. Likewise, regardless of whether one believes that the ultimate justification for government policies is the goal of promoting welfare and minimizing harms or respect for self-determination, one can agree that policies represent commitments to action and hence generate obligations. Moreover, any plausible ethical viewpoint will recognize that when individuals assume roles in organizations they thereby undertake role-derived obligations.

We have already argued that the basic ethical principles that we employ in evaluating experiments are widely accepted and command significant allegiance not only from our contemporaries but also from reflective and morally sensitive individuals and ethical traditions in the past. It would be very implausible to construe any of them as parochial or controversial.

Retrospective Moral Judgment and the Challenge of Relativism

Some may still have reservations about the project of evaluating the ethics of decisions and actions that occurred several decades ago. The worry is that it is somehow inappropriate, if not muddled, to apply currently accepted standards to earlier periods when they were not accepted, recognized, or viewed as matters of obligation. This is an important worry, though one that does not apply to our framework.

The position that the values and principles of today cannot be validly applied to past situations in which they may not have been accepted is called historical ethical relativism. This is the thesis that moral judgments across time are invalid because moral judgments can be justified only by reference to a set of shared values, and the values of a society change over time. According to this view, one historical period differs from another by virtue of lacking the relevant values contained in the other historical period, namely, those that support or justify the particular moral judgments in question. Understood in this way, historical ethical relativism, if true, would explain why *some* retrospective moral judgments are invalid, namely, where the past society about which the judgments are made lacked the values that, in our time, support our judgments. In other words, the claim is that moral judgments made about actions and agents in one period of history cannot be made from the perspective of the values of another historical period.

The question of whether historical ethical relativism limits the validity of retrospective moral judgment is not a mere theoretical puzzle for moral philosophers. It is an eminently practical question, since how we answer it has direct and profound implications for what we ought to do *now*. Most obviously, the position we adopt on the validity of retrospective moral judgment will determine whether we should honor claims that people now make for remedies for historical injustices allegedly perpetrated against themselves or their ancestors. Similarly, we must know whether there is any special circumstance resulting from the historical context in which the responsible parties acted that mitigates whatever blame would be appropriate. We return to this question later in the chapter.

In addition, something even more fundamental is at stake in the debate over retrospective moral judgment: the possibility of moral progress. The idea of moral progress makes sense only if it is possible to make moral judgments about the past and to make them by appealing to some of the same moral standards that we apply to the present. Unless we can apply the same moral yardstick to the past and the present, we cannot meaningfully say either that there has been moral progress or that there has not. For example, unless some retrospective moral judgments are valid, we cannot say that the abolition of slavery is a case of moral progress, moral regression, or either one. More specifically, unless we can say that slavery was wrong, we cannot say that the abolition of slavery was a moral improvement.

For these and other reasons, the acceptance of historical ethical relativism has troubling implications. But even if we were to accept historical ethical relativism as the correct position, it would not follow from this alone that there is anything improper about making judgments about radiation experiments conducted decades ago based on the three kinds of ethical standards the Committee has identified. Two of these kinds of standards—government policies and rules of professional ethics—are standards used at the time the experiments were conducted. Neither of these kinds of standards involves projecting current cultural values onto a different cultural milieu.

We have already argued that basic ethical principles, the third kind of standard adopted by the Committee, are not temporally limited. Although there have been changes in ethical values in the United States between the mid-1940s and the present, it is implausible that these changes involved the rejection or affirmation of principles so basic as that it is wrong to treat people as mere means, wrong to inflict harm, or wrong to deceive people. Thus, the Advisory Committee's evaluations of the human radiation experiments in light of these basic principles is based on a simple and we think reasonable assumption that, even fifty years ago, these principles were pervasive features of moral life

in the United States that were widely recognized and accepted, much as we recognize and accept them today.[5]

Factors That Influence or Limit Ethical Evaluation

Several considerations influence and can limit the ability to reach ethical conclusions about rightness and wrongness and praise and blame. Some of these may be more likely to be present in efforts to evaluate the past, but all can arise when attempts are made to evaluate contemporary events as well. The most important such limitations relevant to the Advisory Committee's evaluations are these:

1. Lack of evidence as to whether ethical standards were followed or violated and if so, by whom, and
2. The presence of conflicting obligations.

The three kinds of ethical standards adopted by the Committee can yield the conclusion that an individual or collective agent had or has a particular obligation. But this conclusion is not by itself sufficient to determine *in any particular case* whether anything wrong was done or whether any individual or collective agent deserves blame.

Lack of Evidence

Sound evaluations cannot be made without sufficient evidence. Sometimes it cannot be determined if anything wrong was done because key facts about a case are missing or unclear. Other times there may be sufficient evidence that a wrong was done, but insufficient evidence to determine who performed the action that was wrong or who authorized the policy that was wrong or who was responsible for a practice that was wrong. This is why the Advisory Committee strove during our tenure to reconstruct the details of the circumstances under which the human radiation experiments themselves took place. However, these records are incomplete, and even the copious documentation we have gathered does not tell as complete a story as sometimes was needed to make ethical evaluations.

Conflicting Obligations

Because we all have more than one obligation, because they can conflict with one another, and because some obligations are weightier than others, a particular obligation that is otherwise morally binding may not be binding in a particular circumstance, all things considered. For example, a government official might be obligated to follow certain routine procedures, but in a time of dire emergency he or she might have a weightier obligation to avert great harm to many people by taking direct action that disregards the procedures. Similarly, a physician is obligated to keep his patient's condition confidential, but in some cases it is permissible and even obligatory to breach this confidence (for example, in order to prevent the spread of deadly infectious diseases). In such cases, the agent has done nothing wrong in failing to do what he or she would ordinarily be morally obligated to do; that obligation has been validly overridden by what is in the particular circumstances a weightier obligation.

The presence of conflicting obligations may limit our ability to make moral judgments when, for example, it is difficult to determine, in a particular case, which obligation should take precedence. At the same time, however, if it can be determined which obligation is weightier, then the presence of this factor does not serve as an impediment to evaluation; rather, it can lead to the conclusion that nothing morally wrong was done and that no one should be blamed.

An example of a potentially overriding obligation that is especially important for the Advisory Committee's work is the possibility that, during the period of the radiation experiments, obligations to protect national security were sometimes more morally weighty than obligations to comply with standards for human subjects research. If the threat were great enough, considerations of national security grounded in the basic ethical principle that one ought to promote welfare and prevent harm could justifiably override the basic ethical principle of not using people as mere means to the ends of others, as well as the more specific rule of research ethics requiring the voluntary consent of human subjects. Had such an overriding obligation to protect national security existed during the period we studied, it also would have relieved responsible individuals of any blame otherwise attributable to them for using individuals in experiments that were crucial to the national defense.

Especially during the late 1940s and early 1950s, and then again in the first years of the early 1960s, our country was engaged in an intense competition with the Soviet Union. A high premium was placed upon military superiority, not only in "conventional" warfare but also in atomic, biological, and chemical warfare. The DOD's Wilson memorandum, when originally promulgated in 1953, declared that it was directed toward the need to pursue atomic, biological, and chemical warfare experiments "for defensive purposes" in these fields.

It would not be surprising, therefore, to discover that, in the government's policies and rules for human subject research, provisions had been made for the possibility that obligations to protect national security might conflict with and take priority over obligations to protect human subjects, and thus that such policies would have included exceptions for national security needs. The moral justification would also not be surprising: that, in order to preserve the American way of life with its precious freedoms, some sacrifices of individual rights and interests would have to be made for the greater good. The very phrase *Cold War* expressed the conviction that we already were engaged in a life-or-death struggle and that in war actions may be permissible that would be impermissible in peacetime. Survival in the treacherous and heavily armed post–World War II era might demand no less, repugnant as those actions otherwise might be to many Americans.

The Advisory Committee did not undertake an inquiry to determine whether during either World War II or the Cold War there were ever circumstances in which considerations of national security might have justified infringements of the rights and protections that would otherwise be enjoyed by American citizens in the context of human experimentation. Our sources for answering this question were limited to materials pertinent to specific human radiation experiments and declassified defense-related memorandums and transcripts. With regard to the experiments, particular cases are reviewed in part II of this report. In those experiments that took place under circumstances most closely tied to national security considerations, such as the plutonium injections (see chapter 5), it does not appear that such considerations would have barred satisfying the basic elements of voluntary consent. Thus, for instance, although the word *plutonium* was classified until the end of World War II, subjects could still have been asked their permission after having been told that subjects in the experiment would be injected with a radioactive substance with which medical science had had little experience and which might be dangerous and that would not help them personally, but that the experiment was important to protecting the health of people involved in the war effort or safeguarding the national defense.

With regard to defense-related documents, in none of the memorandums or transcripts of various agencies did we encounter a *formal* national security exception to conditions under which human subjects may be used. In none of these materials does any official, military or civilian, argue for the position that individual rights may be justifiably overridden owing to the needs of the nation in the Cold War. In none of them is an official position expressed that the Nuremberg Code or other conventions concerning human subjects could be overridden because of national security needs.

Some government officials, military and civilian, may have personally advocated the view that obligations to protect national security were more important than obligations to protect the rights and interests of human subjects. It is, of course, possible that the priority placed on national security was so great in some circles of government that the ability of security interests to override other national interests was implicitly assumed, rather than explicitly articulated. It is a matter of historical record that some initiatives undertaken by government officials at some agencies during this period adopted the view that greater national purposes justified the exploitation of individuals. Notorious examples are the CIA's MKULTRA project and the Army's psychochemical experiments, which subjected unsuspecting people to experiments with LSD and other substances (see chapter 3).[6] How-

ever, even the internal investigation of the Department of Defense into these incidents in the 1970s concluded that these incidents were *violations* of government policy, not recognized legitimate exceptions to it.[7]

During the era of the Manhattan Project, the United States and its allies were engaged in a declared and just war against the Axis powers. Regarding the possibility of a wartime exception, it is well documented that during World War II the Committee on Medical Research (CMR) of the Executive Office of the President funded research on various problems confronting U.S. troops in the field, including dysentery, malaria, and influenza. This research involved the use of many subjects whose capacity to consent to be a volunteer was questionable at best, including children, the mentally retarded, and prisoners.[8] However, when the CMR considered proposed gonorrhea experiments that would have involved deliberately exposing prisoners to infection, the resulting discussion about the ethics of research exhibited a cautious attitude. The conclusion was that only "volunteers" could be used and that they had to be carefully informed about the risks and benefits of participation. In these and other classified conversations, the CMR took the position that care is to be taken with human subjects, including conscientious objectors and military personnel.[9]

It is difficult to reconcile these deliberations with the fact that many subjects of CMR-funded research were not true volunteers. Whether the CMR believed that the needs of a country at war justified the use of people who could not be true volunteers as research subjects is not known.

It would, however, be an error to conclude that, even in contexts where important national security interests are at stake, such as during wartime, a conflict between obligations to protect national defense and obligations to protect human subjects ought always to be resolved in favor of national security. The question of whether any and all means are morally acceptable for the sake of national security and the national defense is a complex one. Even in the case of a representative democracy that is not an aggressor, it would be wrong to assume that there are no moral constraints in time of war. All of the major religious and secular traditions concerning the morality of warfare recognize that there are substantial limitations upon the manner in which even a just war is conducted.[10] The issue of the morality of "total warfare" for a just cause, including the use of medical science, was beyond the scope of the Advisory Committee's charter, deliberations, and expertise.

Distinguishing Between the Wrongness of Actions and Policies and the Blameworthiness of Agents

Factors That Influence or Limit Judgments About Blame

The factors we have just discussed—lack of evidence and the presence of conflicting obligations—place limits on our ability to make judgments about *both* the rightness and wrongness of actions *and* the blameworthiness of the agents responsible for them. Some factors, however, place limits only on our ability to make judgments about the blameworthiness of agents. Even in cases where actions or policies are clearly morally wrong, it may be uncertain how blameworthy the agents who conducted or promulgated them are, or in fact, whether they are blameworthy at all. Some factors make it difficult to affix blame; other factors can mitigate or lessen the blame actors deserve. Four such factors are of particular concern to the Committee:[11]

1. Factual ignorance;
2. Culturally induced ignorance about relevant moral considerations;
3. Evolution in the interpretations and specification of moral principles; and
4. Indeterminacy in an organization's division of labor, with the result that it is unclear who has responsibility for implementing the commitments of the organization.

Factual Ignorance. Factual ignorance refers to circumstances in which some information relevant to the moral assessment of a situation is not available to the agent. There are many reasons that this may be so, including that the information in question is beyond the scope of human knowledge at the time or that there was no good reason to think that a particular item

of information was relevant or significant. However, just because an agent's ignorance of morally relevant information leads him or her to commit a morally wrong act, it does not follow that the person is not blameworthy for that act. The agent is blameworthy if a reasonably prudent person in that agent's position should have been aware that some information was required prior to action, and the information could have been obtained without undue effort or cost on his or her part. Some people are in positions that obligate them to make special efforts to acquire knowledge, such as those who are directly responsible for the well-being of others. Determinations of culpable and nonculpable factual ignorance often turn on whether the competent person in the field at that time had that knowledge or had the means to acquire it without undue burdens.

Culturally Induced Moral Ignorance. Sometimes cultural factors can prevent individuals from discerning what they are morally required to do and can therefore mitigate the blame we would otherwise place on individuals for failing to do what they ought to do. In some cases these factors may have been at work in the past but are no longer operative in the present, because of changes in culture over time.

An individual may, like other members of the culture, be *morally* ignorant. Because of features of his or her deeply enculturated beliefs, the individual may be unable to recognize, for example, that certain people (such as members of another race) deserve equal respect or even that they are people with rights. Moral ignorance can impair moral judgment and hence may result in a failure to act morally.

In extreme cases, a culture may instill a moral ignorance so profound that we may speak of *cultural moral blindness.* In some societies the dominant culture may recognize that it is wrong to exploit people but fail to recognize certain classes of individuals as being people. Some of those committed to the ideology of slavery may have been morally blind in just this way, and their culture may have induced this blindness.

Here it is crucial to distinguish between *culpable* and *nonculpable* moral ignorance. The fact that one's moral ignorance is instilled by one's culture does not by itself mean that one is not responsible for being ignorant; nor does it necessarily render one blameless for actions or omissions that result from that ignorance. What matters is not whether the erroneous belief that constitutes the moral ignorance was instilled by one's culture. What matters is the extent to which the individual can be held responsible for *maintaining* this belief, as opposed to correcting it. Where opportunities for remedying culturally induced moral ignorance are available, a person may rightly be held responsible for remaining in ignorance and for the wrongful behavior that issues from his or her mistaken beliefs.

People who maintain their culturally induced moral ignorance in the face of repeated opportunities for correction typically do so by indulging in unjustifiable rationalizations, such as those associated with racist attitudes. They show an excessive partiality to their own opinions and interests, a willful rejection of facts that they find inconvenient or disturbing, an inflated sense of their own self-worth relative to others, a lack of sensitivity to the predicament of others, and the like. These moral failings are widely recognized as such across a broad spectrum of cultural values and ethical traditions, both religious and secular.

Only if an agent could not be reasonably expected to remedy his or her culturally induced moral ignorance would such ignorance exculpate his conduct. But even in cases in which the individual could not be blamed for persisting in ignorance, this would do nothing to show that the actions or omissions resulting from his or her ignorance were not wrong. *Nonculpable moral ignorance only exculpates the agent; it does not make wrong acts right.*

Evolution in Interpretations of Ethical Principles. There is another respect in which the dependence of our perceptions of right and wrong on our cultural context has a bearing on the Advisory Committee's evaluations. While basic ethical principles do not change, interpretations and applications of basic ethical principles as they are expressed in more specific rules of conduct do evolve over time through processes of cultural change.

Recognizing that more specific moral rules do change has implications for how we judge the past. For example, the current requirement of informed consent is the result of evolution. Acceptance of the simple idea that medical treatment requires the consent of the patient (at least in the case of competent adults) seems to have preceded by a considerable interval the more complex notion that *informed* consent is required.[12] Furthermore, the notion of informed consent itself has undergone refinement and development through common law rulings, through analyses and explanations of these rulings in the scholarly legal literature, through philosophical treatments of the key concepts emerging from legal analyses, and through guidelines in reports by government and professional bodies.[13] For example, as early as 1914, the duty to obtain consent to medical treatment was established in American law: "Every human being of adult years and sound mind has a right to determine what shall be done with his own body; and a surgeon who performs an operation without his patient's consent commits an assault."[14] However, it was not until 1957 that the courts decreed that consent must be informed,[15] and this 1957 ruling was only the beginning of a long debate about what it means for a consent to be informed. Thus it is probably fair to say that the current understanding of informed consent is more sophisticated, and what is required of physicians and scientists more demanding, than both the preceding requirement of consent and earlier interpretations of what counts as informed consent. As the content of the concept has evolved, so has the scope of the corresponding obligation on the part of these professionals. For this reason it would be inappropriate to blame clinicians or researchers of the 1940s and 1950s for not adhering to the *details* of a standard that emerged through a complex process of cultural change that was to span decades. At the same time, however, it remains appropriate to hold them to the general requirements of the basic moral principles that underlie informed consent—not treating others as mere means, promoting the welfare of others, and respecting self-determination.

Inferring Bureaucratic Responsibilities. It is often unclear in complex organizations such as government agencies who has responsibility for implementing the organization's policies and rules. This is particularly common in new and changing organizations, where it is more likely than in stable organizations that there will be interconnecting lines of authority among employees and officials, and job descriptions that are not explicit with respect to responsibility for implementation of policies and initiatives. When policies are not properly implemented in organizations that fit this description, it often is difficult to assign blame to particular individuals. An employee or official of an agency cannot fairly be blamed for a failed or poorly executed policy unless it can be determined with confidence that the person had responsibility for implementing that policy and should have known that he or she had this responsibility.

The Importance of Distinguishing Wrongdoing from Blameworthiness

Judgments of wrongdoing and judgments of blameworthiness have very different implications. Even where a wrong was done, it does not follow that anyone should be blamed for the wrong. This is because there are factors, including the four we have just described, that can lessen or remove blame from an agent for a morally wrong act but that cannot in any way make the wrong act right. If experiments violated basic ethical principles, institutional or organizational policies, or rules of professional ethics, then they were and will always be wrong. Whether and how much anyone should be blamed for these wrongs are separate questions.[16]

The distinction between the moral status of experiments and that of the individuals who were involved with conducting, funding, or sponsoring them also has important implications for our own time. For a society to make moral progress, individuals must be able to exercise moral judgment about their actions. It is important for social actors to be critical about their activities, even those in which they have been engaged for some time. It is important for them to be able to step back and analyze their actions as right or wrong. If we did not distinguish between actions and agents, then people may feel that, once they have perceived their moral error, it is "too late" for them to change their ways, to

object to the ongoing activity, and to try to rally others in support of reform.

For any generation to initiate morally indicated reforms, it must be able to take this critical stance. As we see in part III of this report, even now there are aspects of our society's use of human subjects that should be critically examined. The actions we ourselves have performed do not condemn us as moral agents unless we refuse to open ourselves to the possibility that we have in some ways been in error. As we have said, even if we are exculpated by our own culturally induced moral ignorance, that does not make our wrong acts right. Even if we must accept a measure of blame for our actions, we are free to achieve a critical assessment and to initiate and participate in needed change.

The Significance of Judgments About Blameworthiness

The Committee believes that its first task is to evaluate the rightness or wrongness of the actions, practices, and policies involved in the human radiation experiments that occurred from 1944 to 1974. However, it is also important to consider whether judgments ascribing blame to individuals or groups or organizations can responsibly be made and whether they ought to be made.

There are three main reasons for judging culpability as well as wrongness. First, a crucial part of the Committee's task is to make recommendations that will reduce the risk of errors and abuses in human experimentation in the future, on the basis of its diagnoses of what went wrong in the past. A complete and accurate diagnosis requires not only stating what wrongs were done, but also explaining who was responsible for the wrongs occurring. To do this is likely to yield the judgment that some individuals were morally blameworthy. Second, unless judgments of culpability are made about particular individuals, one important means of deterring future wrongs will be precluded. People contemplating unethical behavior will presumably be more likely to refrain from it, other things being equal, if they believe that they, as individuals, may be held accountable for wrongdoing than if they can assure themselves that at most their government or their particular government agency or their profession may be subject to blame. Third, ethical evaluation generally involves both evaluation of the rightness or wrongness of actions and the praiseworthiness or blameworthiness of agents. In the absence of any explicit exemption of the latter sorts of judgment in our mandate, the Committee believes it would be arbitrary to exclude them.

Having made a case for judgments of culpability as well as wrongness, the Committee believes it is very important to distinguish carefully between judging that an individual was culpable for a particular action and judging that he or she is a person of bad moral character. Justifiable judgments of character must be based on accurate information about long-standing and stable *patterns* of action in a number of areas of a person's life, under a variety of different situations. Such patterns cannot usually be inferred from information about a few isolated actions a person performs in one particular department of his or her life, unless the actions are so extreme as to be on the order of heinous crimes.

APPLYING THE
ETHICAL FRAMEWORK

The three kinds of standards presented in this chapter provide a general framework for evaluating the ethics of human radiation experiments. In this section of the chapter, we revisit those standards in the specific context of human radiation experiments conducted between 1944 and 1974 and what we have learned about the policies and practices involving human subjects during that period.

Basic Ethical Principles

Earlier in this chapter we identified six basic ethical principles as particularly relevant to our work: "One ought not to treat people as mere means to the ends of others"; "One ought not to deceive others"; "One ought not to inflict harm or risk of harm"; "One ought to promote welfare and prevent harm"; "One ought to treat people fairly and with equal respect"; and "One ought to respect the self-determination of others."

These principles are central to our analysis of the cases we present in part II of the report, although not every case we evaluate engages every principle. Two of the principles, however, recur repeatedly as we consider the ethics of past experiments. These are "One ought not to treat people as mere means to the ends of others" and "One ought not to inflict harm or risk of harm." Whether an experiment involving human subjects violates the principle not to use people as mere means generally depends on two factors—consent and therapeutic intent. An individual may give his or her *consent* to being treated as a means to the ends of others. If a person freely consents, then he or she is no longer being used as a *mere* means, that is, as a means *only*. Thus, if a person is used as a subject in an experiment from which the person cannot possibly benefit directly, but the person's consent to that use is obtained, the person *is not* being used as a mere means to the ends of others. By contrast, if a person is used as a subject in such an experiment but the person's consent is not obtained for that use, the person *is* being used as a mere means to the ends of the investigator conducting the experiment and the institutions funding or sponsoring the experiment.

If an action that involves the use of a person is undertaken in whole or in part for that person's benefit, then the person is not being used as a mere means toward the ends of others. Thus, if a person is used as a subject in an experiment that is intended to offer the subject a prospect of direct benefit, then, even if the subject's consent has not been obtained, the subject is not being used as a mere means to the ends of others. This is because the experiment is intended to serve the subject's interests as well as the interests of the investigator and funding agency. It may be wrong not to obtain the subject's consent in this case, but the wrong does not stem from a violation of the principle not to use people as mere means. Instead, the wrong reflects the violation of other basic principles such as the principles enjoining us to respect self-determination and to promote welfare and prevent harm.

These two factors—the obtaining of consent and an intention to benefit—also can transform the moral quality of an act that involves the imposition of harm or risk of harm. One important way to make the imposition of a risk of harm justifiable is to obtain the person's permission for the imposition. The imposition of risk on a person also is more justifiable when the risk is imposed to secure a benefit for that person, although even in the presence of a prospect of offsetting benefit, the imposition of risk on another without that person's consent is morally questionable because it appears to violate the principle of respect for self-determination.[17]

Consider the following example of how the factors of therapeutic intent and consent can transform a morally questionable action into a morally acceptable one. Patients are enrolled in an experiment in which they are given a new drug that is unproven in humans, induces substantial discomfort or even suffering, and may produce irreversible damage to vital organs. There is, however, no effective treatment for the condition from which these patient-subjects suffer, and the condition is life threatening. The drug is theoretically promising compared with related drugs used in similar diseases, and it has proven effective in animals. Further, the opportunity to participate in the experiment is offered to patients while they are lucid, comfortable, and at ease. Under these circumstances the imposition of harm may be transformed into a caring and respectful act.

Policies of Government Agencies

Where agencies of the government had policies on the conduct of research involving human subjects, and where these policies included requirements or rules that are morally sound, these policies constitute standards against which the conduct of the agencies and the people who worked there, as well as the experiments the agencies sponsored or conducted, can be evaluated. Government agencies must be held responsible for failures to implement their own policies. To do otherwise is to break faith with the American people, who have a reasonable expectation that an agency will conduct its affairs in accord with the agency's stated policies. As we noted in chapter 1, it is not always clear, however, whether statements made in letters or memorandums constitute agency policy. When

there is little evidence that a statement by a government official was ever implemented, it is often difficult to determine whether this was an instance of an agency failing to implement its own policies or an instance where a statement by a government official was not perceived as agency policy in the first place.

Among the general conclusions that can be drawn from the discussions about policies during the late 1940s and early 1950s is that the AEC, DOD, and NIH required investigators to obtain the consent of the healthy or "normal" subject, and prior group review was required for risk in research using radioisotopes for all private and publicly financed research (and, in the NIH, for all hazardous procedures). Also, in 1953, the Department of Defense adopted the Nuremberg Code as the policy for research related to atomic, biological, and chemical warfare, and the NIH Clinical Center articulated a consent requirement for patient-subjects in intramural research (see chapter 1).

Two questions that arise at this juncture are whether an experiment was wrong if it violated one of these policies but took place at another government agency, and whether an experiment was wrong if it took place under the auspices of an agency before it promulgated the policy. The answer to both questions is the same: Even if such an experiment was not wrong according to the policy of the agency sponsoring the experiment at the time, the experiment may nevertheless have been unethical based on one or more basic ethical principles or rules of professional ethics.

As is the case today, decades ago government officials had obligations to take reasonable steps to see that policies were adequately implemented.[18] Policies constitute organizational commitments, and organizational commitments generate obligations on the part of the organization and its members. In some cases, however, it is not clear that conditions stated by individual officials rise to a level that all would be comfortable calling "policies." Accordingly, it is not clear whether corresponding obligations to implement can be inferred. The two letters signed by AEC General Manager Carroll Wilson in April and November 1947 are the best examples of this problem. Nevertheless, if it is correct to say that high officials have an obligation to exert due efforts to implement and communicate the rules they are empowered to establish, then they may reasonably be blamed for failures in this regard. Further, if they do not even attempt to articulate rules that are indicated by basic ethical principles and that are clearly relevant to organizational activities that fall under their authority, they are also subject to moral blame.

The mitigating condition of culturally induced moral ignorance does not apply to government officials who failed to exercise their responsibilities to implement or communicate requirements that clearly fell within the ambit of their office and of which they were aware. The very fact that these requirements were articulated by the agencies in which they worked is evidence that officials could not have been morally ignorant of them.

We have observed, however, that, especially with regard to research involving patients, policies were frequently unclear. When this research offered patient-subjects a chance to benefit medically, the widespread discretion granted physicians to make decisions on behalf of their patients is a mitigating factor in judging the blameworthiness of government officials for failing to impose consent requirements on physician-investigators. This failure could be attributed to a cultural moral ignorance concerning the proper limits to the authority of physicians over their patients.

The same cannot be said of government officials for failing to impose consent requirements on physician-investigators who used patient-subjects in research from which the patients could *not* benefit medically. This use of human subjects took place *outside* of the therapeutic context that defines the doctor-patient relationship and therefore also was outside of the authority then ceded to physicians. In this case responsible agency officials had a ready analogy to healthy subjects for whom there was a lengthy tradition of policies and rules requiring the use of "volunteers" and the obtaining of consent. Government officials could and should have perceived the morally identical nature of these cases—that, without consent, both cases involved violation of the principle not to use people as mere means to the ends of others.

Those who were ill should have been granted the same protections as those who were well.

In contrast to requirements for consent, requirements intended to ensure that risks to experimental subjects were acceptable were far more clearly stated. Government officials are blameworthy if they permitted research to continue that was known to entail unusual risks to the subjects, in direct violation of agency policy.

Finally, some lessons that can be drawn from the experience of the human radiation experiments we considered speak to the conduct of government itself as a collective agent, rather than simply to individual government officials. In too many instances, as we saw in chapter 1, we found a lack of clarity about the status within an agency of specific declarations by responsible officials. Particularly when agencies are engaged in activities that may compromise the rights or interests of citizens, it is critically important that agencies be clear about their commitments and policies and that they not remain passive in the face of questionable practices for which they may bear some responsibility. In chapter 3 we saw an effective response to such a situation in the 1960s by the PHS. This example attests to the fact that institutional clarity and active reform measures can succeed and that when they do they can be great forward strides.

Rules of Professional Ethics

Even if the federal government had adopted no formal human research ethics policy whatsoever, the medical profession and its members would still have moral obligations to those who entrust themselves to their care. The successes of modern medical research, regardless of its funding source, are ultimately due to the efforts of talented and dedicated medical scientists. These investigators bear a profound ethical burden in their work with human subjects. Society entrusts them with the privilege of using other human beings to advance their important work. Although society must not discourage them from the pursuit of new information, it also must diligently pursue signs that medical scientists have not exercised their ethical responsibility with the care and sensitivity that society has good reason to expect from them.

Without reference to the policies adopted by federal agencies, what rules of professional ethics were seen by the medical profession during the 1944–1974 period as relevant to the conduct of its members engaged in human subjects research? The answer to this question depends upon which kind of experimental situation is under discussion: an experiment on a *healthy subject;* an experiment on a *patient-subject* without *a scientific or clinical basis for an expectation of benefit to the patient-subject;* or an experiment on a *patient-subject with a scientific or clinical basis for an expectation of benefit to the patient-subject.*

Experiments on Healthy Subjects: By the mid-1940s it was common to obtain the voluntary consent of healthy subjects who were to participate in biomedical experiments that offered no prospect of medical benefit to them. Sophisticated philosophical analysis is not required to reach the conclusion that using a human being in a medical experiment that offers the person no prospect of personal benefits without that person's consent is wrong. As we have already noted, such conduct violates the basic ethical principle that one ought not use people as mere means to the ends of others.

Experiments on Patient-Subjects Without a Scientific or Clinical Basis for an Expectation of Benefit to the Patient-Subject: The Hippocratic tradition of medical ethics inherited by physicians in the 1940s holds that, unless the physician is reasonably sure that his or her treatment is, on balance, likely to do the patient more good than harm, the treatment should not be introduced. The heart of the Hippocratic ethic is the physician's commitment to putting the interests of the patient first. Subjecting one's patient to experimentation that offers no prospect of benefit to the patient without his or her consent is a direct repudiation of this commitment. (If the patient consents to this use, the moral warrant for proceeding with the experiment comes from the patient's permission, not from the Hippocratic ethic.)

Experiments on Patient-Subjects with a Scientific or Clinical Basis for an Expectation of Benefit to the Patient-Subject: Even in Hippocratic medicine it is recognized that physicians should attempt to use unproven or experimental meth-ods to benefit the patient, whether through efforts at cure or palliation, but only so long as there is no efficacious standard therapy available and innovative measures are compatible with the obligation to avoid doing harm without the prospect of offsetting benefit. Interventions in this category should be based on scientific reasoning and conservative clinical judgment. Arguably, so long as these conditions prevailed, it was not thought morally necessary within the medical profession to obtain the patient's consent to such experimentation prior to the 1960s. But the physician assumed a corresponding obligation to base his or her deviation from standard practice on the reasonable likelihood of patient benefit, sufficient to outweigh the risks associated with being in the experiment. This type of reasoning, too, has been available to and accepted by physicians for many years, even though the ability to assess and calculate risks has developed greatly.

* * *

Although the professional ethics of the period thus had relevant moral rules for each of these three experimental situations, compliance with these rules is a separate matter. There may be many reasons for specific failures by physicians to adhere to the requirements of their ethical tradition, some of which may render them nonculpable, and there are various limitations on our ability to assign blame for particular cases of a physician's failure to adhere to professional ethics. However, any use of human subjects that did not proceed in accordance with these rules of professional ethics was wrong in the sense that it was a violation of sound professional ethical standards. Moreover, even if there was then or is now a lack of clarity about the rules of professional ethics, recognition by morally serious individuals of basic ethical principles is enough to identify certain sorts of human experiments as morally unacceptable.

The special moral responsibilities of the medical profession as a whole, whether decades ago or in our own time, deserve careful consideration, especially insofar as previous experience can help formulate lessons for the future. Like the government, the medical profession as a whole must be held to a higher standard than individuals in society. Confidence in the medical profession is important because individuals put their very lives, and the lives of their loved ones, in the hands of those whom the profession has certified as competent to practice. Unlike government officials, members of the medical profession are explicitly bound to a moral tradition in their professional relations, based on which society grants the medical profession the privilege of largely policing itself. This authority is part of what constitutes the medical profession as a profession, but the authority is granted by society on the condition that the profession will adhere to the high moral rules it professes and that, if necessary, the medical profession will reform or encourage the reform of relevant institutions to ensure that those rules will be honored in practice.

Moreover, many of the privileges that devolve on the medical profession are granted on the condition that it is sufficiently well organized to police itself, with minimal intervention by the government and the legal system. Therefore, members of the medical profession are further legitimately expected to engage in organizational conduct that constitutes sound moral practices. Implicit in this arrangement is also the assumption that it will be *self-critical* even about its relatively well-entrenched attitudes and beliefs, so that it will be prepared to undertake reforms. Without this commitment to self-criticism, self-regulation cannot be effective and the public's trust in the professional's ability to self-regulate would be unwarranted.

Today we regard subjects of biomedical research whose consent was not obtained to have been wronged; under conditions of significant risk, the wrong is greater, and in the absence of the potential for offsetting medical benefit, greater still. The historical silence of the medical profession with respect to nontherapeutic experiments was perhaps based on the rationale that those who are ill and perhaps dying may be used in experiments because they will not be harmed even though they will not benefit. But this rationale overlooks both the principle that people should never be used as mere means and the principle of respect for self-determination; it may also provide insufficient protection against harm, given the position of conflict of interest in which the physician-researcher may

find him-or herself. Nevertheless, until the mid-1960s medical conventions were silent on experiments with patient-subjects that offered no direct benefit but which physicians believed to pose acceptable risk. This silence was a failure of the profession.

One defense of the profession in this regard is that it was as subject to the phenomenon we have called cultural moral ignorance as any other group in society at the time, including the arguably excessive deference to physician authority on the part of the government and possibly the public at large. However, the medical profession was in a wholly different position from the others, in several respects. First, it insisted upon and was given the privilege of policing its own behavior. Second, the profession was the direct beneficiary of the deference paid to it. Third, there were already examples of experiments that had involved subject consent that could have served as models of reform. Under these conditions the profession had an obligation to be self-critical concerning the norms and rules it thought appropriate to govern its members' conduct.

The medical profession could and should have seen that healthy subjects and patient-subjects in nontherapeutic experiments were in similar moral positions—neither was expected to benefit medically. Just as physicians had no moral license to determine an "acceptable risk" for healthy subjects without their voluntary consent, they had no moral license to do so in the case of other subjects who also could not benefit from being in research, even if they were patients. The prevailing standards for healthy subject groups could easily have been applied to patient-subjects for whom there was no expectation of medical benefit. The moral equivalence of the use of healthy people and ill people as subjects of experiments from which no subject could possibly benefit directly was perceptible at the time.

This moral equivalence would have made it clear that no one, well or sick, should be used as a mere means to advance medical science without voluntary consent. Thus, this moral ignorance could have and should have been remedied at the time. Indeed, it is arguably the case that physicians could and should have seen that using patients in this way was morally worse than using healthy people, for in so doing one was violating not only the basic ethical principle not to use people as a mere means but also the basic ethical principle to treat people fairly and with equal respect.

American physicians are members of a society that places a high value on these basic moral principles, still more vital than the advancement of medical science. These principles are as easily known to physicians as to anyone else, and it is unacceptable to single oneself out as an exception to these principles simply because one is a member of an esteemed profession. Someone who is ill deserves to be treated with the same respect as someone who is well. Accordingly, a physician who failed to tell a patient that what was proposed was an experiment with no therapeutic intent was and is blameworthy. To the extent that the experiment entailed significant risk, the physician is more blameworthy; where it was reasonable to assume that the experiment imposed no risk or minimal risk or inconvenience, the blame is less.

We argue here that the use of patients in nontherapeutic experiments without their consent was not only a violation of these basic moral principles but also a violation of the Hippocratic principle that was the cornerstone of professional medical ethics at that time. That principle enjoins physicians to act in the best interests of their patients and thus would seem to prohibit subjecting patients to experiments from which they could not benefit. It might be argued that a widespread practice that is not in conformity with a principle of professional ethics invalidates the principle, since the practice shows that the profession was not really committed to the principle in the first place. This is a misunderstanding, however, of what it means for a profession to adopt and espouse a moral principle. Even if many or most physicians sometimes fail or even often fail to comply with the principle, it is still coherent to say that the principle is accepted by the profession, if the principle has been publicly pronounced and affirmed by the profession, as was clearly the case with respect to the Hippocratic ethic.

To characterize a great profession as having engaged over many years in unethical conduct—years in which massive progress was being made in curbing some of mankind's greatest ills—may strike some as arrogant and unreasonable. However, fair assessment indicates that the circumstance was one of those times in history in which wrongs were committed by very decent people who were in a position to know that a specific aspect of their interactions with others should be improved. Wrongs are not less egregious because they were committed by a member of a certain profession or by people who are very decent in their relationships with other parties. It is common for us to look back at such conduct in amazement that so many otherwise good and decent people could have engaged in it without a high level of self-awareness. Moral consistency requires the Advisory Committee to conclude that, if the use of healthy subjects without consent was understood to be wrong at the time, then the use of patients without consent in nontherapeutic experiments should also have been discerned as wrong at the time, no matter how widespread the practice.

It should be emphasized, however, that often these nontherapeutic experiments on unconsenting patients constituted only minor wrongs. Often there was little or no risk to patient-subjects and no inconvenience. Although it is always morally offensive to use a person as a means only, as the burden on the patient-subject decreased, so too did the seriousness of the wrong.

Much the same can be said of experiments that were conducted on patient-subjects without their consent but that offered a prospect of medical benefit. To the extent that such experiments were conducted within the moral environment of the doctor-patient relationship, that is, based on the physician's considered and informed judgment that it was in the patient's best interests to be enrolled in the research, then the less blameworthy the physician was for failing to obtain consent. However, where the risks were great or where there were viable alternatives to participation in research, then the physician was more blameworthy for failing to obtain consent.

It is often difficult to establish standards and make judgments about right and wrong, and about blame and exculpation. Our charge was all the more difficult because the context of the actions and agents we were asked to evaluate differs from our own. In arriving at this moral framework for evaluating human radiation experiments, we have tried to be fair to history, to considerations of ethics, and above all, to the people affected by our analysis—former subjects, physician-investigators, and government officials.

NOTES

1. International declarations of human rights that would otherwise be relevant to an evaluation of human experimentation, such as the Covenant on Civil and Political Rights (1966), were articulated after the human radiation experiments with which we are mainly concerned, with the significant exception of the Nuremberg Code, as discussed in chapter 2.

2. The Advisory Committee is aware that questions such as precisely what ethical principles should be considered "basic," how they are related to those less basic, and how the basic ethical principles are known are among the most controversial and difficult in moral philosophy. For the Advisory Committee's limited purposes, a comprehensive and systematic moral theory is not required and is, in any case, far beyond the scope of this report. We have rather settled on a list of immediately recognizable and widely accepted ethical principles that are not usually thought to require justification themselves and that should be included in any adequate moral theory.

3. Some view promise keeping as a basic ethical principle on a par with the prohibition against deception. It may also be seen as grounded in one or more of the basic ethical principles on our list of six, such as those concerning deception and treating people as mere means.

4. The President's Commission functioned from 1978 to 1983, under the Carter and Reagan administrations, and produced a number of influential reports and recommendations concerning medical ethics and health care policy.

5. It may be argued that historical ethical relativism reduces to *cultural* ethical relativism. On this position, the notion that even basic ethical principles vary by era is part of a more general claim that what is really at stake is different "world views," and these different world views may exist at the same time but in cultures that are different from one another in certain crucial respects. On this analysis, in other words, the temporal factor is not the essential one. However, some find it easier to reject historical ethical relativism than cultural ethical relativism, for they find it plausible that *essentially* the same values operative in, say, the United States in the 1990s were operative in the 1950s, but not that *essentially* the same values that are operative in the United States in the 1990s are also operative in China in the 1990s.

6. In its report on the CIA and Army psychochemical experiments, the U.S. Senate found that

[i]n the Army's tests, as with those of the CIA, individual rights were . . . subordinated to national security considerations; informed consent and follow-up examinations of subjects were neglected in efforts to maintain the secrecy of the tests.

U.S. Congress, The Select Committee to Study Governmental Operations with Respect to Intelligence Activities, *Foreign and Military Intelligence* [Church Committee report], report no. 94-755, 94th Cong., 2d Sess. (Washington, D.C.: GPO, 1976), book 1, 411l. However, even in the light of the Army's own analysis of its LSD experiments, presented in a 1959 staff study by the U.S. Army Intelligence Corps (USAINTC), the operative legal principles should not have permitted the resulting practices to take place:

It was always a tenet of Army intelligence that the basic American principle of dignity and welfare of the individual will not be violated . . . In intelligence, the stakes involved and the interests of national security may permit a more tolerant interpretation of moral-ethical values, *but not legal limits*, through necessity . . . [emphasis added].

USAINTC Staff Study, Material Testing Program EA 1729 (15 October, 1959), 26. The staff study's distinction between the flexibility of "moral-ethical values" and "legal limits" is puzzling.

7. U.S. Army Inspector General, *Use of Volunteers in Chemical Agent Research* (Army IG report) (Washington D.C.: GPO, 1975).

8. David J. Rothman, *Strangers at the Bedside: A History of How Law and Bioethics Transformed Medical Decision Making* (New York: Basic Books, 1991), 32-50.

9. Rothman writes of the CMR's deliberations on the gonorrhea proposal: "It [the CMR] conducted a remarkably thorough and sensitive discussion of the ethics of research and adopted procedures that satisfied the principles of voluntary and informed consent. Indeed, the gonorrhea protocols contradict blanket assertions that in the 1940s and 1950s investigators were working in an ethical vacuum." Ibid., 42-43.

10. Michael Walzer, *Just and Unjust Wars* (New York: Basic Books, 1977).

11. Another factor often important in assessments of blame is duress. All systems of ethics recognize that people cannot be blamed for actions that violate basic ethical principles if they acted under duress. Duress includes manipulation, blackmail, or threats of physical harm. There is no evidence that any particular individual involved in the human radiation experiments functioned under conditions of duress.

12. Ruth Faden and Tom Beauchamp, *A History and Theory of Informed Consent* (New York: Oxford University Press, 1986).

13. For example, the National Commission for the Protection of Human Subjects of Biomedical and Behavioral Research published ten reports. Many of these recommendations were enacted into federal regulation. U.S. Congress, Office of Technology Assessment, *Biomedical Ethics in U.S. Public Policy— Background Paper*, OTA-BP-BBS-105 (Washington, D.C.: GPO, June 1993), 10.

14. *Schloendorff v. Society of New York Hospital*, 211 N.Y. 125 (1914).

15. *Salgo v. Leland Stanford, Jr., University Board of Trustees*, 317 P.2d 170 (1957).

16. In each case we assume that the principles or policies in question were morally sound; if not, anyone who refused to take part in unethical experiments performed in accordance with them acted, in retrospect, in a praiseworthy manner.

17. Again, with regard to the elements of an ethical framework suited to the intentional releases, we note that different justifications are used to evaluate the risks to collectives or communities as against those used to evaluate risks to individuals.

18. Note, however, that the intended scope of the policy was not always clear. Also, if the government or an agency had no policy at all concerning the use of human subjects but did conduct such research, then the absence of a policy would itself be objectionable.

Part II

Case Studies

PART II
OVERVIEW

When we began our work, the Advisory Committee was aware of several dozen human radiation experiments and the thirteen intentional releases in our charter. Soon, however, we found that these represented a fraction of the several thousand government-sponsored human radiation experiments and hundreds of intentional releases conducted from 1944 to 1974.

It was clear that the Committee would have to decide how to proceed in examining the experiments. Our ability to review all of the experiments and releases in detail was limited not only by time and resources, but even more so by the information available. For the majority of experiments identified, only the barest descriptions remained. It appeared that the vast majority of experiments involved trace amounts of radioisotopes, as are routinely used today for the study of bodily processes and the diagnosis of disease. However, where reports or other data were available, they did not routinely provide information needed to assess the precise risks to which subjects were exposed. These reports were even less likely to identify what kinds of people were chosen as subjects and why and how they were selected.

Since the Committee could not review all experiments, we decided to prepare a series of case studies focused on groups of experiments. We quickly found that there was no one right way to organize the experiments for purpose of case study. For example, the case studies could have been defined by the type of radiation to which subjects were exposed. This would likely have yielded groupings of experiments with differing purposes, differing populations, and differing risks and benefits. Likewise, grouping all experiments according to the characteristics of the people who were the subjects of the research would have lumped together experiments with differing purposes, risks, and scientific procedures.

After extensive deliberation, the Committee settled on eight case studies, which together address the charges to and priorities of the Com-

mittee. For example, we were charged to consider both intentional releases of radiation into the environment and the question of whether any former subjects of human radiation experiments would benefit medically from notification of their involvement. In addition, the Committee saw a responsibility to address those experiments that had received significant public attention at the time of the Committee's creation as well as those brought to our attention by members of the public. These experiments either offered no prospect of medical benefit to subjects or they involved interventions alleged to be controversial at the time. We also, however, recognized the importance of considering the far larger group of experiments that received no such attention but that also may have involved no prospect of benefit to subjects. We also placed a priority on experiments that were conducted on behalf of secret programs and for national security reasons; experiments that posed the greatest risk of harm; and experiments in which the subjects selected for experimentation were particularly powerless to resist or exercise independent judgment about participation. Together, these considerations formed the basis for the selection of the case studies.

In chapter 5, we look at the Manhattan Project plutonium-injection experiments and related experimentation. Sick patients were used in sometimes secret experimentation to develop data needed to protect the health and safety of nuclear weapons workers. The experiments raise questions of the use of sick patients for purposes that are not of benefit to them, the role of national security in permitting conduct that might not otherwise be justified, and the use of secrecy for the purpose of protecting the government from embarrassment and potential liability.

In contrast to the plutonium injections, the vast majority of human radiation experiments were not conducted in secret. Indeed, the use of radioisotopes in biomedical research was publicly and actively promoted by the Atomic Energy Commission. Among the several thousand experiments about which little information is currently available, most fall into this category. The Committee adopted a two-pronged strat-

THE ACHRE EXPERIMENTS DATABASE

By Cabinet directive on January 19, 1994, federal agencies were ordered to "establish forthwith an initial procedure for locating records of human radiation experiments conducted by the Agency or under a contract or grant of the Agency." The agencies most closely associated with these activities— the DOD, DOE, DHHS, NASA, CIA, and VA (and later the NRC)—in cooperation with Advisory Committee staff, identified record collections of importance and provided ACHRE with copies of documents potentially containing information on human radiation experiments. The documents were analyzed to identify individual experiments, which were then described according to a protocol developed by ACHRE members and staff, given unique identifiers, and recorded in an electronic database. Experiments were also identified by Advisory Committee staff in the published literature, discovered through a search of the National Library of Medicine databases and bibliographies, and documented by individuals who came forward with information for the Advisory Committee.

The database contains records for approximately 4,000 human radiation experiments. Information was collected, to the extent it was available, on the identity of the experiment (including investigators, location, dates, title, and documentation); funding, program approval and classification; the type and dose of radiation used; various characteristics of the experimental subjects; and the nature of the consent obtained. The experiments were in addition categorized by various themes and characteristics developed by Advisory Committee members and staff to reflect ACHRE research interests.

Documentation for individual experiments varies widely, sometimes including significant primary protocol documentation, often including only a journal article or abstract and, for the greatest number, just an investigator's name, a location, a date, and a title. As a result, although the database and the records it abstracts constitute an impressive and unique collection of information on human radiation experiments, that collection is not a comprehensive information resource on human radiation experiments but really just the best place to start to look for information.

The supplemental volume titled *Sources and Documentation* contains a more extensive and detailed description of the database and its sources.

---

egy to study this phenomenon. In chapter 6, we describe the system the AEC developed for the distribution of isotopes to be used in human research. This system was the primary provider of the source material for human experimentation in the postwar period. In studying the operation of the radioisotope distribution system, and the related "human use" committees at local institutions, we sought to learn the ground rules that governed the conduct of the majority of human radiation experiments, most of which have received little or no public attention. Also in this chapter we review how research with radioisotopes has contributed to advances in medicine.

The Committee then selected for particular consideration, in chapter 7, radioisotope research that used children as subjects. We determined to focus on children for several reasons. First, at low levels of radiation exposure, children are at greater risk of harm than adults. Second, children were the most appropriate group in which to pursue the Committee's mandate with respect to notification of former subjects for medical reasons. They are the group most likely to have been harmed by their participation in research, and they are more likely than other former subjects still to be alive. Third, when the Committee considered how best to study subject populations that were most likely to be exploited because of their relative dependency or powerlessness, children were the only subjects who could readily be identified in the meager documentation available. By contrast, characteristics such as gender, ethnicity, and social class were rarely noted in research reports of the day.

Moving from case studies focused on the injection or ingestion of radioisotopes, chapter 8 shifts to experimentation in which sick patients were subjected to externally administered total-body irradiation (TBI). The Committee discovered that the highly publicized TBI experiments conducted at the University of Cincinnati were only the last of a series in which the government sought to use data from patients undergoing TBI treatment to gain information for nuclear weapons development and use. This experimentation spanned the period from World War II to the early 1970s, during which the ethics of experimentation became increasingly subject to public debate and government regulation. In contrast with the experiments that flowed from the AEC's radioisotope program, the use of external radiation such as TBI did not in its earlier years involve a government requirement of prior review for risk. The TBI experimentation raises basic questions about the responsibility of the government when it seeks to gather research data in conjunction with medical interventions of debatable benefit to sick patients.

In chapter 9 we examine experimentation on healthy subjects, specifically prisoners, for the purpose of learning the effects of external irradiation on the testes, such as might be experienced by astronauts in space. The prisoner experiments were studied because they received significant public attention and because a literally captive population was chosen to bear risks to which no other group of experimental subjects had been exposed or has been exposed since. This research took place during a period in which the once-commonly accepted practice of nontherapeutic experimentation on prisoners was increasingly subject to public criticism and moral outrage.

Chapter 10 also explores research involving healthy subjects: human experimentation conducted in conjunction with atomic bomb tests. More than 200,000 service personnel—now known as atomic veterans—participated at atomic bomb test sites, mostly for training and test-management purposes. A small number also were used as subjects of experimentation. The Committee heard from many atomic veterans and their family members who were concerned about both the long-term health effects of these exposures and the government's conduct. This case study provided the opportunity to examine the meaning of human experimentation in an occupational setting where risk is the norm.

In chapter 11 we address the thirteen intentional releases of radiation into the environment specified in the Committee's charter, as well as additional releases identified during the life of the Committee. In contrast with biomedical experimentation, individuals and communities were not typically the subject of study in these intentional releases. Rather, the releases were to test intelligence equipment, the potential of radiological warfare, and the mechanism of the atomic bomb. While the risk posed by intentional releases was relatively small, the releases often took place in secret and remained secret for years.

The final case study, in chapter 12, looks at two groups that were put at risk by nuclear weapons development and testing programs and as a consequence became the subjects of observational research: workers who mined uranium for the Atomic Energy Commission in the western United States from the 1940s to 1960s and residents of the Marshall Islands, whose Pacific homeland was irradiated as a consequence of a hydrogen bomb test in 1954. While these observational studies do not fit the classic definition of an experiment, in which the investigator controls the variable under study (in this case radiation exposure), they are instances of research involving human subjects. The Committee elected to examine the experiences of the uranium miners and Marshallese because they raise important issues in the ethics of human research not illustrated in the previous case studies and because numerous public witnesses impressed on the Committee the significance of the lessons to be learned from their histories.

Part II concludes with an exploration of an important theme common to many of the case studies—openness and secrecy in the government's conduct concerning human radiation research and intentional releases. In chapter 13 we step back and look at what rules governed what the public was told about the topics under the Committee's purview, whether these rules were publicly known, and whether they were followed.

5

Experiments with Plutonium, Uranium, and Polonium

IN August 1944, at the secret Los Alamos Laboratory in New Mexico, a twenty-three-year-old chemist was trying to learn what he could about the properties of a radioactive metal. One year later, the new "product"—one of several code words for this three-year-old element with a classified name—would power the bomb dropped on Nagasaki. That day the young scientist, Don Mastick, was working with the entire Los Alamos supply of the material, 10 milligrams. It was sealed in a glass vial several inches long and about a quarter inch in diameter. Unknown to Mastick, a chemical reaction was causing pressure to build up inside the vial. Suddenly it burst, firing an acidic solution against the wall from where it splattered into Mastick's face, some of it entering his mouth.[1]

Realizing the importance to the war effort of the plutonium he had just ingested, Mastick hurried directly to the office of Louis Hempelmann, the health director at Los Alamos. Hempelmann pumped Mastick's stomach and instructed the young scientist to retrieve the plutonium from the expelled contents. Hempelmann expressed a concern related to worker safety: there was no way available to determine how much plutonium remained in Mastick's body. He immediately pressed the lab's director, J. Robert Oppenheimer, for authorization to conduct studies to develop ways of detecting

plutonium in the lungs, and in urine and feces, and of estimating the level of plutonium in the body from the amount found in excreta.[2]

Looming over Mastick's accident was the well-known tragedy of the radium dial workers more than a decade earlier. Like Mastick, they had ingested radioactive material through their mouths, as they licked the brushes they used to apply radium paint to watch dials. As time passed, many suffered from a gruesome bone disease localized in the jaw, and some bone cancers developed. Could plutonium cause a similar tragedy? If so, how much plutonium needed to be ingested before harmful effects might arise? How could one tell how much plutonium a person had already ingested? The answers to these questions were crucial, not only in the case of accidents such as Mastick's, but also, in the long run, to establish occupational health standards for the hundreds of workers who would soon be mass-producing plutonium for atomic bombs. Several pounds of radium, handled without recognition of the dangers, had led to dozens of deaths; what might plutonium cause?

A starting point was to examine the available data on radium poisoning, compare the characteristics of the radiation emitted by radium and plutonium, and try to extrapolate from radium to plutonium. However, plutonium had already revealed unexpected physical properties, which were posing problems for the bomb designers. Could plutonium also have unexpected biochemical properties? Extrapolation from radium was a good starting point, but could never be as reliable as data on plutonium itself.

Oppenheimer agreed that this research was critical. In an August 16, 1944, memorandum to Hempelmann, Oppenheimer authorized separate programs to develop methods to detect plutonium in the excreta and in the lung. With respect to biological studies, which Oppenheimer speculated might involve human experimentation, he wrote: "I feel that it is desirable if these can in any way be handled elsewhere not to undertake them here."[3] The reason Oppenheimer did not want these experiments conducted at Los Alamos remains obscure. Nine days later, Hempelmann met with Colonel Stafford L. Warren, medical director of the Manhattan Project, and others. They agreed to conduct a research program using both animal and human subjects.[4]

Mastick, who reported no ill effects from the accident when Advisory Committee staff interviewed him in 1995,[5] was not the first alert to the potential hazards of plutonium. Human experiments to study the metabolism and retention of plutonium in the body had been contemplated from the earliest days of the Manhattan Project. On January 5, 1944, Glenn Seaborg, who in 1941 was the first to recognize that plutonium had been created in the cyclotron at the University of California at Berkeley, wrote to Dr. Robert Stone, health director of the Metallurgical Laboratory in Chicago (a Manhattan Project contractor) and a central figure in efforts to understand the health effects of plutonium:

It has occurred to me that the physiological hazards of working with plutonium and its compounds may be very great. Due to its alpha radiation and long life

it may be that the permanent location in the body of even very small amounts, say one milligram or less, may be very harmful. The ingestion of such extraordinarily small amounts as some few tens of micrograms might be unpleasant, if it locates itself in a permanent position.[6]

Seaborg urged that a safety program be set up. In addition, "I would like to suggest that a program to trace the course of plutonium in the body be initiated as soon as possible. In my opinion such a program should have the very highest priority."[7] Stone reassured Seaborg that human tracer studies "have long since been planned. . . . although never mentioned in official descriptions of the program."[8] The work began at Berkeley with studies on rats conducted by Dr. Joseph Hamilton.[9]

Even as these studies on the biological effects of plutonium were beginning, the amount of plutonium being produced was dramatically increasing. Most of the effort at Oak Ridge was devoted to the separation of isotopes of uranium. However, the X-10 plant at Oak Ridge was a larger version of the very small plutonium-producing reactor developed at the University of Chicago. The X-10 plant began operating on November 4, 1943, and by the summer of 1944 was sending small amounts of plutonium to Los Alamos.[10] By December 1944 large-scale production of plutonium began at the Hanford, Washington, reactor complex.[11]

By late 1944, in the wake of the Mastick accident, the need to devise a means of estimating the amount of plutonium in the body became acute. It seemed that the only way to estimate how much plutonium remained in a worker's body would be to measure over time the amount excreted after a known dose and, from this, estimate the relationship between the amount excreted and the amount retained in the body.[12]

MAXIMUM PERMISSIBLE BODY BURDEN (MPBB) FOR PLUTONIUM

The plutonium injections were part of a larger research project intended to provide data for an occupational safety program riddled with uncertainty. Not only was there

a need for ways to monitor the exposure of personnel—the driving force behind the plutonium injections—but the maximum permissible body burden (MPBB) for plutonium, the maximum amount of plutonium that would be permitted in the bodies of workers, was still under debate.

The concept of "maximum permissible body burden" had begun to develop before the war in light of the known hazards of radium. Just prior to the war, primarily at the request of the Navy, a committee of experts was formed to establish occupational health standards for the factories producing dials illuminated by radium paint. After examining the data on radium dial painters, this committee agreed that 0.1 microgram fixed in the body should be the "tolerance level" for radium: an amount that, in the words of the committee chairman, Robley Evans, would be "at such a level that we would feel comfortable if our own wife or daughter were the subject."[a] After the war the term *maximum permissible body burden* was adopted and defined more precisely as the amount of a radioisotope that, when continuously present inside the body, would produce a dose equivalent to the allowable occupational exposure (the maximum permissible dose). For radioisotopes that, like radium, primarily reside in bone, biological data and mathematical models were used to determine how much of another bone seeker would produce the same dose as the original 0.1–microgram radium standard.

Between 1943 and the spring of 1945, based on the body burden for radium and preliminary results of animal experiments, a tentative MPBB for plutonium of 5 micrograms was adopted by the Manhattan District.[b] This level was derived by direct comparison of the relative energies of plutonium and radium.

By the spring of 1945, differences between the deposition of radium and plutonium in the body were becoming clearer.

Animal data indicated that plutonium deposited in what was called at the time the "organic matrix" of the bone—the part of the bone most associated with bone growth. This was different from radium, which seemed to deposit instead in the mineralized bone. Wright Langham wrote to Hymer Friedell supporting the choice of 1 microgram as an operating limit in lieu of a more formal policy. Langham wrote that with the adoption of this lower limit "the medico-legal aspect will have been taken care of and of still greater importance, we will have taken a relatively small chance of poisoning someone in case the material proves to be more toxic than one would normally expect."[c] This level was adopted and held until the Tripartite Permissible Dose Conference at Chalk River, Canada, in September 1949.

At this conference, representatives from the United States, United Kingdom, and Canada agreed on tolerance doses for many radioactive isotopes, including a maximum body burden of 0.1 microgram for plutonium. This reduced by a factor of 10 the value under which Los Alamos production had been operating. This reduction was based on the results of acute toxicological experiments with animals, which indicated that plutonium was as much as fifteen times more toxic than radium.

On January 20, 1950, Wright Langham wrote to Shields Warren, then the director of the AEC's Division of Biology and Medicine, alerting him to the problems caused by the Chalk River Conference's new "extremely conservative tolerances [which] may have a drastic effect on the efficiency and productivity of the Los Alamos Laboratory. Their official adoption will undoubtedly force major alteration in both present and future laboratory facilities and may add millions of dollars to the cost of construction of the permanent building program now in the

a. Robley Evans, "Inception of Standards for Internal Emitters, Radon and Radium," *Health Physics* 41 (September 1981): 437–448.

b. W. H. Langham et al., "The Los Alamos Scientific Laboratory's Experience with Plutonium in Man," *Health Physics* 8 (1962): 753.

c. Wright Langham, Los Alamos Scientific Laboratory Health Division, to Hymer Friedell, 21 May 1945 ("Since the Chicago Meeting, I am somewhat lost as to what our program should be in the future . . .") (ACHRE No. DOE-113094-B-7), 1.

planning phases."[d] Langham continued with reasons for regarding the Chalk River value of 0.1 micrograms of plutonium as "unnecessarily low." He cited, among other things, differences between acute and chronic toxicity and new analysis of data from the radium watch dial painters.

On January 24, 1950, Shields Warren, Austin Brues of Argonne National Laboratory, Robley Evans, Karl Morgan, and Wright Langham met in Washington. Langham wrote later: "As a result of this meeting, Dr. Shields Warren of the Division of Biology and Medicine authorized 0.5 ug (0.033 uc) of Pu[239] as the AEC's official operating maximum permissible body burden."[e] There were no minutes or transcripts taken of this

meeting. The calculation of this level was again based on the body burden for radium, this time modified by the 1/15 toxicity factor (since experiments had indicated that plutonium was up to fifteen times more toxic than radium), by the relative retention of plutonium and radium in rodents, and by the energy ratios modified by radon retention.

Thus far, the entire debate had occurred behind the closed doors of the AEC. Consideration of all the complex issues applied in setting a permissible body burden had been within a small circle of scientists and administrators. While the MPBB for plutonium accepted at the January 1950 meeting has held until today, its derivation has changed over the years.

By March 1945, there was disturbing news that urine samples from Los Alamos workers were indicating, based on models developed from animal experimentation, that some might be approaching or had exceeded a body burden of 1 microgram.[13] A March 25 meeting led to Hempelmann's recommendation that the Project "help make arrangements for a human tracer experiment to determine the percentage of plutonium excreted daily in the urine and feces. It is suggested that a hospital patient at either Rochester or Chicago be chosen for injection of from one to ten micrograms of material and that the excreta be sent to the laboratory for analysis."[14] The overall program, as it was envisioned by Dr. Hymer Friedell, deputy medical director of the Manhattan Engineer District,

Oppenheimer, and Hempelmann, consisted of three parts: improvement of methods to protect personnel from exposure to plutonium; development of methods for diagnosing overexposure of personnel; and study of methods of treatment for overexposed personnel. On March 29, Oppenheimer forwarded the recommendation to Stafford Warren, with his "personal endorsement."[15]

The accident at Los Alamos was part of the prelude to experiments conducted between 1945 and 1947 in which eighteen hospital patients were injected with plutonium to determine how excreta (urine and feces) could be used to estimate the amount of plutonium that remained in an exposed worker's body. One patient was injected at Oak Ridge Hospital in Oak Ridge, Tennessee; eleven were injected at the University of Rochester, three at the University of Chicago, and three at the University of California.

The results of these experiments contributed to the development of a monitoring method that, with small changes, is still used today. The experimental data were used to develop a model relating body burden to short-term excretion rate. Known as the "Langham model," it was based on short-term excretion data, long-term

---

d. The letter went on to say that "operations of the Los Alamos Laboratory would be curtailed or stopped if such action were necessary to the reasonable and sensible protection of the personnel. The seriousness of this action, however, seems to be adequate reason for requesting that official adoption of the tolerances by the AEC be postponed until they have been carefully reviewed in order to make certain that the values are not unnecessarily conservative." Wright Langham, Los Alamos Laboratory Health Division, to Shields Warren, Director of AEC Division of Biology and Medicine, 20 January 1950 ("Radiation Tolerances Proposed by the Chalk River Permissible Dose Conference of September 29–30, 1949") (ACHRE No. DOE-020795-D-6), 1.

e. W. H. Langham et al., "The Los Alamos Scientific Laboratory's Experience with Plutonium in Man," *Health Physics* 8 (1962): 754.

excretion data that were collected in 1950 from two injection subjects, and worker excretion data. This model has been used almost universally to monitor plutonium workers since 1950, although it has been modified over the years as longer-term and more extensive data were accumulated. While now, fifty years later, not every question concerning the quality of the science or the basis for estimating risk can be answered with precision, there is general agreement among radiation scientists that the experiments were useful.

Although this would be the first time that plutonium would be injected into human beings, the plutonium experiments were not viewed at the time as being extremely risky, and for good reason. Based on experience with other bone-seeking radioisotopes such as radium, the investigators had firm basis for believing, even in the 1940s, that the amount of material to be injected was likely too small to produce any immediate side effects or reactions. No one was expected to feel ill or have any negative reaction to the injection, and apparently no one did. Because acute effects were not expected, the plutonium injections were viewed as posing no short-term risks to human subjects. There was concern, however, about long-term risk. A draft report, written by one of the primary investigators within a few years of the injections, records that "acute toxic effects from the small dose of pu [plutonium] administered were neither expected nor observed." The document also recognized that "with regard to ultimate effects, it is too early to predict what may occur."[16] Based largely on the experience of the radium dial painters, it was recognized that exposure to plutonium could result, perhaps ten or twenty years later, in the development of cancer in a human subject. This was viewed as a significant risk but also as a risk that could be minimized by the use of small doses and wholly avoided if the subjects were expected to die well before a cancer had a chance to materialize.

Even if the plutonium injections had been entirely risk free, an impossibility in human experimentation, they could still be morally problematic. As we discussed in chapter 2, it was not uncommon in the 1940s for physicians to use patients as subjects in experiments without their knowledge or consent. This occurred frequently in research involving potential new therapies, where there was at least a chance that the patient-subjects might benefit medically from being in an experiment. But it also occurred even in experiments—like the plutonium injections—where there was never any expectation and no chance that the experiment might be of benefit to the subjects.

The conduct of the plutonium experiments raises a number of difficult ethics and policy questions: Who should have been the subjects of an experiment designed to protect workers vital to bomb production in wartime? What should the subjects have been told about the risks of the secret substance with which they were being injected? What should they have been told about the purpose of the experiment? What were the subjects told? Did they know they were part of an experiment in which there was no expectation that they would benefit medically?

An inquiry initiated by the AEC commissioners in 1974 investigated some of these questions. That inquiry focused on whether consent was obtained from the subjects, either at the time of the plutonium injections or during 1973 follow-up studies funded by the AEC's Argonne National Laboratory in Chicago, designed to determine the long-term effects of the injections. Sixteen patient charts were examined for evidence of consent at the time of injection; the other two charts had been either lost or destroyed. Of the sixteen charts examined, only one chart—that of the only subject injected after the April 1947 directive of AEC General Manager Carroll Wilson (discussed in chapter 1) that required documented consent—contained evidence of some form of consent. The other fifteen contained no record of consent.[17] According to AEC investigators, oral testimony pointed to failure to obtain consent in the case of the Oak Ridge injection and to some form of disclosure to patients for the California and Chicago experiments. The AEC concluded that testimony was inconclusive for the Rochester experiments.[18] With regard to the follow-up studies conducted with three surviving subjects in 1973, the investigation concluded that two subjects had deliberately not been informed of the purpose of the follow-up and that one subject had actually been misled about the purpose.[19]

As we will see later in this chapter, the AEC's conclusion that consent was not obtained from the surviving subjects for the 1973 follow-up studies was correct. Moreover, additional documentary evidence and testimony suggests that patient-subjects at the Universities of Rochester and California were never told that the injections were part of a medical experiment for which there was no expectation that they would benefit, and they never consented to this use of their bodies.

The rest of this chapter provides a chronological account of the plutonium injection experiments and follow-up studies conducted over the course of many years, assesses the influence of secrecy on the conduct of the experiments, and examines the motivating factors behind the prolonged secrecy of the experiments and the continued deception of surviving subjects. We also consider the conduct of experimentation with uranium and polonium. Finally, we render judgments where we can about the ethical conduct of these experiments.

THE MANHATTAN DISTRICT EXPERIMENTS

The First Injection

A few days after Hempelmann's March 26, 1945, recommendation that a hospital patient be injected with plutonium, Wright Langham, of the Los Alamos Laboratory's Health Division, sent 5 micrograms of plutonium to Dr. Friedell, with instructions for their use on a human subject.[20] The subject, as it turned out, was already in the Oak Ridge Army hospital, a victim of an auto accident that had occurred on March 24, 1945.[21] He was a fifty-three-year-old "colored male"[22] named Ebb Cade,[23] who was employed by an Oak Ridge construction company as a cement mixer. The subject had serious fractures in his arm and leg, but was otherwise "well developed [and] well nourished."[24] The patient was able to tell his doctors that he had always been in good health.[25]

Mr. Cade had been hospitalized since his accident, but the plutonium injection did not take place until April 10. On this date, "HP-12" (the code name HP—"human product"[26]—was later assigned to this patient and to patients at the University of Rochester) was reportedly injected with 4.7 micrograms of plutonium. (It is important here to distinguish between *administered* dose and *retained* dose; not all of the injected dose would remain fixed in the body. It was not known with certainty, however, how much of the 4.7 micrograms of plutonium would remain in his body.)

The small amount of material injected into Mr. Cade would not be expected to produce any acute effects, and there is no indication that any were experienced. However, except for his fractures, Mr. Cade was apparently in good health and at age fifty-three could reasonably have been expected to live for another ten to twenty years. Thus, in Mr. Cade's case, the risk of a plutonium-induced cancer could not be ruled out.

Dr. Joseph Howland, an Army doctor stationed at Oak Ridge, told AEC investigators in 1974 that he had administered the injection. There was, he recalled, no consent from the patient. He acted, he testified, only after his objections were met with a written order to proceed from his superior, Dr. Friedell.[27] Dr. Friedell told Advisory Committee staff in an interview that he did not order the injection and that it was administered by a physician named Dwight Clark, not Dr. Howland.[28] The Committee has not been able to resolve this contradiction.

Measurements were to be taken from samples of Mr. Cade's blood after four hours, his bone tissue after ninety-six hours, and his bodily excretions for forty to sixty days thereafter.[29] His broken bones were not set until April 15—five days after the injection—when bone samples were taken in a biopsy.[30] Although this was several weeks after his injury, during this era when antibiotics were only beginning to become available, it was common practice to delay surgery if there was any sign of possible infection. One document records that Mr. Cade had "marked" tooth decay and gum inflammation,[31] and fifteen of his teeth were extracted and sampled for plutonium. The Committee has not been able to determine whether the teeth were extracted primarily for medical reasons or for the purpose of sampling for plutonium. In a September 1945 letter, Captain David Goldring at Oak Ridge informed Langham that "more bone specimens and extracted teeth will be shipped to you very soon for analysis."[32] It remains unclear whether these additional bone specimens were extracted at the time of the April 15 operation or later.

According to one account, Mr. Cade departed suddenly from the hospital on his own initiative; one morning the nurse opened his door, and he was gone.[33] Later it was learned that he moved out of state and died of heart failure on April 13, 1953, in Greensboro, North Carolina.[34]

The experiment at Oak Ridge did not proceed as planned. "Before" and "after" urine samples were mistakenly commingled, so no baseline data on kidney function was available.[35] Thus, the subject's kidney function would be difficult to assess. In May 1945,[36] Dr. Stone convened a "Conference on Plutonium" in Chicago to discuss health issues related to plutonium, including the relationship between dose and excretion rate, the permissible body burden, and potential therapy and protective measures.[37] Wright Langham spoke about the Oak Ridge injection at the conference, carefully qualifying the reliability of the excretion data obtained from Mr. Cade. Langham observed that "the patient might not have been an ideal subject in that his kidney function may not have been completely normal at the time of injection"[38] as indicated by protein tests of his urine.

The Chicago Experiments

On April 11, the day after the Oak Ridge injection, Hymer Friedell transmitted the protocol describing the experiment on Mr. Cade to Louis Hempelmann at Los Alamos. "Everything went very smoothly," he wrote, "and I think that we will have some very valuable information for you."[39] He then went on to discuss the injection of more patients: "I think that we will have access to considerable clinical material here, and we hope to do a number of subjects. At such time as we line up several patients I think we will make an effort to have Mr. Langham here to review our setup."[40]

Subsequently, between late April and late December of 1945, three cancer patients, code-named CHI-1, 2, and 3, were injected with plu-

tonium. At least two and possibly all three were injected at the Billings Hospital of the University of Chicago. The doses to subjects CHI-2 and CHI-3 were the highest doses administered to any of the eighteen injection subjects—approximately 95 micrograms.[41] However, the amount of material injected was still below what would be expected to produce acute effects. Moreover, unlike Mr. Cade, all three of these patients were seriously ill and at least two of them died within ten months of receiving the injection. That the selection of seriously ill patients was an intentional strategy to contain risk is indicated in a 1946 report on CHI-1 and CHI-2: "Some human studies were needed to see how to apply the animal data to the human problems. Hence, two people were selected whose life expectancy was such that they could not be endangered by injections of plutonium."[42] It remains a mystery why CHI-3 was not included in this report.

On April 26, 1945, CHI-1, a sixty-eight-year-old man who had been admitted to Billings Hospital in March, was injected with 6.5 micrograms of plutonium. At the time of injection he was suffering from cancer of the mouth and lung. The patient reportedly "remained in fair condition until August 1945, when he complained of pain in the chest."[43] His lung cancer had apparently spread, and he died on October 3, 1945.[44]

The next injection took place eight months later. CHI-2 was a fifty-five-year-old woman with breast cancer who had been admitted to Billings Hospital in December 1945 after the cancer had already spread throughout her body. The 1946 report recorded that "the patient's general condition was poor at the time of admission and deteriorated steadily throughout the period of hospitalization."[45] She was injected with 95 micrograms of plutonium on December 27 and died on January 13, 1946.[46]

There is little known about the condition of CHI-3, the other subject who was injected with approximately 95 micrograms. He was a young man suffering from Hodgkin's disease, reportedly injected on the same date as CHI-2.[47] His condition at the time of injection remains unknown, as does his date of death. There is some question whether he was injected at Billings hospital or at another hospital in the Chicago area.[48]

There was no discussion of consent in the original reports on the Chicago experiments. However, a draft report on an interview conducted with E. R. Russell for the 1974 AEC investigation into the experiments (Russell was coauthor of the 1946 report on the Chicago experiments) summarized Russell's description of consent as follows: "[H]e prepared the plutonium solutions for injection and acted together with a nurse as witness to the fact that the patient was or had been informed that a radioactive substance was going to be injected. The administration of this substance, according to what was said in obtaining consent, was not necessarily for the benefit of the patients but might help other people."[49] To say that the injection was "not necessarily" for the benefit of the patient implies that there was some chance these patients might benefit; in fact, there was no expectation that this would occur.

Russell's account was obtained in the context of an official inquiry into his conduct and the conduct of the other investigators and officials involved in the plutonium injections, an inquiry that focused on whether consent was obtained from the subjects. We have no way of corroborating this account or of assessing what Dr. Russell's motivations were in explaining the plutonium injections to the subjects in the way claimed.

The Rochester Experiments

By the time the war began, the University of Rochester, which had a cyclotron, had assembled a group of first-rate physicists and medical researchers who were pioneering the new radiation research. Following the selection of the university's Stafford Warren to head its medical division, the Manhattan Project turned to Rochester for an increasing share of its biomedical research—including, in particular, research needed to set standards for worker safety.[50]

The university's metabolism ward, at what is now the Strong Memorial Hospital, became the central Manhattan District site for the administration of isotopes to human subjects. The two-bed ward, headed by Dr. Samuel Bassett, was part of the Manhattan District's "Special Prob-

lems Division," which worked on the health monitoring of production plants, the development of monitoring instruments, and research on the metabolism and toxicology of long-lived radioactive elements.[51] An experimental plan called for fifty subjects altogether, in five groups of ten subjects each. Each group would receive plutonium, radium, polonium, uranium, or lead.[52] Although the exact number of subjects remains unknown, at least twenty-two patients were administered long-lived isotopes in experiments with plutonium (eleven subjects), polonium (five subjects), and uranium (six subjects).

At the time the experiment was being designed, the main selection criterion for the subjects chosen at Rochester for the plutonium experiment was that they have a metabolism similar to healthy Manhattan Engineer District workers. In a work plan for the plutonium study based on a September 1945 meeting with a representative of Colonel Warren's office and the Rochester doctors, Langham wrote:

The selection of subjects is entirely up to the Rochester group. At the meeting it seemed to be more or less agreed that the subjects might be chronic arthritics [patients with serious collagen vascular diseases, such as scleroderma] or carcinoma patients without primary involvement of bone, liver, blood or kidneys.

It is of primary importance that the subjects have relatively normal kidney and liver function, as it is desirable to obtain a metabolic picture comparable to that of an active worker.

Undoubtedly the selection of subjects will be greatly influenced by what is available. The above points, however, should be kept in mind.[53]

Although this protocol specifies cancer patients as potential subjects, evidently the deliberate choice was made later by the experimenters to select patients without malignant diseases in the hope of ensuring normal metabolism.[54] Thus no cancer patients were included among the plutonium subjects at Rochester. Preference appears to have been given to patients the doctors believed would benefit from additional time in the hospital.[55]

An additional perspective on the selection of subjects for the plutonium experiments is provided in three retrospective reports written by Wright Langham. In a 1950 report on the plutonium project, including the experiments conducted at Rochester, Langham wrote that "as a rule, the subjects chosen were past forty-five years of age and suffering from chronic disorders such that survival for ten years was highly improbable."[56] In subsequent reports, Langham refers to the plutonium subjects as having been "hopelessly sick"[57] and "terminal."[58]

Documents retrieved for the Advisory Committee show that all but one of the plutonium subjects at Rochester suffered from chronic disorders such as severe hemorrhaging secondary to duodenal ulcers, heart disease, Addison's disease, cirrhosis, and scleroderma.[59] One subject, Eda Schultz Charlton, did not have any such condition. According to the draft of the 1950 report, she was misdiagnosed: "a woman aged 49 years may have a greater life expectancy than originally anticipated due to an error in the provisional diagnosis."[60]

Most of the subjects at Rochester were not terminally ill, and at least some of them had the potential to live more than ten years. Three of the Rochester subjects were known to still be living at the time of the 1974 AEC investigation into the plutonium experiments. Whether the inclusion of subjects at Rochester with the potential to live more than ten years is an indication that the investigators were not using Langham's criterion to select subjects or that they erred in their predictions is unclear. Judgments about the life expectancy of the chronically ill are difficult to make and often in error, even today.

The likelihood that long-term risks can be altogether eliminated does exist, however, if the subject is in the terminal stages of an illness and death is imminent. This was recognized by the plutonium investigators, and it led to the observation that the use of a terminal patient permitted a larger dose, which would make analysis easier. The first terminal patient at Rochester was injected toward the end of that series, and the possibility of further injections into terminal patients was discussed explicitly. In a March 1946 letter, Wright Langham wrote to Dr. Bassett, the primary physician-investigator at Rochester:

In case you should decide to do another terminal case, I suggest you do 50 micrograms instead of 5. This would permit the analysis of much smaller samples and would make my work considerably easier. . . . I feel reasonably certain there would be no harm in using larger amounts of material if you are sure the case is a terminal one [as was done in two of the three Chicago injections].[61]

As was the case at Oak Ridge and Chicago, there was no expectation that the patient-subjects at Rochester would benefit medically from the plutonium injections. The Advisory Committee found no documents that bear directly on what, if anything, the subjects were told about the injections and whether they consented. The recollections of at least some of those intimately involved have survived, however, and these recollections all suggest that the patients did not know they had been injected with radioactive material or even that they were subjects of an experiment.

Milton Stadt, the son of a Rochester subject, told the Advisory Committee the following at a meeting in Santa Fe, New Mexico, on January 30, 1995:

My mother, Jan Stadt, had a number, HP-8. She was injected with plutonium on March 9th, 1946. She was forty-one years old, and I was eleven years old at the time. My mother and father were never told or asked for any kind of consent to have this done to them.

My mother went in [to the hospital] for scleroderma . . . and a duodenal ulcer, and somehow she got pushed over into this lab where these monsters were.

Dr. Hempelmann, in an interview for the 1974 AEC investigation, said he believed that the patients injected with plutonium were deliberately not informed about the contents of the injections.[62] Dr. Patricia Durbin, a University of California researcher who in 1968 undertook a scientific reanalysis of the experiments, reported on a visit with Dr. Christine Waterhouse in 1971. Dr. Waterhouse was a medical resident at Rochester at the time of the plutonium injections. Durbin wrote the following regarding the Rochester subjects who were still alive:

She [Dr. Waterhouse] believes that all three persons would be agreeable to providing excretion samples and perhaps blood samples, but they are all quite old—in their middle or late 70's and cannot travel far. More important, they do not know that they received any radioactive material.[63]

In notes on a 1971 telephone conversation with Wright Langham, Dr. Durbin wrote: "He is, I believe, distressed by . . . the fact that the injected people in the HP series were unaware that they were the subjects of an experiment."[64] This recollection is even more troubling than the recollections of Drs. Waterhouse and Hempelmann, as it indicates not only that the subjects did not know that they were being injected with plutonium or a radioactive substance, but also that they did not know even that they were subjects of an experiment.

Even the doctors in charge of some of the injections at Rochester may not have known what they were injecting into patients. In 1974, Dr. Hempelmann suggested that the physician who actually injected the solution quite possibly did not know of its contents.[65]

Further evidence suggesting that the patient-subjects were never told what was done to them comes from 1950 correspondence between Langham and the physicians at Rochester. These physician-investigators were looking for signs of long-term skeletal effects in follow-up studies with two of the subjects at Rochester. Langham wrote to Rochester that he was "very glad to hear that you will manage to get follow-ups on the two subjects. The x-rays seem to be the all-important thing, but please get them in a completely routine manner. Do not make the examination look unusual in any way."[66]

Moreover, a letter from Langham to Dr. Bassett discussed the undesirability of recording plutonium data in the Rochester subjects' hospital records:

I talked to Col. [Stafford] Warren on the phone yesterday and he recommended that I send copies of all my data to Dr. [Andrew] Dowdy where it would be available to you and Dr. [Robert M.] Fink to observe. He thought it best that I not send it to you because he wanted it to remain in the Manhattan Project files, instead of taking a chance on it finding its way into the hospital records. I think this is probably a sensible suggestion.[67]

Uranium Injections at Rochester

Under the Manhattan Engineer District program, physicians at the Rochester metabolism ward also injected six patients with uranium (in the form of uranyl nitrate enriched in the isotopes uranium 234 and uranium 235) to establish the minimum dose that would produce detectable kidney damage due to the chemical toxicity of uranium metal, and to measure the rate at which uranium was excreted from the body. To achieve the first objective, the experimenters used a higher dose with each new subject until the first sign of minimal kidney damage occurred. Damage occurred in the sixth and last subject (at a calculated amount of radioactivity of 0.03 microcuries), indicated by protein tests of his urine. Unlike the plutonium injections, this was an experiment that evidently was designed not only to obtain excretion data but to cause actual physical harm, however minimal. Thus, although the investigators could reasonably view the plutonium injections as an experiment that was extremely unlikely to produce acute effects, this was not true of the uranium experiment, which was intended to produce acute effects. As with the plutonium injections, the uranium injections also posed a long-term risk of the development of cancer. The Committee does not know in this case how long subjects survived after injection; there is no documentation of follow-up with these subjects as there is for some of the subjects of the plutonium injections.

The subjects of this experiment, like some of the plutonium-injection subjects, were not at risk of imminent death, but did suffer from chronic medical conditions such as rheumatoid arthritis, alcoholism, malnutrition, cirrhosis, and tuberculosis. According to Dr. Bassett, again the primary investigator, the subjects "were chosen from a large group of hospital patients. Criteria of importance in making the selection were reasonably good kidney function with urine free from protein and with a normal sediment on clinical examination. The probability that the patient would benefit from continued hospitalization and medical care was also a factor in the choice."[68]

The 1948 report on the experiment did not discuss the question of consent. We were not able to locate any documents that bear on what, if anything, the subjects were told about the uranium injections, nor have any relevant recollections about the experiment survived. Two 1946 documents, however, discussing whether Dr. Bassett should be permitted to give a departmental seminar on the excretion rate of uranium in humans, illustrate the secrecy that surrounded these injections and suggest that the subjects were not informed of the experiment. By the time of this correspondence, the uranium research with animals at Rochester had been declassified. The first document, a letter written by Andrew Dowdy, the director of the Manhattan Department at the University of Rochester, to a Manhattan District Area engineer requesting permission for Bassett to give the seminar, included the following: "I feel that there is no reason why he should not discuss this matter, and I believe that the fact that this information was actually obtained on his own patients is of more concern to himself than to the District."[69] In the second document, an intraoffice memorandum, the area engineer discussed this point, and more:

Dr. Dowdy states that the patients were Dr. Bassett's, but it should be borne in mind that all the work performed by Dr. Bassett was performed at the request of the Manhattan District Medical Section. This seminar is to be conducted for persons who are all Doctors of Medicine and it is doubtful if this information would get out to any of the families of the patients or the patients on whom the experiments were performed. . . .

At the time these experiments were started, this office was given strict orders that the information should not be released to any but authorized persons. Almost all the correspondence and result of experiments were exchanged between Dr. Wright Langham at Santa Fe and Dr. Bassett of the University of Rochester. This rule is still in effect on some of the material that Dr. Bassett is using and knowledge of the experiments is kept from personnel at the Rochester Area.[70]

Polonium Injections at Rochester

In addition to the subjects injected with plutonium and uranium at Rochester, five subjects were chosen for an experiment with polonium.

The purpose of the experiment was to determine the excretion rate of polonium after a known dose, as well as to analyze the uptake of polonium in various tissues. The primary investigator for these experiments was Dr. Robert M. Fink, assistant professor of radiology and biophysics at the University of Rochester. Four patients were injected with the element, and one ingested it.[71] All five patients selected for this study were suffering from terminal forms of cancer: lymphosarcoma, acute lymphatic leukemia, or chronic myeloid leukemia. It is unclear why patients with malignant diseases were chosen as subjects in this experiment but excluded from the subject pools for the plutonium and uranium experiments. There is no discussion in the 1950 final report on the polonium experiments of the possibility that patients with malignant diseases might have abnormal metabolism, and the excretion data were employed right away in the establishment of occupational safety standards.[72]

The final report, unlike other reports on the Manhattan District metabolism studies, briefly discusses the question of consent: "the general problem was outlined to a number of hospital patients with no previous or probable future contact with polonium. Of the group that volunteered as subjects, four men and one woman were selected for the excretion studies outlined below."[73] This statement leaves no clear impression of what the subjects actually were told; like the experiments with plutonium and uranium, the human polonium experiment was a classified component of the metabolism program. Still, this report provides a contrast to the contemporaneous reports on the Manhattan District plutonium and uranium experiments, which make no mention of consent and which do not refer to the patient-subjects as "volunteers."

The California Experiments

While the University of Rochester had been conducting experiments for the Manhattan Engineer District, a related effort was under way at the University of California at Berkeley.[74] Before the war, Drs. Joseph Hamilton and Robert Stone had been exploring medical applications of radioisotopes with the aid of the University of California's cyclotron. Hamilton and his colleagues had pioneered in using radioisotopes to treat cancer, in particular iodine 131 in the 1930s. At the time the United States entered the war, they were investigating another isotope for cancer therapy, strontium 89. Indeed, it was this area of Hamilton's expertise that attracted the interest of the Manhattan Project. While Stone moved to the Chicago Metallurgical Laboratory during the war, Hamilton remained at the University of California's Radiation Laboratory, or "Rad Lab," at Berkeley. A colleague of both men, Dr. Earl Miller, a radiologist at the University of California, reported regularly to Stone on the progress of the Berkeley plutonium project.

Under the Manhattan District contract, Hamilton's studies originally had involved exposing rats to plutonium in an effort to determine its metabolic fate and thereby project the risk to workers at atomic plants. Toward the end of the war, Hamilton began to conduct plutonium studies on humans for the government.[75] Experiments with humans could be handled expeditiously, Hamilton wrote, because of the close relationship between the Rad Lab and the medical school at the University of California at San Francisco.[76] In January 1945, Hamilton confirmed to the Manhattan District that he planned "to undertake, on a limited scale, a series of metabolic studies with [plutonium] using human subjects."[77] The purpose of this work, Hamilton wrote, "was to evaluate the possible hazards . . . to humans who might be exposed to them, either in the course of the operation of the [Chicago] pile, or in the event of possible enemy action against the military and civilian population."[78]

Subsequently, three subjects, two adults and one child (known as CAL-1, 2, and 3), were injected with plutonium. In addition, in April 1947 a teenage boy (CAL-A) was injected with americium, and in January 1948 a fifty-five-year-old female cancer patient (CAL-Z) was injected with zirconium.[79]

On May 10, 1945, Hamilton reported he was awaiting "a suitable patient" for the plutonium experiment.[80] Four days later, fifty-eight-year-old Albert Stevens, designated CAL-1, was injected with plutonium, becoming the first human subject in the California portion of the project.[81] Albert Stevens was chosen in the belief that he was suffering from advanced stomach cancer.[82] Shortly after the injection, however, a biopsy revealed a benign gastric ulcer instead of the suspected cancer. The researchers collected excreta daily for almost one year, analyzing them for plutonium content.[83] Evidently, by two months after the injection, Mr. Stevens was considering moving out of the Berkeley area; this would have prevented further collection of excretion specimens. Dr. Hamilton proposed to Drs. Stone and Stafford Warren that he be permitted to "pay the man fifty dollars per month" in order to keep Mr. Stevens in the area. Hamilton recognized, however, that there were "possible legal and security situations which may present insurmountable obstacles."[84] In response to this request, Dr. Joe Howland (who was reportedly involved with the Oak Ridge plutonium injection) wrote the following to the California area engineer:

Possible solutions to this problem could be:
  a. Pay for his care in a hospital or nursing home as a service.
  b. Place this individual on Dr. Hamilton's payroll in some minor capacity without release of any classified information.
     It is not recommended that he be paid as an experimental subject only.[85]

According to a 1979 oral history of Kenneth Scott, an investigator at Berkeley who evidently was responsible for the analysis of Mr. Stevens's excretion specimens, the patient was paid some amount each month to keep him in the area. However, Dr. Scott also recalled that he never told Mr. Stevens what had happened to him: "His sister was a nurse and she was very suspicious of me. But to my knowledge he never found out."[86]

In addition, an April 1946 report on the experiment records that "several highly important tissue samples were secured including bone."[87] It appears that these tissue specimens, which included specimens of rib and spleen, were removed four days after the injection in an operation for the patient's suspected stomach cancer.[88]

Four months after Mr. Stevens was injected, Dr. Hamilton told the Manhattan District that the next subject would be injected "along with Pu238 [plutonium], small quantities of radio-yttrium, radio-strontium, and radio-cerium." The purpose of this experiment was to "compare in man the behavior of these three representative long-lived Fission Products with their metabolic properties in the rat, and second, a comparison can be made of the differences in their behavior from that of Plutonium."[89] This research would provide data to improve extrapolation from higher-dose animal experiments.

Despite Hamilton's hope to have a second patient by the fall, CAL-2 was not selected until April 1946. Simeon Shaw was a four-year-old Australian boy suffering from osteogenic sarcoma, a rare form of bone cancer, who was flown from Australia to the University of California for treatment. According to newspaper articles at the time, Simeon's family had been advised by an Australian physician to seek treatment at the University of California.[90] Arrangements then were made by the Red Cross and the U.S. Army for Simeon and his mother to fly by Army aircraft to San Francisco. Within days, he had been injected with a solution containing plutonium, yttrium, and cerium by physicians at the university.[91]

Following his discharge on May 25, about a month after his injection, the boy returned to Australia, and no follow-up was conducted. He died in January 1947. In February 1995 an ad hoc committee at the University of California at San Francisco (UCSF) concluded that probably at least part of the motivation for this experiment was to gather scientific data on the disposition of bone-seeking radionuclides with bone cancers.[92]

One piece of evidence indicating that there was a secondary research purpose for the injection of CAL-2 was a handwritten note in the boy's medical record saying that the surgeons removed a section of the bone tumor for pathology and for "studies to determine the rate of uptake of radioactive materials that had been injected prior to surgery, in comparison to normal tissues."[93]

It is likely that the CAL-2 experiment was designed both to acquire data for the Manhattan District and also to further the physicians' own search for radioisotopes that might treat

cancer in future patients. The California researchers themselves noted the dual purpose of their research at the time. Hamilton wrote in a report to the Army in the fall of 1945 that there were "military considerations which can be significantly aided by the results of properly planned tracer research."[94]

As the February 1995 UCSF report on the experiments concluded, however, the "injections of plutonium were not expected to be, nor were they, therapeutic or of medical benefit to the patients."[95] This corresponds with the evidence of a letter, written by Hamilton in July 1946, three months after the injection of CAL-2, to the author of an article on the peacetime implications of wartime medical discoveries:

To date no fission products, aside from radioactive iodine, have been employed for any therapeutic purposes. There is a possibility that one or more of the long list of radioactive elements produced by uranium fission may be of practical therapeutic value. At the present time, however, we can do no more than speculate.[96]

Documentary evidence suggests that consent for the injections likely was not obtained from at least some of the subjects at the University of California. A 1946 letter from T. S. Chapman, with the Manhattan District's Research Division, said the following regarding preparations for injections:

. . . preparations were being made for injection in humans by Drs. [Robert] Stone and [Earl] Miller. These doctors state that the injections would probably be made without the knowledge of the patient and that the physicians assumed full responsibility. Such injections were not divergent from the normal experimental method in the hospital and the patient signed no release. A release was held to be invalid.

The Medical Division of the District Office has referred "P" reports for project 48A to Colonel Cooney for review and approval is withheld pending his opinion.[97]

Chapman does not specify whether the "injections" referred to in this letter were injections of plutonium or of some other substance. It is unclear whether "'P' reports" refers to Hamilton's overall progress reports on his tracer research, which had reported mostly on research with plutonium (but also on research with cerium and yttrium), or whether "P" referred specifically to reports on work with plutonium. As we noted at the outset of this chapter, Chapman's claim that it was commonplace at the time to use patients in experiments without their knowledge and without asking them to sign a "release" is correct.

In the case of Albert Stevens (CAL-1), no documentary evidence that bears on disclosure or consent has been found. Simeon Shaw's (CAL-2's) medical file contains a standard form "Consent for Operation and/or Administration of Anaesthetic." This form, however, was signed by a witness attesting to consent of Simeon's mother one week after the injection and therefore probably applies to a biopsy done a week after the injection, not to the injection itself.[98]

On December 24, 1946, at the prompting of Major Birchard M. Brundage, who was chief of the Manhattan District's Medical Division, Colonel K. D. Nichols, commander of the Manhattan District, ordered a halt to injections of "certain radioactive substances" into human subjects at the University of California.[99] "Such work," Nichols wrote, "does not come under the scope of the Manhattan District Programs and should not be made a part of its research plan. It is therefore deemed advisable by this office not only to recommend against work on human subjects but also to deny authority for such work under the terms of the Manhattan contract." The following week, the civilian AEC took over responsibility for all Manhattan District research and temporarily reaffirmed the Manhattan District's suspension of human experimentation at the University of California.[100] It is unclear why this action was taken.

THE AEC'S REACTION: PRESERVING SECRECY WHILE REQUIRING DISCLOSURE

When the civilian Atomic Energy Commission took over for the Manhattan District on January 1, 1947, the plutonium injections provoked a strong reaction at the highest levels. One immediate result was the decision to keep information on the plutonium injections secret, evi-

dently for reasons not directly related to national security, but because of public relations and legal liability concerns. The other immediate result, as we saw in chapter 1, was the issuing of requirements for future human subjects research as articulated in letters by the AEC's general manager, Carroll Wilson.

In December 1946, as the civilian AEC was about to open its doors, Hymer Friedell, who had been deputy medical director of the Manhattan Engineer District, recommended the declassification of one of the plutonium reports, "CH [Chicago]-3607—The Distribution and Excretion of Plutonium in Two Human Subjects." The report, Friedell argued, "will not in my opinion result in the release of information beyond that authorized for disclosure by the current Declassification Guide."[101]

Friedell's recommendation was soon reversed. Officials with the new AEC had learned of the human injection experiments, and on February 28, 1947, an AEC declassification officer concluded that declassification was out of the question. The reasons are revealed in a previously classified document recently found at Oak Ridge:

The document [CH-3607] appears to be the most dangerous since it describes experiments performed on human subjects, including the actual injection of the metal plutonium into the body. The locations of these experiments are given and the results, even to the autopsy findings in the two cases. It is unlikely that these tests were made without the consent of the subjects, but no statement is made to that effect and the coldly scientific manner in which the results are tabulated and discussed would have a very poor effect on the public. Unless, of course, the legal aspects were covered by the necessary documents, the experimenters and the employing agencies, including the U.S., have been laid open to a devastating lawsuit which would, through its attendant publicity, have far reaching results.[102]

It is not clear to the Advisory Committee on what basis the declassification officer who wrote this comment concluded that it was unlikely that consent was not obtained from the Chicago subjects. This statement could be read as careful bureaucratic language, intended to leave an appropriate paper trail in the event of subsequent legal problems. On the other hand, the statement does support the claim, noted earlier, made by one of the Chicago doctors in 1974 that some form of oral consent for the injections had been obtained from the Chicago subjects. It is clear that there was no documentation of disclosure or consent on which the AEC could rely. As a consequence, secrecy was to be maintained, not as a defense against foreign powers, but to avoid a "devastating lawsuit" and "attendant publicity." Upon further review the report was "reclassified 'Restricted' on 3/31/47."[103] In a March 19, 1947, memorandum, Major Brundage, by that time chief of the AEC's Medical Division, explained:

The Medical Division also agrees with Public Relations that it would be unwise to release the paper 'Distribution and Excretion of Plutonium' primarily because of medical legal aspects in the use of plutonium in human beings and secondly because of the objections of Dr. Warren and Colonel Cooney that plutonium is not available for extra Commission experimental work, and thus this paper's distribution is not essential to off Project[104] experimental procedures.[105]

In July 1947, Argonne National Laboratory's declassification officer, Hoylande D. Young, inquired about possible declassification of this report as well as Hamilton's report on the CAL-1 injection. She stated that the directors of Argonne's Biology and Health Divisions (including J. J. Nickson, one of the authors of the Chicago report on the injections) believed that declassification of these reports would not be "prejudicial to the national interests."[106] The AEC continued to withhold declassification of these reports, however, on the grounds that they involved "experimentation on human subjects where the material was not given for therapeutic reasons."[107] Thus, there was clearly no expectation at the time that the plutonium injections would benefit the patient-subjects but some expectation that the general public might be disturbed by human experimentation in the absence of a prospect of offsetting benefit.

In 1950, Wright Langham and the Rochester doctors undertook to prepare a "Plutonium Report"[108] that would be "the last word on the plutonium situation."[109] It would be the "last word" to only a select few. In 1947, Rochester's

Andrew Dowdy had urged Los Alamos to give advance notice of declassification of the Rochester part of the experiment "because of possible unfavorable public relations and in an attempt to protect Dr. [Samuel] Bassett from any possible legal entanglements."[110] This is likely a reference to the same concern raised in the discussion of Dr. Bassett's seminar about his having experimented upon his own patients, except in this case the context is the plutonium rather than the uranium injections. "We think," Langham wrote to Stafford Warren, "the classification will be 'Secret,' and the circulation limited, depending on Dr. Shields Warren's [the head of AEC's Division of Biology and Medicine] wishes."[111] In August, Shields Warren approved the report for "CONFIDENTIAL classification and limited circulation as [Dr. Langham] requested."[112]

Even though its data and analysis were the basis for widespread plutonium safety procedures, the report remained unavailable to the public until 1971 when, at the urging of Dr. Patricia Durbin, it was downgraded to "Official Use Only."[113] (This categorization means that while the document was not likely to be released to the public absent specific request, it could be disclosed.)

What was it that was so potentially embarrassing about the plutonium experiments? The answer appears to lie in the 1947 letters from General Manager Wilson, discussed in detail in chapter 1. These letters state rules for both the conduct of human experiments and the declassification of previously conducted secret experiments.[114]

In his April 1947 letter, Wilson stated the requirements that there be expectation that research "may have therapeutic effect" and that at least two doctors "certify in writing (made part of an official record) to the patient's understanding state of mind, to the explanation furnished him, and to his willingness to accept the treatment."[115] In his November 1947 letter, Wilson reiterated these terms for human experiments, again calling for "reasonable hope . . . that the administration of such a substance will improve the condition of patient" and this time calling for "informed consent in writing" by the patient.[116] All of the seventeen plutonium injections conducted prior to the letters violated both these terms. As a consequence, they would have to stay secret. The only secret experiments that could be declassified were those that satisfied these requirements; to do otherwise was to risk adverse public reaction. Thus, the decision to keep the plutonium reports secret was itself an example of the way in which the AEC's assertion of conditions for human experimentation was coupled with the decision to keep secret those experiments that evidently did not adhere to these conditions (see chapter 13).

HUMAN EXPERIMENTATION CONTINUES

In March 1947, just as he was declaring that "public relations" required the reclassification of plutonium data, Medical Division chief Major Brundage approved a 1947–48 "Research Program and Budget" for Rochester that provided for metabolism studies with polonium, plutonium, uranium, thorium, radiolead, and radium.[117] The program was put on hold by the AEC soon after.[118]

The future of the metabolism work at Rochester apparently was decided when Shields Warren was named the first chief of the AEC's Division of Biology and Medicine in fall 1947. In his private diary for December 30, 1947, Warren tersely noted: "Ordered abandonment of human isotope program at Rochester."[119] The program at the University of California at Berkeley, however, continued. On December 4, 1947, Shields Warren had met with Hamilton and Stone;[120] the decision to allow the program to continue clearly was not a hasty one. A 1974 recollection of Shields Warren indicates that his decision to allow the program to continue may have been due to Hamilton's assertion in December 1947 that it had been the University of California's practice to obtain some form of (undocumented) consent.[121]

According to Warren, Hamilton had said that subjects were told "they would receive an injection of a new substance that was too new to say what it might do but that it had some properties like other substances that had been used to control growth processes in patients, or some-

thing of that general sort."[122] Warren went on to observe that "you could not call it informed consent because they did not know what it was, but they knew that it was a new and to them unknown substance."[123] Warren's observation does not go far enough, however. If Warren's secondhand account is accurate and this is indeed what the patient-subjects at the University of California were told, then they were more misled than informed. Analogizing plutonium to substances that "control growth processes in patients," even in prospect, might reasonably lead patients to believe that they would be receiving a substance with some hope of treating their cancer. Certainly such a remark would not communicate to patients that the experiment to be performed was not for their own benefit. It would have been appropriate that these patients be told that their participation might benefit future patients with the same conditions. It would have been crucial to distinguish, however, between this legitimate explanation of potential benefit to future cancer patients and misleading the patient into believing the experiment might benefit him or her.

Human Experimentation Continues at the University of California

By the summer of 1947, human experimentation had resumed at the University of California under AEC contract. In June, "CAL-A," a teenage Asian-American bone cancer patient at Chinese Hospital in San Francisco, was injected with americium. An instruction in the patient's file by one of Hamilton's assistants specifies that "we will use the same procedure as with Mr. S,"[124] evidently a reference to Albert Stevens. Dr. Durbin, Hamilton's associate, believes that CAL-A's guardian was informed of the procedure followed in that case.[125] The Advisory Committee received incomplete records for CAL-A that contained no evidence of disclosure or consent; UCSF has told the Committee that records at Chinese Hospital from the 1950s and earlier have been destroyed.[126]

A thirty-six-year-old African-American railroad porter named Elmer Allen, code-named CAL-3, was believed to be suffering from bone cancer and was injected with plutonium at the

University of California in July 1947. His left leg was amputated shortly thereafter. There is a note in his medical chart signed by two physicians, stating that the experimental nature was "explained to the patient, who agreed to the procedure" and that "the patient was in fully oriented and in sane mind."[127] It is likely that this note was intended to fulfill one of the April 1947 conditions for human experimentation, which allowed for such a procedure as documentation of having obtained the patient-subject's consent. It is not clear from the note, however, whether in explaining about the experimental nature of the procedure the physicians told the patient about the potential effects of the injection, as required by the Wilson letter, or that the injection was not intended to be of medical benefit to the patient. On this second point, the injection was *in violation* of the Wilson letter, which also required that there be an "expectation that it may have therapeutic effect."[128] As acknowledged by the February 1995 UCSF report, there was never any expectation on the part of the experimenters that the injection would be of therapeutic benefit to Mr. Allen.

Mr. Allen lived until 1991. According to UCSF's 1995 review of patient-subjects' medical charts, upon biopsy of his tumor a pathologic diagnosis was made of chondrosarcoma, a type of malignant bone tumor. UCSF reported that patients with this type of tumor "frequently surviv[e] many years beyond diagnosis if there is complete excision of the primary tumor."[129] This pathology finding suggests that Mr. Allen was a long-term cancer survivor. A note in his patient chart recorded that the tumor was "malignant but slow growing and late to metastasize. Prognosis therefore moderately good."[130]

On March 15, 1995, Elmerine Whitfield Bell, the daughter of Elmer Allen, told the Advisory Committee in Washington, D.C., that she

continue[s] to be appalled by the apparent attempts at cover-ups, the inferences that the nature of the times, the 1940s, allowed scientists to conduct experiments without getting a patient's consent or without mentioning risks. We contend that my father was not an informed participant in the plutonium experiment.

He was asked to sign his name several times while a patient at the University of California hospital in San Francisco. Why was he not asked to sign his name permitting scientists to inject him with plutonium? Why was his wife, who was college trained, not consulted in this matter?

On January 5, 1948, a fifty-five-year-old woman with cancer was injected with zirconium at the University of California.[131] The patient record for this case has not yet been located, nor have any other documents that might bear on whether this experiment was conducted in compliance with the consent requirements of the Wilson letters. We do know that the injection of zirconium was not expected to benefit the subject herself.[132]

A secret report on the zirconium injection was reviewed by the AEC in light of public relations and liability concerns. In August of that year, the report was denied declassification with the approval of Shields Warren, who wrote, "This document should not be declassified for general medical publication [and] it would be very difficult to rewrite it in an acceptable manner."[133] Warren was responding to a memorandum from Albert H. Holland, Jr., medical adviser at Oak Ridge, which specified that the concern about rewriting had to do with public relations and the fact that the report "specifically involves experimental human therapeutics."[134]

Follow-up of the Patient-Subjects at Rochester

The investigators at Rochester and the AEC were interested in obtaining long-term data from surviving subjects on excretion levels and the distribution of plutonium in various tissues. Follow-up studies at Rochester continued at least through 1953 with two of the subjects in the HP series, Eda Charlton and John Mousso. We have already noted Wright Langham's 1950 instruction to the physicians at Rochester suggesting that they were not to give these patients any indication of the true purpose of the follow-up studies.[135] In addition, Langham sought help in early 1950 to locate Ebb Cade (the man injected at Oak Ridge Hospital) for follow-up excretion studies. Langham asked Dr. Albert Holland at Oak Ridge to try to locate Mr. Cade and to keep his "eyes open for a possible autopsy."[136] It is unclear to the Committee whether follow-up of any kind was ever done with Mr. Cade.

On June 8, 1953, Eda Charlton's rib was removed during exploratory surgery for cancer and analyzed for plutonium. Louis Hempelmann, who by that time had moved from Los Alamos to Strong Memorial Hospital at Rochester, wrote to Charles Dunham of the AEC's Division of Biology and Medicine in advance of the procedure:

The patient in question was brought in for a skeletal survey, and turned out to have a 'coin-like' lesion inside the chest wall. . . . It is undoubtedly an incidental finding, but she must be explored by the chest surgeon here at Strong. In the course of the operation, he will remove a rib which we can analyze. Her films show the same type of minimal indefinite change in the bone that the others have had.[137]

It was standard practice at the time to remove a section of rib incidental to lung surgery. It is clear that the patient was still being followed for long-term effects of plutonium and that some subclinical bone changes of unclear significance had already been observed by this time. Therefore, the examination of this rib segment would have included special tests to determine whether plutonium was present.

On August 31, 1950, an internal DBM memorandum recorded the understanding of some AEC officials that Wright Langham and Rochester doctors were engaged in follow-up studies.[138] In a 1974 interview, however, Shields Warren recalled that he had no knowledge that the patients were the subjects of follow-up studies: "I did not learn of this continuing contact while I was in office at AEC. . . . I had assumed because I had been told that they were incurable patients that they all had died by the time we talked."[139]

Additional Follow-up Studies and the Argonne Exhumation Project

In 1968 Dr. Patricia Durbin undertook an investigation of the plutonium-injection subjects, which included a reevaluation of the original plutonium data. Her goal was to pursue "some elusive information on Pu in man and the information or assumptions about physiology needed to create a believable Pu model for man." She "decided to look at all the old Pu patients

as individuals rather than in a lump. . . . "[140] Durbin was surprised to find in her search for the original experimental data that the University of California data were drawn from three subjects who received plutonium and one who received americium; the data from only *one* plutonium subject from California had previously been reported in the open scientific literature.[141] Durbin asked the original researchers why these data had not been analyzed. She wrote: "I understand from Wright Langham that this problem has been discussed before and discarded as too messy."[142]

In 1972, after the classified report on the experiments had been downgraded to "Official Use Only," she went on to publish "Plutonium in Man: A New Look at the Old Data," a landmark paper in the plutonium story.[143] This was the first review in the open literature to analyze Langham's results in light of the actual medical conditions of the patient-subjects. Because of the prolonged secrecy surrounding the experiments, it was generally not known that two of the three University of California cases had been omitted from the 1950 analysis. The report also revealed in retrospect that all the patients were not hopelessly or terminally ill, as had been suggested in Langham's later public references, that some were still alive, and that some had been misdiagnosed.

In December 1972, Argonne National Laboratory's Center for Human Radiobiology (CHR), to whom Durbin had provided the names of surviving subjects, began a review of the data from all eighteen people who were injected with plutonium between 1945 and 1947. CHR was the national center designated by the AEC to do long-term follow-up of individuals with internally deposited radionuclides, primarily the radium dial painters. Argonne's follow-up plan for the plutonium experiments was to uncover the postinjection medical histories of all the subjects, obtain biological material from those still living, and exhume and study the bodies of those deceased in order to "provide data on the organ contents at long times after acquisition of plutonium."[144]

In 1973, three patients—Eda Charlton, John Mousso, and Elmer Allen—were admitted to the University of Rochester's metabolic ward for more excretion studies paid for by CHR. Elmer Allen had first been brought to Argonne, where an unsuccessful attempt had been made to detect plutonium by external counting techniques. In the course of his examination, however, CHR found subclinical bone "changes" that an Argonne radiologist characterized as "suggestive of damage due to radiation."[145]

Again there was no disclosure to the subjects that they were now being followed because they had been subjects of an experiment that had been unrelated to their medical care, an experiment in which there was continuing scientific interest. The 1974 AEC investigation concluded that, in the case of the surviving Rochester subjects, Dr. Waterhouse, who conducted the follow-up studies with these patients for Argonne, had not told them the purpose of the studies in 1973 because she believed "that disclosure might be harmful to them in view of their advanced age and ill health."[146] This suggests that Dr. Waterhouse had well-intentioned motivations for not being straightforward with the Rochester subjects. It also suggests that these subjects had not been told the truth about the experiments at the time the injections occurred, or that they had forgotten. According to Dr. Waterhouse, the studies were feasible without the subjects' knowledge of the true purpose of the research since these two patients "were accustomed to participating in clinical studies, unrelated to this matter, involving the collection of excretion specimens."[147] Elmer Allen's physician was told by CHR that the purpose of bringing Mr. Allen to Argonne's CHR and the University of Rochester for follow-up was interest in the treatment he received at the University of California in 1947 for his cancer.[148] This use of the term *treatment* in the information provided Mr. Allen's physician, which he presumably relayed to Mr. Allen and his family, was deceptive and manipulative; it implied that the injection Mr. Allen received had been given as therapy for his benefit.

The second component of this follow-up study was research on the exhumed bodies of deceased subjects. The 1974 AEC investigation concluded that the families were not informed that plutonium had been injected. Instead, they were told that "the purpose of exhumation was to examine the remains in order to determine the microscopic distribution of residual radioactivity from past medical treatment" and that the subjects had received an "unknown" mixture of radioactive isotopes.[149] The investigation concluded that such disclosure "could be judged misleading in that the radioactive isotopes were represented as having been injected as an experimental treatment for the patient's disease."[150] Thus, the families of the deceased subjects as well as those subjects still surviving were deceived by officials of the AEC.

A December 1972 intralaboratory memorandum, written by an Argonne investigator, instructs that "outside of CHR we will *never* use the word *plutonium* in regard to these cases. 'These individuals are of interest to us because they may have received a radioactive material at some time' is the kind of statement to be made, if we need to say anything at all."[151] Robert E. Rowland, the author of this memorandum, told Advisory Committee staff in 1995 that he had written this after he had been instructed earlier that month by Dr. James Liverman, director of the AEC's Division of Biomedical and Environmental Research, that "I could not tell the individuals that they were given plutonium. I protested that they must be given a reason for our interest in them, and I was told to tell them that they had received an unknown mixture of radioisotopes in the past, and that we wanted to determine if it was still in their bodies. Further, we were not to divulge the names of the institutions where they received this unknown mixture."[152] Dr. Rowland said he had received these instructions during a trip to Washington, D.C., to obtain approval and funding for the study.[153] Dr. Liverman told Advisory Committee staff that he has "no recollection of discussions with anyone in which some stricture would have been placed on what could be discussed with the patients. That is a medical ethics issue which would have been left to the physicians."[154]

This study was not brought to the attention of the Argonne Human Use Committee until November 1973, even though it had been established in January 1973. (See chapter 6 for a discussion of human use committees.) In a briefing for the 1974 AEC investigation, Dr. Liverman attributed this failure to bring the study before the Human Use Committee to the following factors: "(1) [Argonne's] opinion that the studies came under the scope of a protocol approved by that Committee in 1971. (2) The nature of the studies was to be suppressed to avoid embarrassing publicity for AEC."[155]

In 1974 the AEC informed at least two of the four living subjects—Eda Charlton and John Mousso—of the plutonium injections and had them sign documents to this effect. These documents did not provide any information on possible effects of the injections, although they did describe the purpose as having been "to determine how plutonium, a man-made radioactive material, is deposited and excreted in the human body."[156] One living patient, Jan Stadt, was not told, because it was her attending physician's opinion that her condition was precarious and that disclosure in this case would be "medically indefensible."[157] This judgment, like that of Dr. Waterhouse's, exemplifies how physicians of the time commonly managed the information they shared with their patients. Physicians typically told patients only what they thought it was helpful for them to know; if in the physician's judgment information might cause the patient to become upset or distressed, this was often considered reason enough to withhold it.[158] The judgment also suggests that Ms. Stadt, like Ms. Charlton and Mr. Mousso, had not been told the truth about the experiments at the time the injections occurred or that she had forgotten.

The AEC recommended that exhumations continue, but only with full disclosure to the subjects' next of kin.

The Boston Project
Uranium Injections

Human experiments conducted to measure the excretion and distribution of atomic weapons materials did not stop with the last of the injections at the University of California. The Boston Project human uranium-injection experiments were conducted from 1953 to 1957 at Massachusetts General Hospital (MGH) as part of a cooperative project between the hospital and the Health Physics Division of Oak Ridge National Laboratory. Eleven patients with terminal conditions were injected with uranium, although

data obtained from three of these subjects were never published.[159] The ORNL and the AEC undertook the Boston Project to obtain better data for the development of worker safety standards. One of the investigators wrote that the Boston Project would provide "a wonderful opportunity to secure 'human data' for the analysis and interpretation of industrial exposures."[160] The occupational standards for uranium at the time were based on animal data and on the experiment conducted at Rochester in the 1940s. No autopsy data were obtained from this earlier experiment at Rochester, however, since none of the patients had terminal diseases. Thus, wrote a Boston Project investigator, "the uncertainty, in so far as the distribution of uranium was concerned, was not reduced [by the Rochester experiment] or could not even be determined."[161]

The Boston Project involved a second purpose—the search for a radioisotope that would localize in a certain type of brain tumor—called glioblastomas—and destroy them when activated by a beam of neutrons. This had long been the research interest of Dr. William Sweet at MGH; at the time, these tumors were clearly diagnosable and 100 percent fatal, and there was no effective treatment. This research involved many radioisotopes over the years, most notably isotopes of boron and phosphorus. It is unclear whether Dr. Sweet would have tested uranium without ORNL's involvement—or whether it would have been made available to him by the AEC. Dr. Sweet has indicated to the Committee that he was interested in the potential of uranium as a therapeutic agent prior to being approached by the AEC about the possibility of conducting a joint project.[162]

The Boston Project produced data on the distribution of uranium in the human body that the earlier Manhattan District uranium studies had not provided. The data obtained indicated that uranium, at least at the dose levels used in the Boston Project, localized in the human kidney at higher concentrations than small animal data had predicted and that therefore the maximum permissible levels for uranium in water and air might be unsafe. Recommendations made by the investigators of the Boston Project for more conservative occupational standards were apparently not heeded, however. The accepted occupational levels for uranium became less rather than more conservative over the years, despite the findings of the Boston Project.[163]

Hopes that uranium would localize sufficiently in brain tumors to be of potential therapeutic use were unfulfilled. In a 1979 interview, Robert Bernard, one of the health physicists at ORNL most intimately involved with the study, was asked if during the experiment uranium was showing any promise as a treatment. "No, it concentrated in the kidney just like Rochester said back in the '40's. . . . They got brain tumor samples. There was very little uranium present, but Sweet was still wondering: maybe [it was] not a high enough dose."[164]

In a 1995 interview, Karl Morgan, head of the Health Physics Division of ORNL at the time of the Boston Project, indicated that the project was ultimately discontinued in 1957[165] because of the concerns of an ORNL health physicist:

He felt that the patients were given very large doses of uranium which our data had indicated—that is, the data we collected [at ORNL] in setting permissible doses—would be very harmful. . . . I immediately cancelled our participation in the program. Apparently, they were given doses that were many times the . . . permissible body burden.[166]

In their application to their radioisotope committee, MGH investigators clearly recorded that the proposed dose of 2.12 rem per week "exceeds maximum permissible exposure rate of 0.3 rem/week but [patients] are terminal."[167]

At least one of the subjects was selected for the distribution part of the study only. Reports describe the patients as "virtually all" having malignant brain tumors; newly available documents indicate that at least one patient injected with uranium did not have a brain tumor at all. An unidentified male, identity and age still unknown at the time of his death, became Boston Project subject VI when he "was brought to the Emergency Ward after being found unconscious. . . . No other information was obtainable."[168] According to his autopsy report, this patient was suffering from a subdural hematoma—a severe hemorrhage—on his brain. There was clearly no benefit intended for this patient from the injection of uranium, but there is evidence of harm attributable to the injection. His autopsy report records clinical evidence of mild kidney failure[169]

and pathological evidence of kidney nephrosis (damage to the kidney tubules) from the chemical toxicity of uranium metal.[170] The report also records that "the liver, spleen, kidneys and bone marrow showed evidence of radiation."[171]

Even for the patient-subjects with brain cancer, there was no expectation on the part of investigators that the experiment would benefit the subjects themselves. The object of the experiment was to test whether uranium would localize sufficiently in brain tumors to be of therapeutic value in the future. In order for uranium to have had therapeutic potential for patient-subjects, exposure to a reactor's neutron beam would have been necessary to then activate the uranium, *if* it had localized sufficiently in the tumors, which it did not. There was, however, no plan to expose these particular patient-subjects to a neutron beam; the goal was to see whether the concentration would justify further research that would involve exposure to a neutron beam. Most of the subjects were already comatose and "in the terminal phase of severe irreversible central nervous system disease."[172]

The doses used in the Boston Project were high; the lowest dose was comparable to the highest used in the earlier Rochester uranium experiment—a dose that had caused detectable kidney damage in one of the Rochester subjects. One document records that at least two Boston Project subjects, in addition to subject VI, had kidney damage at the time of death, although this document does not directly link this damage to the uranium injections.[173]

There is no discussion of consent in any of the Boston Project reports. It appears that ORNL left such considerations to Dr. Sweet and MGH. In an interim report, ORNL discusses the division of responsibility in the experiment: "It was agreed that the Y-12 Health Physics Department [at Oak Ridge] would prepare injection solutions and perform the analytical work associated with this joint effort. Massachusetts General Hospital agreed to select the patients, perform the injections, and care for the patients during the period of study."[174]

Dr. Sweet told the Advisory Committee in 1995 that it was his practice to obtain consent from patients or from their families and "scrupulously to give a patient all the information we had ourselves."[175] The Committee has not been able to locate any documents that bear on questions of disclosure or consent for this experiment.[176] The case of the Boston Project subject who was brought into the hospital after being found unconscious, and who, according to his autopsy report, was never identified and never regained consciousness, indicates that this rule was not applied universally.

CONCLUSION

From 1945 through 1947 Manhattan Project researchers injected eighteen human subjects with plutonium, five human subjects with polonium, and six human subjects with uranium to obtain metabolic data related to the safety of those working on the production of nuclear weapons. All of these subjects were patients hospitalized at facilities affiliated with the Universities of Rochester, California, and Chicago or at Oak Ridge. Another set of experiments took place between 1953 and 1957 at Massachusetts General Hospital, in which human subjects were injected with uranium. In no case was there any expectation that these patient-subjects would benefit medically from the injections.

At fifty years' remove, it is in some respects remarkable that so much information has survived that bears on the question of what the patient-subjects and their families were told. Particularly for the Manhattan Project plutonium experiments information is available, in large part because of the 1974 AEC inquiry in which interviews with principals of these experiments were conducted and records of these interviews maintained. At the same time, however, there are *significant* gaps in the record for all the experiments. Particularly where the evidence is skimpy, it is possible that some of the patient-subjects agreed to be used in nontherapeutic experiments. But the picture that emerges suggests otherwise. This picture is bolstered by the historical context. As we discussed in chapter 2, it was not uncommon in the 1940s and 1950s for physician-investigators to experiment on patients without their knowledge or consent, even where the patients could not benefit medically from the experimental procedures. This

context is referenced in a 1946 letter about the University of California injections: "These doctors state that the injections would probably be made without the knowledge of the patient. . . . Such injections were not divergent from the normal experimental method in the hospital. . . ."[177]

Here we present our conclusions about the ethics of these experiments, first for the set of experiments conducted between 1945 and 1947 and then for the experiment conducted from 1953 to 1957. Because the facts appear to be different in the different institutions at which these experiments took place, we summarize what we have learned about risk, disclosure, and consent at each location. We also analyze the ethical issues the experiments raise in common. In our analysis, we focus on whether the subjects consented to being used in experiments *from which they could not benefit medically,* and the extent to which the subjects were exposed to risk of harm. We also focus on the particular ethical considerations raised when research is conducted on patients at the end of their lives. All but one member of the Advisory Committee believe that what follows is the most plausible interpretation of the available evidence in light of the historical context.

With one exception, the historical record suggests that these patients-subjects were not told that they were to be used in experiments for which there was no expectation they would benefit medically, and as a consequence, it is unlikely they consented to this use of their person.

In the case of the plutonium experiments, there was no reason to think that the injections would cause any acute effects in the subjects. This was not true, however, in the case of the Rochester uranium experiments. Both the plutonium and the Rochester uranium experiments put the subjects at risk of developing cancer in ten or twenty years' time. In some cases, this risk was eliminated by the selection of subjects who were likely to die in the near future. The selection of subjects with chronic illnesses was also an apparent strategy to contain this long-term risk of cancer. However, some of these subjects lived for far longer than ten years, and some were misdiagnosed altogether. On the basis of available evidence, we could not conclude that any individual was or was not physically harmed as a result of the plutonium injections. There is some evidence that there were observable, subclinical bone changes of unclear significance in at least two surviving subjects who were followed up in 1953 and 1973 and in one deceased subject who was exhumed in 1973. The uranium injections at Rochester were designed to produce minimal detectable harm—that was the endpoint of the experiment. Such minimal damage is reported to have occurred in the sixth patient of the series.

In the case of Mr. Cade at Oak Ridge, a physician claiming to have injected Mr. Cade reported that his consent was not obtained. An apparently healthy man in his early fifties, Mr. Cade was put at some (probably small) risk of cancer by the plutonium injection.

At the University of Chicago, the only evidence that bears on disclosure and consent comes from an interview with a Chicago investigator conducted as part of the AEC's 1974 inquiry. The investigator was recorded as saying that in obtaining consent patients were told that the radioactive substance to be injected "was not necessarily for the benefit of the patients but might help other people."[178] This statement is misleading. It suggests that there was some chance these patient-subjects might benefit when there was no such expectation. At the same time, however, this statement suggests that the subjects at Chicago were told something. These subjects also were all apparently terminally ill and thus at no risk of developing plutonium-induced cancer; at least two of the three were known to have died within one year of the injection.

Misleading language was purportedly also used with subjects at the University of California, where a secondhand account suggests that subjects were told they were to be injected with a new substance that "had some properties like other substances that had been used to control growth processes in patients."[179] Language in a 1946 letter suggests that at least some of the injections at the University of California may have occurred altogether without the knowledge of the patients. In the case of Mr. Allen, one of the California subjects, two physicians attested that the experimental nature of the procedure had been explained to Mr. Allen and that he had consented. And yet Mr. Allen's physician was

subsequently informed that the follow-up studies were in relation to *treatment* Mr. Allen had received at the University of California. This suggests that, while Mr. Allen may have been told the procedure was experimental, it is not likely that he was told that the procedure was part of an experiment in which there was no expectation that he would benefit medically. Both Mr. Allen and Mr. Stevens survived long enough after injection to be at risk of plutonium-induced cancer.

All the available evidence suggests that none of the subjects injected with either plutonium or uranium at Rochester knew or consented to their being used as subjects in experiments from which they could not benefit. This evidence comes from recollections of some of the individuals who were involved with the plutonium injections, as well as documents about seminars and follow-up studies in the early 1950s suggesting that information about the experiments should be concealed from the subjects. Most of the subjects at Rochester had serious chronic illnesses. It is unclear how likely it was at the time that these patients would not survive more than ten years. A few of these subjects were still alive more than twenty years after the injections. None of the plutonium subjects but all of the uranium subjects were put at risk of acute effects from the experiment.

The purpose of the 1973 follow-up studies was withheld from two surviving subjects. Also, both Elmer Allen's physician and family members of deceased subjects were misled by AEC officials about the purpose of the follow-up studies. They were told that the follow-up was in relation to past medical treatment, which was not true.

It is unlikely that AEC officials would have lied about or otherwise attempted to conceal the purpose of the follow-up studies if at the outset the subjects had known and agreed to their being used as subjects in nontherapeutic experiments. It is also relevant that when the Atomic Energy Commission succeeded the Manhattan Project on January 1, 1947, officials decided to keep the plutonium injections secret. It appears that this decision was based on concerns about legal liability and adverse public reaction, not national security. The documents show that the AEC responded to the possibility that consent was not obtained in the plutonium experiments, as well as their lack of therapeutic benefit, by stating requirements for informed consent and therapeutic benefit for future research, while still keeping the experiments secret. As a result of the decision to keep the injections secret, the subjects and their families, as well as the general public, were denied information about these experiments until the 1970s.

The one likely exception to this picture of patients not knowing that they were used as subjects in experiments that would not benefit them is the polonium experiment conducted at Rochester. This is the one instance in which the patient-subjects are said to have volunteered after being told about "the general problem." Although there is no direct evidence that these subjects were told that the experiment was not for their benefit, the language of volunteering suggests a more forthright disclosure was made, more in keeping with the conventions in nontherapeutic research with healthy subjects than in research with patients (see chapter 2). We cannot reconcile the account of the polonium experiment with the historical record on the other injections.

The Advisory Committee is persuaded that these experiments were motivated by a concern for national security and worker safety and that, particularly in the case of the plutonium injections, they produced results that continue to benefit workers in the nuclear industry today.[180] However, with the possible exception of the polonium experiments, we believe that these experiments were unethical. In the conduct of these experiments, two basic moral principles were violated—that one ought not to use people as a mere means to the ends of others and that one ought not to deceive others—in the absence of any morally acceptable justification for such conduct. National security considerations may have required keeping secret the names of classified substances, but they would not have required using people as subjects in experiments without their knowledge or giving people the false impression that they or their family members had been given treatment when instead they had been given a substance that was not intended to be of benefit.

The egregiousness of the disrespectful way in which the subjects of the injection experiments and their families were treated is heightened by the fact that the subjects were hospitalized patients. Their being ill and institutionalized left them vulnerable to exploitation. As patients, it would have been reasonable for them to assume that their physicians were acting in their best interests, even if they were being given "experimental" interventions. Instead, the physicians violated their fiduciary responsibilities by giving the patients substances from which there was no expectation they would benefit and whose effects were uncertain. This is clearest at Rochester where at least the uranium subjects, and perhaps the plutonium subjects, were apparently the personal patients of the principal investigator.

Concern for minimizing risk of harm to subjects is evident in several of the planning documents relating to the experiments, an obligation that many of those involved apparently took seriously. At Chicago, for example, where the highest doses of plutonium were used, care was taken to ensure that all the subjects had terminal illnesses. In those cases where this concern for risk was less evident and subjects were exposed to more troubling risks, the moral wrong done in the experiments was greater. Where it was not reasonable to assume that subjects would be dead before a cancer risk had a chance to materialize, or in the case of the uranium injections at Rochester where acute effects were sought, the experiments are more morally offensive.

Consideration for the basic moral principle that people not be put at risk of harm is apparently what animated the decision to give higher doses to only "terminal" patients who could not survive long enough for harms to materialize. A person who is dying may have fewer interests in the future than a person who is not. This does not mean, however, that a dying person is owed less respect and may be used, like an object, as a mere means to the ends of others. There are many moral questions about research on patients who are dying; the desperation of their circumstances leaves them vulnerable to exploitation. At a minimum, nontherapeutic research on a dying patient without the patient's consent or the authorization of an appropriate family member is clearly unethical.

Uranium was also injected in eleven patients with terminal conditions at Massachusetts General Hospital in an experiment conducted jointly by the hospital and Oak Ridge National Laboratory from 1953 to 1957. ORNL's purpose was to obtain data for setting nuclear worker safety standards. A second purpose was to identify a radioisotope that would localize in brain tumors and destroy them when activated by a neutron beam. Although all but one of the patient-subjects had brain cancer, the limited purpose of the experiment—to establish whether uranium would localize sufficiently—meant that there was no expectation that patient-subjects might benefit medically from the uranium injections.

The uranium doses in the Boston experiment were comparable to or higher than the one that caused measurable physical harm in the Rochester subject. Boston subjects were apparently subjected to brain biopsies, presumably solely for scientific purposes. At least three Boston subjects showed kidney damage at the time of death. In one of these cases, a trauma victim who was found unconscious, the autopsy report recorded clinical evidence of some amount of kidney failure and pathological evidence of kidney damage due to the chemical toxicity of uranium.

The only evidence available about what the Boston subjects were told comes from 1995 testimony of one of the investigators, Dr. William Sweet, who said it was his practice to "give a patient all the information we had ourselves." Presumably this would have included that the injections had no prospect of benefiting the patient. The Boston Project was an instance in which high doses were given to dying patients. Some of these patients were comatose or otherwise suffering from severe, irreversible central nervous system disease. Unless these patients, or the families of comatose or incompetent patients, understood that the injections were not for their benefit and still agreed to the injections, this experiment also was unethical. There was no justification for using dying patients as mere means to the ends of the investigators and the AEC. In at least one case, this disrespectful treatment clearly occurred. The trauma victim who arrived at the hospital unconscious was used as a subject despite the fact that his identity was never known. Presumably he was not accompanied by any family or friends who might have authorized such a use of his body.

Only extraordinary circumstances can justify deception and the use of people as mere means by government officials and physicians in the conduct of research involving human subjects. In the case of the injection experiments, we see no reason that the laudable goals of the research could not have been pursued in a morally acceptable fashion. There is no reason to think that people would not have been willing to serve as subjects of radiation research for altruistic reasons, and indeed there is evidence of people writing to the AEC to volunteer themselves for just such efforts (see chapter 13).

That people are not likely to live long enough to be harmed does not justify failing to respect them as people. Concerns about adverse public relations and legal liability do not justify deceiving subjects, their families, and the public. Insofar as basic moral principles were violated in the conduct of the injection experiments, the Manhattan Engineer District, the AEC, the responsible officials of these agencies, and the medical professionals responsible for the injections are accountable for the moral wrongs that were done.

NOTES

1. Don Mastick, telephone interview with Steve Klaidman (ACHRE), 23 July 1995 (ACHRE No. IND-072395–F), 1.

2. L. H. Hempelmann, Los Alamos Laboratory Health Division Leader, to J. R. Oppenheimer, Director, Los Alamos Laboratory, 16 August 1944 ("Health Hazards Related to Plutonium") (ACHRE No. DOE-051094–A-17), 1.

3. J. R. Oppenheimer, Director, Los Alamos Laboratory, to L. H. Hempelmann, Los Alamos Laboratory Health Division Leader, 16 August 1944 ("Your memorandum of August 16, 1944") (ACHRE No. DOE-051094–A-17), 1.

4. L. H. Hempelmann, Los Alamos Laboratory Health Division Leader, to J. R. Oppenheimer, Director of the Los Alamos Laboratory, 29 August 1944 ("Medical Research Program") (ACHRE No. DOE-051094–A-17), 1.

5. Interview with Mastick, 23 July 1995, 1.

6. Glenn Seaborg, head of Chemistry Section C-1 of the Metallurgical Laboratory, to Robert Stone, Health Director of the Metallurgical Laboratory, 5 January 1944 ("Physiological Hazards of Working

with Plutonium") (ACHRE No. DOE–070194–A–3), 1.

7. Ibid.

8. Robert Stone, Health Director of the Metallurgical Laboratory, to Glenn Seaborg, Head of the Chemistry Section C-1 of the Metallurgical Laboratory, 8 January 1944 ("Hazards of Working with Plutonium") (ACHRE No. DOE–070194–A-4), 1.

9. Seaborg suggested that several milligrams of the first shipment of plutonium from Oak Ridge be sent on to Dr. Hamilton at Berkeley. A minute amount of plutonium was sent to Hamilton, who began his studies on rats in February 1944. Next came more animal work at Chicago, focusing on the toxic effects of plutonium, as well as its distribution in various tissues. These studies showed that plutonium, like radium, was a "bone-seeking" element, the potential deadly consequences of which radium had already demonstrated. Furthermore, these studies demonstrated that in rats, plutonium distributed itself in bone in a potentially more hazardous way than radium. J. Newell Stannard, *Radioactivity and Health: A History* (Oak Ridge, Tenn.: Office of Scientific and Technical Information, 1988), 1424.

10. Richard Rhodes, *The Making of the Atomic Bomb* (New York: Simon and Schuster, 1986), 547–548.

11. Ibid., 560.

12. The most likely route of worker exposure to plutonium would be inhalation. Hempelmann and others wrote to Oppenheimer in March 1945 that "the very important and difficult problem of detection of alpha active material in the lungs has been studied only at this project and here only on a very limited scale. This problem should be given much higher priority here and at other projects." L. H. Hempelmann, Los Alamos Laboratory Health Division Leader et al., to J. R. Oppenheimer, Director of the Los Alamos Laboratory, 15 March 1945 ("Medical Research of Manhattan District concerned with Plutonium") (ACHRE No. DOE–051094–A-17), 1. Inhalation experiments with rodents were undertaken, starting in 1944, at the University of California's Radiation Laboratory and the University of Chicago's Metallurgical Laboratory, although these studies did not result in extensive analysis of data until the latter half of the 1940s. W. H. Langham and J. W. Healy, "Maximum Permissible Body Burdens and Concentrations of Plutonium: Biological Basis and History of Development," in *Uranium–Plutonium–Transplutonic Elements,* eds. H. C. Hodge et al. (New York: Springer-Verlag, 1973), 576. Wright Langham wrote in 1945 that "if a limited amount of human tracer data are to form the basis of a method of diagnosing internal body contamination," it would be necessary "to assume that [plutonium] is metabolized in the same way regardless of the route of absorption or administration." Wright Langham, Los Alamos Laboratory Health Division, 28 July 1945 ("Report of Conference on Plutonium—May 14th and 15th")

(ACHRE No. DOE–051094–A-427), 29. Since the time of the experiments, it has become clearer that the deposition of plutonium in the body can differ in cases of chronic inhalation exposure versus other types of exposures.

13. Langham and Healy, "Maximum Permissible Body Burdens and Concentrations of Plutonium," 576.

14. L. H. Hempelmann, Los Alamos Laboratory Health Division Leader, to J. R. Oppenheimer, Director of the Los Alamos National Laboratory, 26 March 1945 ("Meeting of Chemistry Division and Medical Group") (ACHRE No. DOE–051094–A-17), 1.

15. J. R. Oppenheimer, Director, Los Alamos Laboratory, to Colonel S. L. Warren, 29 March 1945 ("We are enclosing a record of discussions . . .") (ACHRE No. DOE–051094–A-17), 1.

16. Samuel Bassett [attr.], undated ("Excretion of Plutonium Administered Intravenously to Man. Rate of Excretion in Urine and Feces with Two Observations of Distribution in Tissues") (ACHRE No. DOE–121294–D-10), 29.

17. Division of Biomedical and Environmental Research and Division of Inspection, AEC, 13 August 1974 ("Disclosure to Patients Injected with Plutonium") (ACHRE No. DOE–051094–A-586), 11.

18. Ibid.

19. Ibid., 10.

20. Wright Langham, Los Alamos Laboratory Health Division, to Hymer Friedell, Executive Officer of the Manhattan District's Medical Section, 6 April 1945 ("Although we sent you directions for the 49 experiment along with the material . . .") (ACHRE No. DOE–120894–E-1), 1.

21. Wilson O. Edmonds, AEC Resident Investigator, to Jon D. Anderson, Director, Division of Inspection, 15 July 1974 ("Division of Biomedical and Environmental Research, Headquarters—Request to Locate Mr. Ebb Cade") (ACHRE No. DOE–051094–A-611), 2.

22. Undated document ("Experiment I on P. 49+4") (ACHRE No. DOE–113094–B-5), 1.

23. The Committee uses names of subjects in this chapter only where the names were already a matter of public record.

24. "Experiment I on P. 49+4," 1.

25. Ibid.

26. Hannah E. Silberstein, University of Rochester, to Wright Langham, Los Alamos Laboratory Health Division, 25 October 1945 ("This letter is to report the injection on the second human *product* subject, HP-2 . . .") (ACHRE No. DOE–121294–D-19), 1.

27. W. H. Weyzen, 25 April 1974 ("Visit with Dr. Joe Howland, Chapel Hill Holiday Inn, April 24, 1974") (ACHRE No. DOE–121294–D-18), 1.

28. Hymer Friedell, interviewed by Steve Klaidman and Ron Neumann (ACHRE), transcript of audio recording, 23 August 1994 (ACHRE Re-

search Project Series, Interview Program File, Targeted Interview Project), 49–50.

29. "Experiment I on P. 49+4," 3.

30. Ibid.

31. Ibid., 2.

32. Captain David Goldring, Medical Corps, to Wright Langham, Los Alamos Laboratory Health Division, 19 September 1945 ("Enclosed is a brief resume of E. C.'s medical history . . .") (ACHRE No. NARA-082294-A-47), 1.

33. Karl Morgan, interviewed by Gil Whittemore and Miriam Bowling (ACHRE), transcript of audio recording, 6 January 1995 (ACHRE Research Project Series, Interview Program File, Targeted Interview Project), 147.

34. Edmonds to Anderson, 15 July 1974, 3.

35. "Experiment I on p. 49+4," 3.

36. On 7 May 1945 Germany had surrendered to the Allied forces. The Manhattan Engineer District continued on with the building and testing of the first atomic bomb (the first test was scheduled for July of that year).

37. Robert Stone, Health Director of the Metallurgical Laboratory, to Stafford Warren, Hymer Friedell et al., undated ("On Monday, May 14th, we plan to have an all day meeting dealing with plutonium . . .") (ACHRE No. NARA-082294-A-51), 1.

38. Wright Langham, Los Alamos Laboratory Health Division, 28 July 1945 ("Report of Conference on Plutonium—May 14th and 15th") (ACHRE No. DOE-051094-A-427), 29.

39. Colonel Hymer Friedell, Executive Officer of the Manhattan District's Medical Section, to L. H. Hempelmann, 11 April 1945 ("Enclosed is a protocol of the clinical experiment as we intend to carry it out . . .") (ACHRE No. DOE-121294-D-1), 1.

40. Ibid.

41. J. J. Nickson to R. S. Stone, 23 January 1946 ("Abstract of Monthly Report for January, 1946") (ACHRE No. DOE-051094-A), 1.

42. E. R. Russell and J. J. Nickson, 2 October 1946 ("The Distribution and Excretion of Plutonium in Two Human Subjects") (ACHRE No. DOE-051094-A-370), 1.

43. Ibid.

44. Ibid.

45. Ibid., 2.

46. Ibid.

47. Nickson to Stone, 23 January 1946, 1.

48. Sidney Marks, 3 May 1974 ("Interview with Dr. Leon Jacobson . . . by Marks and Miazga at about 1:30 p.m. on 4/16/74") (ACHRE No. DOE-121294-D-15), 2.

49. W. H. Weyzen, 25 April 1974 ("Visit with Edwin R. Russell, Savannah River Plant, April 23, 1974") (ACHRE No. 121294-D-17), 1.

50. Andrew H. Dowdy, Director of AEC Rochester Project ("Proposed Research Program and Budget: July 1, 1947—July 1, 1948") (ACHRE No. DOE-061794-B-16).

51. William F. Bale, Head of Special Problems Division, undated ("Contributions of the Division of Special Problems to the Manhattan Project") (ACHRE No. DOE-113094-B), 1.

52. L. H. Hempelmann and Wright H. Langham, undated ("Detailed Plan of 'Product' Part of Rochester Experiment") (ACHRE No. 121294-D-2), 5.

53. W. H. Langham, undated ("Revised Plan of 'Product' Part of Rochester Experiment") (ACHRE No. DOE-121294-D-3), 2.

54. The choice not to use subjects suffering from malignant conditions is discussed retrospectively in a partial draft version of the 1950 report (probably written by Dr. Bassett). This discussion was not included in the final version of the report:

The individuals chosen as subjects for the experiment were a miscellaneous group of male and female hospital patients for the most part with well established diagnoses. Preference was given to those who might reasonably gain from continued residence in the hospital for a month or more. . . . Patients with malignant disease were also omitted from the group on the grounds that their metabolism might be affected in an unknown manner.

Bassett, "Excretion of Plutonium Administered Intravenously to Man," 2.

55. Ibid.

56. Wright Langham et al., 20 September 1950 ("Distribution and Excretion of Plutonium Administered Intravenously to Man") (ACHRE No. DOE-070194-A-18), 10.

57. Wright Langham, 27 September 1957 ("Proceedings of the Second Annual Meeting on Bio-Assay and Analytical Chemistry: October 11 and 12, 1956—Panel Discussion of Plutonium") (ACHRE No. DOE-120894-C-1), 80.

58. W. H. Langham et al., "The Los Alamos Scientific Laboratory's Experience with Plutonium in Man," *Health Physics* 8 (1962): 755.

59. Addison's disease is an endocrine disease produced by adrenal gland failure. Today this disease is treated with steroid therapy that was developed in the 1940s and that was extremely expensive at the time of the experiments. HP-6, diagnosed with Addison's, was given steroid treatment as part of his care at the University of Rochester; he lived until 1984.

Scleroderma is a collagen-vascular disease that can produce extreme pain, especially in the hands; can affect eating and swallowing if the esophagus is involved; and eventually leads to organ failure and death. Steroids are the treatment of choice today, but if this disease is not well controlled it can still be fatal. HP-8, who was diagnosed with scleroderma, lived until 1975.

60. Bassett, "Excretion of Plutonium Administered Intravenously to Man," 2. Her provisional diagnosis according to this report was mild hepatitis and malnutrition. Ibid, 18. Her medical records indicate, however, that she had symptoms related to nutritional

deficiencies, which appear to have been alleviated with proper diet and rest. Strong Memorial Hospital, 20 December 1945 ("Discharge Summary Form") (ACHRE No. DOE-051094-A-612), 1.

61. Wright Langham, Los Alamos Laboratory Health Division, to Samuel Bassett, Head of Metabolism Ward of Strong Memorial Hospital, 13 March 1946 ("Your letter of February 27 regarding Hp 11 was startling, to say the least . . .") (ACHRE No. DOE-121294-D-4), 1.

62. Document dated 17 April 1974 ("Comments on Meeting with Dr. Hempelmann on April 17, 1974") (ACHRE No. DOE-121294-D-16), 1.

A 1955 letter from Dr. Hempelmann to the AEC's Division of Biology and Medicine (discussed in more detail in chapter 13) indicates Hempelmann's belief that, in general, patients could be easily deceived about the true research purpose of a medical intervention. In this letter, Hempelmann (who was by then professor of experimental radiology at Rochester) is proposing that researchers present themselves as life insurance agents to AEC workers as a ruse, in order to conceal the true purpose of follow-up medical examinations. He observes that it would be more difficult to deceive workers than it would be to mislead patients in a hospital:

If you feel that the physical examinations are vital to the survey, then, perhaps, you could offer to pay the people to compensate them for the time and effort that they will spend on the part of your alleged survey for the insurance company. They would think they were getting something for nothing and might not feel that you were worried or they were seriously ill. I don't know if these ideas are helpful at all. It is more difficult to find an excuse for these individual workers than it is in the case of patients who were treated for something or other at a hospital.

Louis Hempelmann, University of Rochester, to Charles Dunham, Director, AEC Division of Biology and Medicine, 2 June 1955 ("I did not have an opportunity . . .") (ACHRE No. DOE-092694-A), 1.

63. Patricia Durbin, 9 December 1971 ("Report on Visit to Rochester") (ACHRE No. DOE-121294-D-12), 1.

64. Patricia Durbin, 10 December 1971 ("Dr. Wright Langham, of the Los Alamos Scientific Laboratory, was the biochemist who performed the Pu analyses . . .") (ACHRE No. DOE-121294-D-13), 1.

65. "Comments on Meeting with Dr. Hempelmann on April 17, 1974," 1.

66. Langham further instructed Rochester to look for the following longer-term "symptoms" in the examination of the patients: "Judging from the recent observations that Robley Evans has made, a generalized osteitis with rarefaction of the bones of the feet, the jaw and the heads of the long bones with coarsening of the trabeculae are the most likely symptoms." Wright Langham, Los Alamos Health Division, to

Dr. Joe Howland, Chief of University of Rochester's Division of Medical Services, 2 October 1950 ("I am very glad to hear that you will manage to get follow-ups on the two subjects . . .") (ACHRE No. DOE-121294-D-11), 1.

67. Wright Langham, Los Alamos Laboratory Health Division, to Samuel Bassett, Head of Metabolism Ward of Strong Memorial Hospital, 25 October 1946 ("I just received a shipment of samples which I am sure are the ones you collected on HP-3 . . .") (ACHRE No. DOE-121294-D-5), 1.

68. Samuel Bassett et al., 19 July 1948 ("The Excretion of Hexavalent Uranium Following Intravenous Administration II. Studies on Human Subjects") (ACHRE No. CON-030795-A-1), 8.

69. Andrew H. Dowdy, Director, Manhattan Department, University of Rochester, to the Area Engineer, Rochester Area, 22 October 1946 ("Clearance of Material for Seminar") (ACHRE No. DOE-120994-A-4), 1.

70. Madison Square Area Engineer, 24 October 1946 ("Uranium Studies in Humans") (ACHRE No. DOE 120994-A-4), 1.

71. Robert M. Fink ("Biological Studies with Polonium, Radium, and Plutonium") (ACHRE No. CON-030795-A-2), 122.

72. K. Z. Morgan, Oak Ridge National Laboratory Health Physics Division, to R. S. Stone, Health Director of the Metallurgical Laboratory, 5 May 1945 ("Tolerance Values for Polonium Used at Clinton Laboratories") (ACHRE No. DOE-113094-B-6), 2.

73. Fink, "Biological Studies with Polonium, Radium, and Plutonium," 122.

74. A supplemental volume contains a chapter on the development of human subject research at the University of California at Berkeley and San Francisco.

75. Hamilton's work with plutonium had begun in 1942 with support from the Office of Scientific Research and Development; it was later supported by the Manhattan Engineer District.

76. Joseph Hamilton, Radiation Laboratory of University of California at Berkeley, to Colonel E. B. Kelly, 28 August 1946 ("Summary of Research Program for Contract #W-7405-eng-48-A") (ACHRE No. DOE-113094-B-8), 2.

77. Joseph Hamilton, 11 January 1945 ("Proposed Biochemical Program at University of California") (ACHRE No. IND-071395-A-14), 2.

78. Ibid.

79. At least eleven patients were injected with columbium (later renamed niobium) or zirconium between 1948 and 1950. These experiments appear to have been outside the federal effort.

80. Joseph Hamilton, 10 May 1945 ("Progress Report for Month of May 1945") (ACHRE No. DOE-072694-B-65), 4.

81. Joseph Hamilton, 14 June 1945 ("Progress Report for Month of June 1945") (ACHRE No. DOE-072694-B-66), 4.

82. Ibid.

83. Joseph G. Hamilton, Radiation Laboratory, University of California, Berkeley, to Captain Joe W. Howland, 23 April 1946 ("The problems of the research program . . .") (ACHRE No. DOE-120894–E-40), 2.

84. Joseph G. Hamilton, Radiation Laboratory, University of California, Berkeley, to Robert Stone, Metallurgical Laboratory, 7 July 1945 ("I am writing concerning our experimental subject . . .") (ACHRE No. IND-071395–A), 1.

85. Joe W. Howland, First Lieutenant, Medical Corps, to the Area Engineer, California Area, 12 July 1945 ("Status of Experimental Subject") (ACHRE No. IND-071395–A), 1.

86. Kenneth Scott, interviewed by Sally Hughes (University of California Oral History Project), transcript of audio recording, 17 December 1979, 49–50.

87. Ibid.

88. Hamilton, "Progress Report for Month of June 1945," 4.

89. Joseph Hamilton, 14 September 1945 ("Progress Report for Month of September 1945") (ACHRE No. DOE-072694–B-67), 5.

90. "Mercy Flight Brings Aussie Boy Here: Suffering From Rare Bone Ailment, He Seeks U.S. Treatment," *San Francisco Examiner*, 16 April 1946, 1.

91. In addition to this injection, which was not performed for his benefit, the child also received superficial external radiation (five doses of 250 rad over five days) for palliation of his pain. A 1995 report written by an ad hoc committee at the University of California at San Francisco (UCSF) described the child's prognosis as having been "grave with palliation the only option." With that in mind, superficial irradiation was performed to reduce the patient's pain, not to destroy the sarcoma of the right leg. University of California at San Francisco, February 1995 ("Report of the USCF Ad Hoc Fact Finding Committee on World War II Human Radiation Experiments, February 1995, Appendix 19: Summary of the medical record of CAL-2") (ACHRE No. UCSF-022495–A-6), 3.

92. UCSF, "Report of the USCF Ad Hoc Fact Finding Committee," 27.

93. Loren J. Larson, Assistant in Orthopedic Surgery, University of California Hospital, 11 June 1946 ("To Whom It May Concern . . .") (ACHRE No. DOE-051094–A-605), 2.

94. Joseph Hamilton, Radiation Laboratory of the University of California at Berkeley, to Samuel K. Allison, 11 September 1945 ("Plans for Future Biological Research") (ACHRE No. IND-071395–A-2), 3.

95. UCSF, "Report of the USCF Ad Hoc Fact Finding Committee," 27.

96. Joseph Hamilton, Radiation Laboratory of the University of California at Berkeley, to John Fulton, Historical Library, Yale University Medical Center,

19 July 1946 ("Inasmuch as both the Lawrence brothers are away at the moment, I thought it best that I answer your letter of July 16, 1946, to John . . .") (ACHRE No. DOE-122294–A-3), 1.

97. T. S. Chapman, Chief of Operations Branch, Research Division, to Area Engineer, Berkeley Area, 30 December 1946 ("Human Experiments") (ACHRE No. DOE-112194–D-3), 1.

98. Form dated 2 May 1946 ("Consent for Operation and/or Administration of Anaesthetic") (ACHRE No. DOE-051094–A-604), 1.

99. Colonel K. D. Nichols, Corps of Engineers, to the Area Engineer, California Area, 24 December 1946 ("Administration of Radioactive Substances to Human Subjects") (ACHRE No. DOE-113094–B-2), 1. This order followed a renewed request to the Army by Hamilton for additional plutonium, "to be used for certain human studies," and a further progress report on the injection of Albert Stevens.

100. John L. Burling, AEC Legal Division, to Edwin E. Huddleson, AEC Deputy General Counsel, 7 March 1947 ("Clinical Testing.") (ACHRE No. DOE-051094–A-468), 1.

101. Undated document ("CH-3607 . . . Excerpts from statements of reviewers") (ACHRE No. 113094–B-9), 1.

102. Ibid.

103. Ibid. For discussion of classification levels, see chapter 13.

104. "Off Project" probably refers to work not sponsored by the AEC.

105. Major B. M. Brundage, Chief, Medical Division, to Declassification Section, 19 March 1947 ("Clearance of Technical Documents") (ACHRE No. DOE-113094–B-4), 1.

106. Hoylande D. Young, Argonne National Laboratory, to Charles A. Keller, 25 July 1947 ("Declassification has been refused for the following reports . . .") (ACHRE No. NARA-050995–A-6), 1.

107. Carroll Wilson, AEC General Manager, to Robert Stone, University of California Medical Center, 12 August 1947 ("Declassification of Biological and Medical Papers") (ACHRE No. DOE-061394–A-111), 1.

108. Wright Langham, Los Alamos Laboratory Health Division, to Stafford Warren, University of California, 1 July 1950 ("Dr. Bassett has been here and helped me finish the semi-final draft of the Plutonium Report . . .") (ACHRE No. DOE-082294–B-72), 1.

109. Wright Langham, Los Alamos Laboratory Health Division, to Joe W. Howland, Chief, Division of Medical Services, University of Rochester School of Medicine and Dentistry, 15 April 1950 ("I am curious to hear your reaction to the names that I sent you . . .") (ACHRE No. DOE-082294–B-73), 1.

110. Andrew H. Dowdy, Director of the Manhattan Department, University of Rochester, to Norris E. Bradbarry [*sic*], Director of the Los Alamos Labo-

ratory, 18 February 1947 ("Dr. Wright Langham and Dr. Samuel Bassett were discussing with me today the technical details relative to writing the report . . .") (ACHRE No. DOE-121294-D-6), 1.

111. Langham to Warren, 1 July 1950, 1.

112. Walter D. Claus, Acting Chief, Biophysics Branch, AEC Division of Biology and Medicine, to Wright Langham, Los Alamos Laboratory Health Division, 30 August 1950 ("You will be pleased to learn that Dr. Shields Warren has approved your report for CONFIDENTIAL classification . . .") (ACHRE No. DOE-082294-B-2), 1.

113. It is not clear when CH-3607, the report Dr. Friedell recommended for declassification in December 1946, was declassified. The copy retrieved by the Committee bears a 31 December 1946 declassification date and no indication of subsequent reclassification. Russell and Nickson, "The Distribution and Excretion of Plutonium in Two Human Subjects," 1. In 1956 Dr. Langham made a brief reference to fifteen experimental subjects at an unclassified technical conference. Langham, "Proceedings of the Second Annual Meeting on Bio-Assay and Analytical Chemistry," 80. In 1951, a report, based on Metallurgical Laboratory Memorandum MUC-ERR-209 ("Distribution and Excretion of Plutonium") appeared in a volume of the public Manhattan District research history.

114. While the Wilson letters do not expressly reference the plutonium experiments, the context seems to leave little question that the policies stated in the letters were arrived at with the plutonium experiments in mind. In 1974, when asked what steps had been taken when the plutonium injections had been brought to the attention of the AEC, Shields Warren, who became director of the AEC's Division of Biology and Medicine in late 1947, said that it had been decided "that the rules [should be] properly drawn up by the . . . Human Applications Isotope Committee . . . so that use without full safeguards could not occur, and that . . . nothing of the sort could happen in the future." Shields Warren, interviewed by L. A. Miazga, Sidney Marks, and Walter Weyzen (AEC), transcript of audio recording, 9 April 1974 (ACHRE No. DOE-121294-D-14), 10.

115. Carroll Wilson, AEC General Manager, to Stafford Warren, University of California, 30 April 1947 ("This is to inform you that the Commission is going ahead with its plans to extend the medical research contracts . . .") (ACHRE No. DOE-051094-A-439), 2.

116. Carroll Wilson, AEC General Manager, to Robert Stone, University of California Medical School, 5 November 1947 ("Your letter of September 18 regarding the declassification of biological and medical papers was read . . .") (ACHRE No. DOE-061395-A-112), 1.

117. Dowdy, "Proposed Research Program and Budget: July 1, 1947–July 1, 1948," 25.

118. A December 1947 memorandum from Dr. Bassett recorded:

In the autumn of 1945 the Section on Human Metabolism was activated under your direction at the request of the Manhattan Engineer District to carry out certain tracer studies with long-lived isotopes. As you know, this program was discontinued in the spring of 1947 under a directive from the Atomic Energy Commission although we were instructed to keep the personnel of the section intact. When this directive was received, it was anticipated that follow-up studies on the several subjects of the original investigation would provide occupation for the employees of the section.

Samuel H. Bassett, Section on Human Metabolism, University of Rochester, to William F. Bale, Head of Special Problems Division, University of Rochester, 2 December 1947 ("Proposal of Work for Metabolism Section") (ACHRE No. DOE-121294-D-7), 1.

Dr. Bassett proposed an interim activity for the employees of the section—a study of certain aspects of radiation injury. This was approved by Bale until "the project research program of the Metabolism Section . . . with regard to tracer studies with heavy elements is clarified." William F. Bale, Head of Special Problems Division, University of Rochester, to Andrew H. Dowdy, Director of AEC's Rochester Project, 3 December 1947 ("Program of Work for Metabolism Section") (ACHRE No. DOE-121294-D-8), 1.

119. Gilbert Whittemore, 3 March 1995 ("Shields Warren Papers: A Cumulative Update of Excerpts") (ACHRE No. BU-030395-A-1), 3.

120. Ibid.

121. Interview with Warren, 9 April 1974, 11.

122. Ibid.

123. Ibid. According to Dr. Durbin, it is likely that the "other substances" referred to were probably phophorus 32 and strontium 89, which were used at the University of California between 1941 and 1944 as experimental tracers or for palliation of pain. Dr. Patricia Durbin, telephone interview with Miriam Bowling (ACHRE), 2 August 1995 (ACHRE No. ACHRE-081095-A), 1.

124. Undated note in medical record of CAL-A from "K.G.S." (Ken G. Scott [attr.]) ("The day after solution is injected . . .") (ACHRE No. UCLA-111094-A-1), 1.

125. Telephone interview with Durbin, 2 August 1995, 1.

126. Lori Hefner; telephone interview by John Kruger (ACHRE), 6 July 1995 (ACHRE No. IND-070695-A), 1.

127. Note in medical record of CAL-3 dated 18 July 1947 ("Elmer Allen Chart") (ACHRE No. DOE-051094-A-615), 2.

128. Wilson to Warren, 30 April 1947, 2.

129. UCSF, "Report of the USCF Ad Hoc Fact Finding Committee, Appendix 20: Summaries of the medical record of CAL-3," 3–4.

130. Ibid., 4. If the diagnosis was correct, surgi-

cal amputation would have been appropriate treatment at the time to completely excise the tumor.

131. B. V. Low Beer et al., Radiation Laboratory, University of California, Berkeley, 15 March 1948 ("Comparative Deposition of Zr-95 in a Reticulo-Endothelial Tumor to Normal Tissues in a Human Patient") (ACHRE No. DOE-101194–B-4), 4.

132. Ibid. The test dose was administered to the patient just twenty-four hours prior to the midthigh amputation of her leg for cancer.

133. Shields Warren, Director of AEC's Division of Biology and Medicine, to Albert H. Holland, Jr., AEC Medical Adviser, 19 August 1948 ("Review of Document") (ACHRE No. DOE-101494–B), 1.

134. Albert H. Holland, Jr., AEC Medical Adviser, to Shields Warren, Director of AEC's Division of Biology and Medicine, 9 August 1948 ("Review of Document") (ACHRE No. DOE051094–A), 1.

135. Langham to Howland, 2 October 1950, 1.

136. Wright Langham, Los Alamos Laboratory Health Division, to Albert H. Holland, AEC Director of Research and Medicine, 20 March 1950 ("It seems that I really fouled up regarding my promise to you at the Washington meeting . . .") (ACHRE No. NARA-082294–A-155), 1.

137. L. H. Hempelmann, University of Rochester, to Charles Dunham, AEC Division of Biology and Medicine, 23 May 1953 ("There are several things on my mind that I would like to bring to your attention . . .") (ACHRE No. DOE-041495–A-1), 1.

138. Walter D. Claus, Acting Chief of the Biophysics Branch, AEC Division of Biology and Medicine, to Charles L. Dunham, Chief, Medical Branch, 31 August 1950 ("Physical Examinations at Rochester") (ACHRE No. DOE-051094–A-471), 1.

139. Interview with Warren, 9 April 1974, 8.

140. Patricia W. Durbin, University of California, to William E. Lotz, AEC Division of Biology and Medicine, 13 September 1968 ("You will never guess what I found today . . .") (ACHRE No. DOE-051094–A-606), 1.

141. Ibid.

142. Ibid.

143. Patricia Durbin, 1972 ("Plutonium in Man: A New Look at the Old Data") (ACHRE No. DOE-051094–A-160), 469.

144. R. E. Rowland, Argonne National Laboratory's Center for Human Radiobiology, 8 November 1973 ("Plutonium Studies at the Center for Human Radiobiology [CHR]") (ACHRE No. DOE-051094–A-608), 4.

145. I. E. Kirch, Radiological and Environmental Research Division, Argonne National Laboratory, 13 June 1973 ("Center for Human Radiobiology: Radiologist's Report") (ACHRE No. DOE-051094–A-616), 1. The report records: "In the proximal portions of both humeri as well as in the adjacent acromions, there are some changes in the trabeculae which are consistent with findings in early radium deposition, but not yet completely specific.

The mandible shows abnormal trabeculae, suggestive of damage due to radiation."

Subclinical bone changes were also observed in a deceased subject who was exhumed for the Argonne study. The same radiologist summarized that an "abnormality is present, namely, that there are very many very small very dense deposits on the surfaces of a number of the bones, and other such deposits in the soft tissues very close to the bone surfaces. This abnormality is attributed to the plutonium which has been administered during the subject's life. The radiographic pattern is unique." I. E. Kirch, Radiological and Environmental Research Division, Argonne National Laboratory, 15 November 1974 ("Center for Human Radiobiology: Radiologist's Report") (ACHRE No. DOE-051094–A-618), 1.

146. AEC Division of Biomedical and Environmental Research and Division of Inspection, 13 August 1974 ("Disclosure to Patients Injected With Plutonium") (ACHRE No. DOE-051094–A-586), 10.

147. Ibid.

148. Ibid.

149. Ibid.

150. Ibid.

151. Robert E. Rowland, Argonne National Laboratory, to H. A. Schultz, 21 December 1972 ("Plutonium Cases") (ACHRE No. DOE-080795–A), 1.

152. Robert E. Rowland to Miriam Bowling (ACHRE Staff), 7 August 1995 ("Attached is the memo of December 21, 1972 . . .") (ACHRE No. DOE-080795–A), 1.

153. Ibid.

154. James L. Liverman to Miriam Bowling (ACHRE Staff), 20 August 1995 ("With your fax of August 9 was included . . .") (ACHRE No. IND-082095–A), 1.

155. James L. Liverman, 29 April 1974 ("Briefing on Plutonium Project by Dr. James L. Liverman on April 29, 1974") (ACHRE No. DOE-051094–A-196), 8. The 1971 protocol referred to in this briefing had covered a follow-up project involving the radium dial painters. Although the procedures for the two follow-up studies were similar, the original conditions of exposure were quite different. The radium dial painters, unlike the plutonium-injection subjects, had not been chosen as subjects in a carefully planned medical experiment organized by the government. They had been exposed either occupationally as dial painters or therapeutically as patients receiving one of a variety of prewar radium treatments.

156. Signed form dated 28 August 1974 ("Acknowledgement of Disclosure") (ACHRE No. DOE-051094–A-619), 1.

157. Document dated 24 May 1974 ("Patients Injected with Plutonium [Draft Report of 5-24-74]") (ACHRE No. DOE-051094–A-607), 1.

158. There is some evidence suggesting that at least one subject had a serious emotional reaction to the news, many years after the fact, that she had been injected with plutonium. This suggests that physicians

involved in the follow-up had cause to be concerned about how at least some patients might respond to knowledge of the injections.

159. K. F. Eckerman to Barry A. Berven, 7 January 1994 ("The Boston-Oak Ridge Uranium Study") (ACHRE No. DOE-051094-A-425), 1.

160. John C. Gallimore, Associate Health Physicist, to Dr. W. H. Sweet, Massachusetts General Hospital, 22 March 1954 ("First Results of Uranium Distribution and Excretion Study") (ACHRE No. NARA-082294-A-35), 1.

161. S. R. Bernard, "Maximum Permissible Amounts of Natural Uranium in the Body, Air and Drinking Water Based on Human Experimental Data," *Health Physics* 1 (1958): 288–305.

162. According to the 1957 interim report on the study, it was Harold Hodge of the University of Rochester's Atomic Energy Project, who had been involved with the MED metabolism work at Rochester, who ultimately coordinated the beginning of the joint research. S. R. Bernard and E. G. Struxness, 4 June 1957 ("A Study of the Distribution and Excretion of Uranium in Man: An Interim Report") (ACHRE No. DOE-051094-A-369), 3.

163. Bernard, "Maximum Permissible Amounts of Natural Uranium in the Body, Air and Drinking Water Based on Human Experimental Data," 296–298; *Standards for Protection Against Radiation*, 9 C.F.R. 20 (1958–1994).

164. Robert Bernard, interviewed by J. Newell Stannard, transcript of audio recording, 17 April 1979 (ACHRE No. DOE-061794-A), 8.

165. A continuation of the study *at lower doses* was proposed by the ORNL in 1958; it is unclear whether this project went forward. Karl Morgan, Director of ORNL's Health Physics Division, to William Sweet, Massachusetts General Hospital, 16 July 1958 ("Your help in our cooperative study on the distribution and excretion of uranium in man has been of great value to us . . .") (ACHRE No. DOE-021695-A-1), 1. A study similar to the one proposed by the ORNL in 1958 may have taken place during the mid-1960s at Argonne Cancer Research Hospital. K. Z. Morgan to W. H. Jordan, 3 September 1963 ("Proposed Study of Distribution and Excretion of Enriched Uranium Administered to Man") (ACHRE No. DOE-051094-A-620), 1.

166. Interview with Morgan, 6 January 1995, 118–119.

167. Form dated 3 November 1953 ("Application for Approval of Radioactive Isotopes: Massachusetts General Hospital") (ACHRE No. MGH-030395-A-1), 4.

168. Leonard Atkins, M.D., 26 June 1954 ("Necropsy No. ___: June 26, 1954 at 12:30 p.m.") (ACHRE No. DOE-050895-D-1), 6.

169. Ibid., 1.

170. Ibid. The "Anatomic Diagnoses" include "Uranium nephrosis, acute."

171. Ibid., 5.

172. Bernard and Struxness, "A Study of the Distribution and Excretion of Uranium in Man: An Interim Report," 6.

173. Undated document ("#1 Cloudy swelling of the epithelium of proximal and distal convoluted tubules . . .") (ACHRE No. DOE-050895-D-2), 1. The document records a diagnosis for the two additional patients as "acute nephrosis," and for subject VI, as "severe subacute nephrosis."

174. Bernard and Struxness, "A Study of the Distribution and Excretion of Uranium in Man: An Interim Report," 4.

175. William Sweet, interviewed by Gil Whittemore (ACHRE), transcript of audio recording, 8 April 1995 (ACHRE Research Project Series, Interview Program File, Targeted Interview Project), 46.

176. By the end of the Committee's deliberations, MGH had not yet completed its search for the patient records of the Boston Project subjects.

177. Chapman to Area Engineer, Berkeley Area, 30 December 1946, 1.

178. Weyzen, "Visit with Edwin R. Russell, Savannah River Plant, April 23, 1974," 1.

179. Interview with Warren, 9 April 1974, 11.

180. The relatively small population that has been exposed to substantial levels of plutonium precludes definitive conclusions about risks to humans, but the available evidence clearly suggests that an epidemic of cancer of the magnitude that afflicted the radium dial painters from an earlier era has not occurred in plutonium workers. In the case of the radium dial painters, the unprotected handling of only a few pounds of radium led to hundreds of deaths; in contrast, studies of plutonium workers suggest that to date there is no definite excess mortality in this population. A forty-two-year follow-up of twenty-six Manhattan Project workers who worked with plutonium found a total of seven deaths, including three cancers (two lung and one osteogenic sarcoma), a substantially lower mortality rate than expected based on the U.S. population. The authors concluded that "the diseases and physical changes noted in these persons are characteristic of a male population in their 60s." G. L. Voelz and J. N. Lawrence, "A 42-year Medical Follow-up of Manhattan Project Plutonium Workers," *Health Physics* 61 (1991): 181–190. A larger study of 15,727 LANL workers followed through 1990, some of whom had plutonium exposures, found no cause of death significantly elevated among the plutonium-exposed workers compared with unexposed workers, although there was a nonsignificant 78 percent elevation in lung cancer (a site that is directly exposed) and a single osteogenic sarcoma, a rare cancer that has been associated with plutonium exposure in animal studies. L. D. Wiggs, E. R. Johnson, C. A. Cox-DeVore and G. L. Voelz, "Mortality Through 1990 Among White Male Workers at the Los Alamos National Laboratory: Considering Exposures to Plutonium and Ionizing Radiation," *Health Physics* 67 (1994): 577–588. Another study

of 5,413 workers at the Rocky Flats Nuclear Weapons Plant found elevated risks for various cancers comparing workers with body burdens of 2 nanocuries (nCi) or greater, but with wide uncertainties; no excesses were seen for bone or liver cancers. The authors concluded that "these findings suggest that increased risks for several types of cancers cannot be ruled out at this time for individuals with plutonium body burdens of $\geq$ 2 nCi. Plutonium-burdened individuals should continue to be studied in future years." G. S. Wilkinson et al., "Mortality Among Plutonium and Other Radiation Workers at a Plutonium Weapons Facility," *American Journal of Epidemiology* 125 (1987): 231–250.

The AEC Program
of Radioisotope Distribution

AT the dawn of the atomic age, many people hoped for dramatic advances in medicine, akin to the new miracle drug penicillin. Many of these hopes have been fulfilled. Radioisotopes have become remarkable tools in three areas. First, as their travels within the body are "traced," radioisotopes provide a map of the body's normal metabolic functions. Second, building on tracer research, diagnostic techniques distinguish between normal and abnormal functioning. Finally, radioisotopes, carried by the body's own processes to abnormal or cancerous cells, can deliver a lethal dose of radiation to those undesirable cells. By supplying radioisotopes and supporting their use, the Atomic Energy Commission (AEC) actively promoted the research needed to achieve this progress.

The growth in the applications of radioisotopes involved thousands of experiments using radioisotopes. No feasible method was found to review in detail the vast number of individual radioisotope experiments in the Advisory Committee's database. This was due not only to the large number of experiments, but also to the scarcity of information about many of the individual experiments. Both consent and exact dose levels were often not discussed in published work; no federal repository was found that had collected records documenting these aspects of experiments. Given the decentralized structure

of American medicine, it is not surprising that the Committee found that records on consent and exact dose, if they exist, would still be held at the local institutions conducting research or perhaps even in the private papers of physicians and scientists. Even when records were found at the local level, there was little documentation about consent.

Thus, for the largest group of human radiation experiments, little documentation remains, and a meaningful examination of all such experiments was not possible. The Committee instead chose to focus its energies in two directions: examining the overall system of oversight created by the federal government and examining small subsets of radioisotope experiments that posed significant ethical issues. The first effort led to this chapter, an overview of the system created by the federal government to monitor radioisotope experiments. The second effort led to the case study on experiments involving children (chapter 7) since those raised questions of both additional biological risk and justification for doing nontherapeutic research on minors.

The AEC's isotope distribution program was faced with three essential ethical questions. The most immediate question concerned the allocation of a scarce resource. Given the likelihood that demand for radioisotopes would exceed supply, how should priorities be set? The question involved not simply the choice among competing proposals for "human uses" (including experimentation, treatment of disease, and diagnosis), but between human uses and other kinds of uses (for example, basic scientific research or industrial uses).

Another immediate question was the safety with which this new material would be used. Since the government was actively promoting the use of radioactive isotopes, it had an obligation to ensure their safe use. Harm to patients, physicians, and others involved could arise from inexperienced and untrained users of radioisotopes. When properly used in trace amounts, radioisotopes posed risks well below those deemed acceptable in occupational settings. Balancing risks versus benefits—and seeking means to decrease risks and increase benefits as the field developed—was a major ethical obligation.

Finally, there was the question of the relationship between researcher and subject—more precisely, the question of the authorization for use in humans and the process of disclosure and consent, if any, to be followed. These uses can be divided into (1) therapeutic/diagnostic uses, (2) therapeutic/diagnostic research, and (3) nontherapeutic research.

As we shall see, great attention was paid initially to the question of resource allocation; but supply soon proved far greater than expected, and the need for this attention evaporated. The control of the risk posed by the use of AEC-provided radioisotopes was also a source of intense focus from the outset and remained so as the program grew. By contrast, notwithstanding the 1947 declarations by AEC General Manager Carroll Wilson on the importance of consent, the matter of consent received only limited attention in the early years of the program.

ORIGINS OF THE
AEC RADIOISOTOPE
DISTRIBUTION PROGRAM
IN THE MANHATTAN PROJECT

The medical importance of radioisotopes was recognized before World War II, but distribution was unregulated by government. The postwar program for distributing radioisotopes grew out of the part of the Manhattan Project that had developed the greatest technical expertise during the war: the Isotopes Division of the Research Division at Oak Ridge.[1] Production of useful radioisotopes required extensive planning for both their physical creation and their chemical separation from other materials. Plans to distribute radioisotopes to medical researchers outside the Manhattan Project were developed in the final year of the Project.

In June 1946, the Manhattan Project publicly announced its program for distributing radioactive isotopes. The new world of radioisotope research was to be shared with all. Most research would be unclassified.[2] An enthusiastic *Science* magazine reported: "Production of tracer and therapeutic radioisotopes has been heralded as one of the great peacetime contributions of the uranium chain-reacting pile. This use of the pile will unquestionably be rich in scientific, medical, and technological applications."[3] An article in the *New York Times Magazine* told readers that "properly chosen atoms can become a powerful and highly selective weapon for the destruction of certain types of cancer."[4] Until now, "the doctors and biologists have had to plea for samples of isotope material from their brothers in the cyclotron laboratories. . . . Now the picture has changed in a revolutionary way. The Government has adapted one of the Oak Ridge uranium piles to the mass production of radioactive 'by-product material.'"[5]

Extensive planning led up to this public announcement. Although the initial expectations were that basic research would precede extensive medical applications, from the very beginning officials planned for "clinical investigation" with humans. In doing so, they recognized that the "administration to humans places extreme demands, both moral and legal, upon the specifications and timing of the radioisotope material supplied."[6] The recognition of special moral and legal aspects of human experimentation and reliance on the professional competency of those administering radioisotopes formed the cornerstones of the radioisotope distribution system's oversight of experiments. Significantly, however, the system was not designed to oversee consent from subjects prior to the administration of radioisotopes.

Radioisotopes could not simply be ordered from the Manhattan Engineer District; each purchase had to be reviewed and approved. For human applications, each application was reviewed by a special group of experts: the Advisory Subcommittee on Human Applications of the Interim Advisory Committee on Isotope Distribution Policy of the Manhattan Project. According to one of the initial planners, "The chief reason for setting this group up as a separate entity from the Research group [another subcommittee] is that of medico-legal responsibility involved in the use or treatment of humans, experimentally or otherwise."[7] (When the AEC began its work, this subcommittee continued but was renamed the "Subcommittee on Human Applications of the Committee on Isotope Distribution of the AEC." In 1959 it was absorbed into the "Advisory Committee on Medical Uses of Isotopes."[8] In 1974, the AEC's responsibilities were transferred to the Nuclear Regulatory Commission.) Coupled with this review was a requirement that those wishing to purchase radioisotopes demonstrate the special competence required for working with radioactive materials. This mechanism for centralized, nationwide review was unusual at the time it was begun.

The breadth of the subcommittee's purview can be seen in the range of proposals examined. Although the Advisory Committee is concerned primarily with medical research, the AEC subcommittee review extended well beyond this realm. Apparently, the subcommittee reviewed all proposed uses for radioisotopes that might result in the exposure of humans to radiation. These included, for example, using cobalt 60 in nails in wooden survey stakes (probably to assist in later locating them), sulfur 35 in firing underground coal mines, and yttrium 90 as a tracer in gasoline in simulated airplane crashes.[9] (Its jurisdiction was limited to by-product material, however, and did not extend to fissionable materials such as plutonium and uranium.)

Soon after the Manhattan Project's public announcement, both the radioisotope distribution system and its oversight structure began operation. On June 28, 1946, the Subcommittee on Human Applications held its first meeting. Attending as members were Dr. Andrew Dowdy, chairman, and biophysicist Gioacchino Failla. Dowdy was director of the University of Rochester's Manhattan Project division, while Failla was a professor at Columbia University and consultant to the Metallurgical Laboratory in Chicago. Not attending was the third member of the subcommittee, Dr. Hymer Friedell, executive officer of the Manhattan Project's Medical Section. Attending as nonvoting secretary was Paul Aebersold, in charge of the production of radioisotopes at Oak Ridge (later to head the AEC's Isotopes Division). His efforts to promote the use of radioisotopes later earned him the nickname "Mr. Isotope." Also attending as advisers from Oak Ridge were W. E. Cohn, the author of the original memorandum proposing a system for distributing radioisotopes, and Karl Morgan, director of Health Physics at Oak Ridge, who would, over the years, become a leading figure in the establishment of occupational exposure limits for radioisotopes.[10]

Although the primary task of the subcommittee was to oversee safety, at the time, many expected a shortage of radioisotopes. Thus, much of this first meeting was taken up with a discussion of priorities for allocation.[11] (As it happened, supply exceeded demand within one year.) It was in the context of this discussion of allocation, not a discussion of safety or ethics, that a system of local committees was suggested. Each local committee (also called "local isotope committee" at this meeting) would include "(a) a physician well versed in the physiology and pathology of the blood forming organs; (b) a physician well versed in metabolism and metabolic disorders; (c) a competent biophysicist, radiologist, or radiation physiologist qualified in the techniques of radioisotopes."[12] The main advantages of a system of local committees were administrative efficiency and delegation of prioritization for scarce isotopes.[13] The primary functions of each local isotope committee were coordination, allocation, and safety. Evidently no mention was made of overseeing subject consent.

At this first meeting, the subcommittee had before it no actual requests to evaluate. Even so, members did agree on the general principles on which they would deny a request:

a. The requestors are not sufficiently qualified to guarantee a safe and trustworthy investigation.

b. Insufficient knowledge exists to permit a safe application of the material in the proposed human cases.[14]

There was no elaboration of crucial terms such as *qualified, safe and trustworthy, insufficient knowledge,* and *safe application.* Although no standards of adequate consent were mentioned, this degree of oversight was unusual in medical research during this time and even later.

Although it had no specific requests before it, the subcommittee did consider the anticipated uses of some isotopes. The uses of some isotopes were apparently rejected, not only because of the hazards of radiation, but also because of chemical toxicity and the availability of less-hazardous alternatives.[15] For others, specific limits were set. For example, the subcommittee was especially cautious concerning isotopes of strontium because it concentrated in bone, as did radium, which was known to be hazardous from the prewar experience of the dial painters. The subcommittee set a specific exposure limit: "the Sr 90 (and Y 90 daughter) should not contribute in excess of 1% to the total rate of beta disintegration."[16]

Such general guidelines have little effect unless a procedure is established for their implementation. At its first meeting, the subcommittee set out in detail the mechanism for its own future operation. What the subcommittee would be reviewing were requests to purchase isotopes for any use in human beings. Only after the subcommittee approved a request would the isotope be sold and shipped to the researcher. The need for speed in responding to requests for human uses was recognized.[17]

Details of the procedure for purchasing isotopes were disseminated to potential users through a brochure issued in October 1946 by the Isotopes Branch at Oak Ridge.[18] Most of the brochure concerned the paperwork, which, among other things, ensured that the Subcommittee on Human Applications would actually be notified of all applications for human use.

The last stage of the purchase procedure indicates the underlying concern with legal liability. Although Manhattan Project approval was required, the actual purchase was from the private contractor-operator of the Clinton Laboratories (later designated the Oak Ridge National Laboratory) in Oak Ridge, at that time

Monsanto Chemical Company. The final purchase agreement contained a clause relieving both the government and the private contractor from any responsibility for "injury to persons or other living material or for any damage to property in the handling or application of this material. . . ."[19] The Manhattan Project also required the purchaser to file with the Isotopes Office a statement required by section 505(i) of the Federal Food, Drug, and Cosmetic Act. However, the Advisory Committee found no evidence of direct involvement by the FDA at that time in the planning or operation of the radioisotope distribution program.[20]

By October 1946, the distribution program was well under way: 217 requests had been received. Of these, 211 had been approved. Human use requests totaled 94, of which 90 had been approved.[21]

THE AEC ASSUMES RESPONSIBILITY FOR RADIOISOTOPE DISTRIBUTION

When the AEC took over responsibility for the program on January 1, 1947, the structure of the radioisotopes distribution system remained intact. The Subcommittee on Allocation and the Subcommittee on Human Applications remained as standing subcommittees of the Interim Committee on Isotopes Distribution Policy, which became known as the Advisory Committee on Isotope Distribution Policy. The forms developed by the Manhattan Project were reissued as AEC forms without substantial revision. The system of application from private users, review, purchase, and distribution continued to operate.

At first, there appears to have been some confusion over the responsibility of the AEC for its own research program and for its program to distribute radioisotopes to private researchers. As discussed in chapter 1, two 1947 letters from AEC General Manager Carroll Wilson describe strong consent requirements. The April letter to Stafford Warren was expressly directed to the terms on which research conducted by AEC contractors (including universities) would be approved. The November letter was sent to

Robert Stone. As we have discussed, those clear statements to contract researchers do not seem to have been made to those applying for radioisotopes. This confusion about the relationship between contract research and isotope distribution is discussed in a September 26, 1947, memorandum from J. C. Franklin, manager of Oak Ridge Operations, to Carroll Wilson.[22] Other correspondence also indicates confusion over whether the AEC's own labs (which were themselves often operated by contractors) were to follow the procedures for the radioisotope distribution program, which would have placed their human use requests before the Subcommittee on Human Applications.

Initially, requests for by-product materials from within the AEC used a form that did not specify whether the radioisotope was to be used on humans.[23] By August 1949, Shields Warren, director of the AEC's Division of Biology and Medicine, had directed that human use by AEC laboratories be subject to review by the Subcommittee on Human Applications.[24] However, when regulations governing radioisotope distribution were first promulgated, AEC-owned facilities were specifically exempted from all such regulations.[25] Warren's goal was achieved instead by a memorandum from Carroll Wilson in July 1950. This memorandum discontinued use of the earlier form and directed that all requests use the same form used by outside purchasers, which directed human use requests to the Subcommittee on Human Applications.[26]

The AEC Subcommittee on Human Applications

At the heart of overseeing the expansion of the use of radioisotopes was the Subcommittee on Human Applications of the AEC's Advisory Committee on Isotope Distribution. Applications had to have been approved by a local isotope committee before even being considered by the subcommittee.[27] The subcommittee itself conducted most of its reviews by mail. Unfortunately, only fragmentary records of this correspondence have been found.

The subcommittee formally met only once a year to discuss general issues. By its second meeting, in March 1948, membership had grown to

four. Dowdy was no longer on the subcommittee; Joseph Hamilton and A. H. Holland had been added. Hamilton was, as described in chapter 5, a physician-investigator with the University of California's Radiation Laboratory in Berkeley. Holland was a physician-investigator who became medical director at the AEC's Oak Ridge Operations in late 1947. (As we shall see in chapter 13, he played a central role in the question of the declassification of secret experiments.) As the subcommittee continued to "examine each case on its own merits" it began to generate principles for "general guidance." In doing so, it began to categorize experiments, apparently according to the degree of hazard posed.

One category was tracer studies in "normal adult humans" using beta and gamma emitters with half-lives of twenty days or fewer. Applications needed to include information on biodistribution and biological half-life of the radioisotope, based on either animal studies or references to the literature.[28]

A second category was studies in "normal children." In 1948 the subcommittee did not issue detailed guidelines, but instead simply stated that such applications "would be given special scrutiny by the Subcommittee on Human Applications."[29] In 1949 it issued more detailed guidelines, which indicate that the concern was with minimizing risk, not requiring or overseeing consent:

In general the use of radioisotopes in normal children should be discouraged. However, the Subcommittee will consider proposals for use in important researches, provided the problem cannot be studied properly by other methods and provided the radiation dosage level in any tissue is low enough to be considered harmless. It should be noted that in general the amount of radioactive material per kilogram of body weight must be smaller in children than that required for similar studies in adults.[30]

Coupled with the children's category in 1949 were studies on pregnant women: "The use of radioactive materials in all normal pregnancies should be directly discouraged where no therapeutic benefit is to be derived."[31] Although not specifically mentioned in the minutes, such a policy may, like research in "normal children," have been waived for "important researches" that could not otherwise be undertaken.

One recurring difficulty was the problem of deciding when an application could be considered "safe." There was no simple, mechanical process for making such a judgment. This can be seen in the subcommittee's detailed consideration of an application for phosphorus 32 to be used in a blood volume study of children. The amount of radioactivity proposed ranged from 1/4 to 1 microcurie per kilogram of body weight. Initially, three of the four members approved the application and the allocation was made. However, the fourth member, replying late, reopened the question. Following reconsideration by the entire subcommittee, three of the four members concluded the original application for use on children should be turned down and the investigator asked to revise the application to "state the importance of making the study in children" and to keep the amount of activity less than 1/2 microcurie per kilogram.[32] The reduction in allowable amount of activity illustrates both the diligence with which the subcommittee pursued its task and the inherent difficulties in making judgments about what constituted "safe" practices in a rapidly developing field of research.

The subcommittee's task was made a bit easier when considering applications with adults, where it could draw upon occupational guidelines. Requests for "long-lived radioisotopes" were placed in a third category, defined as those with a biological half-life greater than twenty days. In contrast with experiments on children, here the subcommittee was willing to set a general dose limit: "The dosage in the critical tissue should be such as to conform to the limitations stated by the National Committee on Radiation Protection."[33] (The NCRP, now the National Council on Radiation Protection and Measurements, is an independent organization that publishes occupational radiation protection guidelines based on expert reviews of contemporary scientific knowledge.) As with children, such applications "must be reviewed separately." The subcommittee did not wish this limit to be ironclad: "In special cases, however, the Subcommittee on Human Applications may permit the use of radioisotopes in higher dosages."[34] At this point the subcommittee appears to have been establishing general principles; no specific radioisotopes or particular research proposals are mentioned.

A final category was applications using radioisotopes with long half-lives in patients with short life expectancies. The term *moribund* was used in correspondence by Paul Aebersold prior to the second meeting of the subcommittee in March 1948. He wrote to the subcommittee members explaining that the item was on the agenda because requests for such work had been received. He referred to a written request from a physician at Massachusetts General Hospital to use calcium 45 and an oral request from a staff member at Presbyterian Hospital in Chicago to use testosterone labeled with carbon 14. Aebersold did not provide any details as to the purposes of the proposed research. The issue was what policy to adopt when the patients were predicted not to live long enough for long-term hazards to develop. Aebersold told the subcommittee that "this office feels that such requests should be allowed if a satisfactory mechanism for determining the 'moribundness' of the patients in question is established. We believe that this question should be decided by a group of doctors and written evidence signed by the group filed with the Isotopes Division prior to use of the material."[35]

The subcommittee had no objection to the basic principle of applying larger doses to patients with short life expectancies, but its language was more oblique than Aebersold's letter: "It is recognized that there may be instances in which the disease from which the patient is suffering permits the administration of larger doses for investigative purposes."[36] Safeguards were to be provided by reliance on the judgment of local physicians, not a precise definition of *moribund*. Indeed, the subcommittee did not even use the term. Applications would be approved providing:

1. Full responsibility for conduct of the work is assumed by a special committee of at least three competent physicians in the institutions in which the work is to be done. This will not necessarily be the local Radioisotope Committee.
2. The subject has given his consent to the procedure.
3. There is no reasonable likelihood of producing manifest injury by the radioisotope to be employed.[37]

No further explanation was given of how the second requirement, giving consent, would be fulfilled by a "moribund" patient, nor was additional guidance provided to clarify the third criterion.

One instance in which this policy was applied took place at the Walter E. Fernald State School in Massachusetts (see chapter 7). Correspondence between the researchers and the AEC indicates that the AEC allowed the administration of 50 microcuries of calcium 45 (fifty times the amount the AEC allowed the researchers to administer to other subjects in the study) to a ten-year-old patient with a life expectancy of a few months, suffering from Hurler-Hunter syndrome (a degenerative disease of the nervous system). In applying for the radioisotope, Dr. Clemens Benda, the researcher, noted that "permission for the use of higher doses administered to moribund patients has been granted by you to other investigators . . . ."[38] This subject was part of a study of calcium metabolism approved by the superintendent of the school. Students had been described as "voluntarily participating" in a letter sent earlier to the parents asking if they objected, but that did not mention the use of radioactive tracers. Lack of response from a parent was presumed to be approval.[39] The subject with Hurler-Hunter syndrome was found to have abnormal calcium metabolism, but died before the study could be completed.[40]

Even as it developed procedures for unusual cases, the subcommittee recognized that some existing uses were becoming routine and did not need to be continuously reviewed by the subcommittee itself. The subcommittee delegated the review of such requests to the Isotopes Division, setting out the criteria to be applied:

Such applications should be justified by:
   a) A commensurate increase in patient load.
   b) An expanded research program.
   c) Provision of adequate storage and handling facilities.
   d) Assurance that personnel protection and supervision are adequate for the larger amounts requested.[41]

An additional simplification occurred with the introduction in 1951 of "general authorizations," which delegated more authority to the local radioisotope committees of approved institutions.[42] These authorizations enabled research institutions to obtain some radioisotopes for approved purposes after filing a single application each year, therefore eliminating the need to file a separate application for each radioisotope order. As such, they also reduced the oversight of the AEC's Subcommittee on Human Applications, as each order was no longer reviewed individually. However, at first the general authorizations did not apply to human use, and when they were expanded to human use in 1952, they were limited to certain radioisotopes for clinical use and excluded radioisotopes in cancer research, therapy, and diagnosis.[43]

Both the AEC and the subcommittee reacted strongly when proper bureaucratic procedures were not followed. One example was a private industrial lab that used iodine 131 for a human study that had not been properly reviewed. Even though no one was harmed, the AEC threatened to suspend shipments of all radioisotopes, not just iodine 131; such action would have put the company out of business.[44] Aebersold, at the direction of the subcommittee, notified the company president that while the incident "did not lead to any unfortunate results from the standpoint of radiation hazard . . . a recurrence of this type of violation should result in cessation of all shipment of radioactive materials to Tracerlab, Inc."[45] For his part, the company president reacted by notifying employees that such action would be grounds for automatic dismissal.[46]

Thus, as it proceeded in its work of evaluating individual applications, the subcommittee developed more general principles such as categories of human uses based upon risk and updating of criteria based upon developing knowledge. The goal, as the AEC's director of research, K. S. Pitzer, stated in 1950, was "to make radioisotopes as nearly as possible ordinary items of commerce in the technical world."[47] For example, cancer researchers initially received radioisotopes at no charge.[48] This free program was changed to an 80 percent discount program in 1952[49] and ended in July 1961.[50]

AEC Regulations and Published Guidelines

An important step toward making the use of radioisotopes a component of medical practice routine was formally enacting regulations governing the use of isotopes. The first regulations were enacted in 1951.[51] These early regulations essentially promulgated facility and personnel require-ments without establishing dose limits or men-tioning the consent requirement established in 1949 for administering larger doses to very sick patients. Throughout the 1950s, changes in the regulations dealt with administrative procedures. Other concerns about radioisotope use, such as consent requirements, were disseminated through circulars, brochures, and guides of the Isotopes Division. In 1948 the circular describing medi-cal applications was only three pages long; by 1956 it had been replaced by a twenty-four-page guide that provided detailed requirements for many different applications of isotopes.[52]

This greater precision can be seen, for ex-ample, in the guidelines for terminal patients. By the time of the 1956 guide, the use of radio-isotopes with half-lives greater than thirty days ordinarily would not be permitted without prior animal studies establishing metabolic properties, unless patients had a short life expectancy. The judgment of local physicians was now to be guided by a more exact definition: exceptions would be "limited to patients suffering from diseased conditions of such a nature (life expect-ancy of one year or less) that there is no reason-able probability of the radioactivity employed producing manifest injury."[53] However, while a more precise definition of *terminal* was now provided, there was no longer explicit mention of a specific requirement for consent from these patient subjects, as had been made earlier.

Consent was required, though, in the section of the 1956 guide on the "use of radioisotopes in normal subjects for experimental purposes." (Pre-sumably, "normal" here means "healthy.") This section included the earlier provisions that the tracer dose not exceed the permissible body bur-den and that such experiments not normally be conducted on infants or pregnant women. It also, however, included a new provision that such ex-periments were to be limited to "volunteers to whom the intent of the study and the effects of radiation have been outlined."[54] The term *volun-teer* would seem to imply a requirement that con-sent be obtained following a disclosure of infor-mation to potential subjects. The disclosure requirement does not include, however, all of the elements of information that today are included in duties to obtain informed consent.

This 1956 consent requirement now governed

all radioisotope experiments in normal subjects, a substantial expansion of the earlier require-ment of consent only from terminal patients receiving larger-than-usual doses. It also explic-itly required that both the purpose and effects of radiation be explained. It is unclear whether the failure to mention consent in the section on terminal patients was an oversight in drafting or a deliberate distinction between patients and "normal" subjects. The Advisory Committee has not found documents revealing the history of this provision, nor any explanation of the choice to limit the broad consent requirement to "nor-mal" subjects.[55]

This broad requirement continued over the next decade as part of AEC policy. In 1965, the AEC published the "Guide for the Preparation of Applications for the Medical Use of Radioiso-topes." The guide described the application pro-cess and specific policies for the "Non-Routine Medical Uses of Byproduct Material." This policy statement reiterated the exclusion of pregnant women and required that subject characteristics and selection criteria be clearly delineated in the application. Another requirement stated that ap-plications should include "confirmation that con-sent of human subjects, or their representatives, will be obtained to participate in the investigation except where this is not feasible or in the investigator's professional judgment, is contrary to the best interests of the subjects."[56]

During the 1960s, the entire system of over-sight of radioisotope research began to change as the Food and Drug Administration began developing a more active role in supervising the development of radiopharmaceuticals.[57] The regulatory history of this shift in authority is complex and beyond the scope of this report. Suffice it to say that by the mid-1960s the regu-lation of radioisotope research was beginning to merge with the regulation of pharmaceutical research in general.

LOCAL OVERSIGHT: RADIOISOTOPE COMMITTEES

From its inception, the AEC distribution system required each local institution to establish a "local radioisotope committee," later termed a "medical isotopes committee." Initially, the primary purpose was to simplify the allocation process by having local institutions establish their own priorities before applying to the AEC.[58] Soon after the program began, supply increased and no dramatic new uses developed, so allocation was no longer a major issue. These local committees also took on responsibilities for physical safety, usually working closely with radiation safety offices. By October 1949 this requirement also applied to the AEC's national labs.[59] When "general authorizations" were issued in 1951, granting broader discretion to qualifying local institutions, local isotope committees assumed greater responsibility.[60]

By 1956, the functions of the local radioisotope committees included reviewing applications, prescribing any special precautions, reviewing reports from their radiological safety officers, recommending remedial action when safety rules were not observed, and keeping records of their own activities. The basic focus on radiological safety remained, although in reviewing applications a local medical isotopes committee could also consider "other factors which the [local medical isotopes] Committee may wish to establish for medical use of these materials."[61] Exactly what these "other factors" might be was not specified.

These local committees together reviewed thousands of applications over the next decades. Although not federal agencies, they were required by the AEC, and their proper functioning was an important part of the oversight system envisioned by the AEC. To fully assess whether this system fulfilled its goals would be an enormous task, requiring the retrieval and examination of thousands of local records. However, to make a preliminary assessment of whether the system as a whole generally appeared to function as planned, the Advisory Committee did examine the records of several public and private institutions: the Veterans Administration (VA), the University of Chicago, the University of Michigan, and Massachusetts General Hospital (MGH).[62] Doing so provided us with an understanding of the techniques of risk management used at the local level on a day-to-day basis. We specifically examined whether local radioisotope committees in fact were established as directed and what techniques they developed to monitor consent and ensure safety.

Establishment of Local Isotope Committees

Overall, the federal requirement seems to have been an effective means of instituting a reasonably uniform structure across the nation for local radioisotope committees. The AEC's requirements for local committees were followed in all the institutions studied, and there is no reason for believing they were exceptional. One local radioisotope committee, that of Massachusetts General Hospital, was established in May 1946, prior to the AEC requirement.[63] The other institutions established a local radioisotope committee when required to do so by the AEC.

Local committees also could have broader tasks than those required by the AEC. For example, the Radiation Policy Committee at the University of Michigan regulated all radioactive substances used on campus, not just those purchased from the AEC. These included reactor products, transuranic elements, and external sources of radiation.[64]

The Veterans Administration added another level of oversight in the form of a systemwide Central Advisory Committee.[65] In 1947 the VA embarked on a radioisotope research program that would take place within newly established radiation units in the hospitals that would be the recipients of AEC-supplied isotopes.[66] Among early research projects were the treatment of toxic goiter and hyperthyroidism with iodine 131 and treatment of polycythemia rubra vera (overproduction of red blood cells) with phosphorus 32 at Los Angeles, radioactive iron tracers of erythrocytes at Framingham, and sodium 24 circulatory tracers in Minneapolis.[67] By the end of 1948, radioisotope units had been established in eight VA hospitals.[68] Each of the eight was asked to establish a radioisotope committee (as required by the AEC) to be appointed by the Dean's Committee of each hospital, while representatives from affiliated universities agreed to serve as consultants in the various units.

Local Monitoring of Consent

Generally, although local institutions created clear procedures to monitor safety, these local radioisotope committees did not establish procedures to monitor or require consent.[69] (See part I for discussion of the broader historical context of consent in medical research.) The standard application form to the Massachusetts General Hospital committee, as of 1953, had no place to describe an informed consent procedure. This does not, of course, resolve the question of whether consent was given. According to one prominent neurosurgeon interviewed by the Advisory Committee staff, William Sweet, at that time, in the case of brain tumor patients, oral consent was obtained from both the patient and, since mental competency could later be an issue, the next of kin.[70]

Similarly, no mention of the 1947 AEC requirements stated in General Manager Wilson's letters is contained in the advice Shields Warren gave in 1948 to the VA, even though Warren, as director of the AEC's Division of Biology and Medicine, must have known of discussions about consent requirements. An issue that arose before the VA Central Advisory Committee was whether patient-subjects should sign release slips. This issue posed the question of whether the radioisotope units in the VA hospitals were treatment wards or clinical research laboratories. If wards, patients need not sign consent forms, for they were simply being treated in the normal course of an illness. Shields Warren agreed with this presumption and felt that there was no need for the patients to sign release slips: "The proper use of radioisotopes in medical practice is encompassed in the normal responsibilities of the individual and of the institution or hospital."[71] In addition, he felt that the practice would draw "undue and unwholesome attention to the use of radioisotopes."[72]

Movement toward more formal consent requirements gradually arose at the local level. In 1956 the University of Michigan's own Human Use Subcommittee directed that in an experiment using sodium 22 and potassium 42, each "volunteer would be required to sign a release indicating that he has full knowledge of his being subjected to a radiation exposure." Since the local committee was concerned about what it termed "unnecessary" radiation, the volunteers presumably were healthy subjects not otherwise receiving radiation for treatment or diagnosis. The committee appended a recommended "release" form to its minutes:

I, the undersigned, hereby assert that I am voluntarily taking an injection of _____ at a dose level which I understand to be considered within accepted permissible dose limits by the University of Michigan Radio-isotope Human Use Sub-Committee.[73]

By 1967, the Michigan subcommittee also required that the subject explain the experiment to the researcher to clarify any doubts or misunderstandings. The following statement was incorporated into all applications to the university's Human Use Subcommittee:

The opinion of the Committee is that INFORMED CONSENT is the legal way of describing a "meeting of the minds" in a contract. In this situation it means that the subject clearly understands what the experiment is, what the potential risks are, and has agreed, and without pressure of any kind, elected to participate. The best way to ascertain that the consent is informed, is to have the subject explain back fully to the interviewer, exactly what he thinks he is submitting to and what he believes the risks might be. This facilitates clarification of any doubts, spoken or unspoken. The content of this discussion will be recorded in detail below.[74]

During the 1960s, as explained in chapter 3, concern was growing over the adequacy of consent from subjects. Although not intended by the AEC to monitor the obtaining of consent from subjects, over the years the local radioisotope committees may have come to take on this task. By requiring such local committees, the AEC had, probably unwittingly, provided an institutional structure that allowed later concern for informed consent to be implemented at the local level.

Local Monitoring of Risk

This local and informal approach to consent is in sharp contrast to the detail and documentation with which risk was assessed. As discussed earlier, monitoring risk was the major task of the AEC's Subcommittee on Human Applications. The local committees mirrored this task, examining in detail the various experiments presented to them. As with the AEC subcommittee, local committees developed a variety of methods, none especially surprising, to ensure what they believed was adequate safety.[75]

The basic dilemma facing local committees was to allow exploration of new territory while attempting to guard against hazards that, precisely because new territory was being explored, were not totally predictable. This dilemma was apparent at the local level, as well as at the level of the AEC's Subcommittee on Human Applications. For example, in the minutes of the Massachusetts General Hospital local radioisotope committee in 1955, during a discussion of new and experimental radiotherapies for patients, one member of the committee declared that the safety of the patient was of "paramount importance."[76] Yet, other members suggested that a risk-benefit analysis needed to be an integral component of such a policy decision. The committee as a whole concluded merely that it was a complicated issue and that "it is not wise in any way to inhibit investigators with ideas, and yet the safety of the patient must come first."[77]

Requiring prior animal studies was a basic method of assessing risk. For example, the twenty-two studies reviewed by the University of Chicago's local committee in 1953 included multiple therapeutic and tracer studies involving brain tumors, the thyroid gland, metastatic masses, and tissue differentiation. Those the Chicago committee viewed as involving any risk to the patient were preceded by extensive animal studies.[78]

Animal studies were usually tailored to each project and also raised the question of the differences between how humans and animals might respond to a particular radioisotope. A more uniform standard directly applicable to humans was the system of dose limits established by the National Committee on Radiation Protection for occupational purposes: the maximum permissible dose for each isotope. In addition, although no national system existed for reporting their decisions, local committees drew upon their knowledge of what had been approved at other institutions.[79] At least one local committee issued its own dose limits. The Massachusetts General Hospital local committee in 1949 issued a seven-point policy on human use of beta- and gamma-emitting radioisotopes.[80] By 1956, the Michigan committee provided explicit limits for exposure of volunteers.[81]

At other times, the condition of subjects who were patients was accepted as justification for higher doses. For example, in 1953 the Chicago committee approved a tracer study using mercury 203 "to study uptake by malignant renal tissue." Although admitted to be unusual, it was approved as potentially efficacious in patients suffering hypernephroma (a kidney cancer). Total dose would not exceed 10 milligrams of ionic mercury, a high dose for most tracer studies, which was approved as reasonable given the illness of the patients.[82] Similarly, the Harvard Medical School committee in 1956 stipulated that "the risk of incurring any type of deleterious effect due to the radiation received should be comparable to the normal everyday risks of accidental injury." For seriously ill patients receiving experimental treatment, however, the committee stated, "the estimated deleterious effect from radiation should be offset by the expected beneficial effects of the procedure."[83]

In addition to setting limits, local committees encouraged the use of technical methods to reduce risk. Use of different detection techniques could reduce the dose required. In 1955, for example, the Michigan committee considered an application to administer to normal volunteers up to 30 microcuries of sodium 22 and up to 350 microcuries of potassium 42, resulting in internal radiation doses of up to 300 millirem per week. (The purpose was to study sodium-potassium exchanges.) The committee asked itself: "Is it justifiable to subject the volunteers to an exposure in excess of the maximum permissible? This Committee did not resolve this question but came forward with the suggestion that more-sensitive counting techniques might permit this investigation at lower dose levels."[84]

Another method of reducing risk was to restrict the type of subjects to those whose life expectancy was too short for long-term effects to appear. This has already been seen regarding terminal patients. Another variation of the same technique was to restrict the use of volunteers to those over a certain age. At Michigan, age restrictions on who would be acceptable as a volunteer began appearing in the 1960s.[85]

When a worthwhile experiment also involved novel risks, another method to control risk was to require additional monitoring by the local committee as the experiment proceeded. At

times, the Michigan committee required preliminary reports before allowing experiments
to proceed further.[86] In another instance, the
Michigan committee required the researcher to
obtain long-term excretion data because of concern that "the usual biologic half-life data might
not be sufficient."[87] Similar additional oversight
was required at the University of Chicago in
1953. A proposal was made to use tritium-
labeled cholesterol to study steroid-estrogen
metabolism in women. The question of the distribution of estrogenic hormones in humans was
unexplored at the time and deemed worthy of
research. While the risk appeared low, the committee ultimately approved the study for the first
round of the experiment only for nonpregnant
women who were sterile or pregnant women
who planned to be sterilized postabortion. If
data from the first round suggested minimal risk
to the women and the fetuses, the program could
be expanded.[88]

Thus, in establishing a system of local radioisotope committees, the AEC effectively increased the detail with which each proposed
experiment was reviewed. Often, it appears, experimental protocols were revised at the local
level before being approved and sent on to the
AEC. Thus, the system created by the AEC did
some of its most effective risk management out
of sight of direct federal oversight.

GENERAL BENEFITS
OF RADIOISOTOPE RESEARCH

The system for distribution of radioisotopes
worked well and encouraged researchers to explore new applications. There are two striking
aspects of the application of radioisotopes to
medicine since World War II: rapid expansion
and complexity. Practices that at the end of the
war were limited to fewer than four dozen practitioners have now become mainstays of modern medicine.[89] The second major aspect of the
field is its complexity. Just as nature at times is
best regarded as a seamless web, not unconnected
scientific fields, knowledge nurtured in one field
often provides unexpected benefits in another.
A few examples can illustrate how some of the
hopes at the dawn of the atomic age have actually been realized.[90]

Improved Instrumentation
to Detect Radiation

Improved instruments, the basic tools for all
biological research using radioisotopes, were
developed through the interaction of biology
and medicine with physics and engineering.
Improvements not only provide greater precision, they also allow the same amount of information to be gathered with lower doses of
radiation, thereby reducing the risk.

Perhaps the best-known example is the application of the "whole-body counter" to biological
problems. The device was originally developed
as a tool for physics, enabling measurements of
minute amounts of radiation by combining sensitive detectors with extensive shielding to eliminate extraneous radiation. The result was similar
to placing a sensitive microphone in a sound-
proofed room, allowing lower levels of radioactivity to be detected than was previously possible. For some research, no radioisotope at all
was administered; the counters could measure
naturally occurring radioisotopes. Whole-body
counters also greatly simplified metabolic studies. In some studies, subjects who previously
would have had to reside continuously in a metabolic ward could now schedule visits to the
whole-body counter for their natural radioactivity to be measured on an outpatient basis.[91] This
device was later adapted for whole-body counting after administration of tracer amounts of
radioisotopes and is the basis for a number of
fundamental nuclear medicine tests.

In the early 1970s, computerized tomographic scanning (CT) was introduced. This
technique was first applied to x-ray imaging
by taking multiple x-ray "slices" through a
region of the body, then programming a computer to construct a three-dimensional image
from the information. Thus, internal structures
of the body may be imaged noninvasively.
Newer types of tomographic scanning include
positron emission tomography (PET), in which
various metabolites or drugs are labeled with a
very short half-life positron-emitting radioisotope, such as fluorine 18, and the passage of the
labeled material is tracked throughout the body
by taking multiple images over several minutes
or hours.

Diagnostic Procedures

The first medical application of any radiation was the use of x rays for diagnostic purposes, such as locating broken bones inside the patient. Radioisotopes later opened another window into the body. The natural tendency of certain organs to preferentially absorb specific radioisotopes, coupled with ever-improving detection techniques, allowed radioisotopes to be used to increase the contrast between different parts of the body. X rays could distinguish between hard and soft tissues because of their different densities. Radioisotopes could go one step further and distinguish different kinds of tissues from one another based upon their metabolic function, not merely their physical density.

Radioisotopes also could go beyond detecting different types of tissues. Since they were distributed throughout the body by the body's own metabolism, their location provided a picture not only of structure, but also of processes. Tracing radioisotopes was a means of observing the body in action. The earliest success was using radioiodine to measure the activity of the thyroid. The gland cannot distinguish between radioactive and nonradioactive forms of iodine and therefore preferentially absorbs all isotopes of iodine. Thus, the activity of the gland can be assessed by observing its absorption of radioiodine. Largely as a result of these advances, the thyroid gland is arguably the best understood of all human endocrine organs, and its hormones the best understood of all endocrine secretions. Since the incidence of thyroid disease is second only to diabetes mellitus among human endocrine diseases, this understanding is basic to therapy in large numbers of patients.[92]

Because the brain is a crucial and delicate organ, techniques for diagnosing brain tumors without surgery were vital. In 1948 radioactive isotopes were applied to this task. Using radiotagged substances that were preferentially absorbed by brain tumors, physicians could more accurately detect and locate brain tumors, allowing better diagnosis and more precise surgery. Similar "scanning" techniques were later developed for the liver, spleen, gastrointestinal system, gall bladder, lymphomas, and bone.

As mentioned, a recently developed technique is PET scanning, which is especially helpful in studying the human brain in action. Glucose is the primary food for the brain; by tagging a glucose analog with fluorine 18, investigators can identify the actively metabolizing portions of the brain and relate that to function. This technique has opened a new era of studies of the brain. Outwardly observable functions, such as language, object recognition, and fine motor coordination can now be linked with increased activities in specific areas of the brain.

Radioisotopes allow investigators to increase the sensitivity for analyzing biological samples, such as tissue and blood components, especially when separating out the material of interest using chemical processes would be difficult. Because instruments to measure radioactivity are so sensitive, radioisotopes are frequently used in tests to detect particular hormones, drugs, vitamins, enzymes, proteins, or viruses.

Therapeutic Techniques

Radioisotopes are energy sources that emit one or more types of radiation as they decay. If radioisotopes are deposited in body tissues, the radiation they emit can kill cells within their range. This may be harmful to the individual if the exposed cells are healthy. However, this same process may be beneficial if the exposed cells are abnormal (cancer cells, for example).

The potential for radiation to treat cancer had been recognized in the early days of work with radiation, but after World War II the effort to develop radiation therapy for cancer increased. Iodine 131 treatment for thyroid cancer was recognized as an effective alternative to surgery, both at the primary and metastatic sites. Cancer is not the only malady susceptible to therapy using radioisotopes. The use of radioiodine to treat hyperthyroidism is perhaps the most widespread example. It illustrates the progression from using a radioisotope to measure a process (thyroid activity) to actually correcting an abnormal process (hyperthyroidism).[93]

Not all experimental applications of radioisotopes are successful. Some experiments end in blind alleys, an important result because this prevents widespread application of useless or even harmful treatments. Negative results also help researchers to redirect their efforts to more promising areas. The importance of negative results is sometimes not appreciated because they do not lead to effective treatments. Negative results may range from simply not obtaining an anticipated beneficial effect to the development of severe side effects. Such side effects may or may not have been anticipated; they may occur simultaneously with beneficial effects, such as the killing of cancer cells. Occasionally negative results include earlier-than-anticipated deaths of severely ill subjects. An example is the experimental use of gallium 72 in the early 1950s on patients diagnosed with malignant bone tumors.[94]

Another radioisotope, cobalt 60, has been used successfully to irradiate malignant tumors, but in this case the radioisotope is not administered internally to the patient; rather, the cobalt 60 forms the core of an external irradiator, and the gamma radiation emanating from the radioisotope source is focused on the patient's tumor. Although cobalt 60 irradiators have been largely replaced by linear accelerators, they were developed under AEC sponsorship and were responsible for many advances in radiation therapy.

Recent efforts to utilize radioisotopes in cancer diagnosis and treatment are based on the ability of antibodies to recognize and bind to specific molecules on the surface of cancer cells and the ability of biomedical scientists to custom-design and manufacture antibodies, thus improving their specificity. These fields are now contributing to a hybrid technique: cloning antibodies and tagging them with radioactive isotopes. As the antibody selectively binds to its target on the surface of the cancer cell, the radioactive isotopes attached to the antibody can either tag the cell for detection and diagnosis or deliver a fatal dose of radiation to the cancer cell. The Food and Drug Administration recently approved the first radiolabeled antibody, to be used to diagnose colorectal and ovarian cancers.[95]

Even in the case of widespread metastases where cure is no longer possible, radiation treatments will often produce tumor regression and ease the pain caused by cancer. Phosphorus 32 has been used to ease (palliate) the bone pain caused by metastatic prostate and breast cancers.

Recently, the FDA approved the use of strontium 89 for similar uses.[96]

Metabolic Studies

Studies of the basic processes within the body may not have any immediate application in diagnosis or therapy, but they can indirectly lead to practical applications. One example is in the study of the metabolism of iron in the body. Iron is an important part of hemoglobin, which carries oxygen from the lungs to all cells in the body. Studies using radioactive iron established the pathway iron takes, from its ingestion in food to its use in the blood's hemoglobin and its eventual elimination from the body; these studies had practical applications in blood disease, nutrition, and the importance of iron metabolism during pregnancy.

Radioisotopes have also been used to study how the weightlessness of space travel affects the human body. Radioisotopes have allowed more precise observation of effects of space travel on blood plasma volume, total body water, extracellular fluid, red cell mass, red cell half-life, and bone and muscle tissue turnover rates.

Other uses of radioisotopes are in studies of the transport and metabolism of drugs through the body. New drugs for any clinical application, whether diagnostic or therapeutic, must be understood in detail before the FDA will approve them for general use. One method for readily determining how a drug moves through the blood to various tissues, and is metabolically changed in structure, is to incorporate a radioactive isotope into the structure of the drug.

Unexpected results from an experiment can at times have widespread consequences. An example is how the work of Rosalyn Yalow and Solomon Berson of the Bronx VA Medical Center opened up the field of radioimmunoassay. In the early 1950s, it was discovered that adult diabetics had both pancreatic and circulating insulin. This appeared odd; previously, it had been believed that all diabetics lacked insulin. To explain the presence of diabetes in people with pancreatic insulin, Yalow and Berson decided to study how rapidly insulin disappeared from the blood of diabetics. To do this, they synthesized radioiodine-labeled insulin. This would act as a radioactive tag, making it much easier to measure the presence of insulin in blood. To their surprise, they found that insulin disappeared more slowly from diabetic patients than from nondiabetic people.[97]

Their work had an impact beyond the study of diabetes, however. In the process of studying the plasma of patients who had been injected with insulin, they discovered that the radioactively tagged insulin was bound to an antibody, a defensive molecule that had been produced by the patient's body and custom-designed to attach itself to the foreign insulin molecule. This was a surprise, since prevalent doctrine held that the body did not produce antibodies to attack small molecules such as insulin. To study the maximum binding capacity of the antibodies, they did saturation tests, using fixed amounts of radiolabeled insulin and of antibody to measure graded concentrations of insulin. With this technique Yalow and Berson realized they could measure with great precision the quantities of insulin in unknown samples. They thus developed the first radioimmunoassay. This technique, for which Rosalyn Yalow was awarded the Nobel Prize in Medicine in 1977, has become a basic tool in many areas of research. Radioimmunoassay revolutionized the ability of scientists to detect and quantify minute levels of tissue components, such as hormones, enzymes, or serum proteins, by measuring the component's ability to bind to an antibody or other protein in competition with a standard amount of the same component that had been radioactively tagged in the laboratory. This technique has permitted the diagnosis of many human conditions without directly exposing patients to radioactivity.

No discussion of the impact of radioisotopes on biomedical science would be complete without a recognition of their fundamental importance in basic biological investigations. The ability of radioisotopically labeled metabolites to act like, and therefore trace, their nonradioactive counterparts has allowed scientists to follow virtually every aspect of metabolism in cells of bacteria, yeasts, insects, plants, and animals, including human cells. Among the benefits of such studies are (a) an understanding of the similarities in metabolism of organisms throughout the evolutionary scale, (b) identification of sometimes subtle differences in cell structure and function between organisms and thus the ability of drugs to kill bacteria, fungi, or insects without harming humans, and (c) elucidation of the fundamental properties of genetic material (DNA). The last of these examples has important implications today, as the human genes controlling many important bodily functions are being identified and cloned and gene therapy is just beginning to find its way into clinical application. Many benefits of understanding the human genetic code have already been realized, and others will likely accrue in the next few years. These benefits are the result of fundamental advances in genetics and molecular biology of the past half century, which in turn depended heavily on studies with lower organisms and with radioisotopically labeled materials. Thus, human health is benefiting from both human and nonhuman research with radioisotopes.

The grandest dream of the early pioneers—a simple and complete cure for cancer—remains unfulfilled. Promising paths at times proved to be dead ends. However, the AEC's widespread provision of radioisotopes, coupled with support for new techniques to apply them, laid the foundation stones for much of modern medicine and biology. This section has only skimmed the field of nuclear medicine, with its vast array of diagnostic and therapeutic techniques, and the use of radioisotopes in many areas of basic research.

AN EXAMPLE OF HOPES UNFULFILLED: THE GALLIUM 72 EXPERIMENTS

Human experiments with gallium 72, as discussed in the section titled "General Benefits of Radioisotope Research," were conducted at the Oak Ridge Institute of Nuclear Studies in the early 1950s. The experiments used gallium 72 because of its short half-life (14.3 hours) and because an earlier animal study indicated it concentrated in new bone, making it useful as a tumor marker and possibly for therapy.[a] The 1953 published report stated that the purpose of the study was "to investigate the therapeutic possibilities in human tumors involving the skeletal system."[b] In 1995 one of the original researchers stated to Advisory Committee staff a somewhat broader purpose: "to exploit to the fullest possible extent *any* possible use of this isotope as a bone seeking element rather than to seek a cure for a specific malignant bone tumor, such as osteogenic sarcoma.... While the Gallium-72 studies did include osteogenic sarcomas, they only represented less than half (9/21), 43%, of all the other primary and metastatic skeletal malignancies studied."[c]

Patients were chosen who had been diagnosed with "ultimately fatal neoplasms not amenable to curative surgery or radiotherapy."[d] The diagnosis later proved to be accurate in all but one of the fifty-five subjects.[e] In one part of the study, thirty-four patients were given trace amounts of gallium. Both external radiation measurements and a variety of excreta, blood, and tissue samples were analyzed to determine the localization of gallium. In another part of the study, twenty-one other patients were given

doses that the researchers hoped would be in the therapeutic range. Total doses ranged from 50 to 777 microcuries.[f] The gallium was administered in fractionated doses biweekly. According to the medical investigators, these patients "were, in general, in a more advanced stage of disease and were completely beyond even palliation from conventional forms of therapy."[g] For these patients, "doses which were believed to be moderate were given and gradually increased to toxic level."[h] The conclusion of the report notes that "most of the patients in whom gallium therapy was attempted were given maximum amounts of the isotope. Only the hopelessness of their prognoses justified a trial of doses so damaging to the hematopoietic tissues."[i]

A major difficulty was lack of knowledge about both the chemical toxicity of stable (that is, nonradioactive) gallium and the radiation toxicity of gallium 72. Calculations and small animal studies indicated that dosimetry techniques used for other radioisotopes would "be of little value."[j] During the study, close monitoring was done of many bodily functions to observe toxic effects as soon as they began to appear. Blood tests revealed changes that "were prominent and were usually of primary importance in determining when the treat-

a. Herbert D. Kerman, M.D., FACR, to Dan Guttman, Executive Director, Advisory Committee on Human Radiation Experiments, 19 May 1995 ("It has come to my attention . . ."), 2. Dr. Kerman cites as the preceding study: H. C. Dudley and G. E. Maddox, "Deposition of radiogallium (Ga-72) in skeletal tissues," *Journal of Pharmacology and Experimental Therapeutics* 96 (July 1949): 224–227.
b. Gould A. Andrews, M.D., Samuel W. Root, M.D., and Herbert D. Kerman, M.D., "Clinical Studies with Gallium-72," 570, in Marshall Brucer, M.D. (ed.), Gould Andrews, M.D., and H. D. Bruner, M.D., "Clinical Studies with Gallium-72," *Radiology* 66 (1953): 534–613.
c. Kerman to Guttman, 19 May 1995, 2. Dr. Kerman presumably was referring to the twenty-one subjects who received doses in the therapeutic range, not the thirty-four who received trace doses.
d. Andrews, Root, and Kerman, "Clinical Studies with Gallium-72," 570.

e. A patient was diagnosed with osteogenic sarcoma in his leg, which was amputated. X rays also revealed dense nodules in his lung, which were diagnosed as inoperable but typical pulmonary metastases. He was discharged after the gallium study. When he later returned to the hospital, an operation revealed that the nodules were not typical metastases, but unidentifiable lesions "not characteristic of any specific lesion." This could not have been known prior to the study, when only x rays were available for diagnosis. Ibid., 585.
f. The researchers reported that these doses were equivalent to 8.5–89.2 mg/kg of body weight. Ibid., 574–577.
g. Ibid., 570.
h. Ibid., 571.
i. Ibid., 587.
j. The investigators wrote that "[n]ormal tissue and whole-body tolerances for amounts of radiogallium necessary to produce a significant effect upon malig-

ments should be discontinued."[k] Other effects included drowsiness, then anorexia, nausea, vomiting, and skin rash.

One problem was determining whether these effects were due to chemical toxicity, radiation toxicity, or a combination. Due to technical difficulties in separating out pure gallium 72, the radioactive gallium was injected with larger amounts of stable gallium, so both chemical and radiation effects could be present. To distinguish them, one patient was administered an amount of stable gallium equal to a therapeutic dose, but with only an insignificant amount of radioactive tracer (to determine localization). Observed toxic effects in this patient did not include bone marrow depression. The researchers concluded, therefore, that the "profound bone marrow depression is characteristic of radiation damage and is probably chiefly caused by radiation, though an element of stable metal toxicity may also be contributory."[l]

Bone marrow depression gradually ended after gallium injections were stopped. While it lasted, bone marrow depression led to greater susceptibility to infection and bleeding. Two subjects died sooner than anticipated, one from infection and bleeding and the other from infection, while their bone marrow was still depressed. "These two patients died in spite of antibiotics, blood transfusions, and toluidine-blue therapy."[m] The researchers reported that "in two patients our estimates of safe dosage limits were in error and radiogallium is believed to have hastened death."[n] One researcher,

writing in 1995, stated that "since 'safe dosage' levels were only estimates and seven other patients had survived with even higher dosages, our choice of language [citing the preceding quotation] was unfortunate. It must be emphasized that this portion of the study must be likened to a current clinical Phase I trial where in a limited fashion [a] broad range of toxicity levels may at best be only estimated."[o]

The major conclusion of the experiment was that hopes for gallium therapy were unfulfilled. Even though the maximum tolerated doses had been administered, the researchers reported that "we were impressed with the almost complete lack of any clinical improvement following gallium treatment, even in patients who showed evidence of striking differential localization of gallium in tumor tissue."[p]

Concerning patient consent, the published study says nothing, which was normal for scientific articles at that time. Near the end of the Advisory Committee's deliberations, ORINS reportedly found consent forms signed by subjects in the gallium study.[q] One of the researchers in 1995 did offer his recollections regarding consent to the Committee:

Forty-five years ago all of our patients and their families were given a booklet of information

---

nant tissues were unknown. Preliminary calculations and small animal experiments had indicated that accepted radiation dosimetry as applied to other isotopes would be of little value in calculating radiation dosage to tissues. It was therefore necessary to utilize the hematologic picture to assess the damaging effects of whole-body irradiation, and clinical and roentgenographic experience in evaluating a therapeutic response." Ibid., 571.

k. Ibid., 573.

l. Ibid., 575.

m. Ibid., 573. Neither had suffered from osteogenic sarcoma; one had suffered from adenocarcinoma of the kidney with lytic bone metastases and another from cancer of the prostate with metastatic skeletal involvement. Kerman to Guttman, 19 May 1995, 3.

n. Andrews, Root, and Kerman, "Clinical Studies with Gallium-72," 571.

o. Ibid.

p. Ibid., 587. Researchers reported evidence of concentration in tumors as being one of the following: "no data," "none," "little," "moderate," or "pronounced." Ibid., 574.

q. Dr. Shirley Fry, telephone interview with Dan Guttman (ACHRE), 30 August 1995, 1. The Advisory Committee did not have enough time to review the forms and related file materials once they were identified, which, because ORINS deemed them privacy-protected material, would have required review at Oak Ridge.

explaining how radioisotopes were used in medicine and more specific information about their own involvement including the possible known risks. Signed applications for admission and waiver and release forms were demanded for all patients. When, as in the ongoing gallium studies, toxicity or enhanced risks were encountered, these were immediately made clear to the patients and their families if they were known in that time frame. Very often toxicity is only apparent after review of the clinical data. In the gallium studies, when on review of the data it was determined that no therapeutic benefit had occurred, the study was immediately terminated.[r]

CONCLUSION

At the end of World War II, radioisotopes were regarded as the most promising peacetime application of our new knowledge of the atom. Venturing into new fields carried with it substantial risks: risks due to our ignorance of what lay ahead, and risks due to the lack of training of many would-be explorers. The AEC consistently accepted and acted upon its responsibility to manage this risk. An extensive administrative system was created to oversee the safety of human radiation experiments that used radioisotopes supplied by the AEC. At the heart of the system was the AEC's Subcommittee on Human Applications of the Advisory Committee on Isotopes Distribution Policy. This system regulated the types of uses allowed according to their hazard and the extent of our knowledge of the risks. It required and provided training of those who would use radioisotopes. It required the establishment of local radioisotope committees, which not only reviewed proposals but suggested changes at the local level in experimental design to reduce risk.

While extensive measures were taken to minimize risk, few measures were taken to ensure that all the explorers, subjects as well as researchers, were fully informed and willing members of the expedition. No evidence has yet been found that the standards for documented consent, articulated by AEC General Manager Carroll Wilson in 1947, were applied by the AEC Isotopes Distribution Division. A limited consent requirement was instituted only for the administration of larger-than-usual doses to very sick patients. Only in the late 1950s did a consent requirement for normal volunteers appear in the AEC guidelines.

Based on the records examined by the Advisory Committee, the adjunct system of local radioisotope committees appears to have functioned as planned. The records of local institutions indicate that they established their own local radioisotope committees, as required by the AEC, and that these local committees closely assessed the risks of experiments. At times, this system went beyond what the AEC had planned. Some local committees had jurisdictions that extended to all radiation-related work, not merely to radioisotopes supplied by the AEC. The local committees also provided, probably unintentionally, a ready-made vehicle for administering greater oversight of consent practices, as concern developed in the 1960s. Requirements for consent on a federal level changed only in the late 1960s, as part of a governmentwide concern.

NOTES

r. Kerman to Guttman, 19 May 1995, 3. The booklet, "ORINS Patient Information Booklet" (*circa* May 1950), is discussed in chapter 1. ORINS hospital was known to be dedicated to experimental work with radiation and radioisotopes. Patients were admitted to the hospital only if they were willing to be experimental subjects. It is not as clear, however, whether the details of any particular experiment were always explained adequately to patients.

1. The first complete proposal for radioisotope distribution is contained in a memo dated 3 January 1946. Radioisotope Committee of Clinton Laboratories (Oak Ridge, Tennessee) to Colonel S. L. Warren, Medical Director of the Manhattan Project, 3 January 1946 ("Specific Proposals for the National Distribution of Radioisotopes Produced by the Manhattan Engineer District") (ACHRE No.

NARA-082294-A-31). This memo, in turn, was derived from a more extensive document prepared by Waldo Cohn, a member of the lab staff. W. E. Cohn, 3 January 1946 ("The National Distribution of Radioisotopes from the Manhattan Engineer District") (ACHRE No. DOE-051094-A-317).

2. The press release announcing the program noted that, in addition to technical qualifications of researchers, "An additional qualification will require all groups using the isotopes for fundamental research or applied science to publish or otherwise make available their findings, thereby promoting further applications and scientific advances." Headquarters, Manhattan District (Oak Ridge, Tennessee), 14 June 1946 ("For Release in Newspapers Dated June 14, 1946") (ACHRE No. NARA-082294-A-31), 1.

3. "Availability of Radioactive Isotopes: Announcement from Headquarters, Manhattan Project, Washington, D.C.," *Science* 103 (14 June 1946): 697-705.

4. Harry H. Davis, *New York Times Magazine* (typescript), 22 September 1946 ("The Atom Goes to Work for Medicine") (ACHRE No. DOE-051094-A-408), 2.

5. Ibid., 6.

6. Cohn, 3 January 1946, 10.

7. Ibid., 14.

8. R. E. Cunningham, 20 February 1971 ("Historical Summary of the Subcommittee on Human Applications") (ACHRE No. NRC-012695-A), 6.

9. AEC Subcommittee on Human Applications of the Committee on Isotope Distribution, 13 March 1949 ("Revised Tentative Minutes of March 13, 1949, Meeting of Subcommittee on Human Application of the Committee on Isotope Distribution of U.S. Atomic Energy Commission: AEC Building, Washington, D.C.") (ACHRE No. NARA-082294-A-62), 7.

10. Advisory Subcommittee on Human Applications of the Interim Advisory Committee on Isotope Distribution Policy, 11 July 1946 ("Minutes of Initial Meeting—Held June 28, 1946; Oak Ridge, Tennessee") (ACHRE No. NARA-082294-A-84), 1.

11. Ibid., 2-8.

12. Ibid., 5.

13. Ibid., 6.

14. Ibid., 10.

15. For example, proposals to study possible therapeutic uses of UX1/ UX2, a "naturally radioactive pair [that] behaves chemically as UX1, a thorium isotope (Th 234). . . . Aside from the danger of bone damage, the material would have to be used with much caution because of likely kidney damage. No advantage could be seen in the use of radiothorium over the use of certain other beta ray emitting radioisotopes which deposit in bone." Ibid., 9.

16. Ibid.

17. "In general, there is more of a need for <u>speed</u> in handling requests for human applications than for others because: (1) therapeutic action may be needed urgently, (2) the case may be an exceptionally good one for some purpose and may only be available for study immediately (for example, the chance to obtain tracer samples resulting from a special operation)." Ibid., 10.

18. Isotopes Branch, Research Division, Manhattan District, Oak Ridge, Tennessee, 3 October 1946 ("Details of Isotope Procurement") (ACHRE No. NARA-082294-A-31).

19. Isotopes Branch, Research Division, Manhattan District, Oak Ridge, Tennessee, 3 October 1946 ("Agreement and Conditions for Order and Receipt of Radioactive Materials") (ACHRE No. NARA-082294-A-31), 2.

20. The statement read:

This is to certify that the undersigned has adequate facilities for the investigation to be conducted by him as proposed in the 'Interim Period Request for Radioelement, Form 313,' Serial Number _____, and that such drug will be used solely by him or under his direction for the investigation, unless and until an application becomes effective with respect to such drug under section 505 of the Federal Food, Drug and Cosmetic Act, Isotopes Branch, Research Division, Manhattan District, Oak Ridge, Tenn.

Isotopes Branch, Research Division, Manhattan District, Oak Ridge, Tennessee, 3 October 1946 ("Certificate . . . EIDM Form 465") (ACHRE No. NARA-082294-A-31), 1.

21. Isotopes Branch, Research Division, Manhattan District ("Report of Requests Received to July 31, 1946," "2nd Report of Request Received August 1 to 31, 1946," "3rd Report of Request Received September 1 to 30, 1946," "4th Report of Requests Received October 1 to 31, 1946") (ACHRE No. NARA-082294-A-31).

22. Franklin asked:

What is the relationship of the Atomic Energy Commission Medical Division to the Isotopes Branch and the medical and biological aspects of the isotope distribution program?

(1) Will allocations for human administration be subject to medical review and what control will be exercised?

(2) What responsibilities does the Atomic Energy Commission bear for the human administration of isotopes (a) by private physicians and medical institutions outside of the Project, and (b) by physicians within the Project? This latter category includes contractor personnel employing Atomic Energy Commission funds (indirectly) to perform tracer research, some of which is of no immediate therapeutic value to the patient. What are the criteria for future human tracer research?

(3) What responsibilities does the Atomic Energy Commission bear for the safe handling by the recipient of the more hazardous radioisotopes?

(4) What responsibilities does the Atomic Energy Commission bear for radioactive waste disposal outside the Project?

J. C. Franklin, Manager, Oak Ridge Operations, to Carroll Wilson, AEC General Manager, 26 September 1947 ("Medical Policy") (ACHRE No. DOE-113094-B-3), 2.

23. Research Division, Manhattan District, 3 October 1946 ("Isotope Request, For Manhattan Project Use Only . . . EIDM Form 558") (ACHRE No. NARA-082294-A-31), 1.

24. In a 5 October 1949 memorandum to Carroll Tyler, Manager of Los Alamos, Paul Aebersold, Chief of the Isotopes Division, noted that "Dr. [Shields] Warren instructed that such allocations would be made by the Isotopes Division only after review and approval by the Subcommittee on Human Applications of the Commission's Committee on Isotope Distribution. It should be emphasized that the instruction applies even though the radiomaterial is produced in the laboratory where it is to be used.

"Since this procedure has not been uniformly followed in the past, we are writing to acquaint you with the appropriate details." Paul Aebersold, Chief, Isotopes Division, to Carroll Tyler, Manager, Los Alamos, 5 October 1949 ("Use of Radioisotopes in Human Subjects") (ACHRE No. DOE-021095-B-4), 1. An identical memo was also sent to the manager of the AEC's New York office regarding requirements for Brookhaven National Laboratory. Paul Aebersold, Chief, Isotopes Division, to W. E. Kelley, Manager, New York, 5 October 1949 ("Use of Radioisotopes in Human Subjects") (ACHRE No. DOE-012795B).

25. Presumably codifying existing practice, 10 C.F.R. 30.10 (1951 supplement to 1949 edition) states:

The regulations in this part do not apply to persons to the extent that such persons operate Commission-owned facilities in carrying out programs on behalf of the Commission. In such cases, the acquisition, transfer, use, and disposal of radioisotopes are governed by the contracts between such persons and the Commission, and internal bulletins, instructions and directives issued by the Commission.

26. Carroll L. Wilson, AEC General Manager, to Principal Staff, Washington, and Managers of Operations, 7 June 1950 ("Bulletin GM-161, Procedure for Securing Isotopic Materials and Irradiation Services") (ACHRE No. NARA-122994-B), 1.

27. Subcommittee on Human Applications, 13 March 1949, 1.

28. Ibid., 3.

29. Ibid. These minutes include a review of the minutes of the 22-23 March 1948, meeting.

30. Ibid., 10-11.

31. Ibid., 10.

32. Ibid., 12-13.

33. Ibid., 4.

34. Ibid.

35. Paul Aebersold, Chief, Isotopes Division, Oak Ridge Operations, to Hymer Friedell, G. Failla, Joseph G. Hamilton, and A. H. Holland, Jr., 9 March 1948 ("Meeting of Subcommittee on Human Applications in Washington, March 22 and 23") (ACHRE No. NARA-082294-A-17), 2.

36. Subcommittee on Human Applications, 13 March 1949, 5-6.

37. Ibid.

38. Clemens Benda, Director of Research and Clinical Psychiatry, to AEC Subcommittee on Human Applications, 29 September 1953 ("This letter is written in order to elicit your permission to administer a dose of 50 uc Ca45 to a moribund gargoyle patient now hospitalized in our institution . . ."), 1. Reproduced at appendix B-27, Task Force on Human Subject Research, to Philip Campbell, Commissioner, Commonwealth of Massachusetts Executive Office of Health and Human Services, Department of Mental Retardation, April 1994 ("A Report on the Use of Radioactive Materials in Human Subject Research that Involved Residents of State-Operated Facilities within the Commonwealth of Massachusetts from 1943 to 1973") (ACHRE No. MASS-072194-A).

39. Clemens Benda, Clinical Director, to "Parent," 28 May 1953 ("In previous years we have done some examinations in connection with the nutritional department of the Massachusetts Institute of Technology . . ."), 1. Reproduced at appendix B-23, Task Force on Human Subject Research, to Philip Campbell, April 1994.

40. Task Force on Human Subject Research, to Philip Campbell, April 1994, 16.

41. Subcommittee on Human Applications, 13 March 1949, 8.

42. AEC Isotopes Division, "General Authorizations for Procurement of Radioisotopes," *Isotopics: Announcements of the Isotopes Division* 1 (April 1951): 1-3.

43. AEC Isotopes Division, "General Authorizations for Clinical Use of Radioisotopes," *Isotopics: Announcements of the Isotopes Division* 2 (April 1952): 1-2.

44. Subcommittee on Human Applications, 13 March 1949, 11.

45. Paul C. Aebersold, Chief, AEC Isotopes Division, to William E. Barbour, Jr., President, Tracerlab, Inc., 11 April 1949 ("Violation of 'Acceptance of Terms and Conditions for Order and Receipt of Byproduct Materials [Radioisotopes]'") (ACHRE No. NARA-082294-A-4), 1.

46. William Barbour, President, Tracerlab, Inc., to Employees, April 1949 ("Violation of AEC Regulations") (ACHRE No. NARA-082294-A-4), 1. Barbour stated that a recurrence would

mean cessation of all radiochemical operations of the Company. In turn this would jeopardize the investments of several thousand new stockholders who have placed great faith in the integrity and ability of the management. A violation of a specific agreement with the AEC would be a breach of that faith and could only result in the automatic dismissal of anyone contributing to such a violation.

47. AEC Isotopes Division, 23 March 1950 ("Meeting of the Advisory Committee on Isotope Distribution, March 23 and 24, 1950, Washington, D.C., Minutes") (ACHRE No. NARA-122994-B-1), 4.

48. AEC Isotopes Division, September 1949 ("Supplement No. 1 to Catalogue and Price List No. 3") (ACHRE No. DOD-122794-A-1), 1.

49. Paul C. Aebersold, Director, Isotopes Division, to T. H. Johnson, Director, Division of Research, 2 November 1954 ("Providing Radioisotopes at Reduced Prices for Medical, Biological, or Other Research Uses") (ACHRE No. TEX-101294-A-4), 1.

50. 10 C.F.R. 37 (1961).

51. A conscious decision was made not to include detailed standards in the regulations. The discussion is summarized in Advisory Committee on Isotope Distribution, 23 March 1950, 7-8. The regulations were first promulgated in 10 C.F.R. 30.50 (1951 supplement to 1949 edition).

52. AEC Isotopes Division, 6 December 1948 ("Isotopes Division Circular D-4: Radioisotopes for Use in Medicine") (ACHRE No. DOE-101194-A-5); Isotopes Division, "Supplement No. 1," September 1949; Isotopes Extension, Division of Civilian Application, U.S. Atomic Energy Commission, "The Medical Use of Radioisotopes: Recommendations and Requirements by the Atomic Energy Commission," RC-12 (February 1956).

53. Isotopes Extension, February 1956, 14.

54. Ibid., 15.

55. R. E. Cunningham, "Historical Summary," 5.

56. AEC Division of Materials Licensing, "Non-Routine Medical Uses of Byproduct Material," *A Guide for the Preparation of Applications for the Medical Use of Radioisotopes* (November 1965), 47-48.

57. See, for example, Bryant L. Jones, Division of Oncology and Radiopharmaceuticals, Bureau of Medicine, Food and Drug Administration, 18 May 1967 ("FDA Responsibility in Radiopharmaceutical Research") (ACHRE No. DOE-051094-A-236).

58. Advisory Subcommittee on Human Applications, 11 July 1946, 6.

59. This requirement is stated in Aebersold's memo of 5 October 1949, quoted earlier in endnote 24, which notified AEC labs that their applications for human use would now be reviewed by the Subcommittee on Human Applications of the AEC's Committee on Isotope Distribution. Concerning local isotope committees, the memo states: "It should be emphasized that each application should be accompanied by a formal, written endorsement, signed by the Chairman of the local "Isotopes Committee," the recommended membership of which is outlined on pages 30 and 31 of the catalog." Paul Aebersold, Chief, Isotopes Division, to Carroll Tyler, Manager, Los Alamos, 5 October 1949 ("Use of Radioisotopes in Human Subjects") (ACHRE No. DOE-021095-B-4); Paul Aebersold, Chief, Isotopes Division, to W. E. Kelley, Manager, New York, 5 October 1949 ("Use of Radioisotopes in Human Subjects") (ACHRE No. DOE-012795-B).

60. AEC Isotopes Division, *Isotopics* 1, 1.

61. Isotopes Extension, February 1956, 7. The full description of the functions of the Medical Isotope Committee is:

1. Formation of a Medical Isotopes Committee. The Medical Isotope Committee shall include at least three members. Membership should include physicians expert in internal medicine (or hematology), pathology, or therapeutic radiology and a person experienced in assay of radioisotopes and protection against ionizing radiations. It is often appropriate that a qualified physicist be available to the Committee, at least in consulting capacity. It is recognized that the composition of local isotope committees may vary from institution to institution depending upon the individual interests of a particular medical facility.

2. Duties of the Medical Isotopes Committee

Generally, the Committee should have the following responsibilities:

a. Review and grant permission for, or disapprove, the use of radioisotopes within the institution from the standpoint of radiological health safety and other factors which the Committee may wish to establish for medical use of these materials.

b. Prescribe special conditions which may be necessary, such as physical examinations, additional training, designation of limited area or location of use, disposal methods, etc.

c. Review records and receive reports from its radiological safety officer or other individual responsible for health-safety practices.

d. Recommend remedial action when a person fails to observe safety recommendations and rules.

e. Keep a record of actions taken by the Committee.

62. The Advisory Committee also reviewed materials from the AEC's Oak Ridge, Los Alamos, Argonne, and Brookhaven laboratories, the Air Force School of Aviation Medicine, and the University of California. The development of research at the University of California at Berkeley and San Francisco is the subject of a case study appearing in a companion volume to this report.

63. N. W. Faxon, Director, Massachusetts General Hospital, to Drs. Aub, Moore, Shulz, and Rawson, 3 May 1946 ("At the meeting of the General Executive Committee held on May 1, 1946, consideration of the use of radioactive isotopes was discussed . . .") (ACHRE No. HAR-100394-A-1), 1.

64. "It should be emphasized that the University

Radiation Policy Committee was established to deal with all types of radiation problems at the University and was not limited to the scope of 'radioisotope committees' suggested by the AEC for radioisotope procurement. In fact this Committee predated the earliest suggestions of the AEC by almost a year." W. W. Meinke, Chairman, University of Michigan Radiation Policy Committee, to I. Lampe, 27 February 1956 ("On October 13, 1950, the President of the University of Michigan established the Radiation Policy Committee . . .") (ACHRE No. MIC-010495-A-2), 1.

65. Consisting of Hugh Morgan (Vanderbilt University), Stafford Warren (University of California at Los Angeles), Hymer Friedell (Case Western Reserve University), Shields Warren (AEC Division of Biology and Medicine), and Perrin Long (Johns Hopkins University).

66. There was some debate at the beginning as to the name of the units. With "radioactive" still a charged word for much of the population, an early memo suggested that "it could to advantage be called a Metabolism Ward." Veterans Administration, 15 September 1948 ("Minutes of the Meeting, Central Advisory Committee on Radioisotopes, U.S. Veterans Administration") (ACHRE No. UCLA-100794-A), 23.

67. The chairman listed the already-achieved benefits to thyroid gland research and blood volume diagnosis, and claimed, "It is not an overstatement to say that progress can be expected to be rapid and on a wide front as greater use is made in medical and biological research when this new tool is applied in attempts to solve such problems." Ibid., 3.

68. Framingham, Massachusetts; Bronx, New York; Cleveland, Ohio; Hines, Illinois; Minneapolis, Minnesota; Van Nuys, California; Los Angeles, California; and Dallas, Texas.

69. Joseph C. Aub et. al., to the Executive Committee, Massachusetts General Hospital, 17 June 1946 ("The Radioactive Isotope Committee had its first meeting on June 15th . . .") (ACHRE No. HAR-100394-A-2), 1-2.

70. William Sweet, interviewed by Gilbert Whittemore (ACHRE), transcript of audio recording, 8 April 1995 (ACHRE Research Project Series, Interview Program File, Targeted Interview Project), 20.

71. VA Central Advisory Committee on Radioisotopes, 15 September 1948, 26.

72. Ibid.

73. University of Michigan Subcommittee on Human Use of Isotopes, 10 December 1956 ("Minutes, Meeting of the Subcommittee on Human Use of Isotopes") (ACHRE No. MIC-010495-A-3), 1.

74. William H. Beierwaltes to Edward A. Carr, Chairman, University of Michigan Subcommittee on Human Use of Radioisotopes, 20 May 1968 ("Enclosed are our calculations to date on our first two patients studied in the Clinical Research Unit . . .") (ACHRE No. MIC-010495-A-6), 3. The form in-

cludes space for a signature by a witness as well as the patient.

75. In an effort to develop an overall assessment of the possible harm from radioisotope experiments conducted in the past, the Advisory Committee extracted dose data from our Experiment Database, whenever available, in order to perform risk analyses using contemporary standards. Unfortunately, most of the data recovered by the Committee was fragmentary and did not provide a sufficient basis for an analysis of possible harm in most cases.

76. Massachusetts General Hospital Radioactive Isotope Committee, 15 March 1955 ("Meeting of the Massachusetts General Hospital Radioactive Isotope Committee") (ACHRE No. HAR-100394-A-4), 1.

77. Ibid.

78. One proposal, for example, involved saturating gelfoam with silver 111 or yttrium 90, and then implanting the gelfoam into the tumor. Preliminary work had been done on animals in the previous year on normal brain tissue. After extensive animal testing, the procedure was to be attempted on those humans who already suffered brain cancer and had undergone surgery. Theodore Rasmussen, 29 May 1952 ("Local Application of Beta Ray Isotopes to Brain Tumors") (ACHRE No. DOE-122194-A).

79. For example, in 1953 the Chicago committee approved a proposal to use tritium and C-14-labeled acetate to trace the development of adrenal cholesterol in advanced cancer patients as well as a control group. The committee noted that the doses "are smaller than have been used in human studies at other institutions and in no case involve amounts which will produce internal radiation in excess of maximum permissible dose." George V. LeRoy, Chairman, Radioisotope Committee, 24 February 1953 ("Minutes of the Radioisotope Committee Meeting") (ACHRE No. DOE-122194-A), 1.

80. This included recommendations for using the minimum amounts of isotopes possible, a limitation of 1 rep [roentgen equivalent physical] for tracers, mandatory blood tests before administration and forty-eight hours after, and a listing of dose recommendations. The policy on patients and children was specific: "Adult humans who are ill and who are expected to receive benefit from the procedure, shall not receive tracer doses of radioactive material giving off radiation in excess of a total of 4 rep. Children (all patients below 15 years of age) shall not receive more than a total of 0.8 rep." J. C. Aub, A. K. Solomon, and Shields Warren, Harvard Medical School, 7 May 1949 ("Tracer Doses of Radioactive Isotopes in Man") (ACHRE No. HAR-100394-A-3), 1.

81. The committee stated that all volunteers receiving Na-22 and K-42 should be subjected to doses no more than 100 millirads for the whole body, nor more than one-third the maximum permissible values to a specific organ. University of Michigan Subcommittee on Human Use of Isotopes, 10 December 1956, 1.

82. W. F. to University of Chicago Radioisotope Committee, 28 September 1953 ("Permission is requested to administer intravenously 500 microcuries, or less, of radio-mercury to a patient . . .") (ACHRE No. DOE-122194-A-2), 1.

83. Harvard Medical School Committee on Medical Research in Biophysics, August 1957 ("Tracer Doses of Radioactive Isotopes in Man") (ACHRE No. HAR-100394-A-5), 2.

84. University of Michigan Subcommittee on Human Use of Isotopes, 27 September 1955 ("Minutes of Human Use Committee Meeting") (ACHRE No. MIC-010495-A), 2.

85. A 1963 memorandum indicates the committee's unwillingness to allow a procedure involving selenium 75-labeled methionine for parathyroid scanning limited to use in patients over forty years old, while in a 1966 letter Carr stated that he was "strongly inclined to refuse to permit the use of radioisotopes in all volunteers below the age of 21, unless there are special mitigating circumstances approved by the whole subcommittee." Ronald C. Bishop, Acting Chairman, Subcommittee on Human Use, 13 August 1963 ("Dr. E. A. Carr has asked me to act as chairman of the Subcommittee on Human Use in his absence . . .") (ACHRE No. MIC-010495-A-4), 1; Edward A. Carr to Dr. Bishop, 3 September 1966 ("To Members of the Subcommittee on Human Use of Radioisotopes") (ACHRE No. MIC-010495-A-5), 1.

86. In 1968 the committee approved a proposal for an experiment that involved doses of NM-125 labeled with I-131 or I-125 for patients with melanomas or a reasonable clinical suspicion of melanoma for thirty patients, and then wished to see results before approving of further administration. Likewise, the committee gave approval to a closely related experiment involving use of the same substances in patients with lung cancer. For that regime, the committee demanded feedback after fifteen patients. For a further related matter involving the same substances in patients with pulmonary carcinoma, the committee limited the work to five patients. In each case the dose was to exceed 2 millicuries per patient. Edward A. Carr, Chairman, Subcommittee on Human Use, to William H. Beierwaltes, Director, Nuclear Medicine, 27 September 1968 ("This is to inform you that the Sub-committee on Human Use of Radioisotopes, at its meeting of September 26, 1968, approved the use of a single dose of NM-113 . . .") (ACHRE No. MIC-010495-A-6), 1.

87. A researcher had applied to use sodium 22 in a tracer procedure with several patients. The committee was concerned that "a small but significant fraction of one of the radioisotopes might remain localized in the body for a long period of time . . ." Edward A. Carr, 3 June 1968 ("Sub-committee on Human Use of Radioisotopes, Minutes of the Meeting of June 3, 1968") (ACHRE No. MIC-010495-A-7), 1.

88. George V. LeRoy, 3 November 1953 ("Minutes of the Radioisotope Committee Meeting") (ACHRE No. DOE-122194-A-3), 1.

89. Paul C. Aebersold, Chief, Isotopes Division, to John Z. Bowers, Assistant to Director, Division of Biology and Medicine, 18 March 1948 ("Investigation of Patients Who Have Received Radioactive Isotopes") (ACHRE No. DOE-061395-E-1), 1.

90. A comprehensive history of the application of radioisotopes is well beyond the scope of this chapter and would needlessly duplicate substantial histories already written. See, for example, J. Newell Stannard, *Radioactivity and Health: A History* (Springfield, Va.: Office of Scientific and Technical Information, 1988).

91. An example is Konstantin N. Pavlou, William P. Steffee, Robert H. Lerman, and Belton A. Burrows, "Effects of Dieting and Exercise on Lean Body Mass, Oxygen Uptake, and Strength," *Medicine and Science in Sports and Exercise* 17 (1985): 466-471. The study was conducted at the Boston University Medical School and the Boston VA Medical Center.

92. There is a vast literature on radioiodine and the thyroid. Government studies specifically noted by the Veterans Administration as significant are the following: H. C. Allen, R. A. Libby, and B. Cassen, "The Scintillation Counter in Clinical Studies of Human Thyroid Physiology Using I-131," *Journal of Clinical Endocrinology and Metabolism* 11 (1951): 492-511; B. A. Burrows and J. A. Ross, "The Thyroid Uptake of Stable Iodine Compared with the Serum Concentration of Protein-Bound Iodine in Normal Patients and in Patients with Thyroid Disease," *Journal of Clinical Endocrinology and Metabolism* 13 (1953): 1358-1368; S. A. Berson and R. S. Yalow, "Quantitative Aspects of Iodine Metabolism: The Exchangeable Organic Iodine Pool, and the Rates of Thyroidal Secretion, Peripheral Degradation and Fecal Excretion of Endogenously Synthesized Organically Bound Iodine," *Journal of Clinical Investigation* 33 (1954): 1533-1552; M. A. Greer and L. J. DeGroot, "The Effect of Stable Iodide on Thyroidal Secretion in Man," *Metabolism* 5 (1956): 682-696; K. Sterling, J. C. Lashof, and E. B. Man, "Disappearance from Serum of I-131 Labeled I-Thyroxine and l-Triiodothyronine in Euthyroid Subjects," *Journal of Clinical Investigation* 33 (1954): 1031; K. Sterling and R. B. Chodos, "Radiothyroxine Turnover Studies in Myxosema, Thyrotoxicosis, and Hypermetabolism Without Endocrine Disease," *Journal of Clinical Investigation* 35 (1956): 806-813.

93. See, for example, J. F. Ross, "Cooperative Study of Radioiodine Therapy for Hyperthyroidism," *Bulletin of the Committee on Veterans Medical Problems* (National Academy of Sciences) (1952): 576-578.

94. Gould A. Andrews, M.D., Samuel W. Root, M.D., and Herbert D. Kerman, M.D., "Clinical Studies with Gallium-72," 570-588 in Marshall

Brucer, M.D. (ed.), Gould Andrews, M.D., and H.D. Bruner, M.D., "Clinical Studies with Gallium-72," *Radiology* 66 (1953): 534-613.

95. OncoScint, developed by Cytogen, was approved by the FDA for diagnosis of colorectal and ovarian cancers on 29 December 1992, Product License Application no. 89-0601, with Amendment no. 90-0278. The use of monoclonal antibodies to treat cancer is discussed in Oliver W. Press, M.D., Ph.D., et al., "Radiolabeled-Antibody Therapy of B-Cell Lymphoma with Autologous Bone Marrow Support," *New England Journal of Medicine* 329 (21 October 1993): 1219-1224. Progress in the field is reviewed in an accompanying editorial, Robert C. Bast, Jr., M.D., "Progress in Radioimmunotherapy," *New England Journal of Medicine* 329 (21 October 1993): 1266-1268.

96. Strontium 89, commercially available as Metastron from Amersham-Mediphysics, was approved on 18 June 1993, New Drug Application no. 20134. One of its therapeutic uses is described in an article by Arthur T. Porter. M.D., and Lawrence P. Davis, M.D., "Systemic Radionuclide Therapy of Bone Metastases with Strontium-89," *Oncology* 8 (February 1994): 93-96.

97. R. S. Yalow and S. A. Berson, "Assay of Plasma Insulin in Human Subjects by Immunological Methods," *Nature* 184 (1959): 1648.

7

Nontherapeutic Research
on Children

I N the late 1940s and again in the early 1950s, Massachusetts Institute of Technology scientists conducting research fed breakfast food containing minute amounts of radioactive iron and calcium to a number of students at the Walter E. Fernald School, a Massachusetts institution for "mentally retarded" children.[1] The National Institutes of Health, the Atomic Energy Commission, and the Quaker Oats Company funded the research, which was designed to determine how the body absorbed iron, calcium, and other minerals from dietary sources and to explore the effect of various compounds in cereal on mineral absorption.

In 1961, researchers from Harvard Medical School, Massachusetts General Hospital, and Boston University School of Medicine administered small amounts of radioactive iodine to seventy children at the Wrentham State School, another Massachusetts facility for mentally retarded children. With funding from the Division of Radiologic Health of the U.S. Public Health Service, the scientists conducting this experiment used Wrentham students to test a proposed countermeasure to nuclear fallout. Specifically, the study was meant to determine the amount of nonradioactive iodine that would effectively block the uptake of radioactive iodine that would be released in a nuclear explosion.

Recently, these two studies have received con-siderable media attention, and an official Massachusetts state task force has reported on both episodes in some detail.[2] Although they represent special cases because they involve institutionalized children, the Fernald and Wrentham experiments nonetheless are the most widely known examples of a category of research that raises particular concerns for the Committee: nontherapeutic experimentation on children.

Experiments involving children are important to the Committee for two reasons. First, children are more susceptible than adults to harm from low levels of radiation, and thus as a group they are more likely than adults to have been harmed as a consequence of their having been subjects of human radiation experiments. Second, an evaluation of research with children is critical to determining whether any former subjects of radiation experiments should be notified in order to protect their health, one of our specific charges.[3] Subjects who were children at the time of their exposure are more likely than adults to be candidates for such notification, both because of their increased biological sensitivity and because they are more likely to still be alive. (See chapter 18 for the Committee's recommendations with respect to notification and follow-up.)

We elected to focus on pediatric research that offered subjects no prospect of medical benefit, so-called nontherapeutic research, because it is this kind of research that has generated the most public concern and is the most ethically problematic. This is not to say, however, that experiments on children in which the children stand to benefit medically never raise ethical issues; such research certainly can and does. But in deciding how to allocate our limited resources, we chose to concentrate where the issues are mostly sharply drawn. Also, because most nontherapeutic research with children involved tracer doses of radioisotopes, focusing on this work allowed us a window into radio-isotope research generally.

We begin the chapter by setting the context for nontherapeutic radiation experiments on children. We review those factors that make nontherapeutic research on children ethically problematic and how such research has been viewed historically. We next consider what the practices and standards were for research on children in the 1940s, 1950s, and 1960s. This is a continuation of the discussion in chapter 2, which focused on professional standards and practices for human research.

The next three sections address human radiation experiments in terms of the central ethical issues raised by nontherapeutic research involving children—level of risk, authorization for the involvement of children, and selection of subjects. To address the question of risk, we analyzed twenty-one nontherapeutic radiation experiments with children conducted during the 1944–1974 period. The focus of this analysis is whether it is likely that any of the subjects of these experiments was harmed or remains at risk of harm attributable to research exposures. A table summarizing these experiments and our risk estimates can be found at the end of this chapter. The twenty-one experiments were selected from eighty-one pediatric radiation research projects identified by the Committee from government documents and the medical literature. Although these eighty-one by no means constitute all the pediatric radiation research conducted during this time, they include what are likely fairly typical examples of such research. Of the eighty-one, thirty-seven studies were judged to be nontherapeutic, and twenty-one of these were conducted or funded by the federal government and thus fell under

the charge of the Committee. Included within these twenty-one studies were the two nutrition experiments conducted at the Fernald School and one fallout-related study conducted at the Wrentham School discussed in the introduction to this chapter. All twenty-one studies employed radioisotopes to explore human physiology and pathology.

We turn next to a consideration of how authorization for the inclusion of the children in these experiments was obtained and who these children were. Unfortunately, for most of these experiments, little is known about either of these issues. The last section of the chapter focuses specifically on the experiments at the Fernald School where, thanks to the work of the Massachusetts Task Force on Human Subject Research, relevant information is available. Throughout the chapter, we focus only on research in which children could not have benefited medically. The Committee did not have the resources to pursue two related areas of research—nontherapeutic research on pregnant women and therapeutic research on children. We include two capsule descriptions of examples of these types of research at the end of this chapter.

THE CONTEXT FOR NONTHERAPEUTIC RESEARCH WITH CHILDREN

Children as Mere Means

In both law and medical ethics, it has long been recognized that children may not authorize medical treatment for themselves, except in special circumstances.[4] Instead, authorization must be sought from the parent. Historically, the source of this respect for parental authority rested upon the view that children were the property of their parents, and thus parents had the right to determine how their "property" was to be treated. Today, we still speak of parental rights, although the justification for these rights no longer rests on an analysis of children as property. Instead, respect for the rights of parents is viewed as a mechanism for valuing and fostering the institution of the family and the freedom of adults to perpetuate family traditions and commitments. Another important line of justification for respecting the authority of parents relies not on a recognition of parental rights but on the view that the interests of the child are generally best served by ceding decisional authority to the parent. The parent is thought not only to be in the best position to determine what is in the interests of the child but is also thought to be generally motivated to act in the child's best interests.[5]

When research involving children offers a prospect of medical benefit to the child-subject, the application of the above analysis is straightforward. Parents are generally thought to have the authority to determine whether their children should be made subjects of such research. Certainly today, any use of a child in research would not be ethically acceptable or legally permissible without the parent's permission.[6] Where the research does not offer any prospect of benefit to the child, however, the legitimacy of the parent as authorizer is less clear.

Respect for the authority of parents to make decisions for their children and otherwise control their children's lives is not without bounds. The law recognizes several exceptions, designed primarily to protect the child from what society at large considers to be unacceptable or unjustifiable harm or risk of harm.[7] Laws against the physical abuse of children are perhaps the most obvious example of such limitations on parental authority. In the context of research, the question arises of whether a parent has the authority to permit a child to be put at risk of harm in an experiment from which the child could not possibly benefit medically. In this case, the child is to be used as a means to the ends of others. Children are not in a position to determine for themselves whether they wish to agree to such a use and thus cannot themselves render the use morally acceptable. Should parents have such authority? Should anyone?

This question was resolved as a matter of public policy in the 1970s through the work of the National Commission for the Protection of Human Subjects of Biomedical and Behavioral Research and the subsequent adoption, in 1983, of federal regulations governing research involving children that were guided by the recommendations of the National Commission.[8] These regulations state that children can participate in federally funded research that poses greater than minimal risks to the subject if a local review committee (an institutional review board, or IRB) finds that the potential risk is "justified by the anticipated benefit to the subjects"; "the relation of the anticipated benefit to the risk is at least as favorable to the subjects as that presented by available alternative approaches"; and "adequate provisions are made for soliciting the assent of the children and permission of their parents or guardians."[9] The word *consent* is purposely avoided in these regulations to distinguish parental *permission* and minor *assent* from the autonomous, legally valid *consent* of a competent adult.

Federal regulations do allow nontherapeutic research on children if an IRB determines that the research presents "no greater than minimal risk" to the children who would serve as subjects, although no clear definition of what constitutes minimal risk is provided.[10] As with therapeutic pediatric research, parents or guardians must grant "permission" and children who are deemed capable must offer "assent."

The regulations also allow for nontherapeutic research with children that *does* present more than minimal risk, again with parental permission and assent of the child (as appropriate), *but only if* the risk represents a minor increase over minimal risk, the procedures involved are commensurate with the general life experiences of subjects, and the research is likely to yield knowledge of "vital importance" about the subjects' disorder or condition.[11] Research with children that is not otherwise approvable may be permitted, but only under special, and presumably rare, circumstances. In addition to local IRB review, such research must withstand the special scrutiny of the secretary of the agency sponsoring the research, who is to be advised by a special IRB.[12] The secretary must also allow the opportunity for "public review and comment" on a proposed nontherapeutic research project that is not otherwise approvable.

The regulations thus draw a sharp distinction between therapeutic and nontherapeutic research. Nontherapeutic research, while severely restricted, is not banned. The decision to permit parents to authorize the use of their children in nontherapeutic research reflects both the recognition that some important advances in pediatrics could come only from research with children that was of no benefit to them and the recognition that we all—as parents, as potential future parents, and as members of society—share in the interest of advancing the health of the young. At the same time, however, parental authority to permit such use of a child is generally restricted to research judged to pose little risk; as important as it is to promote the welfare of children (as a class), this interest justifies only minor infringements of the principle not to use people as mere means to the ends of others.

These 1983 regulations, and the reasoning behind them, were the culmination of considerable public debate and scholarly analysis, much of which occurred in the 1970s. To situate properly the experiments of interest to the Committee, it is necessary to examine the social and professional roots of the issues and arguments that ultimately led to the federal regulations.

Public Attitudes, Professional Practices

Attitudes and Practices Prior to 1944

There was significant research interest in infants and children as early as the eighteenth century, as scientists began to experiment with vaccines and immunization. Children were particularly valuable subjects for this type of research because in general, they were less likely than adults to have been exposed to the disease being studied.[13] A child's response to immunizations was also of great interest because most immunizations are performed during childhood.

During the nineteenth century, the Industrial Revolution greatly increased the number of child laborers, and the public began to acknowledge the need for laws to protect children from abuse.[14] Physicians started to specialize in pediatrics, studying specifically the health problems and diseases that afflicted children. Simultaneously, as social reformers were creating a wide range of institutions for children, such as orphanages, schools, foundling homes, and hospitals, scientists recognized the value of research conducted in these types of institutions. In the

late nineteenth and early twentieth centuries, Alfred F. Hess, the medical director of the Hebrew Infant Asylum in New York City, conducted pertussis vaccine trials and undertook extensive studies of the anatomy and physiology of digestion in infants at the asylum. According to Advisory Committee member and historian Susan Lederer, Hess sought to take advantage of the conditions in the asylum as they approximated those "conditions which are insisted on in considering the course of experimental infection among laboratory animals, but which can rarely be controlled in a study of infestation in man."[15]

Although many shared Hess's laudable goal of improving the health of asylum children, many people drew the line at the pediatrician's investigations of scurvy and rickets. In order to study the disease, Hess and his colleagues withheld orange juice from infants at the asylum until they developed lesions characteristic of scurvy. Responding to the public discussion of the ethics of using children in such nontherapeutic experiments, the editors of one American medical journal insisted that such investigations gave the children an opportunity to repay their debt to society, even as they conceded that experimentation on human beings should be limited to "children as may be utilized with parental consent."[16]

Hess's work was not the only case in which experiments involving children attracted negative public opinion. In 1896, for example, American antivivisectionists attacked a Boston pediatrician, Arthur Wentworth, who performed lumbar punctures on infants and children in order to establish the safety and utility of the procedure. The antivivisectionists were particularly alarmed because this procedure, which caused pain and discomfort, did not confer any benefits to the subjects. John B. Roberts, a physician from Philadelphia, labeled Wentworth's procedures "human vivisection," saying that "using the children in the hospital without explaining his plan to their mothers or gaining their permission intensified public fear of hospitals."[17]

The twentieth century brought new drugs and advanced technologies, which allowed for increased research on children. The conduct of this experimentation, however, was largely left to the individual investigator. When his experimental gelatin injections provoked "alarming symptoms of prostration and collapse in three normal children (including a 'feeble-minded' four-year-old girl), the physician Isaac Abt stopped his pediatric experiments and began experimenting on rabbits."[18] Meanwhile, legislation was being proposed throughout the country to protect children and pregnant women from experimenting physicians. Two proposals were introduced in the U.S. Senate in 1900 and 1902; proposals "'to prohibit such terrible experiments on children, insane persons and pregnant women . . . ,' and to ensure 'that no experiment should be performed on any other human being without his intelligent written consent' were introduced in the Illinois legislature" in 1905 and 1907; in 1914 and 1923, the New York legislature considered bills that prohibited experimentation with children.[19] Although these bills did not become law, it is clear that some unease concerning nontherapeutic research on children existed among the public and elected officials.

Reaction to the polio vaccine trials conducted during the 1930s further demonstrated the growing discomfort over pediatric experimentation as thousands of American children were involved in what some considered at the time to be premature human trials of the polio vaccine. Although it appears that parental consent was obtained for a number of these studies, the controversy over these trials stalled polio vaccine research for almost two decades and generally made investigators ambivalent about the use of human subjects.[20]

Although there are no legal cases that bear directly on nontherapeutic research with children during this period, an appellate court ruling in 1941, *Bonner v. Moran*, involving the performance of a nontherapeutic medical procedure on a child without parental consent, suggests how such a case might have been decided.[21] John Bonner, a fifteen-year-old African-American boy from Washington, D.C., had undergone an experimental skin graft for the benefit of Clara Howard, a cousin suffering from severe burns. When he discovered that John Bonner had the same blood type as the burn victim, Howard's plastic surgeon, Robert Moran,

persuaded Bonner to allow him to fashion "a tube of flesh" by cutting from the boy's "arm pit to his waist line."[22] This procedure, however, was conducted without the consent of a parent, as "his mother, with whom he lived, was ill at the time and knew nothing about the arrangement."[23] Moran then attached the free end of Bonner's flesh tube to Clara Howard, hoping that the flesh-and-blood link would bring benefit to the burned girl. Due to poor circulation in the tube, the procedure did not help the burn patient and put the healthy boy, who was required to stay in the hospital for two months, at significant risk (and left him with permanent scars). Bonner's mother brought suit against Moran for assault and battery.

The appellate court based its ruling against Moran on what it perceived as a disturbing combination of a lack of direct benefit for John Bonner and a lack of permission from the boy's mother:

[H]ere we have a case of a surgical operation not for the benefit of the person operated on but for another. . . . We are constrained, therefore, to feel . . . that the consent of the parent was necessary.[24]

The court did not refer to the episode as an instance of experimentation, but the parallels between this novel procedure performed for the benefit of another and a nontherapeutic medical experiment are quite powerful.[25]

Attitudes and Practices 1944–1974

As best the Committee can establish, there were no written rules of professional ethics for the conduct of research on children prior to 1964. Taken literally, the Nuremberg Code, which requires that all subjects of research "have legal capacity to give consent," precludes all research with children.[26] There is no reason to believe, however, that the judges at Nuremberg meant to impose such a prohibition, and the Nuremberg Code did not result in a ban on research with children.

Pediatric research flourished after World War II, as did biomedical research in general. What is less clear is how this research was conducted, and on whom. One source of evidence about legal thinking on pediatric research, if not actual practice, is the writings of Irving Ladimer, a lawyer who, in 1958, was completing a doctoral degree in juridical science at the same time he was employed as an administrator at the National Institutes of Health. Ladimer concluded his doctoral dissertation, "Legal and Ethical Implications of Medical Research on Human Beings," with an appendix devoted to the issues surrounding "Experimentation on Persons Not Competent to Provide Personal Consent," whom he defined broadly as minors and mental incompetents.[27] Ladimer argued that it was "permissible to employ minors and incompetents as subjects of medical investigations . . . where there is informed consent by a parent or guardian (including the state) for procedures which also significantly benefit or may be expected to benefit the individual."[28] Ladimer was less sanguine, however, about nontherapeutic research with these populations. He expressed particular concern about the use of institutionalized children—even with proxy permission—in research that did not hold the possibility of personal benefit: "Permission given by parents or the state to utilize institutionalized children, without any suggestion of benefit to the children, may well be beyond the ambit of parental or guardianship rights."[29]

Ladimer did, however, leave open a window for the use of legally incompetent subjects in nontherapeutic research, but he clearly harbored great discomfort with his own suggestion:

[T]he availability of certain persons, not able to consent personally, may constitute a strategic resource in terms of time or location not otherwise obtainable. It must be remembered, however, that the Nazis hid behind this rationalization in explaining certain highly questionable or clandestine medical experiments. Such justification should not even be considered except in dire circumstances. If ever employed, it should not be assimilated into the concept of personal benefit, else there may be no legal or ethical control for the protection of both prospective subject and investigator and their individual integrity.[30]

As part of the Committee's Ethics Oral History Project, we interviewed two pediatricians who were beginning their careers in academic medicine in the late 1940s. One of these respondents, Dr. Henry Seidel, had some research experience with institutionalized children. He noted that "we got access [to the children] very easily," and although his research was merely observational, it was "not hard to imagine" that experimental research with these children could have been conducted.[31] When asked about the studies conducted by Dr. Saul Krugman on institutionalized children at the Willowbrook State School (discussed later in this chapter), Seidel observed, "I didn't have any problem imagining that possibility. In retrospect, I'm sure it could happen, you know. There was something about those reports that rang true. . . ."[32] William Silverman, the other pediatrician interviewed, had clear recollections of how research was conducted in pediatrics at that time. He recalled that, in the 1950s, many pediatricians, including himself, believed that it was not necessary to obtain the permission of parents before using a pediatric patient as a subject in research—even if the research was nontherapeutic (he has since become a strong proponent of the parental permission requirement in pediatric research).[33] He also asserted that performing nontherapeutic experiments on children without authorization from parents was part of a broader "ethos of the time" in which "everyone was a draftee" in a national war on disease.[34] Dr. Silverman's account squares with the picture that emerged in chapter 2 of practices in research with adults, in which it was not uncommon to use adult patients as subjects of research without their knowledge or consent.

Silverman was among the researchers invited by Boston University's Law-Medicine Research Institute (LMRI) to participate in a conference on "Social Responsibility in Pediatric Research" held in May 1961.[35] This meeting was one in a series of closed-door conferences organized by LMRI to investigate actual practices among clinical researchers. The transcripts of the conference provide an important window onto practices and attitudes of the time; in large measure, they confirm Silverman's recollection of his own position some thirty-five years ago. Early in the meeting, Silverman asserted that "there is an unwritten consent by being a living person at this time to participate in this kind of advancement of knowledge [that is, nontherapeutic pediatric research]."[36] Some of the other participants employed the same analogy to the military draft that Silverman recently used to relate his recollections.

However, there was by no means unanimity about the appropriateness of this view:

Dr. A: [Dr. B] says that this [research without consent] is like military conscription.

Dr. C: Not comparable. We voted to do military conscription.[37]

The proceedings of the conference suggest that while it may not have been uncommon for pediatric patients to be used as subjects of nontherapeutic research without the permission of their parents, at least some physician-investigators, including investigators who followed this practice, thought it was morally wrong to do so. Consider, for example, a story relayed by one pediatrician-investigator at the conference who seemed to embrace with particular earnestness the desire of the conference organizers to learn the unvarnished reality of clinical research. In the opening minutes of the meeting, this researcher reminded his colleagues that "the question for us to discuss here today is how we operate on a daily basis."[38] He offered for discussion a provocative case from his personal experience in which he and his associates "wanted [to do] lumbar punctures on newborns."[39] He explicitly noted that "this study [was] not of benefit to the individual; it was an attempt to learn about normal physiology."[40] One of the other conferees asked, "Did you ask [parental] permission?" The researcher responded, "No. We were afraid we would not get volunteers."[41] The case prompted a great deal of discussion at the conference, but perhaps most tellingly this researcher frankly acknowledged toward the end of the discussion—in a meeting that had begun with an assurance of confidentiality from the organizers—that he had "sinned" in carrying out these lumbar punctures in "normal infants" without parental permission.[42]

The proceedings of the conference also suggest that at least some pediatrician investigators routinely obtained the permission of parents before embarking on research with their children. It is perhaps significant that the pediatric researcher who articulated this position at the conference was from Canada—and the conference transcript seems to suggest that he was pro-

viding a general characterization of practices in his country:

Dr. A: Let's ask [Dr. B] from Canada.

Dr. B: We have been quite sticky on consent. If we want a biopsy or a radioactive exposure and the parent says "no" then we don't do it. . . . The question of morals is too valuable.[43]

If this statement represents the sensitivity of Canadian pediatrician-investigators to issues of parental permission (which this single quotation does not prove), there is no obvious explanation as to why many of their colleagues in the United States behaved differently.

The LMRI conference is noteworthy not only for what it reveals about the range of views and practices concerning parental permission for nontherapeutic research, but also for the unanimity expressed about the importance of obligations to prevent or minimize harm to pediatric subjects of research. Minimizing risk was recognized by those at the conference as the most important (and, for some participants, the only) moral duty of pediatric investigators.[44]

Three years after the LMRI conference, in the summer of 1964, the World Medical Association ratified a code of ethics for human experimentation at a meeting in Helsinki. Unlike the Nuremberg Code, this statement, known as the Declaration of Helsinki, recognizes that research may be conducted on people with "legal incapacity to consent."[45] The Declaration distinguishes between two kinds of research: "Clinical Research Combined with Professional Care" and "Non-therapeutic Clinical Research."[46] It permits the use of people with legal incapacity to consent as subjects in both kinds of research, provided that the consent of the subject's legal guardian is procured.

Subjects of the first kind of research are referred to as patients; disclosure to and consent from patient-subjects are required by the Declaration, "consistent with patient psychology."[47] The Declaration does not specify whether considerations of "patient psychology" also could justify not obtaining the consent of the guardian where the subject does not have the legal capacity to consent.

The subjects of "non-therapeutic clinical research" are not referred to as patients but as human beings who must be "fully informed" and whose "free consent" must be obtained.[48] The Declaration also requires that nontherapeutic research be discontinued if in the judgment of the investigators to proceed would "be harmful to the individual."[49] Thus, although the Declaration permits parents to authorize the use of their children as subjects in nontherapeutic research, such research is not intended to be "harmful" to the subjects.

The language and reasoning of the Declaration was unclear and confusing with regard to clinical research, both therapeutic and nontherapeutic, on legally incapacitated individuals. It was revised in 1975, at a time when the ethics of research with human subjects was receiving considerable public attention in the United States (see chapter 3).

Both in the 1960s and early 1970s, public controversies erupted about several cases of research involving human subjects, controversies that led to the establishment of the National Commission and publication of the federal regulations (see chapter 3). One of the most well known of these cases involved research on institutionalized children. During the 1950s and 1960s, Dr. Saul Krugman of New York University conducted studies of hepatitis at the Willowbrook State School, an institution for the severely mentally retarded.[50] To study the natural history, effects, and progression of the disease, Krugman and his staff systematically infected newly arrived children with strains of the virus. Although the investigators did obtain the permission of the parents to involve their children in the research, critics of the Willowbrook experiments maintained that the parents were manipulated into consenting because, at least in the later years of the research, the institution was overcrowded and the long waits for admittance were allegedly shorter for children who were entering the research unit. Henry Beecher, a Harvard anesthesiologist whose impact on the history of research ethics is detailed in chapter 3, condemned Krugman and his staff for not properly informing the parents about the risks involved in the experiment.[51] Beecher also challenged the legal status of parental consent when no therapeutic benefit for the child was anticipated. A New York state senator, Seymour R.

Thaler, criticized the Willowbrook research on the pages of the *New York Times* in 1967, only to come to its defense later in 1971. Also in the early 1970s Willowbrook became the subject of a heated debate in the medical literature.[52]

Interestingly, Dr. Krugman was one of the participants at the LMRI "Social Responsibility in Pediatric Research" conference where he expressed pride that he routinely obtained permission from the parents of the children in his studies. In that group in 1961, Krugman was thus among those pediatric investigators most sympathetic to the position that children could not be used as mere means to the ends of the researcher without the authorization of the parent.

AEC Requirements for Radiation Research With Children

Although in the 1940s and 1950s there were apparently no written rules of professional ethics for pediatric research in general, there were guidelines for the investigational use of radioisotopes in children. In 1949, the Subcommittee on Human Applications of the Atomic Energy Commission's Isotope Division established a set of rules to judge proposals submitted by researchers for the use of radioisotopes in medical experiments with human subjects, including "normal children."[53] These standards appeared in the fall 1949 supplement to the AEC's isotope catalogue and price list. Under the heading "Normal Children" the isotope catalogue offered the following statement:

In general the use of radioisotopes in normal children is discouraged. However, the Subcommittee on Human Applications will consider proposals for such use in important researches, provided the problem cannot be studied properly by other methods and provided the radiation dosage level in any tissue is low enough to be considered harmless. It should be noted that in general the amount of radioactive material per kilogram of body weight must be smaller in children than that required for similar studies in the adult.[54]

These guidelines did not mention consent— of parents, guardians, or children.[55] Instead, this statement simply discouraged nontherapeutic experiments with children. The guidelines did not, however, suggest that the practice was completely inappropriate; the subcommittee asserted that "important" research using "harmless" levels of radiation dosage with children was acceptable. The crucial terms *important* and *harmless* were left undefined.

It seems reasonable to expect that "important" pediatric research would address a significant medical problem affecting children or would explore key aspects of normal human physiology—relevant to health promotion or disease prevention—for which research on children is indispensable. By these standards, the twenty-one nontherapeutic radiation experiments with children whose risks we review in the next section of this chapter could all be said to address important questions relevant to pediatric health care. This judgment is not based on a determination of whether a given study proved important in the subsequent development of a particular field. Such retrospective analysis would place an unreasonable burden on investigators of the past, as research is an inherently speculative enterprise. Many experiments that prove to be of little value in the advance of medical knowledge are, at the time they are implemented, well designed and appropriate attempts to address important research questions.

It is easier to infer what the members of the AEC Subcommittee on Human Applications would have considered "important" research than what the subcommittee would have considered "harmless" radioisotope research. Acute toxicity is not seen following administration of nontherapeutic (tracer) doses of radioisotopes. Thus, the principal potential harm from radiation exposure at lower doses is the subsequent development of cancer. In the 1940s and 1950s, some in the field apparently discounted the risk, while others were wary of a prevailing uncertainty. Dr. John Lawrence, an early radioisotope researcher at the University of California, described how some researchers conducted public demonstrations of tracers, using an "unsuspecting physician out of the audience to act as the guinea pig," presumably to reassure the audience that tracers were innocuous.[56] By contrast, other investigators focused on the tragedy of the radium dial painters, concerned that this might be repeated with man-made radionuclides.

Evidence of how well the AEC enforced its 1949 guidelines with respect to research on chil-

dren is elusive (see chapter 6). AEC correspondence with researchers at the Fernald School suggests that in at least one case there was oversight of research in which children were administered radioisotopes.[57]

RISK OF HARM AND NONTHERAPEUTIC RESEARCH WITH CHILDREN

The Twenty-One Case Examples

During the 1944–1974 period, there was an explosion of interest in the use of radioisotopes in clinical medicine and medical research, including pediatrics. The twenty-one research projects we review here include only a small number of all those that were likely conducted. These twenty-one do include, however, every nontherapeutic study that was funded by the federal government and fell into our original group of eighty-one pediatric radiation experiments. The table that appears at the end of the chapter provides information about the number of children involved in each of the experiments, the radioisotopes used, and risk estimates for cancer incidence. These twenty-one represent a subset of eighty-one studies identified in documents of the Atomic Energy Commission and a review of the medical literature that met the criteria described above.[58]

All twenty-one projects analyzed in detail involve the administration of radioisotopes to children in order to better understand child physiology or to develop better diagnostic tools for pediatric disease. In this respect, the studies supported by the federal government do not differ from those reviewed that had other funding sources. With the exception of the study at the Wrentham school to evaluate protective measures for fallout, none of the twenty-one experiments reviewed was related to national defense concerns. Seventeen of the twenty-one experiments involved the use of iodine 131 for the evaluation of thyroid function.

Three examples of research reviewed by the Committee will help illustrate the nature of the experiments and the risks posed to children. In the first example, investigators at Johns Hopkins in 1953 injected iodine 131 into thirty-four children from ages two months to fifteen years with hypothyroidism and an unknown number of healthy "control" children in order to better understand the cause of this disease.[59] Iodine is normally taken up and used by the thyroid gland for hormone production. In this experiment, a radiation detector was placed over the thyroid to detect the amount of iodine 131 taken up. Most children with hypothyroidism have an under-developed thyroid gland, in which case only very low levels of iodine 131 uptake will occur. Indeed, this is what the investigators found in this experiment, which was one of the first projects to use iodine 131 uptake as a measure of thyroid function in children. Hypothyroidism is a relatively common condition (1 per 4,000 births) that can cause profound mental retardation if untreated. Today, better diagnostic tests for thyroid function including radioimmunoassay and effective thyroid hormone replacement have virtually eliminated hypothyroidism as a cause of mental retardation in the developed world.

A second example of research reviewed by the Committee is an experiment by investigators at the University of Minnesota in 1951 in which four children with nephrotic syndrome were injected with an amino acid labeled with sulfur 35, along with two "control" children hospitalized for other conditions.[60] Nephrotic syndrome is a serious pediatric condition in which protein is excreted by the kidneys in large quantities. There was controversy at the time over whether children with nephrotic syndrome have low blood protein levels solely because of renal losses or whether they also have impaired protein production. This experiment looked at the incorporation of the radioisotope-labeled amino acid into protein, and the results suggested that the protein production in children with nephrotic syndrome is normal.

A third example of research reviewed by the Committee is a study of iodine 125 and iodine 131 uptake by eight healthy children performed at the Los Alamos Laboratory in 1963.[61] The purpose of the study was to evaluate the use of radioisotopes in very small doses (nanocurie levels) as a measure of thyroid function. The study demonstrated that the technique was scientifically valid and exposed the children to smaller radiation doses than earlier methods.

Estimating Risk

How can the risks posed to children in these types of experiments be estimated? The primary risk posed by the administration of radioisotopes is the potential development of cancer years, even decades, after the exposure. As will be discussed further, the risk of cancer following external radiation exposure was not well documented until the late 1950s and the early 1960s. Thus, the published reports of research projects prior to that time rarely discuss the issue of long-term risks.

The principles of risk assessment for radioisotopes are laid out in "The Basics of Radiation Science" at the end of "Introduction: The Atomic Century."[62] To review: the increased risk of cancer is generally assumed to be proportional to the dose of radiation delivered to the various organs of the body. This dose depends upon a number of factors, including the amount of radioactivity administered, its chemical form (which determines which organs will be exposed), and how long it stays in the body, which in turn depends upon the radioactive decay rate and the body's normal excretion rate for that substance. For many radioisotopes, the overall personal dose can be derived by the "effective-dose method," in which the doses to the ten most sensitive organs are computed and added together, weighting the various organs in proportion to their radiosensitivity. Thus, this effective dose can be thought of as producing the same excess risk of cancer (all sites combined) as if the whole body had received that amount as a uniform dose. This risk is then computed by multiplying the effective dose by established risk estimates per unit dose for various ages. For this chapter, the Advisory Committee has adopted the effective doses and risk estimates tabulated by the International Commission on Radiation Protection and the National Council on Radiation Protection.[63] The lifetime-risk estimate used in this chapter is 1/1,000 excess cancers per rem of effective dose for children and fetuses exposed to slowly delivered radiation doses, like those from radioactive tracers.

The risks of thyroid cancer following exposure to radioactive iodine (generally I-131) represent a special case for three reasons. First, use of the effective-dose method is inappropriate because the dose is much greater to the thyroid than for other organs, and the lifetime risk is therefore dominated by the thyroid cancer risk. Therefore, risk is best calculated using only the thyroid dose and its associated risk. Second, the thyroid cancer risk varies even more by age than for other cancers. Third, the risk for iodine 131 has not been measured directly, but several lines of evidence suggest that it may be substantially lower than for external radiation. For this chapter, the Advisory Committee has adopted estimates provided by three follow-up studies of external irradiation of the thyroid by x rays or gamma rays in childhood: 2,600 children who received x-ray treatment for enlarged thymus glands in the first year of life;[64] 11,000 children who were treated by x rays in Israel for ringworm under age ten;[65] and Japanese atomic bomb survivors under age twenty.[66] The risk estimates from these studies were divided by three to convert them to internal iodine 131 exposures.[67] The estimates from these studies are for cancer incidence; for mortality we have divided them by 10, since 90 percent of thyroid cancers are curable. The resulting estimates are summarized in table 1. These are the same estimates used by the Massachusetts Task Force, which investigated the Fernald and Wrentham experiments.[68]

We can use data from the previously described Johns Hopkins iodine 131 study as an example. In this study, the amount of radioactivity administered was 1.75 microcuries per kilogram body weight; equivalent to 44 microcuries in a seven-year-old child weighing 25 kilograms. Based on interpolation of the tables in ICRP 53, and assuming a 13 percent thyroid uptake, this would produce a thyroid dose of 115 rem to a child aged seven. In this age range (5–9), the lifetime risk of developing thyroid cancer would be calculated by multiplying this dose by 20 per million person rems to produce an estimate of 2.3 cases per 1,000 exposed individuals, or 0.23 percent for a particular child. The risk of dying of thyroid cancer would be one-tenth of this, or 0.023 percent.

The twenty-one experiments subjected to the Committee's detailed risk analysis included approximately 800 children. Eleven of the studies produced estimates of average risk of cancer incidence within the range of 1 and 0.1 percent; eight studies ranged within 0.09 and 0.01 percent, and the remaining two studies produced average risk estimates of 0.001 percent. The maximum potential risk estimate was 2.3 percent in a few children aged one to two years at the time of exposure. The average risk of cancer incidence for the Fernald radioiron and radiocalcium studies were 0.03 percent and 0.001 percent respectively, and for the Wrentham fallout (iodine 131) study, 0.10 percent. All of the highest-risk experiments involved iodine 131, and hence the risks of dying of cancer would be about ten times smaller. (See table 2 at the end of this chapter for further details.)

Based on the average risk estimate for each of the twenty-one experiments, we would estimate an excess cancer incidence of 1.4 cases for the entire group of 792 subjects. However, given the uncertainties built into the risk analysis, it is also possible that no excess cases resulted. Furthermore, since most of that excess would have been thyroid cancer, it is particularly unlikely that any cancer deaths would have been caused. Finally, as thyroid cancer does occur in the general population, it would be difficult to attribute these cases to an individual's involvement in research. In addition, any cases of thyroid cancer among former subjects attributable to their participation in research conducted in the 1940s and 1950s are likely to have occurred already, although there is little long-term follow-up data to know for certain what the ultimate lifetime risk might be.

How do these risk figures compare with what is acceptable in nontherapeutic research today? As noted earlier in this chapter, the contemporary regulatory standard permits children to be involved in nontherapeutic research if the research poses no more than "minimal risk" to the subjects. "Minimal risk" is defined by analogy only: "A risk is minimal where the probability and magnitude of harm or discomfort anticipated in the proposed research are not greater, in and of themselves, than those ordinarily encountered in daily life or during the performance of routine physical or psychological tests."[69] The regulations also allow for nontherapeutic research with children that *does* present more than minimal risk, *but only if* the risk represents

Table 1. Summary of Risk Estimates for Thyroid Cancer from Iodine 131

EXPOSURE AT VARIOUS AGES

Age	0–4*	5–9†	10–14‡	15–19§
Lifetime risk‖ of cancer incidence per million exposed per rem				
Males	27	13	6.7	1.9
Females	53	27	13	3.7
Both	40	20	10	2.8
Lifetime risk of cancer morality per million exposed per rem				
Males	2.7	1.3	0.7	0.2
Females	5.3	2.7	1.3	0.4
Both	4.0	2.0	1.0	0.3

* From R. E. Shore et al., "Thyroid Tumors Following Thymus Irradiation," *Journal of the National Cancer Institute* 74 (1985): 1177–1184, based on 2.9 cases per million person-year-rem.
† From E. Ron and B. Modon, "Thyroid and Other Neoplasms Following Childhood Scalp Irradiation," in J. D. Boice, Jr. and J. F. Fraumeni, Jr., eds., *Radiation Carcinogenesis: Epidemiology and Biological Significance* (New York: Raven, 1984), 139–151, based on the risk in this age group being half that in the 0–4 age group.
‡ From R. L. Prentice et al., "Radiation Exposure and Thyroid Cancer Incidence Among Hiroshima and Nagasaki Residents," *National Cancer Institute Monographs* 62 (1982): 207–212, based on the risk in this age group being one-third of that in the 0–9 age group.
§ Ibid. based on 0.21 per million person-year-rem.
‖ Based on an assumed forty-year period at risk from five to forty-five years after exposure and assuming females have twice the excess risk of males.

a minor increase over minimal risk, the procedures involved are commensurate with the general life experiences of subjects, and the research is likely to yield knowledge of "vital importance" about the subjects' disorder or condition.[70] The regulations do not specify what would count as a minor increase over minimal risk. With this general guidance, it is the obligation of individual institutional review boards (IRBs) to determine whether a nontherapeutic study involving children is acceptable.[71] It is likely that a cancer risk of greater than 1 per 1,000 subjects would be considered by most, if not all IRBs to be unacceptable by a minimal-risk standard, even for nonfatal cancers. It is less clear whether this risk would be considered unacceptable by the "minor increase over minimal risk" standard (assuming the research satisfied the "vital importance" condition). The difficulty of establishing an acceptable level of risk in nontherapeutic radiation research with children is currently being

debated in the medical literature,[72] a debate that will likely continue at least until federal guidelines become more specific.

What Was Known at the Time About Risk in Children

Assuming that any study that posed risks of greater than 1 excess case of cancer per 1,000 subjects would be judged to be more than minimal risk, eleven of the twenty-one research projects reviewed by the Committee exposed children to higher risk than is acceptable today for nontherapeutic experiments. From a moral perspective, a crucial question is whether investigators at the time could or should have known that they were putting their pediatric subjects at greater than minimal risk. If they could have known, then, arguably, these investigators were not conforming to the AEC's requirement permitting nontherapeutic research in children provided that "the radiation dosage level in any tissue is low enough to be considered harmless."

It is clear that the medical community's understanding of the nature and magnitude of risks posed to children by radiation exposure is not what it is today. Researchers did not positively associate prior exposure to external radiation with an increased risk of cancer until the mid to late 1950s. In 1950, Duffy and Fitzgerald raised the question as to whether there might be cause to investigate a possible association between therapeutic thymic irradiation during childhood and subsequent development of thyroid or thymic cancers:

To pose a cause and effect relationship between thymic irradiation and the development of cancer would be quite unjustified on the basis of data at hand when one considers the large number of children who have had irradiation of an "enlarged thymus." However, the potential carcinogenic effects of irradiation are becoming increasingly apparent, and such relationships as those of thymic irradiation in early life and the subsequent development of thyroid or thymic tumors might be profitably explored.[73]

By 1959, several studies had reported an association between radiation exposure and the subsequent development of leukemia.[74] Saenger et al. performed an epidemiologic study of several thousand children in 1960 to evaluate the association between radiation exposure and cancer.[75] They stated:

The question of whether or not radiation can be indicted as the principal causative factor in the induction of neoplasia following radiation exposure for either diagnostic or therapeutic purposes has been of increased interest over the past several years.[76]

In completing their analysis, they concluded: "It remains a fact, indisputable in all respects, that the rate of thyroid cancers in the irradiated group is disproportionately high."[77]

In 1961, Beach and Dolphin prepared a detailed analysis of the literature on the relationship between radiation and thyroid cancer in children.[78] They reported:

The thyroid has always been considered to be an organ comparatively radio-resistant to alteration and subsequent tumor development. Although no definite development of radiogenic tumor has been reported in adults after therapeutic administration of iodine-

131, Jelliffe and Jones (1960) discuss a total of 10 cases of thyroid cancer reported in the literature in persons treated early in life by x-ray irradiation in the neck region. [T]he total of malignant thyroid tumors which develop in children given a dose of x-radiation to the thyroid that is of the same order of magnitude as the incidence estimated for other tumors if a linear dose-response relationship is assumed. No biologic significance is attached to this point, apart from noting the fact that the child's thyroid appears to be more radio-sensitive than an adult's but not more sensitive than some adult tissues.[79]

This lack of appreciation for the potential long-term effects of radiation in children is further reflected in institutional policy development for use of radioisotopes at the time. The Massachusetts General Hospital developed standards for tracer doses of radioisotopes in May 1949. Dr. Shields Warren, director of the AEC Division of Biology and Medicine, assisted in the development of the MGH standard:

Tracer doses in humans will always be kept to the absolute minimum required to make the observation.

Adult humans who are ill and who are expected to benefit from the procedure, shall not receive tracer doses of radioactive material giving off radiation in excess of a total of 4 rep. Children (all patients below 15 years of age) shall not receive more than a total of 0.8 rep.[80]

In any other cases, tracer doses will be limited to radioactive material giving off radiation in an amount less than a total of 1 rep.

In the case of iodine, the thyroid, which retains most of the radioactivity, is radioresistant. In this case, the permitted dosage may be increased by a factor of 100.[81]

Despite the cautious tone of this document, the policy illustrates the complete lack of understanding of the true radiosensitivity of the thyroid gland, especially in the pediatric population. Further allowances must be made with regard to what was known about the distribution of radioisotopes in children at the time. It is evident that investigators using radioisotopes in children were not employing available information on organ weights in children to calculate tissue exposures at least until the mid-1960s. When "standard man" assumptions were used to calculate pediatric exposures before pediatric standards were developed, investigators may have significantly and systematically underestimated effective tissue dosages in children. It is notable that the highest levels of risk posed in the experiments reviewed were to infants administered iodine 131.

Iodine 131 was routinely used for diagnostic procedures in the pediatric population until the 1980s, when it was replaced by I-123, a newly available radioisotope with a significantly shorter half-life, which reduced the thyroid dose markedly. The Wrentham fallout study, performed in 1961, employed doses of iodine 131 that resulted in an average dose of 44 rad to the gland, slightly less than the dose that would have been received for a diagnostic thyroid scan during this time.

Although the doses of radioisotopes subsequently declined during these years for both therapeutic medicine and nontherapeutic research, these guidelines were not based on long-term outcome studies of exposed individuals but rather on conservative extrapolations from high-dose studies and on the dosages necessary to enable detection with the available equipment.

The debate over the potential risks of low-dose exposure continues today, as epidemiological studies of thyroid cancer incidence subsequent to iodine 131 administration in both the diagnostic as well as therapeutic dose range have been largely negative. Risks as a result of iodine 131 exposure are still unclear, and risk analyses for exposure to radioisotopes are thus based on extrapolations from studies involving external irradiation.

In summary, during the period in which children were exposed to the highest levels of risk from nontherapeutic research involving radioisotopes, investigators had a limited understanding of the potential long-term risks of low-dose radiation and of methods to accurately calculate the tissue doses in children. Today, we cautiously assume that any exposure to radiation likely produces some small increase in cancer risk, so that no exposure is absolutely harmless. Instead, the concept of minimal or acceptable risk is commonly used, as discussed earlier. Some of the studies during this period involved risks that would be judged as minimal even today, whereas others would be clearly viewed as unacceptable today. Should the investigators then

have viewed any of these studies as harmless? Though an understanding of the association between exposure to external radiation and subsequent development of cancer was emerging during this time, a similar association had not been made for exposure to low dose levels of radioisotopes. In addition, the relative radio-sensitivity of many pediatric tissues, including thyroid, had not been established, and most researchers during this period subscribed to the "threshold" theory of risk, which assumed that sufficiently low doses were probably harmless. In the face of such widespread factual ignorance, it is difficult to hold these investigators culpable for imposing risks on their subjects that were not appreciated at the time.

BEYOND RISK: OTHER DIMENSIONS OF THE ETHICS OF NONTHERAPEUTIC RESEARCH ON CHILDREN

The level of risk to which children are exposed is critical in evaluating the ethics of nontherapeutic research on children. Also important, however, is whether and how the authorization of parents was solicited, and also which children were selected to be so used. For nineteen of the twenty-one studies reviewed by the Committee, we know almost nothing about whether the permission of parents was sought or what the parents were told about their children's involvement. Two of the studies conducted at the Fernald School were the exceptions, as a result of extensive historical and archival research by the Massachusetts Task Force on Human Subjects Research.

There is a reference to parents in the published literature on only one of the remaining nineteen studies, a 1954 iodine uptake experiment at the University of Tennessee. This paper included the following line: "The procedure was described to the mothers of the infants studied, and the mothers gave consent for the study before the tests were made."[82] (The inclusion of this line is noteworthy for it suggests that at least some investigators thought parental permission was worth mentioning in published reports of their research.)

If the Committee had devoted extensive investigatory resources to these nineteen studies, it is likely we would have learned more about whether or how parental authorization was obtained in at least some cases. It is also almost certain that even the deepest archival digging would have produced no useful information about parental authorization for some of these experiments. The recent experience of the Massachusetts Task Force demonstrates the possibility of both outcomes: for some of the experiments conducted at the Fernald School, the task force's diligent historical research uncovered a variety of documents that shed important light on what both parents and children were told; for the experiments at Wrentham, similar efforts did not produce any significant information on questions of parental authorization.

Again with the exception of the experiments conducted at Fernald and Wrentham, we know almost nothing about who the children were who served as subjects in these experiments. The journal articles on these remaining studies do not describe the sociodemographic characteristics of the subjects. They do sometimes mention whether the subjects had relevant medical conditions and usually that the children, including the "control" subjects, were hospitalized patients. In some of the experiments reviewed by the Committee, the scientific research questions of interest could have been pursued only in children who were ill and hospitalized. In other instances, however, the hospitalized children were likely samples of convenience. This is particularly plausible in the case of control subjects, when a sample of healthy, nonhospitalized children might have made a better control group from a scientific perspective. As we saw in chapter 2, hospitalized patients were often viewed by physician-investigators as a convenient source of research subjects.

Because so little is known, the Committee cannot draw conclusions about the ethics of most of the nontherapeutic studies involving children we reviewed, apart from the important issue of risk of harm to the children involved. We turn now to an analysis of the studies where relevant information about parental authorization, disclosure, and subject selection is available: the studies conducted at the Fernald School.

THE STUDIES AT THE FERNALD SCHOOL

Researchers from the Massachusetts Institute of Technology, working in cooperation with senior members of the Fernald staff, carried out nontherapeutic nutritional studies with radioisotopes at the state school in the late 1940s and early 1950s. The subjects of these nutritional research studies were young male residents of Fernald, who were members of the school's "science club." In 1946, one study exposed seventeen subjects to radioactive iron. The second study, which involved a series of seventeen related subexperiments, exposed fifty-seven subjects to radioactive calcium between 1950 and 1953. It is clear that the doses involved were low and that it is extremely unlikely that any of the children who were used as subjects were harmed as a consequence. These studies remain morally troubling, however, for several reasons. First, although parents or guardians were asked for their permission to have their children involved in the research, the available evidence suggests that the information provided was, at best, incomplete. Second, there is the question of the fairness of selecting institutionalized children at all, children whose life circumstances were by any standard already heavily burdened.

Parental Authorization

The Massachusetts Task Force found two letters sent to parents describing the nutrition studies and seeking their permission. The first letter, a form letter signed by the superintendent of the school, is dated November 1949.[83] The letter refers to a project in which children at the school will receive a special diet "rich" in various cereals, iron, and vitamins and for which "it will be necessary to make some blood tests at stated intervals, similar to those to which our patients are already accustomed, and which will cause no discomfort or change in their physical condition other than possibly improvement." The letter makes no mention of any risks or the use of a radioisotope. Parents or guardians are asked to indicate that they have no objection to their son's participation in the project by signing an enclosed form.[84]

The second letter, dated May 1953, we quote in its entirety:

Dear Parent:

In previous years we have done some examinations in connection with the nutritional department of the Massachusetts Institute of Technology, with the purposes of helping to improve the nutrition of our children and to help them in general more efficiently than before.

For the checking up of the children, we occasionally need to take some blood samples, which are then analyzed. The blood samples are taken after one test meal which consists of a special breakfast meal containing a certain amount of calcium. We have asked for volunteers to give a sample of blood once a month for three months, and your son has agreed to volunteer because the boys who belong to the Science Club have many additional privileges. They get a quart of milk daily during that time, and are taken to a baseball game, to the beach and to some outside dinners and they enjoy it greatly.

I hope that you have no objection that your son is voluntarily participating in this study. The first study will start on Monday, June 8th, and if you have not expressed any objections we will assume that your son may participate.

Sincerely yours,

Clemens E. Benda, M.D.

[Fernald] Clinical Director

Approved:_____

Malcom J. Farrell, M.D.

[Fernald] Superintendent[85]

Again, there is no mention of any risks or the use of a radioisotope. It was believed then that the risks were minimal, as indeed they appear to have been, and as a consequence, school administrators and the investigators may have thought it unnecessary to raise the issue of risks with the parents. There was *no basis,* however, for the implication in both letters that the project was intended for the children's benefit or improvement. This was simply not true.[86]

The conclusion of the Massachusetts Task Force was that these experiments were conducted in violation of the fundamental human rights of the subjects. This conclusion is based in part on the task force's assessment of these letters. Specifically, the task force found that

[t]he researchers failed to satisfactorily inform the subjects and their families that the nutritional research

studies were non-therapeutic; that is, that the research studies were never intended to benefit the human subjects as individuals but were intended to enhance the body of scientific knowledge concerning nutrition.

The letter in which consent from family members was requested, which was drafted by the former Fernald superintendent, failed to provide information that was reasonably necessary for an informed decision to be made.[87]

Fairness and the Use of Institutionalized Children

The Fernald experiments also raise quite starkly the particular ethical difficulties associated with conducting research on members of institutionalized populations—especially where some of the residents have mental impairments. Living conditions in most of these institutions (including Fernald and Wrentham) have improved considerably in recent years, and sensitivity toward people with cognitive impairments has likewise increased. As Fred Boyce, a subject in one of these experiments has put it, "Fernald is a much better place today, and in no way does it operate like it did then. That's very important to know that."[88]

The Massachusetts Task Force describes conditions in state-operated facilities like Fernald, particularly as they bear on human experimentation, as follows:

Until the 1970s, the buildings were dirty and in disrepair, staff shortages were constant, brutality was often accepted, and programs were inadequate or nonexistent. There were no human rights committees or institutional review boards. If the Superintendent (in those days required to be a medical doctor) "cooperated" in an experiment and allowed residents to be subjects, few knew and no one protested. If nothing concerning the experiments appeared in the residents' medical records, if "request for consent" letters were less than forthright, or if no consent was obtained there was no one in a position of authority to halt or challenge such procedures.[89]

Although public attitudes toward people who are institutionalized are admittedly different today than they were fifty years ago, it is likely that this state of affairs would have been troubling to most Americans even then. Historian Susan Lederer has revealed several episodes of experimentation with institutionalized children in America that caused considerable public outcry even before 1940, presaging the concern generated by Willowbrook when this research became a public issue in the 1960s.[90]

The LMRI staff reported in the early 1960s that the pediatric researchers whom they had gathered agreed in principle that the convenience of conducting research on institutionalized children did not outweigh the moral problems associated with this practice:

Several investigators spoke about the practical advantages of using institutionalized children who are already assembled in one location and living within a standard, controlled environment. But the conferees agreed that there should be no differential recruitment of ward patients rather than private patients, of institutionalized children rather than children living in private homes, or of handicapped rather than healthy children.[91]

A particularly poignant dimension of the unfairness of using institutionalized children as subjects of research is that it permits investigators to secure cooperation by offering as special treats what other, noninstitutionalized children would find far less exceptional. The extra attention of a "science club," a quart of milk, and an occasional outing were for the boys at Fernald extraordinary opportunities. As Mr. Boyce put it:

I won't tell you now about the severe physical and mental abuse, but I can assure you, it was no Boys' Town. The idea of getting consent for experiments under these conditions was not only cruel but hypocritical. They bribed us by offering us special privileges, knowing that we had so little that we would do practically anything for attention; and to say, I quote, "This is their debt to society," end quote, as if we were worth no more than laboratory mice, is unforgivable.[92]

Even when a child was able to resist the offers of special attention and refused to participate in the experiment, the investigators seem to have been unwilling to respect the child's decision. One MIT researcher, Robert S. Harris, explicitly noted that "it seemed to [him] that the three subjects who objected to being included in the study [could] be induced to change their minds."[93] Harris believed that the recalcitrant children could be "induced" to join in the study by emphasizing "the Fernald Science Club angle of our work."[94]

From the perspective of the science, it was considered important to conduct the research in an environment in which the diet of the children-subjects could be easily controlled. From this standpoint, the institutional setting of Fernald was ideal. The institutional settings of the boarding schools in the Boston area, however, would have offered much the same opportunity. Although the risks were small, the "children of the elite" were rarely if ever selected for such research. It is not likely that these children would have been willing to submit to blood tests for extra milk or the chance to go to the beach.

The question of what is ethical in the context of unfair background conditions is always difficult. Perhaps the investigators, who were not responsible for the poor conditions at Fernald, believed that the opportunities provided to the members of the Science Club brightened the lives of these children, if only briefly. Reasoning of this sort, however, can all too easily lead to unjustifiable disregard of the equal worth of all people and to unfair treatment.

Today, fifty years after the Fernald experiments, there are still no federal regulations protecting institutionalized children from unfair treatment in research involving human subjects.[95] The Committee strongly urges the federal government to fill this policy void by providing additional protections for institutionalized children.[96]

CONCLUSION

If an ethical evaluation of human experiments depended solely upon an assessment of the risks to subjects as they could reasonably be anticipated at the time, the radiation experiments conducted on children reviewed in this chapter would be relatively unproblematic.[97] During this time, the association between radiation exposure and the subsequent development of cancer was not well understood, and in particular, little was known about iodine 131 and the risk of thyroid cancer. Both researchers and policymakers appear to have been alert to considerations of harm and concerned about exposing children to an unacceptable level of risk.

At the same time, however, the scientific community's experience with radionuclides in humans was limited, and this approach to medical investigation was new. Although the available data about human risk were encouraging and the biological susceptibility of children to the effects of radiation was not appreciated, we are left with the lingering question of whether investigators and agency officials were *sufficiently* cautious as they began their work with children. This is a difficult judgment to make at any point in the development of a field of human research; it is particularly difficult to make at forty or fifty years' remove. Investigators and officials had to make decisions under conditions of considerable uncertainty; this is commonplace in science and in medicine. Although the biological susceptibility of children was not then known, investigators and officials held the view that children should be accorded extra protection in the conduct of human research, and they made what they thought were appropriate adjustments when using children as subjects. If human research never proceeded in the face of uncertainty, there would be no such experiments. How little uncertainty is acceptable in research involving children is a question that remains unresolved. Today, we continue to debate what constitutes minimal risk to children, in radiation and in other areas of research. The regulations governing research on children offer little in the way of guidance, either with respect to conditions of uncertainty about risk or when risks are known.

As best as we can determine, in eleven of the twenty-one experiments we reviewed, the risks were in a range that would today likely be considered as more than minimal, and thus as unacceptable in nontherapeutic research with children according to current federal regulations. It is possible, however, that four of the eleven might be considered acceptable by the "minor increase over minimal risk" standard.[98] In these four experiments, the average risk estimates were between one and two per thousand, the studies were directed at the subjects' medical conditions,

and they may well have had the potential to obtain information of "vital importance."

Physical risk to subjects is not the only ethically relevant consideration in evaluating human experiments. With the exception of the studies at Fernald, we know almost nothing about whether or how parental authorization for the remaining nineteen experiments we reviewed was obtained. And with the exception of the Fernald studies and the experiment at Wrentham, we know very little about the children who were selected to be the subjects of this research. Therefore, we cannot comment on the general ethics of these other experiments.

The experiments at Fernald and at the Wrentham School unfairly burdened children who were already disadvantaged, children whose interests were less well protected than those children living with their parents or children who were socially privileged. At the Fernald School, where more is known, there was some attempt to solicit the permission of parents, but the information provided was incomplete and misleading. The investigators successfully secured the cooperation of the children with offers of extra milk and an occasional outing—incentives that would not likely have induced children who were less starved for attention to willingly submit to repeated blood tests.

One researcher speaking almost thirty-five years ago set out the fundamental moral issue with particular frankness and clarity:

. . . we are talking here about first and second class citizens. This is a concept none of our consciences will allow us to live with. . . . The thing we must all avoid is two types of citizenry.[99]

It might have been common for researchers to take advantage of the convenience of experimenting on institutionalized children, but the Committee does not believe that convenience offsets the moral problems associated with employing these vulnerable children as research subjects—now or decades ago.

THE VANDERBILT STUDY

In an exceptionally large study[a] at Vanderbilt University in the 1940s, approximately 820 poor, pregnant Caucasian women were administered tracer doses of radioactive iron. Vanderbilt worked with the Tennessee State Department of Health, and the research was partly funded by the Public Health Service.[b] Today, most women take iron supplements during pregnancy. This experiment provided the scientific data needed to determine the nutritional requirements for iron during pregnancy.

The radioiron portion of the nutrition study, directed by Dr. Paul Hahn, was designed to study iron absorption during pregnancy.[c] The women, who were anywhere from less than ten weeks to more than thirty-five weeks pregnant, were administered a single oral dose of radioactive iron, Fe-59, during their second prenatal visit, before receiving their routine dose of therapeutic iron.[d] On their third prenatal visit, blood was drawn and tests performed to determine the percentage of iron absorbed by the mother. The infants' blood was then examined at birth to determine the percentage of radio-

iron absorbed by the fetus. The doses to the women were estimated in the study article, using crude dose-estimation methods available at the time, to be from 200,000 to 1,000,000 countable counts per minute.[e] Although the investigators did not estimate doses to the fetuses in the original study, Dr. Hahn later estimated fetal doses to be between 5 and 15 rad. This estimate, however, has been questioned.[f]

There is at least some indication that the women neither gave their consent nor were aware they were participating in an experiment. Vanderbilt study subjects, expressing bitterness at the way they believed they had been treated, testified at an Advisory Committee meeting that the proffered drink, called a "cocktail" by the investigators, was offered with no mention of its contents. "I remember taking a cocktail," one woman said simply. "I don't remember what it was, and I was not told what it was."[g] Although it is not clear what, if anything, the subjects were told, information about the Vanderbilt experiment was available to the general public. In late 1946 news reports appeared in the Nashville press.[h]

---

a. Most of the other tracer studies involving pregnant women and offering no prospect of benefit that were reviewed by the Committee involved fewer than twenty women as subjects.

b. William J. Darby, Director of the Tennessee-Vanderbilt Project et al., *Summary Report, Section B, Tennessee-Vanderbilt Nutrition Project, July 1, 1946 to December 31, 1946* (ACHRE No. CORP-020395-A), 97–110. This nutrition study summary report notes, "Considerable expansion of the program of study of maternal and infant nutrition has been made possible by a grant of $9,000 per year which was made by the U.S. Public Health Service. These funds were available beginning November 1, 1946." Ibid., 99. The summary observes that the grant was to be used for additional personnel, including the appointment of Dr. Richard Cannon, an obstetrics resident, to the staff of the Division of Nutrition beginning 1 January 1947. Dr. Cannon's name subsequently appears as an investigator in the medical report discussing the radioiron portion of the study, along with Dr. Paul Hahn's and others.

c. P. Hahn et al., "Iron Metabolism in Human Pregnancy as Studied with the Radioactive Isotope, Fe-59," *American Journal of Obstetrics and Gynecology* 61 (March 1951): 477–486. The exact years of the radioiron portion of the nutrition study are uncertain. Minutes from a meeting of the nutrition study investigators indicate the study was to begin in September 1945. Tennessee-Vanderbilt Nutrition Project, Nutrition in Pregnancy Study, "Minutes of Meeting for Discussion of Nutrition in Pregnancy Study, August 17, 1945" (ACHRE No. CORP-020395-A), 17A-C.

The radioiron study probably began at approximately that time and appears to have continued until sometime in 1947, based on a review of periodic study summaries.

d. The Advisory Committee has not been able to determine whether Dr. Hahn got the radioactive iron used in the study from a private or government source, or both.

e. Counts per minute is a measure of the radioactivity detected by a specific counting instrument. The sensitivities of counting instruments vary; a specific instrument may not "see" and count all the radiation coming from a particular substance. Thus, the total amount of radiation emitted by a substance may be calculated by considering the sensitivity of the counter.

f. Contemporary estimates of the fetal doses by the Committee and others suggest that the fetal effective dose was a few hundred millirems.

g. Wilton McClure, transcript of audio testimony before the Advisory Committee on Human Radiation Experiments, Small Panel Meeting, Knoxville, Tennessee, 2 March 1995, 182.

h. "Iron Doses with Radioactive Isotopes Aid to Pregnancy, Experiment Shows," *Nashville Banner*, 13

The actual risk to the fetuses in the Vanderbilt experiment has long been a matter of study. In 1963–1964, a group of researchers at Vanderbilt found no significant differences in malignancy rates between the exposed and nonexposed mothers.[i] However, they did identify a higher number of malignancies among the exposed offspring (four cases in the exposed group: acute lymphatic leukemia, synovial sarcoma, lymphosarcoma, and primary liver carcinoma, which was discounted as a rare, familial form of cancer). No cases were found in a control group of similar size, and approximately 0.65 cases would have been expected on Tennessee state rates, compared to which the three observed cases is a marginally significant excess. This led the researchers to conclude that the data suggested a causal relationship between the prenatal exposure to Fe-59 and the cancer. The investigators also concluded that Dr. Hahn's estimate of fetal exposure was an underestimation of the fetal-absorbed dose.

A 1969 study, funded by the AEC and conducted by one of the investigators from the 1963–1964 study, attempted to reconstruct the doses of Fe-59 to the fetuses in the original Vanderbilt study.[j] The investigators observed that the one case of leukemia might have been due to radiation damage, but that the doses in the other two cases were low; therefore, the relationship between the radiation exposure and the cancer in those cases might not have been causal. However, the researchers also noted that due to incomplete data, they could not estimate the dose absorbed by the fetus with confidence and that no definitive conclusions could be drawn from this study as to whether these exposures resulted in damage to the fetus.[k]

The Vanderbilt study raises many of the same ethical issues as the experiments reviewed in this chapter. Like these experiments, the Vanderbilt study offered no prospect of medical benefit to the pregnant women or their offspring, raising the question of the conditions under which it is acceptable to put children at risk for the benefit of others, whether before or after birth. What could the investigators reasonably have been expected to know about the risks to which they put their subjects? Did they exercise appropriate caution in exposing fetuses to radiation? What were the pregnant women told, if anything, and was their permission sought? Who were these women, and how were they positioned relative to pregnant women, generally?

The Committee did not have the resources to pursue these questions in both research in which children were the subjects and research in which children were exposed as fetuses. We did establish that the Vanderbilt study was not the only experiment during this period to expose fetuses in research that offered no prospect of medical benefit to them or their mothers. While the Committee did not conduct an exhaustive review of the scientific literature, we did find twenty-seven human radiation

---

December 1946; "VU to Report on Isotopes," *The Nashville Tennessean*, 14 December 1946 (ACHRE No. CORP-020395-A).

i. The investigators identified the hospital records of 751 exposed mothers and 771 unexposed controls, as well as 719 exposed offspring and 734 unexposed offspring, and mailed them questionnaires. Of the exposed mothers, 90.4 percent responded, as did 91.45 percent of the unexposed mothers, 88.2 percent of the exposed offspring, and 89.2 percent of the unexposed. Ruth M. Hagstrom et al.,"Long Term Effects of Radioactive Iron Administered During Human Pregnancy," *American Journal of Epidemiology* 90 (1969): 1–8.

j. Norman C. Dyer and A. Bertrand Brill, "Fetal Radiation Dose from Maternally Administered Fe-59 and I-131," in *Radiation Biology of the Fetal and Juvenile Mammal: Proceedings of the Ninth Annual Hanford Biology Symposium at Richland, Washington, May 5-8, 1969*, eds. Melvin R. Sikov and D. Dennis Mahlum (Washington, D.C.: GPO, December 1969),

---

78-88. This study was reviewed in detail by the Committee. The study also investigated fetal absorption of radioiodine because that isotope was and is commonly used in diagnosis and therapy, including in pregnant women.

k. Ibid., 85.

studies that included pregnant or nursing women as subjects between 1944 and 1974.[l] Of these studies, eight were considered therapeutic, and nineteen offered no prospect of benefit to the subject. Most of the nineteen were tracer experiments.

These studies were performed in order to examine human physiology during pregnancy or to study the uptake of radioactive substances by fetuses or nursing infants.[m] They generally addressed valid scientific questions that could not be investigated in other populations. Knowledge of fetal exposure to radioiodine, for example, was relevant to issues such as potential harm to the fetus from maternal uptake of radioiodine in diagnostic tests or to estimate the potential effects of environmental exposure to radioiodine on the human fetus. In other studies, radioactive iron was administered to better understand the physiology of maternal and fetal intake of iron during pregnancy.

---

l. All of the nineteen studies reviewed in detail by the Committee were conducted or at least partially funded by the federal government or were supplied with radioisotopes by the AEC. For the earlier years, the Committee relied on the ACHRE experiments database, AEC isotope distribution lists provided by DOE, and relevant biographies. The Committee also consulted relevant medical indexes and computer databases; the isotope distribution lists provided by DOE did not cover these years. While the computer search would have located nontherapeutic tracer experiments for this period as well, very few were identified.

m. Of the nineteen tracer experiments (funded by the government) involving pregnant or nursing women identified by the Committee, only three administered tracer doses to *nursing* women that offered no prospect of benefit; in at least one of the studies the infants were exposed. In one case, six nursing women were given radioiodine to determine excretion in breast milk, the infants were not given the exposed milk. In another case, two infants were intentionally exposed to the breast milk of their mothers, who were given I-131. An I-131 tracer study on the general population, incidentally included two nursing women. The report indicates that both had been nursing their children, and since there is no indication that the mothers were warned to avoid breastfeeding after the exposure, it is quite probable that the infants were exposed.

NASOPHARYNGEAL IRRADIATION

Nasopharyngeal irradiation,[a] introduced by S. J. Crowe and J. W. Baylor of the Otological Research Laboratory at the Johns Hopkins University, was employed from 1924 on as a means of shrinking lymphoid tissue at the entrance to the eustachian tubes to treat middle ear obstructions, infections, and deafness. For this treatment, intranasal radium applicators (sealed ampules containing radium salt) were inserted (at least three insertions per treatment cycle) into the nasopharyngeal area for twelve-

---

a. Nasopharyngeal irradiation was studied in adults as well as children. In the early 1940s, 732 submariners were subjects of a controlled experiment designed to test whether nasopharyngeal radium treatments could be used to shrink lymphoid tissue surrounding the eustachian tubes, thereby preventing and treating aerotitis media in submariners by equalizing external and middle ear pressure. This treatment was successful in 90 percent of the cases. H. L. Haines and J. D. Harris, "Aerotitis Media in Submariners," *Annals of Otology, Rhinology, and Laryngology* 55 (1946): 347–371. In a 1945 journal article, it was noted that a controlled study was considered by the Army Air Forces, but rejected because of the urgent need to treat fliers immediately and keep them flying. However, the published report describes differences between various dose groups, implying an uncontrolled experimental comparison was made. Captain John E. Hendricks et al., "The Use of Radium in the Aerotitis Control Program of the Army Air Forces: A Combined Report by the Officers Participating," *Annals of Otology, Rhinology, and Laryngology* 54 (1945): 650–724. Tens of thousands of servicemen were subsequently given this nasopharyngeal radium treatment.

Relying on the risk estimate developed in the Sandler study, Stewart Farber, a radiation-monitoring specialist with a background in public health, has projected minute periods. [b] The therapeutic effect of the treatments resulted from the penetrating radiation emitted from the radium source (gamma and beta rays), not from the internal deposition of radium itself. Crowe and his colleagues reported that "under this treatment, the lymphoid tissue around the tubal orifices gradually disappeared, marked improvement or complete return of the hear-

ing followed, and in many the bluish discoloration of the tympanic membrane also disappeared." [c] This method was used for more than a quarter century as a prophylaxis against deafness, for relieving children with recurrent adenoid tissue following tonsillectomy and adenoidectomy, and for children with chronic ear infections. Asthmatic children with frequent upper respiratory infections were also often considered for this type of irradiation.

An average of 150 patients a month, mostly children, were given the treatment at the Johns Hopkins clinic over a period of several years. [d] Many children received the treatment more than once as recurrent lymphoid tissue was considered an indication for treatment.

Crowe and his colleagues reported that the results following irradiation of the nasopharynx alone were not only as good as, but often better than, those following removal of tonsils and adenoids. [e] In review articles, they noted that approximately 85 percent of treated patients responded with decreased numbers of infections and/or improved hearing when treated at young ages. They also concluded that "it is effective, safe, painless, inexpensive and has proved particularly valuable for prevention of certain ear, sinus and bronchial condition in children." [f]

Although early articles by Crowe and colleagues indicate that nasopharyngeal radium treatments were accepted as standard procedure for the prevention of childhood deafness, these treatments, like most standard interventions in medicine, had not been subjected to formal scientific evaluation. A controlled study was conducted from 1948

51.4 excess brain cancers over a fifty-year period in the 7,613 servicemen irradiated in the Navy and Army Air Forces studies noted above. Stewart Farber, Consulting Scientist of the Public Health Sciences, to Stephen Klaidman, ACHRE Staff, 8 March 1995 ("Nasopharyngeal Radium Irradiation-Initial Radiation Experiments Performed by DOD on 7,613 Navy and Army Air Force Military Personnel during 1944–45"). Alan Ducatman, M.D., of the University of West Virginia School of Medicine, who coauthored a letter with Farber to the *New England Journal of Medicine* regarding the radium exposure of military personnel, wrote that he found "no convincing evidence of excess cancer in the exposed population." He added, however, "there is also no good evidence for the null hypothesis." Alan Ducatman, West Virginia University School of Medicine, to Duncan Thomas, Member of the Advisory Committee on Human Radiation Experiments, 22 February 1995 ("I'm sorry I could not respond . . .") (ACHRE No. WVU-021795–A).

Han K. Kang, with the Environmental Epidemiology Service of the Veterans Health Administration, is currently conducting a study to assess the feasibility of an epidemiologic study of Navy veterans who received radium treatments. Han K. Kang, Environmental Epidemiology Service, Veterans Health Administration, "Feasibility of an Epidemiologic Study of a Cohort of Submariners Who Received Radium Irradiation Treatment," 23 August 1994. It is not clear, however, that sufficient numbers of treatment-documented personnel can be identified, as a group representing submariners has apparently been able to identify only six former Navy personnel from of a pool of twenty-seven whose records indicate they received radium treatment. (It is not clear whether the data being collected by the VFW with the support of Senator Joesph Lieberman of Connecticut will be from a representative sample of respondents. If, in fact, these data are from a highly nonrepresentative sample, the study may not be considered scientifically valid.) However, the Veterans of Foreign Wars organization apparently is now processing hundreds of surveys filled out by veterans who say they underwent nasopharyngeal radium treatment. Once this task is completed, Senator Lieberman plans to present the data to the Department of Veterans Affairs with a recommendation that an epidemiologic study be conducted.

---

b. Samuel J. Crowe, "Irradiation of the Nasopharynx," *Annals of Otology, Rhinology and Laryngology* 55 (1946): 31.

c. Ibid., 30.

d. Ibid., 33.; Dale P. Sandler et al., "Neoplasms Following Childhood Radium Irradiation of the Nasopharynx," *Journal of the National Cancer Institute* 68 (1982): 3–8.

e. Ibid., 33.

f. Ibid.

to 1953 by Crowe and his colleagues to determine "the feasibility of irradiation of the nasopharynx as a method for controlling hearing impairment in large groups of children associated with lymphoid hyperplasia in the nasopharynx; to draw conclusions concerning the per capita cost of such an undertaking as a public health measure."[g] Crowe et al. wrote in an NIH "Notice of Research" that "the procedure of treatment is not new, as an individual measure; this is the first adequately controlled experiment of sufficient size for accurate statistical analysis."[h]

This work was funded by NIH for the entire period of study. As recorded in an NIH grant application, the study involved approximately 7,000 children screened for hearing impairment.[i] Of those screened, approximately 50 percent were selected for further study based on the chosen criteria for hearing loss. Half of this study group was irradiated with radium, while the other half served as a control group. Crowe and colleagues reportedly concluded from this study (published in 1955) that the radium treatments did shrink swelling of lymphoid tissue and improve hearing.[j] This type of therapy was ultimately discontinued because of newly available antibiotics and the use of transtympanic drainage tubes, as well as awareness of the potential risks of radiation treatment.

In addition to the targeted lymphoid tissue, the brain and other tissues in the head and neck region, including the paranasal sinuses, salivary glands, thyroid, and parathyroid glands are also exposed to significant doses of radiation during the radium treatments, prompting concern that these treated individuals might have been placed at increased risk for radiation-induced cancers at these sites. Dale P. Sandler et al., in their 1982 study of the effects of nasopharyngeal irradiation on excess cancer risk for children treated at the Johns Hopkins clinic, found "a statistically significant overall excess of malignant neoplasms of the head and neck among exposed subjects," based however on only four cases in comparison with 0.57 expected.[k] This excess was accounted for mainly by three brain tumors that occurred in the irradiation subjects. One other malignant tumor, a cancer of the soft palate, was also reported. The Department of Epidemiology at the Johns Hopkins University has undertaken a further follow-up study of the Crowe et al. cohort of children irradiated there, previously studied by Sandler et al.[l] Verduijn et al., in their 1989 study of cancer mortality risk for those individuals (mostly children) treated by nasopharyngeal irradiation with radium 226 in the Netherlands, reported that "the present study has found no excess of cancer mortality at any site associated with radium exposure by the Crowe and Baylor therapy. Specifically, the finding of Sandler et al. of an excess of head and neck cancer was not found in this study group."[m]

Among the Japanese atomic bomb survivors, no excess of brain tumors was found. However, several studies have noted an increased risk of both benign and malignant brain tumors following therapeutic doses of radiation to the head and neck region during childhood.[n] From the Com-

g. S. J. Crowe et al., The Johns Hopkins University School of Medicine and School of Hygiene and Public Health, to Federal Security Agency, Public Health Service, National Institutes of Health, July 1948 ("The Efficiency of Nasopharyngeal Irradiation In the Prevention Of Deafness in Children, Notice of Research Project, Grant No. B-19") (ACHRE No. HHS No. 092694-A).

h. Ibid.

i. Ibid.

j. Ibid.

k. For the combination of benign and malignant neoplasms, there were 23 cases, for a relative risk of 2.08 with a 95 percent confidence interval of 1.12 to 3.91. Sandler, "Neoplasms Following Childhood Radium Irradiation," 5.

l. Jessica Yeh and Genevieve Matanowski, fax to Anna Mastroianni (ACHRE), 7 July 1995 ("Nasopharyngeal Power Analysis"), 1-3.

m. Verduijn et al., "Mortality after Nasopharyngeal Irradiation," *Annals of Otology, Rhinology, and Laryngology* 98 (1989): 843.

n. S. Jablon and H. Kato, "Childhood Cancer in Relation to Prenatal Exposure to Atomic-Bomb Radiation," mittee's own limited risk analysis of these experiments, we concluded that the brain and surrounding head and neck tissues would be put at highest risk and estimated the lifetime risk at approximately 4.35 per 1,000 and an increased relative risk of 62 percent. [o]

The Hopkins nasopharyngeal study raises different ethical issues than those posed by the other experiments reviewed in this chapter, all of which offered no prospect of medical benefit to the children who served as subjects. By contrast, the nasopharyngeal irradiation experiment was designed to determine whether children at risk for hearing loss would be better off receiving radiation treatments or not receiving such treatments.

A central issue here was whether it was permissible to withhold this intervention from "at risk" children. The application of radium was at this point a common, but scientifically unproven, treatment for children at risk of hearing loss; the risks of the treatment were not well characterized. If it was really unknown which was better for children—receiving radium or no intervention—then the medical interests of the children were best served by being subjects in the research because, as a consequence, they would have a 50 percent chance of receiving the better approach. The nasopharyngeal experiment thus belongs to a class of research the Committee did not investigate—therapeutic research with children.

The Lancet, ii (1970): 1000–1003.; M. Colman, M. Kirsch, and M. Creditor, "Radiation Induced Tumors," in *Late Biological Effects of Ionizing Radiation, Vol. 1* (Vienna: International Atomic Energy Agency, 1978), 167–180; R. E. Shore, R. E. Albert, and B. S. Pasternak, "Follow-up Study of Patients Treated by X ray Epilation for Tinea Capitis: Resurvey of Post-Treatment Illness and Mortality Experience," *Archives of Environmental Health* 31 (1976): 17–24; and C. E. Land, "Carcinogenic Effects of Radiation on the Human Digestive Tract and Other Organs," in *Radiation Carcinogenesis,* eds. A. C. Upton et al. (New York: Elsevier, 1986), 347–378.

o. The radiation dose estimate to the head and neck region was calculated according to the following assumptions: (1) Source description: 50 mg of radium, active length 1.5 cm, filtered by 0.3 mm of Monel metal. (2) Average treatment: 60 mg/hrs; based on three 12-minute treatments (radium applicators inserted through both nostrils)= (12x3x50x2)/60 mins per hour= 60 mg-hrs. (3) Dose rate at points in a central orthogonal plane surrounding the source: for distances up to 5 centimeters dose estimated using published data (Quimby Tables, Otto Glasser et al., *Physical Foundations of Radiology,* 3d ed. [New York: Paul Hoeber, Inc., 1961]) for linear radium sources with dose increased by 50% to allow for the reduced filtration provided by the applicator wall and converting roentgen to rad by a multiplication factor of 0.93. For distances greater than 5 centimeters, the dose rate is reduced in accordance with the inverse square

law, with a proportionality constant of 690 rad-cm^2. There was no dose correction for attenuation of the gamma rays by tissue absorbtion, which has been calculated to be about 2%/cm (yielding a dose reduction of about 20% at 10 cm).

The local gamma dose to the head and neck region was assumed to be distributed according to an inverse square law $d(r) = 690/r^2$ rad. The Committee approximated the exposed region of the body by a sphere with radius 10 centimeters. This was felt to be a conservative assumption, because although the dose does not go to zero at the base of the neck, a 10-centimeter sphere would also extend outside the skull. Averaging this dose distribution over the exposed sphere, the average dose to the head was found to be 20.7 rad. The exposed volume is about 4189 cm^3, or 29 percent of the total body, so the average whole body dose is about 6.0 rad. Multiplying this by the BEIR V risk coefficient for children exposed at age five, 1.4/1,000 person-rad, produces a lifetime risk of about 8.4/1,000. This calculation assumes that the brain and other head tissues have average radiosensitivity. BEIR V also gives absolute-risk coefficients for brain cancer ranging from 1 to 9 per million person-year-rad, with 3 being a reasonable average. Applying this figure to an average head dose of 20.7 rad, the Committee estimates a lifetime risk of about 4.35/1,000. The corresponding relative risk coefficients average about 3 percent per rad, so this dose would correspond to an excess relative risk of 62 percent.

Table 2. Summary and Risk Analysis for Studies Examined by the Advisory Committee

Primary Author	Date of Publ.	Title of Study	Isotope	Number of Children	Cancer Risk	Risk Estimate (%) Incidence*
K. M. Saxena	1965	Thyroid Function in Mongolism	I-131	104	Thyroid	0.03 (0.1)
McDougall	1964	Estimation of Fat Absorption from Random Stool Specimens	I-131		Thyroid	0.06
M. A. Van Dilla	1963	Thyroid Metabolism in Children and Adults Using Very Small (Nanocurie) Doses of Iodine-125 and Iodine-131	I-131	8	Thyroid	0.001
R. T. Morrison	1963	Radioiodine Uptake Studies in Newborn Infants	I-131	25	Thyroid	0.15 (0.2)
K. M. Saxena	1962	Minimal Dosage of Iodine Required to Suppress Uptake of Iodine 131 by Normal Thyroid	I-131	63	Thyroid	0.10 (.18)
R. E. Ogborn	1960	Radioactive-iodine Concentration in Thyroid Glands of Newborn Infants	I-131	28	Thyroid	0.25 (0.49)
S. Kurland	1957	Radioisotope Study of Thyroid Function in 21 Mongoloid Subjects, Including Observations in 7 Parents	I-131	24 (17 children)	Thyroid	0.15 (1.0)
L. Oliner	1957	Thyroid Function Studies in Children: Normal Values for Thyroidal I-131 Uptake and PBI-131 Levels up to the Age of 18	I-131	83	Thyroid	0.34 (2.3)
E. E. Martmer	1956	A Study of the Uptake of Iodine (Iodine-131) by the Thyroid of Premature Infants	I-131	70	Thyroid	0.7 (2.2)
F. Bonner	1956	Studies in Calcium Metabolism: Effect of Phytates on 45-Ca Uptake in Boys on a Moderate Calcium Breakfast	Ca-45	57	Total	0.001
A. Friedman	1955	Radioiodine Uptake in Children with Mongolism	I-131	129	Thyroid	0.12 (0.82)
C. E. Benda	1954	Studies of Thyroid Function in Myotonia Dystrophica	I-131	6	Thyroid	0.02
L. V. Middlesworth	1954	Radioactive Iodine Uptake of Normal Newborn Infants	I-131		Thyroid	0.20 (0.27)
S. H. Silverman	1953	Radioiodine Uptake in the Study of Different Types of Hypothyroidism in Childhood	I-131	34	Thyroid	0.13 (0.6)
P. S. Lavik	1952	Use of Iodine-131 Labeled Protein in the Study of Protein Digestion and Absorption in Children with and without Cystic Fibrosis of the Pancreas	I-131	15	Total	0.05

H. W. Scott	1951	Blood Volume in Congenital Cyanotic Heart Disease: Simultaneous Measurements with Evans Blue and Radioactive Phosphorus	P-32	20	Total	0.08
L. M. Sharpe	1950	The Effect of Phytates and Other Food Factors on Iron Absorption	Fe-55 Fe-59	17	Total	0.03
W. A. Reilly	1950	Carrier-Free Radioactive Iodine-131 Thyroid Uptake and Urinary Excretion in Normal and Hypothyroid Children	I-131	16	Thyroid	0.04 (0.3)
V. C. Kelley	1950	Labeled Methionine as an Indicator of Protein Formation in Children with Lipoid Nephrosis	S-35	4	Total	0.02
G. H. Lowrey	1949	Radioiodine Uptake Curve in Humans: II. Studies in Children	I-131	26	Thyroid	0.2 (1.0)
E. H. Quimby	1947	Uptake of Radioactive Iodine by the Normal and Disordered Thyroid Gland in Children	I-131	54	Thyroid	0.16 (2.2)

*Risk estimates are reported as average values for each experiment: maximum values () are reported when available.

NOTES

1. As noted in the report of the Massachusetts Task Force, "many of the people who became residents of the Walter E. Fernald School . . . were not admitted with a diagnosis of mental retardation. Societal and cultural norms of the day permitted persons to be admitted to state-operated institutions for a number of reasons. All were labeled mentally retarded just by virtue of having lived within the facility." Task Force on Human Subject Research, to Philip Campbell, Commissioner, Commonwealth of Massachusetts Executive Office of Health and Human Services, Department of Mental Retardation, April 1994, "A Report on the Use of Radioactive Materials in Human Subject Research that Involved Residents of State-Operated Facilities within the Commonwealth of Massachusetts from 1943 to 1973" (ACHRE No. MASS-072194-A), 1.

2. Task Force on Human Subject Research, April 1994 ("A Report on the Use of Radioactive Materials in Human Subject Research that Involved Residents of State-Operated Facilities within the Commonwealth of Massachusetts from 1943 to 1973"); and the Working Group on Human Subject Research to Philip Campbell, June 1994 ("The Thyroid Studies: A Follow-up Report on the Use of Radioactive Materials in Human Subject Research that Involved Residents of State-Operated Facilities within the Commonwealth of Massachusetts from 1943 through 1973") (ACHRE No. MASS-072194-B).

3. Unfortunately, the published reports of the twenty-one research projects we review in this chapter often provide little or no information that could be used to identify the individual children. Many published reports provide information only about the child's age, weight, and diagnosis. Other reports provide only the child's initials and diagnosis. In either case, it would be difficult or impossible to identify specific individuals from this limited information. An existing chart may or may not confirm a child's involvement in a research project. If the investigators maintained records, those could serve as a key to identify the individuals. Even if the hospital records do exist, however, records for a period of several years prior to publication of the research would have to be reviewed in order to match a set of initials with a diagnosis. However, it is unlikely that research records have been maintained for many of these projects for the past three to five decades. Finally, the identification of an individual would be only the first step in tracking him to his current location.

Many of the children at the Wrentham and Fernald Schools have been located through extensive local efforts. The existence of the research records, as well as the records of these long-term residential institutions, have made these identifications possible.

4. There are a few exceptions to the usual involvement of parents in decisions concerning their minor children. Children who are considered either "emancipated minors" or "mature minors" are generally able to receive routine medical care without any need for parental involvement. Emancipated minors are minor children who have taken on adult responsibilities, such as maintaining financial independence and/or living away from the parents' home. A mature minor, on the other hand, is considered to be decisionally capable under special circumstances because he or she has demonstrated the maturity and ability to decide treatment decisions for himself or herself. Adolescents can be considered emancipated or mature minors and are thereby exempted from parental consent. In addition, if a minor is close to the age of majority (at least fifteen), the treatment clearly benefits the minor and is medically necessary, there is good justification for not obtaining parental consent, and if the procedure is not extraordinary or one involving substantial risk to the child, then practitioners are usually able to deliver medical care without parental permission. A number of states permit minors to give consent to the diagnosis or treatment of venereal disease, drug addiction, alcoholism, pregnancy, or for purposes of giving blood. For more information on this subject, please see: A. R. Holder, *Legal Issues in Pediatrics and Adolescent Medicine* (New Haven, Yale University Press, 1985), 123; and Robert H. Mnookin and D. Kelly Weisberg, *Child, Family, and State: Problems and Materials on Children and the Law* (Little Brown and Company, New York, 1995).

5. Mnookin and Weisburg, *Child, Family, and State*, 536. In addition, parents are considered to be "legally responsible for the care and support of their children," and "the parental consent requirement protects parents from having to pay for unwanted or unnecessary medical care and from the possible financial consequences of supporting the child if unwanted treatment is unsuccessful."

6. In addition to the exceptions given in endnote 4, there are other standard common law and statutory limitations and exceptions to the general parental consent requirement. "These relate to mandatory immunizations and screening procedures (applicable to all children), the neglect limitation (where a court may override a parental decision for an individual child), the emergency treatment of children (where no parental consent is required if the parent is unavailable)." Ibid.

7. "Some medical procedures are required of all children and in this sense represent generally applicable limitations on parental prerogatives. The Supreme Court has held, for example, that a state could impose a compulsory smallpox vaccination law as a 'reasonable and proper exercise of police power.' *Jacobsen v. Massachusetts*, 197 U.S. 11, 35 (1905) quoting *Viemeister v. White*, 72 N.E. 97 (1904). A vaccination requirement may act to protect society from various public health hazards created by communicable diseases where a parental decision may endanger not only a particular child but society at large." Mnookin and Weisburg, *Child, Family, and State*, 551.

8. The National Commission for the Protection of Human Subjects of Biomedical and Behavioral Research, *Research Involving Children: Report and Recommendations* (Washington, D.C.: GPO, 1977), and *Protection of Human Subjects*, 45 C.F.R. § 46, subpart D.

9. *Protection of Human Subjects*, 45 C.F.R. § 46.408.

10. Ibid., § 46.404.

11. Ibid., § 46.406.

12. Ibid., § 46.407.

13. Susan E. Lederer and Michael A. Grodin, "Historical Overview: Pediatric Experimentation," in *Children as Research Subjects: Science, Ethics, and Law*, eds. Michael A. Grodin and Leonard H. Glantz (New York: Oxford University Press, 1994), 4.

14. Ibid., 5.

15. Ibid., 6.

16. "Orphans and Dietetics," *American Medicine* 27 (1921): 394-396.

17. Lederer and Grodin, *Children as Research Subjects*, 11-12.

18. Ibid., 14.

19. Ibid., 12.

20. Ibid., 15.

21. This ruling is summarized in Jay Katz, *Experimentation with Human Beings, the Authority of the Investigator, Subject, Professions, and State in the Human Experimentation Process* (New York: Russell Sage Foundation, 1972), 972-974.

22. Ibid.

23. Ibid.

24. Ibid.

25. This case is also discussed in "Use of Fifteen Year Old Boy as Skin Donor Without Consent of Parents as Constituting Assault and Battery: Bureau of Legal Medicine and Legislation Society Proceedings," *Journal of the American Medical Association* 120 (17 October 1942): 562-563.

26. For more information on the Nuremberg Code, please see *United States v. Karl Brandt, et al.*, "The Medical Case," Trials of War Criminals before the Nuremberg Military Tribunals under Control Council Law No. 10 (Washington, D.C.: GPO, 1949), 2; Jay Katz, "Human Experimentation and Human Rights," *St. Louis University Law Journal* 38 (1993); and George J. Annas and Michael A. Grodin, eds., *The Nazi Doctors and the Nuremberg Code: Human Rights in Human Experimentation* (New York: Oxford University Press, 1992).

27. Irving Ladimer, "Legal and Ethical Implications of Medical Research on Human Beings," (S.J.D. diss., George Washington University, 1958), appendix II, 202-208.

28. Ibid., 207.

29. Ibid., 206.

30. Ibid., 208.

31. Henry Seidel, interview by Gail Javitt (ACHRE), transcript of audio recording, 20 March 1995 (Research Project Series, Oral History Project), 67-68.

32. Ibid.

33. William Silverman, interview by Gail Javitt (ACHRE), transcript of audio recording, 14 February 1995 (Research Project Series, Oral History Project), 26.

34. Ibid.

35. Boston University, Law-Medicine Research Institute, 1 May 1961 ("Conference on Social Responsibility in Pediatric Research")(ACHRE No. BU-062394-A). This was part of a larger LMRI project (which was funded by NIH) to investigate actual practices in clinical research. The project began in early 1960 and continued until 1963, resulting in a lengthy final report, which was never published.

36. Ibid., 5. In this document, speakers are identified by initials. A list of participants found in these same records generally makes identifying particular speakers in the transcripts quite straightforward. In this case, however, a complexity arises because the speaker is identified as "WF." The list of participants reveals no one with these initials, and "WF" appears only once in the transcripts. It is almost certain that "WF" is a typographical error, and given the flow of the transcripts, it is also almost certain that "WF" should have been "WS"—William Silverman.

37. Ibid., 7.

38. Ibid., 3.

39. Ibid.

40. Ibid., 2.

41. Ibid., 6.

42. Ibid., 17.

43. Ibid., 15.

44. Ibid.

45. The Declaration of Helsinki can be found in many sources, but its earliest published appearance was perhaps "Human Experimentation: Code of Ethics of the World Medical Association," *British Medical Journal* 2 (1964): 177.

46. Ibid.

47. Ibid.

48. Ibid.

49. Ibid.

50. Much has been written on the Willowbrook studies; for a short summary of this episode see Ruth R. Faden and Tom L. Beauchamp, *A History and Theory of Informed Consent* (New York: Oxford University Press, 1986), 5, 163-164.

51. Henry Beecher, *Research and the Individual: Human Studies* (Boston: Little, Brown, and Company, 1970), 122-127.

52. There were many exchanges in the medical literature over the hepatitis studies conducted by Saul Krugman at the Willowbrook State School. Stephen Goldby wrote an editorial to *The Lancet*, expressing his outrage over *The Lancet*'s position on Krugman's research, saying that the research was "quite unjustifiable, whatever the aims, and however academically or therapeutically important are the results. . . . Is it right to perform an experiment on a normal or mentally retarded child when no benefit can result to that individual?" The editors of *The Lancet* responded to

Goldby's letter, expressing agreement with his position, stating, "The Willowbrook experiments have always carried a hope that hepatitis might one day be prevented there and in other situations where infection seems almost inevitable; but that could not justify the giving of infected material to children who would not directly benefit." Krugman responded to these editorials by arguing,

Our proposal to expose a small number of newly admitted children to the Willowbrook strains of hepatitis virus was justified in our opinion for the following reasons: 1) they were bound to be exposed to the same strains under the natural conditions existing in the institution, 2) they would be admitted to a special, well-staffed unit where they would be isolated from exposure to other infectious diseases which were prevalent in the institution. . . . Thus, their exposure in the hepatitis unit would be associated with less risk than the type of institutional exposure where multiple infections could occur; 3) they were likely to have a subclinical infection followed by immunity to the particular hepatitis virus; and 4) only children with parents who gave informed consent would be included.

The debate over these experiments continued, as evidenced by editorials by Geoffrey Edsall, Edward Willey, and Benjamin Pasamanick in *The Lancet* and through an editorial in *JAMA* as well. Jay Katz, *Experimentation with Human Beings*, 1007-1010; Geoffrey Edsall, "Experiments at Willowbrook," *The Lancet* (10 July 1971): 95; Edward N. Willey and Benjamin Pasamanick, "Experiments at Willowbrook," *The Lancet* (22 May 1971): 1078-1079; and "A Shedding of Light," *Journal of the American Medical Association* 212 (11 May 1970): 1057-1058.

53. S. Allan Lough, Chief of the Radioisotopes Branch, AEC Isotopes Division, to Drs. Hymer L. Friedell, G. Failla, Joesph G. Hamilton, A. H. Holland, Members of AEC Subcommittee on Human Applications, 19 July 1949 ("Revised Tentative Minutes of March 13, 1949 Meeting of Subcommittee on Human Applications of Committee on Isotope Distribution of U.S. Atomic Energy Commission, AEC Building, Washington, D.C.") (ACHRE No. NARA-082294-A-24). For price list and isotope catalogue, see AEC Isotopes Division, Supplement No. 1 to Catalogue and Price List No. 3, September 1949 (ACHRE No. DOD-122794-A), 3-4.

54. S. Allan Lough, 19 July 1949 ("Revised Tentative Minutes of March 13, 1949 Meeting . . ."), 10.

55. AEC General Manager Carroll Wilson's two 1947 letters that address the consent issue (see chapter 1) did not specifically mention children. The second letter, dated November 1947, required that "the patient give his complete and informed consent in writing, and (c) that the responsible nearest of kin give in writing a similarly complete and informed consent. . . ." It is not clear, however, that Wilson's phrase, "responsible nearest of kin," was written out of concern for children and other patients not capable of

giving "complete and informed consent," as opposed, for example, to adult patients who were too sick to give such consent. Moreover, it is not even clear whether the letter was intended to apply to experiments with healthy subjects, as opposed to sick patients, or to experiments using tracer amounts of radioactive substances. The second letter is specifically focused on "substances known to be, or suspected of being, poisonous or harmful." It is plausible, for example, that tracer amounts of radionuclides were considered "harmless," especially since the Wilson letter expressly prohibited the administration of "harmful" substances unless there was a reasonable hope that "such a substance will improve the condition of the patient." Carroll L. Wilson, General Manager of the AEC, to Stafford Warren, the University of California, Los Angeles, 30 April 1947 ("This is to inform you that the Commission is going ahead with its plans . . .") (ACHRE No. DOE-051094-A-439), 1. Also C. Wilson, General Manager, AEC, to Robert Stone, University of California, 5 November 1947 ("Your letter of September 18 regarding the declassification of biological and medical papers was read at the October 11 meeting of the Advisory Committee on Biology and Medicine.") (ACHRE No. DOE-052295-A).

56. J. H. Lawrence, "Early Experiences in Nuclear Medicine," *The Journal of Nuclear Medicine* 20 (1979): 561. (Publication of speech given in 1955). Dr. Lawrence concludes, however, that "as a matter of fact, in the 20 years since we first used artificially produced radioisotopes in humans, we have not run into delayed effects or complications as some of the skeptics predicted we would." Ibid., 562.

57. This correspondence can be found in Task Force on Human Subject Research, *A Report on the Use of Radioactive Materials*, appendix B, documents 16-18.

58. Citations for the studies for which the Committee performed detailed risk analysis can be found in the supplemental volumes.

59. S. H. Silverman and L. Wilkins, "Radioiodine Uptake in the Study of Different Types of Hypothyroidism in Childhood," *Pediatrics* 12 (1953): 288-299.

60. V. C. Kelley et al., "Labeled Methionine as an Indicator of Protein Formation in Children with Lipoid Nephrosis," *Proceedings of the Society for Experimental Biology and Medicine* 76 (1950): 153-155.

61. M. A. Van Dilla and M. J. Fulwyler, "Thyroid Metabolism in Children and Adults Using Very Small (Nanocurie) Doses of Iodine-125 and Iodine-131," *Health Physics* 9 (1963): 1325-1331.

62. For more information, please see the "Introduction: The Atomic Century," sections entitled "How Do We Measure the Biological Effects of Internal Emitters?" and "How Do Scientists Determine the Long-Term Risks From Radiation?"

63. International Commission on Radiological Protection, *Publication 53: Data for Protection Against Ionizing Radiation from External Sources* (New York:

Pergamon Press, 1973); see also National Council on Radiation Protection and Measurements, *Report 80: Induction of Thyroid Cancer by Ionizing Radiation—Recommendations of the National Council on Radiation Protection and Measurements* (New York: The Council, 1985).

64. R. E. Shore et al., "Thyroid Tumors Following Thymus Irradiation," *Journal of the National Cancer Institute* 74 (1985): 1177-1184.

65. E. Ron and B. Modan, "Thyroid and Other Neoplasms Following Childhood Scalp Irradiation," in J. D. Boice, Jr., and J. F. Fraumeni, Jr., eds., *Radiation Carcinogenesis: Epidemiology and Biological Significance* (New York: Raven, 1984), 139-151.

66. R. L. Prentice et al., "Radiation Exposure and Thyroid Cancer Incidence among Hiroshima and Nagasaki Residents," *National Cancer Institute Monographs* 62 (1982): 207-212.

67. National Council on Radiation Protection and Measurements, *Report 80: Induction of Thyroid Cancer by Ionizing Radiation.* The BIER V report recommends a figure of 0.66 but with a broad confidence interval of (0.14-3.15). National Research Council, Board on Radiation Effects Research, Committee on the Biological Effects of Ionizing Radiation, *Health Effects of Exposure to Low Levels of Ionizing Radiation: BIER V* (Washington, D.C.: National Academy Press, 1990), 5, 298.

68. Task Force on Human Subject Research, "A Report on the Use of Radioactive Materials in Human Subject Research that Involved Residents of State-Operated Facilities within the Commonwealth of Massachusetts from 1943 to 1973."

69. *Protection of Human Subjects*, 45 C.F.R. § 46.102.

70. Ibid., § 46.406.

71. F. P. Castronovo, "An Attempt to Standardize the Radiodiagnostic Risk Statement in an Institutional Review Board Consent Form," *Investigative Radiology* 28 (1993): 533-538.

72. W. L. Freeman, "Research with Radiation and Healthy Children: Greater than Minimal Risk," *IRB: A Review of Human Subjects Research* 5, no. 16 (1994): 1-5.

73. B. J. Duffy and P. J. Fitzgerald, "Thyroid Cancer in Childhood and Adolescence: A Report of Twenty-eight Cases," *Cancer* 3 (November 1950): 1018-1032.

74. R. Murray, P. Heckel, and L. H. Hempelmann, "Leukemia in Children Exposed to Ionizing Radiation," *New England Journal of Medicine* 261 (1959): 585-597.

75. Eugene L. Saenger et al., " Neoplasia Following Therapeutic Irradiation for Benign Conditions in Childhood," *Radiology* 74 (1960): 889-904. For more information on the work of Eugene Saenger, please see chapter 8, "Total-Body Irradiation: Problems When Research and Treatment Are Intertwined."

76. Ibid., 889.

77. Ibid., 901.

78. S. A. Beach and G. W. Dolphin, "A Study of the Relationship Between X-Ray Dose Delivered to the Thyroids of Children and the Subsequent Development of Malignant Tumors," *Physics in Medicine and Biology* 6 (1962): 583-598.

79. Ibid., 583.

80. One rep (roentgen equivalent physical), a unit that is no longer used, is approximately equivalent to one rem (roentgen equivalent man).

81. J. C. Aub, A. K. Solomon, and Shields Warren, Harvard Medical School, 7 May 1949 ("Tracer Doses of Radioactive Isotopes in Man") (ACHRE No. HAR-100395-A). It appears that at least one physician-researcher of the time determined to avoid an unknown risk by not administering radioisotopes in studies with pregnant women and children. In his recent autobiography, Dr. Francis Moore, an eminent Boston-based surgeon, recalled that "in pregnancy, even very small doses of radiation are dangerous to the unborn child, so we did not use radioactive isotopes in studying the body composition in pregnant women or in young children." Presumably Dr. Moore is referring to the 1940s when he began his pioneering research employing radioisotopes to determine the composition of the body, although this is not clear. Whether Dr. Moore's view was informed by dialogue with the relevant pediatric perspectives reviewed here also is unclear. Francis D. Moore, M.D., *A Miracle and a Privilege: Recounting a Half Century of Surgical Advance* (Washington, D.C.: Joseph Henry Press, 1995), 109, 111.

82. L. Van Middlesworth, "Radioactive Iodide Uptake of Normal Newborn Infants," *American Journal of Diseases of Children* 88 (1954): 441.

83. Malcom J. Farrell, Superintendent, Walter E. Fernald State School, to Parent, 2 November 1949 ("The Massachusetts Institute of Technology and this institution are very much interested . . ."), as cited in the Task Force for Human Subject Research, "A Report on the Use of Radioactive Materials," appendix B, document 19.

84. This form states,

To the Superintendent of the Walter E. Fernald State School:

This is to state that I give my permission for the participation of _____ in the project mentioned in your letter of_____

Witnessed by:

_____   _____

                    Signature

Date:_____   _____

                Relationship

Permission form from Parent to the Superintendent of the Walter E. Fernald State School, 2 November 1949 ("This is to state that I give my permission . . ."), as cited by the Task Force on Human Subject Research, in "A Report on the Use of Radioactive Materials," appendix B, document 19.

85. Clemens E. Benda, Director of Research, the Walter E. Fernald State School, to Parent, 28 May 1953, as cited by the Task Force on Human Subject Research, in "A Report on the Use of Radioactive Materials," appendix B, document 23.

86. As stated in the Massachusetts Task Force report, the purpose of the nutritional research studies was to "understand how the body obtained the minerals iron and calcium from dietary sources and to find out whether compounds in cereals affected their absorption . . . the immediate goal of the research was to understand if either of these cereals was preferable from a nutritional point of view." Ibid., 16.

87. Ibid., 43.

88. Fred Boyce, transcript of audio testimony before the Advisory Committee on Human Radiation Experiments, 16 December 1994, 38.

89. Task Force on Human Subject Research, in "A Report on the Use of Radioactive Materials," 33.

90. Susan E. Lederer and Michael A. Grodin, "Historical Overview: Pediatric Experimentation," 12-13.

91. Boston University, Law-Medicine Research Institute, 1 January 1960 to 31 March, 1963, *A Study of the Legal, Ethical, and Administrative Aspects of Clinical Research Involving Human Subjects: Final Report of Administrative Practices in Clinical Research, Research Grant No. 7039* (ACHRE No. BU-053194-A), 34.

92. Fred Boyce, 16 December 1994, 38.

93. Robert S. Harris, Professor of Biochemistry and Nutrition, Massachusetts Institute of Technology, to Clemens E. Benda, 1 May 1953, as cited by the Task Force on Human Subject Research, in "A Report on the Use of Radioactive Materials," appendix B, document 21, 1.

94. Ibid.

95. Children who are considered "wards of the State or any other agency, institution, or entity" can become subjects of research if the research is related to their status as wards and conducted in a setting in which the majority of children involved in the research are not wards. If the research meets these conditions, the IRB must then appoint a special advocate not associated in any way with the research, who will act in the best interests of the child. *Protection of Human Subjects*, 45 C.F.R. § 46.409.

96. There are also no special regulations protecting institutionalized adults. The Committee believes that the federal government should implement public policies to fill this regulatory gap.

97. This conclusion does not hold for people who believe that it is *never* acceptable to use children as subjects in nontherapeutic research, even if the research is risk-free.

98. G. S. Kurland et al., "Radioisotope Study of Thyroid Function in 21 Mongoloid Subjects, including Observations in 7 Parents," *Journal of Clinical Endocrinology and Metabolism* 17 (1957): 552-60; A. Friedman, "Radioactive Uptake in Children with Mongolism," *Pediatrics* 16 (1955): 55; S. H. Silverman and L. Wilkins, "Radioiodine Uptake in the Study of Different Types of Hypothyroidism in Childhood," 288-299; and E. H. Quimby and D. J. McCune, "Uptake of Radioactive Iodine by the Normal and Disordered Thyroid Gland in Children," *Radiology* 49 (1947): 201-205.

99. Boston University Law-Medicine Research Institute, *Final Report of Administrative Practices in Clinical Research*, 34.

8

Total-Body Irradiation: Problems when Treatment and Research Are Intertwined

IN the fall of 1971, a public controversy erupted about the ethics of a research project at the University of Cincinnati College of Medicine funded for more than a decade by the Department of Defense (DOD). In this research, the subjects were cancer patients who underwent external total-body irradiation (TBI); the DOD was funding postirradiation research on the biological effects of this type of exposure to radiation. Critics of the research charged that the physician-investigators were exposing unknowing patients to potentially lethal doses of TBI— not to treat their cancer, but to collect data on the effects of nuclear war for the military—and that numerous patients had died or seriously suffered from the radiation. Defenders asserted that the TBI was reasonable medical treatment for people with incurable cancer and that this treatment was performed in accordance with contemporary professional ethics. Over the next four months, the research was reviewed favorably by ad hoc committees of physicians appointed by the American College of Radiology (ACR), the preeminent professional organization of radiologists, and by University of Cincinnati officials, but critically by an ad hoc committee of junior nonmedical faculty members at the university. Following these reports, the university president rejected further Defense Department funding for the posttreat-

ment data-collection program, and the use of TBI was suspended.

When news reports about human radiation experiments appeared in late 1993, journalists and investigators again focused on this Cincinnati project. Critics charged that the reviews commissioned by the university and the ACR were biased and had been "whitewashes"; supporters countered that the Cincinnati research had been conducted in the open, had been thoroughly and favorably reviewed by the medical community, and was old news. In addition, patients were identified publicly for the first time, leading a number of their family members to file a lawsuit against the university, the physicians, and other parties in federal court.[1] The family members also formed an advocacy group called the Cincinnati Families of Radiation Victims Organization.

The University of Cincinnati was only the last in a line of institutions that received funds to provide data to the government on the effects of total-body irradiation on humans. In this chapter we review thirty years of research supported by the Manhattan Project, the Department of Defense, and the AEC aimed at gathering data on the effects of radiation on hospitalized patients who were medically exposed to total-body irradiation. Much of the record is incomplete, and some of it is contradictory.

We cannot and do not resolve all the inconsistencies and uncertainties in the record. We do, however, focus on the ethical issues that emerged in this research, some of which are still with us today.

The history of TBI research is important to the Committee for several reasons. First, in the other case studies conducted by the Committee, there was *never* any expectation or any claim that subjects, even if they were patients, would benefit medically from their being involved in experiments. By contrast, in the TBI research, the TBI itself was recommended as treatment for incurable cancer, for which the expectation of benefit was low, although possible; chemotherapy, which would be considered "standard" today, was not well established until the mid- to late 1960s. (The postradiation effects studies sponsored by the DOD, however, were not intended to benefit the patients.) As we noted in chapter 4, the presence of an intent to benefit, if that intent is both genuine and reasonable, alters the ethics of the situation. An intent to benefit the patient-subject does not, however, ensure that an experiment is ethically acceptable. Many perplexing questions about the ethics of research involving human subjects that we face today occur at the bedside with patient-subjects who may or may not benefit medically by their participation. The TBI story thus foreshadows important issues we discuss in part III of this report when we focus on contemporary research involving human subjects, much of which involves patient-subjects and the prospect of medical benefit. The core of the ethical problem is straightforward. Whenever the treatment of a patient is intertwined with the conduct of research, the potential emerges for conflict between the interests of science and the interests of the patient. The patient may, for example, be exposed to additional risk or discomfort as a consequence. At the same time, for some patients, participation in research may offer the only chance, or the best chance, of improving their medical condition.

The second reason the history of TBI research is important to the Committee is that although the research was conducted on cancer patients, the government's interest in the research was not to advance the treatment of cancer but to find answers to problems facing the military in the development and use of atomic weapons and nuclear-powered aircraft. It is this disparity that raised questions, both in 1971 and today, about the motivations behind treatment of these patient-subjects with TBI. Whether it matters morally that the government pursued its interests in the effects of TBI on patient-subjects depends in large measure on whether the government's objectives in supporting this research inappropriately compromised the medical care the patient-subjects received. We have just noted that the conjoining of research with medical care necessarily creates a potential for conflict between the interests of the research and the interests of the patient. This is true even where the objective of the research is to find a treatment for the condition from which the patient suffers. A central issue in the case of the TBI research is whether this conflict was exacerbated by the nature of the gap between the interests of the patient and the objectives of the research. A complicating feature of the TBI story is that the DOD did not pay directly for the patients to be administered TBI; the funding by these agencies was restricted to the costs associated with the physiological and psychological measurements taken in conjunction with the TBI, rather than the costs of the TBI itself.

The Committee was also struck with how well the history of TBI research illustrates two very contemporary problems—how to draw boundaries between medical care and medical research, and how to draw boundaries between research with patient-subjects that is "therapeutic" and research that is "nontherapeutic." Was the administration of TBI always an instance of medical research, was it ever standard care, or was it sometimes administered as a departure from standard care outside of research? When TBI was administered in the context of research, was there a basis for believing that there was a reasonable prospect that patients could benefit, or was it the kind of research from which patients could not benefit medically? Because of conflicting and incomplete evidence, these were questions that we could not always answer but that guided our inquiry.

The Committee began our review by seeking to track down TBI research identified in a "Retrospective Study" of TBI exposures conducted in the mid-1960s by the Oak Ridge Associated Universities on behalf of the National Aeronautics and Space Administration (NASA), which collected records on more than 2,000 TBI exposures on both radiosensitive and radioresistant cancers from forty-five U.S. and Canadian institutions.[2] The Committee then focused on approximately twenty research studies that were published between 1940 and 1974 on the use of TBI in the United States. Nine of these twenty studies involved at least some patients with "radioresistant" cancers. Eight of the nine institutions that conducted the studies received funding from either the Manhattan Project or the DOD;[3] the Atomic Energy Commission sponsored one of the studies involving "radiosensitive" cancers at the Oak Ridge Institute of Nuclear Studies (ORINS).[4] In addition, the Committee found only one instance in which nongovernment-funded TBI research involved patients with radioresistant cancers.[5]

In this chapter, we begin with a definition of TBI, including a discussion of the then-contemporary distinction between the use of TBI to treat radiosensitive and radioresistant tumors. The distinction is important to what follows, because patients with radiosensitive cancers (for which TBI was considered most promising medically) were less useful subjects for obtaining the type of information that the military sought—information on the acute effects of radiation on healthy soldiers or citizens during the course of atomic warfare-related activities. In these patients, it would be less clear whether signs such as nausea, vomiting, or other acute effects were due to rapid destruction of cancer cells by the radiation or due to the radiation acting on normal tissue, such as normal blood cells. Similarly, patients with radiosensitive cancers were less useful for research intended to find biological measures of radiation doses ("biological dosimeters"), because this research depended on measuring various cell products in the blood or urine that could also be released by tumor cells that were destroyed. Patients with radioresistant tumors were more desirable because it was more likely that the effects seen were related to radiation effects on normal tissue rather than rapid destruction of their tumor cells.

Following a discussion of TBI itself, we turn to a chronological history of government sponsorship of research related to the effects of TBI with radioresistant tumors. This research began during the Manhattan Project. In 1949 and 1950, as we next discuss, DOD and AEC experts and officials met to consider the need for further TBI human experiments in order to gain information needed in the development of the nuclear-powered airplane. When the decision was made not to proceed with human experiments involving healthy subjects, the military began to fund research on the effects of TBI on patients undergoing treatment for cancer. As we discuss, this program began in 1950 at the M. D. Anderson Hospital for Cancer Research in Houston and continued through the end of the Cincinnati research, in the early 1970s. We conclude our review with a discussion of the AEC-funded TBI research conducted at Oak Ridge between 1957 and 1974, which focused on patients with radiosensitive cancers.

WHAT IS TBI?

Medically administered total-body irradiation, also known as whole-body radiation, involves the use of external radiation sources that produce penetrating rays of energy to deliver a relatively uniform amount of radiation to the entire body. Total-body irradiation was used as a medical treatment long before the 1944–1974 experiments, and it continues to be used today. Soon after doctors began to experiment with radiation, they recognized that radiation had different effects on different types of cancers. They thus began to distinguish between radiosensitive cancers, which generally responded to the radiation treatment, and radioresistant cancers, which most often did not respond. By the 1940s, TBI was recognized as an acceptable treatment for certain radiosensitive cancers that are widely disseminated throughout the body, such as leukemia and lymphoma (a cancer whose origin is in the lymphoid tissue). By the late 1950s, TBI was also being used to assist in conjunction with research on bone marrow transplantations for radiosensitive cancers. During this period, TBI was also explored as a possible palliative treatment (providing relief, but not cure) for disseminated radioresistant cancers, such as carcinomas of the breast, lung, colon, and other organs (carcinoma is a cancer that originates from the cells lining these organs).[6] However, TBI alone did not prove to be of value in treating these cancers because, without support measures to maintain bone marrow function, the doses needed to significantly shrink the tumors were potentially lethal to the patient.

In the 1950s, there were few effective methods for treating radioresistant cancers. Chemotherapy was just being developed; it was risky to use and only marginally effective. With no better alternatives, interest in TBI continued. In addition, the development in the 1950s of high-energy sources of radiation—cobalt 60, cesium 137, and megavolt x-ray sources—represented a significant advance in technology. These new teletherapy units allowed high-energy radiation to penetrate deeper into the body without damaging the overlying skin and soft tissues; thus higher absorbed marrow doses (in rad) could be delivered than with previous equipment. The advent of this new teletherapy encouraged researchers to retest TBI on patients with radioresistant cancers even though prior TBI techniques with older x-ray therapy machines had failed. By the late 1960s, however, chemotherapy began to be recognized as more effective, and interest in TBI waned. During the 1970s, researchers explored the effectiveness of administering TBI without bone marrow transplant through multiple exposures at lower doses

(e.g., 10 to 30 rad), known as "fractionated radiation," to achieve cumulative total body doses of 150 to 300 rad, rather than single exposures of an equivalent total body dose.[7] They also focused much more extensively on partial-body irradiation, because the risk of patient bone marrow failure was lower. Since the 1980s, TBI has again been used to treat certain widely disseminated, radioresistant carcinomas at doses as high as 1,575 rad in conjunction with effective bone marrow transplantation, which became routinely available in the late 1970s.[8]

TBI can cause acute health effects during the first six weeks following an acute (single) exposure. The type and severity of the effects depend, among other things, on the dose, the dose rate, and the individual's sensitivity to radiation.[9] The most serious side effects seen during this period are related to radiation-induced depression of the bone marrow, which can cause a decrease in the number of circulating platelets and white blood cells, which in turn can result in small hemorrhages and infections, possibly leading to death. Moderate bone marrow depression results with doses of about 125 rad. The following table describes the general acute effects that are likely to occur to healthy persons from a single exposure;[10] these effects can be exaggerated and prolonged for people who are ill or have had prior radiation treatments. As with an ordinary diagnostic x ray, the patient feels nothing during the radiation exposure itself. In addition, TBI, like most other forms of radiation exposure, can potentially have long-term effects such as can-

Midline Tissue Dose	Symptoms	Percentage	Time Postexposure
50–100 rad	nausea	5–30	3–20 hours
100–200 rad	nausea	30–70	4–30 hours
	vomiting	20–50	6–24 hours
	death[11]	<5	5–6 weeks
200–350 rad	nausea	70–90	1–48 hours
	vomitting	50–80	3–24 hours
	death	5–50	4–6 weeks
350–500 rad	nausea	90–100	1–72 hours
	vomiting	80–100	3–24 hours
	death	50–99	4–6 weeks
550–750 rad	death	100	2–3 weeks

cer induction; however, most patients who receive TBI do not live long enough to experience most long-term effects.

EARLY USE OF TBI FOR RADIORESISTANT TUMORS: THE MANHATTAN PROJECT EXPERIMENTS ON PATIENTS AND THE SUBSEQUENT AEC REVIEW

In the early 1930s, researchers at Memorial Hospital in New York, a major cancer research center (now known as the Memorial Sloan-Kettering Cancer Research Institute) engaged in an extensive study on the medical effects of total-body irradiation. As part of this study, the researchers attempted to treat a few radioresistant carcinomas. When they published their results in 1942, they noted that "[e]xcept for transient relief of pain in a few cases, the results in generalized carcinoma cases were discouraging. The reason for this is quickly apparent. Carcinomas are much more radioresistant than the lymphomatoid tumors, and by total body irradiation the dose cannot be nearly large enough to alter these tumors appreciably." They cautioned that a cancer-killing dose "will produce deleterious reactions in the bone marrow and general metabolism which may prove lethal to the patient."[12] The equipment used to deliver the TBI during this time was suboptimal because most of the radiation was deposited in superficial structures such as the skin and other soft tissues.

During World War II, Memorial Hospital was one of three medical institutions chosen by the Health Division of the Manhattan Project's "Metallurgical Project" to conduct TBI experiments in order to help understand the effects of radiation; the other two were the Chicago Tumor Institute[13] and the University of California Hospital.[14] All three studies focused on individuals with radioresistant diseases. From the limited records that are currently available, it appears that these three studies achieved little, if any, medical benefit to subjects. In addition, the interest of the military in these studies was classified and kept secret from the patients in order not to reveal the ongoing atomic bomb project.

The first experiment was carried out from 1942 to 1946 at the University of California Hospital in San Francisco to "study the effects of total-body irradiation with x-rays of varying energy on hematologically normal individuals."[15] Twenty-nine patients were treated with total dosages ranging from 100 to 300 R (using a 250-KV machine). The investigators noted that the "treatments were administered as part of the normal therapy of these patients" and reported that "advantage was taken of the fact that patients were receiving such treatment by making numerous blood studies for the Manhattan Project."[16] Little is known, for this and the other two studies, about the treatment of the patients or the issue of patient consent. A number of the patients in the University of California study had rheumatoid arthritis, and the use of TBI for that disease was severely criticized after the war by the Advisory Committee for Biology and Medicine of the newly formed Atomic Energy Commission (see below).[17] In 1948 Dr. Robert Stone, chief of the Manhattan Project's Metallurgical Laboratory Health Division, noted that although "no signed consent was received from the patient . . . the treatment was explained to them by the physicians and they, in full knowledge of the facts, accepted the treatments." At the same time, however, it was admitted that "the fact that Manhattan District was interested in the effects of total body irradiation was kept a secret."[18]

A second Manhattan Project experiment was performed from December 1942 to August 1944 at Memorial Hospital in New York by one of the researchers who had previously written that they were "discouraged" by the use of TBI for radioresistant cancers—Dr. L. F. Craver.[19] Despite his earlier negative results, eight patients were given a total of 300 R (using a 250-KV machine), at various dose rates, in order "to yield some detectable effects on the blood count and to serve as a guide to the clinical tolerance for whole-body irradiation."[20] The patients had to have

metastatic cancer of such an extent and distribution as to render their cases totally unsuitable for any accepted method of surgical or radiological treatment, yet . . . be in good enough general condition so that they might be expected not only to tolerate the exposure to 300 R of total-body irradiation in a period of ten to thirty days, but also to survive the combined effects of their disease and the irradiation for at least six months in order that some conclusions might be drawn as to the later effects of the irradiation.[21]

The report on this experiment makes clear that the primary purpose of this TBI was to obtain data for the military. Dr. Craver essentially acknowledged that there was little prospect of actual medical benefit to the patients in light of the previous failure using the same procedure.[22]

A third TBI study involving fourteen people was performed at the Chicago Tumor Clinic from March 1943 to November 1944; doses up to 120 R were given with a 250-KV machine.[23] None of these individuals had radiosensitive cancers. The use of TBI was justified by the investigators because there were no known treatments for their illnesses, and therefore, "x-ray exposures that were given were as likely to benefit the patient as any other known type of treatment, or perhaps even more likely than any other."[24] This study appears to be the only TBI study that included healthy subjects: three "normals" were each given three doses of 7 R.

After the war, Dr. Stone took on the task of editing an official history of the experiments done for the plutonium project. At one point, he complained to Dr. Shields Warren, chief of the AEC's Division of Biology and Medicine (DBM), that declassifiers were withholding the report of the Chicago TBI experiment on grounds that its release would cause "adverse publicity and even encourage litigation."[25] Stone proposed to solve the problems by carefully rewording the report. The report would make clear that the patients were suffering from incurable illnesses for which radiation was as good, if not better than, any other known treatment. Stone then proposed to deflect the likelihood of adverse publicity or litigation by deleting identifying information so that the patients could never "connect themselves up with the report."[26] The study was declassified and published in the form that Stone proposed.[27]

At about the same time in the fall of 1948, Dr. Alan Gregg, chairman of the AEC's Advisory Committee for Biology and Medicine (ACBM), engaged Stone in an exchange regarding the Manhattan Project TBI experiment on the arthritic patients at the University of California Hospital. Stone, who by this time had returned to the staff of the UC Hospital, had requested funding to monitor these arthritic patients. Gregg told Stone that "I think that I do not misrepresent the opinion of the Advisory Committee [for Biology and Medicine] in saying that we agree with those who believe the x-ray treatment of arthritic patients you have been giving patients is not justified."[28] (Despite Dr. Gregg's concerns, the role of TBI in the treatment of benign autoimmune diseases such as rheumatoid arthritis continues to be explored today.[29]) Gregg stated that the AEC had an obligation to provide a check on overly enthusiastic researchers. While admitting that a conservative consensus against certain treatments is not always correct, Gregg cautioned that "there is plenty of experience that shows that some forms of therapy attract enthusiastic supporters only to be discarded later as unsafe or unjustified."[30]

In response, Stone acknowledged that the military's need for worker safety information during the war was a primary motivation for choosing patients with nonradiosensitive carcinomas and some benign disorders such as arthritis. "At that time I was confronted with the problem of building up the morale of the workers on the new atomic bomb project, many of whom were seriously worried about the effects of prolonged whole body irradiation." But he countered that he and the other researchers did believe that the total-body irradiation would be therapeutic. Moreover, Stone retorted, Gregg's statements threatened researcher and doctor freedom: "Wittingly or otherwise you have dictated how I should treat patients even outside of the Atomic Energy Commission's supported activities."[31] Stone's declaration marked a boundary that government officials (including Stone's fellow medical researchers) would not be eager to cross.

RENEWED INTEREST IN TOTAL-BODY IRRADIATION

In 1949 AEC and Defense Department planners were seeking information on the human effects of a nuclear reactor-powered airplane. The proponents of the so-called NEPA project,[32] which at the time was managed out of Oak Ridge by the Fairchild Engine and Airplane Corporation on behalf of the Air Force and the AEC, needed to know how much external radiation air crews could tolerate. This question was critical because, depending upon the answer, the shielding needed to separate the crew from the aircraft's nuclear reactor might render the project impractical.

Those involved with the NEPA project were primarily interested in the acute effects of total-body exposure over a relatively short time (although they were also concerned about long-term effects of radiation on longevity and reproduction). It was anticipated that NEPA pilots would be exposed to as much as 25 roentgens in the course of a twenty-four-hour flight. How would this amount of radiation affect the crew's abilities to fly the plane and perform their tactical military function? How many such missions could a crew endure before being incapacitated for flight duty, as well as facing a significant risk of developing a life-shortening disease?

In early 1949, the NEPA Medical Advisory Committee was created to research the questions noted above and to advise on the project. Dr. Andrew Dowdy of UCLA was the chairman.[33] Dr. Robert Stone was chosen to head a human experiment subcommittee. At an April 3, 1949, meeting, Stone proposed to the full committee a program of experimentation using total-body irradiation on healthy subjects. In defense of this proposal, Stone noted that experimentation with normal human subjects had been done in the past when there was no other way to obtain necessary data. At the same time, however, Stone discounted the value of the TBI research that had been performed on sick patients.[34] As Brigadier General James P. Cooney, representing the AEC's Division of Military Applications, put it, "We have lots of cases of whole body radiation treatments, but all of them in patients and we have no controls and we don't have anything we can put our finger on. . . . Most of this work was unsatisfactory because the data was poor."[35] However, Shields Warren was not persuaded that experiments on healthy men would provide any more useful information and was concerned about the long-term health consequences. War-

ren noted that "[i]t was not very long since we got through trying Germans for doing exactly the same thing."[36] Nonetheless, General Cooney argued that even if the data would not be statistically valid, "psychologically it would make a lot of difference to the soldier if we were able to tell him that various doses of total-body irradiation were given to a group of people and here are the effects that were discerned."[37]

As we have seen in earlier chapters, the question of medical ethics was considered by the NEPA discussants. Stone urged that the committee approve TBI human experimentation in accordance with three basic principles of the 1946 American Medical Association Judicial Council: (1) "the voluntary consent of the person on whom the experiment is to be performed must be obtained"; (2) "the danger of each experiment must have been previously investigated by animal experimentation"; and (3) "the experiment must be performed under proper medical protection and management."[38] Shields Warren added that the experiments should be unclassified, so that there would be "no suspicion that anything is being hidden or covered up, that it is all being done openly and straightforwardly."[39] MIT's Robley Evans responded that "we don't have to advertise it, but at the same time it doesn't have to be concealed, as Dr. Shields Warren has said."[40] Dr. Hymer Friedell raised the question of whether decisions on these issues could be made by doctors alone: "I am just wondering whether someone else ought not to hold the bag along with us with regard to making such a recommendation. Previously in medical experiments the physicians and doctors have made such recommendations because the problem was primarily a medical one. I think this is something larger than that. It is really not a medical problem alone. It has to do with how critical this is with regard to the safety of the nation."[41]

In January 1950, the NEPA Medical Advisory Board recommended, with the exception of one member (not named), that human experimentation be conducted.[42] Dr. Stone then prepared a January 1950 paper on "Irradiation of Human Subjects as a Medical Experiment" to be presented to the DOD's Research and Development Board (RDB). The paper explained that as long as they kept doses below 150 R, the chances of long-term effects such as "leukemia could be entirely ruled out."[43] (This assertion would prove to be inaccurate; subsequent epidemiological research has shown that radiation doses at such levels will produce approximately a sevenfold increase in leukemia risk and a doubling in the risk of many other cancers.)[44] Accordingly, the experiments were designed only to analyze the acute effects of radiation. Stone extolled the "inestimable value" that would come from being able to tell pilots that "normal human beings had been voluntarily exposed without untoward effects to larger doses than they would receive while carrying out a particular mission." Stone then described a "plan of attack," in which he would start with 25 R total-body irradiation and then gradually increase the dose to 50 R, 100 R, and 150 R if no immediate effects were seen.[45]

The RDB's Joint Panel on the Medical Aspects of Atomic Warfare met in March 1950 and endorsed the NEPA recommendations in Stone's paper.[46] From there, the issue was debated by the RDB's Committee on Medical Sciences (CMS) in May 1950. When one committee member asked whether "you can get answers from people subjected to radiation therapy usually by reason of neoplastic disorders" as an alternative to experiments on healthy persons, Dr. Stone responded that it might be possible, but only if the patients had radioresistant cancers: "you can't pick lymphomas, but [rather] carcinomas [*sic*] types of metastases"[47]—the death of lymphoma cells would release quantities of unknown biologic chemicals and complicate the data collection.

The Defense Department shied away from making a final decision and instead deferred the matter to the AEC on the grounds that NEPA involved "civilian" as well as military problems. Accordingly, the AEC appointed another panel of experts, who met in Washington, D.C., on December 8, 1950.[48] This ad hoc "biological and medical committee," which included a number of participants in the DOD's NEPA advisory committee, addressed four questions:

- Assume that troops are acutely exposed to penetrating ionizing radiation (gamma rays). At what dosage level will they become ineffective as troops?
- What dosage will render an air crew . . . unable to complete a mission during a flight of one to three, four to twelve, and twelve to forty-eight hours?
- How often may an aircraft crew accept an exposure of 25 R per mission and still be a reasonable risk for subsequent missions?
- A submarine crew are receiving 25 R per mission. How many missions should they be allowed to make?[49]

This group of experts concluded, somewhat in contrast to Stone, that the acute effects of doses of 150 R or more would pose "grave risks" of rapidly making troops "ineffective as fighting units," but that doses held below 75 roentgens should be "unimportant in determining the success of a mission provided the crew members had not previously received an appreciable amount of radiation." (Current reports suggest that tolerance levels for acute effects may be a little higher, and that a dose of 125 rad (approximately 200 roentgens) would cause vomiting in approximately 30 percent of those exposed within twenty-four hours, and 200 rad would cause vomiting in 50 percent.)[50] They also said that air and submarine crews could withstand eight missions of 25 roentgens, but that cumulative doses of more than 200 roentgens could "substantially reduce the life expectancy of the irradiated individual." The ad hoc committee based these conclusions on "the results of extensive animal experiments, the response of patients treated for disease by X-ray and radium, observations on the effect of radiations from the atomic bombs detonated over the Japanese cities of Hiroshima and Nagasaki, and accidental exposures within the Manhattan Project and the Atomic Energy Commission."[51] Accordingly, this committee found that additional human experimentation was not needed to come up with reasonable answers.

Dr. Joseph Hamilton, a Manhattan Project physician involved with the plutonium injections, was unable to attend the December 8 meeting and sent a note to Shields Warren explaining his views:

For both politic and scientific reasons, I think it would be advantageous to secure what data can be obtained by using large monkeys such as chimpanzees which are somewhat more responsive than the lower animals. Scientifically, the use of such animals bears the disadvantage of the fact that they are considerably smaller than most adult humans and a critical evaluation of their subjective symptoms is infinitely more difficult. If this is to be done in humans, I feel that those concerned in the Atomic Energy Commission would be subject to considerable criticism, as admittedly this would have a little of the Buchenwald touch. The volunteers should be on a freer basis than inmates of a prison. At this point, I haven't any very constructive ideas as to where one would turn for such volunteers should this plan be put into execution.[52]

Following the ad hoc committee's conclusion, the AEC's Division of Biology and Medicine, headed by Shields Warren, declared "that human experimentation at the present time is not indicated." Moreover, the AEC also stated that such experiments "would have serious repercussions from a public relations standpoint, particularly if undertaken by an agency that has to do a portion of its work in secret." If data were needed, the DBM concluded, they could be obtained from the sources cited by the ad hoc committee.[53] The AEC position spelled the end of the DOD's request to do radiation exposure experiments on healthy people, and roughly coincident with the rejection of this proposal, the DOD contracted to gather data from cancer patients receiving TBI treatments.

POSTWAR TBI-EFFECTS EXPERIMENTATION: CONTINUED RELIANCE ON SICK PATIENTS IN PLACE OF HEALTHY "NORMALS"

In October 1950, the Air Force entered into a contract with the M. D. Anderson Hospital for Cancer Research in Houston, Texas, to provide the DOD with data obtained from TBI studies on cancer patients. Dr. Shields Warren, who seemed to oppose human experimentation on healthy persons during the NEPA debates, did not appear to have any misgivings about this project.[54] By the end of that decade, the DOD would have several contracts with TBI researchers. When, in 1959, a DOD newsletter

announced the renewal of a TBI contract between the Army and the Sloan-Kettering Institute in New York, readers were told: "It is hoped to make this work [by Sloan-Kettering] as well as the work of Baylor University College of Medicine and University of Cincinnati a complete program to provide us with answers on the human whole body radiation effects."[55] The Navy also conducted TBI-related research in conjunction with patient treatments at the Naval Hospital in Bethesda, Maryland. All five of these studies used TBI on many patients with radioresistant cancers.[56] In contrast, physicians at the AEC's hospital in Oak Ridge operated by the Oak Ridge Institute of Nuclear Studies (ORINS), a university-based consortium, chose to perform TBI only on patients with radiosensitive diseases. In each project, the research institutions accepted the dual purposes of treating the patients' illnesses and collecting and analyzing postexposure information for the military.

The DOD-funded experiments would seek to address the three main questions the military wanted answered: How do different doses of radiation correlate with the acute effects? How do different doses of radiation influence psychological effects? And most important, is there a way to find a biological dosimeter to measure how much radiation someone has received? The military was also interested in the diagnosis and treatment of radiation injuries. One reviewer of the initial Cincinnati proposal described the interest in finding a biological dosimeter: if "accurate knowledge of the total dose of radiation received could be determined it would be of inestimable value in case of atomic disaster or nuclear warfare."[57]

When the DOD contracted with medical professionals to perform additional research on their patients receiving TBI, it is not clear whether department officials believed that the TBI itself should be covered by the ethical standards being established for "human experimentation" following the Nuremberg trials. For example, in the NEPA debates, Dr. Robert Stone distinguished *experiments*, which involved healthy persons such as the prisoners, from *studies*, which involved sick patients.[58] Did Stone mean by this that patients receiving treatment did not need to give informed consent, while healthy subjects should? The AEC, for example, took the view that the consent standards should apply to patients. Indeed, as we saw in chapter 1, AEC General Manager Carroll Wilson wrote in a 1947 letter to Robert Stone that "the patient [must] give his complete and informed consent in writing."[59]

In 1953, Secretary of Defense Charles Wilson issued a memo establishing the Nuremberg Code as DOD policy. The Wilson memo required that all experimental subjects sign a statement that explained "the nature, duration, and purpose of the experiment; the method and means by which it is to be conducted; all inconveniences and hazards reasonably to be expected; and effects upon his health or person which may possibly come from his participation in the experiment." Unlike the AEC's 1947 pronouncements, the 1953 Wilson memo did not explicitly refer to patients. In addition, if DOD officials believed that the experiments they were sponsoring did not include the administration of the radiation, but only the collection of postradiation biochemical and psychological data, then they might have interpreted the Wilson memo as applying only to the postexposure testing, not to the radiation treatments.

Although the Wilson memo was classified, its requirements were reiterated by the surgeon general of the Army in a 1954 document that was transmitted to at least some university contract researchers. The Committee found no evidence that this memo was transmitted to the TBI contractors in particular. As discussed in chapter 1, in 1952 Congress had passed legislation that provided for Defense Department indemnification of private contract researchers in cases where human experiments resulted in injury to subjects.[60] As we have seen, the DOD appears to have linked the requirements of this statute to contractor adherence to the principles stated by the 1954 Army surgeon general memo, including written consent of the subject. For example, in a March 1957 letter to the University of Pittsburgh, which was proposing to use medical student-volunteers in a (nonradiation) experiment, the Army stated that the indemnification provision in the contract was "contingent upon your adhering to the following [March 1954 Office of the Surgeon General]

principles, policies, and rules for the use of human volunteers in performing subject medical research contracts."[61] Although this indemnification provision was in the contract of at least one of the five institutions that conducted DOD-sponsored TBI-effects research,[62] no available information indicates that its inclusion demanded adherence to the principles set forth by the surgeon general. Nonetheless, at least three of the institutions had written forms authorizing the radiation treatment procedure, although the forms did not explicitly spell out all of the risks and benefits of the additional experiments. This chapter will now review what is known about the five DOD-sponsored experiments.

M. D. Anderson Hospital
(Houston, Texas)

The Air Force's School of Aviation Medicine (SAM) contract with M. D. Anderson Hospital for Cancer Research, in association with the University of Texas Medical School, declared that the Air Force was willing to use sick patients for the needed data because "human experimentation" had been prohibited by the military:

The most direct approach to this information would be by human experiment in specifically designed radiation studies; however, for several important reasons, this has been forbidden by top military authority. Since the need is pressing, it would appear mandatory to take advantage of investigation opportunities that exist in certain radiology centers by conducting special examinations and measures of patients who are undergoing radiation treatment for disease. While the flexibility of experimental design in a radiological clinic will necessarily be limited, the information that may be gained from the studies of patients is considered potentially invaluable; furthermore, this is currently the sole source of human data.[63]

The M. D. Anderson TBI-effects study extended from 1951 to 1956 and involved 263 cancer patients.[64] M. D. Anderson had a well-established and ongoing radiation treatment program. The project began at the same time that M. D. Anderson received the first cobalt 60 teletherapy unit developed by the AEC's Oak Ridge Institute of Nuclear Studies (ORINS). The M. D. Anderson study involved three phases beginning with low doses—15 to 75 R—and gradually increasing to a maximum of 200 R. The patients in the first group were "in such a state that cure or at least definite palliation could still be expected from established methods of treatment [in addition to the TBI]."[65] Based on these results, the researchers then moved to the second phase, which involved doses ranging from 100 to 200 R. The researchers noted that this greater possible risk necessitated "the selection of patients whose disease had advanced to such a state that, in general, significant benefit could not be expected from conventional procedures other than systemic ones."[66]

The final phase involved thirty patients, all of whom had radioresistant carcinomas for which "cure by conventional means was regarded as completely hopeless."[67] These thirty received the highest doses, 200 R, and reportedly "knew about the advanced state of their disease and the experimental nature and possible risks of the proposed radiotherapy."[68] The Advisory Committee has found no written documentation on what types of risks were described to and understood by these patients. Beginning in 1953, patients signed a release form authorizing the physicians to administer "x-ray therapy, radium and radioactive isotopes, . . . which in their judgment they deem necessary or advisable, in the diagnosis and treatment of this patient."[69] This form was designed apparently to waive legal liability, but did not inform the patient of the risks and benefits of treatment and thus did not meet the other requirements established by the 1953 Wilson memo.

With respect to the biomedical findings, a 1954 Air Force review noted that M. D. Anderson had obtained positive preliminary results by finding a biological dosimeter in the blood. However, one of the reviewers commented that because "the patients were not normal people the changes could very well be the effect of the radiation on the abnormal tissue."[70] The review noted that an effort earlier in the study to find a marker in patients who received repetitive small doses of radiation, similar to what might occur on repeated NEPA flights, was not successful; accordingly, the researchers looked for it in patients who received larger doses in single exposures.[71]

An additional aspect of the M. D. Anderson study was the mental and psychomotor tests that most of the patients were subjected to before and after receiving TBI. (The patients reportedly participated "by their own consent and judgment of the hospital staff."[72]) They performed three tests related to the skills required for piloting aircraft. But the value of testing the abilities of extremely ill patients as a measure for the performance of highly fit pilots was doubtful to the Air Force.[73] In an attempt to lessen this problem, the investigators sought outpatients who were in reasonably good physical and mental condition.[74] Nonetheless, because patients received TBI radiation doses according to the severity of their disease rather than from an arbitrary experimental protocol, there was difficulty in determining whether the performance changes noted resulted from the underlying disease or the radiation.[75]

The M. D. Anderson researchers found medical benefit in three of thirty patients who received 200 R:[76] "200 [roentgens] whole-body x-irradiation produced a definite transitory amelioration of the disease in 3 cases, and a questionable improvement in several additional patients."[77] The study concluded that "the threshold dose, beyond which in a small percentage of patients severe complications begin to appear, lies somewhere between 150 and 200 r."[78] This conclusion seems to have moved the threshold tolerance level for acute effects slightly higher than the 1950 level; at that time the AEC's ad hoc NEPA committee had decided that doses above 150 R would pose "grave risks" to troops.

There is very little information concerning subject selection. It appears that many of the patients were indigent members of minorities, although no information is available to determine whether the ratio of minorities differed in relation to the general hospital population. In the context of an Air Force discussion about the costs of the study, one report noted that "there is some racial problems [sic] involved with colored patients and the colored out-patient maintenance facilities were located in another part of the city and, therefore, it would be difficult to have them transported back and forth to the hospital for testing. . . . Colonel McGraw stated that if we are paying for the maintenance of indigents of the State of Texas with research funds, and the State is also paying for the maintenance of those patients, there could be some difficulty. . . ."[79] Another report stated that "language barriers, both of degree and kind," caused problems in the testing of cognitive functions as part of the psychomotor study.[80]

Several years later, researchers at the School of Aviation Medicine and the University of Texas issued a report comparing the effects of radiation based on TBI treatment of eleven patients (most of whom had radiosensitive diseases) with the M. D. Anderson group. The researchers used the data to report on the civil defense implications that would result from mass exposures to doses between 150 and 200 R. They concluded that 60 percent of the people would experience varying degrees of disability from acute radiation sickness that would cause fatigue, nausea, and vomiting for the first twenty-four hours.[81]

TBI-Effects Studies at Baylor University College of Medicine, Memorial Sloan-Kettering Institute for Cancer Research, and the U.S. Naval Hospital in Bethesda

Within a few years after the Air Force's M. D. Anderson program began, the Army funded two TBI-effects programs with leading cancer centers, both of which appear to have been using TBI to treat radioresistant cancers even before receiving the Army contract. The studies began before M. D. Anderson had published any of its findings.

From 1954 to 1963, Baylor University College of Medicine in Houston, Texas, performed TBI on 112 patients (54 of whom had radioresistant cancers) during the military study; doses ranged from 25 to 250 R, and a 2-megavolt (MV) machine was used in place of a 250-KV machine after the first two years.[82] The principal researchers, Drs. Vincent Collins and Kenneth Loeffler, again sought a biological dosimeter and data on the acute effects of radiation.[83] The researchers noted that even though a significant amount of data had been amassed on radiation effects, no one had been able "to es-tablish clear and dependable relationships with precise physical data."[84] There is no discussion of consent or peer review in any of the twelve available reports or published papers currently available.

The Baylor researchers recognized the same problem that confronted M. D. Anderson: that seeking data from sick patients who require therapeutic TBI treatments may be in conflict with an optimal experimental design.[85] They also noted the problems with giving "last-resort" treatment of this kind:

When patients are referred as a "last resort," the radiotherapist does not wish to withhold treatment that may offer possible benefit but he cannot be certain that the benefit will outweigh the risk. The risk is not that the patient will die but that the undesirable effects of radiation [i.e., bone marrow suppression] will appear more severe in the terminal cancer patient and that the time of death may be destined to coincide with the undesirable effects of radiation.[86]

They concluded that for patients, radiation sickness may be avoided for doses up to 200 R by administering proper care (the researchers suggested that nausea and vomiting for some patients may have been caused by the power of suggestion).[87] They then hypothesized that "with correct information and proper preparation, normal healthy individuals could tolerate even higher exposures without undue incapacitation."[88] Efforts to find a biological dosimeter were said to be unsuccessful because the pool of patients was too small and many either died or were unable to tolerate the necessary tests.[89]

From 1954 to 1961, Dr. James J. Nickson of the Memorial Sloan-Kettering Institute for Cancer Research in New York City performed TBI on more than twenty patients with doses ranging from 20 to 150 R and participated in a DOD study on the acute effects of radiation on humans.[90] Again, the military aims were to find a biological dosimeter and better understand the effects of radiation.[91] Sloan-Kettering was a leading U.S. cancer research center and had a long history of using and experimenting with TBI. The patients selected at Sloan-Kettering had a variety of radioresistant and radiosensitive cancers and were in "relatively good condition."[92] However, patients with kidney, liver, or bone marrow impairment were deliberately excluded from the study because their conditions would "contaminate" biological dosimeter data (the record does not indicate whether these patients who were excluded still received TBI). There is no mention in the currently available records regarding consent of the patients or any form of peer review of the protocol or the experiments.

Between 1959 and 1960, the Navy treated seventeen patients using TBI for a variety of radioresistant and radiosensitive disorders at the Naval Hospital in Bethesda, Maryland, with a cobalt 60 teletherapy unit.[93] The report on these treatments concluded that "total-body radiation therapy in a dose range of 100–400 [roentgens] [air dose] appears to offer relatively safe and reasonably effective palliative therapy for advanced radiosensitive disease."[94] There was no equivalent success on the radioresistant tumors. Urine from some of the patients was collected and retained for analysis to see if there was any amino acid change that corresponded to the radiation exposure received by patients, as part of another attempt to identify a biological dosimeter in the urine. The investigators of the urine study could not find any direct correlation between the dose of radiation and biochemicals in the urine, and they acknowledged that the poor state of health of the patients, as well as age, nutritional state, and renal function changes, may have contributed to this problem.[95]

Surviving patient records indicate that the Naval Hospital used an authorization form, which states that the patient "hereby consent[s] to the performance . . . of total body radiation therapy. This procedure has been fully explained to me by a staff physician of the Department of Radiology."[96] There is no information available to determine if patient permission was or was not given for the collection of urine for evaluation as a biological dosimeter or if the biological dosimeter project had any effect on the patients' treatment. Neither is any information currently available on whether the patients were informed about the additional military research interest in the project, or whether there was any form of review of the project as required by Navy procedures.

The early postwar TBI researchers, such as those at M. D. Anderson, may have been enthu-

siastic to test the new cobalt 60 teletherapy TBI technology on cancers that resisted older TBI techniques, but by the end of the 1950s the new technology did not appear to be producing any more favorable results on radioresistant cancers. Dr. Shields Warren seemed to confirm this view in a 1959 article in *Scientific American;* he noted that "radioresistant tumors are generally not treated with radiation because the damage to surrounding tissue is too great."[97]

However, in March 1960 the Defense Atomic Support Agency (DASA) sponsored a conference on the effects of whole-body radiation on humans. A DASA summary of the meeting reported: "First, experience at the dosage levels up to 200 r indicates that man is able to tolerate far greater radiation dosages than was predicted in the NEPA report of 1949; second, there is a need for continuation of this work and, more important, investigation and analysis of the radiation syndrome in man up to the 300 r level, is the next logical area of study."[98] Indeed, DASA had just signed a contract with the University of Cincinnati to provide information on the effects from these higher doses of TBI.

The University of Cincinnati College of Medicine—the Last DOD-Sponsored TBI-Effects Study

The University of Cincinnati (Cincinnati) was the last institution that the Department of Defense contracted with to collect information on radiation effects from patients exposed to total-body irradiation. From 1960 to 1971, Dr. Eugene L. Saenger and a team of medical researchers from the university's College of Medicine (referred to in this chapter as the "Cincinnati doctors")[99] conducted TBI and partial-body irradiation (PBI) on approximately eighty-eight cancer patients. Cincinnati was the only nonmilitary research institution in the DOD program that did not have preexisting clinical experience with TBI therapy.[100] It was also the only institution using TBI to focus almost exclusively on patients with radioresistant cancers (except for three children with Ewing's sarcoma, a childhood bone cancer for which widespread irradiation is still considered an accepted form of treatment.)[101] The military contract was, as before, to obtain more information on the acute effects of radiation and to find a biological dosimeter.

The University of Cincinnati experiments came to public attention in 1971, a time in the national debate over the Vietnam War when university associations with the military were being questioned by students, the press, and the public. Research by Roger Rapoport, who subsequently wrote a book entitled *The Great American Bomb Machine,* led to a story in the *Washington Post* on October 8, 1971, that described the Department of Defense contract with the University of Cincinnati to measure radiation effects in humans.

It appears that by 1971 the University of Cincinnati was the only remaining institution doing post TBI-effects studies for the Department of Defense. The publicity prompted the University of Cincinnati to hold a news conference on October 11, 1971, to explain its TBI program. The public attention resulted in three investigations of the Cincinnati experiments, all of which reported their findings in January 1972: (1) a January 3, 1972, American College of Radiology report in response to a request by Senator Mike Gravel (the ACR report),[102] which was generally supportive of the program; (2) a January 1972 Ad Hoc Review Committee of the University of Cincinnati Report to the Dean of the College of Medicine on "The Whole Body Radiation Study at the University of Cincinnati" (the Suskind report),[103] which probed the facts and supported the overall objectives of the study; and (3) a January 25, 1972, "Report to the Campus Community" of the Junior Faculty Association of the University of Cincinnati (the JFA report),[104] which severely criticized the TBI program. Following these reviews, the president of the University of Cincinnati decided not to renew the DOD contract in the spring of 1972. The use of TBI was suspended after that time, and the effects study was ended. As recently as April 1994 in a congressional hearing, an ACR representative reiterated its belief that the Cincinnati project was reasonably conducted based on the standards of the time, even if "one might judge them harshly from a perspective 20 years later."[105]

Because of this public attention, a substantial number of documents concerning the University of Cincinnati experiments were preserved, including the original application and subsequent progress reports by the researchers for the Department of Defense, records of the Faculty Committee on Research (the Cincinnati IRB) review of a midcourse research protocol, relevant medical literature, and certain patient medical records. In addition to reviewing these documents, the Advisory Committee staff also interviewed Dr. Eugene Saenger, the principal investigator of the study; and the Committee and staff met with and heard from numerous patient family members and other critics of the Cincinnati experiments. The Advisory Committee also held a public hearing in Cincinnati on October 21, 1994, where more than thirty family members and other interested parties related their concerns about what they believed was wrong with the Cincinnati TBI experiments, chiefly that informed consent was inadequate. Other family members have appeared before the Advisory Committee at another public hearing in Knoxville, Tennessee, and at the Advisory Committee's meetings in Washington, D.C. The Committee also heard public testimony from Dr. Bernard Aron, a coresearcher of Dr. Saenger, and heard from counsel for Dr. Saenger and others involved in pending litigation.

What Was the Purpose of the University of Cincinnati TBI Program?

The experimenters were supported by the military to find a biological dosimeter and provide additional human performance data of military interest. There is no question that the patients were seriously ill with terminal cancers; indeed, many received other forms of treatment in addition to TBI, including surgery, chemotherapy, and localized radiation. Although there is no indication that the Defense Department had any direct role in patient selection or treatment, there have been questions raised publicly as to whether the military interest influenced or at all compromised the physicians' willingness to objectively present reasonable treatment options (including no treatment at all) to these cancer patients. Thus, the Advisory Committee has sought to determine what effect, if any, the DOD contract requirements had on the actual treatment of patients.

In 1958 Dr. Saenger applied to the Department of the Army for funding of a research proposal entitled "Metabolic Changes in Humans Following Total Body Radiation."[106] (Dr. Saenger had joined the radiology department of the University of Cincinnati College of Medicine in 1949 and became the director of its Radioisotope Laboratory in 1950, serving until 1987. Before and after starting the TBI program, he was a consultant in radiology to the Army, the Air Force, the AEC, DASA, and the PHS.)[107] The primary purpose of his proposal was to determine whether amino acids or other biochemicals in the urine could "serve as an indicator of the biological response of humans to irradiation."[108]

Later, the first of approximately ten progress reports to DASA described the purpose of the research program it was funding as "to obtain new information about the metabolic effects of total body and partial body irradiation so as to have a better understanding of the acute and subacute effects of irradiation in the human."[109] The second progress report added that "this information is necessary to provide knowledge of combat effectiveness of troops and to develop additional methods of diagnosis, prognosis, prophylaxis and treatment of these injuries."[110] The study would focus generally on post-TBI effects in patients with radioresistant carcinomas; those with radiosensitive lymphomas and other hematological diseases were for the most part not included, with the exception of the children with Ewing's sarcoma.[111] Dr. Saenger reported to DASA in 1962 that the further studies would be conducted "so long as the following criteria are fulfilled: 1. There is a reasonable chance of therapeutic benefit to the patient. 2. The likelihood of damage to the patient is no greater than that encountered from comparable therapy of another type. 3. The facilities for support of the patient and complication of treatment offer all possible medical services for successful maintenance of the patient's well being."[112]

Midway into the study, the post-TBI-effects researchers added a program of psychological and psychiatric testing, to determine "whether single doses of whole or partial radiation produce any decrement in cognitive or other functions mediated through the central nervous system."[113] They also recorded data on the incidence of nausea and vomiting from radiation. Within the first three years, the Cincinnati doctors reported to the DOD the information that the March 1960 DASA conference had sought, that "[h]uman beings can tolerate doses of 200 rad (300 r) relatively well as far as combat effectiveness is concerned."[114]

In 1973, two years after their work terminated, the Cincinnati doctors published a journal article describing the purpose of their irradiation study as "to improve the treatment and general clinical management and if possible the length of survival of patients with advanced cancer."[115] Unfortunately, no written research protocol now exists for this treatment study, nor did Dr. Saenger state that they had a written protocol while carrying out the TBI palliation treatment study. This lack of a written protocol is consistent with the confusion doctors had at this time (and to a lesser extent today) distinguishing what constituted research from what constituted innovative treatments (see chapter 2).

The clinical objective of the Cincinnati TBI treatments remains difficult to categorize precisely even now. Dr. Saenger stated in a 1994 interview with Advisory Committee staff that there was no need for an experimental treatment protocol because the TBI treatments were given as a palliative cancer therapy for people for whom there was no better alternative.[116] In contrast, the Suskind and ACR reports seem to have assumed that the TBI treatments were experimental; they both describe them as being in "Phase II" of a standard three-phase experimental process.[117]

Because the Cincinnati doctors recognized that higher doses of TBI (150 rad and above) were causing severe bone marrow suppression in some of the patients, beginning in 1963 they sought to develop countermeasures through the use of bone marrow transplants. Over a six-year period, they instituted a program to remove bone marrow from the patient prior to the radiation and to reinfuse it afterward so as to counter any deleterious effect. In 1966, they submitted a protocol on bone marrow transplantation to the College of Medicine's institutional review board, which provisionally approved it in 1967. However, the use of high-dose TBI continued during this time, with the first successful transplant being administered in 1969.

Critics have suggested that the irradiation of the patients who were subjects of the Cincinnati experiments may not have taken place at all in the absence of DOD funding. The TBI regimen did not begin until after the DOD funds were secured in 1960. The DOD provided a total of $651,482 for the TBI-effects study. In addition to the DOD funds, Dr. Saenger has estimated that the hospital spent $483,222 on patient care.[118] Dr. Saenger has stated that he was very careful to separate the DOD work from the patient care and to make sure that the DOD funding was in no way used for the patient therapy.[119] For this reason, he states that he was never personally involved in patient selection or treatment, in order not to influence the judgment of the attending physicians.[120]

When asked if TBI treatment could have begun before the DOD money arrived, Dr. Saenger said: "No, we had to, we hired some people. We had laboratory equipment to set up. . . . It proceeded as one, as really a sort of a two-pronged investigation."[121] Dr. Saenger stated that the work for the DOD "started when we started [administering TBI]—this [DOD] protocol permitted us to get a technique going in trying to look at whole body radiation in comparison with other forms of palliation."[122] Dr. Saenger also said that "if we had found in the first ten or twelve patients a clear biochemical indicator, we possibly would have done something else. We kept being on the edge of finding what we were looking for so we kept on treating the patients."[123]

In the 1969 proposal to renew the DOD contract, Dr. Saenger wrote that, in light of "world tensions from the possibility of nuclear warfare on any scale . . . it is necessary to pursue with increased diligence the scientific investigations of acute radiation effects and the attendant treatment possibilities in the human being." In outlining a plan to compare total-body, partial-body, and trunk and thorax irradiation, the proposal noted that in most cases bidirectional radiation would be used for each of these treatments, but that "whenever possible unidirectional radiation will be attempted since this type of exposure is of military interest."[124] There is

no available evidence to show that the Cincinnati doctors ever actually used unidirectional radiation.

The military's interest in the onset level for the acute effects of radiation, such as nausea and vomiting, led the Cincinnati doctors to intentionally withhold from the patients, as discussed later in the chapter, any premedication or information about these effects for the first three days after irradiation in order not to induce them via psychological suggestions. No mention was made of nausea or vomiting in any of the consent forms.

To the extent that palliation of cancer symptoms was the goal of the Cincinnati doctors, TBI presumably would have been given either as part of a planned experimental protocol or as conventional clinical therapy. If the former, then the currently available evidence indicates rather poor scientific design, even by contemporary standards; if the latter, then the TBI treatment administered for the vast majority of patients was nonstandard therapeutic practice for patients with radioresistant carcinomas at that time.

Institutional Review

Department of Defense.   The Army Research and Development Command review of Dr. Saenger's proposal in 1958 was limited to an evaluation of the usefulness of the proposed work to the military. One Army medical officer wrote that there are "so few radiobiologists in the country willing to do total body radiation that those that are should be encouraged." The proposal should be approved even though there was "very little hope that [this study] will result in practical data. As is pointed out in the proposal, a number of people have looked at the problem and the levels vary widely and there appears to be no consistency. A great deal of work has been done in animals, again without consistent findings." The reviewer hoped that the researchers "will soon decide that some other phase of the radiation program should be investigated and switch to this."[125]

Another reviewer noted that Saenger's study would "augment work being done by Dr. Collins at Baylor and the Sloan-Kettering Institute who are working with humans."[126] This point was reiterated when the contract was approved, at which point the approving officer declared that "diversification is required to achieve adequate results in a field of whole body radiations [*sic*] in humans."[127] A third reviewer noted that correlating tumor response to total dose of irradiation "would be of great value in the field of cancer . . . [and] in case of atomic disaster or nuclear accident."[128]

There is no indication that the Army reviewers considered whether any therapeutic benefits to the patients outweighed the risks that the TBI treatments might pose. These reviewers seemed to have based their support for funding this proposal on the military's need for collaborative researchers and the reputation of the applicant, rather than on the substance of the science within the application or their knowledge of radiation therapy practice at that time.

There is no evidence that the DOD reviewed the treatment of the patients as the study progressed, even though the University of Cincinnati appears to have been the only federally funded institution in the country that was treating radioresistant carcinomas with total-body irradiation at that time. The Cincinnati doctors administered doses up to 250 rad, and had indicated in their first DASA progress report that they planned to go up to 600 rad,[129] without seeming to raise any concerns within the DOD contract office.

University of Cincinnati. As was customary at the time, there was no formal review of the TBI proposal within the University of Cincinnati when it was initially submitted to the DOD in 1958. The University of Cincinnati established an institutional review board, known as the University of Cincinnati Faculty Committee on Research (FCR), in 1964. (IRBs were just beginning to be formed at this time and did not become formalized in most institutions until several years later, nor were they required as a condition of government funding until 1974.)

Subsequent internal reviews by Cincinnati committees raised several concerns. In March 1966, Dr. Saenger and a colleague submitted a protocol entitled "Protection of Humans with Stored Autologous Marrow" to the University of Cincinnati FCR. This proposal was considered an adjunct to the TBI treatments, which Saenger said he did not consider an experiment, and was therefore not subject to review.[130] Some members of the FCR, however, raised concerns that attended to the underlying TBI treatments. These questions included whether each patient was advised that "no specific benefit will derive to him," the need for a more detailed description of the potential hazards, and whether the irradiation would "influence the morbidity or the mortality in these patients."[131]

The proposal was revised and resubmitted on March 1967 under the title "The Therapeutic Effect of Total Body Irradiation Followed by Infusion of Stored Autologous Marrow in Humans." A five-person FCR subcommittee reviewed the proposal. One member, Dr. George Shields, recommended that the study be disapproved because "the radiation proposed has been documented in the author's own series to cause a 25% mortality. . . . I believe a 25% mortality is too high, (25% of 36 patients is 9 deaths) but this is of course merely an opinion." Shields added that if the study were to be approved, then his concern could be addressed by improving the consent process—that is, by ensuring that "all patients are informed that a 1 in 4 chance of death within a few weeks due to treatment exists, etc."[132] Another member, Dr. Thomas E. Gaffney, initially recommended disapproval for several reasons, including the "considerable morbidity associated with this high dose radiation,"[133] but he subsequently recommended approval along with Dr. Harvey Knowles, Dr. Edward Radford, and R. L. Witt. The proposal was then given "provisional approval" on May 23, 1967. The requirements did not include Shields's recommendation on mortality, but did stipulate that "the protocol should be modified to indicate that the exclusive purpose of the study is to determine the therapeutic efficacy of whole body irradiation."[134] There is no written evidence as to whether the FCR ever re-reviewed and approved the revised protocol or the new consent forms that the investigators produced in response to this review.

In 1970, the FCR reexamined the bone marrow protocol because it had not been reviewed since it received provisional approval in 1967. Following two protocol revisions intended to meet the committee's concerns, the FCR noted that it still could "not find adequate methods of evaluation in this study protocol. . . . The real problem seems to be how are we going to evaluate the effectiveness of marrow transplants in protecting against the side effects of total body irradiation. Secondly, how are we going to evaluate the effectiveness of total body irradiation."[135] Nonetheless, after yet another revision, the protocol was approved on August 9, 1971.

In April 1972, Dr. Edward Silberstein, a colleague of Dr. Saenger, submitted a protocol to the FCR entitled "Evaluation of the therapeutic effectiveness of total and partial body irradiation as compared to chemotherapy in humans with carcinoma of the lung and colon."[136] This presumably was to be the next experimental phase of the ongoing TBI work, for which an NIH grant was also contemplated. But that same month, the president of the University of Cincinnati refused to allow continued DOD funding for the post-TBI treatment data collection and analysis following the negative public attention brought to the study. TBI was suspended pending FCR review. Dr. Silberstein's protocol was approved by the university's FCR in August 1972 as a grant application to the NIH's National Cancer Institute.[137] However, in February 1973 the NIH elected not to fund the proposal.

National Institutes of Health.   The TBI research was incorporated into a general research grant that the NIH funded beginning in 1966. According to Dr. Evelyn Hess, chair of the FCR, writing in 1971:

Background to Grant Approvals: This research had, of course, been submitted to the DOD initially with yearly reports since the initiation of the project. The entire total body irradiation protocol, including all the therapeutic, metabolic, chemical, hematologic, immunologic, and psychologic studies was incorporated as one of the components for the General Clinical Research Center (GCRC) grant submitted to the NIH in 1966. There was a site visit and counsel visit on this grant application and many aspects of the radiation project were presented to the scientific review committees. This NIH grant was given full approval and was funded. It came up for renewal in 1970, and again all aspects of the radiation study were incorporated. This grant also had full approval by the NIH.[138]

However, a 1969 internal FCR memorandum from the then-chairman of the Cincinnati FCR, Dr. Thomas Gaffney, noted that the NIH had rejected a University of Cincinnati grant application for TBI research "on ethical grounds," even though it had been approved by the FCR: "We learned that two applications received by NIH from this institution have been rejected on ethical grounds. Both had been through this committee. As far as I know, neither of the principal investigators involved were notified of the reason of the rejection by the NIH. One of these grants was the total body radiation study in patients with malignancy. . . ."[139]

In 1974, D. T. Chalkley, then chief of the NIH's Office for Protection of Research Risks, vigorously responded to a magazine article criticizing the Cincinnati experiments. Chalkley stated that "none of the patients involved died from radiation sickness. . . . In all instances, death was clearly attributable to the advance of cancer, or to intercurrent disease associated with advanced cancer."[140]

Risk of TBI on Mortality and Morbidity.   The risks associated with total-body irradiation were reported by all of the previous DOD-sponsored TBI research institutions and were known to the Cincinnati doctors. Midway into the program, the Cincinnati doctors acknowledged that the TBI treatments posed a risk of death: "bone marrow suppression was the most life threatening radiation effect at the doses used."[141] In their 1966 report to the DOD, they noted that the general response of the first fifty patients was that their "total white count falls to a low point 25 to 40 days after irradiation. There was lymphopenia [low white blood cell count] which persisted for 40 to 60 days."[142] The same report stated that "severe hematologic depression was found in most patients who expired."[143] Although the efforts by the Cincinnati doctors to employ bone marrow transplantation in response to this problem did not succeed until 1969, they continued to administer TBI without bone marrow transplantation throughout the six-year interim period: thirteen patients received doses of 150 or 200 rad TBI (including five on whom autologous marrow infusion was attempted but did not succeed); nine of these patients died between twenty-five and seventy-four days after being irradiated, and the other four survived longer.[144]

Some relatives of deceased TBI patients contend that their relatives may not have been as seriously ill as the reports claim. While all patients had advanced cancer (indicated either by the presence of metastatic or locally advanced tumors) and thus could be considered "end stage" in terms of unlikely curability, they were clearly not all "near death," in that family members reported some of the patients feeling well enough to carry on normal activities of daily life (e.g., holding down jobs, caring for children) until the day they received the TBI. Patient status reports written by the Cincinnati doctors seem to bear out this view. For example, the first seventeen patients were described as all having "incurable and/or metastatic cancer . . . although *in reasonably good clinical condition* [emphasis added]."[145] Similarly, the 1969 DASA report states that the patients "have inoperable, metastatic carcinoma but are *in relatively good health* [emphasis added]."[146] The 1970 DASA report states that the studies conducted during the prior year "were all performed on ambulatory human subjects . . . [who] were *all clinically stable*, many of them *working daily* [emphasis added]."[147] This report also noted the "comparatively better physical condition of these new subjects" and went on to state that "only three of our 11 new patients died in less than 100 days following irradiation. This was in sharp contrast to the almost 50 percent low survival rate for earlier years in this study,"[148] when lower radiation doses had been administered.

Although the Advisory Committee has received some partial patient hospital records, it has not analyzed the records of every patient, which would be required to determine if any deaths could be attributed to the TBI alone, or if such conclusions could be reached at all from the data currently available. (The Committee did not have the time or resources to review the individual files of every patient from this and the numerous other experiments that it has investigated.) Contemporaneous reports, however, state that TBI treatments may have contributed to the deaths of at least eight and as many as twenty patients. The Suskind report, for

example, said that "19 died within 20–60 days and possibly could have died from radiation alone," but noted that bone marrow failure was found in only eight at the time of death.[149] (An additional death occurred six days after irradiation to a patient under anesthesia in the course of a bone marrow transfusion to support the TBI, bringing the number to twenty.) The Suskind report also stated that "there is absolutely no evidence that whole body radiation shortened the period of survival of the treated patients," referring, apparently, to the statistical "survival rate" of the entire group of patients.[150] Similarly, the 1972 ACR report associated the death of eight patients to the fact that "the bone marrow function was subnormal and thus relatable to radiation syndrome." The ACR report also noted that "it is not possible to determine positively that those patients who died within 60 days of the treatment would not have succumbed to their disease within that period, even though the clinical assessment had been that their disease was stable enough to justify their inclusion in the study."[151]

Similarly, following the completion of their study, the Cincinnati doctors wrote that "if one assumes that all severe drops in blood cell count and all instances of hypocellular or acellular marrow at death were due only to radiation and not influenced by the type or extent of cancer and effects of previous therapy, then one can identify 8 cases in which there is a possibility of the therapy contributing to mortality."[152] In 1994 Dr. Saenger wrote:

It is important to realize that in any given patient it is not possible to determine objectively whether death occurred too soon or was prolonged as a consequence of treatment. The only way that an estimate can be made is to compare the length of survival of a group of patients with the same tumor and extent of tumor treated by radiation to a group of patients with the same tumor and extent of tumor treated by different methods.[153]

The attempt by the Cincinnati doctors in their 1973 article to compare survival rates for their TBI patients with the statistics from other published reports[154] is problematic for a number of reasons: first, the comparisons were not controlled for known prognostic factors, such as age, tumor subtype, and stage; second, comparisons with external and historical comparison groups are easily confounded by unmeasured characteristics, such as differences in patient populations and trends in prognosis over time; and third, survival from time of irradiation (some considerable time after diagnosis) in the TBI series is compared to survival from diagnosis in the published series, without using appropriate survival analysis methods. A more meaningful comparison would have been between subgroups of the patients receiving different doses of radiation, preferably including a concurrent unexposed chemotherapy group, adjusting for the interval from diagnosis to the time of death and other relevant prognostic factors. Although limited statistical data were made available to the Advisory Committee, they were not adequate to allow meaningful statistical analysis, and it was not feasible for us to abstract the necessary data from the charts. The Suskind report stated that "before 1966 the design of the study to measure palliation was unstructured and not uniformly applied, particularly as regards uniform definitions and methods of reporting."[155] The report also noted that "it is uncertain whether this study and similar studies reported in the medical literature are truly comparable in all major factors that influence survival, such as selection of patients and ancillary medical management. Therefore, the significance of comparisons of survival rates is doubtful, unless marked differences are found."[156]

The nature of the DOD-sponsored research raises additional concerns as to whether patients were subjected to unnecessary discomfort without full disclosure of experimental purpose or prior consent. In order to collect data on certain side effects of radiation for the military, the patients were not premedicated or informed of potential acute side effects of TBI such as nausea and vomiting so as not to induce these effects psychologically.[157] Patients were to be treated to relieve their symptoms if they affirmatively requested medication. In contrast, researchers at the City of Hope Medical Center conducting a purely clinical TBI study (from 1960 to 1964) gave antinauseant medication to all patients who received 40 or more rad within one hour prior to being irradiated to alleviate possible side effects.[158]

Informed Consent

There is no indication that the DOD ever informed the Cincinnati doctors about the secretary of defense's 1953 Nuremberg Code directive or any subsequent Army implementation of the directive. It is not clear what patients (or family members) were told about the TBI program in the early years of the experiments, because written consent forms were not standard practice at that time. During the later years of the program, written consent forms were employed, but they have been criticized for not clearly stating all of the risks involved.[159]

Written consent forms were first produced and used in the experiments in 1965, two years before they were required by the University of Cincinnati review committee and NIH;[160] the form was revised twice thereafter. According to the Cincinnati doctors' 1973 article summarizing the study, all patients gave informed consent in accordance with the requirements of the Cincinnati Faculty Committee on Research and the National Institutes of Health. Although by the end of the study the consent forms did describe the TBI procedure and its effects, information about risks associated with TBI—nausea and possible death from bone marrow suppression— was not included in these forms.

The first Cincinnati form, dated May 1, 1965, is entitled "Consent for Special Study and Treatment." It states that the "nature and purpose of this therapy, possible alternative methods of treatment, the risks involved, the possibility of complications, and prognosis have been fully explained to me. The special study and research nature of this treatment has been discussed with me and understood by me."[161] There was no mention in the form about the possible risk of death from bone marrow suppression or of the possible side effects of nausea and vomiting, which the doctors were studying and did not want to induce by suggestion. There is no available documentation on what the patients were told orally about the "risks involved." Because Dr. Saenger was not responsible for recruiting or treating patients, he could not speak to what was actually said to the patients.

In 1981, Dr. Robert Heyssel, director of Johns Hopkins Hospital, discussed the ethical climate before the mid-1960s:

I should say that in the climate of those times . . . that many things were done with human subjects, including the investigator himself, which would no longer be condoned. . . . None of these activities had to be reviewed by anyone else in any formal sense within the institutions. I think this was the situation probably up to and around 1966 in most institutions. I am not suggesting that that was the proper thing; I am simply saying that was the case. In terms of experimental therapeutics, I think an honest effort was made by most investigators to explain to families, to patients, that what was being done was in the range of the untried or experimental, but there were certainly no informed-consent rules that anyone was operating under during that period of time up to the midsixties.[162]

A second consent form went into effect in 1967, following the 1967 FCR review of the bone marrow protocol discussed above. This form listed the risks as these: "The chance of infection or mild bleeding to be treated with marrow transplant, drugs, or transfusion as needed." It also said that consent was for "a scientific investigation which is not directed specifically to my own benefit, but in consideration for the expected advancement of medical knowledge, which may result for the benefit of mankind."[163] One member of the University of Cincinnati FCR had suggested in 1967 that the consent form should inform the patient of a one-in-four risk of death. The 1967 FCR review of the protocol required only that the form make clear "the danger inherent in the method and the steps intended to protect the patient."

In 1971, a third form came into use, following the second Faculty Committee review of the bone marrow protocol. This form expanded on the previous form by explaining that "the bone marrow's ability to make [white] cells will be decreased for four or five weeks after you receive your radiation. If you receive a dose of radiation of 200 rads or more, which your doctor will tell you, your blood counts will fall to levels where infection or bleeding could be a problem." It also refined the previous form by describing the re-

search as "a scientific investigation which is not *only* directed specifically to my own benefit, but *also* in consideration for the expected advancement of medical knowledge, which may result for the benefit of mankind."[164] There was no mention of any risk of death.

Beginning in 1968, patient consent was solicited over a two-day period. Dr. Saenger described this process: "Dr. Silberstein was the person who did all this, in that phase. He would explain to somebody the first day what the problems were, what was going to happen, what the risks were, etc. or what the benefits were. Then he had the patient and a representative come back the next day, the representative could have been the patient's mother or cousin, or some family person, or it could have been the patient's minister. And you go through the whole thing with the minister, and the patient and family were all happy with this desperate situation, and the signature was affixed."[165] To the extent it was employed, this procedure appears to have been innovative and above the standard practice of the time.

Family members of some patients testified to the Advisory Committee that neither the patients nor their families were adequately informed about the nature and risks of the radiation treatments. They claim that this occurred despite multiple and persistent requests by family members to meet and discuss their concerns with the doctors involved in administering these treatments. Family members also told the Advisory Committee that patients were not informed about the source of the program funding (although it should be noted that such disclosures are still not mandatory in most institutions).[166] The Suskind report noted that "this information was not withheld if the patient asked about this matter. The procedure follows the custom of every other research project in this University."[167] The ACR report stated that "in the last few years they were told that the information might have military as well as clinical significance."[168]

Subject Selection

All but five of the patients were referred into the study from either the wards of the Cincinnati

General Hospital or its Out-Patient Tumor Clinic. The remaining five were private patients, three of whom were children treated for Ewing's sarcoma. The Suskind report noted that fifty-one of eighty-two patients were black (62 percent) and most were indigent; the report commented that "this distribution reflects the patient population of the Cincinnati General Hospital."[169] Psychological data from the TBI study suggest that some of the subjects may have been of questionable competence or may have been temporarily incapacitated. However, the meaning and importance of these data have been criticized and are in dispute.[170]

AEC-SPONSORED TBI AT OAK RIDGE

At the same time that the University of Cincinnati was conducting TBI experiments for the DOD, the Medical Division of the AEC's Oak Ridge Institute of Nuclear Studies (ORINS)[171] was also treating patients with selected tumors with TBI; retrospective and prospective analyses of these data were supported by the National Aeronautics and Space Administration.[172] ORINS was established in the late 1940s as a research institution to help advance the field of nuclear medicine through research, training, and technology development.[173] From 1957 to 1974, the ORINS/ORAU hospital treated 194 patients with TBI. In contrast with the DOD-sponsored experiments at Cincinnati and the other institutions, ORINS/ORAU used TBI only to treat patients with radiosensitive cancers.

Indeed, in 1972, the ORAU Medical Program Review Committee issued a report on the ORAU TBI activities in light of the recent revelations about the University of Cincinnati TBI program, noting that the studies were ethically conducted and that survival rates were as good as with other methods of treatment.[174]

Nonetheless, similar questions have been raised about the dual-purpose nature of the Oak Ridge program. As happened at Cincinnati, the Oak Ridge TBI experiments, although known in the national and international medical and scientific communities through presentations and publications, first came to the attention of

the general public through the news media. In September 1981, *Mother Jones* magazine published an article charging that ORINS/ORAU treated its patients with total-body irradiation in order to collect data for NASA.[175] The article focused on one patient in particular—Dwayne Sexton, who suffered from acute lymphocytic leukemia and was treated with TBI and chemotherapy over the course of three years until he died in 1968. That article prompted an investigation and public hearing by the Investigations and Oversight Subcommittee of the House Science and Technology Committee, which was chaired by Representative Albert Gore.[176] Testifying before the subcommittee were patients and patient relatives; administrative officials from Oak Ridge, the AEC, and NASA; the medical staff of ORAU; and two cancer experts: Dr. Peter Wiernik, director of the Baltimore Cancer Research Center, and Dr. Eli Glatstein, who was then chief of radiation oncology at the National Cancer Institute and is now a member of the Advisory Committee on Human Radiation Experiments.

ORINS began treating patients with TBI in 1957. Following a 1958 accident at the Oak Ridge Y-12 production plant, in which eight workers were irradiated and treated by the ORINS hospital, ORINS took a heightened interest in the use and effects of TBI. As William R. Bibb, then director of the Department of Energy's Research Division at Oak Ridge, testified at the Gore Hearing: "In order to provide the best possible care in case of an accident the AEC expected that hematologic data from patients being treated with total body irradiation in addition to being used to benefit other patients would also be used to benefit any radiation accident victim."[177] In 1960, the ORINS hospital completed a newly designed irradiation facility that could deliver a uniform dose to all portions of the body without having to move the patient, known as the Medium Exposure Total Body Irradiator (METBI). The METBI facility delivered approximately 1.5 rad per minute. Several years later, ORINS sought to test the hypothesis that exposure to low doses of radiation over an extended period of time would be more effective than a single administration of a similar total radiation dose to the whole body in treating certain types of diffuse tumors known to be responsive to radiation. Accordingly, it developed the Low Exposure Total Body Irradiator (LETBI) as a "one of a kind" system to test this hypothesis. LETBI, which could deliver a whole-body radiation dose of 1.5 rad per hour, went into operation in 1967 and patients could spend several days or weeks in this facility. AEC sponsored all activities concerned with the construction and operation of the LETBI and its use in patient treatment. The results of this treatment approach, however, were found to be no better than others then available, and the use of the LETBI was discontinued in the early 1970s.

The LETBI project was conceived at approximately the same time that NASA had commissioned ORINS to study the effects of total-body irradiation. NASA was particularly interested in the effects of low dose-rate radiation that the LETBI would produce because astronauts would most likely be exposed to low-dose cosmic radiation. Accordingly, NASA provided approximately $65,000 to the AEC for monitoring equipment and the radiation sources used for the LETBI.[178] At the Gore Hearing, officials from the AEC and NASA testified that the LETBI program was conceived purely for therapeutic purposes and that NASA's interest in the data from LETBI exposures in no way influenced the decision to construct the facility or its use for patients. Dr. Clarence Lushbaugh, who ran the LETBI facility under Dr. Gould Andrews, and succeeded Andrews as director of the ORAU medical division, testified: "First, neither NASA nor AEC program monitors, to my knowledge, ever attempted to become involved directly or indirectly with the treatment of patients at the ORINS/ORAU Medical Division. Second, the ORINS/ORAU NASA study group never influenced the clinicians in their selection of patients or the prescription of the exposure dose and dose rates."[179]

There was little dispute with the view of the 1972 Medical Program Review Committee, expressed above, that, at least in the early years, TBI was a legitimate form of treatment worth exploring for the radiosensitive cancers that ORINS/ORAU was treating. The Review Committee's concern was whether the Oak Ridge medical staff conducted their investiga-

tions in an effective manner and whether the AEC's or NASA's interest in the data compelled the continuation of this modality at a time when other forms of treatment were considered more effective. Dr. Peter H. Wiernik, one of the two expert witnesses, acknowledged, for example, that in the early years it was legitimate to experiment with TBI at the high doses being used to try to improve treatment, because "clearly treatment needed to be advanced in those days."[180]

The record of the 1972 review suggests that the ORINS/ORAU staff did not engage in the type of rigorous, systematic research that would be necessary to evaluate the usefulness of that type of therapy. The Oak Ridge doctors acknowledged that they were not evaluating the long-term effectiveness of single-exposure, high-dose TBI and that fractionated exposures (in which numerous smaller doses are given over a period of several weeks or months) "probably offers a preferable approach for total-body irradiation therapy."[181] Dr. Lushbaugh explained that, because the doctors would administer whatever treatment they thought was "best for each patient," they did not adhere to an established research protocol based exclusively on TBI.[182]

In commenting on the 1972 report before the Gore Committee, Dr. Glatstein questioned the "manner of administration and the uncontrolled nature of the studies." Oncology research, he said, requires "an obsession with time"—the effect that a given treatment has over months or years. Glatstein noted that the reports he reviewed "are interesting in terms of acute radiation effects but really don't have any substance in terms of oncologic practice."[183] Glatstein summarized his view of the ORINS/ORAU TBI research program: "If you are talking about the early 60's I think this is probably fairly representative of protocols that were going on at that time. . . . [B]y the end of that decade I believe this was probably not acceptable."[184]

Both Wiernik and Glatstein criticized Dwayne Sexton's medical (nonradiation) treatment, in particular the decision to withhold maintenance chemotherapy, which was recognized as an effective treatment at that time, in order to attempt a never-before-used experimental procedure.[185] Even if the new treatment was worth pursuing, they argued, it should have been done only as part of a larger protocol and only when the patient was in secondary remission following the failure of more-effective treatments.

All patients accepted into the ORINS/ORAU hospital program signed a "Patient Admittance Agreement" that explained that the hospital operated for the purpose of conducting radiation-related research. The form stated that the patient is being admitted because his physical condition "makes me a suitable patient for a currently active clinical research project," that experimental examinations, treatments, and tests may be prescribed for which the patient hereby gives his or her consent, and that the patient "can remain in the research hospital only so long as I am needed for research purposes." Additional forms were used to establish "Consent for Experimental Treatment," which stated that "the nature and purpose of the treatment, possible alternative methods of treatment, the risks involved, and the possibilities of complications have been explained to me. I understand that this treatment is not the usual treatment for my disorder and is therefore experimental and remains unproven by medical experience so that the consequences may be unpredictable."[186] The form made no mention of the possible risk of death from bone marrow suppression or specific side effects such as nausea or vomiting.

In 1974, the AEC conducted a program review of the Medical Division of ORAU. It recommended that the clinical TBI programs be closed, having found that the METBI and LETBI programs had "evolved without adequate planning, criticism or objectives, and have achieved less in substantial productivity than merits continued support."[187]

At the end of his hearing, Gore noted that the subcommittee would issue a report with conclusions and recommendations. Although no formal report was ever completed, the full committee issued the following statement in January 1983: "The Subcommittee testimony revealed that while many of the conditions at [ORAU] were not satisfactory, particularly when judged by the routine institutional safeguards and medical knowledge of today, the more scandalous allegations could not be substantiated. Given the

standards of informed consent at the time, and the state of nuclear medicine, the experiments were satisfactory, but not perfect."[188]

Perhaps the most striking contrast between philosophies of the Oak Ridge and the University of Cincinnati TBI programs can be gleaned from an exchange that occurred in 1966. That year, the AEC's Medical Program Review Committee suggested that ORAU consider using TBI for treatment of radioresistant cancers (similar to what was being done at Cincinnati).[189] The ORAU physicians responded that they had carefully considered treating such diseases, but had declined to do so:

[W]e are very hesitant to treat them because we believe there is so little chance of benefit to make it questionable ethically to treat them. Lesions that require moderate or high doses of local therapy for benefit, or that are actually resistant (gastroenteric tract) are not helped enough by total body radiation to justify the bone marrow depression that is induced. Of course, in one way these patients would make good subjects for research because their hematologic responses are more nearly like those of normals than are the responses of patients with hematologic disorders.[190]

CONCLUSION

When we began our work, the controversy surrounding the Cincinnati TBI research had been rekindled. There was, however, little public awareness that Cincinnati was the last in the line of many years of sponsorship of similar TBI-related research by the Defense Department and other federal agencies. The ethical issues raised by the Cincinnati case are made more acute by the fact that both the government and the medical community already had had decades of experience with TBI, although comparatively less experience with cobalt 60 as a means to deliver higher doses than had been delivered in the earlier era.

This history provides compelling evidence of the importance of the rules that regulate human subject research today—prior review of risks and potential benefits, requirements of disclosure and consent, and procedures for ensuring equity in the selection of subjects. The history also highlights four issues in the ethics of research with human subjects that are as important today as they were then, issues that are not easily resolved or even addressed by present-day rules. As discussed below, these issues are (1) how to protect the interests of patients when physicians use medical interventions that are not standard care; (2) the effects and attendant obligations of the government when it funds research involving patient-subjects; (3) the impact on patients when research is combined with medical care; and (4) what constitutes fairness in the selection of subjects for research.

The first issue is how best to protect the interests of patients when physicians propose to use medical interventions that are not standard care. Today, when nonstandard interventions are part of a formal research project, the interests of the patient are protected in theory by the institutional review board, which is charged with determining that the risks of the nonstandard intervention are acceptable in light of the available alternatives and the prospect for benefit. Patients are also protected by the requirement of informed consent, which is intended to allow the potential patient-subject to assess whether the balance of risks to potential benefits is acceptable. There is no federally mandated parallel IRB mechanism of review, however, when a medical intervention that is experimental or innovative or even controversial is to be used outside the confines of a research project, although some institutions voluntarily have adopted mechanisms of peer review. The requirement of informed consent remains; the physician is obligated to inform the patient that the proposed intervention is not standard practice, whether it is controversial within the field, and how it compares with alternative approaches, but this requirement provides the patient less protection than would a professional peer review.

At the time of the TBI studies, none of these mechanisms were well developed. During the Cincinnati project, IRBs were in their infancy and the convention of obtaining informed consent from patient-subjects was just emerging. The record is confused and confusing as to whether or when TBI at Cincinnati was viewed as part of a cancer research project and thus properly the subject of IRB review. It is not clear whether the treatment of the Cincinnati patients

with TBI was initially intended to be research. In the practice of medicine there has always been a fine boundary between practices or treatments that are accepted as standard, those that are "innovative," and those that are experimental or the subject of research. The use of TBI at Cincinnati is emblematic of the difficulties inherent in sorting through these categories.

By the mid-1960s, TBI without bone marrow protection was a treatment that had been tried and had not been proven effective for patients with radioresistant cancers. By this time, total-body irradiation was not standard treatment for such cases, nor could it be called innovative treatment; some at the time considered its continued use in patients with radioresistant cancers to be controversial. The history of medicine, however, is replete with instances in which failure is followed by success. The continued use of TBI in patients with radioresistant cancers would not have been unethical if the physicians had established clear benchmarks for determining how much additional use was warranted, and if patients had been informed of the speculative nature of the treatment and the gravity of the risks involved. It is not clear that either of these things occurred.

What is clear is that neither the university's IRB nor the funding agency reviewed the appropriateness of continuing to treat patients with radioresistant cancers using TBI without bone marrow protection, despite mounting evidence casting doubt on the utility of TBI treatment for radioresistant tumors in the absence of bone marrow protection. It is also clear that the consent forms did not disclose that it was by this time at best unconventional to treat patients with radioresistant cancers with TBI and that no other medical centers were engaged in this practice at the time; whether physicians told this to their patients is not known. The system of checks and balances that is usually in place today to protect patients' interests was in its early phase at the University of Cincinnati, and the system did not work well at the time. The responsibility for failure rests at all levels, but it is reasonably clear that patient protection was compromised.

Today, as in the past, there are occasions when nonstandard medical interventions are not subject to human research regulations. In such situations, neither IRB review nor the rigors of scientific design are in place to help determine whether an experimental intervention should continue to be used. Today, for example, many innovations in reproductive technologies and surgery proceed with little oversight and few constraints on the practices of physicians. A physician wishing to use an intervention that other colleagues in the field believe to be ineffective or inferior—as was arguably the case with TBI and radioresistant tumors after several years in the Cincinnati program—will find little standing in his or her way to do so save the fear of malpractice claims and, increasingly, the likelihood that such interventions will not be reimbursed, particularly in managed-care settings. The Cincinnati experience underscores the importance of (1) establishing benchmarks for judging the propriety of continued use and (2) providing for special disclosures to patients in all cases where interventions are not standard—without regard for whether the intervention is deemed "human subject research" or is governed by the Common Rule (see chapter 3).

The question of what role the Department of Defense should have played in reviewing the appropriateness of TBI as medical care for the patient-subjects in its biological dosimetry and radiation-effects research points to the second major issue illustrated by our review of the TBI history. Arguably, the ultimate responsibility for determining that TBI was acceptable medical practice rested with the physicians at Cincinnati and with the university and associated hospitals. At the same time, however, thirty years of government interest in the effects of TBI also arguably had a significant influence on medical practice.

From one vantage, the DOD had little or no obligation to consider the value of TBI to the patients who provided the data it was seeking. The DOD was not paying for the irradiation of the patients. It had reason to assume that the decision about the propriety of the treatment would be made by doctors whose judgment in the matter could be trusted. Yet the TBI experience illustrates that when the government funds research, particularly over a long period,

its funding may well have effects beyond the simple conduct of the science and well beyond the confines of the strict terms stated in the contracts or grants authorizing the research.

Over the course of three decades, there was a substantial coincidence between the use of TBI on patients with radioresistant cancers and funding from the Department of Defense and its predecessor. With the exception of work conducted at the City of Hope Hospital, every journal article in the professional literature on the use of TBI with radioresistant tumors during this period was reporting on work supported by the government for military purposes.

In the case of Cincinnati, Dr. Saenger told the Advisory Committee in 1994 that the irradiation of patients might not have been initiated were it not for funding by the DOD and, once initiated, might not have been continued if the objective sought by the DOD (a biological dosimeter) had been realized early on. As Dr. Saenger explained, while the DOD did not directly pay for irradiation, its funding provided for other items—including laboratory equipment and specialists—that facilitated the initiation and maintenance of the TBI program.

Even where the medical care of patients is peripheral to the interests of a funding agency, so long as the research supported by the agency is to be conducted on patient-subjects, it is likely that the research will affect the care patients receive. This is particularly true when agencies support research programs extending over many years, as was the case with the Department of Defense and TBI. Such programs can motivate physician-investigators to alter their practice and can stimulate the adoption of different approaches to the care of patients. Although there is today a greater appreciation of the impact on medical practice of funding patterns in research, it is not clear even now that funding agencies regularly think through the implications for medical care of the research programs they support or that they monitor the impact on patients of their programs over time.

That the joining of research with medical care can alter what happens to a patient is the third issue in research ethics illustrated by the TBI experience. Each purpose introduced into the clinical setting in addition to the treatment of the patient increases the likelihood that the patient will receive more, fewer, or different medical interventions than he or she would otherwise receive. It is naive to think that, either today or thirty years ago, research can be grafted on to the clinical setting without changing the experience for the patient, now turned subject. When the demands of science alter the standard medical practice by increasing the monitoring of physiological indicators, the additional blood tests or bone scans or biopsies are frequently presented as in the interest of patient-subjects. Sometimes this claim is defensible, and the patient-subjects are indeed advantaged by more careful monitoring of their medical condition; at other times, however, this claim is an insupportable rationalization, and there are no offsetting benefits to patients for the risks and discomforts associated with additional monitoring.

In the case of the Cincinnati experiments, the impact of the research protocol on the care of the patient-subjects cannot be construed as beneficial to the patients; in addition, there is evidence of the subordination of the ends of medicine to the ends of research. The decisions to withhold information about possible acute side effects of TBI as well as to forgo pretreatment with antiemetics were irrefutably linked to advancing the research interests of the DOD. To the extent that this deviated from standard care, and caused unnecessary suffering and discomfort, it was morally unconscionable; to the extent that the standard of care in this area is uncertain, it is morally questionable. As troubling as this is, far more troubling is the evidence, including the testimony of the principal investigator, that TBI might not have been employed as treatment for the patients, or once employed continued, in the absence of the government's funding and research requirements.

Whether the ends of research (understood as discovering new knowledge) and the ends of medicine (understood as serving the interests of the patient) necessarily conflict and how the conflict should be resolved when it occurs are still today open and vexing issues. Increasingly, advocates for patients with serious, chronic diseases such as AIDS and breast cancer maintain that it is often in the interests of patients to participate as subjects in clinical research. These advocates are particularly concerned to ensure fair access to participation in research for people who are politically less powerful, such as the poor, minorities, and women. This contemporary perspective upends the traditional way of viewing the fourth issue in research ethics raised by the TBI experiments—fairness in the selection of subjects.

At both M. D. Anderson Hospital and the University of Cincinnati, almost all the patients were drawn from public hospitals, and many were African-Americans. It was common during this period for medical research to be conducted on the poor and the powerless. In part, this practice reflected a general societal insensitivity to questions of justice and equal treatment. In this case, people who were poor disproportionately bore the burdens of questionable research to which their interests as ill people were subordinated. The practice also reflected the view, however, that poor people were better off being patients at hospitals affiliated with research-oriented medical schools where they were likely to become subjects of research (as well as subject matter for clinical teaching). Such institutions, it was thought, offered poor people their best, and perhaps their only, chance to secure quality medical care. Recently, this kind of reasoning has emerged again, as constraints on access to medical care—from the narrowing of entitlement programs to the narrowing of coverage in managed-care medical plans—have made participation in research, as a route to medical care, more attractive. The question of whether the "side benefits" of being a subject should be weighted in the review of the risks and potential benefits of research remains unresolved today.

These findings highlight the contemporary resonance of the TBI story. The issues discussed above are either not now addressed or not addressed adequately by regulation; neither are they covered by clear conventions or rules of professional ethics. Thus, the history of TBI research sponsored by the government is important not only for what it tells us about our past but also for how it illuminates the present.

NOTES

1. As of September 1995, the lawsuit was still ongoing. In January 1995, the court issued an opinion rejecting the defendants' request to dismiss the case, thus allowing the plaintiffs to proceed with discovery and a possible trial. *In re Cincinnati Litigation*, 874 F. Supp. 796 (S.D. Ohio, 1995). That decision is now on appeal in the U.S. Court of Appeals.

2. A list was published in the 1967 National Academy of Sciences report on "Radiobiological Effects of Manned Space Flight" and is reprinted in *Hearing on the Human Total Body Irradiation (TBI) Program at Oak Ridge before the Subcommittee on Investigations and Oversight of the House Science and Technology Committee* [Gore Hearing], 97th Cong., 1st Sess. (23 September 1981), 355. The Advisory Committee requested information from most of these institutions (except for those performing five or fewer procedures) on their use of TBI. Virtually all of them informed the Advisory Committee that they no longer have any records describing these activities, besides what has been published in the literature; some of the institutions have informed the staff orally that they did "nonexperimental" TBI treatments on patients with leukemia during the 1950s and therefore would have no protocols or review documents of that work.

3. Three were conducted during the Manhattan Project years, and five between 1951 and 1971.

4. Renamed the Oak Ridge Associated Universities (ORAU) in 1966.

5. City of Hope Hospital in Duarte, California. Melville L. Jacobs and Fred J. Marasso, "A Four-Year Experience with Total-Body Irradiation," *Radiology* 84 (1965): 452–456 (using a cobalt 60 teletherapy unit).

6. The terms *radiosensitive* and *radioresistant* are relative terms that appear to have little meaning in current medical parlance, but were widely used at least into the 1970s.

7. J. T. Chaffey et al., "Total Body Irradiation in the Treatment of Lymphocytic Lymphoma," *Cancer Treatment Report* 61 (1977): 1149–1152; M. H. Lynch et al., "Phase II Study of Busulfan, Cyclophosphamide, and Fractionated Total Body Irradiation as a Preparatory Regimen for Allogeneic Bone Marrow Transplantation in Patients with Advanced Myeloid Malignancies," *Bone Marrow Transplant* 15, no. 1 (January 1995): 59–64 (This study combines chemotherapy with doses of 1,200 rad of TBI).

8. R. A. Clift, C. D. Buckner, and F. R. Appelbaum, "Allogeneic Marrow Transplantation in Patients with Chronic Myeloid Leukemia in the Chronic Phase: A Randomized Trial of Two Irradiation Regimes," *Blood* 77 (1991): 1660.

9. The methods of reporting radiation doses have changed over the years. Throughout the 1950s, researchers tended to report the dose in *roentgens* ("R" or "r"), which represented the amount of radiation emanating from the source; from the 1960s through the present, researchers reported the dose in *rad* (also known as *centigrays*), which represent the amount of radiation absorbed by the body. For x rays, the *rem*—for *roentgen equivalent man*—is equivalent to a rad. A given air dose (roentgen) of radiation is generally equivalent to a lesser body dose (rad). For example, 325 R (air dose) was reported in one experiment as approximately equivalent to 200 rad (body dose).

10. Fred A. Mettler, Jr., and Arthur C. Upton, *Medical Effects of Ionizing Radiation*, 2d ed. (Philadelphia: W. B. Saunders Co., 1995), 41, 278 (table adapted from J. T. Conklin).

11. One of the difficulties in reporting the chance of death at a given dose is a confusion between the use of air (or skin) dose as opposed to body (or midline tissue) dose (the former is generally significantly higher than the latter), and whether or not medical care is provided. Many investigators fail to indicate clearly what type of dose they are using. Mettler and Upton report that "[i]t is probable that with appropriate medical treatment, the LD50 [lethal dose for 50 percent of recipients] skin dose may be in the range of 6 Gy (600 rad) or an MTD [midline tissue dose] of 3.96 Gy (396 rad)." Mettler and Upton, *Medical Effects of Ionizing Radiation*, 278.

12. Fred G. Medinger and Lloyd F. Craver, "Total Body Irradiation, with review of cases," *American Journal of Roentgenology* 48 (1942): 651, 668. The investigators used a 250-kilovolt (KV) machine and doses up to 450 R. The dose that would cause severe bone marrow damage and be potentially lethal was considered to be between 200 and 400 roentgens. This figure can vary depending on the length, frequency, and intensity of the dose; for example, a single dose of 200 R will generally cause more severe reactions than five doses of 40 R spaced over one or two weeks.

13. J. J. Nickson, "Blood Changes in Human Beings Following Total-Body Irradiation," in *Industrial Medicine on the Plutonium Project: Survey and Collected Papers*, ed. Robert S. Stone (New York: McGraw-Hill Book Co., 1951), 337.

14. B. V. A. Low-Beer and Robert S. Stone, "Hematological Studies on Patients Treated by Total-Body Exposure to X-ray," in Stone, *Industrial Medicine*, 338.

15. Ibid.

16. Ibid., 338–39.

17. Alan Gregg, M.D., Chairman of the AEC Advisory Committee on Biology and Medicine, to Dr. Stone, 20 October 1948 ("The secrecy with which some of the work . . .") (ACHRE No. UCLA-111094-A-24), 1.

18. Robert Stone, M.D., to Allen Gregg, M.D., Chairman of the AEC Advisory Committee on Biology and Medicine, 4 November 1948 ("The candor of your letter of October 20th . . .") (ACHRE No. UCLA-111094-A-25), 3.

19. Medinger and Craver, "Total Body Irradiation," 668.

20. L. F. Craver, "Tolerance to Whole-Body Irradiation of Patients with Advanced Cancer," in Stone, *Industrial Medicine*, 485.

21. Ibid., 486.

22. Ibid.

23. See Nickson, "Blood Changes in Human Beings Following Total-Body Irradiation," in Stone, *Industrial Medicine*. After World War II, Dr. Nickson continued to engage in TBI research at Sloan-Kettering.

24. Ibid., 309. "The persons who were subjected to radiation during this study were divided into three general groups. The first group consisted of eight persons who had neoplasms that could not be cured but still were not extensive enough to influence general health. . . . The second group consisted of three persons who had illnesses that were generalized and chronic in nature [two had arthritis]. . . . The third group consisted of three normal volunteers from among the personnel of the Metallurgical Laboratory." Ibid.

25. Robert Stone, M.D., to Shields Warren, M.D., 6 October 1948 ("I have recently been shown a letter from Mr. Keller . . .") (ACHRE No. DOE-120994–A-27), 1.

26. Ibid.

27. The chief of the AEC's Insurance Branch supported declassification: "It is conceivable that if it became a matter of common knowledge that experiments on human beings such as those described in this document were being carried on by the Commission it might result in some adverse publicity and perhaps encourage litigation. However, we feel that this objection is outweighed by the advantages to be gained by making this information available to technically trained personnel." Clyde E. Wilson, Chief of the AEC Insurance Branch, to Anthony Vallado, Deputy Declassification Officer, AEC Declassification Branch, 10 September 1948 ("Review of Document") (ACHRE No. DOE-120894–E-42).

28. Gregg to Stone, 20 October 1948, 1.

29. For example, M. Soden et al., "Lymphoid Irradiation in Intractable Rheumatoid Arthritis: Longterm Follow-up of Patients Treated with 750 rad or 2,000 rad, . . ." *Arthritis and Rheumatism* 15, no. 3 (May 1989): 577–582.

30. Ibid.

31. Stone to Gregg, 4 November 1948, 2.

32. NEPA stood for Nuclear Energy for the Propulsion of Aircraft.

33. Other participants included Robley Evans, Hymer Friedell, Robert Stone, Shields Warren, and Stafford Warren, all of whom were often called on for other radiation research advice in the late 1940s and 1950s by the AEC, the DOD, and other agencies. NEPA Advisory Committee on Radiation Tolerance of Military Personnel, proceedings of 3 April 1949 (ACHRE No. DOE-120994–B-1).

34. In a subsequent paper, Stone cited Jenner's experiments with smallpox, Walter Reed's experiments with yellow fever, and World War II experiments with malaria on prisoners as examples. Robert S. Stone, paper of 31 January 1950 for the NEPA project ("Irradiation of Human Subjects as a Medical Experiment") (ACHRE No. NARA-070794–A-9). Stone also sought to justify the use of normal humans on national security grounds: "The information desired is sufficiently important to the safety of the U.S.," and the doses would be "relatively low in relation to the lethal doses." The "safety of the U.S." language was later dropped by the full committee. NEPA Advisory Committee, 3 April 1949, 43–48.

35. Advisory Committee for Biology and Medicine, transcript (partial) of proceedings of 10 November 1950 (ACHRE No. DOE-012795–C-1), 6.

36. Ibid., 29.

37. ACBM, transcript of proceedings of 3 April 1949.

38. Ibid., 39–40. Stone noted that "volunteering exists when a person is able to say Yes or No without fear of being punished or of being deprived of privileges due him in the ordinary course of events." Ibid., 39.

39. Ibid., 40.

40. Ibid., 42.

41. Ibid., 41.

42. NEPA Medical Advisory Committee, 5 January 1950 ("Radiation Biology Relative to Nuclear Energy Powered Aircraft") (ACHRE No. DOE-060295–C-1), 4.

43. Robert S. Stone, 31 January 1950, 3. "The extremely small hazard of undetectable genetic effect, undetectable effect on the life span and possibly slight effect on the blood picture are the extremely small hazards that must be weighed against the value of having actual experience with exposure of humans." Ibid., 4.

44. See Committee on the Biological Effects of Ionizing Radiation, National Research Council, *Health Effects of Exposure to Low Levels of Ionizing Radiation: BEIR V* (Washington, D.C.: National Academy Press, 1990), 168–169, 171–175, 242–253. See also Donald A. Pierce and Michael Vaeth, "Cancer Risk Estimation from the A-Bomb Survivors: Extrapolation to Low Doses, Use of Relative Risk Models and Other Uncertainties," in *Low Dose Radiation: Biological Bases of Risk Assessment*, eds. K. F. Baverstock and J. W. Stather (London: Taylor & Francis, 1989), 54, 58–59. The BEIR V leukemia model includes both a linear and quadratic term in dose, the latter dominating the risk above 90 rad. For a single exposure to 10 rad, the BEIR report gives an estimated excess relative risk of 15 percent (lifetime, averaging over ages at exposure). Extrapolating this figure linearly to 150 rad would produce an excess relative risk of 2.25, but when the quadratic term is included, the total relative risk becomes 7. This estimate is in perfect agreement with the observed and fitted risks among the atomic bomb survivors at 150 rad, as plotted by Pierce and Vaeth. For exposures at particular

ages and expressed at particular follow-up times, the relative increase can be either larger or smaller than this figure.

45. Ibid., 3–4.

46. W. A. Selle, Secretary, NEPA Medical Advisory Committee, to Dr. Richard Meiling, Director, Medical Services Division, 22 March 1950 ("As indicated to you in person on March 8 . . .") (ACHRE No. NARA-070794-A-9).

47. DOD Research and Development Board, Committee on Medical Sciences, proceedings of 23 May 1950 (ACHRE No. DOD-042994-A-15), 10.

48. Marion W. Boyer, AEC General Manager, to Robert LeBaron, Chairman, Military Liaison Committee, 10 January 1951 ("As you know, one of the important problems that would confront us . . .") (ACHRE No. DOE-040395-B-1), 3–4.

49. Participants were Alan Gregg (ACBM chairman), Dr. Austin Brues of the Argonne National Laboratory (operated by the University of Chicago), Dr. Simon Cantril of Swedish Hospital in Seattle, Dr. Andrew Dowdy (who had chaired the NEPA Medical Advisory Committee), Dr. Louis Hempelmann of Rochester, Dr. Robert Loeb of Columbia, Dr. Curt Stern of the ACBM, and Dr. Shields Warren of the DBM. Ibid.

50. Mettler and Upton, *Medical Effects of Ionizing Radiation*, 278.

51. Marion W. Boyer, AEC General Manager, to Robert LeBaron, Chairman, Military Liaison Committee, 10 January 1951 ("As you know, one of the important problems that would confront us . . .") (ACHRE No. DOE-040395-B-1), 3–4.

52. Joseph Hamilton to Shields Warren, 28 November 1950 ("Unfortunately, it will not be possible . . .") (ACHRE No. IND-071395-A-9).

53. Armed Forces Medical Policy Council to the Secretary of Defense, 30 June 1951 ("Annual Report") (ACHRE No. DOD-091694-A-1), 158.

54. Interview of Colonel John Pickering by Dr. John Harbert and Gilbert Whittemore (ACHRE staff), transcription of audio recording, 2 November 1994 (ACHRE Research Project Series, Interview Program File, Targeted Interview Project), 14ff.

55. Wilson F. Humphreys, Colonel, USMC, Assistant Chief of Staff, to Commanding Officers, 16 November 1959 ("DASA Base Commanders' Weekly Bulletin") (ACHRE No. DOD-082694-A-1), 2.

56. At least four of the five DOD institutions used the new, high-energy radiation sources—cobalt 60 or megavoltage x rays. (The available data did not make clear what type of unit Sloan-Kettering used.)

57. John A. Isherwood, Chief, Army Radiological Service, to Assistant Chief, U.S. Army Medical Research and Development Command, 22 October 1958 ("1. Recommend approval . . .") (ACHRE No. DOD-042994-A-16).

58. Executive Panel of the NEPA Medical Advisory Committee, proceedings of 8 July 1949 (ACHRE No. DOD-042994-A-17), 77 (emphasis added).

59. Carroll Wilson, AEC General Manager, to Robert Stone, 5 November 1947 ("Your Letter of September 18 . . . ") (ACHRE No. DOE-052295-A-1).

60. P.L. 82–557, sec. 5, 66 Stat. 725 (16 July 1952).

61. Max Brown to Vice Chancellor, University of Pittsburgh, 12 March 1957 ("This in reply . . .") (ACHRE No. NARA-012395-A-6).

62. Department of Defense, Defense Atomic Support Agency, Contract DA-49-146-XZ-029 (contract with the University of Cincinnati), Modification no. 1, ASPR. no. 7-203.22, 28 February 1961 (ACHRE No. DOD-042994-A-23).

63. Lieutenant Lando Haddock, USAF, 19 October 1950 ("Negotiation of Cost-Reimbursement") (ACHRE No. DOD-062194-B-3).

64. The results were published in Lowell S. Miller, Gilbert H. Fletcher, and Herbert B. Gerstner, to the School of Aviation Medicine, report of May 1957 ("Systemic and Clinical Effects Induced in 263 Cancer Patients by Whole Body X-Irradiation with Nominal Air Doses of 15 to 200 R") (ACHRE No. DOD-102594-A-1); and Lowell S. Miller, Gilbert H. Fletcher, and Herbert Gerstner, "Radiobiologic Observations on Cancer Patients Treated with Whole-Body X-Irradiation," *Radiation Research* 4 (1958): 150–165.

65. Miller, Fletcher, and Gerstner, "Systemic and Clinical Effects," 1. "Colonel McGraw stated that he thought that the cases studied were terminal cases. He was answered in the negative." Air Force Research Council, proceedings of 14 January 1954 (ACHRE No. DOD-092894-A-1), 7.

66. Miller, Fletcher, and Gerstner, "Systemic and Clinical Effects," 2.

67. Ibid. The report also states that "[a]t the time of irradiation, these patients were still able to walk and perform light physical tasks."

68. Ibid.

69. Attachment, Lester J. Peters, Division of Radiotherapy, M. D. Anderson Cancer Center, to Steve Klaidman (ACHRE), 22 December 1994 (ACHRE No. CORP-010995-A-1).

70. School of Aviation Medicine (SAM) Research Council, proceedings of 14 January 1954 (ACHRE No. DOD-092894-A-1), 6–7.

71. Ibid.

72. Robert B. Payne to the School of Aviation Medicine, report of February 1963 ("Effects of Acute Radiation Exposure on Human Performance") (ACHRE No. DOD-121994-C-1), 10.

73. One Air Force reviewer stressed that "the patients cannot be considered as normal people." SAM Research Council, proceedings of 14 January 1954, 6. Even the study investigator warned that "the application of these results to operational problems should be made with cautious regard for the medical status of the subjects and the limited relevance of experimental criteria." Payne, "Effects of Acute Radiation," 12.

74. SAM Research Council, proceedings of 14 January 1954, 7.

75. SAM Research Council, proceedings of 29 August 1955 (ACHRE No. DOD-092894–A-2), 6.

76. Miller, Fletcher, and Gerstner, "Systemic and Clinical Effects," 20. The researchers noted that their paper deals only with "those aspects of the problem which are of general radiobiological interest; a strictly clinicotherapeutic evaluation will be given elsewhere." Ibid., 1. It is not known if such a therapeutic evaluation was ever completed, and none has been located to date.

77. Ibid., 7. The report states that 30 percent claimed subjective improvement, but that this was possibly due to psychological rather than clinical factors. Ibid.

78. Ibid., 20. They cautioned, however, that the condition of terminally ill patients may increase their sensitivity to both acute and longer-term radiation symptoms. Ibid.

79. SAM Research Council, proceedings of 14 January 1954, 7.

80. Colonel Robert B. Payne, "Effects of Acute Radiation Exposure on Human Performance," Review 3–63, USAF School of Aerospace Medicine, Aerospace Medical Division, February 1963, 3.

81. William C. Levin, Martin Schneider, and Herbert B. Gerstner, paper of 1960 for Air University, School of Aviation Medicine ("Initial Clinical Reaction to Therapeutic Whole-Body X-Irradiation") (ACHRE No. DOD-072794–B-18).

82. The original contract was for a "Study of the Effects of Total and Partial Body Radiation on Iron Metabolism and Hematopoiesis"; it was later known as "The Effect of Total Body Irradiation on Immunologic Tolerance of Bone Marrow and Homografts of Other Living Tissue." (The Advisory Committee has eight progress reports.) At least three of the patients had arthritis. Baylor University College of Medicine to the Armed Forces Special Weapons Project (AFSWP), report of 1 January 1954 (ACHRE No. BAY-101794–A-1). This condition was not among the diseases listed as having been treated in the Baylor University College of Medicine to Defense Atomic Support Agency (DASA), report of 1 February 1963–31 January 1964 (ACHRE No. BAY-101794–A-2), 6.

83. The "fundamental problem has been to define effect of irradiation and quantitate effect with amount of radiation exposure." Baylor University College of Medicine to DASA, report of 1 February 1963–31 January 1964, 1. Collins and Loeffler published preliminary findings in 1956. Vincent P. Collins and R. Kenneth Loeffler, "The Therapeutic Use of Single Doses of Total Body Radiation," *American Journal of Roentgenology* 75 (1956): 546. Collins appears to have first conducted DOD-sponsored TBI research in 1953 while at Columbia University. See Joint Panel on the Medical Aspects of Atomic Warfare, proceedings of 7 January 1953 (ACHRE No. DOD-072294–B-1), item 10.

84. Baylor University College of Medicine to AFSWP, report of 1 September 1955–31 January 1956 ("A Study of the Effects of Total and Partial Radiation on Iron Metabolism and Hematopoiesis") (ACHRE No. DOD-090994–D-2), 6.

85. Baylor University College of Medicine to DASA, report of 1 February 1963–31 January 1964, 1; Baylor University College of Medicine to Defense Atomic Support Agency, report of 1 February 1961–31 January 1962 (ACHRE No. BAY-101794–A-3), 2.

86. Baylor University College of Medicine to DASA, report of 1 February 1963–31 January 1964, 7–8.

87. Ibid., 5–6.

88. Ibid., 6.

89. Ibid., 12.

90. Because the Advisory Committee did not receive all of the progress reports on this study, the total number of patients cannot be determined. Nor could the Advisory Committee determine what type of teletherapy unit the investigators used. Memorial Hospital was the site of the second Manhattan Project experiment under Dr. Craver. Dr. Nickson was the author of the report on the third Manhattan Project experiment at Chicago; he came to Sloan-Kettering after World War II.

91. Sloan-Kettering Institute for Cancer Research to AFSWP, report of 1 March 1958 ("Quarterly Report-Study of the Post-Irradiation Syndrome in Humans") (ACHRE No. DOD-062194–A-9).

92. Sloan-Kettering Institute for Cancer Research to AFSWP, report of 1 May 1956 ("Annual Report") (ACHRE No. DOD-060794–A-1), abstract page.

93. Captain E. Richard King, "Use of Total-Body Radiation in the Treatment of Far Advanced Malignancies," *Journal of the American Medical Association* 177 (2 September, 1961): 86–89.

94. Ibid., 613.

95. Ralph R. Cavalieri, Milton Van Metre, F. W. Chambers, and R. Richard King, "Taurine Excretion in Humans Treated by Total-Body Radiation," *Journal of Nuclear Medicine* 1 (1960): 186, 187, 190. Taurine is an amino acid that is excreted in the urine. A 1962 request for funding document from the Navy Bureau of Medicine and Surgery proposed to continue the biological dosimeter project through 1967 by collecting "daily urine specimens from patients exposed to total body irradiation for therapeutic purposes." U.S. Navy, Bureau of Medicine and Surgery, 1 June 1962 ("Biological Dosimeter of Radiation Injury") (ACHRE No. DOD-090994–C-3), 1–4. The Navy informed the Advisory Committee that there is no evidence that this project was ever funded.

96. Standard Form 522, revised August 1954 ("Authorization for Administration of Anesthesia and for Performance of Operations and other Procedures") (ACHRE No. DOD-020695–B-1). Standard Form 522 was modified for TBI procedures. A second form was sometimes used for "consent to drastic radiation/

chemical therapy," which states that "[t]he possibility that drastic radiation therapy may be useful . . . , the results expected, and the consequences likely to ensue, have been explained to me and I hereby give my consent. . . ." NHBETH Form 27B, August 1959 ("Consent to Drastic [Radiation] Therapy") (ACHRE No. DOD-020695–B-2).

97. Shields Warren, "Ionizing Radiation and Medicine," *Scientific American*, September 1959, 165.

98. AG Central Files, January-May 1962 ("Conference, Meetings, Military, Naval and All Divisions") (ACHRE No. DOD-081994–A-1), 2.

99. Over the course of the eleven-year DOD study, ten researchers participated in the project and contributed to the DASA reports: Eugene L. Saenger, M.D.; Edward B. Silberstein, M.D.; Ben L. Friedman, M.D.; James G. Kereiakes, Ph.D.; Harold Perry, M.D.; Harry Horwitz, M.D.; Bernard S. Aron, M.D.; I-Wen Chen, Ph.D.; Carolyn Winget, M.A.; and Goldine C. Gleser, Ph.D. Dr. Saenger was the principal investigator of the study, but not the attending physician or the administering radiologist for any of the patients.

100. Dr. Saenger had developed an interest in studying the "effect of whole body radiation on the patient suffering from cancer" while serving in the military as chief of the radioisotope laboratory at Brooke General Hospital, Fort Sam Houston, Texas. Raymond Suskind et al. to Dean of the College of Medicine, University of Cincinnati, January 1972 ("The Whole Body Radiation Study at the University of Cincinnati") (ACHRE No. DOD-042994–A-2) (hereafter "Suskind Report"), 40. He also "had treated occasional private patients with leukemia and lymphoma in my office using whole body radiation and had been impressed with its potentiality." Saenger's response to "Questions from the [Suskind] Committee" (undated) (ACHRE No. DOD-042994–A-24), 2.

101. Philip A. Pizzo et al., "Solid Tumors of Childhood," in *Cancer: Principles and Practice of Oncology*, 4th ed., eds. Vincent T. Devita, Samuel Hellman, and Steven A. Rosenberg (Philadelphia: J. B. Lippincott, Co., 1993), 1782–1783.

102. Robert W. McConnell, President of the American College of Radiology, to U.S. Senator Mike Gravel, 3 January 1972 ("This letter represents our response . . .") (ACHRE No. DOD-042994–A-7). The ACR is the principal organization serving radiologists, with programs that focus on the practice of radiology and the delivery of comprehensive radiological health services. The stated purposes of the ACR are to advance the science of radiology, improve radiologic service to the patient, study the economic aspects of the practice of radiology, and encourage improved and continuing education for radiologists and allied professional fields. The ACR committee consisted of Drs. Henry Kaplan (Stanford), Frank Hendrickson (Chicago Presbyterian-St. Lukes), and Samuel Taylor (also Presbyterian-St. Lukes).

103. Suskind report (released in February 1972). The Suskind Committee was appointed by the dean of the medical school at the University of Cincinnati. It was constituted of eleven physicians from the University of Cincinnati; one of the members, Dr. Bernard S. Aron, had worked on the TBI program.

104. Junior Faculty Association, 25 January 1972 ("A Report to the Campus Community") (ACHRE No. DOD-042994–A-8). The JFA was a group of untenured arts and sciences faculty who organized at the University of Cincinnati in the late 1960s to protect each other's rights to speak out on social issues and to work for fair tenure procedures. In the fall of 1971, a committee of this association, chaired by Martha Stephens, currently professor of English at the university, obtained the reports submitted to the DOD by the Cincinnati doctors and studied the TBI project's history of consent, as well as the blood counts and survival times of the subjects. The JFA then held a press conference on 25 January 1972, charging that "many patients in the project paid severely for their participation and often without even knowing they were part of an experiment" and asking that the study be terminated by the president of the university (which did in fact take place the following March). The JFA report was covered widely in the press and in a number of subsequent studies.

105. James M. Cox, past Chairman of the ACR Commission on Radiation Oncology, statement at Hearing on Radiation Experiments Conducted by the University of Cincinnati Medical School with Department of Defense Funds before the Subcommittee on Administrative Law and Governmental Relations of the House Judiciary Committee, 103d Cong., 2d Sess., 11 April 1994 (ACHRE No. IND-091594–A-1).

106. Dr. Eugene Saenger, University of Cincinnati, to Department of the Army, Office of Surgeon General, 25 September 1958 ("Application for Research Project: Metabolic Changes in Humans Following Total Body Irradiation") (ACHRE No. UCIN-103194–A-1).

107. Dr. Saenger is also a long-standing member of the American College of Radiology and has served on the National Council on Radiation Protection and Measurements (NCRP) and the BEIR (Biological Effects of Ionizing Radiation) Committee in 1972. In 1963, he wrote the AEC's handbook on *Medical Aspects of Radiation Accidents* and regularly consulted for the AEC on radiation accidents and workers' claims of radiation exposure. Saenger has written almost 200 articles in the medical literature on radiology and other topics. Eugene L. Saenger, statement at Hearing on Radiation Experiments Conducted by the University of Cincinnati Medical School with Department of Defense Funds before the Subcommittee on Administrative Law and Governmental Relations of the House Judiciary Committee, 103d Cong., 2d Sess., 11 April 1994 (ACHRE No. IND-091594–A-1).

108. Saenger to the Department of the Army, 25 September 1958, 3. Dr. Saenger's proposal to the

DOD was restricted to a description of the post-TBI metabolic studies for which he was requesting funding, and it made no mention of a primary clinical purpose for the TBI such as a medical therapy or therapeutic research.

109. University of Cincinnati College of Medicine to the Defense Atomic Support Agency, report of February 1960–October 1961 ("Metabolic Changes in Humans Following Total Body Irradiation") (ACHRE No. DOD-042994–A-1), 1.

110. University of Cincinnati College of Medicine to the Defense Atomic Support Agency, report of 1 November 1961–30 April 1963 ("Metabolic Changes in Humans Following Total Body Irradiation") (ACHRE No. DOD-042994–A-1), 3.

111. See, for example, ibid., 3 ("Patients with solid neoplasms which are not radiosensitive are sought."). Three children with Ewing's sarcoma were also treated and included in the study. In addition, the original grant proposal stated that they would compare one group of patients "with relatively radio-resistant lesions (e.g., stomach, bowel, brain)" with a second group "with highly radio-sensitive tumors (lymphomas)." Saenger to the Department of the Army, 25 September 1958, 4–5. The records indicate that the latter group was never used.

112. Eugene Saenger and Ben Friedman, 14 November 1962 ("An appraisal of Human Studies in Radiobiological Aspects of Weapons Effects") (ACHRE No. DOD-091894–A-1).

113. University of Cincinnati College of Medicine to the Defense Atomic Support Agency, report of February 1960–30 April 1966 ("Metabolic Changes in Humans Following Total-Body Irradiation") (ACHRE No. DOD-042994–A-1), 2.

114. University of Cincinnati College of Medicine to DASA, report of 1 November 1960–30 April 1963, 19.

115. Eugene L. Saenger et al., "Whole Body and Partial Body Radiotherapy of Advanced Cancer," *American Journal of Roentgenology* 117 (1973): 670–685. The article makes no mention of the Defense Department-related funding or studies that were performed as part of this program.

116. Eugene Saenger, M.D., interview by Ron Neumann, M.D., Gary Stern, and Gilbert Whittemore (ACHRE staff), transcript of audio recording, 15 September 1994 (ACHRE Research Project Series, Interview Program File, Targeted Interview Project), 50–51.

117. Suskind report, 12; ACR report, 10.

118. Calculated from 3,804 patient days at approximately $114 per day. Eugene L. Saenger, statement at Hearing on Radiation Experiments Conducted by the University of Cincinnati Medical School with Department of Defense Funds before the Subcommittee on Administrative Law and Governmental Relations of the House Judiciary Committee, 103d Cong., 2d Sess., 11 April 1994 (ACHRE No. IND-091594–A-1), 7.

119. Interview with Saenger, 15 September 1994, 50; Saenger testimony, 11 April 1994, 7.

120. Interview with Saenger, 15 September 1994, 54.

121. Interview with Saenger, 20 October 1994, 22–23.

122. Interview with Saenger, 15 September 1994, 64.

123. Ibid., 94.

124. Eugene L. Saenger to Dr. Steven Kessler, DASA Project Officer, 19 February 1969 ("Enclosed are eight copies . . ."), item 9 ("Dosimetry").

125. Lieutenant Colonel James B. Hartgering, Director of the Army Division of Nuclear Medicine and Chemistry, to Lieutenant Colonel Arthur D. Sullivan, Assistant Chief of the Army Medical Research and Development Command, 7 November 1958 ("Application for Research Contract") (ACHRE No. DOD-042994–A-18) (reviewing application submitted by Dr. Eugene Saenger).

126. Lieutenant Colonel Arthur D. Sullivan, Assistant Chief of the Army Biophysics and Astronautics Research Branch, to Colonel Hullinghorst, 12 November 1958 ("Application for Research Contract") (ACHRE No. DOD-042994–A-19).

127. Captain David Lambert, Deputy Chief of DASA Weapons Effects and Tests, to DASA Director of Logistics, 29 October 1959 ("Negotiation of Contract") (ACHRE No. DOD-042994–A-20).

128. Colonel John A. Isherwood to Assistant Chief of the Army Biophysics and Astronautics Research Branch, 22 October 1958. Isherwood also described Dr. Saenger as "well qualified to conduct such research."

129. University of Cincinnati College of Medicine to DASA, report of February 1960–October 1961, 1.

130. Dr. Saenger informed the Advisory Committee that he did not believe that he was required to submit the marrow proposal to the FCR, but elected to do so on his own. Interview with Saenger, 20 October 1994, 6.

131. Dr. Edward A. Gall, FCR Chairman, to Dr. Clifford G. Grulee, Dean of the University of Cincinnati College of Medicine, 6 May 1966 ("This relates to a request . . .") (ACHRE No. DOD-042994–A-11).

132. Dr. George Shields to Dr. Edward A. Gall, FCR Chairman, 13 March 1967 ("Protection of Humans with Stored Autologous Marrow") (ACHRE No. DOD-042994–A-11). Shields also indicated that he was withdrawing from the FCR subcommittee "for reasons of close professional and personal contact with the investigators and with some of the laboratory phases of this project." Ibid.

133. Thomas E. Gaffney to Dr. Edward A. Gall, FCR Chairman, 17 April 1967 ("I cannot recommend approval . . .") (ACHRE No. DOD-042994–A-11). In a subsequent letter, dated 18 May 1967, Gaffney indicated to Gall his approval subject to a proviso (ACHRE No. DOD-042994–A-11).

134. Dr. Clifford G. Grulee, Dean of the College

of Medicine, to Dr. Ben I. Friedman, 23 May 1967 ("The Research Committee has reported . . .") (ACHRE No. DOD-042994–A-11).

135. Dr. Evelyn V. Hess, FCR Chairman, to Drs. Edward B. Silberstein and Eugene L. Saenger, 22 July 1971 ("The Therapeutic Effect of Total Body Irradiation Followed by Infusion of Autologous Marrow in Humans") (ACHRE No. DOD-042994–A-11), 2.

136. Dr. Edward B. Silberstein to Dr. Evelyn Hess, 4 April 1972 ("Enclosed is the protocol . . .") (ACHRE No. DOD-042994–A-11[5]).

137. Dr. Evelyn Hess, FCR Chairman, to Clifford G. Grulee, Jr., Dean, College of Medicine, 28 August 1972 ("Evaluation of the Therapeutic Effectiveness of Wide-Field Radiotherapy . . .") (ACHRE No. 042994–A-11[25]).

138. Evelyn Hess, 20 December 1971 ("Historical Review of the Total Body Irradiation and the Faculty Research Committee Reviews") (ACHRE No. CORP-080195–A-1).

139. FCR Chairman to the Members of the Faculty Committee on Research, 18 April 1969 ("NIH Review . . .") (ACHRE No. DOD-042994–A-21), 1 (describing meeting with Dr. Mark Connor, NIH representative from the Institutional Relations Division).

140. D. T. Chalkley, Ph.D., Chief of the NIH Office for Protection of Research Risks, to U.S. Senator Sam Nunn, 9 December 1974 ("Thank you for your notes . . .") (ACHRE No. DOD-042994–A-12[5]),1–2. In his letter to Senator Nunn, Chalkley states that "[w]hole-body radiation at levels of a few hundred rads is lethal only when it destroys the blood building cells of the bone marrow. In the treatment of these patients who had widespread metastatic cancer, a large part of the marrow was first removed, the patient was then treated and the marrow returned. None of the patients involved died from radiation sickness." However, the marrow transplants that Chalkley refers to were not successfully performed until 1969 and were done on only eight of the eighty-eight patients. See Saenger et al., "Radiotherapy of Advanced Cancer," 682.

141. University of Cincinnati College of Medicine to the Defense Atomic Support Agency, report of 1 May 1966–30 April 1967 ("Metabolic Changes in Humans Following Total-Body Irradiation") (ACHRE No. DOD-042994–A-1).

142. University of Cincinnati College of Medicine to DASA, report of February 1960–30 April 1966, 31.

143. Ibid., 17.

144. Saenger, et al., "Radiotherapy of Advanced Cancer," 682.

145. University of Cincinnati College of Medicine to DASA, report of February 1960–October 1961, 20 (emphasis added). Note that the Sloan-Kettering study sought patients in similar condition.

146. University of Cincinnati College of Medicine to the Defense Atomic Support Agency, report of 1

May 1968–30 April 1969 ("Radiation Effects on Man: Manifestations and Therapeutic Effects") (ACHRE No. DOD-042994–A-1), 1 (emphasis added).

147. University of Cincinnati College of Medicine to the Defense Atomic Support Agency, report of 1 May 1969–30 April 1970 ("Radiation Effects on Man: Manifestations and Therapeutic Effects") (ACHRE No. DOD-042994–A-1), 1 (emphasis added).

148. Ibid., 34. Over the course of the study, twenty-eight patients lived more than one year after being irradiated; seventeen survived for more than two years, at least three people lived more than five years.

149. Suskind report, 65.

150. Ibid., 63.

151. McConnell to Gravel, 3 January 1972, 8. Autopsies were performed on only eight patients, so bone marrow biopsies were used for this evaluation.

152. Eugene L. Saenger et al., "Radiotherapy of Advanced Cancer," 677.

153. Eugene L. Saenger, 20 October 1994 ("How Can the Quality and Length of Life of a Cancer Patient Be Determined?") (ACHRE No. IND-021095–B-2).

154. The article reported that the median survival time for the three cancers for which there was a statistically large-enough group—colon, cancer, breast—was longer than comparable groups receiving no treatment, and almost as long as for patient groups receiving chemotherapy. Saenger et al., "Radiotherapy of Advanced Cancer," 672–676.

155. Suskind report, 59.

156. Ibid.

157. University of Cincinnati College of Medicine to the Defense Atomic Support Agency, report of February 1960–October 1961 ("Metabolic Changes in Humans Following Total Body Irradiation") (ACHRE No. DOD-042994–A-1), 3–4. When asked by ACHRE staff whether information on these side effects should have been withheld in the absence of the DOD funding, Dr. Saenger replied in the first interview that "we would not have [had that period of silence]." Interview with Saenger, 15 September 1994, 74. In the second interview he said that he probably would still have withheld the information. Interview with Saenger, 20 October 1994, 12. A number of contemporary articles discuss the use of compazine in the treatment of radiation sickness. See, for example, Joseph H. Marks, "Use of Chlorpromazine in Radiation Sickness and Nausea from Other Causes," *New England Journal of Medicine* (June 1954): 999–1001; M. J. Solan, "Prochlorperazine and Irradiation Sickness," *British Medical Journal* (21 November 1959): 1068–1069; G. H. Berry, W. Duncan, and Carol M. Bowman, "The Prevention of Radiation Sickness: Report of a Double Blind Random Clinical Trial Using Prochlorperazine and Metopimazine," *Clinical Radiology* 22 (1971): 534–537.

158. Melville L. Jacobs and Fred J. Marasso, "A

Four-Year Experience with Total-Body Irradiation," *Radiology* 84 (1965): 452–456 (twelve of fifty-two patients still experienced some degree of nausea and vomiting). This was the only known U.S. TBI-effects study that performed TBI on patients with radioresistant cancers that does not appear to have been funded by the DOD. Sixteen of the fifty-two patients had radioresistant carcinomas and were chosen because they had life expectancies of less than one month. Ibid.

159. Three main consent forms were used: the first beginning in 1965; the second beginning in 1967; and the third beginning in 1971 (ACHRE No. DOD-042994-A-22).

160. Dr. Saenger stated that he was prompted to begin the use of written consent when he received a letter in 1964 from the DASA requiring all DOD components and contractors to obtain written consent for the use of "investigational drugs in any manner, including research programs." Saenger said he reasoned that there was little difference between drugs and radiation in this context and therefore applied the same standard. Interview with Saenger, 15 September 1994, 77.

161. A second form went into effect at the same time for bone marrow aspiration and storage and is basically the same as the above.

162. Gore Hearing, 35–36.

163. University of Cincinnati Medical Center, Faculty Committee on Research, 1967, "Voluntary Consent Statement."

164. University of Cincinnati, "Consent Form," 1971.

165. Interview with Saenger, 20 October 1994, 7–8. This procedure is also described by Dr. Silberstein in "Extension of Two-part Consent Form," *New England Journal of Medicine* 291 (1974): 155–156.

166. Advisory Committee on Human Radiation Experiments, transcript of Cincinnati Small Panel Meeting, 21 October 1994, 136.

167. Suskind report, 50.

168. ACR report, 7.

169. Suskind report, 28–29. It also notes approximately 2 percent of all patients (both in- and outpatient) at the hospital were private patients. Ibid., 29.

170. In 1969, several of the University of Cincinnati researchers reported on the effects of TBI and PBI on cognitive and emotional processes based on a study involving sixteen of the patients. The published article notes that the "relevant intellectual characteristics of the patient sample were as follows: a low-educational level (ranging from 0 to eight years of education with a mean of 4.2 years), a low-functioning intelligence quotient (ranging from 63 to 112 on the full-scale Wechsler-Bellevue with a mean of 84.5), and a strong evidence of cerebral organic deficit in the baseline (preradiation) measure of most of the patients." Louis A. Gottschalk, Eugene L. Saenger et al., "Total and Half Body Irradiation: Effect on Cogni-

tive and Emotional Processes," *Archives of General Psychiatry* 21 (November 1969): 574, 575. Although these findings suggest that there may have been serious issues about competence among the Cincinnati subjects, questions have been raised about this interpretation. For example, Martha Stephens, former chair of the JFA, has argued as follows: "These citizens were not necessarily dumb or defective, and nothing whatever, in my view, can be judged from the batteries of tests of the psychologists; in the very face of them, they are unconvincing and contradictory. . . . These individuals were in a bad place in life—cancer was more frightening then than it is now, and so were public hospitals. To find people coming to tumor wards (or in the hospital for evaluation of their cancers) to be depressed and upset and somewhat disoriented, and not particularly interested in answering irrelevant questions . . . would be quite normal, I would guess. . . . It seems to be best not to lead the public to believe that what happened in Cincinnati could only have happened to people who weren't smart enough to protect themselves, were virtually retarded." Martha Stephens to Gary Stern, Advisory Committee staff, 3 June 1995 ("I want to thank you for the documents . . .") (ACHRE No. IND-060595-A), 10–11.

171. ORINS was renamed the Oak Ridge Associated Universities (ORAU) in 1966. In 1991, ORAU became the contractor of the Oak Ridge Institute for Science and Education (ORISE).

172. Following standard retention schedules, NASA destroyed funding and procurement records on the Oak Ridge project in 1980. Additional administrative, technical monitoring, and contractor reports still exist. Medical records associated with this project were never in the possession of NASA and always resided with ORINS/ORAU.

173. ORINS had training courses in the handling and use of radioisotopes; it also helped develop the supervoltage cobalt 60 teletherapy machine.

174. Gore Hearing, 110.

175. Howard L. Rosenberg, "Informed Consent: How the Space Program Experimented with Dwayne Sexton's Life," *Mother Jones*, September/October 1981, 31–44. On 13 March 1994, *60 Minutes* aired a story based on the *Mother Jones* article.

176. Gore Hearing, 110.

177. Gore Hearing, 144.

178. Gore Hearing, 161 (statement of Andrew J. Stofan, Acting Associate Administrator for NASA Office of Space Sciences).

179. Ibid., 200. See also page 151 (testimony of William R. Bibb, Director of the Research Division, Oak Ridge Operations Office) and 159–63 (testimony of Stofan).

180. Ibid., 290.

181. G. A. Andrews et al., paper of December 1970 for Oak Ridge Associated Universities ("Hematologic and Therapeutic Effects of Total-Body Irradiation (50 R-100 R) in Patients with Malignant

Lymphoma, Chronic Lymphocytic and Granulocytic Leukemias, and Polycythemia Vera"), 2, reprinted in Gore Hearing, 49.

182. Gore Hearing, 264.

183. Ibid., 291.

184. Ibid., 293.

185. Ibid., 294–295. Following the Gore Hearing, Dr. Helen Vodopick, one of the ORAU physicians who testified, wrote a memo to the file stating that "the therapy given was different from other therapy given that had been tried at that time. However, the other therapies that were being investigated were also radically different. . . . All of these various approaches were tried since nothing had worked before and certainly something new and innovative had to be tried to try to improve the survival rate of acute leukemia in children." Comments to the file prepared by Helen Vodopick, M.D., following the Gore Hearing, 6.

186. Reprinted in Gore Hearing, 32–33.

187. U.S. Atomic Energy Commission, report of 16 April 1974 ("ORAU Review"), reprinted in Gore

Hearing, 186. William Bibb testified that this report was written for the purpose of shutting down the hospital, which had outlived its purposes and could no longer be justified as a necessary AEC program. Accordingly, he stated, some of the statements in the report were "overstatements in order to accomplish what we felt should be accomplished, knowing full well that closing down any Government hospital is hard, closing down that hospital was extraordinarily hard." Gore Hearing, 182.

188. H. R. Resolution 1010, 97th Cong., 2d Sess. 186 (1983). The Committee noted that a later hearing was held on the state of radiation epidemiology within DOE on 19 May 1982. Ibid.

189. It had recommended that ORAU compare the effect of TBI with chemotherapy "for a variety of other solid tumors such as carcinoma of the breast, carcinoma of the gastroenteric tract, the urogenital tract, etc., as well as for lymphomas." U.S. Atomic Energy Commission, report of 16 April 1974, reprinted in Gore Hearing, 247.

190. Ibid., reprinted in Gore Hearing, 252.

9

Prisoners: A Captive Research Population

IN July 1949 a medical advisory panel met in Washington, D.C., to discuss psychological problems posed by radiation to crews of a then-planned nuclear-powered airplane. During the meeting an Air Force colonel noted that crewmen were concerned about anything physically harmful, but especially anything seen as a threat to what he delicately called, using a euphemism of that gentler era, the "family jewels."[1] The nuclear-powered airplane was never built, but concern about radiation hazards to testicular function in space flight, weapons plants, nuclear power plants, and on an atomic battlefield remained.

This concern provides some of the context for a brace of almost identical experiments carried out between 1963 and 1973 in which 131 prisoners in Oregon and Washington submitted to experimental testicular irradiations with national security and other societal goals, but no potential for therapeutic benefit for the subjects. The studies were directed by Carl G. Heller, M.D., a leading endocrinologist of his day, and by Dr. Heller's protégé, C. Alvin Paulsen, M.D. Perhaps because they involved irradiation of the testicles, they have caused great public concern. They were also noted briefly among the thirty-one experiments Representative Edward J. Markey of Massachusetts publicized in his 1986 report on radiation research on

human subjects.[2] Both studies were funded solely by the Atomic Energy Commission. Drs. Heller and Paulsen were interested in the effects of radiation on the male reproductive system, especially the production of sperm cells. The government was interested in the effects of ionizing radiation on workers, astronauts, and other Americans who might be exposed, in a nuclear attack for example.

Both doctors viewed prisoners as ideal subjects. They were healthy, adult males who were not going anywhere soon. In 1963 few if any researchers had moral qualms about using them as subjects, although there seems to have been a consensus in the research community on the rules that should govern such experimentation. By 1973, however, some ethicists, researchers, and others, such as the investigative journalist Jessica Mitford, pointed out that incarcerated people were not well placed to make voluntary decisions. In 1976, the National Commission for the Protection of Human Subjects of Biomedical and Behavioral Research recommended the banning of almost all research on prisoners. Prison experimentation effectively came to an end in this country a few years after the commission offered its recommendations.

The Heller and Paulsen experiments were groundbreaking scientifically, and they were conceived as having an important government purpose—protecting Americans engaged in building the nation's high-priority nuclear and space programs. But looking back through the lens of history, there appears to be an inconsistency between the way human subjects were treated in this research and the standards intended to govern their treatment. Although both Dr. Heller and Dr. Paulsen showed sensitivity to some ethical issues, in both cases the researchers themselves and some of those charged with oversight at both the federal and state levels did not completely live up to what appear to have been well-understood standards applicable to their research. In this failure they were no different from many if not most of their contemporaries. Times were changing, however, and in the end, state officials shut down both sets of experiments, bringing practice more into line with the standards already on the books of some government agencies and private research organizations.

Among researchers who used prisoners as subjects, as early as 1958 the Nuremberg Code was recognized as a model set of rules for conducting human subject research.[3] It is equally clear that the work in the Oregon and Washington prisons did not carefully follow all these rules. Moreover, the funding agency, the Atomic Energy Commission, had its own rules for the conduct of research with human volunteers, which were not fully observed in these experiments. As discussed in chapter 1, in 1956 the AEC's Isotope Division program provided that where healthy subjects were used for research, they needed to be volunteers "to whom the intent of the study and the effects of radiation have been outlined." A 1966 memorandum from the AEC's office of general counsel to the director of the Division of Biology and Medicine sheds some light on the agency's standards at that time, and why it had them. The specific experiments referred to in the memo—plutonium and promethium injections or ingestion—appear not to have been carried out, but the "use of human volunteers in experiments" is addressed in general terms. The memo calls for "volunteer[s]" to sign a written, witnessed agreement attesting to their sound mental state and free will, to their understanding of the purposes and risks of the planned experimentation, and

that the experiment was not being done for their benefit. The relevant paragraph concludes: "Assuming complete understanding and no unequal bargaining factors (e.g. pressure on prisoners to submit), such an agreement would protect against liability for unauthorized invasion of the person."[4]

Finally, those attending a 1962 conference on research using prisoners as subjects reached a consensus on a higher standard for subject selection and informed consent than was typically observed in Oregon and Washington. For example, the conferees argued that potential prisoner subjects should have enough information to avoid their being deceived and that inducements to prisoners should not be so high as to invalidate consent.

The surviving researchers disagree somewhat about the genesis of the testicular irradiation experiments, which the available documentary evidence does not completely resolve. What follows is a version based on and consistent with both the Heller and Paulsen accounts.

Early in 1963 the AEC held a conference in Fort Collins, Colorado, for investigators who were using radiation in studies of reproduction in animals. Dr. Heller was invited. In a bedside deposition taken after he suffered a stroke in 1976, he recounted what happened:

The whole conference finally focused on man. A given group at Fort Collins was working on mice and another group was working on bulls, and then they concluded, what would happen to man[?] They extrapolated the data from bulls or mice to man. I commented one day to Dr. [Paul] Henshaw, who was then . . . with the AEC, that if they were so interested in whether it was happening to man, why were they fussing around with mice and beagle dogs and canaries and so on? If they wanted to know about man, why not work on man[?][5]

According to Dr. Heller, that remark stimulated the AEC to solicit a research proposal from him to study the effects of radiation on the male reproductive system.

Dr. Paulsen, however, recalled a different scenario in a 1994 interview by Committee staff at his office in Seattle.[6] He said he was invited to the AEC's Hanford, Washington, facility in 1962 to act as a consultant after three workers were accidentally exposed to radiation. Like Dr.

Heller, Dr. Paulsen had no previous experience with radiation exposure. He said he was brought in because of a chapter he had written on the testes in an endocrinology text. As a result of that experience, Dr. Paulsen said, he became interested in doing work on the effects of radiation on testicular function, discussed his idea with colleagues, and contacted the AEC to see if the agency would be interested in funding his work.

Whether or not Drs. Heller and Paulsen initiated their projects separately, the practical result was that both received AEC funding and carried out their research projects during the 1960s and early 1970s in the Oregon and Washington state prisons, respectively. Although the two studies were very much alike in their methods and objectives, there were small differences. They used different consent forms, different levels and means of irradiation, and different subject-selection procedures.

This chapter provides accounts of the Washington and Oregon experiments that focus on the failure of these two research projects to live up fully to ethical standards of their time; the Committee's analysis of the risk to subjects in the two experiments; capsule descriptions of a number of other radiation experiments using prisoners as subjects; and a general ethical analysis of radiation experiments using prisoners as subjects.

THE OREGON AND WASHINGTON EXPERIMENTS

Oregon

In 1963 Carl Heller was an internationally renowned medical scientist, a winner of the important Ciba Prize. In the field of endocrinology, he was a preeminent researcher, so it is not surprising that when the AEC decided to fund work on how radiation affects male reproductive function, they would turn to him. He designed a study to test the effects of radiation on the somatic and germinal cells of the testes, the doses of radiation that would produce changes or induce damage in spermatogenic cells, the amount of time it would take for cell production to recover, and the effects of radiation on hormone excretion.[7] To accomplish this

he had a machine designed and built that would give a carefully calibrated, uniform dose of radiation from two sides. The subject lay face down with his scrotum in a small plastic box filled with warm water to encourage the testes to descend. On either side of the box were a matched set of x-ray tubes. The alignment of the x-ray beams could be checked through a system of peepholes and mirrors. Subjects were required to agree to be vasectomized because of a perceived small risk of chromosomal damage that could lead to their fathering genetically damaged children. To carry out this work Dr. Heller was to receive grants totaling $1.12 million over ten years.

Mavis Rowley, Dr. Heller's former laboratory assistant, who was interviewed by Advisory Committee staff in 1994, said that the AEC "was looking for a mechanism to measure the effect of ionizing radiation on the human body. . . ." She said testicular irradiation was promising because the testes have "a cell cycle and physiology which allows you to make objective measurements of dosimetry and effect without having to expose the whole body to radiation."[8]

Although official documentation is fragmentary, it is clear from other evidence such as interviews and contemporary newspaper articles that the concerns cited above—worker exposures, potential exposures of the general population as a result of accidents or bomb blasts, and exposures of astronauts in space—were of interest to the AEC.

In the case of the astronauts, the National Aeronautics and Space Administration has been able to find no evidence of direct involvement in Dr. Heller's project. Yet Ms. Rowley remembers with clarity that NASA representatives, even astronauts themselves, attended meetings with their research team. In her 1994 interview, she said, "NASA was also very interested in this. . . . There was a section of activity which was devoted to what effect would the sun flares and so forth, which give out significant radiation have on the astronauts. And so there were meetings that went on which actually included some of the astronauts attending them. . . ." Rowley explained that the astronauts were concerned that reduced testosterone production might make them lose muscle function, which could compromise their mission, but, belying the comment of the colonel in the 1949 nuclear-powered airplane meeting who said that crewmen were concerned about anything physically harmful, she said they seemed altogether unconcerned "about their own health."[9] During his 1976 deposition, Dr. Heller remarked: "What we would like to supply the medical community with is what happens when you give continual very small doses such as might be given to an astronaut."[10] Moreover, in 1965, Dr. Heller served as a consultant to a Space Radiation Panel of the National Academy of Sciences-National Research Council. And finally, Harold Bibeau, an Oregon subject, recalls that Dr. Heller told him when he signed up for the program that NASA was interested in the results.[11]

At the time the Oregon experiment got under way, using prisoners as research subjects was an accepted practice in the United States. And in this particular study Oregon law was interpreted by state officials as permitting an inmate to give his consent to a vasectomy, which they appear to have seen as analogous to consenting to becoming an experimental subject. However, important ethical concerns of today such as balancing risks and benefits, the quality of informed consent, and subject-selection criteria appear, on the whole, not to have been carefully addressed or not addressed at all by the investigators or those responsible for oversight.

With respect to the health risks associated with the testicular irradiations, there was very little reliable "human" information at the time about the long-term effects of organ-specific testicular exposure to radiation. Hiroshima and Nagasaki bomb data, however, which of course were not organ specific, suggested that the likelihood of inducing cancers with the amount of radiation Dr. Heller planned to use was small. By way of comparison, today's standard radiotherapy of the pelvis, for prostate cancer for example, often results in doses to the testicles in the ranges encountered in these experiments.

So what did Dr. Heller tell subjects about the chronic risk? The answer appears to have been nothing in the early years and, later on, perhaps a vague reference to the possibility of "tumors" but not cancer. In a deposition taken in 1976 a subject named John Henry Atkinson said he was never told there was a possibility of getting cancer or any kind of tumors as a result of the testicular irradiation experiments. Other subjects deposed in 1976 also said they had not been warned of cancer risk, and when asked by one subject about the potential for "bad effects," Dr. Heller was reported to have said, "one chance in a million."[12] When asked in his own deposition what the potential risks were, Dr. Heller said, "The possibility of tumors of the testes." In response to the question "Are you talking about cancer?" Dr. Heller responded, "I didn't want to frighten them so I said tumor; I may have on occasion said cancer."[13]

The acute risks of the exposures included skin burns, pain from the biopsies, orchitis (testicular inflammation) induced by repeated biopsies, and bleeding into the scrotum from the biopsies. Using consent forms and depositions as a basis for determining what the subjects were told, it appears that they were adequately informed about the possibility of skin burns; sometimes informed, but perhaps inadequately, about the possibility of pain; informed about the possibility of bleeding only from 1970 on; and never informed of the possibility of orchitis.

As far as the quality of consent is concerned, the evidence suggests that many if not most of the subjects might not have appreciated that some small risk of testicular cancer was involved. It is also not clear that all subjects understood that there could be significant pain associated with the biopsies and possible long-term effects.

In selecting subjects, Dr. Heller appears to have relied on the prison grapevine to get out the word about a project he apparently believed the Atomic Energy Commission did not want publicized. In a 1964 memorandum he was paraphrased as saying "at Oregon State Penitentiary, the existence of the project is practically unknown."[14] In a 1966 letter to the National Institutes of Health describing the review process at the Pacific Northwest Research Foundation, a respected, free-standing research center, Dr. Heller and two colleagues wrote that "the inmates are well informed by fellow inmates regarding the general procedures concerned (i.e., collecting seminal samples, collecting urines for hormone studies, submitting to testicular biopsies, receiving medication orally or by injection, and having vasectomies . . .)."[15] If the volunteers were healthy and normal they were accepted for a trial period during which they donated semen samples. If all went well, in a matter of weeks they were accepted into the radiation program, as long as the prison's Roman Catholic chaplain certified that they were not Roman Catholics—because of the church's objection to their providing masturbated semen samples—and they could pass what appears to have been a cursory psychological screening designed to ensure they had no underlying objections to the required vasectomy. A copy of a form titled "Psychiatric Examination" provided by Harold Bibeau and signed with the initials of the examining psychiatrist, WHC for William Harold Cloyd, says in full:

11–4–64 Seen for Dr. Heller—Never married, quite vague about future. Feels he doesn't want children—shouldn't have any. I agree. No contraindication to sterilization.

As far as potential health benefits to the subjects are concerned, there were none, and the inmates who volunteered for the research were told so. The benefits were in the form of financial incentives. A review of applications for Dr. Heller's program, and depositions of prisoners who sued Dr. Heller, various other individuals, and the state and federal governments for violation of their rights, clearly indicates that money was in most cases the most important consideration in deciding to volunteer. In prison industry inmates were typically paid 25 cents a day. For participating in the Heller program they received $25 for each testicular biopsy, of which most inmates had five or more, plus a bonus when they were vasectomized at the end of the program, which appears to have been an additional $25. Some inmates indicated that they were grateful for an opportunity to perform a service to society. An obvious ethical question is whether the money constituted a coercive offer to prisoners.[16]

During the course of his study between 1963 and 1973 Dr. Heller irradiated sixty-seven inmates of the Oregon State Prison. Nominally, three institutions had some oversight responsibility for Dr. Heller's work—the Oregon Department of Corrections, the Atomic Energy Commission, and the Pacific Northwest Research Foundation, where Dr. Heller was employed. Practically speaking, however, it appears that Dr. Heller conducted his research independently. As an example of his independence, as recounted by Ms. Rowley, the AEC requested that Dr. Heller begin irradiating subjects at 600 rad and work upward, but he refused and in the end set 600 rad as an upper limit.[17] (It is not clear whether Dr. Heller was concerned about risk to the subjects' health or other research criteria.) Dr. Heller also was a member of the committee at Pacific Northwest Research Foundation that had responsibility for overseeing his research, giving him a voice in the oversight process. This committee was authorized under a foundation regulation titled "Policy and Procedures of the Pacific Northwest Research Foundation With Regard to Investigations Involving Human Subjects." In a section on ethical policy, the document says: "Since 1958 the investigators of this Foundation have conducted all research under the ethical provisions of the Nuremburg [*sic*] Code, modified to permit consent by parents or legal guardians."[18]

In January 1973, in a rapidly changing research ethics environment, the Oregon irradiations were terminated when Amos Reed, administrator of the Corrections Division, ordered all medical experimentation programs shut down essentially because he concluded that prisoners could not consent freely to participate as subjects. It is not known exactly what was behind the timing of Reed's decision, but according to *Oregon Times Magazine*, he had recently read Jessica Mitford's article in the *Atlantic Monthly* titled "Experiments Behind Bars" and an article in *The (Portland) Oregonian* headlined "Medical Research Provides Source of Income for Prisoners."[19]

In 1976, a number of subjects filed lawsuits effectively alleging poorly supervised research and lack of informed consent. In their depositions they alleged among other things that prisoners had sometimes controlled the radiation dose to which they were exposed, that an inmate with a grudge against a subject filled a syringe with water instead of Novocain, resulting in a vasectomy performed without anesthetic, and that the experimental procedures resulted in considerable pain and discomfort for which they were not prepared.[20] These suits were settled out of court in 1979. Nine plaintiffs shared $2,215 in damages.[21]

For the last twenty years all efforts to put in place a medical follow-up program for the Oregon subjects have been unsuccessful. Dr. Heller and Ms. Rowley explicitly favored regular medical follow-up. During the period between 1976 and 1979, the pending lawsuits might have been the reason for the state's reluctance to initiate a follow-up program, but it is less clear why during other periods such efforts have also failed. Two possible reasons suggested by state officials are the cost of such a program and the difficulty of finding released convicts. Other possible reasons are that a follow-up program would not provide a significant health benefit to former subjects and that it would not provide significant new scientific knowledge. According to Tom Toombs, administrator of the Corrections Division of the State of Oregon at the time of the lawsuits, the Corrections Division wrote to the AEC's successor (the Energy Research and Development Administration) in early 1976 recommending medical follow-up for the subjects. Mr. Toombs said there was no record of a response to this request.[22] In 1990, James Ruttenber, an epidemiologist at the Centers for Disease Control, designed a follow-up program for Oregon, but it has not been implemented. In an interview with Advisory Committee staff, Dr. Ruttenber said state officials told him that Oregon does not have sufficient funds to carry out his plan.[23]

Washington

C. Alvin Paulsen was a student of Carl Heller at the University of Oregon in the late 1940s, and in the early 1950s he was a fellow in Heller's lab. But by 1963 he was ready to direct a substantial research program on his own. His chance came when he was called to Hanford to consult on an accidental radiation exposure of three workers. The upshot of this experience was a $505,000 grant from the Atomic Energy Commission to study the effects of ionizing radiation on testicular function. Dr. Paulsen remarked in the 1994 interview with Advisory Committee staff that the main research questions he was trying to answer were what would constitute "a reasonably safe dose" of ionizing radiation to the testes as well as what dose "would cause some change in sperm production and secondly, to determine the scenario of recovery."[24] He recalled a 1962 letter to the Washington State Department of Institutions in which he wrote that he would like to find out "the maximum dose of radiation that would not alter spermatogenesis" and "the maximum dose of radiation that affects spermatogenesis, but only temporarily."[25] Dr. Paulsen said in a 1995 telephone interview, however, that for reasons he can no longer remember, he limited dosage to 400 rad, not enough to test a maximum-dose thesis.[26]

In the 1994 interview, Dr. Paulsen said:

When I recognized a tremendous void of information relative to human exposure, and space travel had started and there was the question of solar explosions and ionizing radiation exposure in space, the nuclear power plants were going in then, a few men throughout the world were exposed . . . I then contacted the Atomic Energy Commission to determine . . . whether they would entertain receiving an application.[27]

Obviously, Dr. Paulsen too was interested in the space applications of his research. In 1972 he and a colleague published their work titled "Effects of X-Ray Irradiation on Human Spermatogenesis" in the proceedings of the National Symposium on Natural and Manmade Radiation, a NASA-sponsored symposium. And Dr. Paulsen said that when he explained his research to potential subjects, one of the things he referred to was concern about exposures in space.[28] An August 1, 1963, article in the *Oregonian* about the Washington experiments said, "Although one of the primary benefits of the research will be in space exploration, the findings are also expected to be of value to an atomic industry where an occupational hazard might exist."[29]

One major difference between the Heller and Paulsen projects was that from the outset Dr. Paulsen planned to eventually move from x rays to neutron irradiation, which, among other things, is more analogous than x rays with the radiation encountered in space.[30] A neutron generator was purchased, calibrated, and shielding was developed. However, the work took years to complete, and this part of the research was never carried out. Dr. Paulsen has expressed the belief on a number of occasions that one reason his project was terminated by the state of Washington in 1970 was concern about the possibly greater risks of exposing subjects to neutrons. Another difference was that Dr. Paulsen used a standard General Electric x-ray machine, which he says he believed would deliver as precise and well-targeted a dose of radiation as Dr. Heller's specially designed machine.[31]

Still another difference was that at a certain stage of the Washington study, Dr. Paulsen used the prison bulletin board to advertise for volunteers. Under the headline "Subject: Additional Volunteers for Radiation Research Project," a notice said in part:

The project concerns effects of radiation on human testicular function and the results of the project will be utilized in the safety of personnel working around atomic steam plants, etc. . . . It is possible that those men receiving the higher dosages may be temporarily, or even permanently, sterilized. It should be understood that when sterilized in this manner, a man still has the same desires and can still perform as he always has. . . . Submit to surgical biopsy. (This is a simple procedure performed under local anesthesia. It is not a very painful procedure.)[32]

According to a March 9, 1976, report prepared for then-Governor Daniel J. Evans by Harold B. Bradley, director of Washington state's Adult Corrections Division, neither Dr. Paulsen's 1963 outline of his research project nor the November 1964 announcement to inmates mentioned a requirement to undergo a vasectomy at the end of the experiment to ensure that subjects would not father genetically damaged children.[33] Dr. Paulsen said he did not recall precisely when in the recruitment process the vasectomy requirement was conveyed to subjects, but he pointed out that once it was they had the option of dropping out of the project without penalty.[34]

Dr. Paulsen's review process and consent procedures are less well documented than Dr. Heller's, but he says his research application, including provisions for subject selection and consent, was approved by what he described as a "human experimentation committee" at the University of Washington. He said the process was "very informal," noting that it was done over the phone. Paulsen added that "somewhat later" his work was also reviewed by a "radiation safety committee."[35] His recollection of both processes is vague. The minutes of a December 10, 1969, meeting of a University of Washington Research and Clinical Investigations Committee at the U.S. Public Health Service Hospital in Seattle includes a recommendation that Dr. Paulsen's consent form be modified to indicate that "a risk of carcinoma of the testes exists although it is extremely small."[36] According to Mr. Bradley's report, his department's records show that Dr. Paulsen's project was reviewed and approved on two occasions—March 1963 and June 1966— by the University Hospital Clinical Investigation Committee. The report shows no state Department of Institutions review until mid-1969.[37]

The Bradley report and related correspondence from 1970 show that at that time some state officials had a sharp concern for research ethics. In mid-1969 a review of all experimentation in the prison system was undertaken by Dr. Audrey R. Holliday, chief of research for the Department of Institutions. At this time Dr. Holliday took steps to temporarily halt the irradiation phase of the project. After investigating the origins of Dr. Paulsen's research, Dr. Holliday asked the University of Washington to conduct a new review of the study, emphasizing her concern about the state's responsibility to safeguard human rights. The university stood by its initial findings allowing the research to continue, although at about the same time it turned down Dr. Paulsen's request to move into the neutron-irradiation phase of his project.[38]

Dr. Holliday then debated the issue with Dr. William Conte, director of the Department of Institutions, who was disposed to allow the project to continue. On March 18, 1970, she wrote a letter to Dr. Conte noting,

. . .There is no question but what the Federal Government has made considerable investment in this project. The Federal Government, however, as a reading of any newspaper will show, has supported a number of projects over which there have been many moral-ethical questions (both large and small) raised, e.g., nerve gasses, toxins, etc. I remind you that the

Federal Government is not responsible for the care, safety and safeguarding of human rights of populations under the purview of the Department of Institutions. This is a responsibility we must discharge, regardless of the amount of money that the Federal Government is willing to invest in a project. . . .

There is no doubt but what the prison setting is an ideal setting for this type of research. . . . I suppose concentration camps provided ideal settings for the research conducted in them. . . . If, in fact, non-inmates were to volunteer in the substantial numbers of persons Dr. Paulsen needs, then I would have less qualms about offering up a captive population for this research, i.e., I would have some evidence, assuming the volunteers were, in fact, normal, that non-captive populations might make the same decision as a captive population. . . .

I am not against high risk research. I have engaged in some myself. I am not against federally sponsored research. I have engaged in some myself. However, the risk should be commensurate with the probable benefits to be received by the population or others like it to follow. I don't think we can argue that in this case.

Neither am I opposed to use of a prison population on a volunteer basis for research projects that may not be of direct benefit to the population, but which are of clear benefit to society or mankind. I don't think we can argue that in this case either.[39]

Dr. Holliday also argued that the study should have been done on "lower order primates" and that if the state allowed Dr. Paulsen's study to continue it would forfeit its right to speak out on behalf of human rights relating to future research proposals.[40]

While favoring continuation of Dr. Paulsen's research, Dr. Conte authorized a review by the Department of Institutions's Human Rights Review Committee. The committee recommended that the study be shut down, noting that the Paulsen project "seems clearly inconsistent with the standards laid down by the Nuremberg Code" for the protection of human subjects with respect to freedom of choice and consent. The recommendation went on to say that "within the context of Dr. Paulsen's project, it is largely irrelevant whether or not a volunteer declares his 'desire to undergo vasectomy' since there is no assurance that his real reasons would be ethically-morally acceptable or that his reasons (whatever they may be) will stand the test of reality after

release." It specified that the money paid for participation and the expectation of privileges, "real or imagined," could constitute undue inducements.[41]

This review, according to the report, "recommended that Dr. Paulsen's request for continuation of his study be rejected as it was found to be inconsistent with standards for the protection of the individual as a research subject. The essential issue raised by departmental personnel was that of informed consent." On March 23, 1970, Dr. Holliday wrote to Dr. Paulsen to inform him that his project was over.[42] The Bradley report added that "so far as is known to departmental personnel, no ill effects have been reported by subjects of the experiments."[43] In 1994, however, a former Washington state inmate named Martin Smith told Karen Dorn Steele of the Spokane Spokesman-Review that ever since participating in the experiment he has suffered testicular pain.[44] Dr. Paulsen notes, however, that Smith was a control and therefore not actually irradiated, although he did have one testicular biopsy.[45]

There has been less debate than in Oregon on the subject of medical follow-up. This may be in part because Dr. Paulsen has taken the position, based on his conversations with inmates, that the subjects of the Washington experiments want their privacy protected, and he has refused to disclose their names. A December 1975 AEC memorandum from Nell W. Fraser, a government contract administrator, to Oscar J. Bennett, director of the Contracts and Procurement Division, paraphrases Dr. Paulsen as saying that a follow-up program was not medically indicated and "a follow-up program would be harmful because most of the prisoners wish to disassociate themselves with the prison experience."[46] According to the memorandum, Dr. Paulsen also noted that his medical malpractice insurance would apply in the event that litigation resulted from his radiation study.[47] In recent years, however, a handful of former subjects have told reporters such as Karen Dorn Steele that they would like to be followed up.[48] In late 1994 state officials said they would seek federal funds to carry out a follow-up program or ask the Department of Health and Human Services to mount such a program.

The Advisory Committee conducted its own analysis of the risks incurred by the Oregon and Washington testicular irradiation subjects based on a 600–rem dose, which was the maximum testicular exposure of any subject in either state. For purposes of this analysis we assumed that the testicles have average radiation sensitivity; that there is a linear relationship between cancer incidence and dose, and that there is a linear relationship between the risk of cancer and the amount of tissue exposed. Using these assumptions, we calculated that it would take more than double the dose received by any prisoner-subject to yield an effective dose of 1 rem. This means that the predicted increase over the expected cancer rate for the individuals who received the greatest exposure would be less that four-hundreths of 1 percent. For those who received smaller doses of radiation, the risk would, of course, be smaller, too.[49]

OTHER RADIATION EXPERIMENTS

There is no comprehensive list of radiation experiments with prisoners as subjects, but in the course of the Advisory Committee's historical research a handful of such experiments other than those in Oregon and Washington has been identified. In many cases there is only fragmentary information available, which the Committee has not always been able to verify. To provide a sense of what else might have been going on at the time (which may or may not have been representative), consider the following:

• A former prison administrator in Utah has confirmed that experiments were conducted on prisoner subjects in the late 1950s or early 1960s in which blood appears to have been removed, irradiated, and returned to the body. Prisoners at the time who were interviewed by the *Deseret News,* a Salt Lake City newspaper, said they believed that about ten prisoner-volunteers were studied in this way. One subject said, "They told us nothing about the tests. They just said it wouldn't bother us."[50] In a 1959 confidential report to the president of the University of Utah, Lowell A. Woodbury, the radiological safety officer said: "One group of medical experimenters with authorization for human experimentation was administering isotopes to volunteers at the state prison. This was in direct violation of the terms of their license and while not an extremely serious violation was apt to result in a citation [from the Atomic Energy Commission]."[51]

• Experiments were conducted at the Medical College of Virginia in the early 1950s under the sponsorship of the Army and possibly the Public Health Service using radioactive tracers. The goal was to study the life cycle of red blood cells. As discussed in more detail in chapter 13, Dr. Everett I. Evans, in a letter to the superintendent of the state penitentiary, quoted from a letter from Colonel John R. Wood of the Army surgeon general's office, which provided that no information related to research being conducted for the Army surgeon general be released without review by the Public Information Office of the Defense Department. Dr. Evans said the reason for this was that "the problem of the use of prisoner volunteers is not yet clarified."[52]

• During the 1960s "prison volunteers" in the Colorado State Penitentiary were used as subjects in an experiment designed to determine the survival time and characteristics of red blood cells during periods of rapid red cell formation and during periods of severe iron deficiency. Red cells transfused into normal recipients were tagged with either radioactive iron or radioactive phosphorus.[53] In a 1976 report on the study, which used five subjects, the investigators wrote:

The rights of the prisoners were respected in conformance with the Helsinki Declaration of the World Health Organization and the Nuremberg Code. Approval was obtained from the Governor, Attorney General, and Director of Institutions of the State of Colorado, the warden and psychiatrist of the Colorado State Penitentiary, and the nearest of kin of each volunteer.[54]

It is not clear from this publication or other documents available to the Committee precisely what use was made of the principles stated in the Nuremberg Code and the Declaration of Helsinki in obtaining the consent of the prisoner-subjects in this experiment. However, if the investigators did accept Nuremberg and Helsinki as standards for consent in the 1960s it adds weight to other evidence (for example, the citation of Nuremberg by the Human Rights Review Committee of the Department of Institutions in the Washington testicular irradiation experiment) that these standards were considered relevant to research on prisoners in the 1960s.

- Other federally sponsored experiments on prisoner volunteers appear to have been conducted in Pennsylvania (Holmesburg State Prison, the effects of radiation on human skin), Oklahoma (Oklahoma State Penitentiary, routine metabolic studies of experimental drugs using tracer amounts of radionuclides), Illinois (Stateville Prison, measurements of radium burden received from drinking water), and California (San Quentin, tracking movement of iron from plasma to red blood cells using a radioactive marker).[55]

HISTORY OF PRISON
RESEARCH REGULATION

Dr. Paulsen reported in a recent interview that he had "asked a lot of people" in 1963 about the use of prisoners as research subjects. He went on to say that at that time "no one said no" to the use of such subjects in his research. However, Dr. Paulsen explained in the same interview that he had started to sense a shift in public opinion around 1970. In particular, he pointed to comments critical of prison experimentation that he had heard at a New York Academy of Sciences conference, "New Dimensions in Legal and Ethical Concepts for Human Research," which he attended in the spring of 1969.[56] Of course, we cannot rely solely on Dr. Paulsen's recollections to provide historical context for experiments in which he was so intimately involved—

and which have now become controversial. But ample evidence suggests that Dr. Paulsen was essentially correct in his impression that testicular irradiation experiments in Washington and Oregon bridged a transitional period in the history of human experimentation generally and particularly in the history of experimentation in American prisons.

Isolated incidents of prison-based research before World War II formed the foundation for a practice that would become firmly embedded in the structure of American clinical research during World War II. Perhaps the most significant wartime medical research project in which American scientists employed prisoners as research subjects was centered in Illinois's Stateville Prison. Beginning in 1944, hundreds of Illinois prisoners submitted to experimental cases of malaria as researchers attempted to find more effective means to prevent and cure tropical diseases that ravaged Allied forces in the Pacific Theater.[57] In 1947, a committee was established by the governor of Illinois to examine the ethics of using state prisoners as research subjects. The committee was chaired by Andrew Ivy, a prominent University of Illinois physiologist and the chief expert witness on medical ethics for the prosecutors at the Nuremberg Medical Trial, where prison research was a salient topic (see chapter 2). The committee pronounced the wartime experiments at Stateville Prison "ideal" in their conformity with the newly adopted rules of the American Medical Association concerning human experimentation. The AMA rules, which Ivy had played a key role in developing, included provisions stipulating voluntary consent from subjects, prior animal experimentation, and carefully managed research under the authority of properly qualified clinical researchers.[58] Perhaps most significantly, the findings of Ivy's committee were announced to the American medical community when the group's final report was reproduced in the *Journal of the American Medical Association*.[59] The appearance of this report in the nation's leading medical journal both represented and reinforced the sentiment that prison research was ethically acceptable.

Publicly aired assertions that experimentation on prisoners relied on exploitation or coercion were extremely rare in the United States before the late 1960s. One criticism of medical research behind bars did, however, emerge with some frequency: prisoners who participated in research were somehow escaping from their just measures of punishment. Inmates were usually offered rewards in exchange for their scientific services, ranging from more comfortable surroundings, to cash, to early release. Perhaps the most powerful statement of the concern that convicts should not receive special treatment because they had participated in an experiment came from the AMA. In 1952, this organization formally approved a resolution stating its "disapproval of the participation in scientific experiments of persons convicted of murder, rape, arson, kidnapping, treason, or other heinous crimes." The AMA was alarmed that some such criminals "have not only received citations, but have in some instances been granted parole much sooner than would otherwise have occurred."[60] (In the Oregon testicular irradiation experiments it appears that this recommendation against using inmates accused of "heinous crimes" was not always observed.)

It should be noted that the use of prisoners as research subjects seems to have been a uniquely American practice in the years following World War II. The large-scale successes of prison experimentation during World War II—and the authoritative pronouncement of the Ivy Committee that prison research *could* be conducted in an ethical fashion—seem to have given the practice a kind of momentum in this country that it did not have elsewhere. In other countries it seems that the first clause of the Nuremberg Code was interpreted to preclude the use of prisoners in experimentation.[61] This clause begins with the assertion that the only acceptable experimental subjects are those who are "so situated as to be able to exercise free power of choice."

It is difficult to overemphasize just how common the practice became in the United States during the postwar years. Researchers employed prisoners as subjects in a multitude of experiments that ranged in purpose from a desire to understand the cause of cancer to a need to test the effects of a new cosmetic. After the Food and Drug Administration's restructuring of drug-

testing regulations in 1962, prisoners became almost the exclusive subjects in nonfederally funded Phase I pharmaceutical trials designed to test the toxicity of new drugs. By 1972, FDA officials estimated that more than 90 percent of all investigational drugs were first tested on prisoners.[62]

It appears that throughout the history of medical experimentation on American prisoners many inmates have valued the opportunity to participate in medical research. One must quickly add that such an observation points to the paucity of opportunities open to most prisoners. The common perception among inmates that participating in a medical experiment was a good opportunity has had an important impact on the racial aspects of prison experimentation. Because of the large numbers of African-Americans in prison (and the overt racial exploitation of the notorious Tuskegee syphilis study, in which black men with syphilis were observed but not treated), it might be assumed that minorities predominated as research subjects in prisons. The opposite has generally been true; white prisoners have usually been overrepresented in the "privileged" role of research subject. In most prison studies before and during World War II, it seems that all of the research subjects were white.[63] In 1975, the National Commission for the Protection of Human Subjects of Biomedical and Behavioral Research carefully examined the racial composition of the research subjects at a prison with a major drug-testing program. The commission found that African-Americans made up only 31 percent of the subject population, while this racial "minority" formed 68 percent of the general prison population.[64]

The shift in public opinion against the use of prisoners as research subjects, which began in the late 1960s, was no doubt tied to many other social and political changes sweeping the country: the civil rights movement, the women's movement, the patients' rights movement, the prisoners' rights movement, and the general questioning of authority associated with the anti-Vietnam War protests. But, as has been common in the history of human experimentation, scandal galvanized public attention, brought official inquiry, and resulted in significant change. A major scandal in prison experimentation came when the *New York Times* published a front-page article on July 29, 1969, detailing an ethically and scientifically sloppy drug-testing program that a physician had established in the state prisons of Alabama.[65] Even more sensational was Jessica Mitford's January 1973 *Atlantic Monthly* article. In this article, Mitford portrayed experimentation on prisoners as a practice built on exploitation and coercion of an extremely disadvantaged class.[66] When the article reappeared later in 1973 as a chapter in her widely read book critiquing American prisons, she had come up with an especially provocative and suggestive title for this section of the book: "Cheaper than Chimpanzees."[67] Mitford, and most of the growing number who condemned experimentation on prisoners during the 1970s (and after), offered two arguments against the practice. First, prisoners were identified as incapable of offering voluntary consent because of a belief that most (some argued, *all*) prisons are inherently coercive environments. Another line of argument was based on a principle of justice that stipulated that one class—especially a disadvantaged class such as prisoners—should not be expected to carry an undue burden of service in the realm of medical research.

A few months after the publication of Mitford's article, Senator Edward M. Kennedy of Massachusetts held hearings to investigate human experimentation. Kennedy was primarily fired into action by the revelations of the Tuskegee syphilis study, which made headlines in 1972, but he devoted one full day of his hearings to the issue of prison experimentation.[68] The chief outcome of Kennedy's hearings was the formation of the National Commission for the Protection of Human Subjects of Biomedical and Behavioral Research, which, among other topics, was specifically charged with investigating experimentation on prisoners (see chapter 3).

The eleven commissioners, including Advisory Committee member Patricia King—with the assistance of twenty staff members—gathered a wealth of data on prison medical research, made site visits to prisons, held extensive public hearings, and engaged in long debates among themselves.[69] After their deliberations, the commission concluded that it was "inclined toward protection as the most appropriate expression of respect for prisoners as persons."[70] But the commission did not call for an absolute ban on the use of prisoners in medical research. A steadfast minority on the commission held to a belief that prisoners should not arbitrarily be denied the opportunity to participate in medical research. An excursion to the State Prison of Southern Michigan, where Upjohn and Parke-Davis pharmaceutical companies had cooperatively built and maintained a large Phase I drug-testing facility, served to reinforce the opinions of this contingent. In candid conversations with the visiting commissioners, randomly selected inmates spoke in convincing terms about their support for the drug-testing program in the Michigan prison.[71]

The commission's final report reflected this hesitancy to call for a complete halt to the use of prisoners in nontherapeutic experimentation. The commission recommended that prisoners could be considered ethically acceptable experimental subjects if three requirements were satisfied: (1) "the reasons for involving prisoners in . . . research [were] compelling," (2) "the involvement of [the] prisoners . . . satisfie[d] conditions of equity," and (3) subjects lived in a prison characterized by a great deal of "openness" in which a prisoner could exercise a "high degree of voluntariness." The final requirement involved a detailed prison accreditation scheme intended to ensure the possibility of voluntary consent.[72]

The National Commission derived its primary power from the fact that the secretary of the Department of Health, Education, and Welfare (DHEW) was legally compelled to respond to the commission's findings and to justify the rejection of any commission recommendations.[73] Joseph Califano, DHEW secretary in the Carter administration, spent nearly a year formulating his response regarding the use of prisoners in medical research. Califano explored the possibility of an accreditation scheme as suggested by the commission. However, in a letter to the commission, Califano reported that the American Correctional Association, "the one currently qualified [prison] accrediting organization," had no interest in "accrediting correctional institutions as performance sites for medical research." "On the contrary," Califano went on to explain, the ACA had recently decided it "would not fully accredit any institution which permitted research on prisoners."[74] After his interchange with the ACA, Califano ultimately decided to issue regulations that, for almost all intents and purposes, brought an end to federally funded nontherapeutic medical research in American prisons.[75]

In the interest of uniform federal regulations, Secretary Califano also "directed" the FDA to issue similar rules governing the use of prisoners in "research that the FDA accept[ed] to satisfy its regulatory requirements."[76] The FDA published final rules in the spring of 1980 that were intended, on the planned effective date of June 1, 1981, to eliminate prisons as acceptable sites for nontherapeutic pharmaceutical testing.[77] However, in July of 1980, almost a year before the FDA's regulations were scheduled to take effect, a group of prisoners at the State Prison of Southern Michigan filed suit against the federal government. These inmates claimed that the impending FDA regulations threatened to violate their "right" to choose participation in medical research. The case was settled out of court when FDA attorneys decided to reclassify the agency's prison drug-testing regulations as "indefinitely" stayed. The FDA's regulations still exist in this bureaucratic limbo.[78]

But even before the FDA issued its proposed regulations on the use of prisoners in drug testing, pharmaceutical companies had already largely abandoned a practice that had been so widespread only a few years earlier. Most significantly, pharmaceutical researchers, along with other medical scientists, had discovered that sufficient numbers of experimental subjects could be found beyond prison walls. Students and poor people proved to be especially viable alternative populations from which to draw participants for nontherapeutic experiments—if the cash rewards were sufficient. The growing controversy surrounding the use of prisoners as research subjects, combined with the realization that they could find enough alternate subjects for their needs, led drug companies to make decisions that were based not so much on ethics as expediency. The comments of an administrator associated with an Eli Lilly testing operation at an Indiana prison are revealing and provide a fitting conclusion to this brief historical analysis: "The reason we closed the doggone thing down was that we were getting too much hassle and heat from the press. It just didn't seem worth it."[79]

ETHICAL CONSIDERATIONS

It is quite clear that all of the radiation experiments that have come to the Advisory Committee's attention in which prisoners were employed as research subjects would have been in violation of federal standards as they exist today. Federal regulation stipulates an extremely limited range of permissible medical research in prison populations. Only four types of investigations can currently receive approval: (1) low-risk studies of "the possible causes, effects, and processes of incarceration, and of criminal behavior"; (2) low-risk studies of "prisons as institutional structures or of prisoners as incarcerated persons"; (3) "research on conditions particularly affecting prisoners as a class (for example, vaccine trials and other research on hepatitis which is much more prevalent in prisons than elsewhere . . .)"; and (4) research that has "the intent and reasonable probability of improving the health or well-being of the subject." Almost certainly, none of the various episodes of radiation research on prisoners treated in this chapter would have fallen into any one of these categories.

But as noted above, widespread concern about coercion and exploitation of prisoner-subjects—which brought about these restrictive federal regulations—arose relatively recently in this country. For the period before roughly 1970, it is almost certainly unfair to condemn, in retrospect, a research project as unethical *solely* because researchers employed prisoners as subjects; historical sensitivity demands some appreciation for what seems to have been a genuine lack of widespread professional or public concern for the ethical problems of prison research that came to the fore during the 1970s. Only in the case of the Washington and Oregon testicular irradiation experiments do we know enough to make any legitimate claims about the extent to which researchers conformed with reasonable contemporary standards for the ethical conduct of prison experimentation. And, even for these relatively well-known studies, the individual complexities of each series of experiments have grown hazier with time.

One of the first known efforts to examine the ethics of using prisoners as research subjects was organized by the Law-Medicine Research Institute (LMRI) of Boston University. The conference was called "The Participation of Prisoners in Clinical Research," and it opened on February 12, 1962. The conference was part of a larger LMRI project to study and report on "the actual practices, attitudes, and philosophies currently being applied in the legal and ethical aspects of clinical investigation" (see chapter 2). LMRI's conference on prison research was one of several "invitational work conferences" organized to gather information on several important topics in human experimentation (other conferences were devoted to "the concept of consent," pediatric research, and pharmaceutical testing). The participants at each conference received an agenda and briefing book in advance of the meetings, but discussions tended to be free-ranging. Those who attended the conferences understood that their words were being recorded, but they tended to speak in a frank and revealing fashion because LMRI pledged to preserve their anonymity when reporting on the meetings.[80]

A copy of the list of participants at the conference on "The Participation of Prisoners in Clinical Research," which survives at Boston University, confirms the following characterization of those who attended:

[T]he thirty-six invited participants comprised two main categories. The first was composed of clinical research administrators and clinical investigators with a variety of academic, commercial, and governmental affiliations, who have had experience in conducting medical studies with prisoners as subjects. The second category consisted of prison administrators and prison medical officers with various federal, state, and municipal correctional programs. Also participating in the conference were representatives of various related fields such as behavioral science, criminal law, organized medicine, pharmaceutical manufacturing, and the military services.[81]

Unfortunately, a copy of the actual meeting transcript has not survived. However, the lengthy unpublished "Analytic Summary," which contains many (anonymous) transcript excerpts, seems to be a fair representation of the daylong meeting.[82] It is relatively easy to extract several important points of agreement about the proper conduct of experimentation in prisons from this report. And, given the broad cross section of those involved in prison experimentation who attended this 1962 conference, it seems reasonable to employ the standards enunciated at this conference as evidence of prevailing interpretation of ethical standards for prisoner experiments that began in 1963.

First, the "conferees generally agreed that experimental risks must be balanced against benefits." In the case of research that was not intended to be of potential direct benefit to the subject, which was generally the case in prison experiments, most meeting participants believed that the social or scientific value of new knowledge that might result from an experiment should be weighed as a benefit.[83] However, when "confronted with the direct question of whether or not a relatively high degree of risk can ever constitute a legitimate reason for the use of prisoner subjects, the conferees were almost unanimous in rejecting this position."[84] Interestingly, those at the conference believed that the *general public* was *less* inclined to worry about subjecting prisoners to high levels of experimental risk. Two brief transcript excerpts are revealing:

When the public hears that inmates are [participating in a seemingly very hazardous study], they rationalize, "Well, I wouldn't do it, but it's all right with prisoners."[85]

[T]he public will allow the investigator to go a lot further, with regard to risks taken with prisoners, than the investigator would go himself.[86]

The conferees spent a large portion of their day together discussing the matter of consent. They reached agreement that meaningful consent should be both voluntary and informed, provided the reach of these terms is carefully circumscribed. The report stated,

[T]he legal prerequisites of consent are, first, not absolute free will, but sufficient free choice to avoid coercion or duress; and, second, not absolutely perfect knowledge, but enough information to avoid fraud or deceit.[87]

The conference participants "unanimously agreed that rewards offered to prisoner volunteers should not be so high as to invalidate their consent to participate as research subjects."[88] There seems to have been considerable disagreement about exactly where to draw the line between ethically acceptable and unacceptable rewards to prisoners for service as experimental subjects, but there was a general desire to "minimize rewards" because it was "consistent with the penological desirability of maximizing prisoners' 'opportunity for altruism.'"[89] As for sentence reductions, some thought that small amounts of "good time" credits were appropriate, but all agreed that "maximum rewards of this type, i.e., definite promises of pardon or parole, should not be given."[90] There seems to have been little discussion of the possibility that the authoritarian structure of prison life was in itself coercive and therefore limited a prisoner's ability to make an autonomous decision.

The disclosure component of consent received extensive attention at the conference. The following was offered as a summation of what the conferees perceived as the "essential content and emphasis" of the information that should be conveyed to "prospective prisoner-subjects":

The explanation of a clinical research project . . . should describe completely the procedures entailed and should stress the possible consequences of these procedures. Even though it may be necessary to "stop somewhere short of full revelation when you reach intricacies a layman would never comprehend," there should be no omission of any adverse consequences, detriments, or risks.[91]

To strive toward this level of communication, the conference participants cited procedures that were "usually" followed in most prison experiments: a general announcement of the research project to the inmates (usually by notices posted on bulletin boards or printed in prison newsletters); a general explanation of the project (often in an auditorium) to groups of prisoners who expressed initial interest in an experiment; and, finally, one-on-one meetings between prospective participants and research personnel.[92] Conferees who had administered or conducted prison experiments also reported that prisoner-subjects "usually sign[ed] some type of 'consent

agreement.'"[93] (Generally speaking the provisions specified above were followed in the Washington and Oregon experiments, but the information provided was often inadequate.)

Even with all of these measures, some meeting participants asserted that the "ideals of comprehension, evaluation, and decision on the part of prisoners were seldom attained in practice." They pointed to two general difficulties in achieving these ideals. First, "the lack of intelligence, education, or 'medical sophistication' among many prisoners." Second, they cited "various 'motives or pressures which so often stand in the way of objective understanding.'"[94] The participants in the conference also recognized that the consent forms used in prison experiments were often less than perfect. They understood that the "waiver or release" components of many forms were probably inappropriate. They also recognized that reasonably predictable risks of an experiment were not always carefully listed on consent forms, but at the same time they "agreed that 'no serious' risk should ever be disguised or concealed" on these forms.[95]

In sum, the records from this conference suggest that even apart from formal, federal rules for experimentation on prisoners, ethical conditions for the conduct of prison research were articulated in the early 1960s. Now, with these conditions in mind, let us turn to a more detailed analysis of the Washington and Oregon testicular irradiation experiments.

As we have noted, the Committee's ability to assess the quality of consent obtained from a research subject thirty or forty years earlier can be confounded in a thousand ways. To begin with, the records are invariably incomplete; then, the investigators are either no longer alive or their memories have grown hazy or selective with time; the same is true of subjects; and, of course, there are confidentiality considerations, which limit the availability of records, the concern of researchers for their reputations, and so on. All of these considerations, to greater or lesser degrees, apply to the Oregon and Washington experiments.

With respect to these experiments, however, we believe we have a clear-enough picture of the standards and practices of the time to evaluate the conduct of the research against them without reference to the standards and practices of today.

In both Oregon and Washington, some subjects were not warned, warned only after enrolling in the experimental program, or inadequately warned that there was potential risk, albeit small, of testicular cancer. While it might not have been uncommon at the time for physicians to avoid using the word *cancer* with sick or even terminally ill patients for paternalistic reasons, such avoidance is harder to justify, even by the standards of the time, in the case of healthy subjects who are participating in research that offers them no direct benefit.[96]

As far as acute effects are concerned, the pain of testicular biopsy may have been understated in both programs, and the risk of orchitis from repeated biopsies seems to have been ignored. Some former subjects have complained of long-term pain, sexual dysfunction, and skin rashes. It is not clear whether these conditions were caused by the experiments, nor is it certain that long-term medical follow-up can answer this question.

Subjects in both sets of experiments were required to have a vasectomy at the end of the program because of concerns about possible chromosomal damage. In both cases the vasectomy consent forms signed by the subjects, and their wives if they were married, adequately described the procedure, its consequences, and the small possibility it could be reversed. However, appropriate questions have been raised about the reasons inmates might agree to vasectomy in the circumstances of prison research, and the possibility, as actually occurred in a number of cases, that in the end the subject would refuse to undergo the procedure.

Finally, there appears to be little doubt that the financial incentives offered for participation were the main reason most inmates volunteered. Payments totaling more than $100 could be seen as unduly influencing the judgment of potential volunteers. While money also is a powerful incentive for research participation outside prison walls, we believe that the conditions of confinement *can* magnify the perceived value of the reward. Whether the payments offered to participants in these programs constitute an unfair inducement to participate in research may vary from inmate to inmate.

While the prison experiments were unethical with respect to current requirements for disclosure of risk and noncoercion, the researchers functioned during a period of rapid evolution of the interpretation of ethical principles in the prison context. Their actions, however, were less than fully consistent with the existing AEC requirements, especially concerning the information the prisoner-subjects were provided.

NOTES

1. Colonel Don Flickinger, NEPA Medical Advisory Panel, Subcommittee IX, Washington, D.C., 22 July 1949 (ACHRE No. DOE-121494–A-2), 17.

2. Subcommittee on Energy Conservation and Power, Committee on Energy and Commerce, House of Representatives, *American Nuclear Guinea Pigs: Three Decades of Radiation Experiments on U.S. Citizens*, 99th Cong., 2d Sess., 3.

3. Pacific Northwest Research Foundation, undated ("Policy and Procedures of the Pacific Northwest Research Foundation with Regard to Investigations Involving Human Subjects") (ACHRE No. IND-011195–A-1).

4. Bertram H. Schur to Dr. Charles L. Dunham, 13 May 1966 ("Use of Human Volunteers in Biomedical Research") (ACHRE No. DOE-051094–A-138).

5. Deposition of Mavis Rowley and Carl G. Heller, 19 July 1976, Poulsbo, Washington (ACHRE No. CORP-013095–A-2), 18.

6. C. Alvin Paulsen, interview by Steve Klaidman (ACHRE), 8 September 1994, Seattle, Washington, transcript of recording (ACHRE Research Project Series, Interview Program File, Targeted Interview Project), 10–11.

7. Pacific Northwest Research Foundation, proposal for Atomic Energy Commission, Division of Biology and Medicine, February 1963 ("Effects of Ionizing Radiation on the Testicular Function of Man") (ACHRE No. DOE-122994–A-2); Carl Heller, Pacific Northwest Research Foundation, 27 April 1967 ("Fifth Yearly Proposal, June 1, 1967–May 31, 1968") (ACHRE No. DOE-122994–A-2); Carl Heller, Pacific Northwest Research Foundation, May 1972 ("Effects of Ionizing Radiation on the Testicular Function of Man: 9 Year Progress Report") (ACHRE No. DOE-122994–A-2); Mavis Rowley, Division of Nuclear Medicine, Tumor Clinic, Swedish Research Hospital, undated ("The Effect of Graded Doses of Ionizing Radiation on the Human Testis: Progress Report, October 1, 1975–September 30, 1976") (ACHRE No. DOE-011895–B-3). The following is a staff-prepared abstract of Dr. Heller's research based on annual reports, the final report, and his research proposal:

I. OBJECTIVES

To determine the nature of the cytological changes, both somatic (Sertoli cell) and germinal (spermatogonia) induced by acute irradiation.

- To determine the dosage required to produce these changes, as well as the dose to induce permanent damage to spermatogenic cells.
- To determine recovery time.
- To determine radiation-produced alteration of testicular parameters, such as total gonadotropin, interstitial-cell hormone excretion, estrogen excretion, and androgen excretion.

II. METHODOLOGY

Subjects received varying doses of X-irradiation to both testes from 8– to 600–rad single dose. Testicular effects were determined by histological (light microscopy) examination of pre- and serial postirradiation biopsy specimens. Sperm counts, motility, morphology, and seminal fluid volume were monitored in serial postirradiation ejaculates. Hormonal excretion was to be monitored by serial urine and plasma analyses.

- Radiation exposure was controlled by a specially constructed device that assured uniform (plus or minus 5%) irradiation at a dose rate of 100 r/min, approximately 140 kVp with 5 mA tube current, and 2 mm Al filter.
- Some subjects received 10 mCi 3H-thymidine injected intratesticularly to assess (via autoradiography) effects of radiation on incorporation into-spermatogonial DNA as a measure of chromosome replication.

III. RATIONALE FOR THE USE OF HUMAN SUBJECTS

- To determine radiosensitivity of germinal elements in man. According to Dr. Heller, man is unique among commonly studied species in being able to submit to serial testicular biopsy without damage and biopsy-induced testicular artifacts. (Mavis Rowley has pointed out that improved techniques have made it more practical to do biop-sies on large animals.)
- To determine germinal cell recovery, thereby allowing prognosis in cases of accidental irradiation.

IV. FINDINGS

- Sperm count reduction and recovery of sperm count are both dose related. At 400–600 rad, sperm count was zero at 156 weeks.
- By autoradiographic studies of 3H-thymidine uptake into spermatocytes in nonirradiated subjects, it was shown that there are approximately four cycles of spermatogenesis of approximately sixteen days each, so that the complete evolution of spermatogonia to mature sperm is approximately sixty-four days. This is approximately the same as other mammalian species.
- Urinary and plasma gonadotropins rose in proportion to testicular dose and fell with germinalrecovery. Plasma FSH and LH also rose. Urinary estrogen remained unchanged. Urinary testosterone fell slightly after irradiation.
- Histologically, spermatogonia were the most radio sensitive. Spermatocytes were damaged above 200–300 rads. Spermatids showed no overt damage.
- Germinal cell recovery time increased as radiation dose increased. Complete recovery occurred within nine to eighteen months for doses of 100 rad and below. Complete recovery required five or more years for doses of 400–600 rad. Germinal tissue appears to be somewhat more radiosensitive in humans than other studied species.

V. FINANCIAL SUPPORT

Contract AST (45–1) 1780, U.S. Atomic Energy Commission.

8. Mavis Rowley, interview with ACHRE staff, 8 September 1994 (ACHRE No. ACHRE-051795–B) 6.

9. Ibid., 12–13.

10. Deposition of Mavis Rowley and Carl Heller, 19 July 1976, 32.

11. Harold Bibeau, telephone interview with ACHRE staff, 11 August 1994 (ACHRE No. IND-081194–A).

12. Depositions of John Henry Atkinson, 54; Ivan Dale Herland, 22, 68; Donald Eugene Mathena, 94; taken 14 October 1976 in *Donald Mathena et al. v. Amos Reed et al.* Civil nos. 73–326, U.S. District Court, Dist. Oregon (ACHRE No. CORP-013095–A).

13. Deposition of Carl Heller, 19 July 1976.

14. L. C. Wertz to Warden C. T. Gladden, 10 July 1964 ("Dr. Heller stopped in the office . . .") (ACHRE No. IND-061594–A-1).

15. William B. Hutchison, M.D., Joseph E. Primeau, and Carl G. Heller, M.D., Ph.D., to Dr. John C. McDougall, Assistant Director for Operations, National Institute of Child Health and Human Development, National Institutes of Health, 12 May 1966 ("This letter is in response . . .") (ACHRE No. DOE-082294–B-70).

16. C. T. Gladden, Warden, to Mark Hatfield, Governor, 8 May 1963 ("I am glad to provide . . . ") (ACHRE No. IND-061594–A), 2; C.T. Gladden, Warden, to Hon. Robert Thornton, Attorney General, 9 September 1963 ("Carl Heller, Medical Research Programs") (ACHRE No. IND-061595–A); undated ("Consent and Release") (ACHRE No. IND-110994–A).

17. Mavis Rowley, interview with ACHRE staff, 8 September 1994, 36.

18. Pacific Northwest Research Foundation, undated ("Policy and Procedures of the Pacific Northwest Research Foundation with Regard to Investigations Involving Human Subjects").

19. William Boly, "The Heller Experiments," *Oregon Times Magazine*, November 1977, 45.

20. *Robert Case v. State of Oregon et al.*, Civil no. 76–500; *Paul Tyrell v. State of Oregon et al.*, Civil no. 76–499.

21. Tom Toombs, Administrator of Corrections Division, undated testimony before Oregon legislature (ACHRE No. IND-101294–D-1), 7.

22. Ibid., 11.

23. James Ruttenber, telephone interview with Steve Klaidman (ACHRE staff), 20 July 1995 (ACHRE No. IND-072095–E).

24. C. Alvin Paulsen, proposal to Atomic Energy Commission, undated ("Study of Irradiation Effects on the Human Testis: Including Histologic, Chromosomal and Hormonal Aspects") (ACHRE No. IND-110994–A-2). The following has been abstracted by staff from Dr. Paulsen's research proposal and annual progress reports:

I. OBJECTIVE

- To determine the dose-dependent relationship between external irradiation and cell kill and inhibition of mitosis in spermatogenic cells. The cells in question are spermatagonial stem cells, and the dose response would be expected to differ from other kinds of cells.

II. METHODOLOGY

- Subjects with normal ejaculates received 7.5–400 rad to both testes. The details of irradiation are not specified.
- Weekly seminal fluid was examined for the endpoint response of azoospermia. Duration was not specified.
- Subjects and some controls received periodic unilateral testicular biopsies. Number not specified.
- Irradiated subjects agreed to be vasectomized at the completion of the experiment.

III. RATIONALE FOR THE USE OF HUMAN SUBJECTS

One cannot directly relate animal data to the human male with security. Among other things, the rate of spermatogenesis in man is different from that in various animal species.

IV. FINDINGS

- The average presterile period was 142 days.
- The maximum sterile period was 501 days.
- Spermatogenesis in man is more radiosensitive than in rodents and recovery time is longer. Man is more radiosensitive to complete sterility than rodents.
- Testicular biopsy by itself can reduce seminal fluid sperm concentration.

V. FINANCIAL SUPPORT

AEC contracts AT (45–1) 1781 and AT (45–1) 2225.

25. C. Alvin Paulsen, telephone interview with Steve Klaidman (ACHRE), 20 July 1995 (ACHRE No. IND-072095–D).

26. C. Alvin Paulsen, telephone interview with Steve Klaidman (ACHRE), 7 March 1995 (ACHRE No. ACHRE-030995–A).

27. C. Alvin Paulsen, interview with ACHRE staff, 8 September 1994, 9.

28. T. W. Thorslund and C. Alvin Paulsen, "Effects of X-Ray Irradiation on Human Spermatogenesis," *Proceedings of the National Symposium on Natural and Manmade Radiation in Space*, ed. E. A. Warman (NASA Document TM X-2440, 1972), 229–232.

29. "Prison Inmates Sought in Prison Experiment," *The (Portland) Oregonian*, August 1963.

30. C. Alvin Paulsen, proposal to Atomic Energy Commission, undated ("Study of Irradiation Effects on the Human Testis: Including Histologic, Chromosomal and Hormonal Aspects").

31. C. Alvin Paulsen, interview with ACHRE staff, 8 September 1994, 56–57.

32. C. E. Heffron, M.D., Prison Physician, to All Inmates Interested, 2 November 1964 (ACHRE No. WASH-112294–A-1).

33. Harold Bradley, Director, Adult Corrections Division, to Hon. Daniel J. Evans, Governor, Washington State, 9 March 1976 ("Secretary Morris has asked . . .") (ACHRE No. WASH-112294–A-2).

34. Paulsen, telephone interview with ACHRE staff, 20 July 1995.

35. Ibid.

36. University of Washington, Research and Clinical Investigations Committee, proceedings of 10 December 1969 (ACHRE No. WASH-112294–A-3), 4.

37. Bradley to Evans, 2.

38. George Farwell, Vice President for Research, University of Washington, to John Totter, Division of Biology and Medicine, 16 July 1969 ("Thank you very much for your prompt response") (ACHRE No. DOE-082294–B-71).

39. Audrey Holliday, Research Administrator, Department of Institutions, to William Conte, Director, Department of Institutions, 18 March 1970 ("I received the review . . .") (ACHRE No. WASH-112294–A-4), 2.

40. Ibid.

41. Research Review Committee, Department of Institutions, to Audrey Holliday, Research Administrator, Department of Institutions, 13 March 1970 ("Disposition of Division Review Committee in Regard to Irradiation Project of Dr. C. Alvin Paulsen at the State Penitentiary") (ACHRE No. WASH-112294–A-5), 2.

42. Audrey Holliday to C. Alvin Paulsen, 23 March 1970 ("The Department of Institutions received copies . . .") (ACHRE No. WASH-112294–A-6).

43. Bradley to Evans, 9 March 1976, 2.

44. Karen Dorn Steele, "Experiments A Life Sentence," *Spokane Spokesman-Review*, 19 June 1994, 1.

45. C. Alvin Paulsen, telephone interview with ACHRE staff, 7 March 1995 (ACHRE No. ACHRE-030795–A).

46. Nell Fraser to Oscar Bennett, 23 December 1975 ("Contracts AT[45–1]-1780–1781, Irradiation

of Prison Volunteers") (ACHRE No. DOE-082294–B).

47. Ibid.

48. Karen Dorn Steele, "State Agrees to Find Victims of Experiments," *Spokane Spokesman-Review*, 16 December 1994, 1.

49. The Advisory Committee calculated the risk from the testicular irradiation study as follows:

The radiation dose to the testicles ranged from 7.5 to 600 rem. The Committee's risk analysis was based on a 600-rem dose and the following three assumptions:

1. The testicles have average radiation sensitivity.
2. The risk of cancer is linearly related to dose.
3. The risk of cancer is linearly related to the amount of tissue exposed.

Based on these assumptions, the Committee calculated the maximum risk expected to any of the prisoner subjects using the following two steps:

1. Calculate effective dose by multiplying a 600-rem testicular dose by the proportion of the body exposed: (2 x 25 grams/70 kilograms), or (50/70,000) x 600 rem = 429 mrem.
2. Calculate the risk (assuming average radiosensitivity) by multiplying this effective dose by the age-specific risk for males age 25: (0.429 x 921/1,000,000 person rem), or a risk of about 0.4/1,000 for males age 25.

50. Lee Davidson, "Did Secret Radiation Tests on Inmates Doom Offspring?" *Deseret News*, 10 November 1994, A1.

51. Lowell A. Woodbury, Radiological Safety Officer, to Dr. A. Ray Olpin, President of the University of Utah, 9 July 1959 ("Resume of Activities While Acting as Health Physicist and Radiological Safety Officer to the University of Utah Isotope Committee") (ACHRE No. UTAH-111394–A-2). A 1964 article titled "The Kinetics of Granulopoiesis in Normal Man" appears to describe the experiment. The article compares methods of labeling white blood cells with various radioisotopes and formulates a concept of forming white cells in normal man based on information obtained using a DFP32 (diisopropylfluorophosphate) label. G. E. Cartwright, J. W. Athens, and M. M. Wintrobe, "The Kinetics of Granulopoiesis in Normal Man," *Blood* 24, no. 6 (December 1964).

52. Everett Evans, Professor of Surgery, Medical College of Virginia, to W. F. Smyth, State Penitentiary, 13 December 1951 ("We continue to enjoy and appreciate . . .") (ACHRE No. VCU-012595–A-17).

53. Matthew Block to John Lawrence, 10 April 1969 ("I have met a very serious . . .") (ACHRE No. DOE-121294–B-7).

54. Matthew Block, *Text-Atlas of Hematology* (Philadelphia: Lea & Febiger, 1966), 503.

55. Henry De Bernardo, "University of Pennsylvania and the Holmesburg Connection," *Philadelphia News Observer*, 12 October 1994, 12; Shelby Thompson, Director (Acting), Division of Information Services, AEC, to H. C. Baldwin, Information Officer, Chicago Operations Office, 21 August 1953 ("Information Guidance on Any Experimentation Involving Human Beings") (ACHRE No. DOE-051094–A-473); "10 San Quentin Felons Used for Atom Tests," source unknown, 12 April 1949; Oak Ridge National Laboratory Human Studies Review Taskforce, University of Oklahoma Human Studies Reviews, undated ("Findings for License 35–03176–01") (ACHRE No. NRC-012695–A).

56. C. Alvin Paulsen, M.D., interview with ACHRE staff, 8 September 1994. The meeting Paulsen referred to took place on 19–21 May 1969. The proceedings of the conference are reported in the *Annals of the New York Academy of Sciences* 169 (21 January 1970): 293–593. Paulsen is not listed among those who offered formal presentations.

57. Published accounts of this research include Alf S. Alving et al., "Procedures Used at Stateville Penitentiary for the Testing of Potential Antimalarial Agents," *Journal of Clinical Investigation* 27, no. 3 (part 2) (1948): 2–5; Joseph E. Ragen and Charles Finston, *Inside the World's Toughest Prison* (Springfield, Ill.: Charles C. Thomas, 1962), 391–395; Nathan F. Leopold, Jr., *Life Plus 99 Years* (Garden City, N.Y.: Doubleday, 1958), 305–355. David J. Rothman, in *Strangers at the Bedside: A History of How Law and Bioethics Transformed Medical Decision Making* (New York: Basic Books, 1991), gives some considerable attention to this tropical disease research in Illinois in a chapter entitled "Research at War," 30–50.

58. These rules, as approved by the AMA House of Delegates on 11 December 1946, read as follows:

1. The voluntary consent of the individual on whom the experiment is to be performed must be obtained; 2. The danger of each experiment must be previously investigated by animal experimentation, and 3. The experiment must be performed under proper medical protection and management.

From "Minutes of the Supplemental Session of the House of Delegates of the American Medical Association, Held in Chicago, December 9–11, 1946," *Journal of the American Medical Association* 133 (4 January 1947): 35. For more background on the development of the AMA standards for human experimentation, see chapter 2 of this report.

59. "Ethics Governing the Service of Prisoners as Subjects in Medical Experiments: Report of a Committee Appointed by Governor Dwight H. Green of Illinois," *Journal of the American Medical Association* 136 (14 February 1948): 457–458.

60. "Abstract of the Proceedings of the House of Delegates Meeting, Denver, 2–5 December 1952," *Journal of the American Medical Association* 150 (27 December 1952): 1699. The Illinois delegation to this meeting introduced the resolution. It is likely that the Illinois group was motivated by the possibility that Nathan Leopold, who had participated in a highly

publicized kidnapping and murder that had been dubbed by the press as "the crime of the century," might be paroled as a result of his participation as a subject in the wartime tropical disease research at Stateville Prison.

61. Martin Jaffe and C. Stewart Snoddy, "An International Survey of Clinical Research in Volunteers," in *Appendix to Report and Recommendations: Research Involving Prisoners*, National Commission for the Protection of Human Subjects of Biomedical and Behavioral Research (Washington, D.C.: DHEW, 1976).

62. Aileen Adams and Geoffrey Cowan, "The Human Guinea Pig: How We Test New Drugs," *World* (5 December 1971): 20.

63. Jon M. Harkness, "Vivisectors and Vivishooters: Medical Experiments on American Prisoners before 1950," paper presented at "Regulating Human Experimentation in the United States: The Lessons of History," Columbia College of Physicians and Surgeons, New York, 23 February 1995.

64. National Commission for the Protection of Human Subjects of Biomedical and Behavioral Research, *Report and Recommendations: Research Involving Prisoners* (Washington, D.C.: DHEW, 1976), 36.

65. Walter Rugaber, "Prison Drug and Plasma Projects Leave Fatal Trail," *New York Times*, 29 July 1969, 1, 20–21.

66. Jessica Mitford, "Experiments Behind Bars: Doctors, Drug Companies, and Prisoners," *Atlantic Monthly* 23, January 1973, 64–73.

67. Jessica Mitford, *Kind and Usual Punishment: The Prison Business* (New York: Alfred A. Knopf, 1973), 138–168.

68. For an analysis of the chain of events leading up to Senator Kennedy's hearings, see Mark S. Frankel, "Public Policymaking for Biomedical Research: The Case of Human Experimentation" (Ph.D. diss., George Washington University, 1976), 190–192. For the transcripts of the actual hearings, see U.S. Congress, Senate, Committee on Labor and Public Welfare, Subcommittee on Health, *Hearings on Quality of Health Care—Human Experimentation*, on S. 974, S. 878, and S. J. Res. 71, 93d Cong., 1st Sess., part 3, 7 March 1973 (Washington, D.C.: GPO, 1973).

69. A record of the National Commission's work can be found in a complete set of the commission's papers in the archives of the National Reference Center for Bioethics Literature, Kennedy Institute of Ethics, Georgetown University. For a useful (and critical) overview of the commission's work with regard to prisoners see Roy Branson, "Prison Research: National Commission Says 'No, Unless . . . ,'" *Hastings Center Report*, February 1977, 15–21.

70. The National Commission for the Protection of Human Subjects of Biomedical and Behavioral Research, *Report and Recommendations: Research Involving Prisoners*, 8.

71. Branson, "Prison Research," 17; National Commission, Staff Paper, "Biomedical and Behavioral

Research Involving Prisoners," 5 March 1976, 12–13, Archives, National Reference Center for Bioethics Literature, Kennedy Institute of Ethics, Georgetown University, National Commission Papers, Box 6; form letter sent to randomly selected prisoners for permission to conduct an interview, 3 November 1975, Archives, National Reference Center for Bioethics Literature, Kennedy Institute of Ethics, Georgetown University, National Commission Papers, Box 22.

72. National Commission, *Report and Recommendations: Research Involving Prisoners*, 16–19.

73. Frankel, 402; see also the enabling legislation for the commission, the *National Research Act*, P.L. 93–348.

74. Joseph Califano to Kenneth J. Ryan, Chairman of the National Commission for the Protection of Human Subjects of Biomedical and Behavioral Research, 2 May 1978; see also a memorandum from Julius P. Richmond, Assistant Secretary for Health, to Califano, 26 July 1977. Both documents can be found in the Office of the Director (OD) Files, National Institutes of Health, Central Files, "Human Subjects" folders.

75. For the proposed DHEW regulations see the *Federal Register* 43 (5 January 1978), 1050–1053; the final regulations can be found in the *Federal Register* 43 (16 November 1978), 53652–53656. These regulations remain essentially unchanged today and can be found at 45 C.F.R. part 46, subpart C. These regulations have also been adopted by other federal agencies that have any concern with human experimentation as part of the so-called Common Rule, with the exception of the FDA (see below).

76. *Federal Register* 43 (5 January 1978), 1051.

77. *Federal Register* 45 (30 May 1980), 36386–36392.

78. *Henry Fante et al. v. Department of Health and Human Services et al.*, U.S. District Court, Eastern District of Michigan, Southern Division, Civil Action No. 80–72778. The records from this case are now kept at the National Archives, Great Lakes Regional Archives, Chicago, Accession No. 21–88–0016, Location No. 331792–332283, Box No. 269. FDA officials announced the decision to stay indefinitely the regulations in the *Federal Register* 46 (7 July 1981), 35085. The current "stayed" status of the FDA prison research regulations can be found at 21 C.F.R. part 50, subpart C.

79. Charles Miller, as quoted in Stephen Gettinger and Kevin Krajick, "The Demise of Prison Medical Research," *Corrections Magazine* 5 (December 1979): 12.

80. On 1 January 1960, NIH awarded $97,256.00 to the LMRI of Boston University to carry out this study. Irving Ladimer, who had completed a doctor of law dissertation at George Washington University in 1958 on the legal and ethical aspects of human experimentation, served as the project's principal investigator through June 1962, when he left Boston University. Ladimer was replaced as principal investigator by

his chief assistant, Donald A. Kennedy, an anthropologist by training, who saw the project through to completion. The first characterization of the purpose of the project is taken from page 1 of Kennedy's preface to the final report: "A Study of the Legal, Ethical, and Administrative Aspects of Clinical Research Involving Human Subjects: Final Report of Administrative Practices in Clinical Research, Research Grant No. 7039," Law-Medicine Research Institute, Boston University (1963) (hereafter cited as LMRI final report); both chapter and page numbers will be provided because pages within chapters are numbered separately. The second, and lengthier purpose statement is taken from page 1 of chapter 1 of the LMRI final report, "Focus of the Inquiry." This unpublished report is in the collections of the Mugar Memorial Library, Boston University (ACHRE No. BU-053194–A). This report, which is more than 360 typewritten pages, is a wealth of information that has remained largely untapped by recent scholars interested in the development of research ethics in this country. The few citations of the project that do appear in the published literature almost all refer to the summary that appears in William J. Curran, "Governmental Regulation of the Use of Human Subjects in Medical Research: The Approach of Two Agencies," *Daedalus* 98 (spring 1969): 546–548. In this very brief reference to the project, Curran makes no mention of the "invitational work conferences," which the project staff identified as the investigational technique that "yielded the most valuable information." This characterization appears on page 8 of chapter 2, which is devoted to research method; pages 3–5 of the same chapter provide more details on the specific methodology employed in these conferences.

81. LMRI final report, chapter 8, Roger W. Newman, LL.B. [of the project staff], "The Participation of Prisoners in Clinical Research: Analytic Summary of a Conference," 1–2. A collection of documents related to the project is located in the files of the Center for Law and Health Sciences, School of Law, Boston University. This entity is the successor to the Law-Medicine Research Institute (ACHRE No. BU-062394–A).

82. Newman's "Analytic Summary" is more than 100 pages of typescript and seems to cover the conference in considerable detail. Also, comparisons between the transcripts of other LMRI "invitational work conferences" that have survived with the related summaries produced for the final report reveal a skillful and fair rendering of the meetings.

83. LMRI final report, chapter 8, 27.

84. Ibid., 18.

85. Ibid., 31.

86. Ibid.

87. Ibid., 85.

88. Ibid., 71.

89. Ibid., 72.

90. Ibid., 74.

91. Ibid., 88–89.

92. Ibid., 85–86.

93. Ibid., 93.

94. Ibid., 89–90.

95. Ibid., 96.

96. Deposition of Carl Heller, 19 July 1976, 32. Heller said he avoided the word *cancer* because "I didn't want to frighten them [the prisoners]." Dr. Paulsen said in a telephone interview on 12 September 1995 that he explained to the inmates that data from Hiroshima and Nagasaki "showed no additional incidence of testicular cancer." Undated consent forms from the Washington experiment differ. Some specify an "extremely small" risk of testicular cancer, and others do not specifically mention cancer. C. A. Paulsen, interview with ACHRE staff, 12 September 1995 (ACHRE No. ACHRE-091295–A).

10

Atomic Veterans:
Human Experimentation
in Connection with
Atomic Bomb Tests

IN 1946, the United States conducted Operation Crossroads, the first peacetime nuclear weapons tests, before an audience of worldwide press and visiting dignitaries at the Bikini Atoll in the Pacific Marshall Islands. In 1949 the Soviet Union exploded its first atomic bomb, and in December 1950, shortly after the United States entered the Korean War, President Truman chose Nevada as the site for "continental testing" of nuclear weapons. Testing of atomic bombs in Nevada began in January 1951 and continued throughout the decade. Further testing of atomic, then hydrogen, bombs took place in the Pacific. By the time atmospheric testing was halted by the 1963 test ban treaty, the United States had conducted more than 200 atmospheric tests and dozens of underground tests.[1]

The rules governing nuclear weapons tests were not spelled out by law or handed down by tradition. They had to be created in ongoing interplay between the new Atomic Energy Commission and the new Department of Defense.

The tests were important to many governmental agencies but, of course, critical to the AEC and the DOD. The AEC, as the source of weapons design expertise, was interested in the performance of new bomb designs and, along with DOD, in the effects of the weapons. The DOD, and each of the armed services, had particular interests in the use of the tests to learn

how atomic wars could be fought and won, if, as seemed quite possible at midcentury, they had to be. Along with "civilian agencies," such as the Public Health Service, the Veterans Administration, and the Department of Agriculture, they shared an interest in civil defense against the use of the bomb in wartime and the impact of the bomb's use—in peacetime tests as well as war—on the public health and welfare. The bomb tests inevitably involved risk and uncertainty; safety was a basic and continued concern, and the development of radiation safety practices and understanding was therefore an essential part of the test program.

At its core, the test program was established to determine how well newly designed nuclear weapons worked; but officials and researchers quickly saw the need and opportunity to use the tests for other purposes as well. More than 200,000 people, including soldiers, sailors, air crews, and civilian test personnel, were engaged to staff the tests, to participate as trainees or observers, and to gather data on the effects of the weapons.

The Committee was not chartered to review the atomic bomb tests or the experience of the troops present at the detonations. However, early in our tenure we heard from veterans who participated in the tests, and their family members, who urged that we include their experiences in our review. In testimony before the Advisory

Committee, "atomic vets" and their widows stated forcefully that all those who participated in the bomb tests were in a real sense participants in an experiment. It also was argued that biomedical experiments involving military personnel as human subjects took place in connection with the tests. The interest among atomic veterans and their families in the activities of the Advisory Committee and the government's commitment to investigating human radiation experiments was intense. When the Department of Energy established its Helpline for citizens concerned about human radiation experiments, for example, bomb-test participants and their family members were the single largest group of callers among the approximately 20,000 calls received.

That the bomb tests were in some sense experiments is, of course, correct. The tests of new and untried atomic weapons were, wrote the chief health officer of the AEC's Los Alamos lab, "fundamentally large scale laboratory experiments."[2] At the same time, although there was a real possibility that human subject research had been conducted in conjunction with the bomb tests, the tests were not themselves experiments involving human subjects.

The Committee reviewed the historical record to determine if human experiments had taken place in connection with the tests. We found that somewhere in the range of 2,000 to 3,000 military personnel at the tests did serve as the subjects of research in connection with the tests. In most cases, these research subjects were engaged in activities similar to those engaged in by many other service personnel who were not research subjects. For example, some air crew flew through atomic clouds in experiments to measure radiation absorbed by their bodies, but many others flew in or around atomic clouds to gather data on radiation in the clouds. The Defense Department generally did not distinguish such research from otherwise similar activities, treating both as part of the duties of military personnel. The experience of the atomic veterans illustrates well the difficulty in locating the boundary between research involving human subjects and other activities conducted in occupational settings that routinely involve exposure to hazards.

The more the Committee investigated the human research projects conducted in conjunction with the bomb tests, the more we found ourselves discussing issues that affected all the service personnel who had been present at the tests, and not just those who also had been involved as subjects of research. This occurred both because of the boundary problem just described and because critical decisions about initial exposure levels and follow-up of veterans were generally not made separately for research subjects and other personnel present at the tests. Legislation passed in 1984 and 1988 that provides the basis for compensation to some atomic veterans similarly does not distinguish between those veterans who were research subjects and the vast majority who were not.

In this chapter we present what we have learned about human experimentation conducted in conjunction with atomic bomb testing as well as some observations about the experience of the atomic veterans generally. In the first section of the chapter we focus on research involving human subjects. We begin by a review of the 1951–1952 discussions in which DOD biomedical advisers considered the role of troops at the bomb tests and the need for biomedical research to be conducted in conjunction with them. We then look at a research activity that was given the highest priority by these advisers, the psychological and physiological testing of troops involved in training maneuvers at bomb tests and of officers who volunteered to occupy foxholes in the range of one mile from ground zero. We next turn to the so-called flashblindness experiments conducted to measure the effect on vision of the detonation of an atomic bomb. Finally, we look at research in which men were used to help measure the radiation absorbed by protective clothing, by equipment that humans operated, and by the human body. We note at the outset that while the studies all took place in the context of the atomic bomb, and therefore involved some potential exposure to radiation, none of them were designed to measure the biological effects of radiation itself (as opposed to the levels of exposure). A basic reason this was so was the determination of the DOD and the AEC to keep exposure levels of test participants below those at which acute radiation effects were likely to be experienced (and therefore measurable).

In the second section of the chapter we discuss issues of concern to the Committee that affected all the atomic veterans. We review how risk was considered by AEC and DOD officials at the time the tests were being planned, the creation and maintenance of records related to bomb-test exposure, and what is now known about the longer-term risks of participation in the tests. We also discuss the legacy of distrust among atomic veterans and their families that stems, in part, from the failure to create and maintain adequate records. Finally, we conclude with a discussion of what the atomic bomb-test experience tells us about the boundary between experimental and occupational exposures to risk and some lessons that remain to be learned from the experience of the atomic veterans.

HUMAN RESEARCH AT THE BOMB TESTS

The Defense Department's Medical Experts: Advocates of Troop Maneuvers and Human Experimentation

As we saw in the introduction, in 1949, when AEC and DOD experts met to consider the psychological problems connected to construction of the proposed nuclear-powered airplane, the NEPA project, there was a consensus that America's atomic war-fighting capability would be crippled unless servicemen were cured of the "mystical" fear of radiation.[3] When routine testing of nuclear weapons began at the test site in Nevada in 1951, the opportunity to take action to deal with this problem presented itself. DOD officials urged that troop maneuvers and training exercises be conducted in connection with the tests. Whole military units would be employed in these exercises, and participation, as part of the duty of the soldier, would not be voluntary. DOD's medical experts simultaneously urged that the tests be used for training and "indoctrination" about atomic warfare and as an opportunity for research. The psychological and physiological testing of troops to address the fear of radiation was the first of the research to take place;

this testing was largely conducted as an occupational rather than an experimental activity.

In a June 27, 1951, memorandum to high DOD officials, Dr. Richard Meiling, the chair of the secretary of defense's top medical advisory group, the Armed Forces Medical Policy Council, addressed the question of "Military Medical Problems" associated with bomb tests.[4] The memorandum made clear that troops should be placed at bomb tests not so much to examine risk as to demonstrate relative safety.

"Fear of radiation," Dr. Meiling's memorandum began, "is almost universal among the uninitiated and unless it is overcome in the military forces it could present a most serious problem if atomic weapons are used." In fact, "[i]t has been proven repeatedly that persistent ionizing radiation following air bursts does not occur, hence the fear that it presents a dangerous hazard to personnel is groundless." Dr. Meiling urged that "positive action be taken at the earliest opportunity to demonstrate this fact in a practical manner."[5]

He continued, a "Regimental Combat Team should be deployed approximately twelve miles from the designated ground zero of an air blast and immediately following the explosion . . . they should move into the burst area in fulfillment of a tactical problem." The exercise "would clearly demonstrate that persistent ionizing radiation following an air burst atomic explosion presents no hazards to personnel and would effectively dispel a fear that is dangerous and demoralizing but entirely groundless."[6]

Dr. Meiling's proposal to put troops at the bomb tests in order to allay their fears may well have been an echo of what the military already had in mind. The Army's 1950 "Atomic Energy Indoctrination" pamphlet, a primer for soldiers, showed that the military was concerned that misperception of the effect of an air burst could be damaging in combat. "[L]ingering radioactivity will be virtually nonexistent in the case of the normal air burst,"[7] it reassured the soldiers. The greater danger, it told them, was the probability that "an unreasoning fear of lingering radioactivity" would take "an unnecessary toll in American lives."[8]

While the tests provided an opportunity to allay fears, they simultaneously provided the opportunity to gather data. In this regard, Dr. Meiling appeared to be ahead of his military colleagues in expressing concern that the military was not taking adequate advantage of the bomb tests as an opportunity for "biomedical participation." In February 1951, in fact, following tests in Nevada, he had urged the DOD to incorporate "biomedical tests" into plans for future bomb tests.[9]

Meiling's suggestion that planning for biomedical tests be undertaken wound its way through the secretary of defense's research and development bureaucracy and fell into the lap of the civilian-chaired Joint Panel on the Medical Aspects of Atomic Warfare.[10] Under the chairmanship of Harvard's Dr. Joseph Aub, the Joint Panel was the gathering place for the small world of government radiation researchers and their private consultants. Its periodic "Program Guidance Reports" laid out the atomic warfare medical research agenda, summarizing work that was ongoing and that which remained. At its meetings, participants heard from the CIA on foreign medical intelligence, debated the need for human experimentation, and learned of the latest developments in radiation injury research, of the blast and heat effects of the bomb, and of instruments needed to measure radiation effects.

In September 1951 the Joint Panel considered a draft report on "biomedical participation" in bomb tests.[11] "It is, of course obvious," the report noted, "that a test of a new and untried atomic weapon is not a place to have an unlimited number of people milling about and operating independently." Planning was therefore in order. There were, the document explained, basic criteria for "experimentation" at bomb tests. For example, "Does the experiment have to be done at a bomb detonation; is it impossible or impractical in a laboratory?"[12]

The document turned to "specific problems for future tests." The list of twenty-nine problems was not intended to be all-inclusive, but was "designed to show the types of problems which should be considered as a legitimate basis for biomedical participation in future weapons tests." The term *human experimentation* was not used, and most of the items could be performed without humans.[13] However, the list included several examples of research involving human subjects:

11. Effects of exposure of the eye to the atomic flash . . .
24. Measurements of radioactive isotopes in the body fluids of atomic weapons test personnel . . .
27. The efficiency and suitability of various protective devices and equipment for atomic weapons war . . .
28. Psychophysiological changes after exposure to nuclear explosions.
29. Orientation flights in the vicinity of nuclear explosions for certain combat air crews.[14]

By the end of the decade, human research would be conducted in all these areas.[15]

At the same September meeting, the Joint Panel also considered a "Program Guidance Report" on the kinds of atomic warfare-related research that needed to be conducted, in the laboratory as well as in the field. The areas singled out for immediate and critical attention included the initiation of "troop indoctrination at atomic detonations" and "psychological observations on troops at atom bomb tests."[16]

A section on "Biomedical Participation in Future Atomic Weapons Tests" concluded that the next step should be

4.1 To complete present program and plan for participation in future tests in light of results from Operation GREENHOUSE [a prior atomic test series]. *These plans should include studies on the effect of atomic weapons detonations on a troop unit in normal tactical support* [emphasis added].[17]

Thus, while it was well known at the time that troops participated at the bomb tests and were subjected to psychological testing, it now is evident that the DOD's medical advisers advocated the presence of the troops at the tests for both training and research purposes. The doctors were not alone in attaching high priority to such research. The Joint Panel's September guidance punctuated, perhaps echoed, the Armed Forces Special Weapons Projects's midsummer 1951 call for a "systematic research study . . . [to] provide a sound basis for estimating troop reaction to the bomb experience and . . . the indoctrination value of the maneuver."[18]

The HumRRO Experiments

Just two months later, in November 1951, at a bomb test in the Nevada desert, the Army conducted the first in a series of "atomic exercises."[19] This exercise was designed primarily to train and indoctrinate troops in the fighting of atomic wars. The exercise also provided an opportunity for psychological and physiological testing of the effects of the experience on the troops.

Desert Rock was an Army encampment in Nevada adjacent to the nuclear test site. At the exercise named Desert Rock I, more than 600 of the 5,000 men present would be studied by psychologists from a newly created Army contractor, the Human Resources Research Organization (HumRRO). HumRRO's research was directed by Dr. Meredith Crawford, who was recruited by the Army from a deanship at Vanderbilt University.[20] The identity of all the participants involved in the "HumRRO experiments," and the further DOD research discussed later in this chapter, is not known. The numbers of those who participated must be reconstructed from available reports.[21]

The highly publicized bomb test was well attended by military and civilian officials. "Las Vegas, Nevada," *Time* magazine reported, "had not seen so many soldiers since World War II. . . . The hotels were jammed with high brass. . . . [o]ut on the desert, 65 miles away 5,000 hand-picked troops were getting their final briefing before Exercise Desert Rock I—the G.I.'s introduction to atomic warfare."[22] The detonation, Representative Albert Gore (father of the current Vice President), told the *New York Times,* was "the most spectacular event I have ever witnessed. . . . As I witnessed the accuracy and cataclysmic effect of the explosion, I felt the conviction that it might be used in Korea if the cease-fire negotiations broke down."[23]

To render the experience more realistic, the observers and participants were told to imagine that aggressor armies had invaded the United States and were now at the California-Nevada border. An atomic bomb would be dropped, with the troops occupying a position seven miles from ground zero. After the detonations they would "attack into the bombed area."[24]

At their home base, two groups of troops—a control group that would stay at home base and an experimental group that would go to Nevada—had listened to lectures and seen films intended to "indoctrinate" them about the effects

of the bomb and radiation safety. Both groups were administered a questionnaire to determine how well they had understood the information provided. Dr. Crawford explained in a 1994 interview that "indoctrination," which today has a negative connotation, was not intended to suggest misrepresentation of fact, but "had more to do with attitude, feeling and motivation."[25] At Desert Rock, the experimental group was given a further "non-technical briefing." They were "reminded that no danger of immediate radiation remains 90 seconds after an air burst; that they would be sufficiently far from ground zero to be perfectly safe without shelter; and that with simple protection they could even be placed quite close to the center of the detonation, with no harm to them."[26]

After the blast, a questionnaire was again administered to most of the experimental subjects, and physiological measurements including blood pressure and heart rates were taken. The questionnaire was designed to test the success of the "indoctrination."[27] For example, questions included (answers in parentheses were those the HumRRO report stated were correct):

1. Suppose the A-bomb were used against enemy troops by exploding it 2000 feet from the ground and suppose all enemy troops were killed. How dangerous do you think it would be for our troops to enter the area directly below the explosion *within a day?* (Not dangerous at all). . . .

6. If an A-bomb were exploded at 2000 feet, under what conditions would it be safe to move into the spot directly below, *right after* the explosion? (Safe if you wore regular field clothing.)[28]

These answers were not correct. Answers to questions like the above depend on weather conditions, the yield of the weapon, and the assumptions about the degree of risk from low levels of exposure. For example, while an airburst (where the fireball does not touch the ground) may result in little fallout in the immediate area of the blast, it does not result in none; if rain is present, a substantial amount of fallout may be localized.

Similarly, whereas the 1946 Bikini bomb tests at Operation Crossroads in the Pacific had caused contamination so severe that many of the surviving ships were scrapped, the question and answer provided said:

Some of the ships in the Bikini tests had to be sunk because they were too radioactive to be used again. (False).[29]

In a 1995 review of the 1951 questionnaire, the Defense Department found that "changes/corrections/clarifications" would be in order for nine of the thirty questions.[30]

In January 1952, the Army surgeon general expressed "continuing interest in the conduct of psychiatric observations," offering funds for "Psychiatric Research in Connection with Atomic Weapons Tests Involving Troop Participation."[31] In March 1952, however, the Army and the Armed Forces Special Weapons Project (AFSWP), which coordinated nuclear weapons activities for the DOD, provided critical reflections on Desert Rock I. "[O]ne is inevitably drawn to the conclusion," the Army reported, "that the results, though measurable, were highly indeterminate and unconvincing. The limitations of evaluation were inherent in the problem. Handicapped by a preconceived notion that there would be no reaction, it took on the form of a gigantic experiment whose results were already known. No well controlled studies could be undertaken which could presume even superficial validity. . . ."[32] In a letter to the AEC, the AFSWP went further. Owing to the "tactically unrealistic distance of seven miles to which all participating troops were required to withdraw for the detonation,"[33] troops might get the wrong impression about nuclear warfare.

In 1994, Dr. Crawford reflected on the logic of testing for panic in an environment that was thought to be too safe. "No troops," Dr. Crawford recalled, "were exposed anywhere where anybody thought there was any danger, so you might ask the question, so what? I've asked that question myself and I've thought about it. It was the first HumRRO project. It was really pretty well agreed upon before I got up here from Tennessee . . . so we did what we could."[34]

Despite the reservations about the 1951 study, on May 25, 1952, the Army conducted its second HumRRO experiment at the exercise called Desert Rock IV. It was similar in methodology to the first experiment and involved about 700 soldiers who witnessed the shot and 900 who served in the control group as nonparticipants.[35] This time, to add more realism, the troops witnessed the blast, an 11-kiloton weapon that was set off from the top of a tower, from four miles from ground zero. By the end of the second research effort, there was even more reason to question the utility of the experiments. HumRRO's report on Desert Rock IV stated that while knowledge increased as a result of the indoctrination, the actual maneuver experience did not produce significant improvement in test scores and decreases were actually reported on some questions.[36]

In both Desert Rock I and Desert Rock IV, the Army hoped that the troops who witnessed the blasts would disseminate information to the troops who stayed at home base. However, the troops who participated in the exercises were warned that discussion of their experiences could bring severe punishment, and the researchers found that communication was at a minimum.[37] Moreover, those who stayed home, HumRRO found, "showed no evidence of great interest, of extensive discussion, or of any important benefit from dissemination as a result of the atomic maneuver."[38] Meanwhile, the experience that the participants had been warned not to discuss and that was of little interest to their comrades was front-page news throughout the country. "When they returned to camp," *Time* reported of the first Desert Rock exercise, "the men were quickly herded into showers. Some were given test forms to fill out. Did you sweat? Did your heart beat fast at any time? Did you lose bladder control? Most of the answers were no."[39]

Without any direct comment on the results of the Desert Rock I and IV experiments, in September 1952, the Joint Panel urged that the psychological research continue:

It is possible that inclination to panic in the face of AW [atomic warfare] and RW [radiological warfare] may prove high. It seems advisable, therefore, to increase research efforts in the scientific study of panic and its results, and to seek means for prophylaxis. . . . The panel supports the point of view that troop participation in tests of atomic weapons is valuable. As many men as possible ought to be exposed to this experience under safe conditions. Psychological evaluation is difficult and results can be expected to appear superficially trivial, but the matter is of such extreme importance that the research should be persisted in, utilizing every opportunity.[40]

Indeed, a third set of experiments was carried out in April 1953, at Desert Rock V; this time, the number of participants is unknown.[41]

The final HumRRO bomb test study was conducted in 1957 at Operation Plumbbob.[42] No formal report was prepared, but the experience was recorded in a personalized recollection by a HumRRO staffer.[43] Weather-related delays, the departure of HumRRO staff, the continued redesign of the exercises, and the failure of a fifth of the troops to return from a weekend pass in time for the events took their toll. The researchers were not given the script used in the indoctrination lectures to the troops. Thus, it was impossible for the researchers to know whether incorrect responses were due to "lack of inclusion of the topic in the orientation or to ineffective instruction."[44] The research was to include exercises such as crawling over contaminated ground.[45] But, yet again, the researchers found that the safety rules in force precluded important study: "shock . . . and panic . . . would not be observed."[46]

There is no question that HumRRO activities were research involving human subjects; the projects involved an experimental design in which soldier-subjects were assigned either to an experimental or a control condition. Available evidence suggests, however, that the Army did not treat HumRRO as a discretionary research activity but as an element of the training exercise in which soldiers were participating in the course of normal duty. The HumRRO subjects were apparently not volunteers. Dr. Crawford in 1994 said of the HumRRO subjects, "Whether they were requested to formally give their consent is pretty unknowable because in the Army or any other military service people generally do what they're asked to do, told to do."[47] Indeed, as HumRRO's initial report stated, the primary purpose of the atomic exercise was training; "research was necessarily of secondary importance."[48] However, Dr. Crawford felt confident that the risks were disclosed. Because of the "number and intensity of briefings . . . [n]o soldier, to our knowledge, went into the test situation with no idea about what to expect. They were adequately informed."[49]

We now know that in 1952 the Defense Department's medical experts were simultaneously locked in discussion of the need for psychological studies and other human research at bomb tests and, as we saw in chapter 1, the need for a policy to govern human experimentation related to atomic, biological, and chemical warfare. In October of that year, the Armed Forces Medical Policy Council recommended that the Nuremberg Code be adopted, as it was by Secretary Charles Wilson in 1953. What is still missing is information that might show how, as seems to be the case, the same experts could have been having these discussions without communicating the essence of them to those responsible for conducting the human research at the tests. There is no evidence that the investigators responsible for HumRRO were informed about the Wilson memo. Dr. Crawford, for example, when queried in 1994, reported that he did not know of the 1953 Wilson memorandum. It is possible that HumRRO was not viewed as being subject to the requirements stated in the Wilson memo despite the fact that it was human research relating to atomic warfare. Although the experimental variable was participation at a bomb test, arguably, the troops would have been present at the test in any event, along with many thousands of other soldiers who were not subjects in the HumRRO research.

Atomic Effects Experiments

At the same time that the third set of HumRRO experiments was being conducted, in April 1953 at Desert Rock V, the Army called on several dozen "volunteers for Atomic Effects Experiments."[50] According to the Army, all were officers familiar with the "experimental explosion involved" and were able to personally judge "the probability of significant variations in [weapon] yield." They were instructed to choose the distance from ground zero they would like to occupy in a foxhole at the time of detonation, as long as it was no closer than 1,500 yards. If the surviving documentation is the measure, these officers, and perhaps officer volunteers in the subsequent Desert Rock series, were the only subjects of bomb-test research who signed forms saying that they were voluntarily undertaking risk.[51] The exposures were meant to set a standard for developing "troop exposure programs and for confirming safety doctrine for tactical use of atomic weapons."[52]

An Army report on the volunteers at Desert Rock V concluded that there would be "little more to be gained by placing volunteer groups in forward positions on future shots."[53] An April 24, 1953, Army memorandum recommended termination of the program "as little will be gained in repeatedly placing volunteers in trenches 2000 yards from ground zero."[54] However, officer volunteers were called on again at the next Desert Rock exercises at the 1955 nuclear test series called Operation Teapot. Following Teapot, the Army recommended that further experiments be conducted in which the volunteers would be moved closer to ground zero, "until thresholds of intolerability are ascertained."[55] This "use of human volunteers under conditions of calculated risk," the Army told the AFSWP, "is essential in the final phase of both the physiological and psychological aspects of the overall program."[56]

In response, the AFSWP pointed out that the injury threshold could not be determined "without eventually exceeding it."[57] The Army was essentially proposing human beings be exposed to the detonation's blast effects to the point of injury. The proposal, an AFSWP memo explained, would not pass muster under the rules of the Nevada Test Site and was otherwise unacceptable:

In particular, it is significant that the long range effect on the human system of sub-lethal doses of nuclear radiation is an unknown field. Exposure of volunteers to doses higher than those now thought safe may not produce immediate deleterious effects; but may result in numerous complaints from relatives, claims against the Government, and unfavorable public opinion, in the event that deaths and incapacitation occur with the passage of time.[58]

If the Army wanted more data on blast effects, AFSWP declared, it should proceed with laboratory experiments, for which money would be made available. The AFSWP was not opposed to the kinds of activities that had previously taken place at Desert Rock. But those activities, AFSWP's memo observed in passing,

"cannot be expected to produce data of scientific value."[59]

The Desert Rock experience was apparently repeated, again with officer volunteers, in the next Nevada test series, the 1957 Operation Plumbbob. Although the total number of officers involved in all of the "officer volunteer" experiments is not known, it is probably fewer than one hundred.

The Flashblindness Experiments

Beginning with the 1946 Bikini tests, experiments with living things became a staple of bomb tests. At Operation Crossroads, animals were penned on the decks of target ships to study the effects of radiation. In the 1948 Sandstone series at the Marshall Islands Enewetak Atoll, seeds, grains, and fungi were added. In 1949, the AEC and the DOD began to coordinate the planning of the biomedical experiments at tests and set up a Biomedical Test Planning and Screening Committee to review proposals.[60] Presumably, the human experiments at bomb tests should have been filtered through this or some other review process designated to consider experiments. Yet, in only one case—flashblindness experiments—did this happen.

With Dr. Meiling's 1951 call for renewed DOD effort at biomedical experimentation came a revival of the DOD-AEC joint biomedical planning. From the start, the AEC doubted DOD's willingness to cooperate.[61] In a January 1952 letter to Shields Warren, Los Alamos's Thomas Shipman complained that the committee was limited to reviewing proposals from civilian groups, and not the military: "[I]f," he wrote, the "AEC can not exercise a measure of control in this matter, they might better withdraw from the picture completely and permit the military to continue on its own sweet way without the somewhat ludicrous spectacle of an impotent committee's snapping its heels like a puppy dog."[62] In retrospect, Shipman wrote to Warren's successor in June 1956, the military's refusal to participate "reduced that committee to impotence."[63]

Whatever its effectiveness, in 1952 the biomedical research screening group did consider at least one of the military's flashblindness experiments.[64] Flashblindness—the temporary loss of vision from exposure to the flash—was a serious problem for all the armed services, but particularly for the Air Force. Pilots flying hundreds of miles an hour in combat could not afford to lose concentration and vision even temporarily.[65]

The flashblindness experiments began at the 1951 Operation Buster-Jangle, the series that included Desert Rock I, with the testing of subjects who "orbit[ed] at an altitude of 15,000 feet in an Air Force C-54 approximately 9 miles from the atomic detonation. . . ."[66] The test subjects were exposed to three detonations during the operation, after which changes in their visual acuity were measured.[67] Although these experiments were conducted at bomb tests that potentially exposed the subjects to ionizing radiation, the purpose of the experiment was to measure the thermal effects of the visible light flash, not the effects of ionizing radiation.

When another experiment was proposed for Operation Tumbler-Snapper, the 1952 Nevada tests, the AEC sought a "release of AEC responsibility" on grounds that "there is a possibility that permanent eye damage may result."[68] It is not clear how the military responded, but the experiment proceeded. Twelve subjects witnessed the detonation from a darkened trailer about sixteen kilometers from the point of detonation.[69] Each of the human "observers" placed his face in a hood; half wore protective goggles, while the other half had both eyes exposed.[70] A fraction of a second before the explosion, a shutter opened, exposing the left eye to the flash.[71] Two subjects incurred retinal burns, at which point the project for that test series was terminated.[72] The final report recorded that both subjects had "completely recovered."[73]

At the 1953 tests, the Department of Defense engaged in further flashblindness study.[74] During this experiment, "twelve subjects [dark-adapted] in a light-tight trailer were exposed to five nuclear detonation flashes at distances of from 7 to 14 miles."[75]

The flashblindness experiments were the only human experiments conducted under the biomedical part of the bomb-test program and the only human experiments where immediate injury was recorded. They were also the only experiments where there is evidence of any connection to the 1953 Wilson memorandum applying the Nuremberg Code to human experimentation.

Recently recovered documents show that upon a 1954 review of a report showing the injuries at the 1952 experiment, AFSWP medical staff immediately declared that "a definite need exists for guidance in the use of human volunteers as experimental subjects."[76] Further inquiry revealed that a Top Secret policy on the subject existed. That policy detailed "very definite and specific steps" that had to be taken before volunteers could be used in human experimentation. But, the AFSWP wrote, "No serious attempt has been made to disseminate the information to those experimenters who had a definite need-to-know."[77]

Nonetheless, some form of consent was obtained from at least some of the flashblindness subjects. In a 1994 interview, Colonel John Pickering, who in the 1950s was an Air Force researcher with the School of Aviation Medicine, recalled participating as a subject in one of the first tests where the bomb was observed from a trailer, and his written consent was required. "When the time came for ophthalmologists to describe what they thought could or could not happen, and we were asked to sign a consent form, just as you do now in the hospital for surgery, I signed one."[78] There is no documentation showing whether subsequent flashblindness experiments, which followed upon the issuance of the secretary of defense's 1953 memorandum, required informed and written consent. However, given the recollection of Colonel Pickering and the military tradition of providing for voluntary participation in biomedical experimentation, this may well have been the case. (A report on a flashblindness experiment at the 1957 Plumbbob test uses the term *volunteers*;[79] a report on 1962 "studies" at Dominic I provides no further information.)[80]

In early 1954 the Air Force's School of Aviation Medicine reported that animal studies and injuries at bomb tests (to nonexperimental participants) had shown that potential for eye damage was substantially worse than had been understood.[81] Studies of flashblindness with humans continued in both field and laboratory tests through the 1960s and into the 1970s. These experiments tested prototype versions of eye protection equipment, and the results were used to recommend requirements for eye protection for those exposed to atomic explosions.[82]

Research on Protective Clothing

In late 1951, following the first Desert Rock exercise, the government conducted Operation Jangle, a nuclear test series that detonated two nuclear weapons, one on the surface and one buried seventeen feet underground. The two Jangle shots were tests where the weapon's fireball touched the ground. When a nuclear weapon's fireball touches the ground it creates much more local fallout than an explosion that bursts in the air. Consequently, these tests posed some potential hazard to civilians who lived near the test site and to test observers and participants.

Two weeks before Jangle the DOD requested an additional 500 observers at each of the Jangle shots, to acclimate the troops to nuclear weapons. The AEC advised against the additional participants, declaring that "[t]his [the first detonation] was an experiment which had never been performed before and the radiological hazards were unpredictable." In the AEC's view, no one should approach ground zero for three or four days after the surface shot.[83]

The AEC seems to have been successful in persuading the Department of Defense not to include the extra observers, but the DOD did not agree to the AEC's suggestion on approaching ground zero. Four hours after the first shot, the DOD conducted research involving troops who were accompanied by radiation safety monitors.[84] Eight teams of men walked over contaminated ground for one hour to determine the effectiveness of protective clothing against nuclear contamination.[85] Similar tests were conducted after the second shot at Jangle, but this time after a longer period. Five days after the shallow underground shot, men crawled over contaminated ground, again to determine the effectiveness of protective clothing.[86] Other men rode armored vehicles through contaminated areas to check the shielding effects of tanks and to check the effectiveness of air-filtering

devices.[87] According to the final report, the protective clothing was "adequate to prevent contact between radioactive dust and the skin of the wearer."[88]

The information on this research is limited. The only mention of the subjects in the report reads, "The volunteer enlisted men, too numerous to mention by name, who participated in the evaluation of protective clothing were of great assistance which is gratefully acknowledged."[89] It is likely that at the time these men were not viewed as subjects of scientific research but rather as men who had volunteered for a hazardous or risky assignment. We know nothing about what these men were told about the risks or whether they felt they could have refused the assignment if they had an interest in doing so.

The Jangle activities are a good illustration of difficulties in drawing boundaries in the military between activities that are research involving human subjects and activities that are not. Although the Jangle evaluation was likely not considered an instance of human research at the time, it has many similarities to ground-crawling activity conducted several years later, not in conjunction with a nuclear test, that was treated as research involving human subjects. In 1958 ninety soldiers at Camp Stoneman, in Pittsburg, California, were asked to perform "typical army tactical maneuvers" on soil that had been contaminated with radioactive lanthanum.[90] The soldiers were then monitored for their exposure to study beta contamination from this nonpenetrating form of radiation. In 1963 soldiers were again asked to maneuver on ground contaminated with artificial fallout, this time at Camp McCoy in Wisconsin.[91]

The plans for the 1958 maneuvers, which were administered by the Navy's Radiological Defense Laboratory, had been submitted for secretarial approval, as was required for biomedical experiments involving Navy personnel.[92] In accordance with the Navy rules, the soldiers signed "written statements of voluntary participation."[93] During the 1963 experiments the Army processed the activity under its 1962 regulation on human experimentation (AR 70–25).[94] This rule, a public codification of the secretary of defense's 1953 Nuremberg Code rule, also required secretarial review and written consent.[95]

Cloud-Penetration Experiments

What are the dangers to be encountered by the personnel who fly through the cloud?—How much radiation can they stand?—How much heat can the aircraft take?—Can the ground crews immediately service the aircraft for another flight?—If so, what precautions are necessary to insure their safety?[96]

The Air Force felt that it was essential to answer these questions. To do so, it carried out experiments, including some with animals and a few with humans.

· At the first atomic tests the military used remote-controlled aircraft, called "drones," to enter and gather samples from atomic clouds in order to estimate the yield and learn the characteristics of the weapon being tested. Military pilots did, however, "track" mushroom clouds, gathering information and plotting the cloud's path in order to warn civilian aircraft. During a 1948 test, a cloud tracker piloted by Colonel Paul Fackler inadvertently got too close to a cloud. But after the accident, Colonel Fackler commented, "'No one keeled over dead and no one got sick.'"[97] Colonel Fackler's experience, an Air Force history later recorded, showed that manned flight through an atomic cloud "would not necessarily result in a lingering and horrible death."[98]

Some of the trackers had "sniffers" on their aircraft to collect small samples. The Air Force conducted experimental sampling missions at 1951 tests and later permanently replaced the drones with manned aircraft because drones were difficult to use, and they often did not get the quality samples of the atomic cloud that Atomic Energy Commission scientists desired. By Operation Teapot (1955), the AEC considered the testing of a nuclear device "largely useless" unless sampler aircraft were used to obtain fission debris that would be used to estimate the nuclear weapon's performance.[99]

As the sampling mission became routine, a new mission in the clouds began. At Teapot the Air Force performed the first manned "early cloud penetration." The phrase was used by the Air Force to refer to missions conducted as soon as minutes after detonation of the test weapon. The main purpose was to discover the radiation and turbulence levels within the cloud at early times after detonation.

Like the first sampling missions, the first early cloud-penetration missions were conducted by unmanned drone aircraft. In 1951 Colonel (now General) E. A. Pinson, an Air Force scientist who had earlier conducted tracer experiments on himself and other scientists, placed mice aboard a drone aircraft; in 1953 he flew mice, monkeys, and instrumentation in drone aircraft through atomic clouds. Pinson concluded that the radiation risk from flying manned aircraft through atomic clouds could be controlled by monitoring the external gamma dose.[100] But the Air Force was not convinced and asked Pinson to follow up the animal experiments with studies with humans during Operation Teapot (1955) and Operation Redwing (1956) to confirm the results. This research appears to have involved a small number of subjects, perhaps in the range of a dozen or so.

Pinson designed the human experiments to "learn exactly how much radiation penetrates into the human system"[101] when humans flew through a mushroom cloud. The Air Force had pilots swallow film contained in small watertight capsules. The film was attached to a string held in their mouths, so that it could be retrieved at the end of the mission.[102] When the film was retrieved, the researchers compared the exposures measured inside the human body with those measured on the outside. They found that the doses measured outside the body were essentially identical to the doses inside the body; this was a critical finding, because it meant that surface measurements would be "representative of the whole-body dose."[103]

For the experiment, the AEC test manager for Teapot waived the AEC's test-exposure limit of 3.9 roentgens and permitted four Air Force officers to receive up to 15 roentgens whole-body radiation.[104] The exemption was "based on the importance of [the project] to the Military Effects Test program and the fact that radiation up to 15 R may be necessary for its successful accomplishment."[105] When the air crews entered the atomic clouds, they measured dose rates of radiation as high as 1,800 rad per hour. Since the crews were in the cloud for such a short period of time, however, the actual doses were much lower than 1,800 R.[106] The maximum reported dose received on a single mission was 17 R,[107] higher than the 15 R authorized for the project. Since the air crews flew on several missions, two of the crew members received more than 17 R.[108]

A year later, at Operation Redwing, where the atomic and hydrogen bombs were tested, the Air Force conducted another series of experimental cloud penetrations. Part of the Redwing experiment was to measure the hazard from inhaling or ingesting radioactive particles while flying through a mushroom cloud. When mice and monkeys were flown through clouds during earlier tests they were placed in ventilated cages to determine the hazard from inhaling radioactive particles. The studies found that the hazard from inhalation was less than 1 percent of the external radiation hazard. As General Pinson put it, "In other words, if the internal hazard were to become significant, the external hazard would be overwhelming."[109] To confirm this finding, Pinson undertook a similar experiment with humans, and again, as with the Teapot experiment, Pinson was a subject as well as a researcher. To perform the experiment, no filters were installed in the penetration aircraft.[110] Again, it is estimated that about a dozen subjects were involved.

The military this time set the authorized dosage (the maximum dosage to which Pinson could plan to have people exposed) at 25 R and a limiting dosage (in which case a report had to be filed) at 50 R.[111] During the experiment "maximum radiation dose rates as high as 800 r/hr were encountered, and several flights yielded total radiation doses to the crew of 15 r."[112] (To measure the internal dose of radiation the scientists analyzed urine samples and used wholebody counters.)

The project, as Pinson's final report noted, marked the transition from animal experimentation to human measurement:

Although a considerable amount of experimentation had been done with small animals which were flown through nuclear clouds, the early cloud-penetration project of Operation Redwing was the first instance in which humans were studied in a similar situation.[113]

The results confirmed those of the animal experiments. The internal hazard of radiation was insignificant relative to the external hazard. Consequently, the researchers recommended "that no action be taken to develop filters for aircraft pressurization systems nor to develop devices to protect flight crews from the inhalation of fission products."[114]

Experimental Purpose: Military Tactics, Money, and Morale

Why was the Air Force interested in showing that atomic clouds could be penetrated soon after a detonation?

Most important, the military wanted to assure itself that it was safe for combat pilots to fly through atomic clouds, if need arose during atomic war. But the research did not make much of a scientific contribution. Researchers had already established the levels of radiation in atomic clouds by flying drone aircraft through them, and there was nothing pathbreaking about humans being exposed to levels of radiation under 25 R. General Pinson later noted, "there are no research people that I know of that gave a damn [about manned early cloud-penetration experiments], because this is . . . a negligible contribution to research and scien[ce]—scientifically, you know, this contributes less than I suspect anything I've ever done . . . its only virtue is the practical use of it."[115]

From the scientific perspective the data would not likely be of great use; from an immediate practical perspective human data were felt to be essential for reassurance. Should the Air Force have been satisfied with the wealth of data it had from the drone experiments? In retrospect Pinson found the question difficult. "There's reason to say, 'Well, you should have been satisfied with the data that had been gathered with the drones.' But, you know, these are hardnosed, practical people that—that put their life on the line and in military combat . . . where the hazards are far greater than in this modest exposure to radiation."[116]

The budget also played a key role in cloudpenetration research, as well as the related decontamination experiments, which will be discussed shortly. The Defense Department declared that the knowledge gained through its cloud-penetration experiments would save "the taxpayers thousands upon thousands of dollars" because there would be no need to develop special protective clothing or equipment, which had been thought to be necessary.[117]

As in the case of the HumRRO experiments and the troop maneuvers, indoctrination and morale were important forces behind the experimentation. "Perhaps the most important problem of all," a popular men's magazine of the day wrote about the Teapot experience, "might be a psychological resistance of combat pilots and crews flying into the unknown dangers of hot, radio-active areas."[118] The press, therefore, depicted the Teapot experiment as a message to the world—pilots can fly through atomic clouds safely.

Research, Consent, and Volunteerism

Like the HumRRO experiments, the cloud flythrough experiments were treated as occupational, rather than experimental, activities. None of the participants signed consent forms, and waivers to dose limits were sought, and approved, under the process followed for the nonexperimental flythrough activities. In 1995 General Pinson said that he had not been aware of the ethical standards declared in the 1953 secretary of defense memorandum. If he had been, he "would have gotten written consent from the people that were involved in this."[119]

A 1963 Air Force history of the cloud-sampling program does not describe the process of crew and pilot selection, but does provide a perspective:

The Strategic Air Command pilots picked to fly the F-84G sampler aircraft were pleased to learn that they were doing something useful, . . . not serving as guinea pigs as they seriously believed when first called upon to do the sampling.[120]

Did the personnel understand the risks? Some of them surely did. The aircraft carried airmen and scientific observers. Because the scientific observers were the very scientists who designed the experiments, they certainly understood the radiation risks as well as anyone could be expected to. In this way, the cloud flythrough experiments exemplified the ethic of researcher self-experimentation. As Pinson recalled in

1995, "If you are going to do something like this and you think it's safe to do it, then you shouldn't ask somebody else to do it. The way you convince other people that at least you think it's all right, is do it yourself."[121]

The nonscientists were briefed and informed that the risks from their radiation exposure would be minimal.[122] A pilot in the cloud-tracking activities recalled one of the briefings: "The scientists line up at a briefing session and tell you there's no danger if you will follow their instructions carefully. In fact, they almost guarantee it."[123]

But many of the pilots seemed to have been neither worried at the prospect of risk nor excited at the prospect of glory. Pinson, for example, described the attitude of the pilot who flew his aircraft as "matter of fact."[124] And at Operation Teapot, Captain Paul M. Crumley, project officer for the early cloud penetrations, stated, "We consider these flights routine. Neither the pilots nor observers are unduly concerned over the fact that no one else has flown into an atomic cloud so soon after detonation."[125]

Decontamination Experiments

In conjunction with the Teapot cloud flythrough experiment, the military also conducted an experiment on ground crews "to determine how soon these same aircraft could be reserviced and made ready to fly again."[126] The Air Force used the contaminated aircraft from the early cloud-penetration experiment.[127] The research sparked a debate between the Air Force and the AEC over the costs and benefits of safety measures, a debate that was itself resolved by further experimentation.

In one part of the "experimental procedure," personnel (the number involved is not reported) rubbed their gloved hands over a contaminated fuselage, and in another part "the bare hand was also rubbed over a surface whose detailed contamination was known and a radioautograph of the hand surfaces [was] made."[128] None of the "survey team" exceeded the AEC's gamma exposure limit of 3.9 R.[129] Concluding that aircraft did not need to be "washed down" or decontaminated after they flew through the atomic clouds, Colonel William Kieffer, deputy commander of the Air Force Special Weapons Center, proposed

that decontamination procedures be eliminated except in extreme circumstances. This change in procedures might cause overexposures, Kieffer wrote, but they would be acceptable as long as "dangerous" dosages would be avoided.[130]

The proposal was not warmly received by the AEC. Los Alamos's Thomas Shipman complained that the goal should be to reduce exposures to zero.[131] Harold Plank, a Los Alamos scientist who was in charge of the cloud-sampling project and who rode along on many of the cloud-sampling missions, said, "Kieffer simply could not understand the philosophy which regards every radiation exposure as injurious but accepts minimum exposures for critical jobs."[132]

Kieffer suggested a compromise; test the proposal with only one or two sampler aircraft.[133] Plank objected, but the AEC test manager promised to "do everything possible to obtain a waiver of AEC operating radiological safety requirements."[134] The Air Force carried out the study during the 1957 Operation Plumbbob. An additional plane was flown through the atomic clouds created by five "events" to determine the hazard from the Air Force's proposed procedures.[135] The study showed that decontamination would be necessary to prevent overexposures at test sites.[136] In the end, the Air Force was unsuccessful in its attempt to change the decontamination procedures for sampler aircraft.

We do not know how the Air Force viewed this activity. Given that it did not treat the cloud flythroughs as an experiment, it is unlikely that the Air Force considered the ground personnel activity to be an experiment. There is no record of what the ground personnel were told or whether they were volunteers.

THE BOMB TESTS: QUESTIONS OF RISK, RECORDS, AND TRUST

In this chapter, the Advisory Committee reviewed six different activities that were conducted in conjunction with bomb tests that today we would consider research involving human subjects.[137] Only two of the six—the "atomic effects experiments" conducted on officer volunteers and the flashblindness experiments—were clearly treated as instances of

human research at the time. The six human research projects likely included no more than 3,000 of the more than 200,000 people who were present during the bomb tests.[138] Some of the research subjects, perhaps as many as several hundred, were placed at greater risk of harm than the other bomb-test participants who were not also research subjects. However, most of the research subjects were not. At this point, we turn to a consideration of several issues that affect all atomic veterans, regardless of whether they were also research subjects. These include how at the time the DOD and the AEC determined what exposures would be permitted, issues of record keeping, and what is known today about long-term risks and participation in the bomb tests.

AEC and DOD Risk Analysis for Exposure at Bomb Tests

In counseling human subject research at bomb tests, the Joint Panel on the Medical Aspects of Atomic Warfare stated that the research had to be performed under "safe conditions." What "safe" meant for all those exposed, both experimental subjects and other military participants at the bomb tests, was subject to arrangements between the AEC and the DOD.[139] While the military, of course, is responsible for the safety of its troops, the AEC had responsibility for the safe operation of the Nevada and Pacific sites at which the weapons were tested. "Secrecy," summarized Barton Hacker, a DOE-sponsored historian of the bomb tests, "so shrouded the test program . . . that such matters as worker safety could not then emerge as subjects of public debate."[140]

As we have seen in the case of the cloud flythrough research, by the mid-1950s the AEC and the Defense Department had arrived at a method of operation through which waivers to the basic radiation safety standards for the tests would be granted for particular activities. In the early 1950s, in the context of the Desert Rock exercises, the AEC and the DOD established the precedent for departure from the standards that the AEC relied on for its own bomb-test work force.

At this time the AEC was the main source of expertise on radiation effects. Its guidepost for

its own workers (at the Nevada Test Site and elsewhere) was the 3 R per thirteen-week standard established for occupational risk by a private organization (the National Committee on Radiation Protection). This level, it may be recalled from the debates on nuclear airplane experimentation (discussed in chapter 8), was well below that at which the experts assumed acute radiation effects, such as would limit combat effectiveness, could occur.[141]

In 1951, the Los Alamos Laboratory, the AEC's right hand in weapons test management, called on the Division of Biology and Medicine's director, Shields Warren, for "official but unpublicized authority to permit exposures up to 3.9r" for AEC test personnel.[142] Warren granted the request, counseling that "this Division does not look lightly upon radiation excesses."[143]

As we have seen, the DOD shortly thereafter determined to use the tests for troop maneuvers and did so at Desert Rock I, keeping the troops at seven miles distance during the detonation. In early 1952 the DOD asked the AEC to endorse its request to station troops at Desert Rock IV as close as 7,000 yards from ground zero (approximately four miles), far closer than the seven-mile limit the AEC permitted its own test-site personnel. The AEC's Division of Military Applications was willing to concur. Shields Warren, however, dissented on grounds of safety.[144] The dispute was settled when AEC Chairman Gordon Dean advised DOD that "the Commission would enter no objection to stationing troops at not less than 7000 yards from ground zero," provided that proper precautions were taken.[145]

Even so, an internal review of the Desert Rock IV exercise by the Division of Military Applications, generally supportive of DOD's request for troop maneuvers, raised questions about the wisdom of deviation from the AEC standard—and the potential for "delayed" casualties.[146]

Determined to proceed, DOD called for "a study to be made to determine the minimum distance from ground zero that should be permitted in a peacetime maneuver."[147] A December 1952 report recommended that dosages for Army personnel be above the limit set by the AEC for its personnel. The soldiers, by compari-

son with the AEC personnel, would be exposed "very infrequently." The report summarized the state of knowledge:

There is no known tolerance for nuclear radiation, that is, there is no definite proof that even small doses of nuclear radiations [*sic*] may not, in some way, be harmful to the human body. On the other hand, there is no evidence to indicate that, within certain limits, nuclear radiation has injured personnel who have been exposed to it.[148]

In response to the DOD's proposal to assume full responsibility for physical and radiological safety of troops and troop observers within the Nevada Test Site, the AEC stated that general safety practice and criteria at the Nevada Proving Grounds was, and must continue to be, the responsibility of the AEC. The AEC did, however, "accept the proposal that the DOD assume full responsibility for physical and radiological safety of troops and all observers accompanying troops within the maneuver areas assigned to Exercise Desert Rock V, including establishment of a suitable safety criteria." The AEC further explained that

The Atomic Energy Commission adopts this position in recognition that doctrine on the tactical use of atomic weapons, as well as the hazards which military personnel are required to undergo during their training, must be evaluated and determined by the Department of Defense.

The Atomic Energy Commission has, however, established safety limits. . . . We consider these limits to be realistic, and further, are of the opinion that when they are exceeded in any Operation, that Operation may become a hazardous one. So that we may know in which particulars and by how much these safety standards are being exceeded, we desire that the Exercise Director transmit to the Test Manager a copy of his Safety Plan. . . .[149]

For the spring 1953 Desert Rock V exercises, the DOD deemed the permissible limit for the troops (for a test series) to be 6 R.[150] In the case of the officer volunteers, a 10 R test limit was agreed to, with the proviso that "it is not intended that these exposures result in any injury to the selected individuals."[151] The Army's limit at Desert Rock was well below the level understood to potentially cause acute effects and far below the recommendation of Brigadier General James Cooney that the military depart from the

"infinitesimal" industrial and laboratory limits and accept 100 roentgens for a single-exposure limit.[152] But the level was not only higher than the AEC level but also above the 0.9 R per week being urged by the British and Canadians, partners in U.S. testing.[153] (The AEC itself objected that a 0.9 R-per-week limit would make testing at Nevada impractical.)[154]

Interestingly, in 1952 the Navy, also faced with the need for more-realistic training exercises, considered spraying radioactive materials on ships during training exercises. The Navy's Bureau of Medicine (BuMed) rejected the proposal. BuMed told the chief of naval operations that while it "fully appreciates" the need for more "realistic radiological defense training," it could not approve the use of radioisotopes in a form other than "sealed sources commonly used in basic training . . . since such use might produce an internal radiation hazard serious enough to outweigh the advantages of area contamination for training purposes."[155]

By the mid-1950s, AEC test health and safety staff were continually concerned about radiation safety at the tests and the failure to reduce them to a predictable and assuredly safe routine. "There are," Los Alamos Health Division leader Thomas Shipman wrote to the AEC Division of Biology and Medicine's Gordon Dunning in 1956, "two basic facts . . . which must never be lost sight of. The first of these is that the only good exposure is zero. . . . The second fact is that once the button for a bomb detonation is pushed you have to live with the results no matter what they are. . . ."[156] In fact, while the AEC had set a limit of 50 kilotons (more than twice the power of the Hiroshima and Nagasaki bombs) for Nevada tests, this limit had already been exceeded by 10 kilotons in 1953.[157] "It is all very nice," Shipman wrote in another 1956 memorandum, "to have a well-meaning Task Force commander who by a stroke of the pen can absolve our radiologic sins, but somehow I do not believe that overexposures are washed away by edict."[158] Shipman's comments illustrate an acute awareness among experts at the center of the testing program of the real and continuing element of risk and uncertainty in the attempt to define and control exposures at the bomb tests.

The Aftermath of Crossroads: Confidential Record Keeping to Evaluate Potential Liability Claims

In the midst of the Korean and Cold Wars, researchers and generals were focused on the short-term effects of radiation, not effects that might take place years later. Thus, the benefits from knowledge about the bomb, or training of troops in its use, loomed large, and the risks from long-term exposure likely seemed distant and small. Government officials undertook to guard against acute radiation effects; the surviving documentation indicates that they were remarkably successful. Of the more than 200,000 service participants in the tests, available records indicate that only about 1,200 received more than today's occupational exposure limit of 5 rem, and the average exposure was below 1 rem.[159] But there was no certainty that lower exposures were risk free.

During the summer of 1946, the contamination of ships at the Crossroads tests put officials and medical experts on alert to the radiation risk posed to participants at atomic bomb detonations. "[D]ifficult and expensive medico legal problems," Crossroads medical director Stafford Warren feared, "will probably occur if previously contaminated target ships are 'cleared' for constant occupancy or disposal as scrap."[160] A "Medico-Legal Advisory Board" sought to deal with these questions,[161] and the Navy created a research organization dedicated to the study of decontamination and damage to ships.[162]

Concern for long-term liability stimulated by Crossroads led to more steps to guard against the legal and public relations implications if service personnel exposed to radiation filed disability claims.

In the fall of 1946, General Paul Hawley, administrator of the Veterans Administration, "became deeply concerned about the problems that atomic energy might create for the Veterans Administration due to the fact that the Armed Services were so actively engaged in matters of atomic energy."[163] In August 1947 Hawley met with representatives of the surgeon general's offices of the military services and the Public Health Service.[164] The meeting was also attended by former Manhattan Project chief General Leslie Groves,[165] (Groves reportedly was "very much afraid of claims being instituted by men who participated in the Bikini tests.")[166] An advisory committee was created, which included Stafford Warren and Hymer Friedell, Warren's deputy on the Manhattan Project medical team. The committee was given the name "Central Advisory Committee," as "it was not desired to publicize the fact that the Veterans Administration might have any problems in connection with atomic medicine, especially the fact that there might be problems in connection with alleged service-connected disability claims."[167]

The committee recommended the creation of an "Atomic Medicine Division" of the VA to handle "atomic medicine matters" and a radioisotope section to "implement a Radioisotope Program."[168] The committee further recommended that "for the time being, the existence of the Atomic Medicine Division be classified as 'confidential' and that publicity be given instead to the existence of a Radioisotope Program."[169]

This history is contained in a 1952 report presented by Dr. George Lyon to the National Research Council.[170] The 1952 report records that "General Hawley took affirmative actions on these recommendations and it was in the manner described that the Radioisotope Program was initiated in the Fall of 1947."[171] Lyon, who had worked with Stafford Warren at Crossroads, was appointed special assistant to the VA's chief medical director for atomic medicine, and through 1959 served in a variety of roles relating to the VA's atomic medicine activities. Dr. Lyon's 1952 report recounts that he was present at the August 1947 meeting and involved in the deliberations of the Central Advisory Committee, as well as subsequent developments.[172]

Working with the VA and the Defense Department, we sought to retrieve what information could be located regarding the Atomic Medicine Division and any secret record keeping in anticipation of potential veterans' claims from radiation overexposures. Among the documents found was a Confidential August 1952 letter to the attention of Dr. Lyon, in which the Defense Department called for comment on the Army's proposal to "eliminate the requirement for maintaining detailed statistical records of radiological exposures received by the Army personnel."[173] The requirement, the letter recorded, "was originally conceived as being necessary to protect the government's interest in case any large number of veterans should attempt to bring suit against the government based on a real or imagined exposure to nuclear radiations during an atomic war."[174]

In 1959 Dr. Lyon was recommended for a VA "Exceptional Service Award."[175] In a memo from the VA chief medical director to the VA administrator, Dr. Lyon's work on both the publicized and confidential programs was the first of many items for which Dr. Lyon was commended. Following a recitation of the 1947 developments similar to those stated by Dr. Lyon in his 1952 report, the memo explained:

It was felt unwise to publicize unduly the probable adverse effects of exposure to radioactive materials. The use of nuclear energy at this time was so sensitive that unfavorable reaction might have jeopardized future developments in the field . . . [Dr. Lyon] maintained records of classified nature emanating from the AEC and the Armed Forces Special Weapons Project which were essential to proper evaluation of claims of radiation injury brought against VA by former members of the Armed Forces engaged in the Manhattan project.[176]

The Advisory Committee has been unable to recover or identify the precise records that were referred to in the documents that have now come to light. An investigation by the VA inspector general concluded that the feared claims from Crossroads did not materialize and that the confidential Atomic Medicine Division was not activated.[177] However, the investigation did not shed light on the specific identity of the records that were kept by Dr. Lyon, as cited in the 1959 memo on behalf of his commendation.[178] While mystery still remains, the documentation that has been retrieved indicates that prior to the atomic testing conducted in the 1950's, the government and its radiation experts had strong concern for the possibility that radiation risk borne by servicemen might bear longer-term consequences.

Looking Back: Accounting for the Long-Term Risks

Civilians, a UCLA psychologist observed during a 1949 NEPA meeting convened to consider the psychology of radiation effects, question "whether the medical group have actually discovered thus far all the effects of radiation on human beings . . . that is going to be one of the most insidious things to combat. . . ."[179] "[W]hen you talk about probable delayed effects possible, unknown, and so forth," Dr. Sells, of the Air Force, asked, "what is the proper evaluation of the ethical question as to how to treat the possible or probable unknown effects?"[180] While not answering the question, he observed that "certainly we can create more anxiety by being scientifically scrupulous than if we simply treated these matters as we are inclined to treat other matters in our every-day life."[181]

This may have been the case following Crossroads. "Now we are very much interested in long-term effects," a military participant in a 1950 meeting of the DOD Committee on Medical Sciences stated, "but when you start thinking militarily of this, if men are going out on these missions anyway, a high percentage is not coming back, the fact that you may get cancer 20 years later is just of no significance to us."[182]

Decades following the 1946 Crossroads tests, researchers began to study the longer-term effects of the bomb on test participants.

In 1980 the Centers for Disease Control (CDC) reported a cluster of 9 leukemias among the 3,224 (then identified) participants of shot Smoky at the Nevada Test Site in 1957.[183] A later report[184] increased the count of leukemias to 10 compared with 4.0 expected on the basis of U.S. rates, but found no excess cancers at other anatomical sites (the total observed was 112, compared with 117.5 expected). The Smoky test was the highest-yield tower shot ever conducted at the Nevada Test Site; however, the measured doses for the Smoky participants as a group were too low to explain the excess. Whether this cluster represents a random event, an underestimation of the doses for the few participants who got leukemia, or some other explanation remains unclear.

In light of the CDC research, the National Academy of Sciences (NAS) thereafter undertook an enlarged study of five series of nuclear tests totaling 46,186 (then identified) participants.[185] The 1985 NAS report confirmed the excess of leukemia at the Smoky test but found no such excess at any of the test series (as opposed to individual tests) and no consistent pattern of excesses at other cancer sites. Later, however, the NAS study was found to be flawed by the inclusion of 4,500 individuals who had never participated and the exclusion of 15,000 individuals who had participated in one or more of the five series, as well as incompleteness of dosimetry.[186]

The belated discovery that thousands of test participants had been misidentified punctuated the deficiencies in record creation and record keeping faced by those who seek to reconstruct, at many years' remove, the exposures of participants at the tests.

Documents long available, and those newly retrieved by the Committee, provide further basis for concern about the data gathering at test series in which human subject research took place. At the 1953 Upshot Knothole series, which included the Desert Rock V HumRRO research, 1994 DOD data show that only 2,282 of the 17,062 participants are known to have been issued film badges to serve as personal dosimeters.[187] At Desert Rock V, the Army surgeon general's policy that one-time exposure need not be reported led to a determination that maneuver troop units would be issued one film badge per platoon, and observers would be issued one per bus.[188] An AFSWP memo recorded that the Radiological Safety Organization did not have enough pocket dosimeters for efficient operation.[189] A recently declassified DOD memo records that "[a]lthough film badges on the officer volunteers [at Desert Rock V] indicated an average gamma dose of 14 roentgens, best information available suggests that the true dose was probably 24 rem initial gamma plus neutron radiation."[190]

In a 1995 report, the Institute of Medicine found that the dose estimates that were proposed for use in the NAS follow-up study were unsuitable for epidemiologic purposes, but concluded that it would be feasible to develop a dose-reconstruction system that could be used for this purpose. Nonetheless, there are some further studies that are of direct relevance.[191]

Recently, Watanabe et al.[192] studied mortality among 8,554 Navy veterans who had participated in Operation Hardtack 1 at the Pacific Proving Grounds in 1958. This is, to date, the only study of U.S. veterans to include a control group of unexposed military veterans. Overall, the participant group had a 10 percent higher mortality rate, but the cancer excess was significant only for the combined category of digestive organs (66 deaths compared with 44.9 expected, a 47 percent increase). On average, the radiation doses were low (mean 388 mrem), but among the 1,094 men with doses greater than 1 rem, there was a 42 percent excess of all cancers. No categories of cancer sites showed a significant excess or clear dose-response relationship, but the number of deaths in any category was small.

Two sets of foreign atomic veterans have been studied. In a study of 954 Canadian participants,[193] no differences with matched controls were found, but only very large effects would have been detectable in such a small study. In contrast, a large study of British participants of test programs in Australia found higher rates of leukemia and multiple myeloma than in a matched control group (28 vs. 6).[194] However, the cancer rates among the exposed veterans were only slightly higher than expected based on national rates, whereas those in the control group were much lower than expected, and there was no dose-response relationship. No excess was found at any other cancer site. Although the difference between the exposed and unexposed groups was quite significant, the interpretation of this result is unclear. Does it mean that for some unknown reason, soldiers are less likely than the general population to get cancer (the "healthy soldier effect," which is usually not thought to be so large for cancer), or is it an indication of some unexplained methodological bias? This point has never been resolved.

These observed effects need to be put in the context of what might reasonably be expected based on current understanding of low-dose radiation risks and the doses the atomic veterans are thought to have received. Approximately 220,000 military personnel participated in at least one nuclear test. The film badges for those monitored (thought to be roughly representative of all participants) average 600 mrem.[195] As summarized in "The Basics of Radiation Science" section of the Introduction, the consensus among scientific experts is that the lifetime risk of fatal cancer due to radiation is approximately 8 per 10,000 person-rem. On this basis, one might anticipate approximately 106 excess cancer deaths attributable to participation in the nuclear tests. Not only is this a number with considerable uncertainty, it is small in comparison with the total of about 48,000 cancer deaths that are eventually anticipated in this population.

Such a small overall excess would be virtually impossible to detect by epidemiologic methods. In some subgroups, however, the relative increase above normal cancer rates could be large enough to be detectable. Leukemia, for example, is proportionally much more radiosensitive than other cancers and the largest excess occurs fairly soon after exposure, when natural rates are low. Focusing on those with highest exposure would also enhance the relative increase, albeit with many fewer people at risk. The Defense Nuclear Agency estimates that about 1,200 veterans received more than 5 rem (mean 8.1 rem).[196] On this basis, about eight excess cancer deaths would be anticipated. These factors may have contributed to the observed leukemia excess among participants of shot Smoky, for example.

Although these numbers represent the best estimate currently available of the expected cancer excess, there are uncertainties in both the real exposures received by the participants and the magnitude of the low-dose risk. As described in "The Basics of Radiation Science" section, there is roughly a 1.4 uncertainty in the low-dose radiation risk coefficient simply due to random variation in the available epidemiologic data, with additional uncertainties of unknown magnitude about model specification, variation among studies, extrapolation across time and between populations, unmeasured confounders, and so on. These uncertainties are hotly contested, although the majority of radiation scientists believe the figures quoted above are unlikely to be seriously in error. If low-dose radiation risks were indeed substantially higher than this, then there would be a serious discrepancy to explain with the effects actually observed at higher doses. The uncertainties in the doses received by participants are perhaps more substantial, but given the limitations in the dosimetry and record keeping, it may be difficult ever to resolve them.

As is clear from the epidemiologic data available today, there is no consistent pattern in increased cancer risk among atomic veterans, although there are a number of suggestive findings, most notably the excesses of leukemia among shot Smoky and British test participants, the causes of which are still unclear. The low recorded doses, the small size of the expected excesses, and problems in record keeping and dosimetry make it very difficult to resolve whether atomic veterans as a group are at substantially elevated cancer risk and whether any such excess can be attributed to their radiation exposures. The Advisory Committee debated at some length the merits of further epidemiologic studies and concluded that the decisions to conduct such studies should be made by other appropriately constituted bodies of experts.

Looking Back: The Legacy of Distrust

The chain of events set in motion by the CDC research, and renewed interest in the fate of the "atomic vets," led to congressional enactment of legislation that provides veterans exposed at atmospheric tests with the opportunity to obtain compensation for injury related to radiation exposure.

The Veterans Dioxin and Radiation Exposure Compensation Standards Act of 1984 provides for claims for compensation for radiation-related disabilities for veterans exposed at atmospheric tests. The Radiation Exposed Veterans Compensation Act of 1988 provides that a veteran who was exposed to radiation at a designated event and develops a designated disease may be entitled to benefits without having to prove causation.[197]

Notwithstanding the passage of this legislation, the Committee heard from many atomic veterans, and their widows, who complained that the records that were created and maintained by the government—records on which veterans' claims may stand or fall—were inadequate, missing, or wrong.[198] Atomic veterans also stated that the laws and rules do not adequately reflect the kinds of illnesses that may be caused by radiation, that they do not provide for veterans who were exposed to radiation in settings other than atmospheric tests, and that the practical difficulties—in time and resources—of pursuing their rights under the laws are often excessive. The Committee heard from many who told of the time, expense, and difficulty of getting information on the full circumstances of bomb-test exposures. They told of their continued efforts, over the course of the years, to reconcile what they have learned from government sources with that which they have been told by other test participants, that which they recovered from the private letters of test participants to family members, and their own further research.

For numerous atomic veterans, the testimony was not simply that the bomb tests themselves had been large experiments, but that they had been put at risk in the absence of planning to gather the data and perform the follow-up studies needed to ensure that the risks of the unknown, however small, would be measured and adequately accounted for.

CONCLUSION

The story of human research conducted in connection with nuclear weapons tests illustrates the difficult questions that are raised when human research is conducted in an occupational setting, especially a setting, such as the military, where exposure to risk is often part of the job. The story illustrates that it may often be difficult to discern whether or not an activity is a human experiment. By the same token, it also illustrates the importance of guarding participants against unnecessary risks, whether or not the activity is a human experiment.

Human experiments at atomic bomb tests were undertaken by the military, which had a long tradition of requiring voluntary consent from participants in biomedical experiments. The need for written consent in experiments related to atomic, biological, and chemical warfare was clearly stated in the secretary of defense's 1953 memorandum. That memorandum also required the approval in writing of the appropriate service secretary and precluded experiments that did not adhere to its further requirements. The 1953 memorandum, however, does not appear to have been transmitted to those involved in human research at bomb tests, although the tenet of voluntary consent was followed in some cases. In addition to consent, the 1953 memorandum contained other significant ethical requirements, including that research be reasonably likely to produce useful scientific results and that proper precautions be taken to minimize risk.

The bomb-test research illustrates the significance of the position that bad science is bad ethics. Unless a research project is scientifically defensible, there is no justification for imposing on human subjects even minimal risk or inconvenience. For example, the DOD's biomedical advisers advocated the conduct of psychological and physiological research on troops participating in bomb tests with an awareness that the likelihood of scientifically useful results was small and that the effort would be part of a larger exercise in indoctrination and training. Having done so, they had an obligation to at least review continued research efforts to determine if the research design was developing useful information. In the case of the psychological and physiological testing, the evidence indicates that early results showed that the research design was not likely to produce useful scientific information, if only because the military, the researchers, and perhaps even the subjects did not view the setting as sufficiently realistic.

At the same time, this question of ethics and science is irrelevant if the HumRRO activities did not entail research involving human subjects. An activity that has a poor research design would not be an ethical human experiment. However, the same activity might be ethical if conducted as a training activity whose essential purpose is to provide reassurance. Similarly, to the extent that research was intended solely to provide reassurance, ethical questions arise that might not be present if the activity were not experimental.

Just what makes something an instance of research involving human subjects? The answer to this question is not discoverable; instead, it is fashioned by people in particular contexts for particular purposes. Today, we would likely consider all the activities reviewed in the first part of this chapter—the HumRRO testing, the "atomic effects experiment," the flashblindness experiments, the cloud flythroughs, and the protective clothing and decontamination tests—to

be cases of research involving human subjects to which the current federal regulations and the current rules of research ethics would apply. Some of these activities are, nevertheless, more paradigmatically instances of human research than others. Depending on the context, for example, the protective clothing and decontamination tests might be considered within the normal course of duty for military personnel.

One of the reasons it is important to be able to distinguish research involving human subjects from other activities is that military policy clearly states that service personnel may not be ordered to be human subjects. In contrast to much else in military service, participation in research is a discretionary activity that service personnel are permitted under military policy and federal regulation to refuse. Thus, in the military as elsewhere, human subjects are supposed to be volunteers whose valid consent has been obtained.

Human subject research is not the only activity in the military, however, for which consent is a requirement. The military also often asks for volunteers in settings where the risk is unusually great. For example, the testing of equipment may often be hazardous, may involve the use of volunteers, but may not be considered human re-search. Thus, in the case of test pilots, there may be significant risk, volunteers may be called for, but the activity might not be considered research with human subjects and thus would not be thought subject to human use research regulations.

Conversely, a requirement of consent may not necessarily mean that subjects have some measure of control over the risks to which they are to be exposed. Even under today's rules, informed consent in the HumRRO tests would be limited to the psychological and physiological testing, and not required for participation in the bomb test itself.

Whether the activity is research involving human subjects or an unusually risky assignment that is not considered human subject research, how free are military personnel to accept or refuse offers (as opposed to orders) put to them? Dr. Crawford, when asked to comment in 1994 on consent in his HumRRO research, responded by observing that "military service people generally do what they're asked to do, told to do." He was speaking of an army that included many conscripts; today's all-volunteer military is doubtless different in many respects that bear on questions of voluntariness. Nevertheless, the culture of the military, with its emphasis not only on following orders but on the willingness to take risks in the interests of the nation, surely influences and in some circumstances may restrict how service personnel respond to such offers.

Because in the military volunteering is often seen as a matter of duty and honor, and the boundaries between experimental and occupational activities may not be clear, the importance of minimizing risk emerges as a central concern. Above all, the activities discussed in this chapter confirm that the ethical requirement that risks to service personnel be minimized should not depend on whether an activity is characterized as an experiment or occupational. In the case of the atomic veterans, the risks run were usually no different for those who were subjects of research and those who were not.

The military took precautions, with great success, to preclude exposure to radiation at levels that might produce acute effects. However, bomb-test participants were exposed to lesser, long-term risks without adequate provision for (1) the creation and maintenance of records that might be needed, in retrospect, to determine the precise measure of risks to which military personnel were exposed; and (2) the tracking of those exposed to risk, so that follow-up and assurance, as needed, could be efficiently undertaken.

It might be argued that, at the time, there was no awareness of a potential for long-term risk, or that the potential was understood to be nonexistent. But, while the possibility of long-term risk from low exposures was seen as low, it was not seen as nonexistent. Following the 1946 Crossroads tests, officials and experts connected with the DOD, AEC, and VA thought action was needed to collect data in secret to evaluate potential disability claims.

Since the bomb tests, the Defense Department has come to recognize the importance of providing for an independent risk assessment when service personnel may be exposed to new weapons—regardless of whether the exposure is classed as experimental or occupational.[199]

However, for the numerous atomic veterans (and their family members) who spoke to the Committee, a continuing source of distress is not simply that the government put service personnel at risk but that, having undertaken to do so, the government did not undertake to collect the data and perform the follow-up that might provide them knowledge and comfort in later years. The Advisory Committee agrees. When the nation exposes servicemen and women to hazardous substances, there is an obligation to keep appropriate records of both the exposures and the long-term medical outcomes.

From listening to those who appeared before us, and from reflection on the laws that are already in effect, the Committee came to appreciate that there are several reasons record keeping is important. First, those who served, and their widows and surviving family members, have a great interest in knowing the facts of service-related exposures. We repeatedly heard from veterans and family members whose inquiries into the circumstances and details of exposures has spanned many years. Second, information may provide basis for scientific analysis that may shed light on the relation between exposure to risk and subsequent disability or disease. Third, where disability or disease appears to be a possible result of exposure, data are needed to provide the basis for a fair and efficient system of remedies.

The experience of the bomb-test participants indicates that several different kinds of records or data should be of use. First, of course, there are data about the exposure of individual service personnel to particular potential hazards. In the case of the atomic bomb tests, the potential that radiation would be a hazard was, of course, obvious. In addition, radiation is a phenomenon that is almost uniquely susceptible to measurement. In other settings faced by service personnel, the precise nature of the hazard may not be easily anticipated or, even if anticipated, readily measurable. Second, there are data concerning the location of service personnel. In the case of the bomb tests, as we have seen, data on the identity and location of all test participants (so that their position in relation to the putative hazard can be retrospectively reconstructed, if need be) were not readily available. Even if the hazard cannot be anticipated, such data can be useful in later efforts to reconstruct the nature of the hazard and its effect. Third, the maintenance of complete medical records, including linkages where multiple sets of records exist, is essential. Records suitable for use in epidemiologic studies of long-term medical consequences of military actions would be valuable for both medical science and service members.

But having heard from many atomic vets and their family members, the Advisory Committee does not believe that, but for the inadequate record keeping and lack of follow-up, there would be no anger or disappointment among atomic veterans and their families. The real offense to many is the belief that the risk was unacceptable and that they or their loved ones may suffer illness unnecessarily as a consequence. Proper attention to record keeping should provide some basis for gaining and assuring the trust of those who are exposed to risk in the future and, perhaps, scientific results that may be of real value to them, but it is hardly a guarantee against perceptions of abuse or unfairness.

If nothing else, our experience makes us appreciate the difference between technical, analytic data and the reality of the human experience. The available data, as we have discussed, indicate that the *average* amount of radiation to which bomb-test participants were exposed was low. But those who believe they have suffered as a consequence of these exposures do not believe these risks to have been as slight as the data indicate. When we review this decades later, we rely on numbers; the atomic veterans and their family members who have appeared before the Committee associate, in a "cause and effect" way, the exposure with some kind of result that they have personally experienced or witnessed. The emotions and concerns expressed to the Committee by these citizens (and those downwind from atomic tests and intentional releases) were very, very real. Both the public and the scientific community must understand that, when data indicate that risks are low, the risks are not necessarily zero; and it is possible for a rare event to occur. The risk analysis may only indicate that it is unlikely that such events will occur with significant frequency or probability.

NOTES

1. See Department of Energy, *Announced United States Nuclear Tests: July 1945 Through December 1992* (Springfield, Va.: National Technical Information Service, May 1993); Department of Energy, *Expanded Test Information for Nuclear Tests With Unannounced Simultaneous Detonation* (Springfield, Va.: National Technical Information Service, 20 June 1994).

2. Thomas L. Shipman, Los Alamos Laboratory Health Division Leader, to Dr. Harry G. Ehrmentraut, Committee on Medical Sciences, Research and Development Board, Department of Defense, 25 July 1951 ("Dr. Robert Grier has passed on to me . . .") (ACHRE No. DOE-033195–B), 2.

3. See the July 1949 transcript of a meeting convened by the NEPA [Nuclear Energy for the Propulsion of Aircraft] Medical Advisory Panel to discuss the "Psychological Problem of Crew Selection Relative to the Special Hazards of Irradiation Exposure," 27. NEPA Medical Advisory Panel, Subcommittee IX, proceedings of 22 July 1949 (ACHRE No. DOD-121494–A-2).

4. Richard L. Meiling, Chairman, Armed Forces Medical Policy Council, to the Deputy Secretary of Defense et al., 27 June 1951 ("Military Medical Problems Associated with Military Participation in Atomic Energy Commission Tests") (ACHRE No. DOD-122794–B), 1.

5. Ibid.

6. Ibid.

7. Department of the Army, September 1950 ("Atomic Energy Indoctrination") (ACHRE No. DOD-020395–D), 73.

8. Ibid.

9. Richard L. Meiling, Chairman, Armed Forces Medical Policy Council, to the Chairman, Research and Development Board, 23 February 1951 ("Department of Defense Biomedical Participation in Atomic Weapons Tests") (ACHRE No. NARA-071194–A), 1.

10. The Joint Panel was created in 1949 by the Committee on Medical Sciences and the Committee on Atomic Energy, which were committees of the Research and Development Board. (See the Introduction and chapter 1 for further discussions of the Joint Panel.)

11. The agenda noted that while civilians were polled in the preparation of the draft, "very few" responded. The draft was therefore " offered not as a proposed statement, to be accepted after only minor revisions, but as a general guide to the type of paper which is expected of the Joint Panel." Joint Panel on the Medical Aspects of Atomic Warfare, 20 September 1951 ("Agenda, 8th Meeting, Item 3—Preparation of Statement on Biomedical Participation in Future Weapons Tests") (ACHRE No. DOD-072294–B), 1–2.

12. Joint Panel on the Medical Aspects of Atomic Warfare, 20 September 1951 ("Biomedical Participation in Future Atomic Weapons Tests [Attachment to Agenda, 8th Meeting]") (ACHRE No. DOD-072294–B), 2. The quoted language appears to have come from Dr. Thomas Shipman of Los Alamos. See Thomas Shipman, Los Alamos Laboratory Health Division Leader, to Shields Warren, Director, AEC Division of Biology and Medicine, 15 September 1951 ("Permissible Exposures, Test Operations") (ACHRE No. DOE-120894–C).

13. The draft stated a concern that "actual animal exposures should be limited as much as possible," but did not expressly address human experimentation. Joint Panel on the Medical Aspects of Atomic Warfare, 20 September 1951, ("Biomedical Participation in Future Atomic Weapons Tests"), 3.

14. Ibid., 5–7.

15. We discuss the data gathering on radioisotopes in the body fluids in chapter 13, in the context of a discussion of secret human data gathering on fallout.

16. Joint Panel on Medical Aspects of Atomic Warfare, 20 September 1951 ("Program Guidance Report") (ACHRE No. DOD-072294–B), 23.

17. Ibid., 20. A further section on "Psychological Studies" recommended the following:

4.1.3 Continue studies in psychology of panic.

4.1.4 Seek technics [sic] for reducing apprehension and for producing psychologic resistance to fear and panic, especially in presence of radiation hazard (emotional vaccination).

4.1.5 Spread knowledge of radiation tolerance, technics [sic] of avoidance, and possibility of therapy through military and civilian populations and measure their acceptance.

4.1.6 Prepare to make psychologic observations at and after bomb tests.

Ibid., 14.

18. Colonel Michael Buckley, Acting Deputy Assistant Chief of Staff, Research and Development, to Chief of Army Field Forces, Fort Monroe, Virginia, 20 August 1951 ("Proposed Study of Behavior of Troops Exposed to A-Bomb") (ACHRE No. NARA-013195–A), 1.

19. Peter A. Bordes et al., February 1953 ("DESERT ROCK I: A Psychological Study of Troop Reactions to an Atomic Explosion [HumRRO-TR-1]") (ACHRE No. CORP-111694–A), 3.

20. Dr. Meredith Crawford, interview by Dan Guttman and Patrick Fitzgerald (ACHRE), transcript of audio recording, 1 December 1994 (ACHRE Research Project Series, Interview Program File, Targeted Interview Project), 6–7. Dr. Crawford was recruited to head the new HumRRO by psychologist Harry Harlow, an Army adviser who was famed for his work with monkeys. HumRRO was a private contractor created at the Army's behest and initially

affiliated with George Washington University. In the 1951 experiments, HumRRO worked with the Operations Research Organization (ORO), which was affiliated with Johns Hopkins University.

21. In 1994, Dr. Crawford prepared a retrospective memorandum titled "HumRRO Research During Four Army Training Exercises." Based on the 1953 report, "A Psychological Study of Troop Reactions to an Atomic Explosion," Dr. Crawford estimated that 633 service personnel were involved in the maneuvers at Desert Rock I. Meredith P. Crawford to William C. Osborn, 27 January 1994 ("HumRRO Research During Four Army Training Exercises Involving Atomic Weapons—1951–1957") (ACHRE No. CORP-112294–B), 8. In addition, hundreds of additional troops were involved as the "nonparticipant" group (which did not attend the test maneuvers, but was given psychological tests). The "experimental paradigm" for the HumRRO tests is described in this 1994 memorandum. Ibid., 4.

22. "Armed Forces: Exercise Desert Rock," *Time*, 12 November 1951, 21–22.

23. *New York Times*, 1 November 1951, 4.

24. Bordes et al., "DESERT ROCK I: A Psychological Study of Troop Reactions to an Atomic Explosion," 6.

25. Interview with Crawford, 1 December 1994, 57.

26. Bordes et al., "DESERT ROCK I: A Psychological Study of Troop Reactions to an Atomic Explosion," 5.

27. Ibid., 103.

28. Ibid., 107–108.

29. Interestingly, the troops evidently did not buy the "correct" answer; only about 40 percent of the troops at the maneuver were reported to have been correctly indoctrinated. Bordes et al., "DESERT ROCK I: A Psychological Study of Troop Reactions to an Atomic Explosion," 130.

30. The Committee asked the DOD to review the 1951 questionnaire and comment on whether the information presented regarding the effect of an airburst is, based on DOD's current expert understanding, still correct. DOD provided "changes/corrections/clarifications" on nine items. In the case of items 1 and 6, quoted in the text, DOD commented:

1) As stated, the answer is wrong. The ground zero hazard 1 day after an atomic explosion depends on the yield. At 20 kt, there would be no fallout for a burst at 2000 feet, but there would be induced activity. . . .

6) There is the same problem with this answer as with 1, above.

In one case the DOD reported that the 1951 questionnaire erred on what might be called the side of caution; where a 1951 answer stated that a posited detonation would not kill anybody beyond the range of three miles, the answer today would be one mile.

Department of Defense, Radiation Experiments Command Center, 26 April 1995 ("ACHRE Request 032795–A, HumRRO Questionnaire and Air Burst Material") (ACHRE No. DOD-042695–A), 1.

31. Colonel R. G. Prentiss, Executive Officer, Office of the Army Surgeon General, to Chief, Army Field Forces, Fort Monroe, Virginia, 9 January 1952 ("Psychiatric Research in Connection with Atomic Weapon Tests Involving Troop Participation") (ACHRE No. DOD-080594–A), 1. The memo recorded:

l. For your information in connection with planning for future exercises and operations in which atomic weapons tests will be used and troops will participate, this office has a continuing interest in the conduct of psychiatric observations regarding the effects of the weapons on the participating troops.

2. Funds for the conduct of psychiatric observations which may be approved for future atomic weapons tests will be made available through the Surgeon General.

The memorandum bears concurrences from the "Medical Research and Development Board," "Medical Plans and Operations," "Fiscal Division," and "Chief, Psychiatry and Neurology Consultants Division." It is not clear what role Army psychiatrists (i.e., medical doctors) played in the implementation of the "psychological" experiments.

32. Major P. C. Casperson, for the Chief of Army Field Forces, to First Army et. al., 7 March 1952 ("Extracts, Final Report Exercise DESERT ROCK I") (ACHRE No. NARA-013195–A), 122. In an age of "polls and questionnaires," the report suggested, the overpsychologized troops may have been putting the psychologists on:

The psychological evaluators, of whom there were many and various, were perhaps too obvious and eager. This is an era of polls and questionnaires and here was a new and untried field with unlimited possibilities. The ultimate response, finally, was a humorous and deliberate program in the troops to confuse the psychological people with fictitious reactions.

Ibid.

33. Brigadier General A. R. Luedecke, AFSWP, to Director, AEC Division of Military Application, 7 March 1952 ("Reference is made to your letter of 28 December 1951 . . .") (ACHRE No. NARA-010495–A), 2.

34. Interview with Crawford, 1 December 1994, 12–13.

35. Dr. Crawford's 1994 reconstruction of events estimated that 672 soldiers witnessed the shot, and 914 served in the control group as nonparticipants. Crawford to Osborn, 27 January 1994, 10.

36. Motivation, Morale, and Leadership Division, Department of the Army, August 1953 ("Desert Rock IV: Reactions of an Armored Infantry Battalion to an Atomic Bomb Maneuver [HumRRO-TR-2]") (ACHRE No. CORP-111694–A), *ix*, 17.

37. Benjamin W. White, 1 August 1953 ("Desert Rock V: Reactions of Troop Participants and Forward Volunteer Officer Groups to Atomic Exercises") (ACHRE No. CORP-111694–A), 10.

38. Department of the Army, "Desert Rock IV: Reactions of an Armored Infantry Battalion to an Atomic Bomb Maneuver [HumRRO-TR-2]," 72.

39. "Armed Forces: Exercise Desert Rock," *Time,* 12 November 1951, 22. At Desert Rock I, physiological testing, including the use of a polygraph, sought to measure anxiety before and after D-Day. Bordes et al., "DESERT ROCK I," chapter 6. At Desert Rock IV, before and after "sweat tests" measured troops' hand sweating as a possible index of fear. Department of the Army, "Desert Rock IV: Reactions of an Armored Infantry Battalion to an Atomic Bomb Maneuver [HumRRO-TR-2]," 10.

40. Joint Panel on the Medical Aspects of Atomic Warfare, 9 September 1952 ("Minutes: 9, 10, 11 and 12 September 1952, Los Alamos Scientific Laboratory") (ACHRE No. DOD-072294–B), 3–4. The panel's statement was in the form of a motion to be transmitted to the DOD Research and Development Board's Committee on Human Resources, to which the advisory role on the HumRRO effort was being turned over.

41. The available research reports do not indicate the numbers of participants in the research.

42. Defense Nuclear Agency, 8 August 1995 ("Atmospheric Test Series/Activities Matrix") (ACHRE No. DOD-081195–A).

43. Robert D. Baldwin, March 1958 ("Staff Memorandum: Experiences at Desert Rock VIII") (ACHRE No. CORP-111694–A). Also at Plumbbob was an experiment to test the efficiency of fallout shelters. Sixteen men were confined in four shelters to collect fallout samples, so that their ability to collect samples could be studied and so that they could be studied for the psychological effect of confinement. The study concluded that the shelters were well suited for both manned stations at nuclear weapons tests and for single-family fallout shelters. J. D. Sartor et al., 23 April 1963 ("The Design and Performance of a Fallout-Tested Manned Shelter Station and Its Suitability as a Single-Family Shelter [USNRDL-TR-647]") (ACHRE No. CORP-032395–A). See also Nevada Test Organization, Office of Test Information, 24 July 1957 ("For Immediate Release") (ACHRE No. DOE-033195–B); Nevada Test Organization, Office of Test Information, 15 July 1957 ("For Immediate Release") (ACHRE No. DOD-030895–F).

44. Baldwin, "Experiences at Desert Rock VIII," 39.

45. Ibid., 12. The troops were not to be told the amount of contamination present, which would depend upon actual fallout amounts. The course was

marked by radiation hazard markers, which might or might not reflect the actual fallout. Ibid., 23.

46. Ibid., 7.

47. Interview with Crawford, 1 December 1994, 52.

48. Bordes et al., "DESERT ROCK I: A Psychological Study of Troop Reactions to an Atomic Explosion," 20.

49. Crawford to Osborn, 27 January 1994, 15.

50. CG Cp Desert Rock to CG Sixth Army, 28 March 1953 ("Reference message G3, OCAFF No. 423") (ACHRE No. NARA-013195–A), 1. The office volunteers participated in three detonations in the 1953 "Upshot-Knothole" series—shots Nancy, Badger, and Simon. "DNA Fact Sheet Operation Upshot-Knothole," January 1992.

51. Captain Robert A. Hinners, USN, Headquarters, Armed Forces Special Weapons Project, 25 April 1953 ("Report of Participation in Selected Volunteer Program of Desert Rock V-7") (ACHRE No. DOE-033195–B), 2.

In an 11 February 1953 letter, the Army informed the Congressional Joint Committee on Atomic Energy of the "steps being taken by the Army in connection with exposure of troops at tests of atomic weapons." Lieutenant General L. L. Lemnitzer, Deputy Chief of Staff for Plans and Research, to Honorable Carl T. Durham, House of Representatives, 11 February 1953 ("The Secretary of the Army has asked that the Joint Committee . . .") (ACHRE No. NARA-112594–A), 1. The Army explained that deployment in foxholes at as close to 1,500 yards was needed to confirm that commanders could risk troops at this distance. The Army assured the committee that experts deemed it "highly improbable that troops will suffer any injury under these conditions." Ibid., 2.

Assurance was given to Congress that no more than twelve volunteers would be used at one shot. G3 DEPTAR, to CG Cp Desert Rock, 15 April 1953 ("Reference your msg AMCDR-DPCO 0498") (ACHRE No. NARA-013194–A), 1.

52. Brigadier General Carl H. Jark, for the Assistant Chief of Staff, Organization and Training Division, to Distribution, 20 February 1953 ("Instructions for Positioning DA [Department of Army] Personnel at Continental Atomic Tests [Attachment to 20 February 1953 memo]") (ACHRE No. NARA-120694–A), 2.

53. White, "Desert Rock V: Reactions of Troop Participants and Forward Volunteer Officer Groups to Atomic Exercises," iii.

54. CG Sixth Army Presidio of SFran Calif, to OCAFF Ft Monroe Va, 24 April 1953 ("Reference Desert Rock msg AMCDR-CG-04237") (ACHRE No. NARA-013195–A), 1.

55. Major R. C. Morris, for the Commanding General, to Chief of Research and Development, 15 November 1955 ("Amendment to Proposed Project Regarding Blast Injury Evaluation") (ACHRE No. DOD-030895–F), 1.

56. Major Benjamin I. Hill, for the Director, Terminal Ballistics Laboratory, to Chief, Armed Forces Special Weapons Project, 13 December 1955 ("Amendment to Proposed CONARC Project Regarding Blast Injury Evaluation") (ACHRE No. DOD-030895–F), 1.

57. Colonel Irving L. Branch, for the Chief, AFSWP, to Chief of Research and Development, Department of the Army, 20 January 1956 ("Annex 'A' to 2nd Endorsement: Detailed Explanation of AFSWP Comments on Feasibility of Human Volunteer Program") (ACHRE No. DOD-030895–F), 1.

58. Ibid.

59. Colonel Irving L. Branch, for the Chief, AFSWP, to Chief of Research and Development, Department of the Army, 20 January 1956 ("Amendment to Proposed Project Regarding Blast Injury Evaluation") (ACHRE No. DOD-030895–F), 1–2.

60. National Military Establishment, Military Liaison Committee, to the Atomic Energy Commission, 24 March 1949 ("Planning for 1951 Atomic Bomb Tests") (ACHRE No. DOE-120894–C).

61. Howard Brown to Shields Warren, 20 August 1951 ("Larry Tuttle advised that he had learned from his agents in AFSWP . . .") (ACHRE No. DOE-040395–A), 1.

62. Thomas L. Shipman, Los Alamos Health Division Leader, to Shields Warren, Director, AEC Division of Biology and Medicine, 20 January 1952 ("Since Wright's return from the meeting in Washington . . .") (ACHRE No. DOE-120894–C), 2.

63. Thomas Shipman, Los Alamos Laboratory Health Division Leader, to Charles Dunham, Director, AEC Division of Biology and Medicine, 9 June 1956 ("This is a rather belated reply . . .") (ACHRE No. DOE-120894–C), 1. In response to the suggestion that Los Alamos participate in another effort, Shipman urged that the committee should

either be given some real responsibility or will at least be able to speak in a loud, strong voice against any proposed program which appears to be poorly or inadequately planned . . . or which appears to be an out and out waste of the taxpayers' money.

Ibid.

64. T. L. Shipman to Alvin Graves, 9 August 1952 ("Meeting of Biomedical Test Planning and Screening Committee") (ACHRE No. DOE-120894–C), 1. DOD records show flashblindness research at Buster-Jangle (1951), Tumbler-Snapper (1952), Upshot-Knothole (1953), Plumbbob (1953), Hardtrack II (1958), and Dominic I (1962), in Defense Nuclear Agency, 8 August 1995, "Atmospheric Test Series/Activities Matrix."

65. The topic of the bomb's effect on vision merited instruction. The 1951 HumRRO questionnaire included: "Watching an A-bomb explode five miles

away can cause permanent blindness. (False)." Bordes et al., "DESERT ROCK I: A Psychological Study of Troop Reactions to an Atomic Explosion"), 109. In a 1995 comment on this question, DOD noted that "[i]n the strictest sense the correct answer is 'true'. Some permanent retinal damage will occur, but complete vision loss will not." Department of Defense, Radiation Experiments Command Center ("ACHRE Request 032795–A, HumRRO Questionnaire and Air Burst Material"), 1.

66. Colonel Victor A. Byrnes, USAF (MC), 15 March 1952 ("Operation BUSTER: Project 4.3, Flash Blindness") (ACHRE No. DOD-121594–C-4), 2.

67. The objectives were

(a) To evaluate the visual handicap which might be expected in military personnel exposed, during daylight operations, to the flash of an atomic detonation.

(b) To evaluate devices developed for the purpose of protecting the eye against visual impairment resulting from excessive exposure to light.

Ibid., 1.

68. J. C. Clark, Deputy Test Director, to Colonel Kenner Hertford, Director, Office of Test Operations, 5 March 1952 ("Attached is an outline of approved Project 4.5 . . .") (ACHRE No. DOE-020795–C), 1. The letter noted that at Buster-Jangle the AEC had sought and received "release of AEC responsibility" in the event of such damage and requested the same release for Tumbler-Snapper.

69. Defense Nuclear Agency, 1952 ("Operation Tumbler-Snapper") (ACHRE No. DOD-102194–C), 92.

70. Ibid.

71. Colonel Victor A. Byrnes, March 1953 ("Operation Snapper, Project 4.5: Flash Blindness, Report to the Test Director") (ACHRE No. DOD-121994–C), 12.

72. Ibid., 15.

73. Ibid. The DOD reported that it does not have the ability to retrieve the names of experimental subjects. Thus, the long-term outcome of those involved in flashblindness tests (estimated by DOD to approximate 100) is not known to the Committee.

74. Colonel Victor A. Byrnes, USAF (MC), et al., 30 November 1955 ("Operation Upshot-Knothole, Project 4.5: Ocular Effects of Thermal Radiation from Atomic Detonation—Flashblindness and Chorioretinal Burns") (ACHRE No. DOD-121994–C), 3.

75. Ibid.

76. Colonel Irving L. Branch, USAF, Acting Chief of Staff, to Assistant Secretary of Defense (Health and Medicine), 5 March 1954 ("Status of Human Volunteers in Bio-medical Experimentation") (ACHRE No. DOD-042595–A), 2.

77. Ibid., 3.

78. Colonel John Pickering; interview by John Harbert and Gil Whittemore (ACHRE), transcript of audio recording, 2 November 1994 (ACHRE Research Project Series, Interview Program File, Targeted Interview Project), 55. DOD did not locate any documents showing written consent.

79. Defense Atomic Support Agency, 15 August 1962 ("Operation Plumbbob: Technical Summary of Military Effects, Programs 1–9") (ACHRE No. DOD-100794–A), 137.

80. Defense Nuclear Agency, 1962 ("Operation Dominic I: Report of DOD Participation") (ACHRE No. DOE-082294–A).

81. John R. McGraw, Deputy Commandant, USAF, to Director, AEC, 20 March 1954 ("Examination of the *Retina* of Individuals Exposed to Recent Atomic Detonation") (ACHRE No. DOE090994–C). The memorandum stated that it "can be assumed that all persons who viewed the actual fireball" of a recent hydrogen bomb test "without eye protection have received permanent chorio-retinal damage." The memorandum went on to recommend that "[p]opulations and observers within an approximate radius of 100 miles from ground zero should be surveyed."

82. See, for example, Byrnes, "Operation Snapper, Project 4.5," 16–17.

83. Roy B. Snapp, Secretary, AEC, minutes of meeting no. 623, 6 November 1951 (ACHRE No. DOE-033195–B), 526.

84. Defense Nuclear Agency, 23 June 1982 ("Shots Sugar and Jangle: The Final Tests of the Buster-Jangle Series") (ACHRE No. DOE-082294–C), 46.

85. John R. Hendrickson, July 1952 ("Operation Jangle, Project 6.3–1: Evaluation of Military Individual and Collective Protection Devices and Clothing") (ACHRE No. DOE-121594–C-14), 5.

86. Ibid.

87. Ibid., 5, 20.

88. Ibid., 19.

89. Ibid., v.

90. U.S. Naval Radiological Defense Laboratory, 26 May 1958 ("Supplement [1] to AEC-313 [2–57] USNRDL") (ACHRE No. DOD-091494–A), 1.

91. Lieutenant Colonel Gordon L. Jacks, CmlC Commanding, to TSG, DA, 12 April 1963 ("Beta Hazard Experiment Using Volunteer Military Personnel") (ACHRE No. DOD-122294–B), 1.

92. Commanding Officer and Director, U.S. Naval Radiological Defense Laboratory, to Secretary of the Navy, 26 May 1958 ("Authorization for use of radioisotopes on human volunteers, request for") (ACHRE No. DOD-091494–A), 1.

93. Ibid.

94. Jacks to TSG, 12 April 1963, 1.

95. "Research and Development: Use of Volunteers as Subjects of Research," AR 70–25 (1962).

96. Office of Information Services, Air Force Special Weapons Center, to Headquarters, Air Research Development Command, 27 January 1956 ("Early Cloud Penetration") (ACHRE No. DOE-122894–B), 1.

97. Air Force Systems Command, January 1963 ("History of Air Force Atomic Cloud Sampling [AFSC Historical Publication Series 61–142–1]") (ACHRE No. DOD-082294–A), 23.

98. Ibid., 229.

99. Air Force Systems Command, "History of Air Force Atomic Cloud Sampling," 121.

100. E. A. Pinson [attr.], 1956 [attr.] ("Gentlemen: this morning I will discuss the following topics . . .") (ACHRE No. DOE-033195–B), 3.

101. Harold Clark, "I Flew Through an Atomic Hell," *Argosy,* December 1955, 63.

102. J. E. Banks et al., 30 April 1958 ("Operation Teapot: Manned Penetrations of Atomic Clouds, Project 2.8b") (ACHRE No. DOE-111694–A), 18.

103. The researchers found: "There appears to be no significant difference between the dose received inside and outside of the body. This indicates that the radiation which reaches the body surface is of sufficiently high energy that it is not greatly attenuated by the body. If this is the case, then measurements made on the surface of the body are representative of the whole-body dose." Ibid.

104. James Reeves, Test Manager, to Colonel H. E. Parsons, Deputy for Military Operations, 11 April 1955 ("Radiation Dosage—Project 2.8, Operation Teapot") (ACHRE No. DOE-122894–A), 1.

105. Ibid.

106. Banks et al., "Operation Teapot, Manned Penetrations of Atomic Clouds Project 2.8b," 5.

107. Ibid.

108. One received 21.8 R and another received 21.7 R. Undated document ("On-Site Personnel Overexposure") (ACHRE No. CORP-091394–A), 6.

109. Pinson [attr.], 1956 [attr.], "Gentlemen: this morning I will discuss the following topics . . . ," 3.

110. "The aircraft were B-57Bs. No special filters were installed in the cockpit pressurization system. The pilots and technical observers were given free choice of the setting of their oxygen controls." Colonel E. A. Pinson et al., 24 February 1960 ("Operation Redwing—Project 2.66a: Early Cloud Penetrations") (ACHRE DOE-122894–B), 41.

111. William Ogle, Headquarters, Task Group 7.1, to Commander Joint Task Force Seven, 8 November 1955 ("Maximum Permissible Radiation Exposure for Personnel Participating in Projects 2.66 and 11.2, Operation Redwing") (ACHRE No. DOE-013195–A), 2.

112. Pinson et al., "Operation Redwing—Project 2.66a: Early Cloud Penetrations," 5.

113. Ibid., 41.

114. Ibid., 51.

115. E. A. Pinson, interviewed by Patrick Fitzgerald (ACHRE), transcript of audio recording, 21 March 1995 (ACHRE Research Project Series, Interview Program File, Targeted Interview Project), 106.

116. Ibid., 121.

117. Office of Information Services to Headquarters, Air Research Development Command, 27 January 1956, 3.

118. Clarke, "I Flew Through an Atomic Hell," 62.

119. Interview with Pinson, 21 March 1995, 94.

120. Air Force Systems Command, "History of Air Force Atomic Cloud Sampling," 66.

121. Interview with Pinson, 21 March 1995, 15.

122. Pinson [attr.], 1956 [attr.], "Gentlemen: this morning I will discuss the following topics . . . ," 8.

123. Raymond Thompson, "A Select Group of ARDC Men Collects Samples from the Mushrooms," *Baltimore Sun—Magazine Section,* 1 May 1955, 17.

124. Interview with Pinson, 21 March 1995, 38.

125. "Center Scientists Fly Through Atom Clouds," *Atomic Flyer,* 29 April 1955 (ACHRE No. DOE-122894–B), 1.

126. Office of Information Services to Headquarters, Air Research Development Command, 27 January 1956, 2.

127. Captain Paul M. Crumley et al., 11 October 1957 ("Operation Teapot—Project 2.8a: Contact Radiation Hazard Associated with Contaminated Aircraft [WT-1122]") (ACHRE No. DOE-111694– A), 9.

128. Ibid., 20.

129. Ibid., 21.

130. Colonel W. B. Kieffer, Deputy Commander, Air Force Special Weapons Center, to K. F. Hertford, Manager, AEC Albuquerque Operations Office, 21 March 1957 ("Recent discussion within the Air Force Special Weapons Center . . .") (ACHRE No. DOE-033195–B), 2.

131. Thomas Shipman, Los Alamos Laboratory Health Division Leader, to Al Graves, J-Division Leader, 29 March 1957 ("Decontamination of Aircraft at Tests") (ACHRE No. DOE-040595–A), 1. Thomas Shipman also argued that the new procedures could compromise the scientific projects.

Without decontamination there will be inevitable migration of contamination carrying activity to other areas where it may be very undesirable. This letter has completely overlooked the fact that people working at tests invariably have neighbors with special requirements.

Ibid., 2.

132. Harold F. Plank, to Alvin C. Graves, Los Alamos Laboratory J-Division Leader, 24 April 1957 ("Col. Kieffer's Proposal for the Decontamination of Sampling Aircraft") (ACHRE No. DOE-040595–A), 2.

133. Colonel W. B. Kieffer, Deputy Commander, Air Force Special Weapons Center, to Colonel Wignall, 22 April 1957 ("Decontamination of Sampler Aircraft at Plumbbob") (ACHRE No. DOE-040595–A), 1.

134. James Reeves, Test Manager, Nevada Test Organization, to Commander, Air Force Special Weapons Center, Attention: Colonel W. B. Kieffer, Deputy Commander, 14 May 1957 ("Reference is made to your letter of March 21, 1957 . . . ") (ACHRE No. DOE-032895–A), 2.

135. First Lieutenant William J. Jameson, 7 October 1957 ("Aircraft Decontamination Study") (ACHRE No. DOE-022395–B), 1.

136. The decontamination experiment had several further components. Lead vests were tested and found to provide a 6.0 percent reduction in exposure levels for air crews. In addition, the experiment tested the consequences of using a fork lift to remove air crews from contaminated planes versus the consequences of letting them climb out with a standard ladder. It concluded that the fork lift was unnecessary. Ibid., 5–6.

Also at Plumbbob a project was undertaken "to measure the radiation dose, both from neutrons and gamma rays, received by an air crew delivering an MB-1 rocket." The report on the research states: "The Joint Chiefs of Staff approved the conduct of a test as a part of Operation Plumbbob in order to obtain the necessary experimental measurements." The report indicates that six studies were involved. Captain Kermit C. Kaericher and First Lieutenant James E. Banks, 11 October 1957 ("Operation PLUMBBOB—Project 2.9: Nuclear Radiation Received by Aircrews Firing the MB-1 Rocket") (ACHRE No. DOD-082294–A), 9.

137. The Advisory Committee is also aware of three more research activities involving atomic veterans. As noted, the body fluid sampling research is discussed in chapter 13. In addition, as mentioned in endnotes in this chapter, the Advisory Committee notes experiments involving fallout shelters and the measurement of radiation exposure to air crews delivering the MB-1 rocket. The inclusion of the subjects of these three types of experiments, however, does not change our estimate that human research in connection with bomb tests involved no more than 3,000 subjects.

138. DOD records did not permit the identification of individuals who participated in particular research projects, and remaining reports do not always indicate the number of subjects. The basis for the very rough estimate of 2,000 to 3,000 research subjects in the activities reviewed by the Committee including those noted in endnote 137 is (l) 1,500 to 2,200 test-site subjects in the psychological and physiological testing, based on reports, as cited in this chapter, for three experiments and an estimated maximum of 800 for the fourth; (2) a dozen test-site subjects in the 1955 body-fluid-sampling research, as cited in the report on this research referenced in chapter 13, and an assumed comparable number for the 1956 research, for which no similar figures appear available; (3) about 100 participants in the flashblindness research, an estimate DOD provided to the Committee; (4) in the range of perhaps one dozen or two dozen participants in aircrew experiments, and perhaps a dozen to several dozen participants in decontamination experiments; (5) perhaps several dozen participants in the protective equipment research; (6) sixteen participants in shelter research; and (7) several dozen participants in the officer volunteer

program. See further endnotes for citations related to particular research.

139. The permissible level of risk to which humans could be exposed in connection with bomb tests lay at the balance point of several factors. Radiation was not the only risk at issue; harm from blast and thermal burn were also possible.

140. Barton C. Hacker, *Elements of Controversy* (Berkeley: University of California Press, 1994), 118.

141. Marion W. Boyer, AEC General Manager, to Honorable Robert LeBaron, Chairman, Military Liaison Committee, 10 January 1951 ("As you know, one of the important problems . . .") (ACHRE No. DOE-040395–B-1).

142. Shipman to Warren, 15 September 1951, 1.

143. Shields Warren, Director, AEC Division of Biology and Medicine, to Carroll Tyler, Manager, Sante Fe Operations Office, 11 October 1951 ("Permissible Levels of Radiation Exposure for Test Personnel") (ACHRE No. DOE-013195–A), 1.

144. Warren's concern was not radiation risk, but injury from the blast. Shields Warren, Director, AEC Division of Biology and Medicine, to Brigadier General K. E. Fields, Director, Division of Military Application, 25 March 1952 ("Draft Staff Paper on Troop Participation in Operation Tumbler-Snapper") (ACHRE No. DOE-040395–A), 1.

145. Gordon Dean, Chairman, Atomic Energy Commission, to Brigadier General H. B. Loper, Chief, Armed Forces Special Weapons Project, 2 April 1952 ("Reference is made to letter of March 7, 1952 . . .") (ACHRE No. DOD-100694–A), 2.

146. Captain Harry H. Haight to General Fields, 21 August 1952 ("Exercise—Desert Rock IV") (ACHRE No. DOE-013195–A), 1. According to this review of Desert Rock activities, "The military importance of permitting major personnel exposures or decreases in drifting distances is not evident from the report. For the Commission to prescribe one limitation for the test personnel and allow greater latitude for the DOD would seem to be unwise and unnecessary. The Commission should strongly object to any special dispensation to the DOD which could possibly result in personnel casualties whether immediate or delayed." Ibid.

147. Colonel John C. Oakes, by direction of the Chief of Staff, to Assistant Chief of Staff, G-3, 5 June 1952 ("Indoctrination of Personnel in Atomic Warfare Operations") (ACHRE No. NARA-112594–A), 1.

148. C. D. Eddleman, Assistant Chief of Staff, G-3, 15 December 1952 ("Complete Discussion" [attachment to "Positioning of Troops at Atomic Weapons Tests"]), 1. In a 1953 memorandum to an AEC committee created to study the Nevada Test Site, Division of Biology and Medicine Director John Bugher similarly wrote:

While it may be stated with considerable certainty that no significant injury is going to result to any individual participating in test operations at the levels mentioned [3.9 R], and while it is true that the same thing would probably have to be said were the limits to be set two or three times as high, it nevertheless is true that there is no threshold to significant injury in this field, and the legal position of the Commission at once deteriorates if there is deliberate departure from . . . the accepted permissible limit.

John C. Bugher, Director, AEC Division of Biology and Medicine, to Members of the Committee to Study NPG, 8 September 1953 ("Interpretation of the Standards of Radiological Exposure") (ACHRE No. DOE-040395-A), 3–4.

149. M. W. Boyer, AEC General Manager, to Major General H. B. Loper, Chief, AFSWP, 8 January 1953 ("Reference is made to letter from Chief . . .") (ACHRE No. DOE-121594–C-8), 2.

150. Jark to Distribution, 20 February 1953, "Instructions for Positioning DA [Department of Army] Personnel at Continental Atomic Tests," 1.

151. Ibid., 2–3.

152. General Cooney presented this view at a July 1951 conference on Past and Future Atomic Tests. Major Sven A. Bach, Development Branch, Research and Development Division, 12 July 1951 ("Conference at OCAFF, Fort Monroe, Virginia, re Past and Future Atomic Weapons Tests") (ACHRE No. NARA-042295–C), 1.

153. Atomic Energy Commission, minutes of meeting no. 862, 13 May 1953 (ACHRE DOE-013195–A), 2.

154. Ibid.

155. Chief, Bureau of Medicine and Surgery, to Chief of Naval Operations, 14 February 1952 ("Radiological Defense Training, comments and recommendations on") (ACHRE No. DOD-080295–B), 1. The proposal would have limited "the dosage of all personnel to 0.3 roentgens per week." Chief of Naval Operations to Chief, Bureau of Medicine, 23 January 1952 ("Atomic Defense Training") (ACHRE No. DOD-080295–B), 1. The proposal originated with the Pacific Fleet. See Commander, Mine Force, U.S. Pacific Fleet, to Commander in Chief, U.S. Pacific Fleet, 17 December 1951 ("Radiological Defense Training") (ACHRE No. DOD-080295–B), 1. In counseling against the use of "area contamination," BuMed solicited advice from the AEC on an isotope that "would have such characteristics that the internal hazard involved would be minimized even though amounts to be used would produce as much as 10 mr/ hr, gamma radiation, three feet from the surface of the contaminated area." Chief, Bureau of Medicine and Surgery, to Director, AEC Division of Biology and Medicine, February 1952 ("Radiological Defense Training, use of radioisotopes in") (ACHRE No. DOD-080295–B), 1.

156. Shipman's comments were specifically directed at the establishment of standards for exposure to the general public. Thomas L. Shipman, Los Alamos

Laboratory Health Division Leader, to Gordon Dunning, AEC Division of Biology and Medicine, 14 August 1956 ("Thanks for sending the draft concerning exposure . . .") (ACHRE No. DOE-022195–C), 1.

157. Department of Energy, *Announced United States Nuclear Tests: July 1945 Through December 1992* (Springfield, Va.: National Technical Information Service, May 1993), 65 (shot Climax in 1953). In the early days, when entirely new types of experimental weapons were being rapidly developed and tested, it was not uncommon for a particular yield to exceed estimates by 50 percent or more. In an October 1957 memorandum to AEC Division of Biology and Medicine director Charles Dunham, Shipman explained that the unpredictability of weapons effects was making biomedical experimentation increasingly difficult. "All too often preshot estimates of yields etc. are just enough in error to make the results of effects tests useless." Thomas L. Shipman, Los Alamos Laboratory Health Division Leader, to Charles Dunham, AEC Division of Biology and Medicine, 7 October 1957 ("Payne Harris is planning to attend the meeting . . .") (ACHRE No. DOE-120894–C), 2.

158. T. L. Shipman, Los Alamos Laboratory Health Division Leader, to Alvin C. Graves, J-Division Leader, 6 August 1956 ("Permissible Exposures") (ACHRE No. DOE-021095–B), 1.

159. Summary information provided by DOD in August 1995 provides a total of 216,507 participants in atmospheric tests, beginning with Trinity in 1945 and concluding with Dominic II in 1962. This tabulation shows about 1,200 instances of exposure in excess of 5 rem. The "total dose may have been measured by one or more film badges, may have been reconstructed, or may be the sum of both film badge data and reconstruction." Some individuals participated in more than one test. Defense Nuclear Agency, 8 August 1995 ("Summary of External Doses for DOD Atmospheric Nuclear Test Participants as of 24 February 1994") (ACHRE No. DOD-081195–A). See also testimony of Major General Ken Hagemann: Senate Committee on Governmental Affairs, *Human Radiation and Other Scientific Experiments: The Federal Government's Role,* 103d Cong., 2d Sess., 25 January 1994, 49–50.

Coincident with the beginning of epidemiological studies discussed in the text above, and growing congressional and public interest in the atomic vets, the Defense Department undertook an information-gathering effort called the "NTPR" (Nuclear Test Personnel Review). The NTPR includes a database, which seeks to include those who participated at tests in an effort to reconstruct the doses they received at tests, and a multivolume history of the bomb tests, which is available in many libraries.

160. Stafford L. Warren, Radiological Safety Consultant, Joint Task Force One, to Admiral Parsons, 6 January 1947 ("Hazards from Residual Radioactivity on the Crossroads Target Vessels") (ACHRE No. DOE-033195–B), 2.

161. Jonathan M. Weisgall, *Operation Crossroads: The Atomic Tests at Bikini Atoll* (Annapolis, Md.: Naval Institute Press, 1994), 210–214, 270–271. Only fragmentary records of the Medico-Legal Board remain.

162. The Naval Radiological Defense Laboratory, the new research laboratory, was established at the Hunter's Point Naval Shipyard in San Francisco, the port to which some ships contaminated in the 1946 Crossroads tests were sent.

163. George M. Lyon, Assistant Chief Medical Director for Research and Education, to Committee on Veterans Medical Problems, National Research Council, 8 December 1952 ("Medical Research Programs of the Veterans Administration") (ACHRE No. VA-052594–A), 553.

164. Ibid.

165. Ibid.

166. J. J. Fee, Commander, USN, as quoted in Weisgall, *Operation Crossroads*, 273–274.

167. Lyon to Committee on Veterans Medical Problems, 8 December 1952, 554.

168. Ibid.

169. Ibid.

170. Ibid. The report was retrieved by the VA at the time of the Advisory Committee's formation in 1994. In an April 1994 statement to the Committee, VA Secretary Jesse Brown stated his determination to find the facts related to the Confidential Division. To this end the VA reviewed significant amounts of documentary information and called on its inspector general to conduct a further review.

171. Ibid., 554.

172. Ibid., 553–554.

173. Major General Herbert B. Loper, Chief, AFSWP, to the Administrator, Veterans Administration, Attention: George M. Lyon, 8 August 1952 ("This activity has received information . . .") (ACHRE No. DOD-100694–A), 1.

174. Ibid. The specific rule or policy that provided for the record keeping referred to in this letter was not located. Thus, it is not clear whether the record keeping referred only to nuclear war-related exposures or more generally to exposures at bomb tests or other nuclear weapons-related activities as well.

175. William Middleton, VA Chief Medical Director, to the VA Administrator, 13 May 1959 ("Recommendation for Administrator's Exceptional Service Award") (ACHRE No. VA-102594–A), 1.

176. Ibid.

177. "12 January 1995 Review of Effort to Identify Involvement with Radiation Exposure of Human Subjects," Inspector General, Department of Veterans Affairs. The inspector general (IG) found that "an 'Atomic Medicine Division' was discussed as a means to deal with potential claims from veterans as a result of exposure to radiation from atomic bomb testing and to be the focal point for VA civil defense planning and support in case of nuclear war. However, claims did not materialize at that time and evidence indicates that the Division was not activated." Stephen A. Trodden, VA Inspector General, to VA Chief of Staff, 12 January 1995 ("Review of Effort to Identify Involvement with Radiation Exposure of Human Subjects") (ACHRE No. VA-011795–A), 1.

With regard to the 1952 history prepared by Dr. Lyon for the National Research Council, which has been previously quoted in the text, the IG stated that "the reference to the Atomic Medicine Division should not be taken literally as documentation that a Division was ever established." Ibid., 4.

178. In communications with Defense Department officials two alternatives were offered: (1) that the records may have been confidential medical examination data taken from participants in Crossroads, pursuant to a regulation providing for such exams; (2) that the records may have related to exposures of military scientists or technicians who worked at the Manhattan Project and were confidential because they contained weapons design or production-related data.

Navy regulations in 1947 provided that

All personnel, both military and civilian, who may be exposed to radiation or radioactive hazard, will be required to have a complete physical examination prior to commencing such duty. Special medical records separate from the normal individuals' health records will be set up and they will be classified as confidential, until declassification is permitted.

Bureau of Medicine and Surgery, 31 January 1947 ("Appendix B—Current Directives; Subject: Safety Regulations for Work in Target Vessels formerly JTF-1") (ACHRE No. DOD-020795–A), B-22. The Navy was not able to locate the records referred to.

The VA told the Committee that "the volume of classified records that are unaccounted for by the VA is too small to have constituted a defense against liability claims." Susan H. Mather, M.D., M.P.H., letter to Dan Guttman (ACHRE), 17 July 1995. Based on discussions with the VA, the basis for this statement appears to be the fact that there were more than 200,000 test participants, and the safe maintained by Dr. Lyon (in which secret documents would presumably have been kept) was relatively small. In the absence of the documents themselves, the VA's statement appears to be only one of several possible speculative alternatives. For example, the VA also explained that few claims eventuated in the period of Dr. Lyon's service; thus, the magnitude of necessary filekeeping may not have been great. Alternatively, documents kept by Dr. Lyon could have been summary documents, which referred to materials in other files. Finally, the VA's statement is also consistent with the possibility that files were kept but that their contents were deemed inadequate to constitute a defense against potential claims.

179. NEPA Medical Advisory Panel, Subcommittee IX, proceedings of 22 July 1949 (ACHRE No. DOD-121494–A-2), 17–18. The meeting is further discussed in the Introduction.

180. Ibid., 18.

181. Ibid.

182. Department of Defense, Research and Development Board, Committee on Medical Sciences, proceedings of 23 May 1950 (ACHRE No. DOD-080694–A), 10.

183. Caldwell et al., "Leukemia Among Participants in Military Maneuvers at Nuclear Bomb Tests," *Journal of the American Medical Association* 244, no. 14 (1980).

184. Caldwell et al., "Mortality and Cancer Frequency Among Military Nuclear Test Participants, 1957 through 1959," *Journal of the American Medical Association* 250, no. 5 (1983).

185. C. D. Robinette et al., *Studies of Participants in Nuclear Weapons Test: Final Report* (Washington, D.C.: National Research Council, May 1985).

186. See U.S. General Accounting Office, *Nuclear Health and Safety: Mortality Study of Atmospheric Nuclear Test Participants Is Flawed* (Gaithersburg, Md.: GAO, 1992), 4. Helen Gelband, Health Program, Office of Technology Assessment, *Mortality of Nuclear Weapons Tests Participants* (Washington, D.C.: Office of Technology Assessment, August 1992), 4.

187. The data appear in table 1 of Clark W. Heath, Chairman, Institute of Medicine (IOM) Committee on the Mortality of Military Personnel Present at Atmospheric Tests of Nuclear Weapons, and John E. Till, Chairman, IOM Dosimetry Working Group, to D. Michael Schaeffer, Program Manager, DNA Nuclear Test Personnel Review, 15 May 1995 ("A Review of the Dosimetry Data Available in the Nuclear Test Personnel Review [NTPR] Program: An Interim Letter Report of the Committee to Study the Mortality of Military Personnel Present at Atmospheric Tests of Nuclear Weapons") (ACHRE No. NAS-051595–A), 9.

188. Hacker, *Elements of Controversy*, 96.

189. The memo explained that the need had been foreseen, but the request for dosimeters had only been partially filled. The memo recorded that 175 "0–5 R dosimeters" were on hand at the Nevada Test Site, but a minimum of 325 were needed for an operation the size of Upshot-Knothole. Colonel Leonard F. Dow, Acting Director, Weapons Effects Tests, to Manager, AEC Santa Fe Operations, 19 February 1954 ("Rad-Safe Equipment for Nevada Proving Grounds") (ACHRE No. DOE-020795–D), 1.

190. Irving L. Branch, Chief of Staff, AFSWP, to Chief of Research and Development, OCS, Department of the Army, 20 January 1956 ("Annex 'A' to 2nd Indorsement: Detailed Explanation of AFSWP Comments on Feasibility of Human Volunteer Program") (ACHRE No. DOD-030895–F), 2.

191. Clark W. Heath and John E. Till, IOM, to D. Michael Schaeffer, DNA, "An Interim Letter Report of the Committee to Study the Mortality of Military Personnel Present at Atmospheric Tests of Nuclear Weapons," 15 May 1995.

192. K. K. Watanabe, H. K. Kang, and N. A. Dalager, "Cancer Mortality Risk Among Military Participants of a 1955 Atmospheric Nuclear Weapons Test," *American Journal of Public Health* 85 (April 1995).

193. S. Raman, G. S. Dulberg, R. A. Spasoff, and T. Scott, "Mortality Among Canadian Military Personnel Exposed to Low Dose Radiation," *Canadian Medical Association Journal* 136 (1987): 1051–1056.

194. S. C. Darby, G. M. Kendall, T. P. Fell et al., "A Summary of Mortality and Incidence of Cancer in Men from the United Kingdom Who Participated in the United Kingdom's Atmospheric Nuclear Weapon Tests and Experimental Programs," *British Medical Journal* 296 (1988): 332–338.

195. *Human Radiation Experiments: The Federal Government's Role, Hearings before the Committee on Governmental Affairs, United States Senate,* 103d Cong., 2d Sess., 25 January 1994, 160.

196. DNA, "Summary of External Doses for DOD Atmospheric Nuclear Test Participants as of 24 February 1994."

197. These laws are further discussed in the Committee's recommendations. In enacting the 1984 Veterans' Dioxin and Radiation Exposure Compensation Standards Act, Congress, among other items, found

(8) The 'film badges' which were originally issued to members of the Armed Forces in connection with the atmospheric nuclear test program have previously constituted a primary source of dose information for . . . veterans filing claims . . . .

(9) These film badges often provide an incomplete measure of radiation exposure, since they were not capable of recording inhaled, ingested, or neutron doses (although the DNA currently has the capability to reconstruct individual estimates of such doses), were not issued to most of the participants in nuclear tests, often provided questionable readings because they were shielded during the detonation, and were worn for only limited periods during and after each nuclear detonation.

(10) Standards governing the reporting of dose estimates in connection with radiation-related disability claims . . . vary among the several branches of the Armed Services, and no uniform minimum standards exist.

198. For example, Frances Brown, of Southwick, Massachusetts, told the Committee of her late husband's experience as a navigator who flew through clouds at weapons tests. Colonel Brown was assigned the duty and was given no protective clothing; he died of cancer in 1983. Ms. Brown shared with the Committee the story of years of inquiry, and her continuing inability to obtain all documents that might shed light on the duty he undertook in the service of his country.

Nancy Lynch, of Santa Barbara, California, told the Committee of her late husband's involvement in the Desert Rock exercises at Operation Teapot in 1955 and her questions regarding the dose reconstruction that was ultimately provided by the government.

Vernon Sousa, a San Francisco veteran, told of years of government "stonewalling" of his information requests. He explained that the oath of secrecy he had taken limited his own ability to discuss the tests for decades after his time in the service ended.

Charles McKay of Severna Park, Maryland, a Navy diver at Operation Crossroads, recalled that he received no briefing on radiation risks before his participation. Mr. McKay said that he received a very low dose reconstruction report from the government, which he believed to be highly inaccurate because it did not take into account diving experiences on Crossroads wrecks.

Rebecca Harrod Stringer of St. Augustine, Florida, wrote to the Committee about the Navy service of her late father in Operation Dominic I, a nuclear weapons test in the Pacific, and the fifteen years it took to obtain copies of his military and medical records.

Linda Terry of California talked of obtaining information about her late father's experiences at the Buster-Jangle tests in 1951–52. She called for full disclosure of information about the weapons tests "so that families do not have to live in the darkness" of not knowing.

Harry Lester of Albuquerque, New Mexico, testified that he was responsible for cleanup at Operation Castle and that he experienced radiation sickness as a result of his exposure. After his involvement in Castle, he was shipped to an Albuquerque hospital every six months for examinations. He told the Committee that his full records remain to be found.

Langdon Harrison of Albuquerque told the Committee about his experiences in cloud flying activities at Operations Redwing and Plumbbob. He recalled routine carelessness in the handling of the film badges of the pilots of cloud flythroughs and occasions when significantly different dose readings were recorded on film badges and personal dosimeters.

Representatives of "atomic veterans" organizations also shared with the Committee information collected in years of research on behalf of themselves and others. These included Pat Broudy of California, whose late husband died of lymphoma and had served at the occupation of Nagasaki, Bikini, and in three Nevada tests; Dr. Oscar Rosen of Massachusetts, who participated in Crossroads; and Fred Allingham of California, whose father served in the occupation of Nagasaki and died several years later of leukemia.

199. The new rules stemmed from the development of a new howitzer. Late in the development cycle a medical hazards review found that alteration to the firing routine was needed if the weapon was to be employed without injuring U.S. soldiers. The discovery caused a long and expensive delay while biomedical studies of blast overpressure effects were done in animals and man and engineering solutions were sought to reduce the hazard. After this experience, the Army determined to conduct health hazard assessments (HHAs) early in the development of weapons and equipment, so that new material is not brought on line with unnecessarily great health and safety risk to the troops using it.

Relevant DOD directives (DODD) and Army regulations are the following: DODD 5000.1, "Defense Acquisition"; DODD 5000.2," "Defense Acquisition Management Policies and Procedures"; AR 70–1; "Army Acquisition Policy"; AR 602– 1, "Human Factors Engineering Program"; AR 602– 2, "Manpower and Personnel Integration (MANPRINT) in the System Acquisition Process"; AR 385–16, "System Safety Engineering and Management"; AR 40–10, "Health Hazard Assessment Program in Support of the Army Material Acquisition Decision Process"; and AR 70–75, "Survivability of Army Personnel and Material."

11

Intentional Releases:
Lifting the Veil of Secrecy

IN February 1986, officials at the Department of Energy responded to requests from activists by releasing 19,000 pages of documents on the early operations of the world's first plutonium factory, at Hanford, Washington. Combing through these documents, reporters and citizens found references to an event cryptically named the "Green Run," in which radioactive material was deliberately released into the air at Hanford in December 1949.[1]

In the aftermath of the public discovery of the Green Run, Senator John Glenn asked the General Accounting Office, the investigative arm of Congress, to find out if there were other instances in which radioactivity had been intentionally released into the environment without informing the surrounding community. In 1993, the GAO reported twelve more instances of such secret intentional releases.[2]

Following additonal research by the DOD and DOE, the number of secret intentional releases has expanded to several hundred, conducted between 1944 and the 1960s. At the Army's Dugway Proving Ground in Utah, dozens of intentional releases were conducted in an effort to develop radiological weapons, some in tests of prototype cluster bombs, others using different means of dispersal; at Bayo Canyon in New Mexico, on the AEC's Los Alamos site, researchers detonated nearly 250 devices, which contained radiolanthanum (RaLa) as a source of radiation to measure the degree of compression and symmetry of the implosion used to trigger the atomic bomb. Other intentional releases were not classified, although not all were made known to the public in advance. At AEC sites in Nevada and Idaho, radioactive materials were released in tests of the safety of bombs, nuclear reactors, and proposed nuclear rockets and airplanes; in still other cases, small quantities of radioactive material were released in and around AEC facilities and in the Alaskan wilderness to determine the pathways such material follows in the environment.[3] Public witnesses from several of these communities told the Committee that they remain deeply disturbed by these releases, wondering whether there is still more information about the secret releases in their communities that they do not know and how much will, at this late date, be impossible to reconstruct.

INTENTIONAL RELEASES AND THE CHARTER THIRTEEN

The Advisory Committee is authorized by its charter to examine "experiments involving intentional environmental releases of radiation that (A) were designed to test human health effects of ionizing radiation; or (B) were designed to test the extent of human exposure to ionizing radiation." The charter also called for the Committee to "provide advice, information, and recommendations" on the following thirteen experiments and similar experiments identified by the Interagency Working Group:

(1) the experiment into the atmospheric diffusion of radioactive gases and test of detectability, commonly referred to as "the Green Run test," by the former Atomic Energy Commission (AEC) and the Air Force at the Hanford Reservation in Richland, Washington;

(2) two radiation warfare field experiments conducted at the AEC's Oak Ridge office in 1948 involving gamma radiation released from non-bomb point sources or at near ground level;

(3) six tests conducted during 1949–1952 of radiation warfare ballistic dispersal devices containing radioactive agents at the U.S. Army's Dugway, Utah, site; [and]

(4) four atmospheric radiation-tracking tests in 1950 at Los Alamos, New Mexico. . . .

Tests of nuclear weapons, intentional environmental releases of radiation in amounts greatly in excess of any of the releases identified above, were not included in the charter. As discussed in chapter 10, the Committee did seek to investigate human subject research conducted in connection with these tests.

This chapter reports on what we found as we sought to retrieve what we could about the releases identified in our charter, determine the nature and number of further intentional releases, identify the ethical standards by which these activities can be evaluated, and determine what lessons can be learned from the past.

Because of the secrecy surrounding these releases—as opposed to atmospheric nuclear weapons tests, which were impossible to hide—many of them took place with no public awareness or understanding. The intentional releases were conducted primarily at sites such as Hanford, Los Alamos, and Oak Ridge, in which defense and atomic energy facilities were located, but they were largely unknown to those who lived in surrounding areas.

There is no evidence in any of these cases that radioactive material was released for the purpose of studying its effects on human communities. As we discuss later in the chapter, the public often was exposed to far greater risk from the routine course of operations of the facilities than from the intentional releases themselves.

That the possible health effects from the Green Run and other intentional releases are so slight that they cannot be distinguished from other sources of disease is small comfort to "downwinders" who were put at risk without their knowledge. The Committee heard from many of them and learned that the longer-term costs of secrecy extend well beyond any physical injury that may have been incurred. These costs include, first, the anxiety and sense of personal violation experienced by those who have discovered that they have intentionally and secretly been put at risk, however small, by a government they trusted. But they also include the consequences for that government, and its people, of the attendant distrust of government that has been created. And finally, they also now include the citizen and taxpayer resources that must be expended in efforts to reconstruct long-buried experiences, and determine, as best as can currently be done, the precise measures of the risks involved.

The chapter is divided into two parts. The first and lengthier section reconstructs the history of the three kinds of releases that were in our charter—the Green Run, radiological warfare tests, and the RaLa tests—and includes a discussion of some types of intentional releases that were not expressly identified in the charter. This section concludes with a review of what is known today about the likely risks of all the releases we consider, as well as a review of the science of dose reconstruction by which this knowledge is obtained. In the second part of the chapter, we focus on the ethical and policy issues raised by intentional releases. We examine the rules that currently govern intentional releases in an effort to learn whether secret environmental releases like the Green Run could take place today and, if so, whether, in light of lessons learned from the past, current procedures and protections are adequate.

WHAT WE NOW KNOW

The Green Run

While the other intentional releases addressed in the Committee's charter were part of the effort to develop the U.S. nuclear arsenal, the Green Run was conducted to develop intelligence techniques to understand the threat posed by the Soviet Union. In 1947 General Dwight D. Eisenhower assigned the Air Force the mission of long-range detection of Soviet nuclear tests.[4] Based on observations from Operation Fitzwilliam, the intelligence component of the 1948

Sandstone nuclear test series, the Air Force determined aerial sampling of radioactive debris to be the best method of detecting atomic releases.[5] An interim aerial sampling network was in place in early September 1949 that detected radioactive debris from the first Soviet nuclear test.[6]

Around the same time, Jack Healy of Hanford's Health Instrument (HI) Divisions noticed anomalous radioactivity readings from an air filter on nearby Rattlesnake Mountain. The HI Divisions were responsible for radiological safety, and Healy had set up this filter to test how radioactive contamination varied with altitude. The rapid decay of his radioactive samples led Healy to conclude that they had come from a recent nuclear test.[7] Soon after news of Healy's observation reached Washington, D.C., Air Force specialists arrived and took Healy's samples and data for analysis. It is not clear whether Healy's observation came in time to support President Harry Truman's announcement on September 23 that the Soviet Union had exploded its first atomic bomb,[8] but it did confirm that radioactivity from a nuclear test could be detected on the other side of the globe.

Now that the Soviet Union knew how to make atomic weapons, the United States needed to know how many weapons and how much of the critical raw material plutonium the Soviets possessed. Like nuclear testing, plutonium pro-

---

HANFORD: THE WORLD'S FIRST PLUTONIUM FACTORY

In 1942 General Leslie Groves selected the Hanford site overlooking the Columbia River in southeast Washington state for the Manhattan Project's plutonium factory. The river would provide a large, reliable supply of fresh water for cooling the plutonium-production reactors, and Hanford's relative isolation from major population centers would make it easier to construct and operate the facility without attracting unwanted attention. The nearby towns of Richland, Kennewick, and Pasco soon became boom towns whose economies depended on Hanford.

At Hanford, neutrons converted uranium 238 in the production reactor's nuclear fuel into plutonium 239. Chemical separation plants then separated this plutonium from the fission products and residual uranium in the irradiated fuel elements. The first separation plants, the T and P plants, used acid to dissolve these fuel elements, but this was superseded by the more efficient Redox and Purex processes in the 1950s.

duction released radioactive gases that sensitive instruments could detect, though not at such great distances.[9] To identify Soviet production facilities and estimate their rate of plutonium production, the Air Force now needed to test ways to monitor these gases.[10]

In late 1948 and early 1949, Air Force and Oak Ridge personnel conducted a series of twenty air-sampling flights at Oak Ridge and three at Hanford.[11] The results were disappointing: instruments detected airborne releases of radioactive material at ranges of up to fifteen miles in the hills and valleys near Oak Ridge, but no farther than two miles from Hanford, because of measures taken to reduce radioactive emissions there. At an October 25, 1949, meeting at Hanford, representatives of the Air Force, the Atomic Energy Commission, and General Electric (the postwar contractor for the Hanford site) agreed to a plan to release enough radioactive material from Hanford[12] to provide a larger radioactive source for intelligence-related experiments.[13]

This intentional release took place in the early morning of December 3, 1949, but information about it remained classified until 1986. Two periodic reports of the HI Divisions described a plutonium production run using "green" fuel elements.[14] The story of this "Green Run" has emerged piecemeal since then. The most complete account comes in a 1950 report coauthored by Jack Healy (referred to as the Green Run report), which was declassified in stages in response to requests from the public under the Freedom of Information Act and inquiries by the Advisory Committee.[15]

Although cooling times of 90 to 100 days were common by 1949, the fuel elements used in the Green Run were dissolved after being cooled for only 16 days. This short cooling time meant that much more radioactive iodine 131 and xenon 133 were released directly into the atmosphere, rather than decaying while the fuel elements cooled. Furthermore, pollution control devices called scrubbers normally used to remove an estimated 90 percent of the radioiodine[16] from the effluent gas were not operated.[17]

When these "green" fuel elements were processed, roughly 8,000 curies of iodine131[18]

flowed from the tall smokestack at Hanford's T plant. This stack was built in the early years of Hanford's operation when large quantities of radioactive gases were routinely released in the rush to produce plutonium. Although the Green Run represents roughly 1 percent of the total radioiodine release from Hanford during the peak release years 1945–1947, it was almost certainly larger than any other one-day release, even during World War II.[19]

One clear purpose of the Green Run was to test a variety of techniques for monitoring environmental contamination caused by an operating plutonium-production plant. A small army of workers, including many from Hanford's HI Divisions, took readings of radioactivity on vegetation, in animals, and in water and tested techniques for sampling radioactive iodine and xenon in the air.[20] The Air Force operated an airplane carrying a variety of monitoring devices—the same aircraft used in earlier aerial surveys at Oak Ridge and Hanford—and set up a special air sampling station in Spokane, Washington.[21]

Those operating the equipment encountered numerous technical problems, including a lost weather balloon and failed air pumps. The greatest problem, however, was the general contamination of monitoring and laboratory equipment. The contamination created a high background signal that made it difficult to distinguish radioactivity on the equipment from radioactivity in the environment. The main cause of this contamination was the weather at the time, which led to much higher ground contamination near the stack than expected.[22]

The plans for the Green Run included very specific meteorological requirements. These requirements were designed to facilitate monitoring of the radioactive plume by aircraft, but they were similar to the normal operational requirements, which were designed to limit local contamination:

- A temperature inversion,[23] to keep the effluents aloft, but at a low altitude;
- No rain, fog, or low clouds to impede aircraft operations;
- Light to moderate wind speeds (less than fifteen miles an hour);

- Wind from the west or southwest, so the plane would not have to fly over rough terrain;[24] and
- Strong dilution of the plume before any possible contact with the ground.[25]

Jack Healy reports that he made the decision to go ahead with the Green Run on the evening of December 2, 1949, even though the weather did not turn out as expected. Some have suggested that the Air Force pressed to go ahead with the release in spite of marginal weather conditions, but Healy recalls no such pressure.[26] The plume from the release stagnated in the local area for several days before a storm front dispersed it toward the north-northeast. As a consequence, local deposition of radioactive contaminants was much higher than anticipated.[27] The Green Run report concludes:

Under the worst possible meteorological conditions for such a test, the airborne instruments detected the radioactive gases at a distance better than 100 miles from the stack. Under favorable conditions, it was estimated that with the same concentrations this distance could have been increased by up to a factor of ten.[28]

Despite the contamination of equipment, the monitoring provided a record of the extensive short-term environmental contamination that resulted from the Green Run. Measurements of radioactivity on vegetation produced readings that, while temporary, were as much as 400 times the then-"permissible permanent concentration" on vegetation thought to cause injury to livestock.[29] The current level at which Washington state officials intervene to prevent possible injury to people through the food supply is not much higher than the then-permissible permanent concentration.[30] Animal thyroid specimens showed contamination levels up to "about 80 times the maximum permissible limit of permanently maintained radioiodine concentration."[31]

In spite of this contamination, the public health effects of the Green Run, discussed later in this chapter, were quite limited. However, in 1949, at the time the Green Run was conducted, the most important environmental pathways for human exposure to radioiodine were unknown. (Understanding developed shortly thereafter that environmental radioiodine enters the human body from eating meat and drinking milk from animals that grazed on contaminated pastures.)[32] Thus, the effects of exposure through these pathways could not have been planned for, and it is fortunate that the risks were not higher.

The Control of Risks to the Public from Plutonium Production at Hanford

From the first years of Hanford's operation, its health physicists were aware of the problems of contamination of the site by radioactive wastes, and it quickly became clear that radioiodine posed the greatest immediate hazard.[33] Most fission products would remain in the dissolved fuel, but iodine gas would bubble out of the solution, up through Hanford's tall stacks into the atmosphere and down onto the surrounding countryside. Other radioactive wastes could be stored and dealt with later, and other radioactive gases were chemically inert and would quickly dissipate.

Over the years, Hanford health physicists adopted three main approaches to the iodine problem:

- Choosing meteorological conditions for releases that would prevent air with high iodine concentrations from contaminating the ground near Hanford;
- Letting the irradiated fuel elements cool for extended periods before separating the plutonium, so that most of the iodine 131, which has an eight-day half-life, could decay; and
- Beginning in 1948, using scrubbers or filters to remove iodine from the exhaust emissions.

During World War II, producing plutonium for bombs was an urgent priority and knowledge of both the environmental hazards from iodine and the ways to prevent it were limited. Over the period 1944–1947, Hanford released nearly 685,000 curies of radioiodine into the atmosphere, about eighty times what was released in the Green Run.[34] After the war, an improving understanding of how iodine could contaminate the food supply,[35] evolving techniques to remove iodine from the plants' emissions, and policy decisions to limit the risks to the nearby population led to a marked reduction in iodine emissions.

When the AEC began operation in 1947, it promptly moved to review safety practices at Hanford and other operating facilities, which had operated largely autonomously until then. The advisory panel established for this purpose concluded that "the degree of risk justified in wartime is no longer appropriate."[36] To address the radioiodine problem at Hanford and related problems, the AEC established a Stack Gas Working Group, which met for the first time in mid-1948 to study air pollution from AEC production facilities. The chair of this group noted that the AEC "desires the removal from gaseous effluents of all [radioactive] material insofar as is humanly and economically feasible" and that because of uncertainties in risk estimates "no limit short of zero should be considered satisfactory for the present."[37] By 1949, daily emissions of radioiodine had fallen by a factor of 1,000 from their wartime highs.

The Green Run clearly did not conform to the practices designed to ensure public safety at Hanford in 1949 or even during the rush to produce plutonium for the first atomic bombs. In his monthly report for December 1949, Herbert Parker, Hanford's manager, concluded that the Green Run had posed a "negligible" risk to personnel, but "[t]he resultant activity came close enough to significant levels, and its distribution differed enough from simple meteorological predictions that the H.I. Divisions would resist a proposed repetition of the tests."[38] This suggests that Parker, at least, considered the risks of such releases potentially excessive even for a one-time event, particularly given the degree of uncertainty.

Parker's recognition of the uncertainties surrounding environmental risks from Hanford's radioiodine emissions was appropriate. At the time, it was not known that drinking milk from cows that graze on contaminated pastures is the main source of exposure, especially for children. Jack Healy recently suggested that if Parker had known of the milk pathway, he would have objected strongly to the Green Run.[39] The question remains as to the consideration that was given by the Green Run's planners to the possibility that they might not fully understand the risks that might be imposed on nearby communities.

Benefits of the Green Run

The Advisory Committee attempted to assess of the national security benefits that were expected and actually resulted from the Green Run. A planning memorandum before the Green Run notes, "the possibility of the detection of stack effluents is of great importance to the intelligence requirements of the country."[40] How important the detection of stack effluents was to the security of the nation in 1949 is not something the Advisory Committee was in a position to judge. We did attempt to ascertain, however, the purpose of the Green Run and the extent to which this purpose was served.

The Green Run report focuses primarily on ground-based monitoring of radioactive contamination in the environment, which provided a test for techniques that could be used on the ground in the Soviet Union. The report also describes efforts to track the radioactive plume by aircraft, but their significance is unclear. Aerial monitoring turned out to be the most effective method for detecting atmospheric nuclear tests, and perhaps it was expected to be equally effective for monitoring Soviet plutonium production. Plutonium production releases relatively little radioactivity into the atmosphere, however—too little to detect outside Soviet air space, and flying inside Soviet air space would have been risky. Alternatively, aerial radiation tracking may have been designed to test techniques for use in monitoring nuclear weapons tests. Finally, the Green Run report compares the pattern of the plume's dispersion with theoretical models, but this appears to be an attempt to estimate the pattern of contamination rather than to test the already well-established theory regarding atmospheric diffusion of gases developed in the 1930s.

It is difficult to ascertain how useful the Green Run actually was. The classified histories of the Air Force's atomic intelligence activities contain no references to the Green Run. These histories jump from events that directly preceded the Green Run—the Oak Ridge and Hanford aerial monitoring tests—to later ones, without any mention of the Green Run.[41] Perhaps most telling, a 1952 AEC report entitled "Technical Methods in Atomic Energy Intelligence" does mention the Green Run in the text, but only in a list of occasions on which a particular type of instrument was used. In describing ways of detecting plutonium-production facilities, the report relies on routine reports of environmental surveys from Hanford's routine operations.[42]

Secrecy and Public Risk

The Advisory Committee accepts that there may be conditions under which national security can justify secrecy in intentional releases like the Green Run, even as we recognize that secrecy can increase the risk to the exposed population.

In discussing this question it is important to explain that when we use the term *secret* we can be referring to secrecy regarding the very fact that a risk has been posed, secrecy regarding the purpose behind the risk, or secrecy regarding the means (for example, the science or technology) by which the risk was imposed. These distinctions are important because even if we agree that the undertaking of an activity is required for national security reasons, it does not follow that secrecy should govern all aspects of the activity. Thus, as an obvious example, atomic bomb tests were quintessential national security activities; information on the design of the bomb was secret, as was information on many of the specific purposes of the tests; however, in many (but not all) cases the public was given notice that a hazardous activity was being undertaken. Similarly, in the cases of other environmental releases, it may be that national security requires secrecy for some aspects of the release but does not necessarily preclude public disclosure sufficient to give basic notification of the existence of potential risk. The Committee is not equipped to say whether this was so in the case of the Green Run. However, in the case of radiological warfare, as we will discuss later, there was contemporary argument that some public disclosure was not inconsistent with national security.

If a release is conducted publicly, affected communities have an opportunity to comment and perhaps influence the conduct of the release in ways that serve their interests. Downwinders can be warned, giving them the options of staying indoors with their windows closed, wearing protective clothing, altering their eating habits, or evacuating the area. If the release is conducted in secret, foreign adversaries are less likely to be alerted, but downwinders will be deprived of their options. Of course, evacuation may not be warranted, and other precautions may not be needed, or they may be of limited value. But, as we have learned during the course of our work, secrecy, even where initially merited, has its long-term price.

At Hanford, as we have noted, the Green Run represented only a fraction of the risks (including nonradiation as well as radiation hazard) to which local communities may have been exposed in secret. The delayed legacy of these risks, in uncertainty and distrust, as witnesses from the Hanford community told the Committee, is only becoming apparent as the secret history of early Hanford operations has been made public.

During World War II, officials at Du Pont, the contractor for Hanford at that time, proposed a practice evacuation to prepare for a possible emergency. General Groves turned them down, saying that "any practice evacuation of the Hanford Camp would cause a complete breakdown in the security of the project."[43] As noted in the Introduction, at the onset of the Manhattan Project concern for the effects of Hanford operations on the surrounding environment, including the salmon in the Columbia River, led to a secret program of research on the environmental effects of Hanford's operations.[44]

Secrecy remained the rule at Hanford after the war. In 1946, as recalled years later by an early biologist at Hanford who wrote to radiation researcher and historian Newell Stannard, Hanford researchers resorted to deception simply to collect information about possible iodine contamination in livestock, by having employees pretend to be agricultural inspectors while surreptitiously monitoring iodine levels in animal thyroids. The biologist wrote: "Though the Environmental Study Group at Hanford had been sampling air, soil, water, and vegetation in a wide area surrounding the Hanford site for several years previous to 1946, it was agreed that sampling from farm animals for uptake of fission product plant wastes would be a much more sensitive problem. At the time, the revelation of a regional I-131 problem would have had a tremendous public relations impact and furthermore the presence of other radionuclides . . . was of possible National Defense significance."

He explained that he was called at home and told to report to work at the director's office in downtown Richland. There:

I was introduced to two security agents of the Manhattan Engineer District . . . who were to be my escorts and contact men during the day. They proved to be the best straight faced "liars" I had ever known. I was no longer "Karl Herde of DuPont" but through the day would be known and introduced as Dr. George Herd of the Department of Agriculture. I was to simulate an animal husbandry specialist who had the responsibility of testing a new portable instrument based on an unproven theory that by external readings on the surface of the farm, the "health and vigor" of animals could be evaluated. I was advised not to be alarmed if at times during the conversations with farmers that they appeared critical or skeptical. I was to be very reserved and answer questions as briefly and vaguely as seemed acceptable. They agreed to carry a clipboard . . . I was to concentrate on the high readings (thyroids, of course) and furnish those for recording when not being observed.

That day we visited several diversified farms under irrigation from the Yakima River between Toppenish and Benton City. . . Smooth talk and flattery enabled us to gain one hundred percent cooperation. . . .

I was successful in placing the probe of the instrument over the thyroid at times when the owner's attention was focused on the next animal or some concocted distraction.[45]

In 1948, the AEC prepared a public relations pamphlet entitled *Handling Radioactive Wastes in the Atomic Energy Program.* The Department of Defense objected to the description of Hanford's operations, arguing that any description of the methods used to reduce contamination might be used by the Soviet Union to avoid detection of its plants.[46] The AEC decided at its October 7, 1949, meeting to release the pamphlet, which contained no specific numbers, in order to "dispel and allay possible latent hysteria."[47]

With a major expansion of Hanford's operations under way in 1954, questions arose over whether to publish information about contamination of the Columbia River. Parker warned that it might be necessary to close portions of the river to public fishing, but he and others noted that this could have a substantial public relations impact.[48] At the same time, there was concern that information on river contamination could make it possible to ascertain Hanford's plutonium output.[49] For this combination of public relations and security reasons, Hanford did not release any quantitative information or public warning on contamination of fish in the Columbia River until many years later.

It is difficult to argue with the need for secrecy about the purposes of the Green Run. Making information on U.S. atomic intelligence methods openly available could have led the Soviet Union to develop countermeasures to these methods. The issue remains important today in responding to the potential proliferation of nuclear weapons capabilities around the world.

But the results of the long delay in informing the public about the activities of which the Green Run was only a part are now evident in public anger and distrust toward the government. At the Advisory Committee's public meeting in Spokane on November 21, 1994, Lynne Stembridge, executive director of the Hanford Education Action League, argued that

Information regarding that radiation release was kept secret for almost 40 years. There was no warning. There was no informed consent. Citizens down wind were never advised of measures that could have been taken to safeguard the health of themselves or their children.

Although the Green Run was not as direct as handing a patient orange juice laced with radioactivity, or giving someone an injection, the Green Run was every bit as intentional, every bit as experimental, every bit as unethical and immoral as the medical experiments which have made headlines over the last year.[50]

Among the most damaging dimensions of the legacy of distrust created by the secrecy that surrounded the routine and intentional releases at Hanford is the government's loss of crediblity as a source of information about risk. Now, when the government is attempting to find out what damage these releases actually did and share that information with the people affected, these people question why they should believe what the government says.[51] Federally funded scientists at the Fred Hutchinson Cancer Research Center in Seattle, Washington, are now studying those exposed as children to all of Hanford's iodine emissions—the many routine emissions as well as the Green Run—to see whether any health effects are detectable.[52] Whatever this study concludes, many residents are convinced that they have already seen the effects. Tom Bailie, who grew up and still lives on a farm near Hanford, spoke to the Advisory Committee's meeting in Spokane in November 1994. He pointed on a large map to what he called a "death mile," where "100 percent of those families that drank the water, drank the milk, ate the food, have one common denominator that binds us together, and that is thyroid problems, handicapped children or cancer."[53] It is doubtful that the results of any study supported with federal funds, no matter how impeccably conducted, would be believable to people like Mr. Bailie. Assuming that the Hutchinson Cancer Research Center study is so conducted, and assuming the study finds that at least some outcomes of concern to the community are not attributable to the Hanford emissions, government secrecy will have deprived Mr. Bailie and people like him of an important source of reassurance and peace of mind.

The Green Run, and the far greater number of environmental releases resulting from Hanford's routine operations, raises challenging questions about the balance between openness and secrecy in settings where citizens may be exposed to environmental hazards. Citizens may reasonably ask whether releases have been determined to be necessary in light of alternatives, whether actions have been taken to minimize risk and provide for any harm that might occur, whether disclosure will be made at the earliest possible date, and whether records will be created and preserved so that citizens can account for any health and safety consequences at the time of disclosure. As we will see, these questions were posed with regard to other environmental releases, and they remain with us today.

Radiological Warfare

The first proposed military application of atomic energy was not nuclear weaponry but radiological warfare (RW)—the use of radioactive materials to cause radiological injury. A May 1941 report by the National Academy of Sciences listed the first option as the "production of violently radioactive materials . . . carried by airplanes to be scattered as bombs over enemy territory."[54] It was not until later that year that a calculation by British physicists demonstrated the feasibility of nuclear weapons, and attention quickly turned to their development.

Military interest in both offensive and defensive aspects of radiological warfare continued throughout World War II. In the spring of 1943, when it was still unclear whether the atomic bomb could be built in time, radiological weapons became a possible fallback. Manhattan Project scientific director J. Robert Oppenheimer discussed with physicist Enrico Fermi the possibility of using fission products, particularly strontium, to poison the German food supply. Oppenheimer later wrote to Fermi that he thought it impractical unless "we can poison food sufficient to kill a half a million men." This proposal for offensive use of radiological weapons appears to have been dropped because of its impracticality.[55] At the same time, military officials developed contingency plans for responding to the possible use of radiological weapons by Germany against invading Allied troops.

The peacetime experience of Operation Crossroads in 1946, particularly the contamination of the Navy flotilla from the underwater nuclear test shot labeled Baker, revived interest in radiological warfare. Some, including Berkeley's Dr. Joseph Hamilton, concluded that radiological poisons could be used as strategic weapons against cities and their food supplies.[56] Once absorbed into the body, radioactive materials would cause slow, progressive injuries. Others proposed that RW could be a more humane form of warfare. Using radioactive material to contaminate the ground would render it temporarily unhabitable, but it would not be necessary to kill or injure people.[57]

Although many discussions of radiological warfare took place in classified military circles,[58] the basic notion of radiological warfare was not secret and was a subject of public speculation. But the government's program in radiological warfare remained largely secret, except in its broadest outlines. The postwar interest in radiological warfare spawned competing programs on radiological warfare both in the AEC and in various parts of the Department of Defense.[59] To meld these into a coherent program, the AEC and DOD established a joint study panel in May 1948, chaired by the chemist W. A. Noyes from the University of Rochester and including civilian experts and DOD and AEC officials.

At its first meeting that month, the Noyes panel recommended work in three areas: (1) biological research on the effects of radiation and radioactive materials, to be carried out mainly at the Army Chemical Corps's Toxicity Laboratory, located at the University of Chicago;[60] (2) studies on the production of radioactive materials for use in radiological warfare, carried out mainly by the AEC; and (3) military studies of possible RW munitions, also carried out mainly by the Chemical Corps.

The latter program was the focus of the Advisory Committee's attention because it involved the intentional release of radioactive materials during several dozen tests of prototype radiological weapons at the Chemical Corps's Dugway Proving Ground in the Utah desert. The offensive radiological warfare program field-testing program coincided with the Korean War years. The Noyes panel issued its final report after its sixth meeting, in November 1950,[61] and was revived briefly in 1952 to assess the status of the RW research program.[62]

The first two field tests were conducted at Oak Ridge. These involved sealed sources of radioactive material that were placed in a field in order to measure the resulting radiation levels. These measurements may have helped predict the effectiveness of radiological weapons. The sources were then returned to the laboratory and left no residual contamination in the environment.[63]

Most of the radiological warfare field tests were carried out by the Chemical Corps at the Dugway Proving Ground, using radioactive tantalum produced at Oak Ridge.[64] From 1949 to 1952, the Chemical Corps conducted sixty-five

field tests at Dugway, intentionally releasing onto the ground roughly 13,000 curies of tantalum in the form of dust, small particles, and pellets. These were prototype tests, releasing much smaller quantities of radioactive material than the millions of curies per square mile that an operational radiological weapon would need to render territory temporarily uninhabitable.[65] Furthermore, the field-test programs used tantalum primarily because it could be produced at existing facilities. An operational radiological warfare program required materials that could be produced in greater quantities than tantalum, but this would have meant constructing special production facilities.[66]

In May 1949, the Chemical Corps established a panel of outside experts to provide advice on the safety of its field-testing program. Chaired by Dr. Joseph Hamilton, a strong advocate of the RW research program,[67] the panel was chartered to consider radiological hazards to the civilian population, including hazards to "the water supply, food, crops, animal population, etc." Occupational safety was left to the Chemical Corps.[68]

Under Hamilton's leadership, this panel raised a number of safety concerns but in the end appears to have been satisfied with the safety of the test program. Several months before the first panel meeting, Hamilton himself had objected to the use of the relatively long-lived isotope tantalum 182 (half-life, 117 days) as the radiological warfare agent in these field tests. He proposed using gold 198 instead (half-life, 2.7 days) to eliminate any lingering radiation hazard to the general population.[69]

At its first meeting, on August 2, 1949, the RW test safety panel provisionally accepted the proposed testing program of the Chemical Corps, subject to a radiological safety review of the results of the first two tests. Hamilton's potential opposition clearly was of consequence, and his agreement to proceed was cause for relief.[70]

Other members of the test safety panel, including Karl Morgan, head of health physics at Oak Ridge, raised concerns about the possible hazard posed by radioactive dust at an arid site like Dugway,[71] both on- and off-site. Morgan proposed the use of airborne monitoring equipment developed at Oak Ridge in tests that preceded the Green Run.[72] The use of such aircraft and other monitoring equipment evolved and expanded as the Dugway field tests continued over the next few years. Panel members approved the continuation of the program based in part on the results of these radiological surveys, which showed that contamination of the area was limited in size.[73]

In 1952 the Chemical Corps proposed a significant expansion of the radiological warfare program, with a large test of 100,000 curies planned for 1953 and still larger tests proposed for later. The test safety panel once again raised concerns over the radioactive dust hazard. Hamilton noted that there were several "hot spots"—areas of unusually high radiation—at Dugway and that trucks at one of the target areas were kicking up significant quantities of radioactive dust.[74] A Chemical Corps study in early 1953 concluded that the hazard was relatively slight.[75]

Hamilton favored going ahead with the 1953 tests and was greatly disappointed when they were canceled, and with them the entire radiological warfare test program.[76] The reasons for this cancellation are not entirely clear, but two factors are evident. The next phase of the program would have required the construction of expensive new production facilities, which collided with military budget cuts at the end of the Korean War. Furthermore, by 1953, only the Chemical Corps maintained a strong interest in the radiological warfare program, making it vulnerable to questions about whether it satisfied any unique military need.[77] The radiological warfare program did not end completely, but its focus narrowed to defensive measures, including shielding and decontamination,[78] with atmospheric nuclear tests providing the main opportunity for study.[79]

The radiological warfare test safety panel was an early example of the use of an expert panel to evaluate possible risks of planned government activities. Ideally, such a panel should not be chaired by a proponent of the program in question, although those with such knowledge of, and interest in, the program are of obvious value to a safety effort. Hamilton's evident enthusiasm for radiological warfare research raises questions about his impartiality as head of the panel,[80] but the panel as a whole appears to have dealt with serious public health issues in a responsible manner.

Secrecy in the Radiological Warfare Program

The U.S. radiological weapons-testing program appears to have remained formally secret until 1974 and remained largely unknown to the public until the GAO's report in 1993.[81] There was a recurring tension at the time between those who wanted to release information to allay unwarranted public fears about radiation hazards and those who thought that publicity would create unwarranted attention and public apprehension that could interfere with the successful prosecution of the program. If there was a concern that public knowledge of the general outlines of the program would undermine national security, none of the available documents state this argument explicitly, except through their classification markings.

In May 1948, at its first meeting, the Noyes panel recommended that the entire program be classified Secret, Restricted Data;[82] the Chemical Corps's RW program was classified at this level.[83] At its second meeting, in August, the Noyes panel revised this recommendation to conclude that "[t]he existence of an RW Program should be considered as unclassified information."[84] The Noyes panel was responding to the recommendation by the AEC's ACBM "that the Advisory Committee on Biology and Medicine urge that the broad subject of Radiological Warfare be declassified" on the grounds that "the subject appears in nearly every Sunday supplement in a distorted manner" and that "better work could be done from the scientific and medical standpoint" if the program were declassified.[85]

In February 1949, Defense Secretary James Forrestal, responding to requests for greater public disclosure of U.S. nuclear activities, appointed Harvard University President James Conant to chair a confidential ad hoc committee to make recommendations on "the information which should be released to the public concerning the capabilities of, and defense against, the atomic bomb and weapons of biological, chemical, and radiological warfare."[86] This high-level committee's work ended in October 1949 in deadlock, without making any strong recommendations. Its report to President Truman was quickly forgotten and, if anything, provided the basis for continuing the existing pattern of secrecy.[87]

Among the listed rationales provided by the majority of committee members who opposed the release of further information on the capabilities of atomic weapons was the absence of "public demand" for the information. (The positions taken "by certain well-known and probably well meaning pressure groups," they suggested, "do not spring from any general public sentiment in this regard and should, therefore be ignored.") James Hershberg, in his biography of Harvard University President James Conant, who chaired "The Fishing Party" (as the committee was code-named), has observed:

Notably missing from this list is any indication that they were worried that the Soviet Union might derive military benefit from the release of data under consideration. . . . The observation [of the majority] that the "public would seem to be more concerned lest their officials release too much classified information, rather than too little" may have been accurate, but would the attitude have been the same if it were known the government was hiding the information not from Moscow but from its own people because it did not trust them? How else to explain the fear that "even a carefully reasoned statement . . . might have a very disturbing effect on the general public and could be misinterpreted by pressure groups in support of any extreme position they were currently advocating"?[88]

In May 1949, while Conant's panel deliberated and the Chemical Corps was preparing for the initial Dugway field tests, the Defense Department's Research and Development Board (RDB) addressed the question of releasing information on radiological warfare. The RDB's Committee on Atomic Energy recommended against a public release of information. Soon after, a joint meeting of the Military Liaison Committee and the General Advisory Council

considered, but rejected a drafted letter to the President, also recommending a press release on the RW program. Later that year, on advice from Joseph Hamilton, the Chemical Corps prepared a release regarding munitions tests at Dugway. The Chemical Corps's proposal for a release was discussed with AEC and DOD officials, who rejected it, saying such a release was "not desirable."[89]

At roughly the same time, Defense Secretary Louis Johnson briefed President Truman on the radiological warfare program. The briefing memorandum prepared for Truman said that the planned tests posed a "negligible risk," but argued that "should the general public learn prematurely of the tests, it is conceivable that an adverse public reaction might result because of the lack of a true understanding of radiological hazards." It also noted that "a group of highly competent and nationally recognized authorities is being assembled to review all radiological aspects of the tests before operations are initiated at the test site."[90]

The reference in the briefing memorandum was to the radiological warfare test safety panel, which was being selected at that time. In August, at the first meeting of this panel, Albert R. Olpin, president of the University of Utah, noted the risk that uranium prospectors might stumble onto the site.[91] Citing Olpin's concern, Joseph Hamilton noted,

While the hazards to health for both man and animals can be considered relatively slight, the adverse effects of having public attention drawn to such a situation would be most deleterious to the program. In particular, Dr. Olpin brought up the interesting point that most of Utah is being very carefully combed by a large number of prospectors armed with geiger counters. Needless to say, it is imperative that such individuals be denied the opportunity to survey any region containing a perceptible amount of radioactivity arising from the various radioactive munitions that are to be employed.[92]

Soon after this meeting, Hamilton also proposed a public release of information, perhaps reasoning that a program that was announced, but played down,[93] would attract less attention than one that was discovered accidentally. Hamilton's proposal was refused.[94] Echoing

Hamilton's concerns, the Chemical Corps proposed once more that the tests be made public, again citing the risk of discovery by uranium prospectors.[95] Robert LeBaron, chairman of the DOD's Military Liaison Committee to the AEC, turned down this request, claiming the need for review by the Armed Forces Policy Council.[96]

The official silence about the prospects for radiological warfare prompted some public speculation about the government's activities, including a report appearing in the *Bulletin of the Atomic Scientists*, a journal created following the war to give a policy voice in print to many of the physicists who had worked on the bomb. The journal had some following in the general public as well as the scientific community. The report mirrored much of the analysis of the Noyes panel and concluded that RW had significant military potential.[97]

In September 1949, the AEC's Declassification Branch recommended that certain general information, civil defense problems, and medical aspects of RW be declassified. Details regarding specific agents and methods of delivery, however, should remain secret.[98] These suggestions appear to have been adopted shortly thereafter, as AEC and DOD reports at the end of 1949 and into the early 1950s discuss some aspects of the RW program in very broad terms.[99] The closest thing to an official announcement of the field-test program appears to have come in a report for the first half of 1951.[100] This report briefly noted that "research and development activities in chemical, biological, and radiological warfare were accelerated," and that "Dugway Proving Ground . . . was reactivated, and major field-test programs in offensive and defensive toxicological warfare were started," but provided no details. The 1994 summary of declassification policy by the Department of Energy notes that offensive radiological warfare was declassified in 1951 by the AEC, although the Defense Department appears to have kept this aspect of the program classified until much later.[101]

The secrecy that surrounded the radiological warfare field-test program raises two related questions. The first question is whether concerns over public reaction are a legitimate basis for security classification. Officials at various levels cited fears of "public anxiety," "undue public apprehension," and even "public hysteria" to justify keeping even the most general information secret.

The documents reviewed by the Advisory Committee do not record the actual decisions at various stages to keep the field-testing program secret; they refer only to such decisions being made by others. It may be that those decisions reflected other reasons for secrecy. Or it may be that public reaction was considered a national security issue. This can be a legitimate argument, when the program in question is considered vital to the nation's security. However, the nation has a vital interest in open public participation in representative government, and making exceptions to the rule of openness requires a high standard of national need.

The second question is the same as the one raised for the Green Run: Can potentially important public health information about secret activities be made available to the public without compromising secrecy about the details and purposes of the activity? As described later in this chapter, this remains a live issue today.

The RaLa Tests: Two Decades of Experimentation

From 1944 to 1961, the Los Alamos Scientific Laboratory used lanthanum 140 (also known as radiolanthanum or RaLa) in 244 identified tests of atomic bomb components.[102] These tests were critical to the development of the plutonium bomb, which required a highly symmetrical inward detonation of high explosive—known as implosion—to compress the plutonium fuel and allow a critical chain reaction. The RaLa method (see "What Were the RaLa Tests?") was the only technique available for measuring whether the implosion was symmetrical enough and continued to be used for testing bomb designs until the early 1960s, when technical advances allowed the use of alternative techniques.[103]

WHAT WERE THE RALA TESTS?

Implosion devices use carefully timed detonations of carefully shaped high-explosive charges to generate a spherically symmetrical inward-directed shock wave. This shock wave in turn compresses the nuclear fuel of an atomic bomb—usually plutonium—causing it to "go critical" and undergo a nuclear chain reaction.[a]

In the RaLa tests, the plutonium core was replaced by a surrogate heavy metal with an inner core of lanthanum. Lanthanum 140 has a half-life of forty hours, emitting a high-energy gamma ray in its decay. Some of these gamma rays were absorbed as they passed through the outer components of the implosion device, the degree of absorption depending on how compressed those components were. Radiation measurement devices placed in various directions outside the device would indicate the overall compression and whether that compression was symmetrical or instead varied with direction. The lanthanum sources typically ranged from a few hundred to a few thousand curies, the average being slightly more than 1,000 curies, and were dispersed in the cloud resulting from the detonation.

In 1950 the Air Force flew a B-17 aircraft carrying an atmospheric conductivity apparatus in four radiation-tracking experiments at Los Alamos. These four experiments were identified subsequently by the General Accounting Office[104] and appear in the Advisory Committee's charter.[105] A historical analysis undertaken by the Los Alamos Human Studies Project Team in 1994 identified three of these experiments, in which the environmental release of radiation was incidental to the experiment, as part of the series of 244 intentional releases mentioned above; the presence of the tracking aircraft is all that distinguishes the three in the Advisory Committee's charter from the other 241.[106]

The Los Alamos Scientific Laboratory was established in 1943 as the atomic bomb design center for the Manhattan Project on a mesa overlooking the Rio Grande valley, about forty miles northwest of Santa Fe, New Mexico. The RaLa tests were conducted in Bayo Canyon, roughly three miles east of the town of Los Alamos, which grew up next to the lab. Although radioactive clouds from the RaLa tests occasionally blew back toward the town, the prevailing winds usually blew those clouds over sparsely populated regions to the north and east. Aside from a small construction trailer park and a pumice quarry within three miles, the next nearest population center was the San Ildefonso pueblo, roughly eight miles downwind of the test site in the Rio Grande valley. Several Pueblo Indian and Spanish-speaking communities lie within twelve miles of Los Alamos.

Risks to the Public

Concerns over risks to the public arose at the beginning of the RaLa program. In the early years, Los Alamos planners and health physicists worried that the detonations could cause some contamination in areas outside the test site, such as the construction trailer park and nearby hiking trails.[107]

As the RaLa program continued, several patterns of public safety practices developed. Initially, the principal way to protect people was to keep them out of the immediate test areas, but in later years it became the practice to test only when the weather was favorable, and later still to survey surrounding roads to detect whether contamination had reached hazardous levels.

Perhaps because early atmospheric monitoring had produced only negative results and be-

---

a. Lillian Hoddenson et al., *Critical Assembly: A Technical History of Los Alamos during the Oppenheimer Years, 1943-1945* (New York: Cambridge University Press, 1993), 268–271.

cause surveys in Los Alamos had indicated only minimal levels of contamination,[108] ground contamination was not believed to be a significant problem at first. Environmental surveys after RaLa tests indicated significant contamination at some locations within three miles of the release, but not at greater distances.

This observation, and the opening of a pumice quarry within three miles of Bayo Canyon, led to intensive studies of fallout from the RaLa tests in 1949 and 1950. These studies led Los Alamos to conclude that "any area which is two miles or more from the firing point may be regarded as a non-hazardous area."[109] As a result of these studies, Los Alamos restricted RaLa testing to take place only when the winds were blowing away from the town and laboratory of Los Alamos.[110] Systematic weather forecasting, therefore, began only in 1949, after more than 120 tests had been carried out, and maintaining the capability to forecast wind conditions for these tests remained an important requirement over the years.[111]

The meteorological constraints presumably reduced the radiation exposures in Los Alamos itself; exposures in more distant communities, while probably more frequent, remained lower than Los Alamos. At the Advisory Committee's public meeting in Santa Fe on January 20, 1995, however, Los Alamos activist Tyler Mercier commented that most of the "shots were fired when the wind was blowing to the northeast. At this point in time, that's where most of the population of this region lived. I mean, half of it is Spanish and half of it Native American." Mercier concluded that there "appears to be a callous disregard for the well-being and lives of the Spanish and Native Americans in our community."[112]

The RaLa tests were suspended from July 1950 to March 1952. Routine radiological survey procedures were put into place when testing resumed. Surveyors would drive along roads in three sectors monitoring radiation hazards. Readings were typically below 1 mrad per hour (1 mR/hr), but reached levels of up to 15 mR/hr at nearby locations and up to 3 mR/hr at distances of several miles. Readings in excess of 6 mR/hr required further action, including possible road closure. If the surveyors detected significant levels, they would continue monitoring

in the next canyon downwind. On at least one occasion, ground contamination at relatively large distances from Los Alamos led monitors to extend their survey to a nearby town (Espanola), where they detected no radioactivity.[113]

The RaLa tests were understood from the beginning to be hazardous, but they were also critical to the design of nuclear weapons. Los Alamos officials took significant steps to understand and limit those risks. On at least two occasions—in late 1946 and from 1950 to 1952—they suspended testing amid questions about the continuing need and decided to continue testing.[114] When the RaLa tests finally ended in 1961, an alternative means of obtaining needed information had become available.

Risks to Workers

From the beginning, the RaLa tests also raised concerns over hazards to workers, particularly the chemists, in spite of elaborate measures adopted to limit these chemists' radiation exposures.[115] Lanthanum 140, with a half-life of forty hours, is itself the decay product of barium 140, which was separated from spent reactor fuel at Oak Ridge or Idaho National Engineering Laboratory in later years[116] and transported in heavily shielded containers to Los Alamos. There, chemists would periodically separate out the highly radioactive lanthanum for use in the implosion tests.

Soon after testing began on September 21, 1944, the RaLa program posed a puzzle for radiation safety. On October 16, Louis Hempelmann, director of the Health Division at Los Alamos, wrote to Manhattan Project medical director Stafford Warren about blood changes observed in the chemists working on the most recent RaLa test:[117]

[I]t looks now as though I was too excited about the blood changes, but at that time it seemed to me to be such a clear cut case of cause and effect that I thought the measurements of dosage must have been incorrect. Now I feel reasonably certain of the dosage.... It was a case where risk was taken knowingly and willingly because it seemed necessary for the project. ... It is my feeling that it should be the decision of the Director whether or not risks of this type should be taken....[118]

In August 1946 Hempelmann termed the exposures of personnel in the Chemical Group "excessive" and recommended that no more "RaLa shots" be attempted until "replacements are obtained for each member in this team."[119] The tests were suspended temporarily "because of over-exposure of personnel to radiation."[120] Los Alamos was faced with the alternative of increasing its staff (so that individual exposures could be reduced) or shutting work down until safety measures were installed.

RaLa testing resumed in December 1946, after a review to determine whether it was still necessary,[121] but no documents are available to determine whether safety procedures or staffing were changed. What did change was that researchers began a formal study of the relationship between the radiation exposures and blood counts of the Bayo Canyon chemists. The chemists' depressed white blood counts (lymphopenia), presumably the same changes noted two years earlier, posed a puzzle that continued for at least a decade, resulting in three scientific reports.[122] In 1954, Thomas Shipman, who had replaced Hempelmann as Health Division director, wrote to the AEC that

The blood counts were done with extreme care . . . and we are satisfied that the changes in counts are actual and not imaginary. It is our belief, however, that they don't mean anything; if they do mean anything, we don't know what it is.[123]

The cause of these blood effects remains uncertain. The reported doses of roughly 10 rad per year are well below levels expected to produce any detectable blood changes, a fact that was known by 1950.[124] While it is possible the effect could have been due to undetected internal contamination,[125] a more likely explanation may be that the chemists were exposed to chemical compounds that produced the observed blood changes.[126]

It appears that in the latter part of the 1940s some Los Alamos officials worried about the possible consequences of publicly releasing data on health effects, including those related to the chemists. A 1946 internal Los Alamos memo records that Dr. Oppenheimer asked that "all reports on health problems be separately classified and issued at his request." The author of the memo indicated his belief that the purpose was to "safeguard the project against being sued by people claiming to have been damaged."[127] Two years later, Norman Knowlton, a Los Alamos hematologist, reported on the blood changes in ten workers at the lab. A 1948 memo from the AEC's insurance branch argued that releasing this report on blood counts could have "a shattering effect on the morale of the employees if they became aware that there was substantial reason to question the standards of safety under which they are working" and concluded that "the question of making this document public should be given very careful study."[128] The report was not classified, however, although later reports were stamped "Official Use Only."

While the remaining information on the Los Alamos chemists is fragmentary, the experience raises an enduring question: What are the obligations of the government and its contractors to notify and protect employees whose work may expose them to continuing hazards, even when the risk is known to be small or is uncertain? As is discussed in chapter 12, during the same period, issues of worker protection and notification were raised much more starkly in the case of the uranium miners, who were placed at significant risk, a risk they had not "knowingly and willingly" taken.

Informing the Public

Although many in Los Alamos—those who worked on bomb design—knew of the RaLa program and its potential hazards, there is no indication of any discussion with other workers or local communities. For example, from the mid- 1940s to the mid-1950s many Pueblo people who may not have been informed worked at the lab as day laborers, domestics, and manufacturers of detonators.[129] The first public mention appears to have come in 1963, when the Los Alamos laboratory newsletter printed an article describing the cleanup of Bayo Canyon.[130] Los Alamos reports that its first concerted efforts to tell the Pueblo people about the RaLa program did not occur until 1994, when Los Alamos began its review of the RaLa program.[131]

Representatives of the pueblos near Los Alamos most likely to be affected by the RaLa tests have complained about past and continuing failures of laboratory officials to communicate with Pueblo workers or communities. Recent efforts at Los Alamos to undo this legacy of secrecy have created a continuing sense of frustration; Pueblo representatives state that information and other relations with the lab are still too tightly controlled to be trusted completely.[132]

It is difficult for any outsider to appreciate fully the unique cultural and religious viewpoint from which the Pueblo Indians perceive the effects of environmental releases. In addition to having several holy sites located near Los Alamos, the Pueblo have a deep respect for the land, which appears to have been violated by many of the activities at Los Alamos.[133] The Pueblo continue to rely to some degree for the basic necessities of food, heat, and shelter on plants, animals, and the earth, and they suspect that they may be at added risk of exposure to radioactivity in the environment.[134]

George Voelz, a Los Alamos physician who was at the lab during some of the RaLa tests, told the Advisory Committee, "As far as I know there was not much communication going on with the people in the area. And that, in retrospect was a mistake."[135] As a result of these failures of communication, Los Alamos now faces a difficult challenge, five decades later, of attempting to establish trust with neighboring communities that have become more suspicious because of what they have learned. Here, as in Hanford, credibility is the casualty of silence and secrecy.

Studies of Environmental Risks and Safety

The Green Run and the radiological warfare and RaLa programs were by no means the only government-sponsored experiments in which radioactive materials were intentionally released into the environment. Scientists undertook a wide variety of studies designed to understand the risks of environmental exposure to radioactive materials. For example, tests of experimental nuclear reactors at the National Reactor Testing Station in Idaho and the National Reactor Development Station in Nevada were designed to simulate possible accident scenarios under carefully controlled and isolated conditions. Similarly, tests at the Nevada Test Site were designed to understand the possible effects of an accidental (nonnuclear) explosion of a nuclear weapon.[136]

In addition to intentional releases designed to test the safety of nuclear machinery, safety was also a concern in studies designed to understand the fate of radioactive materials in the environment. Many of these studies simply took advantage of releases that occurred accidentally or were incidental to other projects. In 1943, studies of the exposure of salmon in the Columbia River to the radioactive effluent from Hanford's reactors set in motion the growing and largely public science of radioecology. The environmental analogue of radioisotope tracer studies designed to better understand the workings of the human body, these studies were intended both to follow the course of radionuclides released into the environment during nuclear weapons production and testing, and use radionuclides to trace the basic workings of the environment. The deliberate release of very small quantities of radioactive material provided the opportunity for more-controlled environmental study than those studies that simply observed radionuclides already released into the environment.[137] The Advisory Committee did not attempt to survey the entire field of radioecology, but we have reviewed the following examples in some detail.

Project Chariot

Project Chariot was a component of Project Plowshare, the brainchild of physicist Edward Teller, who helped develop the first hydrogen bomb. Plowshare arose in the late 1950s in response to public protests against atmospheric nuclear testing and was intended to demonstrate that "clean" nuclear explosives would provide safe, peaceful uses of atomic energy.[138]

In 1958, Teller selected a site in northern Alaska for Project Chariot, the proposed excavation of an Arctic seaport using a series of nuclear explosions. The site chosen was near Cape Thompson, roughly thirty miles from the Inupiat Eskimo village of Point Hope. This proposal, which was the subject of public debate, died in 1962 in the face of popular opposition.[139] However, extensive observations of the Alaskan ecosystem were undertaken between 1958 and 1962 to provide a baseline for comparison with results of the planned nuclear explosions. These observations led to the first awareness of the environmental hazards of cesium 137 from distant (primarily Soviet)[140] atmospheric nuclear tests and led to a series of studies on cesium in the food chain and in humans.[141]

Most of the environmental studies in Project Chariot were purely observational, but one series of studies involved the intentional release of small quantities of radioactive materials—a total of 26 millicuries of iodine 131, strontium 85, cesium 137, and mixed fission products.[142] In several studies, researchers from the U.S. Geological Survey spread radioactive materials on the surface of small plots of land and observed their spread across the surface when sprayed with water to simulate rainfall. In another, researchers placed mixed fission products in a small pit and measured their transport through the subsurface clay, and in yet another, researchers studied the spread of radioactivity in a creek contaminated with radioactive soil from Nevada. After these studies, the contaminated soil was removed and buried in above-ground mounds. Although this was a technical violation of regulatory requirements, an AEC memo expressed general satisfaction with the cleanup, noting that burial in the permafrost would have been too difficult.[143]

After the initial cleanup, the site remained dormant for thirty years until 1992, when a researcher discovered correspondence between the AEC and USGS about the tracer studies. In response to public concerns, the Department of Energy undertook to clean up the mounds' potentially contaminated soil. A survey indicated no externally observable radioactivity, and very little, if any measurable, radioactive material was believed to remain. In 1993, the mounds of soil were removed for disposal at the Nevada Test Site.[144] Caroline Cannon, an Inupiat Indian resident of Point Hope, told the Advisory Committee at its public meeting in Santa Fe,

I have lived in Point Hope all my life and eaten the food from the sea and the land and drank the water of Cape Thompson, along with the others. I have to wonder about my health, what impact the poison on

the earth will have all through my lifetime, emotionally, physically, and most of all for my children and my grandchildren.[145]

Although the risk to the population was minimal, residents still wonder whether other experiments might have occurred and remain secret.[146] Here again, government secrecy in the past is undermining government credibility in the present. How much comfort are Ms. Cannon and others like her able to take in reassurances from the government about risks to future generations, a government that they perceive unjustifiably kept them in the dark?

Controlled Radioiodine Releases

A small number of intentional releases involved the deliberate exposure of human subjects to trace quantities of radioisotopes in the environment. The most systematic of these were five of the roughly thirty Controlled Environmental Radioiodine Tests (CERT), carried out at Idaho National Engineering Laboratory (INEL between 1963 and 1968. Small quantities of I-131 were released into the atmosphere under carefully monitored meteorological conditions.[147]

In one study, seven volunteers drank milk from cows that grazed on the contaminated pasture. The quantity of iodine was measured carefully in the air, on the grass, in the milk, and later in the volunteers' thyroids, allowing a quantitative reconstruction of the full environmental pathway.[148] The maximum exposure among these volunteers was reported as 0.63 rad to the thyroid, nearly a factor of 50 below the contemporary annual occupational exposure limits.[149] In four other studies, a total of about twenty volunteers stood downwind at the time of the release; their exposures, from inhaling I-131 in the air, were much lower.[150] Apparently, all these volunteers were members of the INEL staff.[151] Measurements of the radioactivity in their thyroids provided a quantitative reconstruction of the inhalation pathway.

Studies similar to the CERT took place at Hanford in 1962, 1963, and possibly in 1965. The 1963 Hanford test involved human volunteers from Hanford's health physics staff, as did studies of iodine uptake from milk.[152]

The subjects in all these studies are referred to as volunteers in the relevant documents. No evidence is available bearing on what these subjects knew or were told about the experiments or the conditions under which they agreed to participate. The subjects were all staff members of the agency (or its contractors) conducting the research. The documents suggest that these staff members included knowledgeable individuals who participated in these experiments in the spirit of self-experimentation.

Reconstructing, Comparing, and Understanding Risks

Thus far, we have only briefly characterized the risks associated with the intentional releases reviewed in this chapter. Just how risky were those intentional releases and how much of this risk materialized? Although these questions cannot be answered with certainty, the answers can be approximated. Actual and suspected failures to respect public health in the environmental practices of the past have often led to efforts to reconstruct the basic facts and estimate the likely harm from environmental releases of radioactive materials. This process of environmental dose reconstruction has become an essential part of informing the public.

The task of estimating past environmental exposures to radioactive materials is a complex, multistep process. The first step is to collect data from historical records on the amount of material released. The second is to use records on weather, actual measurements of radioactivity in the environment, and computer models to reconstruct where this material went. The third step is to estimate how this distribution of material might result in radiation exposures to humans. Finally, these exposure estimates can be combined with mathematical models of radiation risks to estimate the resulting harm to people who were exposed.

Radioactive materials released into the environment can affect humans in two ways. First, they can be a source of radiation external to the body: beta radiation, which affects the skin, or more penetrating gamma radiation. Second, they can enter the body from contaminated air, food, or water and provide an internal source of radiation. Of these environmental pathways to radiation exposure, the food pathway is by far the most complicated. Radionuclides can enter the food chain at many points, through contaminated air, water, and soil, resulting in contaminated fruits, vegetables, meat, and dairy products.

The hazards from environmental exposures to radionuclides differ in important quantitative ways from those due to medical procedures or participation in biomedical research. The natural dilution of materials in the environment means that individual exposures even from massive releases are often quite small, although the chemical and biological processes involved in exposures through the food chain can lead to effects that counteract this dilution. Finally, many more people may be exposed, with exposures that vary widely from person to person.

Because individual exposures are generally too low to produce any acute effects, the main form of injury possible from environmental radiation exposure is cancer, which may occur many years after the exposure, and the number of cases attributable to such exposures can be expected to be relatively small. Evidence of cancer from exposure to radiation is difficult to separate out from other possible causes of those injuries; for the intentional releases discussed in this chapter, it is essentially impossible. Instead, we must rely on models of risk based on studies of other human radiation exposures.

Increased cancer rates among Japanese survivors of the atomic bombings provide the basis for most current radiation exposure risk estimates.[153] Health effects from the massive accident at Chernobyl and from other sites in the former Soviet Union should also be detectable and eventually may improve our understanding of the risks of chronic, low-level radiation exposure. The uncertainties in these scientific analyses are a major component of the uncertainty in risk estimation from environmental exposures.

In addition to individual exposures, it is important to know how many people were exposed. The population dose—obtained by adding up the individual exposures—provides a measure of the overall risk to the exposed population. According to models used by the Environmental Protection Agency (EPA), we can expect about

Table 1. Magnitude of Radioactive Releases

Event (number)	Location	Year(s)	Curies Released (Total)	Isotope	Risk (fatal cancers)[a]
Chernobyl	Ukraine, Soviet Union	1986	950,000 1,900,000 17,000,000	Cs-134; Cs-137; I-131[b]	17,400 expected/2.9 billion exposed[c]
Household radon	United States	Lifetime	N/A	Ra-222	14,000 per year expected/240 million[d]
Atomic weapons testing (atmospheric)	Worldwide	1945-1980	~26 million(Cs-137);~18 million (Sr-90); ~19 billion (I-131); ~6.5billion (H-3); ~6 million (C-14)	Cs-137; Sr-90 I-131; H-3; C-14	12,000 expected/5 billion[e]
First A-bombs	Hiroshima & Nagasaki, Japan	1945	~250,000,000	Short-lived fission products[f]	300 estimated/76,000 tracked[g]
Early Hanford operations	Hanford, Washington	1945-1947	700,000	I-131[h]	~1.6 cases of thyroid cancer expected/ 3,200[i]
Three Mile Island	Harrisburg, Pennsylvania	1979	15 10,000,000	I-131; noble gases[j]	0.7/2 million exposed[k]
RaLa tests (254)	Los Alamos, New Mexico	1944-1962	250,000	La-140	0.4 cases/10,000 exposed[l]
Green Run	Hanford, Washington	1949	8,000 20,000	I-131; Xe-133	0.04 expected/30,000 exposed[m]
RW field tests (65)	Dugway, Utah	1949-1952	13,000	Ta-182[n]	Unknown[o]

a. For every event but one, this column displays the risk of excess cancer fatalities. For I-131 released during "Hanford early operations," it displays the risk of excess cases of thyroid cancer.

b. United Nations Scientific Committee on the Effects of Atomic Radiation (UNSCEAR), *Sources and Effects of Ionizing Radiations* (New York: United Nations, 1993), 114, basing findings on L. A. Ilyin et al., "Recontamination Patterns and Possible Health Consequences of the Accident at the Chernobyl Nuclear Power Station," *Journal of Radiological Protection* 10 (1990): 3–29. The radioactivity released in the Chernobyl accident would include other fission products, particularly long-lived ones, but isotopes of cesium and iodine posed the greatest health hazard.

c. Lynn R. Anspaugh, Robert J. Catlin, and Marvin Goldman, "The Global Impact of the Chernobyl Reactor Accident," *Science* 242 (1988): 1516.

d. Environmental Protection Agency, Public Health Service, *A Citizen's Guide to Radon* (Washington, D.C.: GPO, May 1992), 2.

e. United Nations Scientific Committee on the Effects of Atomic Radiation, *Ionizing Radiation: Sources and Biological Effects* (New York: United Nations, 1982), 212–226. While the list of fission products released is incomplete, other products do not contribute much in the way of effective doses.

f. This is the rough level of radioactivity remaining one day after each of the explosions, including biologically active and relatively active isotopes. Samuel Glasstone, ed., *The Effects of Atomic Weapons* (Washington, D.C.: GPO, 1950), 220. The level of radioactivity diminished rapidly thereafter. Prompt neutron and gamma radiation from the nuclear explosion, rather than fallout, was responsible for most of the radiation exposures.

g. "Life Span Study," in Hiroshima Radiation Effects Research Foundation [electronic bulletin board] (cited 31 May 1995); available from www.rerf.or.jp; World Wide Web. This is the number of excess cancer fatalities between 1950 and 1985 among the 76,000 for whom doses have been calculated.

h. Sara Cate, A. James Ruttenber, and Allen W. Conklin, "Feasibility of an Epidemiologic Study of Thyroid Neoplasia in Persons Exposed to Radionuclides from the Hanford Nuclear Facility between 1944 and 1956," *Health Physics* 59 (1990): 169.

i. Kenneth Kopecky et al., "Clarification of Hanford Thyroid Disease Study," *HPS Newsletter*, July 1995, 24–25.

one induced fatal cancer for every 1,940 person-rem of radiation exposure.[154] While the risk to any one person may be small, the exposure of a large population can lead to a statistically significant increase in the number of fatal cancers, but it will be impossible to attribute any particular cancer to radiation exposure.

The Committee was not equipped to reconstruct historical doses from intentional releases, but can make some rough judgments based on more formal analyses performed by others.

The Green Run

The Green Run took place after years of routine emissions of radioiodine from the wartime and early postwar operations of the Hanford plant, and it added a relatively small amount to the overall risk (see the accompanying table1, "Magnitude of Radioactive Releases"). In 1987 the Department of Energy established the Hanford Environmental Dose Reconstruction (HEDR) project to provide an estimate of all the exposures that might have resulted and continues to refine its estimates of the resulting radiation doses to people.[155] These exposures, primarily through the food chain, may have produced a measurable excess in thyroid disease. A follow-up study of the exposed population is attempting to ascertain whether excess thyroid disease can indeed be seen.

The Green Run represents only about 1 percent of all the radioiodine releases from Hanford. Fortunately for most nearby residents, it occurred at a time of year when people were not eating fresh garden vegetables or drinking milk from cattle grazing in open pastures. The estimated radiation dose to members of the public from Hanford's operations for all of 1949 probably did not exceed 600 mrad to the thyroid, and doses ten times lower were more typical of the most highly exposed population. The Committee estimates that the Green Run may have increased the expected number of fatal thyroid cancers in the exposed population by 0.04, within broad error margins.[156] This means it is highly unlikely that even one person died as a result of the Green Run. A larger incidence of benign thyroid conditions is likely, but there is no evidence to support a connection between the intentional releases and any other possible medical conditions.

Radiological Warfare

No formal dose reconstruction has been done for the radiological warfare field tests at Dugway. Although the radioactive tantalum used in these tests does not concentrate in the food chain, because of its long half-life there may have been many opportunities for people to be exposed. Weather and vehicle traffic could have spread some of the contamination outside the Proving Ground, and even repeated low-level exposures to uranium prospectors or hikers who regularly wandered onto the site may have been possible.

Whatever public health hazard the RW tests at Dugway may have posed at the time, the

---

j. UNSCEAR, *Sources and Effects of Ionizing Radiation*, 114.

k. *Report of the President's Commission on the Accident at Three Mile Island: The Need for Change: The Legacy of TMI* (New York: Pergamon Press, 1979), 12.

l. This is an upper estimate based upon a preliminary dose reconstruction by staff of the Los Alamos National Laboratory of 1.1 mSV (1.1 rem). "Assuming an individual had been at the Los Alamos site continuously throughout the experiments, the total dose from the 18 year RaLa series was estimated to have been approximately 1.1 mSv." Using the average dose of 0.6 mSv (0.6 rem), the excess cancer risk falls to 0.24. Los Alamos notes, "A somewhat abbreviated approach could be used wherein a static population of 10,000 is assumed to be uniformly distributed across the Los Alamos of the 1950s. The dose as a function of distance could be used to estimate approximate population doses." D. H. Kraig, Human Studies Project Team, Los Alamos National Laboratory, fax to Gilbert Whittemore (ACHRE staff), 14 September 1995 ("Dose Reconstruction for Experiments Involving La140 at Los Alamos National Laboratory, 1944–1962") (ACHRE No. DOE-091495–A).

m. Maurice Robkin, "Experimental Release of I-131: The Green Run," *Health Physics* 62, no. 6 (July 1992): 487–495.

n. See, for example Chemical Corps, 1952 ("Explosive Munitions for RW Agents") (ACHRE No. NARA-112294–A-10); Chemical Corps, 1952 ("Testing of RW Agents") (ACHRE No. NARA-112294–A-7); George Milly, Chemical Corps, 27 June 1952 ("Report of Field Tests 623 and 624 Airburst Test of Two 1,000 Lb. Radiological Bombs") (ACHRE No. DOD-062494–A-16); E. Campagna, Chemical Corps, 18 September 1953 ("Static Test of Full Diameter Sectional Munitions, E83") (ACHRE No. DOD-062494–A-15).

o. The Advisory Committee knows of no dose reconstructions for these releases.

radioactive decay of the tantalum caused the risks to dissipate over time. By 1960, no more than a few millicuries of tantalum remained, dispersed so widely that by this time it posed no conceivable human or environmental hazard.

RaLa Tests

Los Alamos's 1995 report on the history of the RaLa test program contains basic information necessary for an environmental dose reconstruction, including the amount of radioactivity released, a rough indication of the amount of high explosive used in each test, and meteorological and fallout data where available.[157] Advisory Committee staff reviewed the process by which this information was assembled and reported that the historical reconstruction appears to be a reasonably accurate representation of what actually occurred.

Los Alamos is using this historical information to produce an environmental dose assessment, which it is providing to the state of New Mexico and plans to submit for publication in a peer-reviewed journal. The Committee was not in a position to judge the adequacy of the dose reconstruction, but the sources, methodology, and results will be available for review by outside experts.

Individual exposures from the full series of RaLa tests were somewhat higher than for the single release of the Green Run, and the exposed population was somewhat smaller. According to a preliminary dose reconstruction by the Human Studies Project Team at Los Alamos, the total dose for someone living continuously in Los Alamos for all eighteen years of the program was roughly 110 mrem. With a population of approximately 10,000 in Los Alamos County, 0.4 excess cancer deaths might be expected. The average dose would have been 60 mrem for someone living in Los Alamos.[158]

The General Accounting Office noted an Air Force report that a B-17 airplane detected radioactive debris from one of the tests as far as seventy miles away, over the town of Watrous, New Mexico, but it is unlikely that any significant risks extended to this distance. The Human Studies Project Team concluded, however, that the cloud could not have gone as far as claimed at the time of the observation and suggests that the atmospheric conductivity apparatus used by the Air Force was sensitive to effects other than radioactivity.[159]

Los Alamos has not attempted to reconstruct the doses to the Bayo Canyon chemists. Using data from one of the reports, however, it would appear that the total exposure for these chemists was high enough to place these individuals at some increased risk for developing a radiation-induced cancer.[160]

Other Intentional Releases

No risk estimates are available for the other releases the Committee has studied, and aside from DOE's Idaho National Engineering Laboratory, no dose reconstructions have been undertaken. It does appear, however, that the human health risks were small even compared with the minimal risks of the intentional releases discussed above and with other, more familiar exposures to radioactivity in the environment (see the accompanying table, "Magnitude of Radioactive Releases").

POLICIES AND PRINCIPLES GOVERNING SECRET INTENTIONAL RELEASES: THE EFFECTIVENESS OF CURRENT REGULATIONS

Policies and Practices in the Early Years

When the federal government set out to apply atomic energy to national needs, there were no specific rules or policies to govern the deliberate release of radionuclides into the environment. Nonetheless, the declassified record of the releases just reviewed shows that those responsible considered the basic issues that concern us today and that are today the subject of federal regulation. These include the need to limit risks, the question of who should bear those risks, and the extent of the obligation to inform affected citizens.

This record indicates that, for intentional releases as for biomedical experimentation, the government was most concerned with, and placed the highest priority on, limiting human health risks. At Hanford, for example, this was done by establishing limits for the permitted level of radioactive contamination. Some of these guidelines were exceeded, if only temporarily, by the Green Run. For the radiological warfare program, the Department of Defense established a panel of outside experts to safeguard against excessive risks to the general public.

The federal government struggled throughout these early years to clarify its obligations to protect the general public from the risks of radioactive contamination in the environment, particularly from atmospheric nuclear weapons testing (see chapter 10). The 1953 Nevada test series raised serious concerns about whether and how radioactive fallout from the expanding testing program was exposing nearby people and livestock to risk.[161] In an analysis that seems equally apt for intentional releases, Richard Elliott, information director of the AEC's Santa Fe Operations Office, argued at the time that the AEC had the obligation to show that the testing program was "vital to the nation and that it was conducted as safely as possible." He also asserted, however, that the agency had duties in addition to limiting risk, including

(1) To inform concerned publics of the hazards created and of preventive action which may be undertaken; (2) To warn people in advance of potentially hazardous situations, or of situations which may alarm them; (3) To report after the fact not only with reassurances but also with details and interpretations; (4) And, to the extent of the agency's responsibility, to reimburse the public for its losses.[162]

For most of the intentional releases described in this chapter, information was withheld entirely, even when that information might have enabled the public to reduce its risk, however small, of exposure to ionizing radiation.[163] This secrecy appears to have been motivated by legitimate national security needs in the cases of the Green Run and the RaLa program. The radiological warfare field-testing program was kept secret primarily to avoid public awareness and controversy that might jeopardize the program. The extent of secrecy abated in later years, and many of the intentional releases that occurred from about 1960 onward involved relatively low risks and were made known to the public.

Obligations to limit risk, to consider who should bear the risk, and to inform the public, while recognized, were often subordinated to concerns for national security, which were sometimes joined or melded with concerns for public relations. The information that is available indicates that the physical harm from the radiation is probably less than the damage—to individuals, communities, and the government—caused by the initial secrecy, however well motivated, and by subsequent failures to deal honestly with the public thereafter. The legacy of distrust, as described in the histories presented above, is probably more significant than the legacy of physical harm.

Regulating the Levels of Risk the Government May Impose

The past fifty years has seen the development of a body of laws and regulations governing releases into the environment, including releases of radioactive materials. These laws and regulations give legal standing to moral considerations about limiting risk, fairness in the imposition of risk, and disclosure to and involvement of the public. When environmental releases take place today—for example, in the cleanup of the nuclear weapons complex—they are subject to rules that provide procedures for public review and comment on proposed federal actions and to rules that limit the amounts of radiation that can be released into the environment.

Environmental law contains a variety of quantitative standards designed to limit the risk to human health from exposure to environmental hazards. These limits apply both to private companies and to the federal government.

The Atomic Energy Act of 1954 and the Clean Air Act of 1970 impose the most important constraints on intentional releases of radioactivity into the environment.[164] Regulations under both of these laws limit the maximum exposure to any one person. These limits are often supplemented by secondary standards (for example, on concentrations in air and water) designed to prevent exposures from exceeding this limit. This basic form of regulation remains largely unchanged from the early days of radiation protection, although the quantitative limits have been greatly reduced over the years.[165]

The actual limits on radiation exposures to members of the public have dropped dramatically over time. The initial postwar standard was for occupational exposures: 0.1 R per day.[166] If a person were exposed at such levels for his or her entire working lifetime, about fifty years, a rough extrapolation of current risk models would predict that he or she would be more likely than not to die of radiation-induced cancer. In practice, however, it is extremely unlikely that any worker came close to that level of lifetime exposure. Once it was recognized that standards for the general public should be stricter than those for a potentially hazardous workplace, the exposure standard for members of the public was set a factor of ten below the occupational standard. In 1960, when the occupational standard was reduced to 5 rem per year, the standard for exposures to members of the general public was reduced to 500 mrem per year from all artificial environmental sources.[167]

Since that time, the Environmental Protection Agency and the Nuclear Regulatory Commission (NRC) were established as separate regulatory agencies,[168] and radiation protection standards have been tightened further. The DOE and NRC have adopted the stricter limit of 100 mrem per year for general population exposure, and the EPA has proposed adopting a similar standard. The EPA's standard for atmospheric emissions under the Clean Air Act is a factor of ten lower: 10 mrem per year. A lifetime of exposure at this level would produce an expected excess in cancer deaths of a few in 10,000.[169]

By way of comparison, the average human exposure to background radiation from naturally occurring cosmic rays and radioactive materials is roughly 300 mrem per year. Exposure limits that were initially much higher than natural backgrounds have since fallen substantially below those levels. Actual public exposures are much lower still, with average medical exposures of roughly 50 mrem per year and exposures from nuclear power at roughly 1 mrem per year for people living closest to nuclear power plants.[170] Although the risk associated with the maximum allowed exposure from human-controlled sources has fallen over the years, so that it is now below that from natural background levels, it remains higher than that for exposure to chemi-

cal carcinogens, which range from 1 in 10,000 to 1 in 1,000,000.[171]

However, standards based solely on limiting individual exposures would not address the possibility that—as in the case of intentional releases—large numbers of people might be exposed to risk, though likely at low levels. As described above, the population dose, obtained by adding up all the individual doses, provides a measure of the overall risk to a large exposed population. A more universal application of the population dose in the regulatory process would give greater weight to this overall risk.[172]

Under some circumstances, however, the federal government may invoke exceptions to these baseline standards—imposing greater risks on its citizens where national need dictates. Under the Clean Air Act, only the President may invoke such exceptions, and only on the basis of "national security interest." The President must report to Congress on any such exceptions at the end of the calendar year.[173] Under the Atomic Energy Act, however, the Department of Energy is largely exempt from external regulation. When its predecessor, the Atomic Energy Commission, developed regulations for the civilian nuclear power industry, it also committed to operate its own nuclear facilities according to certain safety provisions, but allowed itself an exemption "when over-riding national security considerations dictate."[174] Such an exception under the Atomic Energy Act could still be invoked today. These exemptions clearly allow national security interests to take precedence over public health concerns. The Advisory Committee is concerned that this could occur without adequate consideration or oversight, and without adequate protection of the public's interest in a safe environment and public notice. Once the exemption is invoked, there is no formal limit on the risks to which members of the public may be exposed, although the requirement to report to Congress could deter some actions.[175]

Public Disclosure and Formal Review

Today's environmental laws require public disclosures of the likely environmental impacts of federal government actions, subject to public and EPA review, and EPA oversight of federal compliance with environmental regulations. As we will discuss below, the classification of information for national security purposes requires certain exceptions to the general rules described here.

The National Environmental Policy Act (NEPA) of 1969 requires that the federal government take into account and publicize the environmental impact of its actions.[176] NEPA's requirements serve the dual purposes of informing the public and forcing agencies of the federal government to inform themselves of the environmental impact of their actions. NEPA requires an agency to prepare an environmental impact statement (EIS) for any proposed "major federal action" having a significant impact on the human environment.[177]

As long as an agency has followed the requisite procedures (and rationally explained its choices in the EIS) it may choose whatever course of action it likes, even the alternative that poses greater environmental risks. Nonetheless, the public process can have dramatic effects on the way agencies make decisions. Assessments that are subject to public comment and decisions that are open to public scrutiny force agencies to consider public reaction when they choose policy alternatives. The adequacy of the process is subject to review by EPA and, if members of the public sue, by the courts. However, environmental impact statements may be classified in whole or in part. The EPA is obliged to review and comment on the classified portions.[178]

The EPA is also charged with making sure the federal government complies with the substantive requirements of the Clean Air Act (and other environmental statutes), and shares oversight responsibilities under the Atomic Energy Act with DOE and the NRC. For example, EPA must approve the construction or expansion of a facility, certifying that such action would not exceed the limits of the Clean Air Act. Furthermore, agencies are required to report on their emissions to EPA and are subject to fines if they violate the emissions limits. Under the Federal Facility Compliance Act, EPA must list and review environmental compliance at all federal facilities.

Selection of Sites and Affected Communities

The sites selected for intentional releases, and thus the populations affected, do not appear to have been chosen arbitrarily, but rather for reasons that are arguably defensible, albeit open to a charge of unfairness. Most of the releases took place in and around "atomic energy communities" and military sites, a choice that had several obvious advantages. First, the sites offered the expertise and facilities, both indoors and out, for the evaluation of releases involving radioactivity. Second, the locations of most of these facilities were originally chosen because of their relative, if not complete, isolation from major "civilian" population centers. Residents near these sites were generally accustomed to secret government activities in their midst. The selection of these sites for repeated exposure to releases of radioactivity—whether experimental, accidental, or routine—probably resulted in fewer people being exposed, but it also meant that the same groups were repeatedly exposed to higher than normal risks.

While there is no formal analogue to the research rules regarding fairness in the selection of subjects in the context of environmental releases, the environmental impact process does provide for public review of, and comment on, the rationale for the choice of taking an action in one locale, as opposed to another. In addition, by a 1994 executive order, President Clinton called on decision makers to consider whether actions affecting the environment may have disproportionate impact on the environment of poor or minority populations.[179] When the environmental review and decisions are made in secret, however, opportunities for any group of citizens to make their concerns known are limited.

The Effects of Secrecy on Current Policies and Protections

As we have seen, current law permits the conduct of intentional releases in secret. Secret intentional releases pose two kinds of problems for the interests of the public—loss of assurance that secret releases comply with laws regulating risk exposure and loss of the protections afforded by public disclosure and comment.

Formally, at least, the regulations limiting radiation exposures to the public and requiring official environmental review and oversight of government programs apply equally to classified programs as to public ones. In practice, however, classification creates complications that have yet to be resolved. Efforts are now under way to put procedures into place to better address proper environmental compliance in classified programs.

For example, security classification can interfere with official oversight of environmental compliance. Even in recent times, environmental oversight of classified programs has not been the rule in practice. Until 1994, the Federal Facilities Enforcement Office at EPA, which is charged with environmental oversight of all federal facilities, had no personnel with suitable clearances to oversee "black" programs—programs so highly classified that their existence is not acknowledged.[180]

Lack of oversight creates opportunities for violations of environmental law to go undetected and unpunished. Some have charged that the Department of Defense, as recently as 1993, used secrecy as a cover for violations of environmental law. Recent lawsuits against the Department of Defense and the Environmental Protection Agency allege that (1) illegal open-air burning of toxic wastes took place at a secret Air Force facility near Groom Lake, Nevada, and that (2) EPA has not exercised its required environmental oversight responsibilities for this facility.[181] Responding to the second of these lawsuits, EPA reported that in early 1995 it had seven regulators on staff with Special Access clearance who inspected the Groom Lake facility.[182] The Committee believes that the federal government has a particular obligation to provide environmental oversight of classified programs and that there is no fundamental barrier to environmental oversight in classified programs. Regulators can be granted the appropriate clearances. For example, before its existence was openly recognized, the F-117 Stealth fighter base in Nevada was subject to oversight by Nevada state regulators who had received the necessary clearances.[183] Such oversight is not automatic; it requires active cooperation between the regulatory agencies and the agencies subject to regulation. The Department of Defense has undertaken a

review of environmental compliance in its "black" programs and is working with EPA to establish mechanisms to provide continuing environmental oversight of those programs.[184]

Even when regulators have the appropriate clearances, however, other aspects of secrecy can create barriers to oversight. Providing clearances often entails lengthy background investigations, which can result in delays. Furthermore, it remains unclear what EPA can do if it detects a violation that results in a dispute with the agency in charge of the program. This is a basis for concern about the credibility of environmental oversight that occurs in secret.

The limits on outside oversight are ameliorated by the fact that both DOE and DOD have established environmental and health offices that are largely independent of their respective agencies' operational programs. Under most circumstances these offices can probably provide adequate oversight over their agencies' classified programs. Because of the potential institutional conflict of interest, however, it would be preferable to have further oversight by an independent entity.

The conduct of intentional releases in secret necessarily deprives the public of information to which it would otherwise be entitled. Security classification modifies or eliminates the various requirements for providing public disclosures. The agency states that its normal practice is to send an EPA employee with appropriate clearances to the agency in question to review the classified information; EPA, however, does not keep copies of the reviewed document or any other records of such reviews.[185] Moreover, review by an EPA employee is no substitute for a process open to public comment and scrutiny.

Secrecy, especially to the degree of "black" programs, severely limits or eliminates the ability of the public to influence decisions about environmental health, either through political action or through the courts,[186] and undermines public confidence that officials are carrying out their responsibilities to safeguard public health. As in the secret releases of the past, there are also concerns about whether and what kind of information can be given to the public about environmental and public health effects when releases are classified and if restrictions on information compromise the ability of members of the public to take protective actions.

CONCLUSION

While the intentional releases described in this chapter put people at risk from radiation exposures, with limited exception, they were not undertaken for the purpose of gathering research data on humans. Thus, in contrast with the biomedical experiments studied by the Advisory Committee, they were not intended as human experiments.

Fifty years ago, unlike today, there was no formal and published body of laws that defined and limited the ability of the government to release potentially hazardous substances into the environment. Nonetheless, the duty to limit risk and, by implication, the duty to balance risks against potential benefits was understood by those who engaged in intentional releases. In the case of the Green Run, risk from the intentional release could be gauged against preexisting guidelines for operational releases; in the case of radiological warfare tests, a separate safety panel was established to consider releases.

The intentional releases studied by the Committee often engaged national security interests and were conducted in secret. However legitimate and well-motivated the releases were, security classification prevented any public notice or discussion of the Green Run—an experiment conducted for intelligence purposes—the radiological weapons field tests, or the RaLa experiments testing atomic bomb components. The essentially complete secrecy surrounding these tests prevented any warnings that might have allowed members of the public to protect themselves from whatever risks might have been inherent in the tests.

In retrospect, and with limited information, it is difficult to know whether and how national security interests affected the decisions to conduct these intentional releases. In the case of the Green Run, for example, how did decision makers seek to balance the national security interests in learning about Soviet bomb testing (and the risks of not performing the Green Run

and thus not gaining relevant information) against the potential risks to the local population of the release?

The health and safety risks posed by the intentional releases appear in retrospect to have been negligible (the Green Run, for example, in comparison with other exposures at Hanford). But this does not mean that the intentional releases were without negative consequences. The secrecy that surrounded the conduct of these releases and the failure to deal forthrightly with citizens after the fact has taken a substantial toll. People living in the affected communities have been robbed of peace of mind, and the government has lost the trust of some of its citizens.

Could this happen again? Could there be another Green Run? The answer is a qualified yes.

In fact, an intentional release like the Green Run probably would not be contemplated (because the scientific and strategic value would seem minimal), but actions that raise similar concerns if undertaken in secrecy could still happen. Environmental regulations apply to secret programs, but the oversight procedures are not fully in place to ensure adherence to these regulations. The public review process that is at the heart of current environmental protections could be limited or rendered nonexistent if the government were to invoke exceptions for "national security interest" to avoid these constraints.

Any government action that is conducted in secret is likely to cause suspicion and distrust, even if the risks to members of the public are minimal or nonexistent. Public policy should operate with a strong presumption favoring public disclosure and openness. There doubtless are limited circumstances under which it is justifiable to conduct an intentional release in secret. The lesson of the Green Run and the other intentional releases is, however, that unless great care is taken to preserve and honor the public's trust, the cost to the body politic of such an action is likely to be substantial. The Committee believes that the current regulatory structure does not go far enough in this regard. Provisions must be made for timely public disclosure, and records must be created and maintained capable of satisfying the affected populations that their interests have been protected. And mechanisms need to be developed to approximate the scrutiny of the public when security interests require the classification of environmental impact statements or otherwise limit disclosure of information to the public. Without such protections, the greatest casualty of the Green Run—the distrust it engendered—cannot be prevented in the future; where this happens, official concern that the public cannot be trusted to appreciate sometimes-complex information about health and safety will become an ever-more-corrosive self-fulfilling prophecy.

NOTES

1. The story of the public discovery of the Green Run is recounted in Michael D'Antonio's *Atomic Harvest: Hanford and the Lethal Toll of America's Nuclear Arsenal* (New York: Crown, 1993), 116-145.

2. U.S. Congress, General Accounting Office, *Examples of Post World War II Radiation Releases* (Washington, D.C.: GPO, 1993) (ACHRE No. DOE-042894-B-1).

3. The Committee did not undertake to review in detail the general development of radioecology, which began during the Manhattan Project with research on the radiosensitivity of aquatic life around the Hanford Reservation and extended to research on flora and fauna in and around other AEC sites. For an introductory overview, see "Survey of Radioecology: Environmental Studies Around Production Sites," in J. Newell Stannard, *Radioactivity and Health: A History* (Springfield, Va.: Office of Science and Technical Information, 1988).

4. General Dwight D. Eisenhower, to Commanding General, Army Air Forces, 16 September 1947 ("Long Range Detection of Atomic Explosions") (ACHRE No. DOD-011595-A).

5. Charles A. Ziegler and David Jacobson, *Spying Without Spies: Origins of America's Secret Nuclear Surveillance System* (Westport, Conn.: Praeger, 1995), 133.

6. Ibid., 203-204. R. H. Hillenkoetter, Rear Admiral, USN, Central Intelligence Agency, 9 September 1949 memorandum ("Samples of air masses recently collected over the North Pacific . . .") (ACHRE No. CIA-011895-A).

7. Jack W. Healy, interview by Marisa Caputo (Department of Energy, Office of Human Radiation Experiments), transcript of audio recording, 28 November 1994 (ACHRE No. DOE-120894-D), 7. Nuclear explosions release larger quantities of short-lived fission products than do nuclear reactors, so the radioactive fallout from a nuclear test decays much more rapidly than emissions from a reactor.

8. "Statement by the President on Announcing the First Atomic Explosion in the USSR, 23 September 1949," reprinted in Robert C. Williams and Philip L. Cantelon, eds., *The American Atom: A Documentary History of Nuclear Policies from the Discovery of Fission to the Present, 1939-1984* (Philadelphia: University of Pennsylvania Press, 1984), 116-117.

9. For example, krypton 85 in the atmosphere comes entirely from atmospheric testing and nuclear fuel reprocessing. Chemically unreactive, with a half-life of roughly eleven years, it is well mixed in the atmosphere, so the concentration of Kr-85 measured at random sites on the globe will provide a rough measure of the total production of plutonium production over time. See Frank von Hippel, Barbara G. Levi, and David H. Albright, "Stopping the Production of Fissile Materials for the Weapons," *Scientific American* 253, no. 3 (September 1985): 43, 45.

10. Charles A. Ziegler and David Jacobson, *Spying Without Spies*, 181.

11. F. J. Davis et al., Department of Energy, 13 April 1949, ORNL-341, declassified with deletions as ORNL-6728, 2 September 1992 ("An Aerial Survey of Radioactivity Associated with Atomic Energy Plants") (ACHRE No. DOE-122194-B), 7, 13, 156-157. Pages 136 and 148 refer to the possibility of tracking a "really strong source" from the Hanford stacks.

12. [Deleted] to Commander, Military Air Transport Service, 10 November 1949 ("In furtherance of the research and development program . . .") (ACHRE No. DOD-032395-A); [deleted] to Dr. S. C. Schlemmer, Manager, Hanford Operations Officer, 10 November 1949 ("This letter will confirm the arrangements made . . .") (ACHRE No. DOD-032395-A). The name of the author of these memorandums remains classified.

13. GAO, *Examples of Post World War II Radiation Releases*, 9.

14. H. M. Parker, Department of Energy, 6 January 1950, HW-15550-E DEL ("Health Instruments Divisions Report for the Month of December 1949") (ACHRE No. DOE-050394-A-4); H. J. Paas and W. Singlevich, Department of Energy, 2 March 1950, HW-17003 ("Radioactive Contamination in the Environs of the Hanford Works for the Period October, November, December, 1949") (ACHRE No. DOE-050394-A-6). These two reports were among more than 19,000 pages of documents on Hanford's early operations released by the Department of Energy in 1986.

15. Lieutenant W. E. Harlan, D. E. Jenne, and Jack W. Healy, Hanford Works, Atomic Energy Commission, 1 May 1950, HW-17381 ("Dissolving of Twenty Day Metal at Hanford") (ACHRE No. DOD-121494-C). This document was originally classified Secret, Restricted Data, but has been declassified in several stages; this citation is to the version that was declassified most recently and completely, on 13 December 1994. The GAO report *Examples*

of Post World War II Radiation Releases also relies on interviews with several people connected with the Green Run, including one who was involved directly in the intelligence aspects of the Green Run, but who has since died.

16. Harlan, Jenne, and Healy, "Dissolving of Twenty Day Metal," 26.

17. Ibid., 7.

18. Hanford workers managing the Green Run estimated the release at 7,780 curies. Ibid., 28. Subsequent estimates have ranged from 7,000 curies to 11,000 curies. Maurice Robkin, "Experimental Release of 131I: The Green Run," *Health Physics* 62 (June 1992): 487-495. Roughly 20,000 curies of xenon 133 were also released, but because this gas does not concentrate in the food supply or in the thyroid, it posed relatively little danger. The section "Reconstructing, Comparing, and Understanding Risks" discusses the significance of these numbers.

19. M. S. Gerber, Department of Energy, May 1994, WHC-MR-0452 ("A Brief History of the T Plant Facility at the Hanford Site") (ACHRE No. DOE-112294-A), 25-32.

20. Harlan, Jenne, and Healy, "Dissolving of Twenty Day Metal," 10-11.

21. Ibid., 12-14. See also 8, 32.

22. Healy, interview with Caputo (Office of Human Radiation Experiments), 28 November 1994, 12; Jack Healy, interview with Mark Goodman (ACHRE staff), 8 March 1995 (ACHRE Research Project Series, Interview Program File, Targeted Interview Project), 26-27. Healy has compared the contamination with that resulting from the 1957 Windscale nuclear reactor accident in the United Kingdom. Although Windscale released a greater quantity of radioiodine, the Green Run contaminated an area five to ten times as large. Healy attributed the high levels of contamination to the weather.

23. Normally the temperature of the atmosphere falls with increasing altitude. An inversion occurs when the temperature near the ground rises before falling at higher altitudes. This traps contamination in the lower levels of the atmosphere.

24. F. J. Davis et al., "An Aerial Survey of Radioactivity Associated with Atomic Energy Plants," 112-116. The Air Force had found it difficult to track radioactivity from Oak Ridge's operations in the hills and valleys of Tennessee.

25. Harlan, Jenne, and Healy, "Dissolving of Twenty Day Metal, 5-6.

26. Healy, interview with Caputo (Office of Human Radiation Experiments), 28 November 1994, 13.

27. Harlan, Jenne, and Healy, "Dissolving of Twenty Day Metal," 20-25; Bruce Napier, Battelle Pacific Northwest Laboratory, to John Kruger (ACHRE staff), 18 August 1995.

28. Harlan, Jenne, and Healy, "Dissolving of Twenty Day Metal," 70. Also, W. E. Harlan, interview by Mark Goodman (ACHRE staff), transcript

of audio interview, 10 April 1995 (ACHRE Research Project Series, Interview Program File, Targeted Interview Project). Harlan recalls no plans to track the plume to greater distances.

29. The "permanent tolerance value" at the time was 0.009 microcuries per kilogram (FCi/Kg) of vegatation. Harlan, Jenne, and Healy, "Dissolving of Twenty Day Metal," 3. Readings from the Green Run were as high as 4.3 *m*Ci/kg. The current intervention level is .013 *m*Ci/kg. Al Conklin, Washington Department of Health, 7 November 1994, personal communication with Mark Goodman (ACHRE staff).

30. In mid-1949, the standard was lowered from 0.2 FCi/kg to 0.1 FCi/kg. Manager, Health Instruments Division, Hanford, to AEC, Hanford Operations Office, 8 November 1950 ("Radiation Exposure Data"), which is still ten times higher than described in Harlan, Jenne, and Healy, "Dissolving of Twenty Day Metal," 3. A footnote in the November 1950 report suggests that the lower level (0.01 FCi/kg) was still controversial and considered by the author to be overly cautious.

31. H. M. Parker, "Health Instruments Divisions Report for December 1949," 2; Harlan, Jenne, and Healy, "Dissolving of Twenty Day Metal," 3, 65.

32. Healy, interview with Caputo, 28 November 1994, 8-9. The largest hazard is now known to come from drinking milk from cows that graze on pastures contaminated with radioiodine. The earliest reference from Hanford to the milk pathway is H. M. Parker, "Radiation Exposure from Environmental Hazards," presented at the United Nations International Conference on the Peaceful Uses of Atomic Energy, August 1955, reprinted in Ronald L. Kathren, Raymond W. Baalman, and William J. Bair, eds., *Herbert M. Parker: Publications and Other Contributions to Radiological and Health Physics* (Richland, Wash.: Battelle Press, 1986), 494-499. A reference to concern over milk contamination in Utah from a 19 May 1953 atmospheric test appears in "Transcript of Meeting on Statistical Considerations on Field Studies on Thyroid Diseases in School Children in Utah-Arizona, December 3, 1965, Rockville, MD" (ACHRE No. HHS-022395-A), 4.

33. Herbert Parker, Chief Supervisor, to S. T. Cantril, Assistant Supervisor, 11 December 1945 ("Xenon And Iodine Concentration in the Environs of the T and P Plant") (ACHRE No. IND-120294-A-1); Herbert Parker to File, 17 December 1945 ("Proposed Revision of Tolerances for I 131") (ACHRE No. IND-120294-A-2); Herbert M. Parker, Department of Energy, 14 January 1946 ("Tolerance Concentration of Radio-Iodine on Edible Plants") (ACHRE No. IND-120294-A-3). This was confirmed by later dose reconstructions. The estimated doses range up to several hundred rad (a few tens of rem) to the thyroid. Technical Steering Panel, Department of Energy, 10 February 1990 ("Hanford Environmental Dose Reconstruction Project: Find-

ing the facts about people at risk.") (ACHRE No. DOE-050694-B-3).

34. Estimates from ARH-3026, by J. D. Anderson as cited in the Technical Steering Panel of the Hanford Environmental Dose Reconstruction Project, Department of Energy, March 1992 ("The Green Run") (ACHRE No. ORE-110794-A).

35. Parker, "Tolerance Concentration of Radio-Iodine on Edible Plants," Kathren, Baalman, and Bair, *Herbert M. Parker: Publications and Other Contributions to Radiological and Health Physics,* art. IV-7.

36. "Report of Safety and Industrial Heath Advisory Board," as cited in Daniel Grossman, *A Policy History of Hanford's Atmospheric Releases* (Ph.D. diss., Massachusetts Institute of Technology, 1994), 169.

37. F. A. R. Stainkan to R. S. Ball, "Stack Gas Conference—Washington, D.C.," 8 September 1948, HW-10956. Michele S. Gerber, *On the Home Front: The Cold War Legacy of the Hanford Nuclear Site* (Lincoln: University of Nebraska Press, 1994), 89.

38. Parker, "Health Instrument Divisions Report for Month of December, 1949," 3.

39. Healy, interview with Caputo, 28 November 1994, 8.

40. [Deleted] to Dr. Schlemmer, 10 November 1949, 2.

41. "Green Run," an Air Force official noted in a 1995 comment on this omission, "was beset by numerous technical and meteorological problems that significantly compromised the value of the results obtained. In our view, then, this omission implies the Green Run was not useful, rather than unnecessary." Major Meade Pimsler, USAF, to ACHRE Staff, 19 June 1995 ("Comments on 5th Set of Review Chapters").

42. ACHRE Research Project Series, Mark Goodman Files, 6-21. The device in question was the atmospheric Conductivity Apparatus; it was used in Operation Fitzwilliam, at the radiation survey flights at Oak Ridge and Hanford, at the Green Run, and in radiation survey flights at the Los Alamos radiolanthanum tests described in this chapter.

43. Grossman, *A Policy History of Hanford's Atmospheric Releases,* 230-232 (references 24 and 35).

44. Neal D. Hines, *Proving Ground: An Account of the Radiobiological Studies in the Pacific 1946-61* (Seattle: University of Washington Press, 1962).

45. Stannard, *Radioactivity and Health: A History,* 761-762.

46. Decision on AEC 180/1 and 180/2, as cited in Grossman, *A Policy History of Hanford's Atmospheric Releases,* 244-245.

47. AEC 180/1; Ibid., 245.

48. Herbert Parker, 19 August 1954 ("Columbia River Situation—A Semi-Technical Review") (ACHRE No. DOE-053095-A), 5.

49. William Bale, Advisory Committee for Biology and Medicine, transcript of proceedings of 13-14 January 1950 (ACHRE No. DOE-072694-A).

The ACBM decided at this meeting that it might be possible to publish a sanitized version of the report to aid scientists studying the contamination problem. It is unclear whether the report was published.

50. Lynne Stembridge, Advisory Committee on Human Radiation Experiments, transcript of proceedings of 21 November 1994 (Spokane, Wash.).

51. For the perspective of a government-sponsored expert involved in the reconstruction of the risk at Hanford and other nuclear weapons sites on the necessity of public participation in dose reconstruction activities, see John Till, "Building Credibility in Public Studies," *American Scientist* 83, no. 5 (September-October 1995).

52. Scott Davis, Ph.D., et al., Fred Hutchinson Research Center, 24 January 1995 ("Hanford Thyroid Disease Study: Final Report") (ACHRE No. DOE-061295-A).

53. Tom Bailie, Advisory Committee on Human Radiation Experiments, transcript of proceedings, 21 November 1994 (Spokane, Wash.), 121-122.

54. Quoted in Richard Rhodes, *The Making of the Atomic Bomb* (New York: Simon and Schuster, 1986), 365. See also Henry DeWolf Smyth, *Atomic Energy for Military Purposes* (Stanford, Calif.: Stanford University Press, September 1, 1945), 71. Princeton physicists Henry DeWolf Smyth and Eugene Wigner reported later that year that the fission products produced in one day's operation of a 100-megawatt reactor could render a large area uninhabitable. Eugene Wigner and Henry D. Smyth, National Academy Project, 10 December 1941 ("Radioactive Poisons") (ACHRE No. NARA-033195-A).

55. J. Robert Oppenheimer to Enrico Fermi, 25 May 1943, reproduced in Barton Bernstein, "Oppenheimer and the Radioactive Poison Plan," *Technology Review* 88, no. 14 (May 1985).

56. These included Dr. Joseph Hamilton of Berkeley, who had performed pioneering studies of the fate of radioactive materials in the bodies of animals and humans (see chapter 5). Joseph Hamilton, M.D., to K. D. Nichols, 31 December 1946 ("Radioactive Warfare") (ACHRE No. DOD-010395-C-1); Lee Bowen, U.S. Air Force ("A History of the Air Force Atomic Energy Program, 1943-1953, volume IV: The Development of Weapons") (ACHRE No. SMITH-120994-A-1), 323.

57. Joseph Hamilton to D. T. Griggs, Project Rand, Douglas Aircraft Co., 7 April 1948 ("I wish to thank you . . .") (ACHRE No. DOE-072694-B-34).

58. For example, General Douglas MacArthur proposed in 1950 to lay down a line of highly radioactive cobalt 60 to block a Chinese return to the Korean Peninsula. Bruce Cummings, *The Origins of the Korean War, volume II: The Roaring of the Cataract, 1947-1950* (Princeton, N.J.: Princeton University Press, 1981), 750.

59. The Armed Forces Special Weapons Project, the Air Force, and the Army's Chemical Corps were

interested in both offensive and defensive radiological warfare, while the Naval Radiological Defense Laboratory focused on defense.

60. This research program, led by Dr. Franklin McLean, used animals to test the toxicity of various candidate radiological warfare agents. The Advisory Committee's research has uncovered no evidence that human subjects were used in any of these studies. The Toxicity Laboratory also performed studies using human subjects on the inhalation of aerosols, but available documents do not indicate any use of radioactive material as tracers or otherwise. These studies may have been related to the Chemical Corps's programs in chemical and biological warefare. Frank C. McLean to Shields Warren, Director, Division of Biology and Medicine, 5 October 1948 ("Program of Chicago Toxicity Laboratory") (ACHRE No. DOE-082294-B-18); Walter J. Williams, Acting General Manager, to Major General A. H. Waitt, Chief, Chemical Corps, 12 April 1948 ("For some time we have been considering ways and means of enlarging programs . . .") (ACHRE No. DOE-012595-B-2); Shields Warren to Frank C. McLean, 4 May 1950 ("In light of our conversations of January 25, 1950 . . .") (ACHRE No. DOE-012595-B-1).

61. Joint AEC-NME Panel on Radiological Warfare, 20 November 1950 ("Radiological Warfare Program Status Report: Sixth Meeting of the Joint AEC-NME Panel on Radiological Warfare") (ACHRE No. CORP-010395-A-2). See also, Atomic Energy Commission, Division of Military Application, 26 December 1950 ("Conclusions and Recommendations of the Sixth Meeting of the RW Panel") (ACHRE No. DOE-092694-B-3).

62. Major Thomas A. Gibson, Chemical Corps, Radiological Branch, to Chief of Staff, AFSWP, 23 April 1952 ("A Technical Study Group to Review the Technical Aspects of Radiological Warfare") (ACHRE No. NARA-033195-A).

63. These two tests appear in the Advisory Committee's charter. One test involved three sources of roughly 1,280, 100, and 20 curies of radioactive lanthanum; Karl Z. Morgan and C. N. Rucker, Oak Ridge National Laboratory, 23 July 1948 ("Single Source Lanthanum Test—AHRUU Program") (ACHRE No. DOE-051094-A-122). The other used 156 tantalum sources of roughly 100 millicuries each distributed in an uniform grid; Karl Z. Morgan, Oak Ridge National Laboratory, 11 August 1948 ("Uniformly Distributed Source, AHRUU Program") (ACHRE No. DOE-051094-A-118).

64. R. W. Cook to Brigadier General James McCormack, Division of Military Applications, 4 May 1949 ("Irradiation of Tantalum for RW Tests") (ACHRE No. DOE-120994-A-24).

65. Atomic Energy Commission, Division of Military Application, 26 December 1950 ("Conclusions and Recommendations of the Sixth Meeting of the RW Panel"), 3.

66. AFSWP to Chief, Chemical Officer, Department of the Army, 31 December 1952 ("Re-evaluation of the Research and Development Program on Offensive Radiological Warfare") (ACHRE No. CORP-010395-A-5); this memo makes reference to the proposed production of zirconium and niobium radioisotopes.

67. Joseph Hamilton had written in 1948: "In concluding, I would like to emphasize that all of the potentialities, including the rather repellent concepts of the use of fission products and other radioactive materials as internal poisons, should be explored up to and including a level of pilot experiments on a fairly large scale. I feel very strongly on the point that unless we ourselves learn all we can about the use and possible methods of protection against these agents and a wide variety of their potential applications as military weapons, we shall have failed to explore the necessary measures which may be desperately needed for the protection of our own people." Joseph Hamilton to D. T. Griggs, Project Rand, Douglas Aircraft Co., 7 April 1948 ("I wish to thank you very much for . . .") (ACHRE No. DOE-072694-B-34), 3.

68. C. B. Marquand, Secretary, Test Safety Panel, to Joseph Hamilton, Director, Crocker Laboratory, 24 August 1949 ("Meeting of the Test Safety Panel at Dugway Proving Ground on August 2, 1949") (ACHRE No. DOE-072694-B-29), 3.

69. Joseph Hamilton to C. B. Marquand, Office of the Chief, Chemical Corps Advisory Board, 6 July 1949 ("I am sorry not to have had some more definitive information . . .") (ACHRE No. DOE-072694-B-3).

70. C. B. Marquand, Secretary, Test Safety Panel, and S. C. Hardwick, Assistant Secretary, Test Safety Panel, Atomic Energy Commission, 5 August 1949 ("Preliminary Report of the Test Safety Panel Meeting at Dugway Probing Ground—August 2, 1949") (ACHRE No. CORP-010395-A-1).

71. G. Failla, Columbia University, to Joseph Hamilton, 13 May 1950 ("In answer to your letter of May 10th, I . . .") (ACHRE No. DOE-072694-B-4).

72. Karl Z. Morgan to Joseph Hamilton, 9 September 1949 (ACHRE No. DOE-072694-B-5).

73. Joseph Hamilton to Albert Olpin, President, University of Utah, 10 May 1950 ("Please find enclosed my report and recommendations for . . .") (ACHRE No. DOE-072694-B-7); Failla to Hamilton, 13 May 1950; Albert Olpin to Joseph Hamilton, 17 May 1950 ("It was good to hear from you again . . .") (ACHRE No. DOE-072694-B-55); Joseph Hamilton to C. B. Marquand, 1 June 1950 ("At long last I have received agreement from . . .") (ACHRE No. DOE-072694-B-21); Joseph Hamilton to C. B. Marquand, 4 August 1949 ("This letter is to confirm") (ACHRE No.CORP-010395-A-1).

74. Joseph Hamilton to C. B. Marquand, 18 November 1952 ("Last week, I spent a profitable two days . . .") (ACHRE No. DOE-072694-B-23). See also,

Joseph Hamilton to John Bugher, Director, Division of Biology and Medicine, 25 February 1953 ("In my opinion . . .") (ACHRE No. DOE-072694-B-49).

75. Department of Defense, 11 May 1953 ("An Evaluation of the Airborne Hazard Associated with Radiological Warfare Tests") (ACHRE No. DOE-072694-B-50), iii.

76. Brigadier General William M. Creasy to Chief Chemical Officer, Department of the Army, 24 June 1953 ("Minimum Fund Requirement") (ACHRE No. DOD-030895-F-3); U.S. Army Chemical Corps, 31 December 31 1953 ("RDB Project Card: Progress Report, Project No. 4-12-30-002") (ACHRE No. NARA-112294-A-1); Joseph Hamilton to C. B. Marquand, 23 July 1953 ("I regret to hear that the RW Program has been so drastically reduced . . .") (ACHRE No. DOE-072694-B-51).

77. Lee Bowen, United States Air Force Historical Division, "A History of the Air Force Atomic Energy Program, 1943-1953, vol. 4: The Development of Weapons" (ACHRE No. SMITH-120994-A-1), 331-332.

78. Merril Evans, 3 June 1953 ("RW Decontamination and Land Reclamation Studies") (ACHRE No. DOD-062494-A-11); also Chemical Corps, 1952 ("Testing of RW Material for Detection, Protection and Decontamination") (ACHRE No. NARA-112294-A-5).

79. Lee Bowen, "The Development of Weapons," 333-337.

80. The Chemical Corps recognized Hamilton's support for the radiological warfare program. A 1952 Chemical Corps memorandum, Lieutenant Colonel Truman Cook to Secretariat, Chemical Corps Advisory Council, 7 April 1952 ("Radiological Warfare Test Saftey Panel") (ACHRE No. DOE-072694-B-46), noting the greater risk associated with planned large-scale tests, including possible plutonium contamination from the use of fission products, recommended that the test safety panel should be dissolved and Hamilton and someone chosen by him be retained as a consultant.

81. Joseph Hamilton's papers, including those dealing with the radiological warfare test safety panel, were declassified in 1974, but the GAO report provided the first opportunity for most people to learn about it.

82. Atomic Energy Commission, Ad Hoc Panel on Radiological Warfare, proceedings of 23 May 1948 (ACHRE No. CORP-051894-A-1). Restricted Data are atomic energy secrets as classified by statute under the Atomic Energy Act. Information may also be classified Confidential, Secret, or Top Secret, depending on its importance to national security, under the authority of an executive order by the President. See chapter 13 on secrecy for details.

83. Chemical Corps, 1 January 1948 ("Quarterly Technical Progress Reports") (ACHRE No. NARA-121594-B).

84. Joint NME-AEC Panel on Radiological Warfare, 29 August 1948 ("Radiological Warfare Report—Second Meeting") (ACHRE No. DOD-041295-A), 72.

85. Atomic Energy Commission, Advisory Committee for Biology and Medicine, proceedings of 11-12 June 1948 (ACHRE No. DOE-072694-A). The ACBM reiterated this recommendation at its next meeting on 11 September 1948. Atomic Energy Commission, Advisory Committee for Biology and Medicine, proceedings of 11 September 1948 (ACHRE No. DOE-082294-B-15), 15.

86. The eight members included future President Dwight D. Eisenhower, who was president of Columbia University at the time, and New York lawyer John Foster Dulles, who would serve as secretary of state in the Eisenhower administration. See James Hershberg, *James B. Conant: Harvard to Hiroshima and the Making of the Nuclear Age* (New York: Alfred A. Knopf, 1993), 378-383. The story of the Conant Committee is told in chapter 20 of Hershberg's book.

87. Ibid., 390.

88. Ibid., 383. The members of the majority opposed to further release included Eisenhower and Dulles.

89. Marshall Stubbs to Joseph Hamilton, 30 August 1949 ("Following your suggestion that we prepare . . .") (ACHRE No. DOE-072694-B-1). Stubbs concluded by reporting, "Both Colonel Cooney and Captain Winant reiterated that regardless of the actions noted above, such a release is not desirable."

90. William Webster to Secretary Johnson, Department of Defense, 11 May 1949 ("Memorandum for the President: This memorandum is to inform you of planned activities . . .") (ACHRE No. DOD-071194-A-4).

91. Joseph Hamilton to C. B. Marquand, Secretary, Test Safety Panel, 4 August 1949 ("This letter is to confirm the decisions . . .") (ACHRE No. CORP-010395-A-3).

92. Ibid., 3.

93. The proposed press release falsely described the program as intended only "to determine a proper defense," involving "the distribution of small amounts of radioactive materials on various types of simulated targets in the field." Marshall Stubbs to Joseph Hamilton, 30 August 1949 ("Following your suggestion . . .") (ACHRE No. DOE-072694-B-1), 2.

94. Ibid.

95. Colonel William M. Creasy, Chief, Research and Engineering Division, U.S. Army Chemical Corps, to Director of Logistics, U.S. Army General Staff, 3 October 1949 ("Public Release On R[adiological] W[arfare] Tests at Dugway Proving Ground") (ACHRE No. DOD-071194-A-1). The letter notes the draft press release contains "no reference to the general RW program or to the use of radioactive materials as agents of warfare. It does indicate the use of radioactive materials in the Dugway area for the purpose of formulating defensive doctrine."

96. C. G. Helmick, Deputy Director for Research and Development, to Robert LeBaron, Chairman,

Military Liaison Committee, 3 January 1950 ("Public Release on RW Tests at Dugway Proving Ground") (ACHRE No. DOD-071194-A-1); Robert LeBaron to C. G. Helmick, 19 January 1950 ("Public Release on RW Tests at Dugway Proving Ground") (ACHRE No. DOD-071194-A-1).

97. Louis N. Ridenour, "How Effective Are Radioactive Poisons in Warfare?" *Bulletin of the Atomic Scientists* 6 (1950): 199-202. On the birth of *Bulletin of the Atomic Scientists* and, more generally, the history of the physics community that worked on the bomb, see Daniel J. Kevles, *The Physicists: The History of the Scientific Community in Modern America* (New York: Vintage, 1979), 351.

98. Atomic Energy Commission, Declassification Branch, 30 September 1949 ("Classification Guide for Radiological Warfare") (ACHRE No. DOE-070695-C).

99. U.S. Department of Defense, *Semiannual Report of the Secretary of Defense and the Semiannual Reports of the Secretary of the Army, Secretary of the Navy, Secretary of the Air Force, July 1 to December 31, 1949* (Washington, D.C.: GPO, 1950), 65-69; Samuel Glasstone, executive editor, *The Effects of Atomic Weapons* (Washington, D.C.: GPO, 1950), 287-290.

100. U.S. Department of Defense, *Semiannual Report of the Secretary of Defense and the Semiannual Reports of the Secretary of the Army, Secretary of the Navy, Secretary of the Air Force, January 1 to June 30, 1951* (Washington, D.C.: GPO, 1951) (ACHRE No. DOD-052695-A), 36. In an August 1951 letter to AEC Chairman Dean, Acting Secretary of Defense Rovert Lovett noted: "The Director of Public Information, Department of Defense, has been directed to undertake a program of public information in this field as recommended by the [Noyes] Panel." A further memo to the Director, Office of Public Information, the author of which is unclear, cited the Noyes panel's recommendation that civil defense agencies and the public be apprised "concerning the nature and possibilities of radiological warfare as well as possible countermeasures so as to avoid the possibility of panic should an enemy carry out an attack . . . and that studies be made of the psychological effects to be expected." Robert Lovett to Gordon Dean, 6 August 1951 ("The Final Report of the Joint AEC-NME Panel . . .") (ACHRE No. DOD-081695-A); memorandum to Director, Office of Public Information, undated ("Public Information Program on Radiological Warfare") (ACHRE No. DOD-021095-A).

101. U.S. Department of Energy, Office of Declassification, *Drawing Back the Curtain of Secrecy: Restricted Data Declassification Policy, 1946 to the Present* (RDD-1) (1 June 1994), 82.

102. Human Studies Project Team, Los Alamos National Laboratory, March 1995 ("The Bayo Canyon Radioactive Lanthanum [RaLa] Program [draft]") (ACHRE No. DOE-031095-B-1). This report lists 254 RaLa tests, but 1 planned test was never fired,

and the last 9, conducted for different purposes, did not release radiolanthanum into the environment.

103. D. P. MacDougall to N. E. Bradbury, 22 June 1961 ("RaLa Shots in Bayo") (ACHRE No. DOE-040695-A-13), concludes that the RaLa program should not be dismantled until the replacement procedure, Phermex, was operating. Jane H. Hall to Distribution, 8 February 1963 ("Subject: Rala") (ACHRE No. DOE-040695-A-8), reports that "RaLa may no longer be released into the Bayo Canyon atmosphere."

104. GAO, *Examples of Post World War II Radiation Releases*, 16.

105. The fourth experiment was not an intentional release; like the Oak Ridge radiological warfare experiments, it involved a sealed source of radiation that was later returned to the laboratory. Samuel Coroniti, Los Alamos Scientific Laboratory, 19 July 1950 ("Radiation Test Conducted at Los Alamos, New Mexico on 19 July 1950") (ACHRE No. DOE-051095-B).

106. Human Studies Project Team, Los Alamos National Laboratory, March 1995 ("The Bayo Canyon/Radioactive Lanthanum RaLa Program [draft]"), 6. The GAO report states that the Air Force Cambridge Research Laboratories and Los Alamos jointly performed the explosions; Samuel C. Coroniti, Air Force Cambridge Research Laboratories, 26 May 1950 ("Report on the Atmospheric Electrical Conductivity Tests Conducted in the Vicinity of Los Alamos, Scientific Laboratories, New Mexico") (ACHRE No. DOD-120294-A-1). S. V. Burriss, Los Alamos Scientific Laboratory, to Colonel Benjamin G. Holzman, Research and Development, Pentagon, 11 October 1949 ("Cloud Studies at Los Alamos") (ACHRE No. DOE-060295-B), indicates that the Air Force simply took advantage of releases that occurred for other purposes.

107. L. H. Hempelmann to David Dow, 29 June 1944 ("Safety of Radiolanthanum Experiment in Bayo Canyon") (ACHRE No. DOE-051094-A-15); Los Alamos Scientific Laboratory, Safety Committee, proceedings of 7 March 1945 (ACHRE No. DOE-052395-B-1).

108. Ralph G. Steinhardt, Jr., to Joseph Hoffman, 19 June 1945 ("Summary Report on Health Conditions in RaLa Program") (ACHRE No. DOE-052395-B-2).

109. T. L. Shipman to R. E. Cole, Atomic Energy Commission, Office of Engineering and Construction, through N. E. Bradbury, 4 April 1950 ("Health Hazards—Guaje Canyon and Vicinity") (ACHRE No. DOE-052395-B), 1.

110. If the wind was blowing toward the main access road to the Los Alamos mesa, tests could not be conducted in the late afternoon. T. L. Shipman to Donald Mueller through N. E. Bradbury and Duncan MacDougall, 28 April 1949 ("Precaution for Bayo Canyon Shots") (ACHRE No. DOE-042495-C).

111. Los Alamos Scientific Laboratory, 8 March 1952 ("H-1 Program for Bayo Canyon Shots")

(ACHRE No. DOE-042495-C); Los Alamos Scientific Laboratory, 23 July 1952 ("H-1 Program for Bayo Canyon Shots") (ACHRE No. DOE-042495-C); Los Alamos Scientific Laboratory, 1 April 1958 ("H-1 Program for Bayo Canyon Shots") (ACHRE No. DOE-042495-C); and C. D. Montgomery, D. W. Mueller, R. O. Niethammer, 30 June 1954, revised 15 January 1960 ("Clearance, Firing, and Monitoring Procedures for Bayo Canyon Site") (ACHRE No. DOE-040695-A-14), 8, all describe the continuing requirements for weather forecasting. N. E. Bradbury to Distribution, 8 March 1956 ("Meteorological Forecasting Service") (ACHRE No. DOE-040695-A-12), notes the continuing need for weather forecasting in the context of an Air Force threat to withdraw two meteorologists.

112. Tyler Mercier, Advisory Committee on Human Radiation Experiments, transcript of proceedings of 30 January 1995, Santa Fe, N.M., 35.

113. Glenn Vogt, General Monitoring Section, Los Alamos, to Dean Meyer, Group Leader, 20 April 1956 ("Bayo Canyon Activities, April 12, 16, 18, 1956") (ACHRE No. DOE-041295-C).

114. D. W. Mueller to RaLa Committee, 22 September 1952 ("RALA Shots") (ACHRE No. DOE-071095-B).

115. Steinhardt to Hoffman, 19 June 1945.

116. The National Reactor Testing Station was established near Idaho Falls in 1950, and plans to process Ba-140 for Los Alamos at Idaho Chemical Processing Plant were made on 5 November 1952. Dick Duffey, Atomic Energy Commission, to W. B. Allred, Chief, Reactor Division, Oak Ridge Operations Office, and H. Leppich, Idaho Operations Office, 5 November 1952 ("This will confirm our telephone conversation . . .") (ACHRE No. DOE-040695-A).

117. The third test had taken place two days earlier, on 14 October 1944. Human Studies Project Team ("The Bayo Canyon [RaLa] Program"), appendix A-1.

118. Louis Hempelmann, M.D., to Colonel Stafford Warren, 16 October 1944 ("Enclosed is an excerpt of my report about the health hazards . . .") (ACHRE No. DOE-071494-A-10).

119. Louis Hempelmann to Norris E. Bradbury, 30 August 1946 ("Excessive Exposures at Bayo Canyon") (ACHRE No. DOE-062295-A-1).

120. E. R. Jette, Acting Director, to Technical Board Members, 3 September 1946 ("The main topic for discussion at the Technical Board meeting . . .") (ACHRE No. DOE-062295-A-2) .

121. Ibid.

122. Norman Knowlton, Los Alamos Scientific Laboratory, "Changes in the Blood of Humans Chronically Exposed to Low Level Gamma Radiation," LA-587, 1948 (ACHRE No. DOE-033095-A-2); Robert Carter and Norman Knowlton, "Hematological Changes in Humans Chronically Exposed to Low-Level Gamma Radiation," LA-1092, 1950 (ACHRE No. DOE-030695-A); Robert E. Carter et al., Los Alamos Scientific Laboratory, July 1952, LA-1440 "Further Study of the Hematological Changes in Humans Chronically Exposed to Low-Level Gamma Radiation" (ACHRE No. DOE-033095-A).

123. Thomas Shipman, Health Division Leader, Los Alamos Scientific Laboratory, to Gordon Dunning, Division of Biology and Medicine, 21 July 1954 ("When we finally decided to issue LA-1440 . . .") (ACHRE No. DOE-020795-D-4).

124. Samuel J. Glasstone, ed., *The Effects of Atomic Weapons* (Washington, D.C.: Atomic Energy Commission, 1957), 342.

125. This dosimetry appears to have been a subject of some care. See Louis Hempelmann to Stafford Warren, 16 October 1944 ("Enclosed is an excerpt . . .") (ACHRE No. DOE-071494-A-10), and William C. Inkret, Human Studies Project Team, Los Alamos National Laboratory, to Michael Yuffee, Office of Human Radiation Experiments, Department of Energy, 16 May 1995 ("This letter is a follow-up to the other materials we have sent . . .") (ACHRE Request No. 032995-C).

126. For example, organic solvents such as benzene and toluene can cause depressed white blood cell counts. The first of the Los Alamos reports (Knowlton, LA-587, 1948) notes that the control group was not exposed to organic solvents, suggesting that the researchers were aware of this fact, but does not consider this as a factor in the chemists' blood counts.

127. J. F. Mullaney to N. E. Bradbury, 3 January 1946 ("Biological Effects of July 16th Explosion"), 1. See also Bradbury to Mullaney, 7 January 1946.

128. Clyde Wilson, Chief, Insurance Branch, to Anthony C. Vallado, Deputy Declassification Officer, 20 December 1948 ("Review of Document by Knowlton") (ACHRE No. DOE-120894-E-32).

129. Leon Tafoya, interview with Mark Goodman (ACHRE staff), 10 March 1995 (ACHRE Research Project Series, Interview Program File, Targeted Interview Project).

130. "Bye Bye Bayo Site," *LASL News,* 23 May 1963 (ACHRE No. DOE-051094-A-622), 7.

131. William C. Inkret, Los Alamos Human Studies Project Leader, to Mark Goodman and Dan Guttman (ACHRE staff), 17 April 1995 ("Attached are the answers to questions 6 and 7 of . . .") (ACHRE No. DOE-042495-C), 4-5.

132. Leon Tafoya, interview with Mark Goodman (ACHRE staff), 10 March 1995; and Gilbert Sanchez, interview with Mark Goodman (ACHRE staff), 9 March 1995 (ACHRE Research Project Series, Interview Program File, Targeted Interview Project).

133. Leon H. Tafoya, "Biocultural Dimension of Health and Environment," *Hazardous Waste and Public Health* (1994): 245-252.

134. Sanchez, interview with ACHRE staff, 9 March 1995.

135. Dr. George Voelz, Advisory Committee on Human Radiation Experiments, transcript of proceedings of 30 January 1995, Santa Fe, N.M., 43.

136. Department of Energy, *Human Radiation Experiments: The Department of Energy Roadmap to the Story and the Records* (Springfield, Va.: National Technical Information Service, February 1995), 214-222.

137. The story of this early Hanford research is told in Hines, *Proving Ground.* See also, Stannard, *Radioactivity and Health,* 745-1368.

138. Daniel O'Neill, *The Firecracker Boys* (New York: St. Martin's Press, 1994), 28, 31-46. Some tests were designed to see whether nuclear explosions could stimulate the release of deep deposits of natural gas. Others were conducted in Nevada to test the ability to conduct massive civil engineering projects using nuclear explosions. The possibility of using nuclear explosions to excavate a second Panama Canal received serious theoretical attention.

139. Ibid., 239-257.

140. The Soviet test site at Novaya Zemla lies north of the Arctic Circle and was responsible for most of the fallout in Alaska and other Arctic locations.

141. Wayne Hanson, interview by Daniel O'Neill, 4 May 1988, transcribed by ACHRE staff, 9 March 1995 (ACHRE No. ACHRE-031395-A), 56.

142. Nevada Environment Restoration Project, Department of Energy, *Project Chariot Site Assessment and Remedial Action Final Report* (Springfield, Va.: National Technical Information Service, 1994), 1-5. See also Arthur Piper and Donald Eberlein to John Phelps, Director, Special Projects Division, 9 October 1962 ("Your letter of September 27, 1962, to . . . ") (ACHRE No. DOE-050295-E), 3-4.

143. Ray Emens, Director, Support Division, to John Phillip, Director, Special Projects Division, 10 April 1963 ("Radioactive Waste Mound At Project Chariot Site") (ACHRE No. CORP-013095-A-1), 1.

144. Nevada Environmental Restoration Project, *Project Chariot Site Assessment and Remedial Action Final Report,* 1-2.

145. Caroline Cannon, Advisory Committee on Human Radiation Experiments, transcript of proceedings of 30 January 1995, Santa Fe, N.M., 136.

146. North Slope Borough Science Advisory Committee, April 1994, "A Preliminary Review of the Project Chariot Site Assessment and Remedial Action Final Report" (ACHRE No. DOE-121494-E-2).

147. C. A. Hawley et al., Health and Safety Division, AEC, Idaho Operations Office, "Controlled Environmental Radioiodine Tests at the National Reactor Testing Station," IDO-12035, June 1964 (ACHRE No. DOE-060794-B-37), 6. After 1968, a variety of radioisotopes were used, and the name of the series was changed to the Controlled Environmental Release Tests.

148. Ibid., iii.

149. Ibid., 31. The exposure limits at the time were 30 rem to the thyroid per year. Richard Dickson, Idaho National Engineering Laboratory, to Bill LeFurgy, Office of Human Radiation Experiments, 16 May 1995 ("Comments on Draft Advisory Committee Report").

150. Data on tests 2, 7, 10, and 11 are contained in C. A. Hawley, Jr., ed., Idaho Operations Office, AEC, "Controlled Environmental Radioiodine Test at the Nuclear Rector Testing Station: 1965 Progress Report," IDO-129457, February 1966, (ACHRE No. DOE-060794-B-37); D. F. Bunch, ed., Idaho Operations Office, AEC, "Controlled Environmental Radioiodine Tests: Progress Report Number Two," IDO-12053, August 1966, (ACHRE No. DOE-060794-B-39), 26-30; D. F. Bunch, ed., Idaho Operations Office, AEC, "Controlled Environmental Radioiodine Tests: Progress Report Number Three," IDO-12063, January 1968 (ACHRE No. DOE-101194-B-3), 14.

151. Dr. George Voelz, interview with Marisa Caputo (DOE Office of Human Radiation Experiments), transcript of audio recording, 29 November 1994 (ACHRE No. DOE-061495-A), 16-18. Members of the INEL's Human Radiation Experiments Team state that the identities of these subjects could be determined from records.

152. [Deleted] Senior Engineer to R. F. Foster (PNL-9370), 1 August 1963 ("Monthly Report: July 1963 [handwritten draft]") and PNL-9369-DEL, 23 August 1963 ("Monthly Report: August 1963"). A proposal for a second Hanford iodine 131 field release test was never implemented. E. C. Watson, BWWL-CC-167, 22 July 1965 ("Proposal for a Second Iodine-131 Field Release Test") (ACHRE No. DOE-033095-A-1). A handwritten notation on the cover sheet reads: "This test was not run. D Gydesen. 3/24/86." The DOE interview with Jack Healy includes descriptions of his role in a study involving iodine uptake through milk. Jack Healy, interview with Mark Goodman (ACHRE staff), 8 March 1995, transcript of audio recording (ACHRE No. DOE-050295-A), 32.

153. The exposures at Hiroshima and Nagasaki came primarily from acute exposure to gamma and neutron radiation, rather than from radioactive fallout.

154. U.S. Environmental Protection Agency, *Estimating Radiogenic Cancer Risks,* EPA, 402-R-93-076, June 1994 (DOE-061195-A). One person-rem corresponds to an aggregate dose of 1 rem spread over any number of people. The result from BEIR V is roughly one cancer fatality for every 1,120 person-rem (see chapter 4, "BEIR V"), but this is from a single exposure to gamma radiation.

155. This project has since been transferred to the Centers for Disease Control and Prevention.

156. The Committee has not attempted to estimate the range of uncertainty in this estimate. Some of the relevant figures are the estimated maximum 600-mrad thyroid dose from Hanford emissions in 1949, the more typical 180-mrad dose for residents of Richland and the roughly 30,000 population of the

Richland area at the time, suggesting a total population exposure of roughly 5,400 person-rad to the thyroid. See W. T. Farris et al., Hanford Environmental Dose Reconstruction Project, PWWD-2228 HEDR, April 1994, ("Atmospheric Pathway Dosimetry Report, 1944-1992 [draft]"), C. 6. Using NCRP risk estimates of 7.5 excess cancer deaths per million person-rad to the thyroid, this leads to an estimate of 0.04 excess thyroid cancer deaths. The corresponding estimate for nonfatal thyroid cancer is a factor of 10 higher. There are many uncertainties in this estimate, but they do not consistently overstate or understate the risk. For example, we have ignored the smaller exposures to other population centers and the relatively high doses and risks to children, as well as the offsetting facts that the Green Run represented only about 80 percent of Hanford's I-131 emissions in 1949 and occurred in December when the food pathway was suppressed.

157. Meteorological and fallout data are more or less complete after 1950. A total of roughly 250,000 curies of radiolanthanum were released (remarkably, less than half a gram) from 1944 to 1960, with releases peaking in 1955 and 1956 at roughly 40,000 curies a year. Strontium 90 was a minor contaminant, with total releases of about 200 millicuries. Los Alamos Human Studies Project Team (draft, 9 March 1995) ("Bayo Canyon/[RaLa] Program"), 15, appendix A.

158. D. H. Kraig, Human Studies Project Team, Los Alamos National Laboratory, 1995 ("Dose Reconstruction for Experiments Involving La140 at Los Alamos National Laboratory, 1944-1962") (ACHRE No. DOE-091495-A).

159. General Accounting Office, *Examples of Post World War II Radiation Releases*, 16. Los Alamos Human Studies Project Team, "Bayo Canyon/RaLa Program," 9-10. In tandem with its historical reconstruction, the Human Studies Project Team at Los Alamos is preparing a report estimating radiation exposures to the general population.

160. According to LA-1440, ten workers were exposed to an average exposure of at least 34 R, and the total exposure was at least 340 person-R, corresponding to roughly 0.2 fatal cancers. Knowlton reports that ten men received an average of 16.21 R over a 77-week period. Knowlton, LA-587, 2.

161. Barton C. Hacker, *Elements of Controversy: The Atomic Energy Commission and Radiation Safety in Nuclear Weapons Testing, 1947-1974* (Berkeley: University of California Press, 1994), chapters 4 and 5.

162. Richard Elliott, Director, Public Information Division, San Francisco Office, AEC, to Public Information Officers, Division Offices, AEC, 2 December 1953 ("The Public Relations of Atmospheric Nuclear Tests") (ACHRE No. DOE-030195-C), 2. Hacker, in *Elements of Controversy*, provides some of the background for the discussion of Elliott's paper, including high levels of fallout observed in communities in southern Utah and injuries and death to livestock that had grazed in the fallout area.

163. We should emphasize that public notification does not mean that members of the public would need to or could take precautionary actions that would not otherwise be taken. Given the relatively low risk posed by the intentional releases, evacuation could have had costs greater than the possible benefits. In the case of the Green Run, a warning not to eat certain foods might have been useful; however, the food pathways were not known at the time. On the other hand, the prospectors around the Dugway site and the Pueblo Indians around Los Alamos could have been warned not to wander into certain areas that may have posed some hazard, however small.

164. Both statutes have since been amended by subsequent legislation. The relevant provisions of the Clean Air Act are the National Emission Standards for Hazardous Air Pollutants (42. U.S.C. 7412), and those of the Atomic Energy Act are 42 U.S.C. 2114, 2133. Other environmental statutes either explicitly exempt most radioactive materials (the Clean Water Act) or are less directly relevant to intentional releases (the Safe Drinking Water Act, the Resource Conservation and Recovery Act, and the Comprehensive Environmental Compensation, Response, and Liability Act).

165. As noted in the Introduction, radiation standards were initially established as recommendations by two private advisory bodies: the International Commission on Radiological Protection (ICRP) and the U.S. National Committee on Radiation Protection, now the National Council on Radiation Protection and Measurements (NCRP). Over time federal and state agencies have based regulatory standards on these recommendations.

166. As noted above, this standard was a recommendation by the NCRP, later adopted as policy by the AEC. Carroll Wilson to Lauriston Taylor, 10 October 1947, as cited in Gilbert Whittemore, "The National Committee on Radiation Protection, 1928-1960: From Professional Guidelines to Government Regulation" (Ph.D. diss., Harvard University, 1986), 326-327.

167. This standard took the form of guidance issued by the Federal Radiation Council in 1960, "Radiation Protection Guidance for Federal Agencies," in Fed. Reg. 25, 4402-4403 (1960); and Fed. Reg. 26, 9057-9058 (1961). See also, D. C. Kocher, "Perspective on the Historical Development of Radiation Standards," *Health Physics* 61, no. 4 (October 1991).

168. The EPA was established by President Nixon. The NRC was formed in 1974 under the Energy Reorganization Act to take over the regulatory functions of the AEC. See 42 U.S.C. 5801 et seq.

169. U.S. General Accounting Office, *Consensus on Acceptable Radiation Risk to the Public is Lacking*, GAO/RCED-94-190, summarizes the existing radiation protection standards in the federal government (see especially table 1, p. 5).

170. Committee on the Biological Effects of Ionizing Radiation, *BEIR V*, 18. Table 1-3 provides a comparison on typical exposure to natural and artificial sources of ionizing radiation.

171. 54 Fed. Reg. 51657; 54 Fed. Reg. 51655; 56 Fed. Reg. 33080, as cited in David O'Very and Allan Richardson, unpublished, "Regulation of Radiological and Chemical Carcinogens: Current Steps Toward Risk Harmonization," 1995.

172. Some regulations already take the population dose into account. The DOE and NRC use the population dose in implementing the principle that radiation exposures be made as low as reasonably achievable (a principle that goes by the acronym ALARA), applying cost-benefit analysis to reduce population doses from the operation of a given facility. As another example, releases of Kr-85 from nuclear power plants are limited on the basis of population doses. 40 C.F.R. 190.10(b).

173. The national security interest exemption to the Clean Air Act is provided in 42 U.S.C. 7412(i)(4): "The President may exempt any stationary source from compliance with any standard or limitation under this section for a period of not more than two years if the President determines that the technology to implement such standard is not available and is in the national security interest of the United States to do so." Other environmental statutes have similar exemptions.

174. AEC 132/64, 7 January 1964, cited in J. Samuel Walker, *Containing the Atom* (Berkeley: University of California Press, 1992), 11-12. Except for such circumstances, the AEC declared its intention to ensure that "reactor facilities are designed, constructed, operated, and maintained in a manner that protects the general public, government and contractor personnel, and public and private property against exposure to radiation from reactor operations and other potential health and safety hazards."

175. The ability to delay any report to Congress by as much as a year greatly limits the effectiveness of this reporting requirement. There also remains the possibility that the information provided to Congress would be classified, and so the report would not be made public.

176. See 42 U.S.C. 4321 et seq.

177. The basic requirements for environmental impact analyses appear at 40 C.F.R. part 1500 et seq. As a preliminary step, an environmental assessment may be done to determine whether the "significant impact" threshold is met and a full EIS is necessary. This EIS must include an analysis of the environmental impact alternatives to the proposed action. Normally, a draft EIS must be made available for public information and comment, and the agency must respond to any comments of the public.

178. The regulations implementing NEPA provide that "environmental assessments and environmental impact statements which address classified proposals may be safeguarded and restricted from public dissemination in accordance with agencies' own regulations applicable to classified information." 40 C.F.R. 1507.3(c). This provision for secret procedures does not relieve an agency of the obligation to inform itself of the environmental impacts of its actions, nor does it relieve EPA of the requirement to review those impacts.

179. On 11 February 1994, President Clinton signed Executive Order 12898, "Federal Actions to Address Environmental Justice in Minority Populations and Low-Income Populations," which requires each federal agency to address disproportionate human health or environmental effects of its policies. This includes requirements to assess those impacts and to seek greater public participation in environmental planning and policymaking. Executive Order 12898, 59 Fed. Reg. 7629 (16 February 1994).

180. Richard Sanderson, Director, Office of Federal Activities, EPA, to Donald Weightman (ACHRE Staff), 22 March 1995 ("NEPA Oversight of Classified Documents"). Such programs are offically referred to as Special Access programs.

181. See *Helen Frost et al. v. William Perry, Secretary of the United States Department of Defense et al.*, Civil Action no. CV-S-94-795-PMP (RLH), filed August 15, 1994 (ACHRE No. 5WU-041495-A), and *John Doe et al. v. Carol M. Browner, Administrator, United States Environmental Protection Agency*, Civil Action no. CV-S-94-795-DWH (LRL), both of which were filed in the U.S. District Court for the District of Nevada.

182. Craig Hooks, Associate Director, Federal Facilities Enforcement Office, EPA, to Donald Weightman (ACHRE staff), 11 April 1995 ("Please find enclosed . . .") (ACHRE No. EPA-041395-A).

183. Gary Vest, Deputy to the Assistant Secretary of Defense for Environmental Security, interview with Mark Goodman (ACHRE staff), 13 December 1994, staff notes (ACHRE Contact Database).

184. Mark Hamilton, USAF, telephone interview with Mark Goodman (ACHRE staff), 4 January 1995, staff notes (ACHRE Contact Database).

185. Richard Sanderson, Director, Office of Federal Activities, EPA, to Donald Weightman (ACHRE Staff), 22 March 1995 ("NEPA Oversight of Classified Documents").

186. The Supreme Court has ruled that the question of an agency's compliance with NEPA is "beyond judicial scrutiny" when a trial of the case would "inevitably lead to the disclosure of matters which the law itself regards as confidential, and respecting which it will not allow the confidence to be violated." *Weinberger v. Catholic Action of Hawaii/Peace Education Project*, 454 U.S. 139, 146 (1981).

12

Observational Data Gathering

SUPPLIES of uranium to build atomic bombs; a remote, sparsely inhabited site to test the bombs; information about the health effects of both the raw material and the bomb: these were the Cold War needs that led directly to the events with which this chapter is concerned.

This chapter examines whether the U.S. government wronged or harmed uranium miners in the American West and Marshall Islanders in the mid-Pacific, in both cases by exposing them to radiation hazards: in the case of the miners by failing to inform them about the risk and failing to mitigate it; and in both cases, perhaps to different degrees, by studying them without having obtained adequate consent. Although the mines of the Colorado Plateau and the seas surrounding the atolls of the Marshall Islands were seen by U.S. policy planners as ideal sites for the government's primary missions—mining uranium and detonating atomic and hydrogen bombs—they became laboratories for studying radiation damage to humans. We also touch briefly on a radiation experiment conducted with a view to the natural laboratory in which the subjects were set: in 1956 and 1957 the Air Force administered iodine 131 to Alaskan residents to determine the role of the thyroid gland in adapting to extreme cold.

The uranium mines, the Marshall Islands, and Alaska were not, of course, the first such occa-sions for studying the effects of radiation on people. As has been reported in earlier chapters, radium dial painters were studied, and in the largest epidemiological study of radiation effects ever, the survivors of the Hiroshima and Nagasaki bombs continue to be followed. The Atomic Bomb Casualty Commission (now the Radiation Effects Research Foundation) began its work soon after World War II. The organization's projects include a mortality study, a periodic health examination study, a study of people exposed in utero, and a genetic effects study. Some of the most important data available on long-term radiation risks have come from these studies. These data have also provided the basis for most current radiation exposure standards. The Hiroshima-Nagasaki studies are different from the cases of the uranium miners and the Marshallese, however, because the exposure ended before the epidemiologic study got under way.

While the miners, and the Marshallese after their high initial exposure, were subjected to continuous exposure to radiation—relatively high for the miners, relatively low for the Marshallese—they were not exposed for the purpose of studying the effects of radiation on their health. But the exposures resulting from the mining and bomb tests provided the government an opportunity, and some would say a duty, to collect needed information on radiation effects on human beings. In both cases researchers were interested in determining the health consequences of exposure to specific and quantified forms and levels of ionizing radiation over a long term. For the miners it was radon gas and its radioactive decay products. For the Marshallese it was the fallout products of nuclear explosions such as iodine 131, strontium 90, and cesium 137. Also, in both cases, the United States has provided, and in the case of the miners, continues to provide, financial compensation. In addition, a class-action lawsuit, *Begay v. United States*, was brought on behalf of a group of Navajo miners.

There were, however, major differences between the situation of the miners and that of the Marshallese. In the case of the miners, the research was conducted even though there were data from European studies clearly indicating that uranium miners were at high risk for lung cancer, which could have been substantially mitigated by ventilating the mines. The study of the miners, conducted by the Public Health Service, was epidemiological in nature and unrelated to their clinical care. The Marshallese were the first population exposed to amounts of fallout perceived as acutely dangerous.[1] The long-term effects of exposure to fallout were unknown; therefore it was important to gather data while treating the exposed population. It appears that the medical monitoring of the exposed population was directly integrated with the management of their health care.

To gather information on the health effects of radiation, federal government agencies mounted observational studies, a term indicating that the conditions of exposure are not under the control of the investigator who is studying the health effects.

For a long time, while they were being studied, it seems evident that no one explained to the miners the extent to which their exposure to radiation might be hazardous and, in many cases, lethal. Nor, it appears, were they told that ventilation of the mines could significantly reduce the hazard. And, evidently no one seems to have told the miners the true purposes of the research. With respect to the Marshallese, efforts to explain to them the purpose of the studies and the hazards of their contaminated environment

were inadequate well into the 1960s, and the difference between medical care and treatment-related research was not clearly explained. The Advisory Committee reports here on both studies and concludes with a discussion of the cold-weather experiment in Alaska in which servicemen, Eskimos, and Indians were given tracer amounts of iodine 131. We begin with the uranium miners.

THE URANIUM MINERS

The competition with the Soviet Union to build atomic arsenals spurred a uranium boom. In the late 1940s, there was a perceived need for a large and reliable domestic source of uranium to replace supplies predominantly from the Belgian Congo and, to a lesser degree, Canada. The AEC's announcement in 1948 that it would purchase at a guaranteed price all the ore that was mined set off a stampede on the Colorado Plateau.[2] Hundreds of mines, ranging from mines run by the prospectors themselves to larger corporate operations, were opened in the Four Corners area of Arizona, New Mexico, Utah, and Colorado, and several thousand miners, many of them Navajo, went to work.[3]

Some of the mines were large open pits, but most were underground networks of shafts, caverns, and tunnels, shored up by timbers. Because uranium milling and open-pit mining is conducted above ground, radon levels tend to be quite low, as radon is readily dispersed into the atmosphere. However, millers are exposed to uranium dust and thorium 230, both of which may have chemical or radiological toxicity, as well as additional chemicals used in the extraction process. In the remainder of this chapter, we focus on the underground miners who were exposed to much higher levels of the hazards that are the principal cause of lung cancer in the miners.[4]

The American boom followed centuries of experience with uranium mining in Europe, where a mysterious malady had been killing silver and uranium miners at an early age in the Erzgebirge (ore mountains) on the border between what is now the Czech Republic and Germany. In 1879, two researchers identified the disease as intrathoracic malignancy. They reported that a miners' life expectancy was twenty years after entering the mine, and about 75 percent of the miners died of lung cancer.[5] By 1932, both Germany and Czechoslovakia had deemed the miners' cancers a compensable occupational disease.

In 1942, Wilhelm C. Hueper, a German émigré who was founding director of the environmental cancer section of the National Cancer Institute (NCI), one of the National Institutes of Health, published a review in English of the literature on the European miners suggesting that radon gas was implicated in causing lung cancer.[6] He eliminated nonoccupational factors because excess lung cancer showed up only among miners. He also eliminated occupational factors other than radon because these other factors had not caused lung cancer in other occupational settings.[7] Among Hueper's peers, dissenters, such as Egon Lorenz, also of the NCI, focused on contaminants other than radon in the mine, the possible genetic susceptibility of the population, and the calculated doses to the lung, which seemed too low to cause cancer because the role of radon daughters—which the radioactive polonium, bismuth, and lead decay products of radon gas are known as—was not yet understood.[8]

At the time its own program began, the AEC had many reasons for concern that the experience of the Czech and German miners portended excess lung cancer deaths for uranium miners in the United States. The factors included the following: (1) No respected scientist challenged the finding that the Czech and German miners had an elevated rate of lung cancer; (2) these findings were well known to the American decision makers; (3) as Hueper points out, genetic and nonoccupational factors could be rejected; and (4) radon standards existed for other industries, and there was no reason to think that conditions in mines ruled out the need for such standards. Moreover, as soon as the government began to measure airborne radon levels in Western U.S. uranium mines, they found higher levels than those reported in the European mines where excess cancers had been observed.[9] As Public Health Service (PHS) sanitary engineer Duncan Holaday, who spent many years studying the miners, recalled in 1959 congressional testimony, there was early recognition

that while there were substantial differences between European and American settings, the exposure levels in U.S. mines were high:

> In 1946 our American mines were not as deep as those in Europe. The men did not work long hours. Furthermore, a great many of them were more or less transient miners, in and out of the industry.

> However, our early environmental studies in these early American mines indicated that we had concentrations of radioactive gases considerably in excess of those that had been reported in the literature.[10]

One important hole in Hueper's argument was that the calculated dose of radiation from the radon in European mines did not seem high enough to cause cancer.[11] But when William Bale of the University of Rochester and John Harley, a scientist at the AEC's New York Operations Office (NYOO) who was working toward his doctorate at Renssaelear Polytechnic Institute, were able to show and explain in 1951 the importance of radioactive particles that attached to bits of dust and remained in the lung, the discovery had a tremendous impact.[12] When doses to the lung were recalculated using Bale and Harley's models, they increased 76 times,[13] making them high enough to explain the observed cancer rates.[14] Recognizing the importance of radon daughters also explained why animal experiments using pure radon gas had not caused cancer.[15]

In the absence of Atomic Energy Commission willingness to press for relatively safe tolerance levels for radon in U.S. mines and to institute an effective program of mine ventilation to reduce the hazard, and a mixed, but mainly unsatisfactory response from the states, the stage was set for intergovernmental buck passing and decades of study, a course that resulted in the premature deaths of hundreds of miners. An analysis of eleven underground miners' studies published in 1994 by the National Cancer Institute supports the view that radon daughters are responsible for an even greater number of lung cancers than previously believed.[16]

The Advisory Committee heard from many miners and their families about the devastation wrought by the experience in the mines and the government's ability to prevent it. Dorothy Ann Purley, from the pueblo of Laguna in New

Mexico, told Advisory Committee members at a public meeting in Santa Fe, "Nowadays people come out and say, 'Did you know so and so died of cancer?' 'I have a brother in law who has got cancer. He worked at the mine.'"[17]

Philip Harrison, a spokesman for Navajo miners and their families, told the Advisory Committee that in New Mexico mines "the working conditions were sometimes unbearable. . . . The government knew all along what the outcome would be and . . . initiated studies on the miners . . . without their knowledge and consent."[18]

A Standard for Beryllium, But Not for Uranium

In 1948, Merril Eisenbud, an industrial hygienist, was recruited by the AEC's New York Operations Office to help set up a health and safety laboratory. The NYOO was responsible for all raw materials procurement for the AEC.[19] At the request of the AEC's Raw Materials Division, Dr. Eisenbud and Dr. Bernard Wolf, a radiologist, reported on potential health hazards in the mines to the NYOO field office in Colorado and to AEC headquarters staff.[20] Dr. Eisenbud and the New York Operations Office recommended that the AEC write requirements for health protection into its contracts with the mine operators.[21]

The AEC had used contract provisions in the case of beryllium, another key (but not radioactive) element in bomb production. One month before Dr. Eisenbud filed his report on the uranium mines, the *Cleveland News* reported on a conference convened to discuss cases of beryllium poisoning at plants in Massachusetts and Lorain, Ohio.[22] Among the fatalities in Lorain were five residents living near the Beryllium Corporation plant.[23] The plant owner, Dr. Eisenbud recalled in 1995, was eager to have conditions studied "because he wanted to know what his liability was."[24]

That same month, June 1948, responding to the "considerable publicity . . . given by the press to cases of berylliosis among plant workers and residents," the AEC set a tentative standard for the permissible levels of exposure to beryllium. The NYOO, "with the approval of the Division of Biology and Medicine, has insisted that the AEC-recommended tolerance levels be met in all

plants processing beryllium or beryllium compounds for the Commission."[25] Despite the fact that by September 1949 there had been at least twenty-seven deaths attributed to beryllium in plants where the AEC had contracts (no one became sick with berylliosis after the tolerance limits had been set in place), the DBM objected to AEC "establishment and enforcement of standards or regulations pertaining to health and safety conditions" and wanted to turn the matter over to the states.[26] Nevertheless, the NYOO enforced standards for beryllium.[27]

The uranium and beryllium situations had much in common. In both cases the AEC was the sole or primary purchaser. In both cases the AEC's New York Operations Office sought to control the hazard. And in both cases there were arguments to be made for inaction: The causation mechanism for the disease was poorly understood, and the legal authority of the AEC to regulate private production was questionable. The essential difference between the two cases was that the illness caused by beryllium appeared shortly after exposure and aroused publicity and associated public concern. By contrast, it would take more than a decade before uranium miners would begin to die of lung cancer, and causality would be harder to infer.

The DBM and the AEC Raw Materials Division rejected Dr. Eisenbud's recommendation for health protection, arguing that the Atomic Energy Act did not give the AEC authority over uranium mine health and safety.[28] The New York Operations Office took the same position that it had taken on beryllium: if it was going to procure uranium, it was going to control radon in the mines.[29] The AEC responded by transferring uranium procurement to a newly created section of the Raw Materials Division in Washington.[30] According to Dr. Eisenbud, the director of the New York Operations Office and many of its employees quit over this move, at least some of them because the shift was intended to keep the AEC out of health-related matters in the uranium mining industry.[31]

Eisenbud's perspective was echoed in at least part of the AEC's Washington office. In May 1949, A. E. Gorman, a sanitary engineer at the AEC, wrote a memo for the files in which he reported on a meeting with Lewis A.

Young, director of the Colorado Department of Health's division of sanitation, and Dr. John Z. Bowers, deputy director of the Division of Biology and Medicine. Bowers "indicated that health conditions [on the Colorado Plateau] were not satisfactory," and Mr. Young reported that "conditions under which uranium ore was being mined and processed were not good."[32] Bowers, the memo recorded, said his office did not want to recommend "drastic steps" to require correction of deficiencies, but preferred to gather facts about the hazard and cooperate with mine operators and state agencies to correct unsatisfactory conditions. Gorman, however, recorded:

I expressed the opinion that if the State of Colorado had only two inspectors to cover industrial hygienic conditions in all mines in the state, it would not be realistic to expect very extensive follow up of the hazards problems [sic] involving silicosis and radioactivity; also that since AEC was purchasing a very large percentage of the uranium produced, we had a moral responsibility at least to improve any unsatisfactory condition which was known to exist involving the health of workers. I suggested that this might be taken care of by a clause in our contracts even though it might result in a higher cost of production. I questioned the point that such action might seriously affect the production of uranium.[33]

Gorman's perspective did not win out. By the 1950s occupational standards or guidelines existed not only for radium[34] (a maximum permissible body burden) but also for radon. By 1941 the data from the European mines had been used to establish a radon standard for "air in plant, laboratory, or office [of] 10 picocuries per liter."[35] But when it came to the mines the federal government took nearly two decades to issue enforceable standards and actions to protect all those miners known to be exposed to significant risk. Instead, it debated responsibility for action while it pursued a long course of epidemiological study. The episode, the judge would declare in the *Begay* case decision in 1984, was a "tragedy of the nuclear age."[36]

The PHS Study

On August 25, 1949, the state of Colorado and U.S. Public Health Service officials met to explore radiation safety in the uranium mines and mills.[37] Colorado was home to about half of the U.S. uranium mines. Because many of them were small mines, they employed less than 10 percent of the country's uranium miners. (New Mexico, with much larger mines on average, had a fraction of the mines, but nearly half of the miners.)[38] The Colorado Department of Health established an advisory panel of federal, state, and uranium industry officials to oversee a comprehensive study. The panel advised the health department that more information was needed on the medical hazards of the uranium mines. In August 1949, the health department, along with the Colorado Bureau of Mines and the U.S. Vanadium Company, formally requested a study of the mines and mills, which the PHS agreed to do.[39] The PHS initiated both environmental studies of the mines[40] and epidemiologic studies of the miners.[41] The environmental study ended in 1956, but the epidemiologic study is ongoing.

In 1949, Henry Doyle, a sanitary engineer who was the chief PHS representative in Colorado, began environmental sampling in the mines.[42] Doyle recruited Holaday to direct the study.[43] The health departments of Utah, New Mexico, and Arizona also participated.[44] The environmental part of the study began first, in 1950. Between 1950 and 1954 medical examinations of uranium miners and millers were done on a "hit-or-miss basis,"[45] but in 1954 a systematic epidemiological study of the miners was begun.

Between 1949 and 1951, PHS investigators took environmental measurements of radon levels in the mines. Like Dr. Eisenbud, they detected high levels of radon.[46] In a February 1950 memo to the PHS Salt Lake City office, Holaday reported on a survey of four mines on the Navajo reservation. He declared that while he "anticipated that the samples would show high radon concentrations, the final results were beyond all expectations." The samples disclosed a "rather serious picture," leading Holaday to conclude "that a control program must be instituted as soon as possible in order to prevent injury to the workers."[47]

On January 25, 1951, representatives from the AEC, the PHS Division of Industrial Hygiene, and other branches of PHS convened to discuss in detail the radon concentrations discovered by the PHS study and what could be done about them.[48] The PHS staff explained that the uranium study demonstrated "radon concentrations . . . in the mines high enough to probably cause injury to the miners. . . ."[49] They also said the hazard could be abated by proper ventilation. The group concluded that the radon concentrations should be reduced to the lowest level possible consistent with good mine ventilation practices, but found it "unrealistic" to set a definite level that mine operators should meet.[50] They recommended further research, especially on ventilation techniques.[51] By this route, "the radon concentrations in the mines would be materially reduced in all cases, and valuable information would be yielded as to the effectiveness of standard ventilation practice in the control of radon."[52] It also was noted at this meeting that the acceptable level of radon in manufacturing was only 10 picocuries per liter, one to three orders of magnitude lower than the observed levels in the mines.[53]

The PHS Progress Report for the second half of 1951 explained that because of the "acuteness of the radon problem it was felt that it was necessary to temporarily put aside our full-scale environmental investigation of this industry and concentrate on the control of this contaminant."[54] The PHS met with the mining companies to discuss the hazards and urged them to undertake ventilation measures.[55] In 1979, Duncan Holaday testified to Congress that "by 1940 I do not believe there was any prominent scientist or industrial hygienist in the United States, except one [presumably Lorenz], who was not thoroughly convinced of the dangers, and it had been demonstrated that the radioactive elements could be removed from a closed area and be completely avoided."[56] However, it appears the mining industry lacked the commitment to improve worker conditions.[57]

The PHS distributed its interim report on a "restricted" basis to state and federal government officials and mining companies in May 1952.[58] A June 26, 1952, press release announcing the completion of the interim report began with the statement that "no evidence of health damage from radioactivity had been found."[59] Mining had been going on for only a few years, and lung cancer has a ten- to twenty-year latency period. The introduction to the report itself noted, however, that "certain acute conditions are present in the industry which, if not rectified, may seriously affect the health of the worker."[60]

Meanwhile, as evidence of hazard mounted, Dr. Hueper, now at the National Cancer Institute, reported continued efforts to limit his speech on the risks involved. Dr. Hueper reported that in 1952 he was invited to speak to the Colorado Medical Society, but declined to attend when ordered by the director of the NCI, at the request of the AEC's Shields Warren, to delete references "to the observation of lung cancer in from 40 to 75 percent of the radioactive ore miners in . . . [Europe] although these occupational cancers had been reported repeatedly since 1879."[61] In a 1952 memo to the head of the Cancer Control Branch of NIH, Hueper reported that an AEC representative had objected that references to occupational cancer hazards in the mines were "not in the public interest" and "represented mere conjectures."[62] After the Colorado episode, according to Hueper, Warren wrote to the director of the NCI, asking for Dr. Hueper's dismissal for "bad judgment." Dr. Hueper kept his job, but was, according to Victor Archer, one of the physicians who ran the uranium miner study, forbidden to travel west of the Mississippi for research purposes.[63]

U.S. officials, including those from the PHS, had no independent authority to enter the privately owned mines—as opposed to those owned by the AEC and leased to private operators—without permission of the mine owners.[64] Duncan Holaday testified in court proceedings that in order to gain access to the mines, an oral agreement was made with mine owners not to directly inform those most affected by their findings, the miners.[65] According to Holaday, "this was routine procedure that was followed in every industrial survey I was aware of . . . this went back for many decades." To gain entry to the mines the researchers agreed that the PHS would not "alarm the miners" by warning them of hazardous conditions.[66] In 1983 Holaday testified in *Begay* that "you had to get the survey done and you knew perfectly well you were not doing the correct thing . . . by not informing the workers."[67] A medical consent form from the PHS study dated May 1960 says nothing about the risk of lung cancer or any other health risk associated with working in uranium mines.[68] "[T]here would be no overt publicity," Holaday recalled in a 1985 deposition, "and when we reported the information that we found, it would be done in such a way that the facilities where a particular set of samples were taken would not be identified and that we would not inform the individual workers of what data we found."[69]

Holaday told Stewart Udall, a former secretary of the interior who represented the miners in the *Begay* case, that he did not try to go public because he didn't think that Washington would notice a "little Utah tweet" from him.[70] Eisenbud has suggested that perhaps this was because in the Cold War environment, with nuclear weapons testing under way, no one would pay much attention to the long-term health risks of a small group of miners.[71]

Although the PHS and the AEC already knew the danger of radon in the mines in 1951, and had pressed the states to take action with mixed results, PHS doctors nonetheless began to conduct basic health examinations to collect baseline data against which long-term health effects of radon could be gauged.[72] These medical examinations did not initially find evidence of harm caused from working in the mines. However, one would not have expected to find such effects because few miners had been on the job for more than five years and lung cancer takes ten to fifteen years to appear.

By 1953, the PHS had completed a series of ventilation studies. As early as 1951, federal and state officials meeting with mine owners in Colorado had told them that "ventilation had been tried in other mines and found to be satisfactory."[73] But while some large mines were ventilated during the 1950s and 1960s, most of the small mines were not ventilated until the 1960s or later, and in those mines that had ventilating systems earlier, they were not always properly used.[74]

The uranium miners were discussed at a January 1956 meeting of the AEC's Advisory Committee for Biology and Medicine. The formally secret transcript records that in a "status report on the Colorado plateau," the Division of Biology and Medicine's Dr. Roy Albert stated:

There are no pressing—particularly pressing—problems associated with it now, but there has always been a rumbling of discontent with the status of the health conditions in the uranium mines of the Colorado Plateau because this is a mining industry which is essentially controlled by the Federal Government and by the AEC in terms of how much it can produce and how much it paid for its product.

Albert explained that the tentative decision was to "sit tight" because it would be "an unusual step" for the federal government to enter the mining industry and the AEC could take a "wait and see" approach as the states "took up the cudgel."[75]

Merril Eisenbud responded, to no evident effect, that the federal government should pay to ventilate the mines: "I think here is where our responsibility lies, because I think this industry would not exist except for the fact that we need uranium. If the cost of operating these mines as determined by us does not permit adequate ventilation of those mines, we will have to change the price. It is as simple as that."[76]

In October 1958, LeRoy Burney, the surgeon general of the Public Health Service, wrote to Charles Dunham, director of the AEC's Division of Biology and Medicine, that the "numbers are too small to permit conclusions to be drawn at this time" about whether there were excess lung cancer deaths among the uranium miners. However, he added, "if this proportion of mortality . . . should increase or even continue in the future, then it might be appropriate to conclude that our American experience is not inconsistent" with that in the Czech and German mines. Dr. Burney added:

Although we do not have complete environmental measurements in all mines, it appears that about 1,500 men in some 300 mines are working in uncontrolled or poorly controlled environments. The median level of alpha emitters in the mines of one state is five times the recommended working level, and in some mines the level is exceeded by more than 50 times. . . . It is usually the older, smaller mines in which the workers are still exposed to these high levels.[77]

Burney concluded by suggesting that as the "sole purchaser of ores produced in the mines," the federal government could require mine owners to conform to federal safety standards.

Several months later, Dunham wrote a memo to AEC General Manager A. R. Luedecke, reporting "it is doubtful if the Commission's regulatory Authority could be extended to cover the mines."[78] The same day, March 11, 1959, AEC General Counsel L. K. Olson wrote to Dunham reporting that "there is nothing in the legislative history of the 1954 [Atomic Energy] Act, or the 1946 [Atomic Energy] Act, which indicates that Congress may have intended to permit AEC to regulate uranium mining practices."[79]

Later in 1959, the AEC asked the Bureau of Mines to inspect mines it leased and then made follow-up inspections to see that the bureau's recommendations were followed, closing sections of mines temporarily until corrective measures were completed. In the ten months between July 1959 when the inspections began and May 1960, levels of radon in these mines improved dramatically.[80]

As the judge in the *Begay* decision found, "the AEC concluded that it could enforce health and safety measures in leased mines [as distinct from privately owned mines] pursuant to the leasing provisions of the Atomic Energy Act" and amended its mines' leases "to contain explicit enforcement language and procedures."[81] The states began to enact standards in 1955,[82] but inspection and enforcement came later and varied greatly. New Mexico began enforcement in 1958.[83] Colorado and Utah did not begin serious enforcement until the 1960s,[84] and Arizona, according to Duncan Holaday, did "nothing outside of take air samples."[85]

In late 1959, the miners were provided with the PHS pamphlet that warned them about the hazards of radon exposure. The pamphlet mentioned the possibility of radon causing lung cancer, but said nothing of the experience of U.S. or European miners or the level of risk. It said that "scientists are working hard to get the final answer on how much radon and its breakdown products, known as *daughters,* you can be exposed to safely."[86] It did not tell the miner the "suggested figures," but suggested bringing "enough clean, fresh air to the face to sweep out the radon gas and dust," as well as several other measures to reduce exposures.[87]

All mining is dangerous, and there is no reason to think that any miners went into the uranium mines unaware of this. Whether the uranium miners had an appreciation of the added cancer risk from radon is another matter. The 1959 pamphlet is the first document we could find that indicated that the federal government tried to warn the miners of the radiation hazards. While the pamphlet mentioned the possibility of radon causing lung cancer, it gave no indication of the level of risk.[88] Duncan Holaday told a congressional hearing in 1979, "We, in the Public Health Service, made every effort to communicate with the men the situation that they were in. We put out pamphlets . . . conducted medical examinations . . . we told them what the story was."[89] This statement is hard to reconcile with Holaday's other statements, as quoted earlier, that the researchers had agreed not to warn the miners as the condition for access to the mines. When Senator Orrin Hatch of Utah suggested to Mr. Holaday that some of the miners "just were not capable of understanding or knowing the dangers to which they were subjected," Mr. Holaday responded, "I understand this perfectly well."[90]

In 1960, the PHS presented to the governors of the mining states what it believed to be conclusive evidence from the PHS study of a correlation between uranium mining and lung cancer. The evidence showed that at least four and a half times more lung cancers were observed than would normally be expected among white miners—for whom comparison data were available—and that there was less than a 5 percent chance that such a difference had appeared by chance. The results of a study of 371 mines (the number of miners surveyed was not stated) in 1959 showed that the number of mines with unacceptable levels of radon had increased from 1958.[91] Yet the federal government continued to defer to the states on rule setting and enforcement in the case of the mines that were not AEC property, and the AEC, the PHS, and the states continued studies and discussions.

Finally, in 1967, Secretary of Labor Willard Wirtz announced the first federally enforceable standard for radon and its daughters in uranium mines that supplied the federal government. "After seventeen years of debate and discussions regarding the respective private, state, and federal responsibilities for conditions in the uranium mines," Wirtz told Congress, "there are today (or were when the hearings were called) no adequate health and safety standards or inspection procedures for uranium mining."[92] The standard was set at 0.3 Working Level (WL).[93] Wirtz established this criterion under the 1936 Walsh-Healy Act, which provided for the regulation of health and safety conditions under government contracts.[94] It is not clear why the authority granted the secretary of labor under this 1936 law was not used earlier to control radon in the mines, but it might have been because most of the mines were privately owned and did not operate under federal contacts, which made the applicability of the act questionable.[95]

The Begay Decision

Begay v. United States was filed on behalf of a group of miners in federal district court in Arizona in 1979; the case came to trial in 1983. During the 1950s, according to the court, the PHS found radiation exposures in some mines higher than the level it recommended, and "even higher than the doses received as a result of the atomic bomb explosion in Japan."[96] But on July 10, 1984, the court decided that the United States was immune from suit,[97] although the judge wrote that the miners' situation "cries for redress."[98]

The decision in the *Begay* case poses basic questions regarding the responsibility of the government and its researchers. The court found that the government's actions were motivated by strong national security interests:

The government, in making its decision in this area, was faced with the immediate need of a constant, uninterrupted and reliable flow of great quantities of uranium . . . for urgent national security purposes and as an energy source in the future for the growing peacetime nuclear energy industry. . . . [T]he decision makers had to be concerned that there was adequate data available to justify the standards to be set and that labor and management would have the tools to know when they were in violation. . . .[99]

The court is not clear, however, on why or how a standard for radon in the mines would have interrupted the flow of uranium, damaged

national security interests, or interfered with the development of peaceful uses of nuclear energy. Ventilating the mines would have been relatively inexpensive, and it would have improved working conditions—this was demonstrated in PHS ventilation studies in 1951[100]—making it more rather than less attractive to a potential work force. In 1960 the deputy commissioner of mines of Colorado is reported as having said that 98 percent of the mines would have to suspend work if forced to abide by a working level standard proposed in 1955: 100 picocuries of radon in equilibrium with 300 picocuries of radon daughters.[101] In any event, the federal government did not invoke national security as a basis for its inaction. For example, in 1986 Duncan Holaday responded in the negative when asked in a deposition, "in all [your] years from 1949 until your retirement, did you ever receive directly or receive indirectly, any document [from the] Public Health Service, from the Atomic Energy Commission, or from any other source, indicating you or directing you that you are to pull punches or nothing was to be done because of national security considerations?"[102] As for the federal government's policy of not regulating the mines, this appears to have involved questions of the AEC's understanding of its authority and political questions relating to the traditional relationship between the states and the federal government.

Was the failure to apply the same approach to the uranium miners as to the beryllium workers a matter of the absence of legal authority, as claimed by the AEC, or of reasoned deference to state regulators, as the court suggested? The court's decision did not address the AEC's action to require its beryllium contractors to comply with hazard standards, nor did it address the fact that radiation standards were enforced in industrial settings. Fragmentation of responsibility—both at the federal level and between the states and the federal government—appears to have provided a convenient opportunity for the federal government to pass the buck among agencies and avoid decisive action until long after such action should have been taken.

Under what conditions should researchers enter into a long-term study where there is reason to suspect at the outset that the subjects are, each day, at continuing and largely avoidable and unnecessary risk?

The *Begay* decision states clearly the bargain entered into by the government and its researchers, on behalf of the epidemiological study:

... it was necessary to obtain the consent and voluntary cooperation of all mine operators. To do this, it was decided by PHS under the surgeon general that the individual miners would not be told of possible potential hazards from radiation ... for fear that many miners would quit and others would be difficult to secure because of fear of cancer. This would seriously interrupt badly needed production of uranium. . . . [N]o individual mine, or mines, would be publicly identified in connection with that data. Consequently, the voluntary consent of mine operators was secured to conduct the PHS study.[103]

The *Begay* decision does not address questions such as whether the researchers could have worked more effectively with state agencies that had authority to enter the mines, or whether they could have conducted the study in mines on federal or Navajo land, to which they had access. In any case, there is no obvious national security or other ground on which to justify the continued exposure of miners to the radon hazard.[104]

As to medical examinations of the miners, the court found that the physicians who had conducted them "had the responsibility for dealing only with the examination and the results of that examination."[105] Thus, the court concluded, "it was neither necessary nor proper for those physicians to advise the miners voluntarily appearing for examinations of potential hazards in uranium mines."[106] In the case of the epidemiological study, the court explained:

An epidemiological study deals with group statistics and the conclusions of such a study appropriately cannot be applied to specific participants of a group. . . . The government did not seek volunteers to work in the mines so that they could become part of the study group. . . .[107]

On this point, the Advisory Committee disagrees with the court. In epidemiological studies such as the one under discussion, group conclusions *are* applicable to the members of the population of which the group is intended to be a representative sample. That is, each individual can be told the probability of developing disease based on his level and conditions of exposure. If the study was poorly designed, then such applicability may not hold, but to the Committee's knowledge, no one has argued this about the PHS study. Moreover, the PHS researchers had opportunities to warn the miners face to face because they examined them periodically over more than twenty years. There is some disagreement about whether any miners were warned of the risk of lung cancer, but even Duncan Holaday, who in one instance indicated that some miners received warnings, acknowledged that very likely these warnings were ineffective.

Radiation Exposure Compensation Act

The *Begay* decision concluded that the plight of the uranium miners "cries for redress." Because of the doctrine of sovereign immunity, however, the court declared that it could not provide the appropriate remedy. By 1990, 410 lung cancer deaths had occurred among the 4,100 miners in the Colorado Plateau study group; about 75 lung cancer deaths would normally have been expected in a group of miners such as this.[108] In the same year, Congress responded with legislation, the Radiation Exposure Compensation Act (RECA), which provided $100,000 compensation for miners with lung cancer or nonmalignant respiratory disease, subject to certain conditions. In the case of lung cancer, the act requires that the claimant demonstrate an occupational exposure to radon daughters from 200 WLM (working level months) to 500 WLM, depending upon his age and smoking history, the higher figure applying to smokers and older miners. In the case of nonmalignant respiratory disease, the act also requires documentation of disease by a panel of radiologists certified in assessing x-ray evidence of lung disease. In both cases, records of occupational histories and civil records for next-of-kin claimants (such as marriage certificates) are also required—records that are often nonexistent or difficult to obtain, particularly for Navajo miners.

The most recent and authoritative analysis of risks of lung cancer from radon in uranium mining comes from a 1994 NIH publication[109] that reanalyzed all eleven of the major occupational radon studies worldwide. This analysis considerably extends that undertaken by the National Academy of Sciences BEIR IV Committee,[110] which was available in 1986 prior to the enactment of RECA. This report used similar methods of analysis but more recent and more detailed data on a larger set of studies. The most important conclusions of this report are

- that the risk rises approximately linearly with level of exposure, with an average slope that is similar to that estimated by earlier committees, including BEIR IV;[111]
- that the risk per WLM varies strongly by age, latency, mining cohort, and especially by dose rate or duration, the latter being a relatively recent observation, but one that is now widely accepted;[112]
- that there is little evidence that the proportional increase in lung cancer risks is substantially different for smokers and non-smokers—as a consequence, the probability that a particular lung cancer was caused or contributed to by radon is not materially altered by smoking history;[113]
- that on average more than half of the lung cancers among white miners in the Colorado plateau cohort and the Navajo New Mexico cohort were caused by radon exposures;[114] and
- that there were substantial uncertainties in the actual doses received by miners in different mines.[115]

Thus, the 200 WLM figure that is used in RECA as the criterion for awarding compensation is not unreasonable as a "balance of probabilities" for the miners as an entire group, but (1) is a much higher risk threshold than is required for either the downwinders of the Nevada Test Site or the atomic veterans covered in the same act and (2) ignores substantial variation in age, latency, and other factors and substantial uncertainties in dose estimates for individuals within the group of all miners, so that many miners whose cancers are likely to have been caused by radon would not have attained this criterion. Furthermore, the distinction between smokers and nonsmokers established in the act is not well supported by currently available scientific evidence and tends

to deny compensation to many miners, most of whom are smokers but suffered substantial increases in risk due to the synergistic effect of the two carcinogens.

Clearly some miners have a stronger case for compensation than others, and RECA makes an attempt to make such distinctions. In principle, it would be possible to construct a formula for determining the probability of causation that would better reflect the current state of scientific knowledge and a threshold on this scale of probabilities that would treat the miners more equitably vis-á-vis the other groups covered by the act. However, the case of the uranium miners presents insurmountable obstacles in this regard, including the loss of records pertaining to occupational histories and exposures and variations in cultural practices that have made record-keeping burdens on claimants especially onerous. When the difficulty of meeting such bureaucratic requirements is coupled with the strong link between lung cancer and uranium mining, the scheme unjustly places too great a burden on the individual. The Committee is strongly persuaded to propose an adjustment in the criteria so that the evidence of a minimum duration of employment underground would be sufficient to qualify for compensation. Any compensation scheme is necessarily imperfect, but given the strength of causal connection, and the severity of the injury, the time spent in the mines is a rational and equitable basis for determining exposure levels.

Conclusions About the Uranium Miners

The Advisory Committee concludes that an insufficient effort was made by the federal government to mitigate the hazard to uranium miners through early ventilation of the mines, and that as a result miners died. The Committee further concludes that there were no credible barriers to federal action. While national security clearly provided the context for uranium mining, our review of available records reveals no evidence that national security or related economic considerations were relied on by officials as a basis for not taking action to ventilate the mines. Since most of the mines were not ventilated, the federal government should at least have warned the miners of the risk of lung cancer they faced by working underground. We recognize that the miners had limited employment options and might have felt compelled to continue working in the mines, but the information should have been available to them. Had they been better informed, they could have sought help in publicizing the fact that working conditions in the mines were extremely hazardous, which might have resulted in some mines being ventilated earlier than they were.

The court in the *Begay* decision did not exaggerate when it called the abuse of these miners "a tragedy of the nuclear age."

The Committee believes that after 1951, when William Bale and John Harley's findings on radon daughters established that miners were getting a much larger dose to the lungs than previously suspected, the mine owners, the state governments, and the federal government each had a responsibility to take action leading to ventilation of all mines. There are basic ethical principles to not inflict harm and to promote the welfare of others (as described in chapter 4) under which all the relevant parties ought to have acted to prevent harm to the miners.

The Advisory Committee has found no plausible justification for the failure of the federal government, which is the focus of our inquiry, to adhere to these principles. It is clear that officials of the federal government were convinced by the early 1950s that radon and radon-daughter concentrations in the mines were high enough to cause lung cancer. The federal government's obligation flows from this knowledge and its causal link to the mining activity. Without the federal government to buy uranium, there would have been no uranium mining industry. Since the miners were put at risk by the federal government, a minimal moral requirement would be that the government ensure that the risk was reduced to an acceptable level. Because the federal government did not take the necessary action, the product it purchased was at the price of hundreds of deaths.

The historical record is tangled and incomplete, but legal responsibility for the health and safety of the miners appears to have rested largely, but not exclusively, with the states. At the same time, the resources to implement remedial measures existed mainly within the federal government.

The Atomic Energy Commission, which was the contracting agency of the federal government in its role as sole purchaser of uranium, interpreted the Atomic Energy Act as not providing it with authority over health and safety in the mines. It is not clear to the Committee why the AEC, as in the case of beryllium, could not have made ventilation a requirement of any contract to mine uranium, or, in any event, why the AEC could not have sought clarification of its authority from Congress. The Labor Department appears to have had authority under the 1936 Walsh-Healy Act to ensure safe working conditions in the mines, but for reasons that are again unclear to the Committee, it was not until 1967 that the Department of Labor applied the act.

According to the *Begay* decision, the United States did not recruit miners to work in the mines, nor did it cause the miners to be exposed to hazard or withhold treatment from any individual. None of the considerations, however, detracts from what was for the Advisory Committee an overarching determinative consideration: without the federal government's initiative and its role as sole purchaser, there would not have been an American uranium industry. Because the government played a pivotal role in putting the miners in harm's way, it follows that the government had a moral obligation to ensure that the harm be controlled, at least to a level of risk that was not in excess of those risks normally associated with underground mining, an argument the government used to act in the case of beryllium.

The uranium mines were not ventilated, however, adding particular significance to a second moral issue raised by this case: Why were the miners not warned about the risk to which they were being exposed, particularly as the likely magnitude of the hazard became clear? Although this question can be properly put to all the relevant parties, including the mine owners, the state governments, and the various federal agencies, most attention has focused on the Public Health Service. Investigators of the PHS were the only federal officials in direct contact with miners as they recruited and then followed the miners in the course of their epidemiological studies. Also, it was in the course of these studies that important evidence about the severity of the risk was accumulated.

When the data collected by the PHS indicated the miners were working in an environment where the threat of lung cancer was significant, which was clearly the case after the Bale-Harley findings, and when the PHS observed in the early 1950s that the states and owners were not ventilating the mines to mitigate the hazard, the PHS was obligated to warn the miners about the implications of its research. This research appears to have been conducted, however, under oral understandings with the mine owners that the PHS researchers would not directly warn the miners of the level of hazard.[116]

The question arises, of course, of whether the PHS should have entered into an agreement to study the miners conditioned on not warning them of the hazard to which they were being exposed. The argument for accepting this condition is that it was the only way the PHS researchers could gain entry to the mines and that ultimately the study results would be valuable and likely save some lives. But acceptance of the condition precluded the PHS from dealing in a straightforward manner with the people they were proposing to study and from providing a warning that had the potential, in this case, for saving at least some lives. The Committee is divided on this issue. Some members concluded that the condition was morally objectionable and should have been rejected, even if this meant that the research could not go forward or could go forward only in a limited way.[117] Others argued that a morally acceptable course would have been to accept the condition and, as the results emerged, warn the miners anyway, because in this case the duty of promise keeping was justifiably overridden by the duty to prevent harm.

The PHS's decision to abide by the agreement not to warn the miners is particularly troubling in light of a regulation, as noted by the court in the *Begay* decision, in force from 1951 to 1978, that governed the disclosure of information obtained and conclusions reached for PHS surveys, research projects, and investigations. The regulation said, in part, that information "obtained by the Service under an assurance of confiden-

tiality . . . may be disclosed . . . whenever the Surgeon General specifically determines disclosure to be necessary (1) to prevent an epidemic or other grave danger to the public health. . . ."[118] Certainly at some point the potential and eventually realized lung cancer epidemic qualified under this regulation. The PHS's 1952 interim report is clear that "certain acute conditions are present in the industry which, if not rectified, may seriously affect the health of the worker."[119] So, while the PHS had legal as well as moral standing to breach its confidentiality agreement, it did not do so, although it appears to have made efforts to communicate its findings, their implications, and abatement recommendations to health authorities, the AEC, mine operators and owners, and state agencies.[120]

The agreement between the PHS and the mine owners no doubt also affected what PHS investigators were willing to tell the miners about the purpose of their investigations at the time the miners were recruited to participate. The PHS told the miners little more than that they were studying "miners' health."[121] In fact they were studying (1) the relationship between exposure to radon and other conditions in the mines and miners' health and (2) engineering methods (specifically, ventilation techniques) for controlling radiation hazards.[122] Had miners been told the true purpose of the study then, even in advance of any warnings connected with the progress of the research, it is possible the miners could have used this information to advocate for their interests. Even if the miners were not well positioned to seek employment elsewhere or to advocate for improved working conditions, the principle of respect for the self-determination of others would have required a more straightforward disclosure.

Current guidelines for the ethics of epidemiological research, as well as current practices, would not counsel the original bargain with the mine owners, the minimal disclosure made to workers about the purpose of the research, or the failure to warn the workers as the hazard became clear. For example, the current Council for International Organizations of Medical Sciences (CIOMS) guidelines explain: "Part of the benefit that communities, groups and individuals may reasonably expect from participating in studies is that they will be told of findings that pertain to their health."[123] The CIOMS guidelines also specify a duty not to withhold, misrepresent, or manipulate data.[124] Today, it is widely recognized among epidemiologic researchers that they have an obligation to report findings indicating potential or actual harm, along with the uncertainties of those findings, to the people being studied and to the public at large.

Although the Committee believes that the federal government should have acted to ensure that the mines were ventilated and that the PHS should have informed the miners about the severity of the risk it was investigating, the Committee did not have enough information to assess the moral responsibility of individual AEC and PHS employees and officials for these failures. Some effort was made by some investigators to get the states and mine owners to ventilate the mines, and some warnings may have been given to individual miners. But the ventilation effort was inadequate and the warnings ineffectual. We lack the information to evaluate whether officials such as Duncan Holaday, Henry Doyle, and Merril Eisenbud should have done more than they did to protect the miners, granting that their superiors had ultimate responsibility for decisions not to press for ventilation and warnings. Whistleblowing to avert serious harm is an important moral responsibility, but there are personal prudential considerations unknown to us that must be weighed before judging whether these people failed in their duty.[125]

While federal and state agencies may debate internally and with one another the limits of their authority, from the vantage of those exposed to risk by the government, the government should be reasonably expected to do what is needed to sort out responsibility and to ensure that action is taken to address risk. This did not happen. Perhaps the most remarkable aspect of the uranium miners tragedy is that, notwithstanding the national security context, so much of it took place in the open; so many federal and state agencies were participants, often with some formal degree of responsibility or authority in an unfolding disaster that appears to have been preventable from the outset.

THE MARSHALLESE

Following World War II, the United States selected the Marshall Islands as the site of the Pacific Proving Grounds for testing nuclear weapons. The Marshall Islands are a widely scattered cluster of atolls located just above the equator north of New Zealand. They were designated a trust territory of the United States by the United Nations in 1947. The Marshallese were granted independence under a Treaty of Free Association that went into effect in 1986. The U.S. Department of the Interior oversees relations with the Marshall Islands, with responsibility to ensure that the terms of the Trusteeship Agreement are carried out. According to the 1947 Agreement, the United States as trustee "shall . . . protect the health of the inhabitants."[126]

Testing of nuclear weapons began on July 1, 1946, with Operation Crossroads, two tests at Bikini Atoll. In preparation for this operation, the Bikinians were evacuated in March of that year. Crossroads did not lead to any immediate exposure of the native population. However, the second shot in the series, Baker, was a 21-kiloton underwater blast that contaminated the surviving test ships, posing major decontamination problems for the military participants. It also contaminated the atoll itself, which, along with further testing, delayed the return of the Bikinians, who began returning to the island in 1969. Although some radioactive contamination was still known to linger, it was believed at the time that restrictions on the consumption of certain native foods and provision of imported foods would make Bikini habitable. Unfortunately, these assumptions proved wrong. After the resettlement, the AEC and its successors monitored the internal contamination levels of the Bikinians and observed increases in plutonium, leading to their reevacuation in 1978.[127] Today, the Bikinians remain scattered around the Marshall Islands, while a new radiological cleanup of their atoll is in progress.

In 1954, the Bravo shot of the Operation Castle series was detonated at Bikini Atoll. Bravo was the second test of a thermonuclear (hydrogen) bomb, with a yield of 15 megatons, a thousand times the strength of the Hiroshima bomb.

A change in wind direction carried fallout from the test toward Rongelap and other inhabited atolls downwind of it. The populations of the Rongelap and Utirik Atolls were evacuated, but not until after they had received serious radiation exposure (about 200 roentgens on Rongelap and about 20 on Utirik). What followed was a program by the U.S. government—initially the Navy and then the AEC and its successor agencies—to provide medical care for the exposed population, while at the same time trying to learn as much as possible about the long-term biological effects of radiation exposure. The dual purpose of what is now a DOE medical program has led to a view by the Marshallese that they were being used as "guinea pigs" in a "radiation experiment."

As happened at Bikini, the Rongelapese were resettled onto their atoll, but after an interval of only three years. Again, it was recognized at the time that some radioactivity remained, but U.S. officials concluded that appropriate dietary restrictions would minimize the danger.[128] Unlike the case of the Bikinians, however, the medical follow-up program has continued to the present, reflecting the seriousness of the initial exposure and the added risk of continuing exposure at low levels. Five years after the Bravo shot, Dr. Robert A. Conard, then the director of the AEC's Brookhaven National Laboratory (BNL) medical team, wrote,

The people of Rongelap received a high sub-lethal dose of gamma radiation, extensive beta burns of the skin, and significant internal absorption of fission products. . . . Very little is known of the late effects of radiation in human beings. . . . The seriousness of their exposure cannot be minimized.

Low levels of radioactive contamination persist on Rongelap Atoll. The levels are considered safe for habitation. However, the extent of contamination is greater than found elsewhere in the world and, since there has been no previous experience with populations exposed to such levels, continued careful checks of the body burdens of radionuclides in these people is indicated to insure no unexpected increase.

From these considerations it is apparent that we are obligated to carry out future examinations on the exposed people to the extent that they are deemed necessary as time goes on so that any untoward effects that may develop may be diagnosed as soon as pos-

sible and the best medical therapy instituted. Any action short of this would compromise our responsibility and lay us open to criticism.[129]

These and similar documents discussed below lay out clearly the purposes of the medical program. However, at the fourth meeting of the Advisory Committee, representatives of the Republic of the Marshall Islands presented documents to support their contention that by ignoring forecasts about the weather patterns at the time of the Bravo shot,[130] and by resettling the Rongelapese on their atoll despite knowledge of residual contamination, the U.S. government was using the Marshallese as guinea pigs in a deliberate human radiation experiment.

The Committee heard extensive testimony about the difficulties the Marshallese have had in obtaining information relevant to their health. Their own medical records are only now being made readily available to them. Many other documents describing U.S. government activities conducted on their soil have for too long been shrouded in secrecy or made inaccessible to the Marshallese by bureaucratic obstacles. This inaccessibility of records, combined with a history of inadequate disclosure of hazards known to U.S. researchers, has contributed to a climate of distrust.

In our review of materials that are now becoming available, we found no evidence to support the claim that the exposures of the Marshallese, either initially or after resettlement, were motivated by research purposes. On the contrary, while there is ample evidence that research was done on the Marshallese, we find that most of it offered at least a plausible therapeutic rationale for the potential benefit of the subjects themselves. We have found only two examples of research in the Rongelap and Utirik populations that appear to have been nontherapeutic: this research was intended to learn about radiation effects in this population and offered little or no prospect of benefit to the individual subjects.

There is, of necessity, some tension between data gathering and patient care when the same physician is responsible for both. The Advisory Committee has found no clear-cut instance in which this tension was likely to have caused harm to patients, but some may have been subjected to biomedical tests for the primary purpose of learning more about radiation effects. This inherent tension, coupled with the additional strains of language and cultural differences between the Marshall Islanders and the physicians, appears to have compromised the process of informing the subjects of the purpose of the tests and of obtaining their consent, which has doubtless contributed to their sense of being treated as guinea pigs. Insensitivity to cultural differences, failure to involve the Marshallese in the planning and implementation of the research and medical care program, divided responsibilities for general medical care, and failure to be fully open about hazardous conditions have all contributed to unfortunate and probably avoidable distrust of the American medical program by the Marshallese.

It is of concern to the Advisory Committee that problems arose in explaining to the Marshallese the nature and purpose of the research activities that accompanied their treatment and in obtaining their consent for both research-related interventions (such as bone marrow, blood, and urine tests) and treatment. Both Brookhaven researchers and the Marshallese agree that general medical care provided by the Trust Territory government was inadequate,[131] but this question was outside the scope of the Advisory Committee's investigation. What follows, as best we can piece it together, is the story of how the United States handled its responsibility to provide medical care to citizens of a U.S. trust territory exposed to hazard by a U.S. nuclear bomb test that went awry.

The Bravo Shot

The Bravo shot was detonated on Bikini at 6:45 a.m. on March 1, 1954. Its yield was substantially greater than expected. The radioactive cloud rose to an altitude of about 100,000 feet before blowing east toward the inhabited atolls of Rongelap, Ailinginae, and Rongerik, and still farther east, toward Utirik, Ailuk, and Likiep, instead of north into the Pacific as planned. It was soon clear to the task force command in charge of the shot that evacuations would be necessary and by the evening of March 2 a ship was steaming toward Rongelap to remove the

population. Over the next three days, 236 Marshallese were transported by sea and 28 U.S. servicemen were airlifted from a weather station on Rongerik to Kwajalein Atoll, south of the fallout pattern, and then to a U.S. naval base with medical facilities.[132]

Merril Eisenbud has observed:

There are many unanswered questions about the circumstances of the 1954 fallout. It is strange that no formal investigation was ever conducted. There have been reports that the device was exploded despite an adverse meteorological forecast. It has not been explained why an evacuation capability was not standing by, as had been recommended, or why there was not immediate action to evaluate the matter when the task force learned (seven hours after the explosion) that the AEC Health & Safety Laboratory recording instrument on Rongerik was off scale. There was also an unexplained interval of many days before the fallout was announced to the public.[133]

The Marshallese and Americans were not the only ones exposed to fallout from Bravo. A 100-ton Japanese fishing vessel with a crew of twenty-three called the *Fukuryu Maru (Lucky Dragon)* was sailing some eighty miles from Bikini when the bomb exploded. Within days, crew members suffered from acute radiation sickness. Seven months after the test, one of the crew members died.[134] The others were hospitalized for more than a year, until May 1955. The event received international attention and contributed to a worldwide protest of atmospheric testing of nuclear weapons.

Dr. Victor Bond, a member of the medical team sent from the United States to treat the exposed population immediately after the accident, said in an interview with Advisory Committee staff that "initial statements by Washington officials underplayed the severity of the effects of the exposure."[135] Dr. Eugene Cronkite, who headed the medical team, said he told Lewis Strauss, chairman of the Atomic Energy Commission in 1954, of his concern that the *New York Times* and others had reported a "downright lie" in reporting that the fallout hazard was minimal.[136] Dr. Cronkite recalled Strauss's response: "Young man, you have to remember that nobody reads yesterday's newspapers."[137]

On March 6, the task force command approved a request by the Armed Forces Special

Weapons Project to establish a joint study of the "response of human beings exposed to significant gamma and beta radiation due to high yield weapons."[138] Thus, it appears to have been almost immediately apparent to the AEC and the Joint Task Force running the Castle series that research on radiation effects could be done in conjunction with the medical treatment of the exposed populations.

Medical Follow-up

On March 8, Dr. Cronkite's mission was formally established in a letter to him that was classified Secret and Restricted Data and said, "The objective of this project is to study the response of human beings in the Marshall Islands who have received significant doses due to the fall-out from first detonation of Operation Castle."[139] The project was given the designation 4.1 and titled, "Study of Response of Human Beings Exposed to Significant Beta and Gamma Radiation Due to Fallout from High Yield Weapons."[140] The letter continued: "Due to possible adverse public reaction, you will specifically instruct all personnel in this project to be particularly careful not to discuss the purpose of this project and its background or its findings with any except those who have a specific 'need to know.'"[141]

As Dr. Cronkite understood it, his mission was to "examine and treat the Marshallese and the American servicemen that were exposed."[142] Initial exposure estimates ranged from 15 rad for people on Utirik to 150 rad for those on Rongelap.[143] Dr. Bond, who accompanied Dr. Cronkite on the mission, told Advisory Committee staff that "we were given estimates of dose. But they were poor, and we still don't know very well the effects."[144] The Marshallese were exposed to highly penetrating gamma radiation, which resulted in whole-body exposure, external radiation from deposition of fission products on the skin, internal radiation from consumption of contaminated food and water and, to a lesser extent, from inhalation of fallout particles. During the first few days after Bravo, several of the people from Rongelap were suffering from nausea and vomiting (the first signs of radiation sickness), depressed white blood cell counts, and slight hair loss. Only one of the Marshallese exposed on Ailinginae Atoll had these symptoms, and none from Utirik had them. The American servicemen on Rongerik were asymptomatic, as well.[145]

Although the medical program for the exposed Marshallese was designated a "study," both Dr. Cronkite and his successor, Dr. Robert A. Conard, maintain the project never included nontherapeutic research.[146] Both men assert that the primary goal has always been the treatment of the exposed population and that the data that were collected were always intended first and foremost to benefit the Marshallese. There is no conclusive evidence available to the Advisory Committee to contradict their statements. In examining various studies of the Marshallese that could have been driven by pure research goals, the Advisory Committee has found treatment-related goals that are at least plausible. It appears that in the medical follow-up to the Bravo shot, treatment and research objectives were essentially congruent.

Dr. Cronkite and his team arrived on Kwajalein the same day he received the memorandum establishing their mission. They set up examination and lab facilities in a building adjacent to the living quarters of the Marshallese and began their work. Team members took medical histories with the help of translators, inspected skin to monitor for radiation burns, took body temperatures, drew blood regularly to check white cell counts, platelet levels, leukocytes, and red cells, took urine samples, checked for eye injuries, and monitored pregnancies.[147]

In the Rongelap population, platelet levels fell to about 30 percent of normal by the fourth week, white blood cell counts fell to half of normal by the sixth week, but at the six-week point, when the initial examinations were completed, these blood elements began moving back up toward normal levels.[148] There was substantially less depression of platelet and white cell counts in the other groups, which received significantly lower doses of radiation. Despite the low platelet and white cell counts, there appears to have been little unusual bleeding or increased susceptibility to infection. Dr. Bond, said "There was some . . . excessive menstruation and blood in the urine . . . but nothing that merited strenuous therapy."[149] About ten to fourteen days after exposure, radiation burns began appearing.[150] These burns were much more pronounced among the Rongelap people than those from Ailinginae or the U.S. servicemen on Rongerik, and there were no burns noted in the Utirik group. Often the burns were accompanied by itching and some of the lesions on the top of the feet were described as painful. In two to three weeks the burns began healing.[151] There was some weight loss in the exposed population, and about 90 percent of the children and 30 percent of the adults lost hair.[152]

Dr. Bond told Advisory Committee staff that the exposed Marshallese "seemed to be perfectly healthy people [but] we were well aware of the latent period, and that they might well become ill later." He went on to say:

And quite frankly, I'm still a little embarrassed about the thyroid. [T]he dogma at the time was that the thyroid was a radio-resistant organ. . . . [I]t turned out they had . . . very large doses of iodine . . . to the thyroid.[153]

Dr. Cronkite noted that "there was nothing in the medical literature . . . to predict that one would have a relatively high incidence of thyroid disorders."[154]

In May 1954 the AEC told the DOD that the "Utirik people" could return home following the completion of the current tests, "provided that specimens reveal absence of radioactive materials in quantity injurious to health."[155] On Rongelap, however, radiation levels were considered to be too high. The Rongelapese were moved to Eijit, a small island in Majuro Atoll.[156] The United States continues regularly to followup the exposed Rongelapese and Utirikese. The U.S. servicemen were sent to Honolulu for further examination by Army physicians.[157] But according to Dr. Cronkite, "Somebody at a higher level within DOD decided that they did not want to study the American servicemen and cast them to the wind. Sort of forget them. I think that's a terrible thing to do, but it was done. Medically, it was unacceptable."[158] Dr. Cronkite went on to explain that if an induced cancer had been identified, early diagnosis and treatment might have benefited the exposed serviceman.[159] The DOD

reported to the Advisory Committee that twelve of the twenty-eight servicemen were examined in 1979 by the Veterans Administration as part of a notification and medical examination program for military personnel exposed to radiation. We have not been able to determine whether any of the twenty-eight had any other medical follow-up.[160]

The Ailuk Exposure

According to a report by Lieutenant Colonel R. A. House, based on an aerial survey done within forty-eight hours of the Bravo blast, "The only other atoll which received fallout of any consequence at all was Ailuk [it is not clear to which atolls the word "other" applies]. . . . [I]t was calculated that a [lifetime] dose would reach approximately 20 roentgens," about the same as or slightly higher than the exposure of the Utirik population.[161] Unlike the people of Utirik and Rongelap, however, the 401 people of Ailuk, south of Utirik in the eastern Marshalls, were not evacuated at all. The January 18, 1955, final off-site monitoring report of Operation Castle, however, gave the Ailuk exposure, based on several aerial and ground readings, as 6.14 roentgens. Readings from this report for other exposed atolls were as follows: Rongerik, 206; Rongelap, 202; Utirik, 24; Ailinginae, 6.7; Likiep, 2.19; and Wotje, 2.54.[162] People living on these atolls would be exposed to additional radiation as a result of consuming contaminated food. Based on the initial reading of 20 roentgens, the U.S. task force should have evacuated the people of Ailuk. A 1987 epidemiological study reported in the *Journal of the American Medical Association,* however, shows higher rates of thyroid abnormalities on other atolls to the south and east of the blast site, including Jaluit and Ebon.[163]

By the afternoon of March 4, two ships, both destroyer escorts, seem to have been available to evacuate the 400 or so people on Ailuk.[164] But according to Colonel House, "the effort required to move the 400 inhabitants," when weighed against potential health risks to the people of Ailuk, seemed too great, so "it was decided not to evacuate the atoll."[165] However, evacuation would have reduced the lifetime exposures of the Ailuk population by a factor of three, according to an estimate provided by Thomas Kunkle of Los Alamos National Laboratory.[166] In testimony before the Advisory Committee, Ambassador Wilfred Kendall of the Republic of the Marshall Islands noted that "the United States Government studied with interest the unexpected and dramatic incidence of thyroid disease on Utirik Atoll [but] no effort was made to reassess the health of the population on Ailuk, or Likiep, or other mid-range atolls."[167]

Resettlement of Rongelap

Between March 1954 and mid-1956, the Rongelap population on Eijit was followed medically, with visits from a U.S. medical team at six months, one year, and every year thereafter.[168] According to a preliminary report on the two-year medical resurvey, "There has been little illness among the people [and] none of the clinical entities noted in the Rongelap people appear to be related in any way to radiation effect."[169]

By late 1956, about a dozen radiological surveys of Rongelap and neighboring atolls had been conducted to determine contamination levels.[170] On February 27, 1957, the AEC informed the commander of the Pacific Fleet that resettlement was approved[171] despite lingering residual radiation, most pertinently, in the food supply.[172] This decision, which was consistent with international pressure for resettlement, was made even though in 1954 U.S. medical officers had recommended that the exposed Rongelapese "should be exposed to no further radiation, external or internal with the exception of essential diagnostic and therapeutic x-rays for at least 12 years. If allowance is made for unknown effects of surface dose and internal deposition there probably should be no exposure for rest of natural lives."[173] However, the displaced Rongelapese were eager to return to their home island. In March 1956, Dr. Conard wrote to Dr. Charles L. Dunham, director of the AEC's Division of Biology and Medicine, that "we are committed to return the people to their homes and that is their express wish."[174]

In June 1957, a final resettlement radiosurvey was made from the air. Gordon Dunning, an AEC health physicist, wrote he would have pre-ferred a full survey, but that "it appears we will have to settle for the external readings only."[175] The exposed Rongelap people and 200 other Rongelapese, who were not on the atoll at the time of the Bravo shot, were returned to their home islands at the end of June. The Advisory Committee has not been able to learn why Dunning's advice to carry out a more thorough, land-based survey was not heeded. A 1957 project report notes that while "the radioactive contamination of Rongelap Island is considered perfectly safe for human habitation. . . . The habitation of these people on this island will afford most valuable ecological radiation data on human beings."[176] Nevertheless, the Advisory Committee does not conclude that the resettlement decision was motivated by AEC research goals. From 1954 on, the U.S. researchers recognized the importance of the opportunity that had been presented to gather data on radiation effects. However, we have seen no evidence, including this report, that convincingly demonstrates that research goals took priority over treatment in a way that would expose the populations to greater than minimal risk.

Apart from the radiation deposited by the Bravo shot, there is evidence that later bomb tests also contributed to the overall radiation level on Rongelap. For example, a January 1957 letter from Dr. Edward Held, the director of a University of Washington group conducting ecological studies for the Joint Task Force, said that "activity levels in the water at Rongelap were higher in July 1956 than the levels . . . obtained at earlier visits [and] the best evidence seems to indicate that the increase . . . is due to the recontamination of Rongelap from the 1956 series of weapons tests."[177] The letter goes on to say, "The decay of the newly added radioactivity is such that it will soon be insignificant when compared with that from the 1954 series."[178]

Atmospheric testing of nuclear weapons was ended in 1963 by international agreement.

Post-Resettlement Medical Follow-up

After the population returned to Rongelap in 1957, Dr. Conard visited annually with a medical team from Brookhaven National Laboratory.[179] The team's primary mission, according to Dr. Conard, "was to treat the people. I don't think at any time the motivation . . . was anything other than treatment of the effects of radiation." He added, however, that "we [also] were trying to get as much information as we could into the medical literature. We knew that we were dealing with an area that was unexplored in human beings and we wanted to find out as much as we could about" the effects of radiation exposure resulting from fallout from a nuclear explosion.[180]

After their return to their native island in 1957, the Rongelapese continued to be monitored annually by the Brookhaven teams. On Utirik, exams were carried out every three years, then annually with the appearance of thyroid abnormalities.[181] The examinations included complete physicals; blood tests; examinations of reproductive effects including fertility, miscarriages, stillbirths, observable birth defects,[182] and genetic studies; growth and development studies of children; thyroid function tests and palpation; and studies of absorption, metabolism, and excretion of radioisotopes.[183] In addition to the annual exams conducted in the Marshalls, in 1957 some Marshallese were flown from their islands to Argonne National Laboratory in Chicago, where a whole-body counter and other advanced equipment was available.[184] When Marshallese developed medical problems that required treatment in the United States, such as thyroid nodules requiring surgery, they were sent to Metropolitan General Hospital in Cleveland or to other hospitals.[185] One eighteen-year-old male was treated in 1972 at NIH and at a Western Reserve University teaching hospital for leukemia, which proved fatal.[186]

In our search of documents related to the Brookhaven medical program, the Advisory Committee has found only two examples of studies that were not primarily intended to benefit the individual participants. In one, a "chelating" agent (EDTA), normally administered shortly after internal radiation contamination to remove radioactive material, was administered seven weeks after exposure. The stated rationale was that the agent would "mobilize and make detection of isotopes easier, even though it was realized that the procedure would have limited value at this time."[187] Because there was virtu-

ally no therapeutic benefit envisioned, it appears the primary goal of the study was to measure radiation exposures for research purposes, although the knowledge may have been helpful in the clinical care of the patient. In the second experiment, a radioactive tracer (chromium 51) was used to tag red blood cells in ten unexposed Rongelapese to measure their red blood cell mass. The purpose was to determine whether the anemia that had been observed among Marshallese was an ethnic characteristic or due to their radiation exposures.[188] The tracer dose used would have posed a very minimal risk, but it was clearly not for the benefit of the ten subjects themselves. The data could, however, have benefited Marshallese exposed as a result of the Bravo explosion. No documentation addressing whether consent was sought is available for either experiment.

The AEC was responsible only for continuing studies of the Marshallese to detect radiation effects and for medical care required for radiation-related effects, while the Trust Territory government under the Department of Interior was responsible for general medical care, but this appears to have been a meaningless distinction to the Marshallese. "All they knew," Dr. Cronkite told Advisory Committee staff, "is that something had happened to them and they wanted to be taken care of, very logically."[189] Often, Dr. Cronkite noted, the members of the Brookhaven team did take care of nonradiation-related health problems. "Physicians being what they are," he said, "you see disease and there's something you can do about, you like to take care and help people."[190] The Brookhaven team sometimes included a dentist because severe dental problems had been observed. The dentist mostly did extractions and "a little restoration."[191] According to Dr. Cronkite, the Marshallese appreciated getting dental care because "they were getting something they had never had before in their lives and they liked it."[192] Although the extractions appear to have been done for therapeutic or prophylactic purposes, the extracted teeth were analyzed for radioactive content.

Primary care, however, remained inadequate. There were serious epidemics of poliomyelitis, influenza, chicken pox, and pertussis, all of which, according to Dr. Conard, were imported into the Marshalls by the U.S. medical teams.[193] The epidemics were severe, with high mortality rates, and could have been prevented by the use of available vaccines. The AEC insisted that primary care be left to the Trust Territory, which had neither the personnel nor the equipment to provide adequate services. Dr. Hugh Pratt, who succeeded Dr. Conard in 1977, wrote as late as December 1978, "The Marshall Islands medical 'system' under the Trust Territory is underfinanced. The professional staff is undertrained and overworked. Critical supplies are usually not available."[194]

By 1958, Dr. Conard was aware of Marshallese dissatisfaction with the annual exams and wrote to Dr. Dunham:

I found that there was a certain feeling among the Rongelap people that we were doing too many examinations, blood tests, etc. which they do not feel necessary, particularly since we did not *treat* [emphasis in the original] many of them. Dr. Hicking and I got the people together and explained that we had to carry out all the examinations to be certain they were healthy and only treated those we found something wrong with. I told them they should be happy so little treatment was necessary since so few needed it . . . etc., etc. Perhaps next trip we should consider giving more treatment or even placebos.[195]

Also in 1958, Edward Held, the University of Washington professor involved in environmental surveys of the islands, wrote to Dr. Conard about a meeting he had with Amata, son of a paramount chief of the Marshalls, in which Amata said the Marshallese were "apprehensive about being stuck with needles."[196] Amata, who is now president of the Republic of the Marshall Islands, asked about the need for continued medical examinations, and Dr. Held told him that he should talk to Dr. Conard, but Held also wrote that "there have been medical benefits not connected with radiation which have resulted from the medical surveys." Held added that Amata agreed this was true.[197]

The annual exams given to the people of Rongelap were described by Konrad Kotrady, a Brookhaven physician resident in the islands from 1975 to 1976, from the Marshallese point of view:

[E]ach March a large white ship arrives at your island. Doctors step ashore, lists in hand of things to do, and

people to see. Each day a jeep goes out to collect people for examinations, totally interrupting the normal daily activities. Each person is given a routing slip which is checked off when things are done. They are interviewed by a Marshallese, then examined by a white doctor who does not speak their language and usually without the benefit of a Marshallese man or woman interpreter. Their blood is taken, they are measured, and at times, subjected to body scans.[198]

Eventually, Dr. Conard tentatively arranged for the AEC to pay the Utirik participants $100 each for their inconvenience.[199]

A Marshallese who acted as a translator for the Brookhaven team said that people didn't believe Dr. Conard. According to this man, they began to say, "You people coming back every 2 years to . . . just do the experiments on us like guinea pigs."[200] According to Dr. Pratt, some of the distrust of Dr. Conard, at least among the people of Utirik, was the fact that he predicted that there would be no cases of thyroid carcinoma in this population and one occurred.[201] Dr. Kotrady wrote that "for 22 years, the people have heard Dr. Conard and other doctors tell them not to worry, that the dose of radiation received at the island was too low to cause any harmful effects. . . . However . . . [i]t has been found that there is as much thyroid cancer at Utirik as at Rongelap—3 cases each. . . . The official explanation for the high incidence of thyroid cancer at Utirik is unknown at present. Yet in the people's mind the explanation is that it is a radiation effect despite what the doctors have said for 20 years."[202]

In 1961, Dr. Dunham wrote an open letter to the exposed people of Rongelap in which he explained the need for medical follow-up.[203] Dr. Dunham specified that one reason was the health care of the exposed population, but that the other was "of no direct value to you (the Rongelap population)." This is the only instance we found in which a U.S. official explicitly says research is being conducted that has no direct benefit to the Marshallese population under the care of the Brookhaven doctors. The letter continued: "The [health studies] help us to understand better the kinds of sickness caused by radiation. The United Nations has a special scientific committee to study these things and the information we get from our work here is made available to that committee and to the whole world."[204] This letter was rescinded before it was sent, however. Although it was read once over the radio, the "broadcast probably did not reach the Rongelap people since there are only three radios on the island."[205] Courts Oulahan, the AEC's deputy general counsel, apparently requested the letter be rescinded, although the reason for the request is unclear. The district administrator of the Marshall Islands, William Finale, complied with the request, and the letter was never published.[206]

Many complaints resulted from the fact that the U.S. researchers had difficulty communicating with the Marshallese, most of whom did not speak English. Information about risk, countermeasures, and radiation was not easily explained to the Marshallese,[207] and cultural differences made it difficult for the researchers to appreciate relevant Marshallese practices and customs. According to Dr. Bond, an early member of the medical team, the Brookhaven doctors did not believe that they needed to obtain consent for treatment or to conduct studies related to treatment.[208] The Brookhaven team offered needed medical care; therefore, despite complaints, the Marshallese requested extension of the medical program provided to the Rongelap and Utirik people to include more general medical care and to include other islands and atolls.[209]

Thyroid abnormalities, in addition to the one fatal case of leukemia, have been the most significant late effect of radiation among the Marshallese. These endpoints appear to have received both extensive study and appropriate treatment. As thyroid abnormalities began to appear in the Utirik population, the Brookhaven team felt a need to establish a baseline in an unexposed Marshallese population.[210] Over the years, members of the Ailuk "control" population—at best an imperfect control population because of their exposure—had emigrated or died and had been lost to follow-up. This population was too small to provide an adequate baseline, so the Brookhaven team conducted surveys of 354 people at Likiep and Wotje Atolls in 1973 and 1976. They also examined more than 900 Rongelap and Utirik people who were

not on their home islands during Bravo.[211] It is likely that many if not most of the controls selected had some radiation exposure resulting from the bomb tests.

During the early 1970s there were increasing complaints about and resistance to participation in the medical surveys coupled with the continuing appearance of thyroid abnormalities, including their development in the less-exposed Utirik population.[212] There were also growing numbers of people from Rongelap and Utirik who, as a result of thyroid surgery or reduced thyroid function, needed thyroid medication and indications that those on medication were not adequately complying with their therapeutic regimen.[213]

As a consequence of all these events, Brookhaven expanded its staff and medical care programs in the Marshalls in the mid-1970s, including for the first time primary care for a number of conditions not thought to be radiation related. Full-time resident staff was increased. In 1973, Brookhaven stationed a full-time physician in the Marshalls. "His principal responsibilities included (a) monitoring the thyroid treatment program, (b) visiting Rongelap, Utirik, and Bikini Atolls for health care purposes every 3 to 4 months, and (c) assisting the TT [Trust Territory] medical services with the care of Rongelap and Utirik patients at the hospitals at Ebeye and Majuro."[214]

In 1974, the researchers conducted extensive screening for diabetes, a nonradiation-related condition, in order to determine the impact of diabetes on the population and form the basis for development of a program for treatment and management of this significant problem, which affects 17 percent of the population.[215] In 1976, a new agreement provided for Brookhaven to provide examinations and health care for all Marshallese living on Rongelap and Utirik when they made their visits and for the resident Brookhaven physician to assist in the care of Rongelap and Utirik patients at the hospitals at Ebeye Island in Kwajelein Atoll and Majuro, the capital of the Marshall Islands in the Majuro Atoll.[216] In 1977, an extensive program to diagnose and treat intestinal parasites was carried out.[217]

By 1978, administrative responsibility in the Trust Territory government shifted to the individual island groups. The Marshallese at this point took responsibility for general health care.[218] While the 1947 Trusteeship Agreement provided for health care for the Marshall Islanders, the Department of the Interior carried out this responsibility mainly in an oversight capacity. The Department of Energy carried on the programs of its predecessor agencies for treating radiation-related illnesses in the people of Rongelap and Utirik. During this period the Brookhaven medical team often treated nonradiogenic as well as radiogenic medical conditions.[219]

In 1985, expressing concern that radioactivity in the food chain represented a significant health hazard, the people of Rongelap rejected the Department of Energy's advice that they stay on their island. At their own request they were evacuated on the Greenpeace ship *Rainbow Warrior* to Majetto Island in Kwajelein Atoll, where they remain today. In 1994 the National Research Council published a report that, among other things, reviewed food-chain data collected and analyzed by Lawrence Livermore National Laboratory. According to this report,

On the basis of current radiation dose estimates, there is no expectation that any medical illness due to exposure to ionizing radiation will occur in any members of the resettlement population of the island of Rongelap from either intake of native foods or environmental contact.[220]

However, the report recommended that no categorical assurances be given the people of Rongelap that their annual exposure upon returning would be less than the 100–mrem limit agreed to in a 1992 memorandum of understanding between the Republic of the Marshall Islands and the United States. Moreover, the report recommended an initial diet in which half the food consumed would be from nonnative sources and that no food be gathered from the northern islands of Rongelap and Rongerik Atolls.[221]

In 1986 a Compact of Free Association went into effect between the United States and the Republic of the Marshall Islands.[222] The compact established a $150 million fund to compensate the Marshallese for damage done by the U.S. nuclear testing program.[223] The United States accepted "responsibility for compensation owing to citizens of the Marshall Islands . . . for loss or damage to property and person of the citizens of the Marshall Islands. . . ."[224]

At present there are three separate health care programs for citizens of the Republic of the Marshall Islands. There is a program of general health care for all citizens for which the Marshallese government is solely responsible; there is a Four Atoll Program, which is run by the Marshallese, but funded by the United States at about $2 million a year[225] (the atolls that benefit from this program are Bikini, Enewetak, Rongelap, and Utirik), and there is the continuation of the Brookhaven program, which is responsible for medical monitoring and care related to radiation exposure. The Lawrence Livermore National Laboratory conducts environmental surveys as part of the Brookhaven program, whose total cost is about $6 million a year.[226] The funding for this entire program is discretionary and can be reduced or eliminated by Congress.

Conclusions about the Marshallese

The United States has a special responsibility to care for the radiation-related illnesses of the exposed Marshallese because of its role as trustee and because it caused the exposures. As best the Advisory Committee can determine, it is carrying out this responsibility well. Treatment has been provided as needed for acute effects, monitoring continues to this day, and latent radiation effects have been identified early and treated. The research conducted between 1954 and today consisted mainly of blood and urine tests and procedures to measure radiation with little or no additional risk to the subjects. Overall, these tests seem to have been related to patient care, although two instances of minimal-risk nontherapeutic research have been identified. The Committee found no evidence that the initial exposure of the Rongelapese or their later relocation constituted a deliberate human experiment. On the contrary, the Committee believes that the AEC had an ethical imperative to take advantage of the unique opportunity posed by the fallout from Bravo to learn as much as possible about radiation effects in humans.

Nevertheless, the inherent conflicts posed by combining research with patient care could perhaps have been reduced by clearer separation of the two activities and clearer disclosure to the subjects. For the most part, consent for tests and treatment appears to have been neither sought nor obtained. Although lack of consent for minimal-risk procedures performed on a patient population was not atypical for the time (see chapter 2), the Committee believes efforts should have been made to ensure that the people being monitored and treated understood what was being done to them and why, and their permission should have been sought.

While cultural and linguistic differences made communication with the Marshallese difficult at first, the Advisory Committee believes the situation continued for much too long. As a consequence, dietary differences and other eating habits were not recognized and may have led to higher exposures among some members of the population. Cultural differences may also have resulted in an inadequate accounting of adverse reproductive outcomes. Certainly, differences in pace and lifestyle contributed to a perception by the Marshallese that they were being told what to do rather than asked. The Advisory Committee was unable to determine whether the early medical teams should have been more aware of such cultural differences, but they do appear to have been slow to learn.

The BNL medical team was constrained by instructions from the U.S. government to restrict its activities to treatment and research related to radiation-related illnesses. General medical care was held to be the responsibility of the Trust Territory government. However, there was no adequate medical service available to refer other complaints to, so the BNL physicians were put in an awkward situation where, as doctors, they felt obliged to treat conditions that were presented to them. The lack of clear lines for general medical care in the early years of the program seriously compromised relations with the Marshallese. Since the Marshall Islands were a trust territory, both general medical care and care for radiation injuries were ultimately the responsibility of the United States, and the care of individuals should not have suffered as a result of bureaucratic confusion. Thus the Committee commends the expansion of the BNL program in the 1970s to include general health care, and the U.S.-supported Four Atoll Program that went into effect after the Compact of Free Association was approved in 1986. It may be, depending on factors such as food-chain and other environmental exposure levels, that certain midrange atolls such as Ailuk and Likiep also merit inclusion.

THE IODINE 131 EXPERIMENT IN ALASKA

In 1956 and 1957 the U.S. Air Force's Arctic Aeromedical Laboratory conducted a study of the role of the thyroid gland in acclimatizing humans to cold, using iodine 131. Like the case of the Marshallese, this study is another instance in which research conducted on populations that were unfamiliar at the time with modern American medicine posed special ethical problems and was therefore of interest to the Advisory Committee. The study involved 200 administrations of I-131 to 120 subjects: 19 Caucasians, 84 Eskimos, and 17 Indians,[227] with some subjects participating more than once. Animal studies had suggested the thyroid gland might play a crucial role in adaptation to extreme cold. This experiment was part of the laboratory's larger research mission to examine ways of improving the operational capability of Air Force personnel in arctic regions. The results of the study were published in 1957 as an Air Force technical report by the principal investigator, Dr. Kaare Rodahl, M.D., a Norwegian scientist hired by the U.S. Air Force for his expertise—rare at the time—in arctic medicine.[228] Many observational studies of Alaska Natives were carried out by a variety of researchers in the 1950s and 1960s; most of these did not administer radiation to the natives, but only measured what had already accumulated in their bodies from fallout.[229] The thyroid study discussed here, however, differed in that it actively administered radionuclides to natives, raising more direct questions of consent, risk, and subject selection. The Alaskan I-131 experiment also offered subjects *no* prospect of medical benefit.

This study is the subject of a review by a committee of the Institute of Medicine and the National Research Council. The IOM/NRC committee was mandated by legislation passed by Congress in 1993 and began operation in June 1994, including an on-site investigation of the experiments.[230]

To the extent possible, the IOM/NRC committee has provided the Advisory Committee with information but, in accordance with its own procedures, has kept its own deliberations confidential. The IOM/NRC report was not available to the Advisory Committee, as it had not been completed by the time the Committee had concluded its deliberations. We did not conduct our own on-site investigation of the Alaskan experiments. Instead, we have relied on published materials (primarily Rodahl's 1957 report on the study, "Thyroid Activity in Man Exposed to Cold") and those observations presented to the Committee in testimony by representatives of the IOM/NRC committee, as well as by representatives of the Inupiat villages of the North Slope of Alaska where the research was conducted. More detailed study may always, of course, lead to different factual conclusions. The Advisory Committee was concerned with understanding the experiments well enough to develop general remedial principles to be applied to more detailed factual findings completed by others.

According to Dr. Chester Pierce of Harvard Medical School, chair of the IOM/NRC committee, in 1994 Dr. Rodahl recalled that the base commander at the Artic Aero-medical Laboratory approved the study, and headquarters in Washington knew of the experiment.[231] Participants in the study were asked to swallow a capsule containing a tracer dose of radioiodine. Measurements were then made of thyroid activity, using a scintillation counter, and samples taken of blood, urine, and saliva.[232] The study's overall conclusion was that "the thyroid does not play any significant role in human acclimatiza-

tion to the arctic environment when the cold stress is no greater than what is normally encountered by soldiers engaged in usual arctic service or by Alaskan Eskimos or Indians in the course of their normal life or activities."[233] One minor consequence of the experiment was to have the noniodized salt in the local stores replaced with iodized salt. Follow-up, Dr. Rodahl told the IOM/NRC Committee, was left to the Alaska Native Service, which was already aware of a goiter problem in these communities.[234] Alaska natives testifying in 1994 before the IOM/NRC committee could not recall any follow-up visits by physicians, according to Dr. Pierce.[235]

Risk

The Advisory Committee did not undertake a detailed dose reconstruction or assessment of the scientific quality of the research, since these tasks were already being undertaken by the IOM/NRC committee. The actual capsules of iodine 131 were prepared in continental U.S. laboratories. As was common at the time, the principal investigator, Dr. Rodahl, took a one-week course on the proper handling and administration to humans of iodine 131.[236] He then instructed the other physicians who would be working in the field. Doses were officially reported to range from 9 to 65 microcuries of iodine 131, with most being approximately 50 microcuries. The doses below 50 microcuries were due to the natural reduction in the radioactivity of the ready-made capsules during the long trip to remote regions.[237] (To compensate for the low doses, longer scanning times were used in the field, but in the 1957 report these results were judged to be unreliable.)[238] According to Dr. Pierce, Dr. Rodahl stated in 1994 that the dosage was standard at the time for tracer studies. This was the dose he had been taught in his training course; the dosage was approved by the AEC.[239]

In terms of dosage and risk, the experiment was not significantly different from tracer studies conducted in the continental United States with two exceptions. First, some subjects were used more than once; several Alaska Native subjects reported they received as many as three doses.[240] Second, the subjects included women who were pregnant or lactating. Dr. Pierce reported that testimony at the IOM/NRC hearings in Alaska indicated that at least one subject may have been pregnant at the time; technical reports, he said, state that two female subjects may have been lactating at the time.[241] Although the AEC discouraged the nontherapeutic use of radioisotopes in pregnant women, such research was sometimes conducted. What sets the Alaska experiment apart from other studies conducted on pregnant and lactating women is that this experiment was not investigating a research question about an aspect of pregnancy or lactation.

As discussed in detail in chapter 6, from its mid-1940s inception the AEC's radioisotope distribution program required prior review of "human uses" of radioisotopes to ensure that risks were minimized and safety precautions were followed. (In 1952 the Air Force issued a rule that required prior review for experiments, but the rule was limited to research conducted at Air Force medical facilities.[242]) As discussed in chapter 6, in 1949 the AEC's Human Use Subcommittee expressly discouraged the use of radioisotopes for research with children or pregnant women.

Disclosure and Consent

This experiment offered no prospect of medical benefit to subjects. If the subjects in this experiment did not understand and agree to this instrumental use of their bodies, then they were used as mere means to the ends of the investigators and the Air Force. It was at this time conventional for investigators to obtain the consent of "normal" (healthy) subjects or "volunteers" in nontherapeutic research. This tradition was particularly strong in the military services (see part I). It was also recognized by the AEC at least by February 1956 when the AEC's radioisotopes distribution program explicitly stated that where normal subjects are to be used they must be "volunteers to whom the intent of the study and the effects of radiation have been outlined."[243]

The Committee is not aware of any documents from the time of the experiment that bear on what, if anything, the subjects were told and whether consent was obtained. There are also no documents bearing on whether the Air Force provided the researchers with guidelines on the use of human subjects or requirements for obtaining consent. However, documents available to the Committee indicate that the radioisotopes used by the Arctic Aeromedical Laboratory and Dr. Rodahl were obtained by the Air Force under license from the AEC.[244] The AEC's provision for healthy volunteers, as just quoted, was included in the AEC's publicly available materials and presumably should have been known to—and abided by—those conducting government research programs involving AEC provided radioisotopes.[245]

The only available evidence comes from personal recollections of the principal investigator and a few of the former subjects. Dr. Rodahl recalled in 1994 that he obtained white volunteers through their military commanders and Indian and Eskimo volunteers through the village elders.[246] When a military volunteer came before him, he explained, in the subject's native tongue (English), the purpose of the study and what a subject would do and gave the person the opportunity to decline to participate.[247] When visiting the villages, the physicians could not communicate directly in the native language. They would find an English-speaking village elder and explain the purpose of the study. The elder would then find people to serve as subjects. What communication occurred between the village elder and the prospective subjects is not known. According to members of the IOM/NRC committee, Dr. Rodahl recalled that, although all potential subjects were given the opportunity not to participate, all of the Indians and Eskimos who reported did participate in the experiment.[248]

Dr. Rodahl also reported that he did not use the term *radiation* in his explanation to the English-speaking village elders who then communicated with others in the villages. Interviews in 1994 by the IOM/NRC committee indicated that there is no word for radiation in the native languages. One Alaska Native subject, interviewed by the IOM/NRC committee in 1994, recalled that at the time he worked in a hospital, spoke English, and did know about "radiation." He could not recall any use of the term in the study.[249] In at least one village—Arctic Village—there were no English speakers. Subjects from this village testified in 1994 to the IOM/NRC that they thought they were taking a substance that would improve their own health and that they would not have participated in the study if they had known it required them to take a radioactive tracer.[250]

These accounts raise difficult ethical questions about authorization and consent, questions made the more difficult by an incomplete historical record. It is, for example, unclear whether the village elders were employed solely as translators who were asked to transmit individual requests for permission to potential subjects, or whether Dr. Rodahl was responding to the perceived authority of the village elder who then "volunteered" members of his community. Thus we do not know what the individual subjects were told or whether their individual permission was sought. Today we continue to debate whether, when human research is conducted in cultures where tribal or family leaders have considerable authority over members of their communities, it is ever appropriate to substitute the permission of these leaders for first-person consent.[251]

Even if the procedure used for securing authorization through the tribal leaders was appropriate, the available evidence suggests that the leaders may not have understood, and thus were not in a position to communicate to the subjects, that the experiment was nontherapeutic, that it had a military purpose, or that it involved exposure to low doses of radiation. The ethical difficulties posed by the language barrier were exacerbated by a significant cultural barrier. The Indian and Eskimo villages had little exposure to modern medicine. One village—Point Lay—is described in Rodahl's 1957 report as "relatively little affected" by the modern world.[252] There is a strong likelihood that there was no appreciation for the difference between treatment of a patient and research unrelated to any illness of the subject.

The danger of exploitation was further heightened by the trusting relationship that developed between the native Alaskans and the field researchers. In part, this trust was the customary welcome given to visitors; in part it was due to the desire for medical care. In at least one village, harsh conditions may have increased the need for outside assistance. Rodahl's report states that Point Lay had suffered from semi-starvation the previous year.[253] Dr. Pierce testified to the Advisory Committee that "in the mid-1950s, doctor visits to native villages were quite scarce." Dr. Rodahl said when his plane landed, the villagers would come running to meet him and the other physicians who came with him, and the villagers would immediately want their ailments treated. He said the physicians treated them because they were medical men. He also said "the natives trusted them, and they trusted the natives."[254] Testimony before the IOM/NRC committee included the recollection of one participant that he had been paid $10 for the study; in other testimony it appears some subjects may have believed there was an implicit quid pro quo, trading medical treatment for participation.[255] The testimony suggests that at least some subjects understood that part of what was being done to them was not medical care.

Subject Selection

The selection of Alaskan Indians and Eskimos as subjects for this research was not arbitrary. In order to better understand acclimatization and human performance under conditions of extreme cold, it was reasonable and potentially important to study people who lived under such conditions. At the same time, however, the population chosen was not one familiar with modern medicine, but rather a population for whom the treatments of modern physicians were a strange but valued innovation, and the research activities of modern medicine were totally unknown. As a consequence, the potential for misunderstanding and exploitation was significant. The Committee does not know whether there were at the time other populations also acclimated to cold weather who were better positioned than Alaskan Indians or Eskimos to be genuine volunteers for this nontherapeutic experiment. There has been no evidence that any attempt was made to explain the military purpose of the study to the Indians or Eskimos. Thus, in general, there was no oversight—or even knowledge—of how the village elders recruited participants and explained the nature of the experiment.

CONCLUSION

The three cases discussed in this chapter all raise troubling questions that will stay with us into the future, but they do so in different ways, and with different consequences.

The iodine 131 experiment conducted in Alaska was conventional biomedical research, although, as discussed in chapter 11, the subject population and its environment were also the object of observational study related to the effects of fallout from nuclear weapons. This experiment took place at a time (the mid-1950s) when the government's rules requiring disclosure and consent in the use of radioisotopes with healthy subjects were established and public; the available documented evidence suggests that these rules were not followed. The evidence also suggests that, like the Marshallese, the Eskimos and Indians in Alaska were, in the 1950s, unacquainted with modern medical science and therefore unlikely to understand the nature and purpose of the research.

As a result of the 1954 Bravo shot, the Marshallese (and those exposed American servicemen and Japanese fishermen) experienced the largest peacetime exposures from fallout from detonation of nuclear weapons, and as a consequence of subsequent detonations, they were subjected to further exposures. The biomedical research that was conducted by the United States in the aftermath of Bravo raises basic questions about the obligations of researchers when long-term study is coupled with treatment, particularly in a setting where communication is difficult and the subjects otherwise have inadequate medical care.

Of all those covered in this report, the uranium miners were the single group that was put most seriously at risk of harm, with inadequate disclosure and with often-fatal consequences. The failure of the government and its researchers to adequately warn uranium miners who were continually being studied is difficult to comprehend; but the greater question is why,

with the knowledge that they had, government agencies did not act to reduce risk in the mines in the first place.

NOTES

1. Downwinders at the Nevada Test Site were exposed to lower levels of fallout during the same period as the Marshallese. The residents of Hiroshima and Nagasaki were exposed mainly to neutron and gamma radiation from the bomb's explosion.

2. Peter H. Eichstaedt, *If You Poison Us: Uranium and Native Americans* (Santa Fe, N.M.: Red Crane Books, 1994), 35–36.

3. Undated document ("Radiation Exposure in the United States—Uranium Mining Industry") (ACHRE No. HHS-092694–A), 1.

4. The Advisory Committee also heard extensive testimony from uranium millers and open-pit uranium miners who expressed dissatisfaction that their health problems were not covered by RECA, as were those of the underground miners. The health problems of the uranium millers appear to have been overshadowed by the clearly established problems of the underground miners and have received little attention in the scientific literature: only three articles have been located by the Advisory Committee. These papers show modest increases in certain cancers (notably lung and lymphatic) and nonmalignant respiratory disease and contain recommendations that these problems merit further study. No excess bone cancer, leukemia, or chronic renal disease has been reported, however. The most recent publication found by the Advisory Committee is dated 1983, and we are not aware of any further studies currently under way. Nevertheless, the millers and open-pit miners attest to numerous health problems they associate with their occupational exposures. See V. E. Archer, S. D. Wagoner, F. E. Lundin, "Cancer Mortality Among Uranium Mill Workers," *Journal of Occupational Medicine* 15 (1973): 11–14; A. P. Polednak and E. L. Frome, "Mortality Among Men Employed Between 1943 and 1947 at a Uranium-Processing Plant," *Journal of Occupational Medicine* 23 (1981): 169–178; R. J. Waxweiler et al., "Mortality Patterns Among a Retrospective Cohort of Uranium Mill Workers" in *Epidemiology Applied to Health Physics, Proceedings of the 16th Midyear Topical Meeting of Health Physics Society*, Albuquerque, N.M., 9–13 January 1983, 428–435.

5. Robert N. Proctor, *Cancer Wars: How Politics Shapes What We Know and What We Don't Know About Cancer* (New York: Basic Books, 1995), 186.

6. William C. Hueper, *Occupational Tumors and Allied Diseases* (Springfield, Ill.: C. C. Thomas, 1942).

7. Ibid., 438.

8. Egon Lorenz, "Radioactivity and Lung Cancer,"

Journal of the National Cancer Institute 5 (August 1944): 13.

9. Duncan Holaday to Chief, Industrial Hygiene, 20 November 1950 ("Radon and External Radiation Studies in Uranium Mines") (ACHRE No. IND-091394–B).

10. Duncan Holaday, Chief, Occupational Health Field Station, Public Health Service, "Employee Radiation Hazards and Workmen's Compensation," Joint Committee on Atomic Energy, 86th Cong., 1st Sess. (1959), 190.

11. See Lorenz, "Radioactivity and Lung Cancer."

12. William F. Bale to Files, 14 March 1951 ("Hazards Associated with Radon and Thoron") (ACHRE No. DOJ-051795–A), 3–8.

13. Ibid., 6.

14. J. Newell Stannard, *Radioactivity and Health: A History* (Oak Ridge, Tenn.: Office of Scientific and Technical Information, 1988), 138.

15. See Lorenz, "Radioactivity and Lung Cancer," for a review of animal experimentation, 7–10.

16. National Cancer Institute, *Radon and Lung Cancer Risk: A Joint Analysis of 11 Underground Miners Studies*, January 1994.

17. Dorothy Ann Purley, Advisory Committee on Human Radiation Experiments, small panel meeting, Santa Fe, N.M., proceedings of 30 January 1995 (morning session), 82–83.

18. Philip Harrison, Advisory Committee on Human Radiation Experiments, proceedings of 21 June 1995.

19. Merril Eisenbud, *An Environmental Odyssey* (Seattle: University of Washington Press, 1990), 43.

20. B. S. Wolf, Medical Director, NYOO, to P. C. Loshy, Manager, Colorado Area Office, 19 July 1948 ("Medical Survey of Colorado Raw Materials Area") (ACHRE No. IND-091394–B), 2.

21. *Health Impact of Low-Level Radiation: Joint Hearing before the Subcommittee on Health and Scientific Research of the Senate Committee on Labor and Human Resources and the Senate Committee on the Judiciary*, 96th Cong., 1st Sess. (1979), 40–41.

22. "A-Bomb Metal Affects Lungs, Doctor Reveals," *Cleveland News*, 22 September 1948.

23. Atomic Energy Commission, Manager of the New York Operations Office, 15 September 1949 ("Policy Regarding Special Beryllium Hazards") (ACHRE No. DOE-01295–B), 11.

24. Merril Eisenbud, telephone interview by Steve Klaidman (ACHRE staff), 7 July 1995 (ACHRE No. IND-070795–B), 1.

25. George Hardie to John Bowers, 29 December 1949 ("Statement of Policy on Be.") (ACHRE No. DOE-012595–B), 1.

26. Ibid., 2.

27. Shields Warren, Director, Division of Biology and Medicine, to W. E. Kelley, Manager, New York Operations Office, 17 January 1950 ("Proposed AEC Staff Paper on Beryllium Policy") (ACHRE No. DOE-012595–B), 1–2.

28. Eisenbud, *Environmental Odyssey*, 61. The DBM's position was apparently based on the view that the Atomic Energy Act did not give authority to the AEC until after the ore was mined. "The position of the New York Operations Office," Eisenbud wrote, "was that while the act did not require that the AEC be responsible for uranium mine safety, neither did it prevent the agency from doing so."

While the Committee did not locate the early AEC legal opinions on this question, as discussed in the text, we did find documentation of AEC lawyer reassertion of this position in the late 1950s.

29. Ibid., 62.

30. Ibid.

31. Interview with Eisenbud, 7 July 1995, 1.

32. A. E. Gorman, AEC Sanitary Engineer, to Files, 26 May 1949 ("Visit of Lewis A. Young, Director, Division of Sanitation, Colorado Department of Health") (ACHRE No. DOE-051195–A), 1.

33. Ibid.

34. The radium standard was set in 1941 when the Navy came to Robley Evans, a leading radiation researcher. A committee was established and came up with a standard based on data on twenty-seven human beings who had been exposed to radium, twenty of whom had been injured. Evans went around the room and asked each of the men for a standard they would feel comfortable having their wives or daughters work with and they agreed on 0.1 mCi. Robley D. Evans, "Inception of Standards for Internal Emitters, Radon and Radium," *Health Physics* 14 (September 1981): 441–443.

35. Stannard, *Radioactivity and Health*, 131–132. Stannard adds in a footnote on page 131:

This standard was not intended to be applied to the mines. The Europeans were totally involved with the war. In the Western Hemisphere, there were not yet enough uranium mines per se to worry about exposure standards. Uranium mining in the United States had hardly begun. In 1967, NCRP representative Lauriston Taylor testified at the congressional hearings on Secretary of Labor Williard Wirtz's proposed uranium mine standard that the 1941 standard was meant for "indoor" environments "where it is quite feasible to accomplish any degree of ventilation . . . that might seem indicated."

According to Taylor, the PHS was handling the situation in the mines, so the NCRP stayed out of it. *Radiation Exposure of Uranium Miners, Part One: Hearings before the Subcommittee on Research, Development, and Radiation of the Joint Committee on Atomic Energy*, 90th Cong., 1st Sess. (1967).

36. *Begay v. United States*, 591 Supp. 991 (D. AZ, 1984), 1013.

37. Duncan A. Holaday, August 1994 ("Origin, History and Development of the Uranium Study") (ACHRE No. DOJ-051795–A), 2.

38. *Begay v. United States*, 994.

39. Undated document ("Progress Report [July 1950—December 1951] on the Health Study in the

Uranium Mines and Mills") (ACHRE No. DOJ-051795–A), 3.

40. Duncan Holaday, *Radiation Exposure of Uranium Miners, Part One*, JCAE (1967), 601.

41. Federal Radiation Council, Preliminary Staff Report, No. 8, *Radiation Exposure of Miners, Part One*, JCAE (1967), 1038.

42. Henry N. Doyle, Senior Sanitary Engineer, USPHS, undated ("Survey of Uranium Mines on Navajo Reservations, November 14–17, 1949 and January 11–12, 1950") (ACHRE No. DOJ-051795–A), 1.

43. Duncan A. Holaday, "Origin, History and Development of the Uranium Study," 5.

44. Ibid., 12.

45. Deposition of Duncan A. Holaday, *Barnson v. Foote Mineral Co.*, 9 October 1985, 25.

46. "Progress Report (July 1950–December 1951) on the Health Study in the Uranium Mines and Mills," 4–5, 8.

47. Duncan A. Holaday, Senior Sanitary Engineer, Radiation Unit, Division of Industrial Hygiene, to Chief, Industrial Hygiene Field Station, 21 February 1950 ("Radon Samples in Uranium Mines") (ACHRE No. DOJ-051795–A), 1.

48. Public Health Service, Division of Industrial Hygiene, proceedings of 25 January 1951 (ACHRE No. HHS-092794–A), 1.

49. Ibid.

50. Ibid., 2.

51. Ibid.

52. Ibid.

53. Duncan Holaday to J. W. Hill, General Superintendent, U.S. Vanadium Company, 26 March 1951 ("I'm sorry that Dr. Cralley . . .") (ACHRE No. IND-091394–B).

54. "Progress Report (July 1950–December 1951) on the Health Study in the Uranium Mines and Mills," 11.

55. Ibid.

56. Duncan Holaday, "Radiation Exposure of Uranium Miners," Subcommittee on Research, Development, and Radiation, 1967, 23.

57. A 1975 report written for National Institute for Occupational Safety and Health (NIOSH) and released by the National Technical Information Service provides the following analysis of the behavior of both the industry and the states: "The early uranium mining industry, was unstable, extremely transient and highly speculative. It was both ill-equipped to remedy the mine radiation hazard and resistant to encroachments by the government. . . ." The report also says that in the absence of actual cases of lung cancer "companies, official agencies and miners alike remained unconvinced of the need for preventative measures to control mine radiation." Jessica S. Pearson, "A Sociological Analysis of the Reduction of Hazardous Radiation in Uranium Mines," National Technical Information Service, PB-267 503 (April 1975), 12.

58. Henry N. Doyle, Senior Sanitary Engineer, Division of Occupational Health, to Chief, Division of Biology and Medicine, AEC, 26 May 1952 ("I am pleased to transmit . . .") (ACHRE No. DOE-061395–E), 1. Doyle wrote: "This is a restricted report [An Interim Report of a Health Study of the Uranium Mines and Mills] and is only being circulated to companies engaged in the production of uranium ores, certain federal agencies concerned with the problem, and the Universities of Rochester, Colorado and Utah."

59. Associated Press, "Survey Shows Miners Unhurt by Radiation," 26 June 1952. The lead paragraph reads: "Examinations of more than 1,100 workers in uranium mines and mills have revealed no evidence of health damage from radioactivity." The existence of the press release suggests the report was generally available; however, the May 1952 letter discussing it, as cited immediately above, indicated that it would be available on a "restricted" basis.

60. Federal Security Agency and Colorado State Department of Public Health, May 1952 ("Interim Report of a Health Study of the Uranium Mines and Mills") (ACHRE No. DOE-061395–E), i.

61. Wilhelm C. Hueper, undated, "Organized Labor and Occupational Cancer Hazards" (ACHRE No. HHS-042495–A), 9–10. Wilhelm C. Hueper, "Adventures of a Physician in Occupational Cancer: A Medical Cassandra's Tale" (1976). Unpublished autobiography, Hueper Papers, National Library of Medicine (ACHRE No. HHS-042495–A), 177–178.

62. W. C. Hueper, M.D., to Dr. R. F. Kaiser, Chief, Cancer Control Branch, NIH, 3 April 1952 ("Re.: Cancer Control Grant") (ACHRE No. IND-083094–A), 4.

63. Proctor, *Cancer Wars*, 44.

64. Duncan Holaday testified in 1983 that "the Division of Industrial Hygiene [PHS] had no right of entry to any facility. We had to have the permission of the owner of the facility in order to get on the property." *Begay v. United States*, Civ. 80–982 Pct. WPC, transcript of trial proceedings, 3 August 1983 (ACHRE No. DOJ-051795–A), 114. With respect to AEC-owned mines, E. C. Van Blarcom of the AEC noted that "the Commission ha[d] carried out independent observations in mines under its control." In addition, the AEC requested that the Bureau of Mines conduct its own independent investigation because of "its statutory responsibility to assist, on request, other Federal and State agencies in matters concerning mine safety and health." Department of Health, Education, and Welfare, PHS, 16 December 1960 ("Proceedings of the Governors' Conference on Health Hazards in Uranium Mines") (ACHRE No. DOJ-051795–A), 29.

65. *Barnson v. Foote Mineral Co.*, Consolidated Action Nos. C-80–0119A, C-81–0719W, C-81–0045W & C-81–0715J, deposition taken upon oral examination of Duncan Holaday, 9 October 1985 (ACHRE No. DOJ-051795–A), 12. *Begay v.*

United States, Civ. 80–982 Pct. WPC, transcript of trial proceedings, 3 August 1983, 116–119.

66. *Begay v. United States*, 116–119.

67. Ibid., 119.

68. Department of Health, Education, and Welfare, Public Health Service, Rev. 5–60 ("Uranium Miner Study Record, PHS 2766, Rev. 5–60") (ACHRE No. IND-012395–A), 1.

69. Deposition of Duncan Holaday, *Barnson v. Foote Mineral Co.*, 12.

70. Stewart Udall, *The Myths of August* (New York: Pantheon Books, 1994), 199.

71. Merril Eisenbud, interview by Steve Klaidman (ACHRE) 7 July 1995, 1.

72. Federal Security Agency and Colorado State Department of Public Health ("An Interim Report of a Health Study of the Uranium Mines and Mills") (ACHRE No. DOE-032195–B), 3–5, 6.

73. P. W. Jacoe to Lester Cleere, 28 March 1951 ("Regarding a Discussion with Uranium Producers on Radon Gas Problems in Mines") (ACHRE No. IND-083094–A), 1.

74. Duncan Holaday, Joint Committee on Atomic Energy, Subcommittee on Research, Development and Radiation, 26 July 1967, 90th Cong., 1st Sess., 1213.

75. Advisory Committee for Biology and Medicine, transcript of proceedings of 13–14 January 1956 (ACHRE No. DOE-072694–A), 22, 23–24.

76. Advisory Committee for Biology and Medicine, transcript of January 13–14, 1956 (ACHRE No. DOE 072694–A), 7. Some state regulators and mine owners took the position that the imposition of a strict safety standard would have resulted in the closing of large numbers of small mines. While conceivably the cost to the federal government of ventilating hundreds of small mines could have been prohibitive, the federal government does not appear to have invoked this claim as a basis for inaction. According to Duncan Holaday, the cost of ventilating these mines would have translated to an increase of 50 cents to $1 a pound in the price of uranium, and the average price of fully processed uranium was in the range of $20 a pound. Richard Hewlett, Francis Duncan, and Oscar Anderson, Jr., *Atomic Shield* (Berkeley: University of California Press, 1990), 173. The second volume of a history of the AEC cites a 1948 estimate of about $20 for uranium mined and processed in the United States.

77. L. E. Burney, Surgeon General, to C. L. Dunham, 27 October 1958 ("Since 1950, as you know . . .") (ACHRE No. IND-083094–A).

78. C. L. Dunham to A. R. Luedecke, 11 March 1959 ("Letter from Surgeon General to C. L. Dunham Concerning Radiation Exposure to Miners in Certain Mines") (ACHRE No. DOE-040395–A), 2.

79. L. K. Olson, AEC, General Counsel, to C. L. Dunham, 11 March 1959 ("Health Hazards in Uranium Mines") (ACHRE No. DOE-040395–A).

80. Department of Health, Education, and Welfare, "Proceedings of the Governors' Conference on Health Hazards in Uranium Mines," 29–33.

81. *Begay v. United States*, 591 F. Supp. 991 (D. AZ., 1984), 1002.

82. Duncan A. Holaday, "Origin, History and Development of the Uranium Study," 12.

83. Ibid., 14.

84. Ibid., 16.

85. *Begay v. United States*, Civ. 80–982 Pct. WPC, transcript of trial proceedings (3 August 1983), 152.

86. Department of Health, Education, and Welfare, Public Health Service and U.S. Department of the Interior, Bureau of Mines ("Uranium Miners: Your Ounce of Prevention") (ACHRE No. IND-083094–A).

87. Ibid., 8.

88. Ibid., 4.

89. *Health Impact of Low-Level Radiation*, 24.

90. Ibid., 25.

91. Department of Health, Education, and Welfare, "Proceedings of the Governors' Conference on Health Hazards in Uranium Mines," 17–23.

92. Holaday, "Radiation Exposure of Uranium Miners," 88.

93. A working level is any combination of short-lived radon daughter products per liter of air that releases an amount of energy equal to the energy that would be released by the short-lived daughter products in equilibrium with 100 picocuries of radon per liter of air.

94. Holaday, "Radiation Exposure of Uranium Miners," 89. Wirtz explained: "Since so much of the uranium ore mined in this country is used by mills which have contracts with the Atomic Energy Commission, the Public Contracts Act [Walsh-Healey] authority has clear applicability to the uranium miners situation. This has not been questioned, except with respect to certain details regarding the coverage of 'independent' mines (those not owned or operated by the milling companies). The AEC contracts with the mills contain broadly phrased 'health and safety' stipulations in accordance with the Public Contract Act Requirements."

95. George T. Mazuzan and J. Samuel Walker, *Controlling the Atom* (Berkeley: University of California Press, 1984), 308.

96. *Begay v. United States*, 591 F. Supp. 991, 1007.

97. Under U.S. law, the federal government may be sued only in circumstances in which it waives its sovereign immunity. The Federal Torts Claims Act spells out these circumstances. Under that law, the federal government cannot be sued when the actions complained of are "discretionary functions" of government. Ibid., 1007–1013. The general theory behind this limitation is that the ability of officials to govern would be seriously compromised if their basic decision making were routinely subject to court challenge. The judge in the Begay case concluded that

because "conscious policy decisions based on political and national security feasibility factors" were involved, he had no authority provide a remedy. Ibid., 1012.

98. Ibid., 1013.

99. Ibid., 1011–1012.

100. Federal Security Agency and Colorado State Department of Public Health ("An Interim Report of a Health Study of the Uranium Mines and Mills"), 9.

101. Mazuzan and Walker, *Controlling the Atom,* 317.

102. Duncan Holaday, deposition, 19 March 1986, *Begay v. United States,* Civ. 80–982, and *Anderson v. United States,* Civ. 81–1057 (ACHRE No. IND-091494-A), 102.

103. *Begay v. United States,* 591 F. Supp. 991, 995.

104. On the question of the economic impact of ventilation costs on the price of uranium from U.S. mines (which was already significantly higher than that from the Belgian Congo, but was, of course, a more secure source), Eisenbud noted in 1956: "While it has a big effect on the price of ore, by the time you get it into a reactor or into a bomb that differential is insignificant." ACBM, transcript of proceedings of 13–14 January 1956 (ACHRE No. DOE-012795–C), 35. See also, Victor Archer, interview by Ken Verdoia (KUED-TV, Salt Lake City, Utah), transcript of audio recording, July 1993 (ACHRE No. CORP-122794-A), 14.

105. *Begay v. United States,* 591 F. Supp. 991, 997.

106. Ibid.

107. Ibid.

108. Richard Lemen, Assistant Surgeon, to D. A. Henderson, Deputy Director, National Institute of Occupational Safety and Health, 12 May 1995 ("Populations at Risk: The Ethics of Observational Data Gathering"). This was a response to a draft ACHRE chapter sent to the agency for review.

109. National Cancer Institute, National Institutes of Health, *Radon and Lung Cancer Risk: A Joint Analysis of 11 Undergound Miner Studies,* Publication No. 94–3644 (Washington, D.C.: National Institutes of Health, January 1994).

110. Committee on the Biological Effects of Ionizing Radiation, *Health Risks of Radon and Other Internally Deposited Alpha-Emittters, BEIR IV* (Washington, D.C.: National Academy Press, 1986).

111. The average risk estimate for all eleven mining cohorts is 0.49 percent per WLM, which translates to a "doubling dose"—the dose at which the probability of causation equals 50 percent—of 204 WLM.

112. For example, the doubling dose for the Colorado cohort is 238 WLM, whereas for the New Mexico cohort it is only 58 WLM. The doubling dose is as low as 84 WLM under age fifty, and exposures at different latency periods should be accumulated with different weights. A consistent "inverse dose rate effect" was found, such that a long low-dose-rate exposure is much more hazardous than a short, intense one (this is the reverse of the usual pattern for x rays and gamma rays). Thus, exposures at a dose rate of greater than 15 WL have 1/10 the effect of those at a rate of less than 0.5 WL, or equivalently a 35+ year exposure is 13.6 times more hazardous than one of less than 5 years.

113. The studies differ considerably in the quality of data available on smoking and on the pattern of interactions between smoking and radon found. Because of these limitations, a joint analysis of smoking and the above temporal modifiers was not attempted for all studies. One analysis gives an estimated doubling dose of 97 WLM for nonsmokers and 294 WLM for smokers in all cohorts combined. The latter figure is close to the 300 WLM figure specified for smokers in RECA. However, for specific cohorts, the results are quite different. Neither the Colorado nor the New Mexico cohorts show any significant differences in slope between smokers and nonsmokers, although the estimated slopes appear to vary in opposite directions. In the Colorado cohort, the doubling doses are higher for smokers, whereas in the New Mexico cohort the doubling doses are lower for smokers.

114. The report estimates that 59 percent of the lung cancer deaths in the Colorado cohort and 66 percent of the New Mexico deaths are attributable to radon exposure (87 percent and 47 percent, respectively, among nonsmokers, 59 percent and 74 percent among smokers).

115. These uncertainties (95 percent confidence limits) are typically of the order of sevenfold, with about 90 percent of the estimates being based on extrapolations from other mines or other years in the absence of any actual measurements.

116. The court in the *Begay* decision concluded that the epidemiological study and the conduct of the researchers were consistent with the "medical, ethical and legal standards of the 1940s and 1950s." The researchers "were not experimenting on human beings. They were gathering data to be used for the establishment of enforceable maximum standards of radiation. . . . " *Begay v. United States,* 591 F. Supp. 991, 997–998.

117. The PHS could have conducted its research on only the small number of mines that were not privately owned.

118. 42 C.F.R. § 1.103 quoted in *Begay v. United States,* 591 F. Supp. 991, 1011.

119. Federal Security Agency and the Colorado State Department of Public Health ("An Interim Report of a Health Study of the Uranium Mines and Mills"), i.

120. For example, see the May 1952 "Interim Report of a Health Study of the Uranium Mines and Mills" compiled by PHS and the Colorado State Department of Public Health; "Proceedings of the Governors' Conference" held in 1960; and correspondence between the Industrial Hygiene Field Station

and mining concerns. Duncan A. Holaday, Senior Sanitary Engineer, to J. W. Hill, General Superintendent, U.S. Vanadium Company, 26 March 1951 ("I'm sorry that Dr. Cralley . . .") (ACHRE No. IND-091394–B).

121. Department of Health, Education, and Welfare, undated ("Uranium Miner Study Record, PHS 2766, Rev. 5–60") (ACHRE No. IND-012395–C), 1.

122. Uranium Study Advisory Committee, proceedings of 3 December 1953 (ACHRE No. DOE-012595–B), 1.

123. *International Guidelines for Ethical Review of Epidemiological Study* (Geneva: CIOMS, 1991), 13.

124. Ibid., 18.

125. Judith Sweazey and Stephen Scher, "The Whistleblower as a Deviant Professional: Professional Norms and Responses to Fraud in Clinical Research," Whistleblowing in Biomedical Research, proceedings of a workshop, 21–22 September 1981, 180–2. President's Commission for the Study of Ethical Problems in Medicine and Biomedical Research, 1981.

126. U.S. Department of State, "Trusteeship Agreement," reprinted in *Trust Territories of the Pacific Islands, 1993*, appendix B.

127. According to a 1976 Lawrence Livermore National Laboratory Study: "Bikini Atoll may be the only global source of data on humans where intake via ingestion is thought to contribute the major fraction of plutonium body burden. . . . It is possibly the best available source of data for evaluating the transfer of plutonium across the gut wall after being incorporated into biological systems." W. L. Robison and V. E. Noshkin, 27 September 1976 ("Plutonium Concentration in Dietary and Inhalation Pathways at Bikini and New York") (ACHRE No. DOE-021795–A), 15.

128. Due to cultural differences and the language barriers, however, Marshallese dietary customs were unknown or ignored. For example, differences in the eating habits between men and women may have led to higher exposure in women. The differences of retention of radionuclides by coconut and land crabs also were not recognized by the American doctors. Ambassador Wilfred Kendall, Advisory Committee on Human Radiation Experiments, proceedings of 15 February 1995; Gordon Dunning to A. H. Seymour, 13 February 1958 ("Operational Responsibilities").

129. Robert Conard to L. H. Farr, 2 June 1959 ("Future Marshallese Surveys").

130. See for example, Joint Task Force-7, spring 1954 ("Operation Castle—Radiological Safety, Final Report") (ACHRE No. CORP-063095–A), K-3; and R. A. House, 1 March 1954 ("Memo for the Record").

131. Robert Conard, *Fallout: The Experiences of a Medical Team in the Care of a Marshallese Population Accidentally Exposed to Fallout Radiation* (Upton, N.Y.: Associated Universities, Inc., September 1992) (ACHRE No. DOE-082494–A), 15.

132. Jonathan Weisgall, *Operation Crossroads* (Annapolis, Md.: Naval Institute Press, 1994), 303.

133. Merril Eisenbud, interview with ACHRE staff, 12 September 1995 (ACHRE No. ACHRE-091895–A).

134. There has been a diffference of opinion regarding whether the death, caused by liver disease, was due to radiation exposure or a blood transfusion received after the incident. Stannard, *Radioactivity and Health*, 914.

135. Victor Bond, interview by Gil Whittemore and Faith Weiss (ACHRE staff), 1 December 1994, transcript of audio recording (ACHRE Research Project, Interview Program Series, Targeted Interview Project), 57.

136. Eugene Cronkite, interview by Gil Whittemore and Faith Weiss (ACHRE staff), 1 December 1994, transcript of audio recording (ACHRE Research Project, Interview Program Series, Targeted Interview Project), 42.

137. Ibid.

138. Commander, Joint Task Force-7, to Chief, Armed Forces Special Weapons Project, 6 March 1954 ("Project 4.1 Study. . .") (ACHRE No. DOE-033195–B).

139. E. K. Gilbert, Commander Task Unit 13, USAF, to Commander Eugene P. Cronkite, 8 March 1954 (Letter of Instruction to Cmdr. Eugene P. Cronkite, USN") (ACHRE No. DOE-033195–B).

140. Ibid.

141. Ibid.

142. Interview with Cronkite, 1 December 1994, 37.

143. Gordon Dunning, Biophysics, to John Bugher, Division of Biology and Medicine, 8 June 1954 ("Basis for Estimation of Whole Body Gamma Dose to Exposed Personnel in the Pacific") (ACHRE No. DOE-033195–B).

144. Interview with Bond, 1 December 1994, 36.

145. Naval Station Kwajalein to AEC, 16 March 1954 ("Pastore, Hollifield, and staff . . .") (ACHRE No. DOE-033195–B).

146. Robert A. Conard, telephone interview with Steve Klaidman (ACHRE staff), 29 June 1995; Interview with Cronkite, 1 December 1994, 60.

147. Project Officers for Follow-up Studies on Marshallese to John Bugher, DBM, 20 July 1954 ("Plans for the first follow-up study on the Marshallese") (ACHRE No. DOE-051095–B), 3.

148. Cronkite et al., "Response of Human Beings Accidentally Exposed to Significant Fallout Radiation," *Journal of the American Medical Association* (1 October 1955): 427–434.

149. Interview with Bond, 1 December 1994, 38.

150. Ibid.

151. Cronkite et al., "Response of Human Beings Accidentally Exposed," 433.

152. Ibid.

153. Interview with Bond, 1 December 1994, 42.

154. Interview with Cronkite, 1 December 1994, 67.

155. Francis Midkiff, High Commissioner, Trust Territories of the Pacific Islands, to Major General P. W. Clarkson, Joint Task Force-7, 6 May 1954 ("Dr. John Bugher . . . conferred with my staff . . . ") (ACHRE No. DOE-033195–B).

156. Ibid.

157. Director, Project 1–M-54, to Surgeon General, 5 July 1954 ("Report of 1–M-54 on 30 Servicemen Exposed to Residual Radiation at Operation Castle") (ACHRE No. DOD-092394–C), 2.

158. Interview with Cronkite, 1 December 1994, 46.

159. Ibid.

160. Colonel Claud Bailey, Department of Defense, Radiation Experiments Command Center, to David Saumweber, ACHRE staff, 14 July 1995 ("DNA response to ACHRE Request 070695–B").

161. Lieutenant Colonel R.A House, undated, "Discussion of Off-Site Fallout," in Operation Castle, Radiological Safety, Final Report," vol. 1, spring 1954 (ACHRE No. CORP-063095–A), K-59.

162. Alfred J. Breslin and Melvin E. Cassidy, 18 January 1955 ("Radioactive Debris From Operation Castle Islands of the Mid-Pacific") (ACHRE No. DOE-033195–B).

163. Thomas Hamilton et al., "Thyroid Neoplasia in Marshall Islanders Exposed to Nuclear Fallout," *Journal of the American Medical Association* (7 August 1987): 630.

164. House, "Discussion of Off-Site Fallout," K-59.

165. Ibid.

166. Thomas Kunkle, Los Alamos, to Ellyn Weiss, Office of Human Radiation Experiments, 17 April 1995 ("More Comments on the Draft ACHRE Chapter"), 19.

167. Ambassador Wilfred Kendall, Advisory Committee on Human Radiation Experiments, proceedings of 15 February 1994.

168. Conard, *Fallout*, 15.

169. Robert A. Conard, undated, "Preliminary Report on the Two-Year Medical Resurvey of the Rongelap People" (ACHRE No. DOE-033195–B).

170. Gordon Dunning, Division of Biology and Medicine, November 1956 ("Review of Data: Radioactive Contamination of Pacific Areas from Nuclear Tests") (ACHRE No. DOE-051095–B).

171. Holmes & Narver, Inc., November 1957, "Report of Repatriation of the Rongelap People," prepared for the Albuquerque Operations Office of the AEC (ACHRE No. DOE-033195–B), 1.18.

172. K. E. Fields to Anthony Lausi, Department of the Interior, 4 March 1957 ("The Atomic Energy Commission . . .") (ACHRE No. DOE-033195–B).

173. Eugene Cronkite to Commander, Joint Task Force-7, 21 April 1954 ("Care and Disposition of Rongelap Natives") (ACHRE No. DOE-051995–B).

174. Robert A. Conard to Charles L. Dunham, 28 March 1956 ("The medical team . . .") (ACHRE No. CORP-062295–B).

175. Gordon M. Dunning, Health Physicist, Division of Biology and Medicine, to C. L. Dunham, Director, Division of Biology and Medicine, 13 June 1957 ("Resurvey of Rongelap Atoll") (ACHRE No. CORP-072195–B). In a 13 February 1958 memorandum to Dr. A. H. Seymour of the Division of Biology and Medicine, Dunning characterized the radiological survey as "a poor second alternative" that provided "only a small part of the data we should have obtained." Gordon M. Dunning to A. H. Seymour, 13 February 1958 ("Operational Responsibilities") (ACHRE No. CORP-072195–B).

176. Robert Conard et al., March 1957 ("Medical Survey of Rongelap and Utirik People Three Years After Exposure to Radioactive Fallout") (ACHRE No. DOE-033195–B), 22.

177. Lauren R. Donaldson, University of Washington, to Allyn H. Seymour, Division of Biology and Medicine, AEC, 11 January 1957 ("During a conversation . . .") (ACHRE No. CORP-072195–B).

178. Ibid.

179. Robert Conard, *Review of Medical Findings in a Marshallese Population Twenty-Six Years After Accidental Exposure to Radioactive Fallout* (Upton, N.Y.: Associated Universities, January 1980), 8.

180. Dr. Robert A. Conard, interview by Steve Klaidman, 30 June 1995.

181. Conard, *Fallout*, 15.

182. According to an interview with Marshallese senator Tony deBrum, taboos would have kept Marshallese women from reporting births of severely deformed children to the BNL medical team. Senator Tony deBrum, interview with Steve Klaidman (ACHRE staff), 16 July 1994.

183. Robert Conard, Three Year Report, 22–23.

184. Ibid., 6.

185. Robert Conard, *Fallout*, 26.

186. Ibid., 24.

187. E. L. Cronkite, V. P. Bond, and C. L. Dunham, *Some Effects of Ionizing Radiation on Human Beings* (Washington, D.C.: Atomic Energy Commission, July 1956) (ACHRE No. CORP-062295–A), 75 .

188. Undated, "Evaluation of Total Body Water and Blood Volume Using Marshallese Individuals" (ACHRE No. CORP-062295–B).

189. Interview with Cronkite, 1 December 1994, 59.

190. Ibid.

191. Ibid., 60.

192. Ibid.

193. For example, Conard noted, "Polio was introduced into the Islands by an infected sailor from a visiting ship. A widespread epidemic occurred, with nearly 200 cases of paralysis." Conard, *Fallout*, 14.

194. Hugh Pratt, "Position Paper for the Marshall Islands Study from Brookhaven National Laboratory," 1 December 1978 (ACHRE No. DOE-051094–A).

195. Robert Conard to Charles Dunham, 5 June 1958 ("I sent you a letter . . .") (ACHRE No. CORP-012395–A), 2.

196. Edward E. Held to Robert A. Conard, 16 September 1958 ("We have been back in Seattle . . .") (ACHRE No. CORP-062295–B).

197. Ibid.

198. Konrad Kotrady, "The Brookhaven Medical Program to Detect Radiation in the Marshallese People," 1 January 1977 (ACHRE No. CORP-062295–B), 5.

199. Robert A. Conard to John R. Totter, 4 November 1970 ("I have just returned . . .") (ACHRE No. DOE-052695–A).

200. Ezra Riklon, interview by Holly Barker, transcript of audio recording, 18 August 1994, provided by Marshallese Embassy (ACHRE No. CORP-092694–A).

201. Dr. Hugh Pratt, telephone interview with ACHRE staff, 29 July 1995 (ACHRE No. ACHRE-091895–A).

202. Kotrady, "Brookhaven Medical Program," 8.

203. Charles L. Dunham to the People of Rongelap, 2 February 1961.

204. Ibid.

205. Robert Conard to Courts Oulahan, Deputy General Counsel, AEC, 17 April 1961 ("In regard to our telephone conversation . . .") (ACHRE No. CORP-062295–B).

206. Ibid.

207. Conard, *Twenty-Six Year Report*, vi.

208. Bond, interview with ACHRE staff, 1 December 1994, 80–81.

209. Kotrady, "Brookhaven Medical Program," 13.

210. Conard, *Twenty-Six Year Report*, vi.

211. Ibid.

212. Martin Biles, AEC, Division of Operational Safety, to Julius Rubin Assistant General Manager for Environment and Safety, 13 March 1972 ("Summary of Activities Related to Several Pacific Atolls") (ACHRE No. DOE-033195–B), 2.

213. Conard, *Twenty-Six Year Report*, v.

214. Conard, *Twenty-Six Year Report*, v.

215. Ibid., 44.

216. Ibid., v-vi.

217. Ibid., vi.

218. Jim Beirne, Senate Energy Committee, interview with Steve Klaidman (ACHRE staff), 3 July 1995 (ACHRE No. ACHRE-091895–A).

219. Conard, *Fallout*, 14; Conard, *Twenty-Six Year Report*, vi.

220. National Research Council, *Radiological Assessments for Resettlement of Rongelap in the Republic of the Marshall Islands* (Washington, D.C.: National Academy Press, 1994), 86.

221. Ibid., 6–7.

222. *Compact of Free Association*, 48 U.S.C., sec. 177(c).

223. Ibid.

224. Ibid., sec. 177 (a).

225. Interview with Beirne, 3 July 1995.

226. Ibid.

227. Kaare Rodahl, M.D., and Gisle Bang, D.D.S., "Thyroid Activity in Men Exposed to Cold," Technical Report 57–36 (Alaskan Air Command, Arctic Aeromedical Laboratory, Ladd Air Force Base: October 1957) (ACHRE No. CORP-071294–A), 81. Charts appearing in the report indicate slightly different subject numbers.

228. Loren Setlow, Advisory Committee on Human Radiation Experiments, transcript of proceedings of 16 March 1995, 440.

229. For an extensive bibliography see Robert Fortuine, M.D., et al., "The Health of the Inuit of North America," *Arctic Medical Research* 52, supplement 8 (1953): 86–91. All fifty-one studies listed under "Radiobiology and Radioactive Substances" are either reports on monitoring fallout or survey articles.

230. The IOM/NRC Committee's members are Professor Chester M. Pierce, Harvard Medical School; Dr. David Baines, native Alaskan physician; Professor Inda Chopra, UCLA School of Medicine; Associate Professor Nancy M. P. King, University of North Carolina School of Medicine; Professor Kenneth Mossman, Arizona State University. Administering the committee is Loren Setlow, director of the NRC's Polar Research Board. The committee is examining the I-131 study to determine compliance with contemporaneous guidelines for human subject research; compliance with contemporaneous and modern radiation exposure standards; notification of participants of possible risk; and whether follow-up studies should have been conducted. The IOM/NRC committee will make recommendations to the Department of Defense, which must then report to Congress.

231. Chester Pierce, Advisory Committee on Human Radiation Experiments, transcript of proceedings of 16 March 1995, 429.

232. Rodahl and Bang, "Thyroid Activity in Men Exposed to Cold," 75–77.

233. Ibid., 83.

234. Pierce, Advisory Committee on Human Radiation Experiments, transcript of proceedings of 16 March 1995, 433.

235. Ibid., 433–434.

236. Ibid., 429.

237. Rodahl and Bang, "Thyroid Activity in Men Exposed to Cold," 3.

238. See Loren Setlow, Advisory Committee on Human Radiation Experiments, transcript of proceedings of 16 March 1995, 442. Concerning longer scanning times, see Rodahl and Bang, "Thyroid Activity in Men Exposed to Cold," 32.

239. Pierce, Advisory Committee on Human Radiation Experiments, transcript of proceedings of 16 March 1995, 430–431.

240. Ibid., 430–431.

241. Ibid., 432.

242. Department of the Air Force, "Research and

Development, Clinical Research," AFAR 80–22 (28 July 1952).

243. Atomic Energy Commission, Division of Civilian Application, "The Medical Use of Radioisotopes: Recommendations and Requirements by the Atomic Energy Commission, " February 1956 (ACHRE No. NARA-082294-A-96), 15.

244. See 15 March 1957 letter from Kaare Rodahl, M.D., to Colonel D. M. Alderson, USAF, Deputy Chief, Preventive Medicine Division, Office of the Surgeon General ("In accordance with your letter of June 21, 1956 . . .") (ACHRE No. NAS-072195-A), and 26 April 1957 letter from Cecil R. Buchanan, Assistant Chief, Byproduct Licensing Branch, Isotopes Extension, Division of Civilian Application, to Colonel Jay F. Gamel, Headquarters, Air Material Command, United States Air Force ("License no. 46–50–1") (ACHRE No. NAS-072195-A).

245. Ibid., 1–2.

246. Pierce, Advisory Committee on Human Radiation Experiments, transcript of proceedings of 16 March 1995, 429.

247. Ibid., 434.

248. Ibid., 436.

249. Ibid.

250. Ibid., 437.

251. C. C. Ijsselmuiden and R. R. Faden, "Research and Informed Consent in Africa: Another Look," *New England Journal of Medicine* 326, no. 12 (1992): 830–834.

252. Rodahl and Bang, "Thyroid Activity in Men Exposed to Cold," 15.

253. Ibid., 16.

254. Pierce, Advisory Committee on Human Radiation Experiments, transcript of proceedings of 16 March 1995, 435.

255. Chester Pierce and Loren Setlow, Advisory Committee on Human Radiation Experiments, transcript of proceedings of 16 March 1995, 454.

13

Secrecy, Human Radiation Experiments, and Intentional Releases

W HEN news reports of human radiation experiments sponsored by the government appeared in late 1993, most citizens were startled to learn about such seemingly secret activities. However, some said that there was nothing new or secret; not only had such experiments been the subject of government inquiry in prior years, but they also had been openly published in the medical literature, and even the popular press, at the time they were performed. Not unlike the atomic bomb itself, human radiation experiments were said to be the darkest of secrets and yet no secret at all. What was secret about human experiments and what was not? This chapter, drawing on what we have reported and adding some new material, summarizes what we have learned about both the rules governing secrecy in human subject research and data gathering and the actual practices employed.

To most citizens it is axiomatic that openness in government is a cornerstone of our society. We believe this is so for many reasons. In a democracy, the free flow of information is essential if we are to choose our governmental leaders, understand their policy choices, and hold them accountable. In our society, when the government puts citizens at risk, those citizens reasonably expect to be informed—both in advance about the potential risks and in retrospect about

the consequences. In the tradition of science, as well as that of democracy, secrecy has often been said to be anathema. Good science requires the testing of theories and findings, and the open flow of information is essential to this end.

Yet we also know that the government must keep some secrets for reasons of national security. But national security may not be the only reason the public cannot obtain information about government activities. In the absence of an affirmative requirement that the government must provide the public with access to information—such as the Freedom of Information Act (FOIA), enacted in 1966—much information that is not classified under secrecy laws is, for practical purposes, out of the citizen's reach. Even under FOIA, access can be denied for reasons other than national security.[1] Finally, the government can make information public in a form—such as technical research reports—that is too obscure or costly to be within the practical reach of many citizens. In short, our discussion of secrecy must begin, but not end, with information intentionally concealed through the formal system of classification.[2] It must also cover information that is intentionally concealed through other means and information that may not have been intentionally concealed but remains inaccessible to the public.

The government's use of secrecy is a measure

of its citizens' ability to understand, participate in, and trust government. Because the government must keep some secrets, the measure of public trust, therefore, is not simply whether secrets are kept but the integrity of the rules used to keep them. The question, then, is not simply whether secrets were kept. Were the rules governing secrecy clear and known to all? Were they reasonable? Were they honored in practice?

To answer these questions, we begin by describing the rules of secrecy that governed the AEC and the Defense Department at the beginning of the Cold War. We found that in addition to national security, classification guidelines instructed officials to keep secrets for other reasons, including the protection of the "prestige" of the government.

We begin reviewing the practices of secrecy with the story of a debate within the early AEC over declassification of Manhattan Project human radiation experiments. While publicly professing the need to limit secrecy in science to matters of national security, the AEC kept information on experiments secret for reasons of public relations and liability. We next turn to the practice of secrecy that began roughly in 1950. We have learned that since that time, human subject research (including those that served military purposes) have typically not been classified. Nonetheless, some important information on human radiation experiments was still concealed from the public. After these two sections on the practices of secrecy in biomedical research, we turn to the issue of secrecy in environmental releases of radiation. When radiation was released into the environment, the government concealed information for reasons that included but were not limited to national security. Finally, we look at the government's practice of record keeping. The government records that the Advisory Committee and the Human Radiation Interagency Working Group have retrieved are invaluable, and the history described in this report could not have been told without them. At the same time there are important gaps in the records that limit the public's ability to know about the rules and practices of secrecy, and most important, the activities that were conducted—in whole or in part—in secret.

While the Cold War is over, the choices faced by biomedical officials and researchers from the onset of the period, and the decisions they made, have substantial relevance today. Early AEC leaders and biomedical advisers came from traditions of science and democracy that recognized that while some secrets must be kept, secrecy is corrosive, and over the longer term secrecy itself can endanger national security. At the same time, these individuals were confronted with continued temptation to keep secrets out of concern that public opinion about sensitive matters would itself imperil programs they believed to be important. The boundary between legitimate concern for national security requirements and concern for the consequences of public opinion was continually tested. The problem of defining this boundary, and ensuring its integrity, remains with us today. So, too, does the no less important question of the means of ensuring public trust in cases where secrecy is merited. In what follows we seek to determine what can be learned from the experience of those for whom the question of defining the rules of secrecy and putting them into practice was routine and essential.

NATIONAL SECURITY AND GOVERNMENTAL PRESTIGE: THE LEGAL TRADITION INHERITED BY COLD WAR AGENCIES

To many citizens, the idea of secrecy in government is linked to the idea of "national security secrets" or "classified information." As we have noted, the government also keeps secrets that fit in neither of these categories. The system of classification, nonetheless, occupies a special place in governmental secrecy. Classified information is accessible only to those who have been "cleared" following investigation and who agree to abide by the rules regarding access to this information; the violation of these rules can result in severe criminal penalties.[3]

Today, classification is limited to matters of national security. At the start of the Cold War, however, the legitimate reasons for classification were not so limited. The legal tradition that information can only be classified for reasons of national security was just beginning to displace a tradition that allowed classification for other interests of state.

The authority to classify information derives from legislation and from presidential executive order. In 1917, Congress passed the Espionage Act to address wartime spying,[4] and further legislation providing for military secrets was enacted in 1938.[5] In 1940 President Franklin D. Roosevelt issued the first executive order on classification, which was based on the authorization of the 1938 law enacted to protect military installations and equipment.[6]

The regulations that interpreted the World War I law declared that secrets could be kept not only for national security reasons but also for other reasons. In 1936, for example, the Army issued rules that provided for Secret, Confidential, and Restricted information. The definition of Confidential provided that

A document will be classified and marked "Confidential" when the information it contains is of such nature that its disclosure, *although not endangering our national security,* might be prejudicial to the interests or prestige of the Nation, an individual, or any governmental activity, or be of advantage to a foreign nation [emphasis added].[7]

Similarly, data could be classified Secret where it might endanger national security "*or* cause serious injury to the interests or prestige of the Nation, an individual, or any government activity [emphasis added]."[8]

The Manhattan Project's "Security Manual" followed the Army rules, requiring classification of information as Confidential, and even at the higher level of Secret, in the absence of likely harm to national security.[9] Before the end of World War II, therefore, there was precedent for using the classification system to do more than protect national security.

The era of atomic energy presented the government with unique questions of secrecy. The government built the atomic bomb behind an extraordinary shield of wartime secrecy. The very existence of the newly created communities surrounding AEC laboratories in Los Alamos, New Mexico; Hanford, Washington; and Oak Ridge, Tennessee; was a secret. Children at Oak Ridge schools did not use their full names, and houseguests were introduced as "Mr. Smith."[10] Following the Hiroshima bombing, the govern-

ment faced the somewhat paradoxical task of protecting its single most important military secret while having to inform the public, if not the world, about both the hazards and peacetime spinoffs that the creation of the bomb had engendered—from radiation fallout and waste to nuclear power and radioisotopes for medical research and treatment.

At the war's end, a committee (known after its chair as the Tolman Committee)[11] convened to determine what information from the Manhattan Project should be declassified. In its report, the Tolman Committee concluded that "in the interest of national welfare it might seem that nearly all information should be released at once."[12] But national welfare had to be considered in light of national security. Still, "it is not the conviction of the [Tolman] Committee that the concealment of scientific information can in any long term contribute to the national security of the United States."[13] The progress of science, the committee reasoned, depends on the free flow of information, and long-term national security depends on the progress of science. In the short term, however, the security of the nation required some secrecy. Thus, the Tolman Committee concluded that secrecy could be justified for reasons of national security and then only if "there is a likelihood of war within the next five or ten years."[14] Applying this general philosophy to the question of secrecy in medical research, it recommended that "all reports on medical research and all health studies" be immediately declassified except for those reports that contained information independently classified in the interest of short-term national security.[15]

While the Tolman Committee report generally advocated openness, it also set the precedent for keeping declassification guides secret. The report recommended that "the whole of the Declassification Guide should not, however, be generally distributed since it gives an overall picture of the whole project and makes mention in certain instances of extremely secret matters. The portions of the Declassification Guide needed for the work of anyone concerned with declassification should be made available."[16] By following this recommendation, the AEC, and later the Department of Energy, would keep from the public the ever-accumulating rules governing weapons-related information. Indeed, the first three declassification guides covering information on nuclear weapons, published in 1946, 1948, and 1950, were declassified only in 1995.[17]

In 1946 Congress enacted the Atomic Energy Act, which, in creating the AEC, expressly addressed the protection of atomic energy information. The act provided that all information related to atomic energy was to be considered as Restricted Data (RD) until the AEC reviewed it and decided that it should be unprotected (RD was, therefore, said to be "born secret").[18] The act prohibited the unauthorized disclosure of RD (making it a capital crime to do so in the course of espionage) and prohibited anyone from receiving access to it without first receiving a security clearance. At the same time, however, the act instructed the AEC not to protect information if the AEC did not consider its disclosure harmful to the national security. Thus, the statute defined RD to mean "all data concerning the manufacture or utilization of atomic weapons, the production of fissionable material, or the use of fissionable material in the production of power, *but shall not include any data which the Commission from time to time determines may be published without adversely affecting the common defense and security* [emphasis added]."[19]

As we look back on a Cold War that spanned four decades, the Tolman Committee's view that secrecy could be justified for reasons of national security only if there is a "likelihood of war within the next five or ten years" may seem quaint. In the decades following the Tolman Committee's work, the possibility of nuclear war would loom as a reality, and information on nuclear weapons design and development would be, and remains today, most closely guarded. But, in the immediate postwar period in which the Tolman Committee worked and the Atomic Energy Act was passed, the question of whether information on atomic energy could, as a practical matter, long be kept secret by one nation, or whether international control of atomic energy and atomic energy information was the best course to national security, was itself a subject of highest-level policy discussion. Most notably, this question was addressed in 1946 by a committee appointed by Secretary of State

James F. Byrnes, and chaired by future Secretary of State Dean Acheson. Acheson selected David Lilienthal (soon to be the first chairman of the new AEC) to chair a board of consultants, which included J. Robert Oppenheimer, the Manhattan Project's senior scientist. In early 1946 the "Acheson-Lilienthal Report" proposed international control of atomic energy under an "Atomic Development Authority." The story of how this proposal was overtaken by the dawning of the Cold War is beyond this report's purview.[20] Nonetheless, as we turn to the new AEC's treatment of information on biomedical research, it is important to recall that in the immediate aftermath of Hiroshima and Nagasaki, there was a window in our history in which the most basic questions of the role of secrecy in nuclear weapons development were an open subject of high-level and public debate.

THE PRACTICE OF SECRECY

The AEC Addresses Secret Manhattan Project Experiments

When it began operation in 1947, the AEC was heir to two traditions: one in which official secrets could extend beyond national security to matters of prestige and another in which the interest in promoting openness and limiting secrecy to matters of national security was recognized. In public, AEC biomedical officials and advisers advocated the latter policy. In secret they embraced the former and even expanded it to encompass "embarrassment." Through as late as 1949, the declassification of reports on human experiments involved their review for public relations and legal liability implications. Documents revealing the dual tracks of public policy making and the secret review process did not become public until 1994. Important pieces of the story remain unclear, including the way in which AEC officials and advisers reconciled seemingly contrary principles.

As described in chapter 5, when Manhattan Project medical official Hymer Friedell recommended in late 1946 that one of the reports on the plutonium injection experiments be declassified, officials inside the new AEC reacted strongly. On March 19, 1947, AEC Medical Division chief Major B. M. Brundage countermanded the declassification decision, on grounds of "public relations." The plutonium report produced the strongest reaction, but it was not the only report on human data at issue. Brundage's March 19 memo also stated that further reports ("Studies of Human Exposure to Uranium Compounds" and "Uranium Excretion Studies") should remain classified. On March 21, an AEC declassification officer confirmed the reclassification on the ground that "these documents may involve matters prejudicial to the best interests of the Atomic Energy Commission in that experiments with humans are involved." The memo expressed hope that "a definite policy in this matter will be announced or explained in the near future."[21]

In April 1947 that hope was partly fulfilled when Colonel O. G. Haywood of the Corps of Engineers wrote to H. A. Fidler, an AEC information officer, that "it is desired that no document be released which refers to experiments with humans and might have adverse effects on public opinion or result in legal suits. Documents covering such work should be classified as secret."[22]

Shortly thereafter the AEC seemingly embraced both of the contradictory traditions to which it was heir. In June 1947, the AEC approved the basic policy of the 1945 Tolman report as an "interim policy."[23] In August 1947 General Manager Carroll Wilson publicized that approval in a letter appearing in the *Bulletin of the Atomic Scientists.* The letter indicated that the AEC endorsed the Tolman report, quoting sections that advocated declassification of nuclear weapons information that posed no "danger to our military security."[24]

Also in June 1947, Chairman David Lilienthal's blue-ribbon Medical Board of Review issued its recommendations on the biomedical program. "Secrecy in scientific research," the board declared, "is distasteful and in the long run is contrary to the best interests of scientific progress." The board recommended that "in so far as it is compatible with national security, secrecy in the field of biological and medical research be avoided."[25] The endorsement of the Tolman report and the broad statement of the Medical Board would seem to indicate that high-level AEC officials and biomedical advisers were opposed to secrecy not required by national security.

But these broad statements left unaddressed the specific response to continued requests to declassify Manhattan Project human experiments. In a June 5 response to researcher Robert Stone, General Manager Wilson suggested that any experiments involving "unwitting subjects" should remain classified as they "might have an adverse effect on the position of the Commission" in "the eyes of the American people and the medical profession in general."[26] In an August 12 letter to Stone, Wilson indicated that the Medical Board of Review had considered the question of secrecy and human experiments in mid-June, but the matter had been deferred.[27]

On August 9, John Derry, serving as acting general manager, evidently in Wilson's absence, proposed a set of guidelines that restated the proposition that secrecy could be based on reasons other than national security. The definition of Confidential that he proposed went beyond the Army and Manhattan Project rules:

CONFIDENTIAL: Documents, information or material, the unauthorized disclosure of which, *while not endangering the National security,* would be prejudicial to the interests or prestige of the Nation or any Governmental activity, or individual, or would cause administrative embarrassment, or be of advantage to a foreign nation shall be classified CONFIDENTIAL [emphasis added].[28]

The Derry memo called for review by a classification board assembled from the AEC's regional sites. In September, this board assembled in Oak Ridge. The available documentation does not show that Derry's proposed rules went into effect, but does show that the Classification Board blessed the illustrations of matter that "should be graded" Secret or Confidential. The former category included "certain selected human administration experiments performed under MED [Manhattan Engineer District]."[29] The latter category contained a broad catch-all:

All documents and correspondence relating to matters of policy planning and procedures, *the given knowledge of which might compromise or cause embarrassment to the Atomic Energy Commission and/or its contractors* [emphasis added].[30]

Following the Classification Board's meeting, Oak Ridge officials wrote to Washington headquarters in search of policy guidance on human subject research. Oak Ridge explained that researchers were eager to have their work declassified. "However, there are a large number of papers which do not violate security, but do cause considerable concern to the Atomic Energy Commission Insurance Branch and may well compromise the public prestige and best interests of the Commission." A problem arose, for example, "in the declassification of medical papers on human administration experiments done to date. Again many of these radioactive agents have been of no immediate value to the patient but rather a much needed opportunity for tracer research."[31]

The problem, Oak Ridge pointed out, was not limited to data from human experiments, but also included health risks that radiation posed for workers and for the public:

Papers referring to levels of soil and water contamination surrounding Atomic Energy Commission installations, idle speculation on future genetic effects of radiation and papers dealing with potential process hazards to employees are definitely prejudicial to the best interests of the government. Every such release is reflected in an increase in insurance claims, increased difficulty in labor relations and adverse public sentiment.[32]

Indeed, the Insurance Branch had already reviewed some papers that were slated for declassification. It had advised against publishing papers that suggested health hazards to the public. In the case of one paper, for example, the Insurance Branch wrote in June 1947:

We question the advisability of publishing this document unless the contractor involved is able to establish that the amount of fissionable material leaving the area is in no way a health hazard to the people living down stream.[33]

In an October memo to Washington, Oak Ridge suggested that the Insurance Branch should routinely review declassification decisions for liability concerns:

Following consultation with the Atomic Energy Commission Insurance Branch, the following declassification criteria appears desirable. If specific locations or activities of the Atomic Energy Commission and/or its contractors are closely associated with statements and information which would invite or tend to encourage claims against the Atomic Energy Commis-

sion or its contractors such portions of articles to be published should be reworded or deleted. The effective establishment of this policy necessitates review by the Insurance Branch, as well as the Medical Division, prior to declassification.[34]

Oak Ridge explained that its acting medical adviser, Dr. Albert Holland, Jr. (whose contribution had been praised in the June 1947 report of the Medical Board of Review), would be in Washington on October 11 to discuss the matter further.[35] On that date the Advisory Committee for Biology and Medicine met and concluded that the "important" policy questions raised by Oak Ridge would require "more study."[36]

While the discussion of Oak Ridge's inquiry did not resolve the question of classification, the matter was otherwise addressed at the October 11 meeting. The draft of the secret minutes of the meeting record the discussion of yet another letter from Dr. Robert Stone, regarding the release of "classified papers containing information on human experiments with radioisotopes conducted within the AEC program."[37] The ACBM concluded that the "problem" was addressed by "the recommendations of the Medical Board of Review and that papers on this subject should remain classified unless the stipulated conditions laid down by the Board of Review are complied with."[38]

What were the recommendations of the Medical Board of Review that the ACBM referred to? Recall that its public report did not address human experiments but briefly declared the importance of limiting secrecy. The matter is cleared up by two letters written by General Manager Wilson on November 5—the first to Stone (this is the "second Wilson letter" discussed in chapter 1) and the second to ACBM Chair Alan Gregg.[39] Consistent with the October 11 ACBM minutes, the letter to Stone explained that all classified research not in compliance with certain conditions laid down by the Medical Board would remain classified. These conditions, as we discussed in chapter 1 included written "informed consent" from the patient and the next of kin. This requirement, Wilson further explained, was contained in an "unpublished and restricted" draft report of the Medical Board of Review, which had been read to the

Commission in June. The letter to Gregg, who had served on the Medical Board of Review, indicated that the ACBM need not consider the matter further because the Medical Board of Review's statement was sufficient.

Thereafter, documents show that the AEC continued to review reports for possible public relations and liability consequences and, as Oak Ridge had recommended, called on the AEC Insurance Branch to vet reports for public relations and liability implications.

In 1948 former Manhattan Project researchers pressed the AEC to declassify data from human experiments for inclusion in a history of Manhattan Project medical research as part of a group of publications called the National Nuclear Energy Series, or "NNES." In February 1948, the University of Rochester's Harold Hodge complained about classification officers gutting his chapter on uranium toxicology. "I would like," Hodge wrote, "to advance the argument that Chapter XVI does not report experiments with humans, and should never have been classified on this basis in the first place."[40]

The researchers sought a "final policy" decision on reports regarding plutonium and uranium from the Division of Biology and Medicine and its advisory committee. In a March 15 letter to a participant in the NNES project, Oak Ridge's Holland reported that it was "the feeling" of these groups that the reports should not be declassified. "While I am sure we both fully appreciate the desirability of declassification, I feel certain that the various individuals concerned will also understand and appreciate the reasons for this decision."[41] (The minutes of the March 10, 1948, ACBM meeting, themselves declassified in 1994, do not refer to the policy decision.)

The policy of classifying reports for reasons of public relations and liability was not limited to human experiments conducted under the Manhattan Project; it extended to at least one human experiment conducted under the AEC. In late 1948, Division of Biology and Medicine chief Shields Warren stated his "complete agreement" with Oak Ridge's Holland that a report on a 1948 University of California experiment with zirconium (the research has since become

known as the "CAL-Z" experiment; see chapter 5) had to be kept under wraps.[42] The report had to remain secret because "it specifically involves experimental human therapeutics" and could not be rewritten in a way that "would not jeopardize our public relations."[43]

In addition, data on workers, as well as sick patients, was vetted for labor relations and legal concerns. In chapter 11 we discussed the exposure of Los Alamos workers involved in the "RaLa" intentional releases. In late 1948 the AEC Declassification Branch reviewed a study entitled "The Changes in Blood of Humans Chronically Exposed to Low Level Gamma Radiation." The document, a memo from the Declassification Branch recorded, "has been issued as an unclassified report by Los Alamos, since it clearly falls within the open fields of research." While agreeing with Los Alamos, the Declassification Branch sent the document to the Insurance Branch, at the suggestion of the medical adviser.[44]

In a December 20, 1948, memo to the Declassification Branch, the Insurance Branch recorded its alarm over the study's finding that accepted gamma radiation safety levels "may be too high." In calling for "very careful study" before making the report public the Insurance Branch declared:

We can see the possibility of a shattering effect on the morale of the employees if they become aware that there was substantial reason to question the standards of safety under which they are working. In the hands of labor unions the results of this study would add substance to demands for extra-hazardous pay . . . knowledge of the results of this study might increase the number of claims of occupational injury due to radiation and place a powerful weapon in the hands of a plaintiff's attorney.[45]

While the Insurance Branch reviewed declassification decisions it did not automatically veto the release of all human experimental data. In an October 1947 memo, Holland approved a report ("The Effect of Folic Acid on Radiation Induced Anemia and Leucopenia") for publication "since purportedly the human work was done in the Department of Medicine of the University of Chicago," and not, presumably, an AEC or Manhattan Project facility.[46] Even when publication might result in bad public relations or might encourage litigation, information was sometimes released.[47]

Thus, while the evidence of formal policy-making that can be recovered is fragmentary, it appears that even though the AEC biomedical officials and advisers publicly advocated limiting secrecy to matters of national security, they secretly endorsed a different policy and followed the secret one. The AEC employed the concepts of "prejudicial to the best interests of the government" and "administrative embarrassment" in determining what information to withhold on human experiments. This course was crafted and administered in secret and remained a secret for decades. Its full reach remains unknown.

While our discussion thus far has focused on the AEC, it was not alone in its concerns that data on human radiation exposure could cause public relations or legal liability problems. As we saw in chapter 10, in 1947, former Manhattan Project head General Groves, and the chair of the new AEC's Interim Medical Advisory Committee, Stafford Warren, were evidently among those who counseled the Veterans Administration to keep secret records in anticipation of potential claims from servicemen. In both cases, the impulse to keep such information secret was accompanied by the decision to create a highly publicized program of radioisotope research, which resulted in numerous human radiation experiments that were not secret.

The practice (and any policy) of keeping information secret on grounds of embarrassment or potential legal liability should have ended no later than 1951, and perhaps as early as 1949.[48] In its 1949 "Policy on the Control of Information," the AEC recognized that secrecy must be balanced against not only the value of the progress of science but also the value of a well-run democracy. Limiting secrecy, the AEC said, ensures "that people may be able to judge the action of their representatives and officials and to participate in public policy decisions. Information about a public enterprise of such consequence as the atomic energy program should be concealed only for reasons soundly based upon the common defense and security."[49] In 1951 President Harry Truman issued a new executive

order on classification.[50] While the order expanded government secrecy by giving every department and agency the authority to classify information, it limited the reasons for classification to national security. Today, the governing executive order expressly prohibits classification of information "in order to: (1) conceal violations of the law, inefficiency, or administrative error; (2) prevent embarrassment to a person, organization, or agency; (3) restrain competition; or (4) prevent or delay the release of information that does not require protection in the interest of national security." The order also prohibits classification of "basic scientific research information not clearly related to national security."[51] As we shall see later in this chapter, while the law has long since begun to draw a line against the keeping of classified secrets for reasons other than national security, the boundary between national security and public relations rationales remains murky.

Human Radiation Experiments In the 1950s: Experiments Are Not Classified, but Some Secrets Remain

The 1947–1948 AEC declassification controversy may have taught Shields Warren and other AEC biomedical officials that secrecy and human radiation experimentation were a troubling mix, to be avoided if possible. The search efforts of the Human Radiation Interagency Working Group and the Committee located very few human radiation experiments in the post-Manhattan Project period that were classified secrets. Nonetheless, important information relating to many experiments was still intentionally concealed from the public.

When the AEC and DOD debated the need for human experiments for the proposed nuclear-powered airplane (NEPA) in 1950, Warren and the Advisory Committee for Biology and Medicine counseled the Defense Department that there would be "serious repercussions from a public relations standpoint" if human experiments were conducted by an agency that did some of its work in secret.[52] As we saw in chapter 1, in March 1951, Los Alamos asked Warren to state the policy on human experimentation. In transmitting to Los

Alamos excerpts from General Manager Wilson's November 1947 letter to Stone, which cited the requirement for "informed consent," Warren added further counsel against secrecy. Warren cited the Medical Board of Review's public declaration that secrecy should only be countenanced when required by national security. He then quoted ACBM chairman Alan Gregg: "The secrecy with which some of the work of the Atomic Energy Commission has to be conducted creates special conditions for the clinical aspects of its work in that the public is aware of this necessity for secrecy and of the subsequent difficulty of probing into it."[53] When in 1952 the DOD's Joint Panel on the Medical Aspects of Atomic Warfare called for renewed discussion of human experiments, Warren reportedly advised "that studies of this type under the Joint Panel's purview should be conducted by the Public Health Service or some agency where security restrictions would not lead to misunderstanding."[54]

Thus, Warren and Gregg's statements convey a profound concern for the public's perception of human experiments, particularly where human experiments are conducted by agencies that also conduct activities in secret.

Under Paul Aebersold, the AEC isotope distribution program—the provider of the source material for many hundreds of human experiments—became a showcase for public research (see chapter 6). At the Defense Department as well, biomedical human radiation experiments—even when there was clear military purpose—were typically not classified. For example, post-Manhattan Project total-body irradiation research sponsored in part by the military, in the wake of the controversy that raged when similar human experiments were proposed for the NEPA project, was not conducted in secret (see chapter 8).

But if the experiments themselves were not secret, important decision-making context for them was sometimes secret, and hidden rules or practices may have also limited what the public was told about particular experiments. The ability of the public and the press to probe experiments connected to secret programs was limited, making it difficult for the public to critically assess the practices of its government.

For example, the 1950–1952 meetings in which DOD biomedical officials discussed the

need for an ethical code to govern human experiments were classified.[55] So were the meetings of the Joint Panel on the Medical Aspects of Atomic Warfare. Similarly, meetings of the ACBM were often conducted in part or whole in secret. These meetings, as we have seen from the review of the 1947–1948 secret keeping, included seminal discussions of the ethics of human experimentation and the rules governing declassification of experimental data.[56]

To some degree experiments sponsored by civilian agencies such as the National Institutes of Health were also rooted in this secret context. The 1952 letter that reported Warren's belief that human experiments should be separated from secret programs communicated the willingness of NIH and PHS to cooperate in conducting research needed for military purposes. These civilian agencies were themselves participants in DOD biomedical planning for atomic warfare, and their research was also listed in the secret digests (which included classified and non-classified research) of atomic warfare-related research that the DOD's Committee on Medical Sciences provided to the Joint Panel on the Medical Aspects of Atomic Warfare.[57] Also in 1952, an internal report on "Defense Activity of the National Institutes of Health (1950–52)" noted that "a second major activity of the NIH relating to radiation research has been participation in the medical and biological aspects of atomic bomb tests. A large share of this activity has been borne by the Armed Forces Special Weapons Project. The substance of this work is classified."[58]

The country's research resources should have been available to serve national security needs. But, as Warren and Gregg suggested, when human research and national security are intertwined, care must be taken to ensure that the public has means to separate out secret and nonsecret purposes with confidence. At this time it is not clear what, if any, classified human radiation experiments were conducted by DHHS's predecessors and what was said in secret about otherwise public human radiation experiments.[59]

Similarly, while most AEC biomedical radiation research was not classified, some was. From available records, it appears unlikely that much of the secret research involved humans. But, given the secrecy and the absence of clear records, certainty is impossible.[60]

Moreover, even if little human subject radiation research itself was classified, information about the research could be concealed by less formal means. As we discussed in the Introduction and chapter 10, in July 1949, the NEPA advisory group met with a group of psychologists and psychiatrists to discuss the psychology of radiation risk. The participants were told:

This is not a closed meeting. Some of our advisers . . . have not been cleared. Ordinarily, medical and biological discussions are not, of course classified. We shall ask you, however, to refrain from discussing these matters on the outside, since of course we do not want newspapers to know of these discussions at this time.[61]

Moreover, the determination to render information formally secret could be applied in a manner that was invisible and arbitrary, as illustrated by the following case. At midcentury, the Medical College of Virginia (MCV) performed research on the effect of thermal burns for the Defense Department. MCV's research, conducted with animals, prisoners, and medical students, initially appears to have been a matter of public record. In January 1951, following inquiry by a reporter from the *Richmond Times-Dispatch*, MCV investigator Dr. Everett I. Evans grew alarmed that press reports decrying the use of dogs would "greatly harm the work we are doing on the experimental burn in relation to atomic bomb injuries."[62] Evans called on the chairman of the Army's Medical Research and Development Board to classify the work so that "I would have legal means of preventing a public newspaper discussion of these experiments. . . ."[63]

The Army immediately provided a declaration that all work under the MCV contract "will be classified RESTRICTED."[64] The Army decreed that a bureaucratic obstacle course would have to be overcome before information was released, including "coordination" with the experimenters, and evaluation by "the other branches of the Armed Forces, the Federal Civil Defense Administration, the National Security Resources Board, the Atomic Energy Commission, and the National Research Council."[65]

This rigor was essential because "individual releases may be mistaken for official advice to civil defense groups and result in confusion of training and procedure, the stockpiling of unnecessary or inappropriate materials, etc."[66] Finally, perhaps on the possibility that the local reporter might be uniquely dogged, the Army added that it "is also the policy of the Department of Defense that public releases to the press are made simultaneously to all national news services, and that the releases are not made to individual reporters or newspapers."[67] While the secrecy was prompted by revelations on animal experiments, in late 1951 Dr. Evans invoked it to close the curtain on the use of prisoner volunteers at the state penitentiary.[68] The prison assured Evans that inmates and staff were informed that "no publicity should be given to the experiment being carried on at the Medical College."[69]

In the case of research related to chemical and biological warfare, the military issued a secret edict that published articles be cleansed of any reference to military purpose.[70] In many cases the opportunity to obscure the full purpose of research by careful wording was obvious. As a DOD document put it, "the term 'radiobiology' is so flexible semantically that, depending upon the investigator's point of view, any project could be classified as 'clinical' or 'basic' or 'nuclear weapons effects.'"[71] In 1961, the U.S. Department of Agriculture issued an extensive bibliography of research on strontium and calcium. The preface made clear the publication was relevant to those researching fallout (radioactive strontium being a major fallout concern).[72] However, Advisory Committee staff review of many of the articles on human experiments included in the bibliography revealed few indications of fallout as a purpose for the research.[73]

The difficulty of determining what was secret is compounded because the government sometimes actively deceived or lied. Most remarkably, the AEC continually told inquiring members of the public that it did not perform human experiments—even when its isotope division very publicly supported them. In 1948, for example, the AEC wrote to a member of the public that "there is no possibility, at present or projected, of human experimentation with atomic energy."[74] In 1951, when the press pursued a rumor that the AEC was sponsoring an experiment with prisoners, the AEC's chief public information official assured the Associated Press that the AEC "has never sponsored a medical research project where human beings were being used for experimental purposes."[75] In 1953 the AEC wrote to members of the public that it "does not deliberately expose any human being to nuclear radiation for research purposes unless there is a reasonable chance that the person will be benefited by such exposure."[76] At the same time an internal AEC memo from the public information office noted that "any experimentation on humans has obvious and delicate public relations aspects. Any project involving such experimentation must have careful prior consideration by both the field and Washington, particularly as to content of any public statements."[77]

As we saw in chapter 12, uranium miners were not adequately informed about the purpose of research regarding their exposure to radon in the mines. Above and beyond lack of disclosure, there is evidence that deception was not unusual in data gathering on AEC workers, as illustrated by a 1955 exchange between the University of Rochester's Dr. Louis Hempelmann and the AEC Division of Biology and Medicine regarding a proposed study evidently designed to measure the occurrence of lung cancer among a group of former workers. "You will have to find a good excuse so as not to worry the person you are contacting," Hempelmann wrote to DBM chief Charles Dunham. "This isn't very clever but, perhaps, you could say in some convincing way that you, or rather the person conducting the study, represents a life insurance company studying the health of people employed by the Harshaw Company during a certain period."[78] Dr. Hempelmann apologized for his lack of imagination:

I don't know whether these ideas are helpful at all. It is more difficult to find an excuse for these individual workers than it is in the case of patients who were treated for something or other at a hospital. I think that someone with imagination might come up with a better idea than I have had to date.[79]

This last comment implies that it was not only workers, but also patients, who were deceived about their participation in research, and more easily at that. The statement is particulary striking when it is recalled that Dr. Hempelmann was, as an adviser to Robert Oppenheimer, a proponent of the plutonium injection experiments, and, following the war, became professor of experimental radiology at the University of Rochester, a major AEC biomedical contractor. Thus, if the statement is a reflection of the readiness to deceive patients, it is one made by a doctor at the center of the AEC biomedical community and, indeed, was made directly to the head of the AEC's Division of Biology and Medicine.

Dunham's assistant evidently agreed that workers should be deceived, but "we have racked our brains for any useful subterfuge in carrying out the study but none came to mind which could possibly hold water for any length of time."[80] The AEC opted for subtle deception:

The attack with which we are going to start the study will be to inform the old Harshaw employees that our interest in them is only part of an over-all program to make sure that the safety controls in the atomic energy business are absolutely perfect. To be sure, such an approach might cause some alarm but this should not be too great, I hope, because it is essentially a negative one; namely, the Commission is sure that there will be no injury to its workers but it needs to document this fact for the record.[81]

The AEC official agreed that "routine physical examination would be relatively fruitless since the ultimate objective is to determine the incidence of lung cancer, which can be obtained best with a post-mortem examination. On the other hand," the official noted, "the attitude of the Western Reserve group [with whom the AEC was proposing to contract for the study] is that physicial examinations are a useful means for maintaining close contact with people and will improve the chances of getting post-mortem information."[82]

In sum, after the Manhattan Project the governing presumption, to which the Advisory Committee found little exception, has been that biomedical human radiation experiments should not be classified. But the presumption included important qualifications, some of which were hidden at the time, and others of which may be beyond our ability to retrieve and reconstruct. These qualifications are shortcomings and legitimate cause for public concern, especially when held up to the ideals publically espoused by the AEC's initial leaders.

Human Data Gathering Connected with Bomb Tests and Intentional Releases: National Security, Secrecy, and Public Opinion

The view that a line needed to be drawn to ensure that human radiation experiments were not too closely associated with secret keeping was not easily translated to settings where entire groups of people were placed at risk by environmental releases of radiation. In March 1951, as we have just noted, Shields Warren advised Los Alamos to avoid secrecy in human experimentation. Warren and other AEC officials also told the military of their concern for public repercussions if human experiments were conducted in close proximity to government secret keeping. At the same time, however, Warren and other AEC biomedical experts were called on to advise on nuclear weapons activities that might place entire populations at risk. Here, the question of public disclosure was more difficult to resolve. In May 1951 for example, as discussed in the Introduction, Warren chaired a secret meeting in Los Alamos to consider the safety concerns of the first underground test of a nuclear weapon. The record of the meeting shows that Warren and other experts worried that fallout from the tests could endanger citizens around the Nevada Test Site. The public was not given access to the discussion of testing that the participants were concerned might endanger surrounding communities.[83] Press information stressed the absence of public danger.[84]

As we saw in the discussion of intentional releases (chapter 11), little or no information was contemporaneously made public about the radiological warfare tests at Dugway, the RaLa tests at Los Alamos, or the Green Run at Hanford. National security required some degree of secrecy; but whether more could or should have been disclosed is unclear in retrospect. In the case of at least the Dugway tests, secrecy was fueled by concern that the public might not understand the tests and might question the program.

Atmospheric nuclear weapons tests were, in contrast to the intentional releases and underground nuclear weapons tests, much more difficult to keep secret. In chapter 10 we saw that activities could simultaneously have elements of deep secrecy and appear as front-page news. A then-secret report on the Desert Rock exercises observed, "It was a constant source of amusement at the camp that the newspapers carried accounts of the atomic tests which included information, usually accurate, which the men had been expressly forbidden to reveal."[85] At the same time that the bomb tests were highly publicized, basic information on the risks to participants was not public. "Secrecy," summarized Barton Hacker, author of a DOE-sponsored history of the bomb test program, "so shrouded the test program . . . that such matters as worker safety could not then emerge as subjects of public debate."[86]

Once bomb tests became routine, fallout presented a further opportunity and obligation for the government to sponsor data gathering, including human subject data gathering. It did so on a global scale. As discussed in chapter 12, the research on the Marshall Islanders to measure fallout effects began in secret. "Due to possible adverse public reaction, "the director of the research project was counseled, those involved should limit discussions of the research to those with a "need to know."[87] The Marshall Islands research was only one component of a worldwide data-gathering program that was constructed and operated in substantial secrecy until the latter part of the 1950s. The Advisory Committee was not created to study atomic bomb testing or the related debate about the effects of fallout. However, the human subject research related to bomb-test fallout also presents questions about openness and secrecy in human research and the ethics of human data gathering.

The Fallout Data Network: Projects Gabriel and Sunshine

The study of fallout began with the effects of the first atomic bomb test in New Mexico in 1945.[88]

In 1949 the AEC commissioned Project Gabriel, a study to determine how many atomic weapons could be detonated before radioactive contamination of air, water, and soil would have a long-range effect upon crops, animals, and humans.[89] The AEC soon created a worldwide network for the collection and measurement of fallout (typically by permitting it to fall on a horizontal gummed paper or plastic sheet).[90] By 1954 Gabriel included about seventy investigations supported by the Division of Biology and Medicine, involving 325 person years of labor per year and costing $3.325 million annually.[91]

In the early 1950s the Defense Department created its own fallout research program, under the auspices of the Armed Forces Special Weapons Project. The Public Health Service joined with the AEC and the DOD in monitoring fallout around the Nevada Test Site.[92]

In 1953, under contract to the AEC and the Air Force, the Rand Corporation convened a review of Gabriel.[93] The study was directed by Dr. Willard Libby, a University of Chicago radiochemist who would receive the Nobel Prize in 1960 for the development of the radioactive carbon dating method. The resulting report concluded that strontium 90 (Sr-90) was the most dangerous long-term, global radioactive product of bomb testing and that a global study of strontium 90 fallout was needed.[94]

The report noted how atmospheric testing had, as an unintended side effect, introduced tracers into the world's ecosystem: "Until comparatively recently it would have been extremely difficult, if not impossible, to obtain a measure of a number of the parameters. Today we are afforded the opportunity of doing a radioactive-tracer chemistry experiment on a world-wide scale."[95] The group recommended that "studies then current be supplemented by a world-wide assay of the distribution of strontium 90 from the nuclear detonations which have occurred. This assay has been designated Project Sunshine." The name for the project would be variously attributed to the project's gestation in Santa Monica, California, (where Rand was headquartered) and to the determination to measure the presence of strontium in "sunshine units." Three laboratories were engaged to analyze samples of strontium 90: one at Libby's research center at the University of Chicago, another at the Lamont Geological Observatory of Columbia University, and a third at the New York office of the AEC.

The long-term goals of the full-scale study would be to (1) determine if a hazard had already been created by fallout; (2) determine the number of bombs that could be exploded without creating a hazard; and (3) determine the mechanisms by which radioactive materials might become concentrated.[96]

Secrecy and Deception in Fallout Studies: Project Sunshine's Collection of Human Bones

Project Sunshine was born in secrecy.[97] The decision to keep the existence of the worldwide assay secret "limited the freedom with which suitable combinations of samples might be obtained from foreign countries."[98] For the pilot program, the report suggested that twelve human samples (bone and teeth) be drawn from each of six regions around the world. In addition, samples would be drawn from livestock, foodstuffs, water, and soil.[99] The discussion of collecting individual samples was limited to means of ensuring uniformity in practice, without mention of the ethical relationship between investigators and human subjects. An early effort concerned the collection of baby bones.

In an October 1953 letter to Dr. Libby, Robert A. Dudley of the DBM explained that the collection process would proceed "through personal contacts with foreign doctors" and groups like the Rockefeller Foundation, which had many overseas contacts. Because the chief of the DBM, Dr. John Bugher, advised that "security specifications" needed to be maintained, a cover story would be employed.[100]

The stated purpose of the collection is to be for a survey of the natural Ra [Radium] burden of human bones . . . there are still enough uncertainties regarding threshold dose for injury . . . to provide a plausible explanation for further surveys. . . . As for the emphasis on infants, we can say that such samples are easy to obtain here, and that we would like to keep our foreign collections comparable.[101]

Dudley explained that the AEC wanted to be kept "out of the picture where possible," but to be helpful "I would still be prepared to do all the work except for providing the signature."[102]

One week later Dudley wrote to Shields Warren in Boston. Dudley, noting that the effort was proceeding "pretty much on the lines you suggested," sought Warren's assistance in contacting another Boston doctor who might not be in on the full story. Dudley offered that "while the real purpose will of course remain secret . . . we do expect to make radium analyses on at least some of the samples, so our story is merely incomplete, not false."[103]

On the same day, Dudley wrote to his father, the director of a missionary organization, also in Boston. The letter explained the public purpose of the data gathering and solicited assistance.[104] On November 10, evidently from a referral from his father, the AEC official wrote to the Christian Medical Association in Nadya Pradesh, India, also soliciting assistance. Finally, the DBM sought assistance from civilian organizations that already had well-developed contacts at the local level in foreign countries.[105]

What was the "real purpose" that had to be kept so carefully concealed, even from those who were actually assisting the project? On December 9, Dudley sent a letter to a doctor at the AEC's project at the University of Rochester that explained "for you alone" the AEC's real interest:

This letter will explain in a little more detail than I was able to do over the phone our interest in obtaining infant skeletons from Japan.

The Division of Biology and Medicine is engaged in a project to evaluate the long range radiological hazard which might result from the large scale use of atomic weapons. . . . In order to help in the evaluation of the hazard, we are providing for the direct measurement of the world-wide Sr-90 distribution which has resulted from the 40 or 50 nuclear detonations in the last few years. One type of sample on which we are concentrating is the bones of infants, either stillborn or up to a year or two of age. We have found that stillborn bones are easy to obtain in the United States, and are trying to extend our collection to foreign countries. It appears that the ABCC [Atomic Bomb Casualty Commission] would be a logical contact in Japan. We could use perhaps 6 or 8 skeletons from that area.

It has been decided, for various reasons including public and international relations, to classify this project SECRET for the present. Hence, the unclassified description of our purpose in obtaining these bones is for Ra analyses.[106]

The July 1954 Gabriel report summarized the "human, animal and animal product samples" that had been analyzed.[107] The list included stillborns from Chicago (fifty-five), Utah (one), Vellore, southern India (three), and human legs from Massachusetts (three).[108]

Soon, the DOD was also engaged in fallout data gathering. In the fall of 1954, the Armed Forces Special Weapons Project established a "Fall-out Study Group" following a request for information from the Joint Chiefs of Staff.[109] In 1954 DOD planned a secret project to collect human urine and animal milk and tissue samples following the 1955 Operation Teapot tests in Nevada. The work was coordinated by the Walter Reed Army Institute for Research, with review from researchers at the Harvard Medical School and the National Institutes of Health. The purpose of the effort was to establish a baseline for forthcoming Pacific tests.[110] The military data gathering also involved a cover story. A December 16, 1954, memorandum from the chief of the Armed Forces Special Weapons Project stated, at least in regard to the animal sampling:

The actual data obtained are SECRET and the sample collection should be discreetly handled. It is suggested that a statement be included in the instructions to the effect that these samples are being collected for nutritional studies.[111]

In January 1955 the Gabriel-Sunshine program was the subject of a classified "Biophysics Conference" convened by the Division of Biology and Medicine. The spring 1954 Marshall Islands disaster had, the attendees were told, added new urgency to their task. "I keep reading," noted one participant, "the articles by the Alsops and others [journalists] of the high level groups which are frantically trying to find the answer to how many bombs we can detonate without producing a race of monsters."[112]

The Secret transcript of the conference, declassified from Restricted Data status only in 1995, shows that the AEC and its researchers assigned a high priority to what was referred to as "body snatching." No AEC program, explained Dr. Libby, who had become an AEC commissioner, was more important than Sunshine. There were great gaps in knowledge and human samples were essential to fill them. "[H]uman samples are of prime importance and if anybody knows how to do a good job of body snatching, they will really be serving their country."[113] In the 1953 Rand Sunshine study, Libby recalled, an "expensive law firm" was hired to study the "law of body snatching." The lawyers' analysis showed "how very difficult it is going to be to do it legally."[114]

Nonetheless, "excellent sources" were available from several places, including New York, Vancouver, and Houston. In Houston, said Columbia University's Laurence Kulp, "they intend to get virtually every death in the age range we are interested in that occurs in the City of Houston. They have a lot of poverty cases and so on."[115]

Where good personal relationships with medical sources existed, Dr. Kulp offered, "the men did not require you to tell them anything except that they realized it was something confidential. They could guess, and they probably didn't guess very wrong, but they were willing to cooperate just on the basis that this was an important thing."[116] With a connection "through one of the top medical people who is internationally known, it will not be hard at all to be able to establish the sites that we should establish." The DBM's Dr. Bugher explained that the AEC was exploring the possibility of a special clearance ("L") so that medical professionals who did not want to "fill out any forms" could be briefed on a limited basis. "You are," he stated, "dealing with directors of hospitals and pathologists and persons in general who have an understanding of the seriousness of the project in which we are engaged."[117]

Libby hoped to declassify at least the existence of the Sunshine program. "Whether this is going to help in the body snatching problem, I don't know, I think it will. It is," he said, "a delicate problem of public relations, obviously."[118]

The efforts bore fruit. A report on Sunshine's 1955–1956 operations recorded that during that period hundreds of human bone samples were collected by dozens of stations abroad and by researchers in Boston, Denver, Houston, and New York.[119]

In addition to the Sunshine-related research, the AEC sponsored further efforts to gather human tissue in order to study the effects of ra-

diation on weapons complex workers, as well as fallout on citizens. In a June 1995 report, the General Accounting Office summarized fifty-nine studies, most of which were conducted and terminated in the 1950s and 1960s. While many, probably the great majority, were not secret programs, the GAO found that typically no information can now be located about the consent practices that were followed. Today, the Department of Energy sponsors a program under which those with documented exposures to certain radioactive elements may donate their tissues for research. The operations of the transuranium and uranium registries are subject to review by an institutional review board, and donors must sign a consent form.[120]

In sum, during the 1950s the AEC promoted human tissue sampling for studies on fallout and other research, and its efforts involved secrecy and deception. The AEC evidently considered the legal aspects of "body snatching," but there is no evidence that it sought to consider any independent ethical requirements for disclosure to the families of the subjects (or the subjects themselves, where alive) whose tissue was sampled. While further rationale for keeping the data gathering secret may have existed, in surviving documents concern for public relations emerges as the dominant motivation. At the same time, the AEC recognized that secrecy hampered the conduct of research that it believed central to the public interest.

Secrecy, Public Opinion, and Credibility

On reviewing the transcript of the 1955 Biophysics Conference in 1995, Dr. Merril Eisenbud, a former AEC official who participated in the session, expressed surprise that the document had been classified in the first place.[121] There was, he observed, nothing that merited national security classification; if anything, perhaps it merited the category of Official Use Only, which instructs officials not to publicize the document but is not a category in the formal classification system.[122] As in the case of the AEC's 1947–1948 decision to keep experimental data secret, however, information on fallout data gathering appears to have been classified out of concern that public opinion (in the United States, but also elsewhere) might imperil U.S. weapons development programs.

In November 1954 AEC officials met for lunch with Secretary of Defense Charles Wilson, the signator of the 1953 memorandum discussed in chapter 1, to discuss civilian evacuation in case of atomic warfare and the related question of what the public should be told about fallout. "Secretary Wilson," an AEC record of the meeting summarizes,

stressed the importance of not arousing public anxiety in this country or abroad by public official discussions of the dangers of atomic warfare, particularly with reference to fall-out. He expressed the view that much too much had already been said publicly about fallout, and he urged that before the Government reveals the full extent of the dangers to be expected the Government work out the answers to a lot of questions as to what our citizens could do in the event of an atomic blitz.[123]

"Obviously," records a history of the AEC by AEC and DOE historians, "estimates of the biological effects of fallout on large human populations were more likely to arouse fear and controversy than were small-scale experiments on laboratory animals. Thus, it was not surprising that initial studies of large-scale effects were highly classified and unknown to the public."[124]

Within a very short period, however, much of the secret research was disclosed, but under circumstances where, as the AEC itself came to recognize, its credibility as an information source was seriously impaired.

The Marshall Islands disaster, and the attendant controversy related to the irradiation of the crew of a Japanese fishing boat in the area, marked the beginning of a worldwide debate on fallout that would end with a ban on atmospheric testing.[125] Following this event, ban-the-bomb protests began in Britain.[126] Two years later, in 1956, presidential candidate Adlai Stevenson called for an end to nuclear testing. Soon thereafter, the closely held fallout research began to become public. In October, Libby, addressing the American Association for the Advancement of Science at the dedication of its new headquarters in Washington, reported that the amounts of strontium 90 entering the bodies of children were well below the maximum permissible concentration.[127]

In February 1957 Dr. Kulp and his associates presented the results of their study of 1,500 human bones from around the world. The report made the front page of the *New York Times*.[128] In June, the National Academy of Sciences issued a report noting that strontium 90 and genetic effects were two potentially long-term hazards from nuclear testing.

The public fallout debate was on, pitting scientists against one another. "Test ban advocates," a historian of the fallout controversy recounted, "always stressed the great potential hazard from fallout over a long period of time; their opponents minimized the danger by pointing to similar or greater risks that people routinely accepted, such as luminous wristwatches and medical X-rays."[129]

In May and June 1957, Congress's Joint Committee on Atomic Energy held its first public hearings on the dangers of fallout. The initial 1953 Sunshine report, "Worldwide Effects of Atomic Weapons—Project Sunshine," was apparently declassifed on May 25, two days before the hearings began.[130] Most of the debate focused on the dangers of strontium 90. In June, Commissioner Libby responded to a proposal from an NIH official for the use of children's milk teeth to measure strontium 90. The idea was good, but he advised (in the immediate aftermath of the first highly publicized hearings on fallout), "I would not encourage publicity in connection with the program. We have found that in collecting human samples publicity is not particularly helpful."[131]

In October 1958, a moratorium on nuclear testing began, and in May 1959 the Joint Committee on Atomic Energy held a second series of hearings on fallout. The hearings concluded that the risk was worth the returns to national security; but the public debate continued.[132]

As AEC documents on the fallout debate have become available in the intervening years, it has become clear that the government's effort to manage public opinion was rooted in a sensitivity to its importance. For example, in 1953, following the spring Nevada test series, ranchers in Utah began to report the deaths of their sheep from what, it appeared, might be radiation burns from the tests.[133] The AEC convened a panel to consider the continuation of testing at the Nevada

Test Site. The panel concluded that continued testing was justified by the national interest, although risks were inevitable.[134] The tests to date were relatively safe, but there were serious problems with "public reaction."[135] The panel found that "a sufficient degree of . . . public acceptance has not been achieved."[136] Radiation remained a "mysterious threat."[137] But the government had surrounded the program with an aura of secrecy, its own statements were not clear, and statements by former AEC experts or officials had caused "near-panic concern."[138] The public, "which is expected to accept a certain degree of hazard, has not been adequately informed of the extent and nature of the hazard."[139] An extensive program of public education was called for.[140]

The AEC study found the problem was not only with the public; there was a "lack of agreement and acceptance, first, within AEC and test management, and second, among health, medical, and other scientific individuals and groups." The problem was exacerbated by "lack of knowledge of this new subject, by lack of definition, by the extreme sensitivity of the subject, and by the resulting nervousness of the various levels of management."[141] As shown by the secrecy surrounding the ongoing Project Sunshine, however, the public was not let in on the uncertainty of knowledge or on the steps being taken to answer questions of admitted import to all citizens.

AEC insiders recognized that credibility was a problem. In a December 1954 letter to DBM's director, Charles Dunham, Los Alamos Health Division Leader Thomas Shipman touted the importance of Sunshine and suggested a possible role for Los Alamos. At the same time he lamented the lack of credibility possessed by those too closely associated with the AEC:

There is also the fact that Los Alamos may be regarded as a rather biased institution. Some people may feel that we are interested parties. I certainly am only too well aware of a resistance, particularly in the Press, to accept pronouncements and conclusions coming out of the AEC itself. Strangely enough, they were quite willing to accept the conclusions of the National Academy of Sciences, completely forgetting that the subcommittees were in very large measure composed of AEC or AEC contractor representatives. They were the same guys wearing different hats.[142]

In the late 1950s the AEC itself came to question whether its data-gathering efforts were serving the purposes of scientific knowledge and public understanding, as had been hoped. Sunshine, internal memos recorded, lacked coordination and clear goals, and the confusion of roles cost credibility. "[T]he primary reason," wrote Hal Hollister, an AEC fallout expert, "the AEC is now in the soup with respect to Congress, the public, and the fallout problem is that all three of these relationships with the public (reporting data scientifically, getting it across to the public, and telling official interpretations of it) have been inextricably mixed up. This has continued to be true after the hearings, and the future promises more of the same."[143]

In 1959 President Dwight Eisenhower acted to take responsibility for radiation safety away from the AEC, placing it in the hands of a new Federal Radiation Council, chaired by the secretary of the Department of Health, Education, and Welfare.

By the mid-1960s the possibility that data gathering could only get the AEC into more trouble became an incentive to "not study at all." In 1965 Dwight Ink, general manager of the AEC, advised against conducting proposed studies on the detrimental effects of nuclear testing partly because of liability concerns: "[P]erformance of the above U.S. Public Health Service studies will pose potential problems to the Commission. The problems are: (a) adverse public reaction; (b) law suits; and (c) jeopardizing the programs at the Nevada Test Site."[144]

In his DOE-sponsored history of the AEC and nuclear testing safety, Barton Hacker, laboratory historian at DOE's Lawrence Livermore Laboratory, concludes that while AEC officials did not doubt that testing could be done safely if precautions were taken, there was divergence about what to tell the public, and reassurance won out over information:

[T]he people in the field, those involved in the test program directly, tended to favor telling the public just what the risk was and stressing that whatever risk testing might impose was far outweighed by the national importance of the test program. Openness, they argued, would retain public trust and ensure continued testing.[145]

However:

AEC officials in general, headquarters staff members in particular, mostly preferred to reassure rather than inform. Convinced that trying to explain risks so small would simply confuse people and might cause panic, they feared jeopardizing the testing vital to American security. Their policy prevailed. A formal public relations plan became as much a part of every test as the technical operations plan. Carefully crafted press releases never to my knowledge lied, though they sometimes erred. Yet, by the same token, they rarely if ever revealed all. Choices about which truths to tell, which to omit, could routinely veil the larger implications of a situation. . . .[146]

"Reluctance to acknowledge any risk, the policy that mainly prevailed in the 1950s," Hacker concluded, "undercut the AEC's credibility when the public learned from other sources that fallout might be hazardous."[147]

THE RECORDS OF OUR PAST

The story that we have told in this report could not have been told if the government did not keep records that could be retrieved. By the same token, the story is often disturbingly fragmentary; seemingly contradictory statements of principle or policy abound, and the trail from policy to practice is often hard to discern. The story is complex, but it is also hard to reconstruct because, notwithstanding considerable search efforts of the Human Radiation Interagency Working Group, many documents appear to have been long since lost or destroyed. In each case, we emphasize, any loss or destruction took place prior—most often many years prior—to the Advisory Committee's creation. Federal records management law provides for the routine destruction of older records, and in the great majority of cases it should be assumed that loss or destruction was a function of normal record-keeping practices. At the same time, however, the records that recorded the destruction of documents, including secret documents, have themselves often been lost or destroyed. Thus, the circumstances of destruction (and indeed, whether documents were destroyed or simply lost) is often hard to ascertain.

As Shields Warren and Alan Gregg suggested, where human research is connected to secret programs, the public has a special interest in the adequacy of record keeping needed to ensure the integrity of experimental activity. Regardless of

whether documents that cannot now be retrieved contained further secrets, they would have provided more confidence in our understanding of the rules and practices that governed the boundary between openness and secrecy. In too many cases, however, documents are no longer available.[148] A number of such examples follow.[149]

The CIA, virtually all of whose records are classified, reported that it was unable to retrieve any records of its participation in the midcentury DOD panels that met in secret to discuss, among other things, human experiments. In addition, the CIA's classified records of its secret MKULTRA human experimentation program were, as reported when the program became a public scandal in the 1970s, substantially destroyed at the direction of then-Director of Central Intelligence Richard Helms in 1973. In 1995 the CIA concluded, following a search for remaining records and interviews of those involved, that it did not likely conduct or sponsor human radiation experiments as part of MKULTRA. The Advisory Committee, which was necessarily limited in its abilities to directly review CIA files, did not find evidence to the contrary. As a CIA report on its own inquiry (which was declassified at the Advisory Committee's request) concluded, the circumstances of the CIA's MKULTRA record keeping will likely leave questions in the public's mind.[150]

The DOD provided many documents that shed light on the rules of secrecy. However, some important collections are incomplete, and other important collections (such as the records of the Naval Radiological Defense Laboratory, the Medical Division of the Defense Nuclear Agency, classified records of the Navy Bureau of Medicine relating to Operations Crossroads physical exams, and entire sections of records of the Army surgeon general) appear to have been substantially lost or destroyed.

The DOE could locate only fragments of the records of the Insurance and Declassification Branches, which reviewed human subject research for declassification. The entirety of the files of the AEC Intelligence Division, which likely contained information on intentional releases, research performed by the AEC for other agencies, and secrecy policy and practices, was subject to "purge" in the 1970s, and as late as 1989.[151] Many other significant collections were retrieved. However, there were often gaps, including, for example, multiyear gaps in the Division of Biology and Medicine fallout collection, gaps in the transcripts from the meetings of the Advisory Committee on Biology and Medicine, and limited collections related to the work of the Isotope Distribution Division's Human Use Subcommittee.

The DHHS was able to locate sufficient information to confirm that it conducted classified research on behalf of the military mission, but could not locate information needed to determine the nature and extent of this research.[152] The classified information it once maintained has been substantially destroyed (or lost).

The VA, similarly, was able to provide fragments of information that show that "confidential" files were kept in anticipation of potential radiation liability claims. However, neither the VA, nor the DOE and DOD (who evidently were parties to this secret record keeping), have been able to determine exactly what secret records were kept and what rules governed their collection and availability.[153] VA publications did contain lists of several thousand (nonclassified) human experiments conducted at VA facilities; however, the information was quite fragmentary, and further information could not be readily retrieved (if it still exists) on the vast majority of these experiments.

Thus, in looking for answers to questions about the secrecy of data on human experiments and intentional releases, we find record-keeping practices that leave questions about both what secrets were kept and what rules governed the keeping of secrets.

CONCLUSION

Openness—the public sharing of all information necessary to govern—has long been an ideal in American democracy and politics. Scientists, also, have traditionally embraced openness as the surest guarantee of continued progress. However, the ideal of openness has often competed of necessity with some measure of government-imposed secrecy. This has been particularly the case in a time of national emergency, such as war. But secrecy existed even at the roots of our democracy: the Constitutional Convention itself was conducted out of the public eye.

In the early part of this century, President Woodrow Wilson called for "open covenants openly arrived at," seeking to shed light upon an area—international diplomacy—traditionally shrouded in secrecy. In the half century since the end of World War II, with the growing importance of science and technology in our lives, the proper place of secrecy at the intersection of government, private enterprise, and research has emerged as a question of central and continuing importance to society.

We have focused upon only one of many Cold War settings where secrecy was often a routine consideration. But human radiation experiments and intentional releases of radiation were often closely related to, if not directly a part of, some of the most closely held of secrets; including, most notably, nuclear weapons design and testing. The episodes we reviewed reveal the tensions underlying the necessarily delicate balance between openness and secrecy.

We found that from the onset, leading government biomedical officials and advisers were aware of the costs of secrecy and proclaimed the need to limit its reach. In one important respect, these officials and researchers lived up to their publicly stated ideals. Since about midcentury, there have been very few instances in which the very existence of human subject radiation research has been officially classified. Nonetheless, we also found that practices often fell short of the ideals that were publicly expressed.

We found that decision making related to the secrecy of human subject research considered not only national security, but also other criteria. At its birth in 1947, the AEC determined to keep Manhattan Project experiments secret on the basis of concern for "adverse effects on public opinion" and possible "legal suits," even where national security itself was not expressly invoked. More generally, we also found that decisions to keep information secret were often accompanied by a concern that the public might not understand the information and thus overreact or that the public would understand the information but that its immediate reaction could undermine support for programs deemed essential by policymakers.

Significantly, we found that AEC and DOD discussions of Cold War human research policy were themselves conducted outside the realm of public debate. For example, the 1947 AEC declarations of requirements for human research involving patients were evidently given minimal distribution within the AEC research community itself. Recently retrieved documents now show that in 1947 the requirement of "informed consent" was itself invoked in secret by the AEC's Medical Board of Review, in response to the request for criteria that had to be met when secret experiments could be declassified, and evidently thereafter relied on to keep some experiments secret. Similarly, the discussions underlying the 1953 memorandum by Secretary of Defense Charles Wilson, concerning human experiments done under DOD auspices, were themselves secret, as, of course, was the Wilson memorandum itself.

Even if there is clear and public consensus on what constitutes "national security," its application to the classification of particular information may be a matter of disagreement. In addition, in some cases the boundary between protecting the nation's security and simply avoiding the potential of adverse public reaction may not be so clear. For example, in an intense national crisis, the release of information that might jeopardize successful resolution of the crisis should properly be proscribed. But it is also clear that the assertion that programs will be jeopardized because of embarrassment or potential legal liability (or, worse, because of a lack of confidence in the American public's ability to understand) can be used to limit disclosure of precisely those matters that most affect us all and that would most benefit from informed public discussion.

If the boundary between openness and secrecy is inherently ambiguous, the public trust in those who define it on a daily basis requires a clear explanation of the principles that they will follow. However, we found that some of the basic principles and rules by which this boundary was defined were themselves kept secret from the public. AEC officials, in consultation with biomedical advisers internally invoked public relations and legal liability as bases for keeping secrets, while publicly declaring that secrecy should be limited to national security requirements. As a corollary, we found that where formal criteria for classification were not established, secrecy was nonetheless achieved by other, informal means. Thus, at midcentury, participants in

discussions of defense-related biomedical research were told that while the information in question was not itself classified, it should nonetheless be kept from the press and public.

Since 1951, presidential executive orders have limited the use of classification stamps to matters of national security. Nonetheless, the keeping of secrets with reference to ill-defined reasons such as public relations, continued. Indeed, as recently as the early 1970s, adverse public relations was reportedly invoked as a reason for keeping secret details of the plutonium injections of the 1940s. In some cases, as we look back, the public relations rationale for secrecy appears to be more clearly documented than any national security rationale. For example, we found that in the early 1950s public relations was an express consideration in keeping secrets related to fallout-related human tissue sampling; but we found it more difficult to locate contemporaneous documentation of national security rationales, and in 1995, surviving participants found it hard to reconstruct one as well.

We also found instances where the keeping of secrets was accompanied by deception. The shades of deception ranged from outright denials by the AEC that it engaged in human experimentation, to the use of cover stories in the collection of human tissue, to incomplete information deliberately given participants in government-sponsored biomedical research. In some such cases, such as the use of a cover story in collecting the bones of stillborn infants, those involved rationalized that since partial truths were being told, active deceit was not involved. In others, a rationalization for deception was a desire not to alarm exposed workers or the public. In yet others, such as the AEC's denial that it sponsored human experiments (when its Isotopes Division publicly advertised the success of human subject research) the rationale is hard to discern in retrospect.

In many cases, of course, some degree of secrecy was merited.[154] We found that where secrecy was initially justified by reasons relating to national security, the classifying authority often gave too little attention to the likelihood that there would come a time when such information was no longer sensitive. Immediately prior to the AEC's creation, the Tolman Committee pointed out that in the long run (which that Committee identified in terms of years, not decades) the nation's interest lies in the disclosure of information that needs to be kept secret over the shorter term. Yet, the practical reality was that once information was "born secret" it often simply remained that way.

Similarly, we found that where a national security rationale for secrecy did exist, adequate attention was often not paid to ensuring that sufficient records would be created and maintained so that all affected individuals (and the public at large) could later know the possible health and safety consequences. As a result, "downwinders," as well as knowing participants in nuclear tests, today wonder whether the information given them represents the full story of these events. (Indeed, as we reported in chapter 11, the number of once-secret intentional releases that are publicly known burgeoned from the thirteen reported by the General Accounting Office in late 1993 to the far greater number reported by the DOD and DOE following their more recent search.) When, as we reported in chapter 10, there is evidence that government officials contemplated, and may have kept, secret records to evaluate potential claims from service personnel exposed to government-sponsored radiation risk, the public has a right to expect that the government can readily and unambiguously account for any record keeping that may have taken place. Its inability to do so is very troubling.

Finally, we found that confusion, misunderstanding, and controversy still characterize public understanding of issues at the core of the Committee's work; for example, what is the nature of the risk from radiation? And to what extent can government statements about human radiation experiments and intentional releases be trusted? It is important to reflect on the ways in which this state of affairs may, in part, be a consequence of past secret keeping.

In testimony before the Advisory Committee, numerous witnesses expressed a common feeling—that the government did not give adequate weight to the interests of an informed public. Secrets, some said, were kept from the American public, not the enemy. Even where information may have been rightly classified in the first instance, many pointed out that there is no longer any reason for the absence of documents that provide a clear and full accounting to all those who were put at risk. There are too many cases where we can give no comforting answer to these angry voices.

However, by paying heed to these voices and by trying to understand the past they point to, we may more readily find our path into the future. Perhaps the first step in this direction is a simple recognition that the proper boundary between openness and secrecy will not be immediately obvious in all cases; many cases will not only require judgment, but also the will to avoid the temptation to keep secrets because the benefits of secrecy may be immediate, while the costs are longer term.

A second step is to understand that where secrecy is truly merited, and citizens are put at risk, there must also be precautions to ensure that a timely public accounting will be possible when the information need no longer be kept secret.

As the Cold War recedes further into history, the issues of secrecy and openness it posed will undoubtedly continue to present themselves, although often in new settings. Our review of the past provides the basis for some specific recommendations about the future, but it also points to a more fundamental understanding of the wisdom of those leaders of the day who identified the long-term costs of secrecy and called for policies to minimize them. The shortcomings of past policies and actions confirm that even when principles are articulated by well-intentioned officials, the translation of principles into practice is not automatic and warrants careful attention by the public. At the same time, the present-day legacy of distrust confirms that too much secrecy in the short term will, in the long run, erode the public's trust in government and the government's ability to keep the secrets that must be kept.

NOTES

1. For example, FOIA exempts from disclosure draft documents and other records reflecting deliberations made before a decision.

2. For a discussion of the definition of secrecy see Sissela Bok, *Secrets: On the Ethics of Concealment and Revelation* (New York: Vintage Books, 1989), 5–9.

3. As discussed in the text that follows, the executive orders on national security information are perhaps the most basic source on the definition of national security requirements in the context of security classification. The currently effective order, Executive Order 12958, provides that "'national security' means the national defense or foreign relations of the United States." The precise contours of this definition are, as discussed in what follows, a perennial subject of attention. Executive Order 12958 further provides that "a person may have access to classified information provided that: (1) a favorable determination of eligibility for access has been made by an agency head or the agency head's designee [i.e., a security clearance]; (2) the person has signed an approved nondisclosure agreement; and (3) the person has a need-to-know the information." Executive Order 12958, sec. 4.2(a), 60 Fed. Reg. 19836 (April 20 1995). The nondisclosure agreement referred to will typically refer, in turn, to various statutes that make it a crime to disclose classified information without authorization—for example, the Espionage Act and the Atomic Energy Act, as also discussed in the text.

4. *Espionage Act,* 18 U.S.C. 793–794.

5. *Military Installation and Equipment Protection Act,* 18 U.S.C. 795–797.

6. President, executive order, "Defining Certain Vital Military and Naval Installations and Equipment, Executive Order 8381," *Federal Register 5,* no. 59 (26 March 1940).

7. Army Classification Guide, AR 320–5 (1936).

8. Ibid.

9. Manhattan Engineer District, 26 November 1945 ("Security Manual") (ACHRE No. DOE-050595–B), 21.

10. Noted by Stafford Warren, who came from the University of Rochester to be the medical director of the Manhattan Project. Stafford L. Warren, interviewed by Adelaide Tusler (UCLA Oral History Program), transcript of audio recording, 21 July 1966 (ACHRE No. UCLA-101794–A), 574–575.

11. Committee on Declassification to Major General L. R. Groves, 17 November 1945 ("Report of Committee on Declassification [Tolman Committee report]") (ACHRE No. DOE-120594–D). The committee was chaired by Dr. H. C. Tolman, a professor of physical chemistry and mathematical physics at the California Institute of Technology. Tolman had advised the government on the creation of the wartime Office of Scientific Research and Development, advised Manhattan Project director General Groves, and served on the Target Committee that made decisions on the dropping of the atomic bombs on Japan. Ronald L. Kathren et al., eds., *The Plutonium Story: The Journals of Professor Glenn T. Seaborg: 1939–1946* (Columbus: Battelle Press, 1994), 818–819.

12. Committee on Declassification to Groves, 17 November 1945, 2.

13. Ibid., 3.

14. Ibid.

15. Ibid., 4–5.

16. Ibid., 13–14.

17. Manhattan Engineer District, 30 March 1946 ("Declassification Guide for Responsible Reviewers") (ACHRE No. DOE-050495–B); Atomic Energy Commission, 1 January 1948 ("Declassification Guide") (ACHRE No. DOE-050495–B); Atomic Energy Commission, 15 November 1950 ("Declassification Guide For Responsible Reviewers") (ACHRE No. DOE-052595–B).

18. Defense Department entities can also create and use RD. However, presidential order has been the primary source of DOD classification authority.

19. *Atomic Energy Act* §10(b)(1) (1946) (emphasis added). In 1954 the Atomic Energy Act was amended, and the requirement to declassify RD was strengthened: "The Commission *shall* from time to time determine the data . . . which can be published without undue risk of the common defense and security . . . "42 U.S.C. § 2162. In addition, a new category of classified information, later termed *Formerly Restricted Data,* was created to apply to information concerning the military use of atomic weapons that was no longer RD. 42 U.S.C.§ 2164.

20. For treatments of this complex story, see Gregg Herken, *The Winning Weapon: The Atomic Bomb in the Cold War 1945–1950* (New York: Knopf, 1980); James Hershberg, *James B. Conant: Harvard to Hiroshima and the Making of the Nuclear Age* (New York: Knopf, 1993); Richard G. Hewlett and Oscar E. Anderson, Jr., *The New World: A History of the United States Atomic Energy Commission, Volume I 1939–1946* (Berkeley: University of California Press, 1990); George T. Mazuzan and J. Samuel Walker, *Controlling the Atom: The Beginnings of Nuclear Regulation, 1946–1962* (Berkeley: University of California Press, 1984); and Richard Rhodes, *Dark Sun: The Making of the Hydrogen Bomb* (New York: Simon and Schuster, 1995).

21. Major Richard T. Batson, Declassification Officer, Research Division, to Dr. A. H. Dowdy, 21 March 1947 ("Reclassification of Documents") (ACHRE No. DOE-101394–A), 1. The two reports were chapters in the volume *Pharmacology and Toxicology of Uranium Compounds,* which was to be part of the public history of Manhattan Project research.

22. Colonel O. G. Haywood to H. A. Fidler, 17 April 1947 ("Medical Experiments on Humans") (ACHRE No. DOE-051094–A-62), 1. In May, Fidler noted that for purposes of classification, "the Declassification Section was giving a very strict interpretation to the above term *[human experiment]* and included routine checks performed on plants' personnel who may or may not have been exposed to excess radiation as well as known accidental exposures where plutonium, for instance, was introduced into the body." Fidler noted, however, that "Colonel Cooney stated that only those experiments where actual materials were intentionally introduced into the human system need to be regarded as secret." H. A. Fidler, 14 May 1947

("Memo to the Files on Policy on Medical Reports") (ACHRE No. IND-071395-A), 1.

23. H. A. Fidler to Carroll L. Wilson, AEC General Manager, 26 May 1947 ("Declassification Policy") (ACHRE No. DOE-121294-C), 7; Carroll L. Wilson to H. A. Fidler, 23 June 1947 ("Declassification Policy") (ACHRE No. DOE-121294-C), 1.

24. Carroll Wilson, "Security Regulations in the Field of Nuclear Research," *Bulletin of the Atomic Scientists* 3 (1947): 322. Wilson's 27 August 1947 letter also, however, stated that information policies were under review and that "since the Commission took over the facilities and operations of the Manhattan District on January 1, 1947, the information program has followed the policies inaugurated by the War Department."

25. Atomic Energy Commission, 20 June 1947 ("Report of the Board of Review") (ACHRE No. DOE-051094-A-191), 11.

26. Carroll Wilson, AEC General Manager, to Dr. Robert Stone, 5 June 1947 ("Your letter of May 7, 1947 . . .) (ACHRE No. DOE-061395-A), 1.

27. Carroll Wilson, AEC General Manager, to Dr. Robert Stone, 12 August 1947 ("Declassification of Biological and Medical Papers") (ACHRE No. DOE-061395-A), 1.

28. John A. Derry, AEC Acting General Manager, to Walter J. Williams, Manager, Field Operations at Oak Ridge, 9 August 1947 ("Establishing Criteria for Proper Classification of Information") (ACHRE No. DOE-020795-B).

29. Further items included the following:

(1) All medical records, reports and correspondence which embodies or refers to other technical information classified secret or higher. . . .

(3) All medico-legal and insurance statistics which refer directly to process hazards.

(4) Claims, allegations or reports of injury on 'investigation prohibited' cases where the material or process involved is considered to be classified secret.

(5) All medical reports, references and correspondence dealing with certain special hazard problems, as for example, the medical aspects of criticality accidents.

Albert H. Holland, AEC Acting Medical Adviser, to the Chairman, Classification Board, 12 September 1947 ("Proposed Classification for Unique Operational and Production Hazards, Including Medical Classification") (ACHRE No. DOE-101394-A), 1.

30. Ibid., 2. Other examples of "information or matter which should be graded confidential" included these:

(1) All documents, claims, allegations and medical reports on injury on 'investigation prohibited' cases, including reports of the Advisory Board on Occupational Disease Claims.

(2) All 'programmatic' medical research.

(3) All records of exposure to classified substances.

(4) All documents and correspondence which state, refer to or give information of known medical or public health hazards.

Ibid., 1.

31. J. C. Franklin, Manager, Oak Ridge Operations, to Carroll L. Wilson, AEC General Manager, 26 September 1947 ("Medical Policy") (ACHRE No. DOE-113094-B-3), 2–3. We note that the documentation available does not permit a definitive understanding of the relationship between the rationale employed for keeping data on human radiation experimentation (and other human radiation data gathering) secret and the particular level of security classification to which the data were assigned. As quoted in the text, documents talk in terms of the need to keep secret information that, while not endangering national security, could nonetheless be damaging to the government. As also quoted in the text, the category of Confidential provided for classification on this basis. However, as indicated in the text, documents invoking the "adverse effect on public opinion" language also call for the classification of human radiation experimentation data as Secret, a higher level of classification. Thus, while it is clear that the rules provided for classification on bases other than national security, and it also seems clear that those calling for continued keeping of radiation experiments secret saw a need for secret keeping independent of national security impact, there also may have coexisted the view that an "adverse effect on public opinion" could equate to endangering national security.

32. Ibid., 3.

33. L. F. Spalding, Chief, Insurance Claims Section, to C. L. Marshall, Deputy Declassification Officer, Technical Information Branch, 11 June 1947 ("Document by Cheka and Morgan") (ACHRE No. DOE-070795-C), 1. See also L. F. Spalding, Chief, Insurance Claims Section, to C. L. Marshall, Deputy Declassification Officer, Technical Information Branch, 11 June 1947 ("Document by Cheka") (ACHRE No. DOE-070795-C), 1.

34. Memorandum to Advisory Board on Medicine and Biology, 8 October 1947 ("Medical Policy") (ACHRE No. DOE-051094-A-419), 8.

35. Holland had proposed that the definition of Unclassified include

All medical and biological documents, reports and research not directly relating to experimental human administration, process hazards, contamination hazards or public health hazards, and which will not result in mass hysteria on the part of employees or the public, or in idle speculation or [illegible] adverse claims against the Atomic Energy Commission or its contractors.

Holland to the Chairman, Classification Board, 12 September 1947, 2.

36. Advisory Committee for Biology and Medicine, 11 October 1947 ("Draft Minutes, Advisory Committee for Biology and Medicine: Second Meeting") (ACHRE No. DOE-072694-A), 10. The agenda for the 11 October 1947, meeting of the Advisory Committee on Biology and Medicine also contained "publication of scientific papers" and "secrecy of work in the field of biological and medical research." Carroll L. Wilson, AEC General Manager, to Commissioners and Division Heads, 9 October 1947 ("Meeting of the Advisory Committee for Biology and Medicine") (ACHRE No. DOE-072694-A), 2. The minutes show the topics were discussed; however, there is no specific reference to human experiments data in these discussions.

Regarding restrictions on the publication of scientific papers [General Manager Wilson] expressed the hope that these restrictions [which were not identified] would diminish in time. He pointed out that information on the physical science mentioned in medical and biological papers frequently delays classification.

"Draft Minutes," 11 October 1947, 5–6. In addition, Chairman Gregg

read a memorandum on the publication of scientific papers prepared jointly by Dr. A. F. Thompson, Chief Technical Information Branch, and H. A. Fidler, Chief, Declassification Branch. The memorandum explained the present status of declassification . . . and stated that papers are not declassified when they include information on nuclear constraints for the heavy elements or reference to classified technological papers.

Ibid., 6–7.
37. "Draft Minutes," 11 October 1947, 11.
38. Ibid., 11.
39. Carroll L. Wilson, AEC General Manager, to Robert S. Stone, University of California Medical School, 5 November 1947 ("Your letter of September 18 regarding the declassification of biological and medical papers was read at the October 11 meeting of the Advisory Committee on Biology and Medicine . . .") (ACHRE No. DOE-061395-A), 1. Carroll L. Wilson, AEC General Manager, to Alan Gregg, Director for Medical Sciences, Rockefeller Foundation, 5 November 1947 ("I want to thank you for your letter of October 14 concerning the questions raised by Dr. Stone in his letter to me of October 18 . . .") (ACHRE No. DOE-061395-A), 1.
40. Dr. Harold Hodge, Chief Pharmacologist, University of Rochester, to Brewer F. Boardman, Chief, Technical Information Division, AEC Field Operations, 12 February 1948 ("Thank you for your letter of February 4th . . .") (ACHRE No. DOE-113094-B-4), 1. Following a section-by-section review of the chapter Hodge declared, "I wish to submit the argument that none of this material is human experimentation unless you would class measuring a man's height or recording his weight as human experimentation."

41. Albert H. Holland, Medical Adviser, to Dr. Hoyland D. Young, Director, Information Division, Argonne National Laboratory, 15 March 1948 ("In accordance with your recent request, the following documents were reviewed for reconsideration of their classification . . .") (ACHRE No. DOE-120894-E-4), 1.
42. Shields Warren, Director, AEC Division of Biology and Medicine, to Albert H. Holland, AEC Medical Adviser, 19 August 1948 ("Review of Document") (ACHRE No. DOE-101494-B), 1.
43. Albert H. Holland, AEC Medical Adviser, to Shields Warren, Director, AEC Division of Biology and Medicine, 9 August 1948 ("Review of Document") (ACHRE No. DOE-051094-A), 1.
44. Anthony C. Vallado, Deputy Declassification Officer, Declassification Branch, to Clyde Wilson, Insurance Branch, 8 December 1948 ("Review of Document by Knowlton") (ACHRE No. DOE-120894-E-32), 1.
45. Clyde E. Wilson, Chief, Insurance Branch, to Anthony C. Vallado, Deputy Declassification Officer, Declassification Branch, 20 December 1948 ("Review of Document by Knowlton") (ACHRE No. DOE-120894-E-32), 1.
46. Albert H. Holland, AEC Acting Medical Adviser, to C. L. Marshall, Deputy Declassification Officer, Technical Information Branch, 23 October 1947 ("Declassification of Document") (ACHRE No. DOE-113094-B-4), 1.
47. In February 1948 the Insurance Branch opined that although a report ("Biochemical Studies Relating to the Effects of Radiation and Metals") "might arouse some claim consciousness on the part of former employees we are unable to predict that the Commission's interests would be unjustifiably prejudiced by its publication." Nonetheless, if latent disabilities resulted from the exposures reported, the public relations section would be involved. The Insurance Department, noting that it was conferring with Dr. Holland on "claims similar in nature to some of the exposures" discussed in the report, urged that he be called in. L. F. Spalding, Insurance Branch, to Charles A. Keller, Declassification Officer, Declassification Branch, 5 February 1948 ("Review of Document") (ACHRE No. DOE-113094-B), 1.
48. Documents available to the Committee show that medical research reports were reviewed by Public Relations and Insurance Branch officials prior to declassification at least as late as April 1949. See 27 April 1949 letter from Anthony C. Vallado, Deputy Declassification Officer, Declassification Branch, to Warren C. Johnson, University of Chicago ("Transmittal of Fink Survey Volume ['Biological Studies with Polonium, Plutonium, and Radium'] for Final Declassification Review") (ACHRE No. DOE-032995-A).
49. Atomic Energy Commission, 2 May 1949 ("Policy on Control of Information") (ACHRE No. IND-071395-B), 3.

50. President, executive order, "Prescribing Regulations Establishing Minimum Standards for the Classification, Transmission, and Handling, by Departments and Agencies of the Executive Branch, of Official Information Which Requires Safeguarding in the Interest of the Security of the United States, Executive Order 10290," 3 C.F.R. (1949–1953 Compilation). The executive order provided that:

[i]nformation . . . shall not be classified under these regulations unless it requires protective safeguarding in the interest of the security of the United States. The use of any of the four security classification prescribed herein . . . shall be strictly limited to classified security info.

"Classified security information" was defined as "official information the safeguarding of which is necessary in the interest of national security." The order did, however, provide that it should not be construed "to replace, change, or otherwise be applicable with respect to any material or info protected against disclosure by statute." It would not have required the alteration of embarrassment- or public relations-based criteria if they were supported by an independent statutory basis. Thus, if statutes like the Espionage Act previously provided adequate basis for classification in the absence of national security endangerment, they would continue to do so.

51. President, executive order, "Classified National Security Information, Executive Order 12958, sec. 1.8 (a)-(b)," *Federal Register* 60, no. 76 (20 April 1995).

52. AEC Advisory Committee for Biology and Medicine, minutes of the twenty-third meeting, 8–9 September 1950 (ACHRE No. DOE-072694–A), 28.

53. Shields Warren, Director, AEC Division of Biology and Medicine, to Leslie M. Redman, "D" Division, Los Alamos Scientific Laboratory, 5 March 1951 ("Dr. Alberto F. Thompson, Chief, Technical Information Service, has asked me to reply to your letter . . . ") (ACHRE No. DOE-051094–A-603), 2. Warren's letter attributes Gregg's statement to a September 1948 meeting of the ACBM. While the statement is not reflected in the minutes, a statement by Gregg in a similar vein was contained in an October letter, as discussed in chapter 8. Alan Gregg, Chairman, AEC Advisory Committee for Biology and Medicine, to Robert Stone, University of California Medical School, 20 October 1948 ("The secrecy with which some of the work of the Atomic Energy Commission has to be conducted creates special conditions for the clinical aspects of its work . . .") (ACHRE No. UCLA-111094–A-24), 1.

54. Warren's view is reported in a 1952 letter from a PHS official. James G. Terrill, Acting Chief, Radiological Health Branch, Division of Engineering Resources, to Charles V. Kidd, Chief, Research Planning Branch, National Institutes of Health, 25 September 1952 ("At the September 8–12 meeting of the Panel at Los Alamos, several subjects were discussed that are of general interest to the Public Health Service . . .") (ACHRE No. HHS-092794–A), 1.

55. See Department of Defense, Research and Development Board, Committee on Medical Sciences, 23 May 1950 ("Transcript of Meeting Held on 23 May 1950 . . . The Pentagon, Washington, D.C.") (ACHRE No. DOD-080694–A). (The Advisory Committee's copy of this transcript was classified Confidential and bears a 1994 declassification stamp.) See also Department of Defense, Research and Development Board, Committee on Chemical Warfare, 10 November 1952 ("Transcript of the Fourteenth Meeting Held 10 November 1952 . . . The Pentagon") (ACHRE No. DOD-080694–A). (The Advisory Committee's copy of this transcript was classified Secret and bears a 1994 declassification stamp.)

56. The Committee's copies of the minutes of the first year of meetings of the ACBM bear a 1994 declassification stamp.

57. Committee on Medical Sciences to Chairman and Members, Joint Panel on the Medical Aspects of Atomic Warfare, 15 December 1952 ("Department of Defense Research Program Under the Technical Objective of AW-6") (ACHRE No. NARA-062094–A). The NIH and PHS had representatives who were associate members on the Joint Panel on the Medical Aspects of Atomic Warfare. See Joint Panel on the Medical Aspects of Atomic Warfare, Minutes of the Sixth Meeting held on 31 October–1 November 1950 (ACHRE No. DOD-072294–B), and Joint Panel on the Medical Aspects of Atomic Warfare, Minutes of the Seventh Meeting held on 25–26 January 1951 (ACHRE No. DOD-072294–B).

58. National Institutes of Health, 13 May 1952 ("Defense Activities of the National Institutes of Health [1950–1952]") (ACHRE No. HHS-071394–A), 30.

59. Since DOD and AEC biomedical human subject radiation research was rarely classified, it would seem most likely that the classified HHS research (except in cases, such as the Marshallese, where there was a direct connection to weapons tests) did not involve humans.

60. In a 7 May 1955 letter to AEC chairman Lewis Strauss, Gioacchino Failla, chairman of the Advisory Committee for Biology and Medicine, wrote that the AEC had reviewed the Division of Biology and Medicine's research program "and is pleased to find that less than 5% of the medical program has security classification." Failla to Strauss, 7 May 1955 ("The Advisory Committee for Biology and Medicine has reviewed again the program of research in the Division . . . .") (ACHRE No. DOE-082294–B), 1. In a May 1955 letter to AEC Chairman Strauss, the ACBM recommended that the AEC "continue to sponsor all research relative to the diagnosis and treatment of radiation injury in a wholly unclassified way

except those experiments directly related to weapons testing or development." Ibid. However, in Senate testimony two months earlier, the University of Chicago's George Leroy (who played a key role as an AEC adviser on bomb test research and who, as medical school dean, oversaw AEC-funded research) told Senator Hubert Humphrey that "there is a considerable amount of information which for one reason or another has not been disseminated to the medical profession and scientific profession." Subcommittee on Reorganization of the Committee on Government Operations, *Hearing Held Before Subcommittee on Reorganization of the Committee on Government Operations, S. J. Res. 21, Joint Resolution to Establish a Commission on Government Security, Statement of Dr. George V. LeRoy, M.D.,* 84th Cong., 1st Sess., 14 March 1955, 851.

61. NEPA Medical Advisory Panel, 22 July 1949 ("NEPA Medical Advisory Panel Subcommittee No. IX, Report No. NEPA 1110–IER-20") (ACHRE No. DOD-121494-A-2), 5. A subgroup was convened to assess what was known about the effects of whole-body exposure to radiation. In a 1951 letter to the Air Force's School of Aviation Medicine transmitting the conclusions, the Air Force's surgeon general explained that "[w]hile this information is not classified, it should not be given general publicity." Major General Harry G. Armstrong, Air Force Surgeon General, to Commandant, USAF School of Aviation Medicine, 24 January 1951 ("Data Relative to External Radiation from Radioactive Material") (ACHRE No. DOD-062194-B-14), 3.

62. Everett Evans, Director of the Laboratory for Surgical Research at Medical College of Virginia, to Doctors W. T. Sanger et al., 23 January 1951 ("I think each of you should be informed of a problem . . .") (ACHRE No. DOD-020995-A), 1. Evans added:

There is much about this experiment I do not like but we are doing it in a manner as humane as possible . . . we have simply had to make the choice between this type of study which I hope will bring relief to atomic bombing victims or simply wait for an atomic bomb attack. . . . This issue here is one of national security.

Ibid., 2. It should be noted that the focus of the MCV atomic bomb-related research appears to have been thermal burns, and not the effects of ionizing radiation.

63. Everett Idris Evans, Director, Laboratory for Surgical Research, Medical College of Virginia, to Colonel John R. Wood, Chairman, Army Medical Research and Development Board, 23 January 1951 ("You will find from the attached letter I am having my problems with the local press . . .") (ACHRE No. DOD-020995-A), 1.

64. John Wood, Chairman, Medical Research and Development Board, to Everett I. Evans, Director, Laboratory for Surgical Research, Medical College of Virginia, 25 January 1951 ("Reference your letter of 23 January 1951 . . .") (ACHRE No. DOD-020995-A), 1.

65. Ibid.

66. Ibid.

67. Ibid.

68. Everett Evans, Director, Laboratory for Surgical Research, Medical College of Virginia, to Major W. F. Smyth, Superintendent, Virginia State Penitentiary, 13 December 1951 ("We continue to enjoy all the help you and your staff are giving us . . .") (ACHRE No.VCU-012595-A-17), 1.

69. Major W. F. Smyth, Superintendent, Virginia State Penitentiary, to Everett Evans, Director, Laboratory for Surgical Research, Medical College of Virginia, 19 December 1951 ("I wish to thank you for your letter of December 13 . . .") (ACHRE No. VCU-012595-A-17), 1.

70. W. G. Lalor, Rear Admiral, U.S. Navy (Ret.), Secretary, Joint Chiefs of Staff, to Chief of Staff, U.S. Army et al., 3 September 1952 ("Security Measures on Chemical Warfare and Biological Warfare") (ACHRE No. NARA-012495-A), 2. In the memo to the service chiefs of staff, the Joint Chiefs decreed that "responsible agencies" should "[e]nsure, insofar as practicable, that all published articles stemming from the BW [biological warfare] or CW [chemical warfare] research and development programs are disassociated from anything which might connect them with U.S. military endeavor."

71. Office of the Director of Defense Research and Engineering, Thirtieth Joint Medical Research Conference, minutes of 8 January 1964 (ACHRE No. DOD-062994-A), 3.

72. R. Wasserman and C. Comar, *Annotated Bibliography of Strontium and Calcium Metabolism in Man and Animals* (Washington, D.C.: Agricultural Research Service, 1961), Publication no. 821, 1. The preface states:

Within recent years it has become necessary to understand the metabolism and movement of radioactive strontium in the biosphere. The behavior of strontium in man and animals is closely linked with that of calcium, and it is therefore necessary to consider the factors that govern the behavior of both elements. This annotated bibliography . . . should be useful to national defense workers who are doing research on the strontium-calcium relationship.

73. In some cases, however, the fallout-related purpose of research was publicly stated. See, for example, Robert P. Chandler and Samuel Wider, "Radionuclides in the Northwestern Alaska Food Chain, 1959–1961—A Review," *Radiological Health Data* (June 1963): 317–324.

74. John Bowers, Assistant to Director, Division of Biology and Medicine, to A. H. Gill, 18 February 1948 ("Your letter to David E. Lilienthal . . .") (ACHRE No. DOE-040395), 1.

75. Shelby Thompson, Chief, AEC Public Information Service, to Frank Starzel, Associated Press General Manager, 7 December 1950 ("We have noted in the November 29 *Baltimore Sun . . .* ") (ACHRE No. DOE-051094-A), 1.

76. John C. Bugher, Director, Division of Biology and Medicine, to Jesse Paul Malone, 2 April 1953 ("This is in reply to your letter of March 23rd . . . ") (ACHRE No. DOE-040395-A), 1. See also another April 1953 letter in which the AEC's Argonne Laboratory told a citizen, "We do not conduct experiments on human beings." Harvey M. Patt, Division of Biological and Medical Research, to Mr. Joseph Vodraska, New York City, 14 April 1953 ("Thank you kindly . . .") (ACHRE No. DOE-050195-B).

77. Shelby Thompson, Acting Director, Division of Information Services, to H. C. Baldwin, Information Officer, Chicago Operations Office, 21 August 1953 ("Information Guidance on Any Experimentation Involving Human Beings") (ACHRE No. DOE-040395-A), 1.

78. Louis Hempelmann, University of Rochester School of Medicine and Dentistry, to Charles Dunham, Director, AEC Division of Biology and Medicine, 2 June 1955 ("I did not have an opportunity to speak to Roy Albert in New York . . .") (ACHRE No. DOE-092694-A), 1.

79. Ibid.

80. Roy Albert, Assistant Chief of the Medical Branch of the Division of Biology and Medicine, to Louis Hempelmann, University of Rochester School of Medicine and Dentistry, 23 June 1955 ("Chuck Dunham passed along to me your letter containing suggestion for the Harshaw study . . .") (ACHRE No. DOE-092694-A), 1.

81. Ibid.

82. Ibid.

83. Committee to Consider the Feasibility and Conditions for a Preliminary Radiological Safety Shot for Operation "Windsquall" [later named Jangle], 21 May 1951 ("Notes on the Meeting . . . 21 and 22 May 1951") (ACHRE No. DOE-030195-A).

84. Barton C. Hacker, *Elements of Controversy; The Atomic Energy Commission and Radiation Safety in Nuclear Weapons Tests 1947–74* (Berkeley: University of California Press, 1994), 69.

85. Benjamin W. White, 1 August 1953 ("Desert Rock V: Reactions of Troop Participants and Forward Volunteer Officer Groups to Atomic Exercises") (ACHRE No. CORP-111694-A), 10.

86. Hacker, *Elements of Controversy,* 118. The pattern applied to animal experiments, as well as human data gathering. A recently declassified 1952 DOD history records that "because of anti-vivisection sentiment, release of such information would be detrimental to the testing program. Decision was made that such information fell into a sensitive, though non-classified, category and should not, therefore, be released to the public." Armed Forces Special Weapons Project, undated document ("First History of AFSWP 1947–1954, Volume 5–1952, Chapter 3–Headquarters") (ACHRE No. DOD-120794-A), 3.12.9. Other evidence indicates, however, that animal experiments were publicized.

87. H. K. Gilbert, Commander, USAF, to Commander Eugene Cronkite, USN, 8 March 1954 ("Letter of Instruction to CMR Eugene P. Cronkite, USN") (ACHRE No. DOE-013195-A), 1.

88. In Rochester, New York, an Eastman Kodak researcher, observing the fogging of a batch of Kodak film, traced the film materials to Iowa and deduced that radiation had been transported by air following an explosion. J. Newell Stannard, *Radioactivity and Health: A History* (Springfield, Va.: Office of Scientific and Technical Information, 1988), 884–886.

89. Roy B. Snapp, 14 February 1952 ("Project Gabriel: Note By the Secretary") (ACHRE No. DOE-033195-A), 1; Shields Warren, Director, AEC Division of Biology and Medicine, to General Advisory Committee, 13 February 1952 ("Project Gabriel") (ACHRE No. DOE-033195-A), 1.

90. AEC Division of Biology and Medicine, July 1954 ("Report on Project Gabriel") (ACHRE No. DOE-040395-A), 8.

91. AEC, 19 January 1954 ("Supplementary Information on Gabriel: Report by the Director of Biology and Medicine") (ACHRE No. DOE-013195-A), 1.

92. Stannard, *Radioactivity and Health,* 934–936, 1064–1080.

93. Rand, the quintessential "think tank," was created to advise the Air Force on emerging issues of policy and strategy.

94. Rand Corporation, *Worldwide Effects of Atomic Weapons: Project Sunshine: AECU-3488* (Oak Ridge, Tenn.: U.S. Atomic Energy Commission, Technical Information Service Extension, 1953), v–vii.

95. The report continues to explain in more detail: "The release in the world of several kilograms (kg) of strontium 90 within less than a decade has probably disseminated enough of the contaminant to provide amounts that are probably now detectable in samples of inert and biological materials throughout the world." Ibid., 7.

96. Ibid., 47.

97. Even the initial conference was kept secret. The attendees had been told: "The letter of invitation [to the conference] . . . should be classified . . . or returned to this office by registered mail." Ernest H. Plesset, Nuclear Energy Division, Rand Corporation, to Forrest Western, Biophysics Branch, Division of Biology and Medicine, 31 July 1953 ("We wish to thank you very much for your participation in the conference . . .") (ACHRE No. DOE-013195-A), 1.

98. AEC Division of Biology and Medicine, "Report on Project Gabriel," 2.

99. The six locales were (1) northern Utah or southwestern Idaho, (2) Kansas or Iowa, (3) New England (Boston), (4) South America, (5) England, and (6) Japan. From each would be drawn twelve

human tissue samples, four from each age group: 0–10 years, 10–20 years, over 20 years. Within each age group, two samples would be epiphysial end or rib and two would be teeth. Rand Corporation, *Worldwide Effects*, 51.

100. Robert A. Dudley, Biophysics Branch, Division of Biology and Medicine, to Gertrude Steel c/ o Willard Libby, Professor of Chemistry, University of Chicago, 16 October 1953 ("There are several matters which I would like to bring to the attention of you and Dr. Libby . . .") (ACHRE No. DOE-013195–A), 1.

101. Ibid.

102. Ibid., 2.

103. Robert A. Dudley, Biophysics Branch, Division of Biology and Medicine, to Shields Warren, Director, AEC Division of Biology and Medicine, 26 October 1953 ("We are now starting to make provision for a collection of foreign bones . . .") (ACHRE No. DOE-013195–A), 1.

104. Robert A. Dudley, Biophysics Branch, Division of Biology and Medicine, to Raymond A. Dudley, ABCRM, 10 November 1953 ("Thanks for the information in your letter of November 4 . . .") (ACHRE No. DOE-013195–A), 1.

105. DBM Director Bugher wrote to the Rockefeller Foundation, providing the cover story and asking for help in obtaining specimens "from Brazil, Colombia, Peru, and Chile or Bolivia." John C. Bugher, Director, Division of Biology and Medicine, to Andrew J. Warren, Director, Division of Medicine and Public Health, Rockefeller Foundation, 30 December 1953 ("Herewith I am enclosing a letter to you which might be used to explain the program of bone collections . . .") (ACHRE No. DOE-013195–A), 1.

106. Robert A. Dudley, Biophysics Branch, Division of Biology and Medicine, to James K. Scott, Atomic Energy Project, University of Rochester, 9 December 1953 ("This letter will explain in a little more detail . . .") (ACHRE No. DOE-013195–A), 1.

107. AEC Division of Biology and Medicine, "Report on Project Gabriel," July 1954, 13.

108. Ibid., 38.

109. Armed Forces Special Weapons Project, undated document ("AFSWP History, Latter Period: 1955–58") (ACHRE No. DOD-072594–B), 37.

110. Walter Reed Army Institute of Research, November 1955 ("Recovery of Radioactive Iodine and Strontium from Human Urine—Operation Teapot [WRAIR-IS-55 {AFSWP-893}]") (ACHRE No. DOD-092394–C), 1. The substance of the work was declassified in the late 1950s.

Research continued through the early 1960s, with use of the new body counter technologies (that permitted measurement of body radiation). U.S. Army Medical Research and Development Command to the Chief of DASA, 25 April 1963 ("Metabolism of Fission Products from Fallout") (ACHRE No. DOD-020195–A); U.S. Army Medical Research and Development Command to the Chief of DASA, 26 April 1963 ("Ionizing Radiation Combined with Trauma") (ACHRE DOD-020195–A).

111. Major General A. R. Luedecke, Chief, AFSWP, to the Surgeon General, Department of the Air Force, 16 December 1954 ("Fall-Out Studies") (ACHRE No. DOD-090994–C), 2.

Another contemporary instance of selective disclosure of fallout-related research, although not directly involving human beings, is discussed in a February 1955 letter written in the aftermath of the March 1954 Bravo bomb test. In this letter, Willard Libby, acting AEC chairman, writes to the chairman of the Congressional Joint Committee on Atomic Energy to report on a proposed marine radiobiological survey in the Pacific. Libby explained that it had been determined that the survey itself did not involve Restricted Data, although the results would involve Restricted Data since they could reveal weapons information. Libby further explained:

The classification "Secret" Defense Information has been assigned to the survey in order to avoid, if possible, an unwarranted recrudescence of fears in Japan of radioactive contamination of fish; and because knowledge by unfriendly interests of bomb-originated debris in the vicinity of Formosa might be used effectively to embarrass the United States. The fact of an oceanographic survey in the Pacific, however, is regarded as unclassified so long as purpose, content, and results are not revealed.

W. F. Libby, Acting Chairman, AEC, to Honorable Clinton P. Anderson, Chairman, Joint Committee on Atomic Energy, U.S. Congress, 16 February 1955 ("We would like to inform the Committee of plans . . .") (ACHRE No. NARA-070595–A), 1–2.

112. AEC Division of Biology and Medicine, 18 January 1955 ("Biophysics Conference") (ACHRE No. NARA-061395–B), 60.

113. Ibid., 8.

114. The researchers had come to recognize the difficult sampling problems; not only was the statistical representativeness of individual subjects a question, but the representativeness of particular body parts. Ibid., 12.

115. Ibid., 81. In 1995, Dr. Kulp recalled that the Columbia researchers followed legal processes to obtain cadavers. Some states required a special permit to dispose of human remains outside the state; others required specific approval from relatives for use of certain organs. At the time there were no restrictions in Houston on the use of unclaimed bodies for any scientific purpose. As a result of these policies, Dr. Kulp recalled, "The group supporting our project said they could obtain samples from virtually every body that came under their jurisdiction (not all of Houston!) that met the legal criteria." The reference to "poverty cases" was "meant to convey the fact that among the lower economic group in the city there are many unclaimed bodies." Dr. Kulp recalled: To the best of

my recollection all human bone samples collected for the Columbia University studies were done legally. They came from medical school cadavers, morgues or amputation material. In all cases the sources were either bodies that had been donated for medical use, unclaimed, or residue from necessary amputations. In all cases the material (with or without Project Sunshine) would have been ashed after examination or research use and then the ash discarded. Taking a portion of this ash for the determination of its calcium and strontium-90 concentration (or for any other trace element such as radium, selenium, arsenic etc.) can hardly be a moral, ethical or legal issue under these circumstances." Dr. J. Laurence Kulp, letter to Dan Guttman (ACHRE), 21 July 1995.

116. DBM, "Biophysics Conference," 185–187.

117. Ibid., 187.

118. Ibid., 12.

119. Geochemistry Laboratory, Lamont Observatory, Columbia University, 15 April 1956 ("Project Sunshine: Annual Report, Period March 31, 1955–April 1, 1956") (ACHRE No. DOE-082294–B), 68–70.

120. U.S. General Accounting Office, *Fact Sheet for Congressional Requestors: Information on DOE's # Human Tissue Analysis Work,* GAO/RCED95–109FS (Gaithersburg, Md.: GAO, 1995), 3.

121. Dr. Merril Eisenbud to Dan Guttman (ACHRE), 25 June 1995 ("I appreciate the opportunity you have given me . . . ") (ACHRE No. IND-070395–A).

122. Ibid. From 1948 to 1950 "Official Use Only" was a category in the formal classification system, but since then it has been used as an informal way of connoting that information should be protected even if it is not classified. While much information labeled Official Use Only never makes it to the public, the information does not have to be protected with formal security measures, readers do not have to be cleared in order to see it, and officials cannot be criminally prosecuted for disclosing it to members of the public.

123. Paul F. Foster, Special Assistant to the General Manager for Liaison, to the AEC General Manager, 9 November 1954 ("Discussion in Office of Secretary of Defense on 'Change in National Dispersion Policy'") (ACHRE No. DOE-033195–A), 2. This memo was circulated within the AEC; see W. B. McCool, Secretary, to Distribution, 16 November 1954 ("Atomic Energy Commission National Dispersion Policy: Note by the Secretary") (ACHRE No. DOE-033195–A), 1.

124. Richard Hewlett and Jack Holl, *Atoms for Peace and War: 1953–1961* (Berkeley and Los Angeles: University of California Press, 1989), 264.

125. On the scientific debate, see Carolyn Kopp, "The Origins of the American Scientific Debate over Fallout Hazards," *Social Studies of Science* 9 (1979): 403–422. For a public AEC presentation of fallout at the time, see Gordon M. Dunning, "The Effects of Nuclear Weapons Testing," *Scientific Monthly* 81, no. 6 (December 1955): 265–270. (Dunning was a health physicist with the AEC Division of Biology and Medicine.)

126. Robert A. Divine, *Blowing on the Wind: The Nuclear Test Ban Debate 1954–1960* (New York: Oxford University Press, 1978), 21.

127. Stannard, *Radioactivity and Health*, 982.

128. Harold M. Schmeck, Jr., "Study Discounts Risk In Nuclear Fall-Out," *New York Times*, 8 February 1957, 1. See J. Laurence Kulp, Walter R. Eckelmann, and Arthur R. Schulert, "Strontium-90 in Man," *Science* 125 (8 February 1957): 219–225. Following the initial declassification of Sunshine in the mid-1950s, the details of the Columbia work, including the identity of medical professionals who had provided assistance, became part of the public research report record.

129. Divine, *Blowing on the Wind,* 106.

130. For a contemporary reaction to the declassification see, Ralph E. Lapp, "Sunshine and Darkness," *Bulletin of the Atomic Scientists* 15 (January 1959): 27–29.

131. W. F. Libby, AEC Commissioner, to Herman M. Kalcker, National Institutes of Health, 10 June 1957 ("I think your idea of using children's milk teeth for strontium-90 measurement is a good one . . .") (ACHRE No. DOE-041295–D), 1.

132. The 1957 and 1959 hearings appear as Joint Committee on Atomic Energy, Special Subcommittee on Radiation, *The Nature of Radioactive Fallout and its Effect on Man,* 85th Cong., 1st Sess., 1957; Joint Committee on Atomic Energy, Special Subcommittee on Radiation, *Fallout from Nuclear Weapons Tests,* 86th Cong., 1st Sess., 5 May 1959.

133. The AEC concluded that radiation did not cause any sheep deaths and consequently did not compensate the Nevada ranchers. Some of the ranchers remained unconvinced by the AEC's explanation and sued the government. While the court ruled in favor of the government, controversy over the case continues to this day. Congress held hearings on the subject in 1979 and concluded that the AEC had suppressed evidence during the trial. House Committee on Interstate and Foreign Commerce, Subcommittee on Oversight and Investigations, *The 'Forgotten Guinea Pigs': A Report on Health Effects of Low-Level Radiation Sustained as a Result of the Nuclear Weapons Testing Program Conducted by the United States Government,* 98th Cong., 2d Sess., 1980, Committee Print 96–IFC 53, 15. In 1981, the judge who heard the first case ruled that the AEC fraudulently suppressed evidence in the trial. On appeal, however, this ruling was overturned. For further discussion of the sheep controversy see Philip L. Fradkin, *Fallout: An American Nuclear Tragedy* (Tucson: The University of Arizona Press, 1989), 147–165; Hacker, *Elements of Controversy,* 106–130.

134. Committee to Study Nevada Proving Grounds, 1 February 1954 ("Abstract of Report, Committee to Study Nevada Proving Grounds") (ACHRE No. DOE-040395–B), 1–2.
135. Ibid., 2.
136. Ibid., 46.
137. Ibid.
138. Ibid., 47.
139. Ibid.
140. Ibid., 50.
141. Ibid., 46.
142. Thomas L. Shipman, Los Alamos Laboratory Health Division Leader, to Charles Dunham, Director, Division of Biology and Medicine, 5 December 1956 (ACHRE No. DOE-020795–D-2), 3. On Sunshine, Shipman also wrote, "such a program obviously cannot be carried out with the complete lack of administration which has characterized past efforts." Ibid, 2.
143. Hal Hollister, Environmental Sciences Branch, Division of Biology and Medicine, to Dunham et al., 27 February 1958 ("Reporting Sunshine") (ACHRE No. DOE-012595–B), 2. Other participants in AEC-sponsored biomedical research had a different perspective on the fallout research.

In 1995, Dr. Kulp recalled that, from the perspective of the researchers at Columbia, the goals were clear—"defining the amount of SR90 in the stratosphere to its mechanism of descent in the ground to the movement through the food chain to man."

The work of Sunshine, he recalled, "provided the scientific basis for the International Treaty banning atmospheric tests." J. Laurence Kulp, letter to Dan Guttman (ACHRE), 21 July 1995.

Another perspective is contained in a 1973 letter to Dixie Lee Ray, the last AEC chairman before its separation into agencies responsible for regulating (the Nuclear Regulatory Commission) and promoting (the Energy Research and Development Administration) nuclear energy. Dr. William F. Neuman, director of the Atomic Energy Project at the University of Rochester, suggested that the difficulties leading to the agency's breakup were not limited to the conflict between its responsibilities to promote and regulate atomic energy. In addition, "the AEC (its Division of Biology and Medicine in particular) has been put in the position of providing a biological *justification* for some other agency's political decision." He explained to Chairman Ray:

Some years back, before the Test Ban, the military wished to test various weapon designs. The Eisenhower Administration concurred. Admiral Strauss was instructed to have the AEC provide the basis for public acceptance. This meant of course that the Division of Biology and Medicine was supposed to convince the public that fallout was good for them and environmental Sr-90 contamination was accordingly expressed in 'Sunshine Units' if you recall. This very nearly tore the Division apart and we were rescued from potential disaster only by the timely signing of the big power Test Ban Treaty.

Neuman had been a participant in the fallout debate and was the spokesperson for a panel that included Libby, Eisenbud, Dunham, Langham, and other AEC-connected experts at the 1959 congressional hearings. William F. Neuman, Wilson Professor and Director, Atomic Energy Project, University of Rochester, to Dixie Lee Ray, Chairman, Atomic Energy Commission, 12 November 1973 ("When you visited the Rochester Biomedical Research Project . . .") (ACHRE DOE-011895–B), 1.
144. Dwight Ink, AEC General Manager, to Seaborg, Chairman of the AEC, 9 September 1965, as quoted in House Committee on Interstate and Foreign Commerce, Subcommittee on Oversight and Investigations, *The 'Forgotten Guinea Pigs,'* 15.
145. Hacker, *Elements of Controversy*, 277.
146. Ibid., 278.
147. Ibid.
148. "The worst thing in the world," Harry Truman reportedly once said, "is when records are destroyed." Merle Miller, *Plain Speaking: An Oral Biography of Harry Truman* (New York: Berkley, 1974), 27.
149. The supplemental volumes to this report contain a detailed description of the record collections reviewed by the Advisory Committee.
150. As a 14 February 1995 CIA report concluded:

CIA has found no evidence that Agency offices sponsored human radiation experiments or deliberately exposed anyone to ionizing radiation for operational or experimental purposes. As noted above, at least two Agency-affiliated contractors [deletion] and [Dr.] Geschickter [a Georgetown University researcher]) may have conducted human radiation experiments while working on other matters for the CIA. Some CIA officers probably knew of human radiation tests by other U.S. government agencies, but apparently did not consider these tests particularly relevant to the Agency's mission.

Circumstantial evidence, however, may not suffice to overcome suspicions fueled by CIA's contacts with persons and programs involved in radiation experiments sponsored by other agencies. The fact that MKULTRA held the authority to conduct radiological experiments, combined with the Agency's destruction of the main MKULTRA files in 1973, has already prompted speculation about the Agency's "real" role. These heightened suspicions will not fade any time soon.

Michael Warner, CIA History Staff, 14 February 1995 ("The Central Intelligence Agency and Human Radiation Experiments: An Analysis of the Findings") (ACHRE No. CIA-061295–A), 14.
151. This conclusion was arrived at by DOE following an investigation conducted in response to the Committee's request for the documents. DOE Office of Human Radiation Experiments, 26 August 1994 ("Destruction of the U.S. Atomic Energy Commission Division of Intelligence Files") (ACHRE No. DOE-082994–A). The DOE interviewed DOE

employees who stated that they destroyed documents under direction from supervisors during this period. DOE reported that, shortly after the AEC Division of Intelligence was abolished in 1971, destruction of older file materials began. "This first file 'purge' continued until at least May 1974. Destruction was probably confined to documents dated prior to 1964." Following the DOE's creation in 1977, a second "purge" began, reportedly based on limited storage space, "destroying most surviving files." In 1988, DOE implemented rules requiring that documents classified at the Secret level be inventoried. "Many offices, however, destroyed Secret documents rather than having the burden of inventorying them. Surviving fragments of the AEC Division of Intelligence files may also have been destroyed during this third 'purge.'" Ibid., 2–3. The investigation reported that records that were kept of the documents that were destroyed had themselves been subsequently destroyed in the routine course of business.

152. There was no central location, within agencies, or among them, that routinely kept anything but the most fragmentary records of human experiments sponsored by the agencies. During the 1950s, a central "Bio-Science" information exchange was maintained. Government and nonfederal agencies (such as foundations) formerly registered descriptions of research projects performed or sponsored by the federal government with an office of the Smithsonian Institution variously called the Scientific Information Exchange or the Bio-Sciences Information Exchange. This group, established at the recommendation of and advised by the National Research Council, collected abstracts of research in progress reports for the period 1949–1979. The Department of Commerce's National Technical Information Service began a similar program two years later.

The abstracts submitted to the Exchange were collected in annual reports and are available on microfiche in the Smithsonian Institution Archives. Unfortunately, the indices to the reports are available only on magnetic tape in a 1970s mainframe format that Smithsonian technologists are currently unable to read. For that reason, Advisory Committee staff did not review the Exchange's records.

153. As noted in chapter 10, the VA concluded that a "confidential" division contemplated in relation to secret record keeping was not activated.

154. Although, as discussed in chapter 11, we must be careful to distinguish the need to keep secret information, for example, weapons design or a weapon's purpose, from the need to keep secret a weapons test that may put surrounding populations at risk.

Part III

Contemporary Projects

PART III
OVERVIEW

IN parts I and II of this report, the Advisory Committee attempted to come to terms with the past. We told the history of standards for conducting human subject research in part I, and the history of human radiation experiments through representative case studies in part II. Here in part III of our final report, we attempt to assess whether the current protections for human subjects are better than the prevailing standards and practices during the 1944 to 1974 period to help recommend what changes, if any, ought to be instituted in current policies governing human subject research.

The Advisory Committee's study of contemporary research ethics is three-pronged. It comprises a review of agency policies and oversight practices, a review of documents from recently funded research proposals (the Research Proposal Review Project, or RPRP) to examine the extent to which the rights and interests of the subjects of federally sponsored research appear to be protected, and the Subject Interview Study (SIS) in which the attitudes and beliefs of patients about medical research and their decisions and experiences regarding participation in research are examined. These projects together form the basis of the Advisory Committee's picture of the protections now afforded the subjects of biomedical research and, along with findings regarding radiation experiments during the 1944–1974 period, inform the forward-looking recommendations of the Advisory Committee, found in part IV.

Chapter 14 reviews the current regulatory structure for human subjects research conducted or supported by federal departments and agencies, a structure that has been in place since 1991. This "Common Rule" has its roots in the human subject protection regulations promulgated by the then-Department of Health, Education, and Welfare (DHEW) in 1974. The historical developments behind these regulations are described in chapter 3. Following a summary of the essential features of the Common Rule, chapter 14 discusses several subjects of particu-

lar relevance to the Advisory Committee's work, such as special review processes for ionizing radiation research, protection for human subjects in classified research, and audit procedures of institutions performing human subject research.

Chapter 15 describes the Research Proposal Review Project (RPRP), the Advisory Committee's examination of documents from research projects conducted at institutions throughout the country, including both radiation and nonradiation proposals. Documents utilized in the RPRP were those available to the local institutional review boards (IRBs) at the institutions where the research was conducted. The goals of the RPRP were to gain an understanding of the ethics of radiation research as compared with nonradiation research; how well research proposals address central ethical considerations such as risk, voluntariness, and subject selection; and whether informed consent procedures seem to be appropriate.

The RPRP reviewed documents prepared by investigators and institutions and submitted in IRB applications. This study was complemented by a nationwide effort to learn about research from the perspective of patients themselves, including those who were and were not research subjects. The Subject Interview Study (SIS), described in chapter 16, was conducted through interviews with nearly 1,900 patients throughout the country. The SIS aimed to learn the perspectives of former, current, and prospective research subjects by asking about their attitudes and beliefs regarding the endeavor of human subject research generally and their participation specifically.

The RPRP tried to understand the experience of human subjects research from the standpoint of the local oversight process, while the SIS tried to understand it from the standpoint of the participant. Although the two studies related to different research projects and different groups of patients and subjects, some common tensions in the human research experience emerge in both projects, and they are described in the "Discussion" section of part III. For example, it has long been recognized that the physician who engages in research with patient-subjects assumes two roles that could conflict: that of the caregiver and that of the researcher. The goals inherent in each role are different: direct benefit of the individual patient in the first case and the acquisition of general medical knowledge in the second case. The interviews with SIS participants suggest that at least some patient-subjects are not aware of this distinction or of the potential for conflict. In our review of documents in the RPRP we found that the written information provided to potential patient-subjects sometimes obscured, rather than highlighted, the differences between research and medical care and thus likely contributed to the potential for patients to confuse the two.

To help complete the picture of current human subject research and its regulation and oversight, the Committee also gathered limited information in two areas: (1) the federal system of human subject protection as viewed by those charged with implementing it at the local level, the chairs of IRBs; and (2) the particular review process applied to human subject research involving radiation as viewed by those charged with implementing it at the local level, the chairs of radiation safety committees.*

A letter was written to forty-one chairs of IRBs and forty chairs of radiation safety committees at institutions throughout the country, attempting to gain their perspectives on the current regulatory systems their committees seek to apply. Many of these letters are reproduced in a supplemental volume to this report. Most of the replies from IRB chairs indicated a general approval of the current system, but many also had useful observations and suggestions for improvement.

For example, several expressed concern about what they believed to be a disparity in the procedures of IRBs from one institution to another. The chairs of radiation safety committees, on the other hand, reported a nearly universal confidence in, and approval of, the review process for human subject research involving the use of radiation. The Committee's recommendations, in part IV of this report, address some of the concerns outlined in response to our queries.

As the Committee's work in part III shows, in the discussion section, contemporary human subject research does not suffer from the same shortcomings witnessed in the 1940s and 1950s, but poses different issues that need to be addressed. With a system of human subjects protections comes issues related to implementation and interpretation of rules and regulations. And with a change in the culture of medicine comes a change in the relationship between researchers and subjects. In the historical period of the Committee's review, we found that subjects needed protections to ensure their basic rights to consent to or to refuse participation in research. While this need to protect the right of consent continues, in the current period we found that subjects also need protections to ensure their interests are served in understanding the distinctions between research and therapy and the limits of the benefits research may offer. These findings and conclusions suggest the need for changes in an oversight system designed to address the concerns of an earlier time, and the Committee makes recommendations for such change in part IV of this report.

---

*The Committee also contacted a sample of institutions at which therapeutic human radiation research involving higher doses of radiation, and therefore imposing substantial risk, had recently been conducted according to reports in the medical literature. The Committee was interested in learning whether the research projects reported in these journal articles had been reviewed by an IRB, and if IRB review had depended upon whether the research was supported by federal funds. Information was received from only nine of the sixteen institutions requested. Although the projects about which we were inquiring were sometimes described as clinical investigations in the journal reports, these institutions did not always view them as satisfying the definition of human subject research and thus did not appear to require IRB review for these projects.

14

Current Federal Policies Governing Human Subjects Research

EACH year many thousands of people participate in biomedical and behavioral research projects conducted, sponsored, or regulated by federal agencies. The federal government invests roughly $3.5 billion annually in research that involves human subjects.[1] The Committee wanted to establish what the federal government currently does to protect the rights and interests of these subjects. The answers to this question all emanate from a seminal event in the history of human subjects research, the adoption of what is widely known as the "Common Rule."

A single, general set of regulatory provisions governing human subjects protections was adopted by sixteen federal departments and agencies[2] in 1991; the Common Rule specifies how research that involves human subjects is to be conducted and reviewed, including specific rules for obtaining informed consent.[3] The Common Rule was developed in response to recommendations made by the President's Commission for the Study of Ethical Problems in Medicine and Biomedical and Behavioral Research in 1981 calling for the adoption by all federal agencies of Department of Health and Human Services regulations then in effect for the protection of human subjects of research.[4] In mid-1982 the President's science adviser, the head of the Office of Science and Technology Policy (OSTP), appointed an ad hoc committee that included the federal departments and agencies engaged in research involving human subjects to address these recommendations.[5] Nine years later, the Common Rule was the result of this committee's efforts.

HISTORY OF THE COMMON RULE SINCE 1974[a]

1974   Title II of the National Research Act (P.L. 93-348)
Required codification of DHEW policy in regulations, imposed a moratorium on federally funded fetal research, and established requirements for IRB review of all human subjects research at any institution receiving DHEW funding.

DHEW regulations for the protection of human research subjects,

a. For a brief history of federal protections for human subjects prior to 1974, see chapter 3.

45 C.F.R. 46

Established IRB review procedures in accordance with Title II. Later in the same year DHEW published regulations providing additional protections for pregnant women and fetuses.

1974–1978 National Commission for the Protection of Human Subjects of Biomedical and Behavioral Research.

Issued reports and recommendations on fetal research; on research involving prisoners, psychosurgery, children, and the mentally infirm; on IRBs and informed consent; and, in The Belmont Report, *discussed criteria for distinguishing research from the practice of medicine and ethical principles underlying the protection of subjects.*

1978 Revised DHEW regulations governing protections for pregnant women, fetuses, in vitro fertilization (subpart B of 45 C.F.R. 46), and prisoners (subpart C) published

1980–1983 President's Commission for the Study of Ethical Problems in Medicine and Biomedical and Behavioral Research

Charged with, among other responsibilities, reviewing federal policies governing human subjects research and determining how well those policies were being carried out. Recommended that all federal agencies adopt the DHHS (a successor agency to DHEW) regulations for the protection of human subjects (1981).

1981 DHHS published a revision of 45 C.F.R. 46, responding to recommendations of the National Commission.

The revision set out in greater specificity IRB responsibilities and the procedures IRBs were to follow.

FDA regulations at 21 C.F.R. 50, governing informed consent procedures, and at 21 C.F.R. 56, governing IRBs, revised to correspond to DHHS regulations to the extent allowed by FDA's statute

1982 President's Science Adviser, Office of Science and Technology Policy (OSTP), appointed an interagency committee to develop a common federal policy for the protection of human research subjects

1983 DHHS regulation governing protections afforded children in research (subpart D of 45 C.F.R. 46) published

1986 Proposed common federal policy for the protection of human research subjects published

1991 Final common federal policy published on June 18, codified in the regulations of fifteen federal agencies and adopted by the CIA under executive order.

This common policy, known as "the Common Rule," is identical to the basic DHHS policy for the protection of research subjects, 45 C.F.R. 46, subpart A. Other sections of the DHHS regulation provide additional protections for pregnant women, fetuses, in vitro fertilization (subpart B), prisoners (subpart C), and children (subpart D). Several agencies have adopted these additional provisions as administrative guidelines. The FDA made conforming changes in its informed consent and IRB regulations.

The promulgation of the Common Rule was a significant achievement. The ability of the Common Rule to protect the rights and interests of human subjects is, however, at least partially dependent on how the departments and agencies to which the Common Rule applies implement and oversee its provisions. As a foundation for the Advisory Committee's recommendations concerning contemporary policies and practices regarding human subjects, we asked the sixteen federal agencies and departments that conduct human subjects research to provide us with information on the relevant policies and practices currently in place. In this brief descriptive overview, we focus on six agencies within the scope of the Advisory Committee's charter: the Department of Defense (DOD), Department of Energy (DOE), Department of Health and Human Services (DHHS), Department of Veterans Affairs (VA), National Aeronautics and Space Administration (NASA), and the Central Intelligence Agency (CIA). (Information on the ten other agencies covered by the Common Rule is provided in a supplemental volume to this report.)

The following sections briefly describe the institutional structures, review mechanisms, and policies prescribed by the Common Rule and the variety of ways in which federal agencies attempt to ensure that human subjects are adequately protected in the conduct of research. The chapter closes with a review of an issue of particular importance to the Advisory Committee—the status of protections for human subjects of classified research.[6]

THE FEDERAL POLICY FOR HUMAN SUBJECTS PROTECTIONS (THE COMMON RULE)

The Common Rule applies to all federally funded research conducted both intra- and extramurally. The rule directs a research institution to assure the federal government that it will provide and enforce protections for human subjects of research conducted under its auspices. These institutional assurances constitute the basic framework within which federal protections are effected. Local research institutions remain largely responsible for carrying out the specific directives of the Common Rule. They must assess research proposals in terms of their risks to subjects and their potential benefits, and they must see that the Common Rule's requirements for selecting subjects and obtaining informed consent are met.

As discussed below, central to the process of ensuring that the rights and well-being of human subjects are protected are institutional

review boards (IRBs). The Common Rule requires that a research institution, as a condition for receiving federal research support, establish and delegate to an IRB the authority to review, stipulate changes in, approve or disapprove, and oversee human subjects protections for all research conducted at the institution. IRBs are generally composed of some combination of physicians, scientists, administrators, and community representatives, usually at the local research institution, but sometimes at an agency that conducts intramural research.[7] IRBs have the authority to suspend the conduct of any research found to entail unexpected or undue risk to subjects or research that does not conform to the Common Rule or the institution's additional protections.

A prominent feature of the Common Rule is the informed consent requirement. The informed consent of a competent subject, along with adequate safeguards to protect the interests of a subject who is unable to give consent, is a cornerstone of modern research ethics, reflecting respect for the subject's autonomy and for his or her capacity for choice. Informed consent is an ongoing process of communication between researchers and the subjects of their research. It is not simply a signed consent form and does not end at the moment a prospective subject agrees to participate in a research project.

The required elements of informed consent stipulated by the Common Rule are summarized as follows:

- A statement that the study involves research, an explanation of the purposes of the research, and a description of the procedures to be followed;
- A description of any reasonably foreseeable risks or discomforts to the subject;
- A description of any benefits to the subjects or to others that might reasonably be expected;
- A disclosure of alternative procedures or courses of treatment;
- A statement describing the extent to which confidentiality of records identifying the subject will be maintained;
- For research involving more than minimal risk, an explanation of the availability and nature of any compensation or medical treatment if injury occurs;

- Identification of whom to contact for further information about the research and about subjects' rights, and whom to contact in the event of a research-related injury; and
- A statement that participation is voluntary, that refusal to participate will involve no penalty or loss of benefits to which the subject is otherwise entitled, and that the subject may discontinue participation at any time.[8]

The Common Rule includes several additional elements of consent that may be appropriate under particular circumstances[9] and describes the conditions under which an IRB may modify or waive the informed consent requirement in particular research projects.[10]

When an IRB reviews and approves a research project, it must pay particular attention to the project's plan for obtaining subjects' informed consent and to the documentation of informed consent. The IRB may require changes in the investigator's procedure for obtaining informed consent and in the consent documents. The board also must be allowed to observe the informed consent process if the IRB considers such oversight important in ascertaining that subjects are being adequately protected by that process.[11]

RESEARCH INVOLVING
IONIZING RADIATION

Beyond the strictures of the Common Rule, research involving either external radiation or radioactive drugs usually undergoes additional reviews for safety and risk (including a review of radiation dose) prior to IRB review at the local research institution. Most medical institutions have a radiation safety committee (RSC) responsible for evaluating the risks of medical activities involving radiation, whether for diagnostic, treatment, or research purposes, and limiting the exposure of both employees and subjects to radiation. In addition, research and medical institutions that perform basic research involving human subjects and radioactive drugs must have such studies reviewed and approved by a radioactive drug research committee (RDRC)—a local

institutional committee approved by the Food and Drug Administration (FDA) to ensure that safeguards, including limitations on radiation dose, in the use of such drugs are met.[12] Notwithstanding the prior review and approval of either or both of these radiation committees, the IRB must also assess the risks and potential benefits of the proposed research before approving it.[13]

SCOPE OF PROGRAMS
OF RESEARCH INVOLVING
HUMAN SUBJECTS

The six federal departments and agencies (DHHS, DOD, DOE, NASA, VA, and CIA) all conduct or support research involving human subjects. Each agency's program is distinctive in terms of its scope, organization, and focus, all of which reflect the primary mission of the agency.

DHHS is the largest federal sponsor of research involving human subjects, with approximately $367 million in intramural funding and $2.4 billion in extramural support for clinical research in fiscal year 1992, the latest year for which an estimate of extramural research funding is available.[14] Intramural research is usually conducted by agency staff members at various field sites, while extramural research is conducted outside the agency by contractors or grantees such as universities. Most of this research is biomedical, and some involves the use of radiation in experimental diagnostic and therapeutic procedures or as tracers in basic biomedical research.[15] The U.S. Public Health Service (PHS) is the operating division of DHHS and the principal health agency of the federal government.[16]

The DOD conducts biomedical and behavioral research involving human subjects within each of the military services and through several additional defense agencies, primarily in areas that support the mission of the department. In fiscal year 1994 DOD spent an estimated $77 million on intramural and $107 million on extramural human subjects research.[17]

The VA operates 171 inpatient medical centers, including short-term hospitals, psychiatric and rehabilitation facilities, and nursing homes.

The VA's largely intramural biomedical research program focuses on the health care needs of veterans. The VA spends approximately $114 million annually in appropriated research money on human subjects research, along with another $110 million in staff clinicians' time. Other federal agencies and private entities also support research in VA facilities.[18]

The DOE conducts and supports research, both intramurally and extramurally, involving human subjects that ranges from diagnostic and therapeutic applications in nuclear medicine to epidemiological and occupational health studies. DOE laboratories also receive funding from other federal agencies such as the NIH and from private sponsors of research. DOE spends $46 million annually on human subjects research, more than $20 million of which is devoted to the Radiation Effects Research Foundation (RERF) in Japan, which is charged with studying the health effects of exposure to radiation from atomic weapons.[19]

Both intramurally and extramurally, NASA conducts ground-based and in-flight biomedical research involving human subjects related to space life. In fiscal year 1994 NASA spent approximately $25 million on ground-based human subjects research.[20]

The CIA supports or conducts a small number of intramurally and extramurally conducted studies involving human subjects each year.[21] No figure for the annual dollar amount spent by the CIA was made available to the Advisory Committee.

ADMINISTRATIVE STRUCTURES AND PROCEDURES FOR RESEARCH OVERSIGHT

The following is an overview of the administrative structures and procedures used by the six departments and agencies to ensure compliance with human subjects ethics rules, particularly as they relate to the Common Rule. The Advisory Committee asked each of these agencies to provide the following information on its program of protections for human subjects involved in research:

- The scope of its human subjects research programs;
- The organizational structure of its human subjects protection efforts and the resources devoted to such activities;
- The policy issuances and guidances pursuant to the Common Rule that the department or agency has prepared and provides to subsidiary agencies and research institutions engaged in human subjects research;
- Monitoring and enforcement activities for ensuring that the provisions of the Common Rule are met;
- Sanctions available for noncompliance with human subjects protections;
- The rules governing classified research involving human subjects; and
- The use or potential use of waivers of any of the requirements of the Common Rule or the agency's human subjects regulations.

In a supplemental volume to this report we provide greater detail on the departments' and agencies' responses.

Each federal department structures its program of administrative oversight of human subjects research somewhat differently, despite the fact that all operate under the requirements of the Common Rule.[22] Some departments conduct reviews of research documentation out of one central departmental office, while others rely on local review; some provide detailed interpretive guidance on human subjects protections to subsidiary intramural research offices, contractors, and grantees, while others simply reference the Common Rule; and some departments audit or review IRB performance routinely, while others conduct investigations only when problems emerge.

The Office for Protection from Research Risks (OPRR) at the National Institutes of Health (within DHHS) serves not only as the locus for that department's policies for the protection of research subjects but also as the principal federal agent approving the assurances of research institutions to conduct human subjects research sponsored by any of a number of departments.[23]

Scientific peer review of federally sponsored research is one layer of protection for research subjects. Most federal research programs require that committees of scientists, expert in the particular subject under consideration and often from outside the agency (generally known as "study sections"), review both intramural and extramural research proposals for scientific merit and make recommendations regarding funding. When these committees of subject-matter experts review research proposals, they also consider the risks that may be involved for subjects. They may recommend that the sponsoring agency more closely consider the potential risks or that the principal investigator make specific changes in the research protocol prior to any funding.

Local review is a key component of the oversight system. The Common Rule requires IRB review and approval prior to the granting of federal funding for research on human subjects. Almost all federal agencies that conduct human subjects research within their own facilities have intramural IRBs, whose members include agency staff and at least one member who is not affiliated with the facility.[24] Likewise, extramural research projects must undergo IRB review prior to agency funding, usually by an IRB at the site of the research activity—for example, a university, medical school, or hospital. The IRB is an administrative unit that must itself comply with certain requirements of the Common Rule in terms of its composition, review procedures, and substantive review criteria; it must also direct researchers to comply with other requirements of the rule, such as adequate informed consent and fair subject selection procedures.

A research institution that has assured either OPRR or the federal agency sponsoring the research that it conducts human subjects research in compliance with the Common Rule must delegate to its IRB the authority to preclude or halt the conduct of any federally funded research project that does not conform with federal human subjects protections.[25] This delegation of authority applies to IRBs within federal research institutions for intramural research and to those at nonfederal research institutions as well. This authority extends even to research performed by military organizations, where unit commanders cannot overrule safeguards adopted by military IRBs.[26] Thus the IRB is the enforcing agent of federal protections that is situated closest to the conduct of research. Much of the success or fail-

ure of the federal regulations governing human subjects research depends on the effectiveness of IRBs in carrying out their responsibilities: assessing research proposals prior to their funding; stipulating any changes in the research protocol or informed consent procedure that strengthen the protections afforded the subjects; disapproving inadequate or excessively risky research proposals; minimizing risks to subjects; reviewing ongoing research at least every twelve months to ascertain that the research poses no undue risks to subjects; and taking action quickly to correct any failings in safeguarding subjects' rights and welfare.[27]

In overseeing human subjects research conducted in-house or supported extramurally, federal agencies acquire the following responsibilities: (1) communication of practice guidelines to research institutions and IRBs based on the policies of the Common Rule, (2) establishment of a structure whereby research proposals involving human subjects are peer reviewed for scientific merit as well as for IRB approval and the adequacy of subject protections, (3) negotiation of assurances with research institutions that ensure that adequate protections will be in place for research subjects, (4) verification that institutions, their IRBs, and researchers are complying with the federal human subjects regulations, and (5) investigation of complaints of noncompliance and adverse outcomes for subjects of research.

Table 1, "Human Subjects Research & Protections in Seven Departments and Agencies" (at the end of this chapter), summarizes information received by the Advisory Committee about human subjects research programs in DHHS, DOD, DOE, VA, NASA, CIA, and FDA (a subagency of DHHS). This chart shows each department's or agency's staffing levels for human subjects protection activities. Both the size of the departments' research programs and their investment of staff resources in oversight activities vary widely. A particularly important distinction in oversight programs is the extent to which they investigate the performance of research institutions and IRBs in carrying out their responsibilities under the Common Rule. Some departments rely heavily on the prospective assurances that research institutions make to the funding agency or to OPRR, while others audit research institutions and IRB records periodically.

The method, intensity, and frequency of research oversight and inspection activities depend entirely on how much staff and budget an agency allots them. OPRR negotiates multiple project assurances (MPAs) with large research institutions that perform a significant amount of research funded by DHHS. If an institution is awarded an MPA by OPRR, the federal agency funding the research must accept that institution's assurance of compliance with federal requirements and may not impose additional assurance requirements on the institution. This provision is intended to avoid duplicative and potentially contradictory enforcement of the federal protections.[28]

OPRR, in overseeing human subjects protections for DHHS-funded research and for all institutions to which it has issued an assurance, generally investigates the conduct of research only in cases where a complaint has been filed; where an institution, IRB, or researcher has reported a problem or adverse outcome; or where a problematic audit finding has been referred to it by the FDA.[29] Principal investigators are required to report to the IRB any adverse outcomes to subjects in the course of their research, and the IRB must have procedures to ensure that the appropriate institutional officials and the funding agency are informed as well. The FDA, in its role regulating new drugs, biologics, and devices for marketing, enforces the somewhat different requirements for human subjects protections of the Food, Drug, and Cosmetic Act through periodic onsite investigations of research institutions (e.g., pharmaceutical firms, university-based research facilities funded by pharmaceutical firms, independent testing laboratories) and their IRBs.[30] The DOD conducts on-site audits of its intramural research programs in addition to negotiating assurances. The DOD also reports that it is common practice in DOD-funded research to appoint independent medical monitors—health care providers qualified by training, experience, or both to monitor human subjects during the conduct of research as advocates for safety of the subjects.[31] The DOE is now planning to institute periodic audits of the research programs that it funds in addition to relying on assurances.[32]

Special Issues Arising in DOD Research

Human subjects research conducted by military agencies and within military settings entails considerations for subject protections and research oversight that are unique to the military context. The activities of military research programs may be difficult to distinguish from innovative training programs and medical interventions undertaken for the protection of the troops. In addition to enforcing policies derived from the requirements of the Common Rule, DOD has in place a parallel set of regulations for managing the risks to which military personnel are exposed in the course of these routine duties.[33] Military leaders are responsible for determining whether human experimentation protections, in addition to the more general risk-assessment requirements, apply to particular practices. A further distinction of the military context is the hierarchical and comprehensive nature of its authority structure, which poses special issues with respect to voluntariness in the recruitment of experimental subjects. In some cases, military researchers have excluded unit officers and senior noncommissioned officers from subject recruitment sessions (e.g., in vaccine trials conducted by Walter Reed Army Medical Center).[34] DOD has regulations that require most more-than-minimal-risk research proposals to be subjected to a second level of review by each military medical service at a central oversight office.[35] The Army, for example, requires greater-than-minimal-risk research protocols to undergo a second level of review at the Human Use Review and Regulatory Affairs Division (HURRAD) and the Human Subjects Research Review Board or the Clinical Investigation Regulatory Office (CIRO).[36]

FEDERAL RESPONSES TO VIOLATIONS OF HUMAN SUBJECTS PROTECTIONS

In the event that the Common Rule is violated in the conduct of federally sponsored research involving human subjects, there are various responses that can affect both investigators and grantee institutions, such as withdrawal or restriction of an institution's or project's assurance and, with that action, of research funding and suspension or termination of IRB approval of the research. In addition, an IRB is authorized by the Common Rule to suspend or terminate its approval of research that fails to comply with the IRB's requirements or when a research subject suffers an adverse event.[37] No federal department or agency may continue to fund a project from which IRB approval has been withdrawn or at an institution whose assurance has been withdrawn.[38]

An institution's or investigator's prior performance with respect to human subjects protections may affect future federal funding as well. If human subjects protection regulations are willfully violated, the department secretary or agency head may bar the organization or individual from receiving funding from any federal source.[39] Such debarment must be for a specified length of time and, in some extreme cases, may be permanent.

Federal agencies may also take disciplinary action against employees involved in human subjects research for failure to follow human subjects protection rules. For example, DOD sanctions for noncompliance by intramural researchers include loss of investigator privileges. For military personnel, potential sanctions are letters of reprimand, nonjudicial punishment, and sanctions under the Military Code of Justice; for civilian DOD personnel, sanctions include reprimands, suspension, or termination of employment.

No requirement of the Common Rule can preempt state and local laws governing the conduct of human subjects research that are stricter or provide additional protections for subjects. Of those states with any laws governing research involving human subjects, only California authorizes sanctions for failure to obtain a subject's informed consent.[40] The California statute authorizes monetary awards for negligent failure to obtain a subject's informed consent (up to $1,000), for willful failure to obtain such consent (up to $5,000) and, if a subject is thereby exposed to "a known substantial risk of serious injury either bodily harm or psychological harm," jail terms of up to one year and/or fines of up to $10,000.

PROTECTIONS FOR HUMAN SUBJECTS IN CLASSIFIED RESEARCH

We were advised that the only classified studies involving human subjects currently conducted by the six federal agencies are a small number of projects sponsored by the DOD and the CIA.[41] The Common Rule does not distinguish between classified and unclassified research in terms of the requirements or procedures it imposes to protect human subjects.

The Department of Defense reported that it currently sponsors a small number of classified research studies involving human subjects.[42] When such research is proposed, IRBs review classified protocols in one of two ways. The chair of the IRB may remove the classified portions of the protocol if he or she judges that those classified portions have no effect on the risks imposed on human subjects. Alternatively, the IRB may be composed of people with appropriate security clearances who then review the protocol in its entirety. A person not affiliated with the institution but with appropriate security clearance is included as a voting member of such IRBs.

The CIA indicated that it is currently performing classified human research projects.[43] The agency informed the Advisory Committee that all human subjects are informed of the CIA's sponsorship and of the specific nature of the study in which they are participating, even if the general purposes of the research are classified.[44]

Although DOE has the authority to conduct or support classified human subjects research projects, it reports that it is not currently conducting such research.[45] According to DOE guidelines, IRB review of classified research may take one of two forms.[46] If the chair of the IRB determines that none of the classified information in a proposal is relevant to the protection of human subjects and that the research can be accurately and fully described to the IRB, the proposed research will be reviewed at a regular IRB meeting without disclosure of any classified information. If the proposed research cannot be reviewed in the foregoing manner, however, the IRB must meet in a secure environment. (The Advisory Committee was advised that to date this has not occurred.) To review classified research, each member of the IRB must have the appropriate security clearance. The member of the IRB who is not affiliated with the institution conducting the research must also have security clearance to participate in the review of classified research. DOE guidelines recommend that IRBs expecting to review classified research obtain clearance for their nonaffiliated members so that they are not excluded from such reviews.

DHHS neither conducts nor sponsors any classified research. Some FDA personnel hold security clearances so that they may review classified investigational new drug or device applications submitted by the DOD, if the need to study or use these items in secret arises.[47] The VA does not now conduct any classified research and does not have original classification authority.[48] NASA currently conducts no classified research that involves human subjects and has not in the past. NASA does have classification authority, however, and conducts some classified research that does not involve human subjects.[49]

Research that involves human subjects and is classified for reasons of national security raises special issues for IRB review and for the process of obtaining informed consent, particularly with respect to the level of disclosure and waivers of informed consent. Specifically, the IRB must consider whether the prospective research subject will be adequately informed about the nature of classified research if some aspects of the research will not be disclosed in the informed consent process, whether security clearances are needed for IRB members, and whether information about classified studies must be partitioned from other IRB study reviews. Institutional review boards can determine that some aspect of a classified research project, if only the identity of the research sponsor, is irrelevant to the process of obtaining a subject's informed consent to participate. IRB members can decide that sponsorship information or complete disclosure of the purpose of the research need not be provided to potential subjects (in contrast to information about physical risk).

The Common Rule grants IRBs the authority to approve modifications in, or to waive entirely, informed consent requirements, but only for research involving no more than minimal risk.[50] A separate provision grants an agency head the authority to waive any requirement of the Common Rule for any kind of human subject research as long as advance notice is given to OPRR and the action is announced in the *Federal Register*.[51] As indicated above, the rule makes no distinction between classified and unclassified research, so this latter route to an informed consent exception would appear to pose a tension between duties to disclose and the need to keep information secret.

CONCLUSION

The Common Rule, adopted by the sixteen federal agencies and departments that conduct human subjects research, is another step in the evolution of human subject research protections policies begun in the 1940s. While those protections are crucial, gaps still remain.

With respect to classified research, the current requirement of informed consent is not absolute; if consent is waived, the research may proceed in ways that do not adequately protect the research subject. Also, military research involves special considerations because of the nature of the subject population, whose voluntary participation must be especially guarded. In addition, nonfederally funded research is not subject to the Common Rule, except under the umbrella of an institution's multiple project assurance.

Further, oversight mechanisms generally are limited to audits for cause and review of paperwork requirements. These offer little in the way of assurances that the prospective review process is working and do not give an indication of the quality or consistency of IRB review, either among IRBs or within a single board. An effective system of oversight relies on the detection of violations of policies and the imposition of appropriate sanctions.

The Committee's recommendations for remedying these and other shortcomings are discussed in chapter 18 of the final report. The remaining two chapters of part III report what documents used by IRBs suggest about the protection of human subjects and what patients think about the enterprise of human subject research.

Table 1. Human Subjects Research & Protections in Seven Departments and Agencies

	Specific Statutory Authority for Human Subjects Protection	Annual Spending on Human Subjects Research†	Locus of Human Subjects Research sponsored by Agency†	Staff Resources Devoted to Human Subjects Protection Activities*†	Nature of Research Oversight and Compliance Activities†	Original Classification Authority/Conduct of Classified Research†	Additional Provisions for Special Populations
DHHS	P.L. 93-348 (1974) P.L. 99-158 (1985) P.L. 103-43 (1993)	$367 million intramural, $2.4 billion extramural fiscal year 1992	Intramural and extramural. Some overseas research	19 full-time staff & 38.3 FTEs (excludes FDA)	Negotiates institutional assurances; reviews IRB performance on an exceptions basis only	Is not conducting classified research with human subjects	Pregnant women, fetuses and in vitro fertilization at subpart B, 45 C.F.R. 46; prisoners at subpart C; children at subpart D
FDA	Food, Drug, and Cosmetic Act, sec. 505(i). 507(d) (1963), and 520(g) (1976)	(included in DHHS total)	Intramural and extramural. Drug and device research regulated by FDA: domestic and foreign	6 full-time staff & 26.7 FTEs	Conducts compliance inspections on a three-year cycle, annually if problematic	No classified research. Maintains security clearances for coordination with DOD	Intramural research governed by 45 C.F.R. 46 subparts B, C, and D
DOD	10 U.S.C. 980 (1988). Requires informed consent	$77 million for intramural programs; $107 million for extramural programs, fiscal year 1994	Intramural and extramural. Some overseas research	80 FTEs	Negotiates assurances, relies on OPRR MPAs; Also conducts on-site audits of research programs	Has original classification authority. Conducted one classified human subject study in FY 1994. May conduct other minimal-risk classified studies	Subparts B, C, and D of 45 C.F.R. 46 adopted as DOD directives
DOE	None. Policy derives from Common Rule	$46 million in fiscal year 1994 ($20.4 million of which is for epidemiological studies in Japan); $10 million from other federal agencies	Intramural and extramural. Some overseas research	7–10.4 FTEs. Program director devotes 85 percent of time to human subjects protection	Negotiates institutional assurances, relies on OPRR MPAs. reviews IRB performance on an exceptions basis; plans to conduct periodic audits	Has original classification authority. Is not conducting any classified research with humans	Subparts B, C, and D of 45 C.F.R. 46 adopted as agency guidelines

VA	38 U.S.C. 7331, 7334. Requires informed consent and references Common Rule	$114 million in research funds; $110 million in clinicians' time; $100 million in privately supported research; $170 million funded by other federal agencies	Intramural only	0.5 FTE, central office staff; 51.6 FTEs, field staff	Central office review of IRB minutes and of research protocols	Does not have original classification authority. No classified human subject research	No distinct requirements
NASA	None. Policy derives from Common Rule	$25 million FY 1994 for ground-based research	Intramural and extramural; Some overseas research	0.5 FTE	In-house IRB provides a second-level review for all air/space human research: ground-based research may be reviewed by one or more IRBs	Has original classification authority. Conducts no classified human subject research	No distinct requirements
CIA	None. DHHS regulations applicable under Executive Order 12333 (1981)	Funding is a small portion of research components of general budget	Intramural and extramural	One senior staff physician (four hours per month)	Director, CIA, approves all human subjects research. In-house IRB also reviews extramural projects. Inspector general reviews Human Subject Research Panel	Has original classification authority. Conducted small number of classified human studies FY 1992-1993	Subparts B, C, and D of 45 C.F.R 46 adopted as agency policy

* This estimate includes staff resources devoted to policy development and guidance. negotiating assurances, oversight. and auditing. It excludes the time of agency staff spent on IRB members or staff and the minimal efforts of grant and contracts personnel who track IRB-approval status on research applications. *Full-time equivalent (FTE) effort represents the cumulative efforts of several people, who spend part of their time on oversight of human subjects experiments.*

† Information on current human subjects research programs and practices provided by agencies to Advisory Committee staff.

NOTES

1. Agency data reported to the Advisory Committee. See table 1 at end of this chapter and supplemental volume for the individual agency spending estimates that make up this total figure.

2. The sixteen departments and agencies that adopted a common policy for human subjects protection are the Department of Agriculture, Department of Energy, National Aeronautics and Space Administration, Department of Commerce, Consumer Product Safety Commission, Agency for International Development, Department of Housing and Urban Development, Department of Justice, Department of Defense, Department of Education, Veterans Administration (now Department of Veterans Affairs), Environmental Protection Agency, Department of Health and Human Services, National Science Foundation, Department of Transportation, and pursuant to an executive order, the Central Intelligence Agency. The Food and Drug Administration, a subagency of Health and Human Services, has somewhat different regulations governing human subjects research, based on its distinct statutory authority to regulate research for the licensing of new drugs, devices, and biologics (e.g., vaccines).

3. *Federal Policy for the Protection of Human Subjects; Notices and Rules,* 56 Fed. Reg. 28002 - 28032 (June 18, 1991). Each department and agency subject to the Common Rule incorporated its provisions within the agency's own regulations (e.g., DHHS regulations are reflected in 45 Code of Federal Regulations [C.F.R.] pt. 46, while DOD regulations are reflected in 32 C.F.R. pt. 219). The June 1991 *Federal Register* announcement is the only publication of the Common Rule as such. The Common Rule is not applicable to nonfederally funded research unless the research is performed at an institution whose research is subject to a multiple project assurance (MPA), described later in this chapter.

4. President's Commission for the Study of Ethical Problems in Medicine and Biomedical and Behavioral Research, *Protecting Human Subjects: The Adequacy and Uniformity of Federal Rules and Their Implementation* (Washington, D.C.: GPO, 1981).

5. Ibid., 140.

6. DHHS regulations specify additional protections for research on certain subject populations: pregnant women, fetuses, and subjects of in vitro fertilization research; prisoners; and children. The DHHS regulation is codified at 45 C.F.R. pt. 46 (1991). Subpart A of this regulation is the Common Rule. Subpart B provides additional protections for research involving pregnant women, fetuses, and in vitro fertilization, subpart C for research involving prisoners, and subpart D for research involving children. At their discretion, some of the other federal agencies whose research programs involve subjects in one of these categories have adopted these regulations

as agency guidelines. See table 1 at the end of this chapter for information on the applicability of special protections by agency. Some agencies, such as DOD, impose other safeguards in addition to those of the Common Rule. Information on individual agency policies and oversight practices at the other ten agencies, and greater detail on the policies of the six agencies above, are reported in a supplemental volume to this report.

7. The Common Rule directs that IRBs must include at least one member whose primary concerns are in scientific areas and at least one member whose primary concerns are in nonscientific areas. They must also include at least one member who is not otherwise affiliated with the institution (§ ___.107). (The provisions of the Common Rule are designated as "§ ___.000." The "___" indicates that these sections are reproduced within the regulations of various departments. Thus §___.107 of the Common Rule is codified for DHHS at 45 C.F.R. § 46.107.)

8. *Federal Policy for the Protection of Human Subjects,* §___.116(a).

9. Ibid., §___.116(b).

10. Common Rule, ___.116(d). Under the Common Rule, four requirements must be met in order for an IRB to waive the rule's informed consent requirements: "(1) the research involves no more than minimal risk; (2) the waiver or alteration will not adversely affect the rights and welfare of the subjects; (3) the research could not practicably be carried out without the waiver or alteration; and (4) whenever appropriate, the subjects will be provided with additional pertinent information after participation."

11. Ibid., § ___.109.

12. *Radioactive Drugs for Certain Uses,* 21 C.F.R. § 361.1

13. National Institutes of Health, Office for Protection from Research Risks, *Protecting Human Research Subjects* (Washington, D.C.: GPO, 1993), 5-23 - 5-28.

14. These figures represent the total amount of funds obligated for projects in which any human subjects were involved regardless of how minimal such involvement may be. Since it is virtually impossible to determine, within any given grant, the exact dollar amount that goes to human subject research, only the funding for the entire project could be calculated.

15. Lily O. Engstrom, Office of Extramural Research, NIH, to Wilhelmine Miller, ACHRE, 21 February 1995 ("Response to ACHRE Request No. 013095-E") and 4 April 1995 ("Additional Information in Response to ACHRE Request No. 013095-E").

16. The PHS comprises a number of agencies, including the National Institutes of Health (NIH), the Centers for Disease Control and Prevention (CDC), and the Food and Drug Administration (FDA). The NIH is the world's largest medical research center and conducts biomedical research (both basic science and clinical) dedicated to the improvement of the public's

health. The CDC focus is primarily on health promotion and disease-prevention, in addition to basic research in epidemiology, disease surveillance, laboratory science, and training of disease-prevention officials. The FDA is responsible for regulating and overseeing the safety and effectiveness of food, cosmetic, medical device, and human and veterinary drug industries, in addition to studying and monitoring consumer products and the industries that produce them. Department of Health and Human Services, Public Health Service, Office of the Assistant Secretary for Health, January 1993 ("The U.S. Public Health Service Today") (ACHRE No. HHS-091395-A).

17. Joseph V. Osterman, Environmental and Life Sciences, Office of the Director of Defense Research and Engineering, DOD, to Principal Deputy Assistant to the Secretary of Defense (Atomic Energy), 27 February 1995 ("White House Advisory Committee on Human Radiation Experiments").

18. Richard Pell, Jr., Deputy Chief of Staff, VA, to Jeffrey Kahn, ACHRE, 10 February 1995 ("We have prepared the enclosed fact sheet"), and Richard Pell, Jr., to Wilhelmine Miller, ACHRE, 19 January 1995 ("In response to your request"), enclosure pages 9-10.

19. DOE Database, Fiscal Year 1994, reported by David Saumweber, ACHRE, to Advisory Staff, ACHRE, 17 October 1994 ("DOE Current Research"), and oral communication by Susan Rose, Office of Health and Environmental Research, Office of Energy Research, DOE, to Wilhelmine Miller, ACHRE, 13 January 1995. There is ongoing discussion as to how large the commitment to RERF will be and how it will be administered.

20. J. Stoklosa, NASA Office of Aerospace Medicine, to Wilhelmine Miller, ACHRE, 14 February 1995 ("Enclosed is the response to"), 2. Figures for in-flight biomedical research were not provided.

21. Notes of Gary Stern, ACHRE, regarding 7 March 1994 meeting with CIA inspector general's staff, 8 March 1994. John F. Pereira, CIA, to Gary Stern, Anna Mastroianni, and Sara Chandros, ACHRE, 7 August 1995 ("Information for Committee's Final Report"). No additional information on this issue was made available by the CIA in response to Advisory Committee queries.

22. In addition, several federal agencies have adopted as policy guidelines the additional provisions of the DHHS regulation, 45 C.F.R. pt. 46, for pregnant women, fetuses, and in vitro fertilization; prisoners; and children (subparts B, C, and D, respectively). See table 1 at the end of this chapter for references to such agency policies.

23. *Federal Policy for the Protection of Human Subjects,* §___.103(a).

24. Occasionally, a federal agency may rely on an IRB at an adjacent academic institution to review research projects conducted at the federal facility. This sometimes occurs at VA hospitals that are affiliated with teaching hospitals and is the case for the Environmental Protection Agency, whose own re-

search facility is located at the University of North Carolina.

25. *Federal Policy for the Protection of Human Subjects,* §___.103 and §___.113.

26. James M. Lamiel, Chief, Clinical Investigation Regulatory Office, Consultant to the Army Surgeon General for Clinical Investigation, to Director, Radiation Experiments Command Center, 28 July 1995 ("Revised Chapter Drafts of the Advisory Committee").

27. Ibid., §___.109 and §___.113.

28. *Federal Policy for the Protection of Human Subjects,* §___.103(a).

29. Gary Ellis, Director, OPRR, to OPRR Staff, 7 December 1993 ("Compliance Oversight Procedures"), 1-4.

30. *FDA Protection of Human Subjects,* 21 C.F.R. pt. 50 and *Institutional Review Board Requirements,* 21 C.F.R. pt. 56 (1995), § 56.115 and § 56.120. See supplemental volume for further discussion of the FDA's distinctive policies and oversight practices.

31. Medical monitors are not permitted to be investigators involved in the protocol. Medical monitors have the authority to terminate an individual volunteer's participation in the study or suspend the study for review by the IRB. Lamiel, "Revised Chapter Drafts of the Advisory Committee."

32. DOE, Office of Health and Environmental Research, *Progress Report: Protecting Human Research Subjects* (Washington, D.C.: DOE, November 1994), A2.

33. See for example, Department of Defense Instruction 5000.2: "Defense Acquisition Management Policies and Procedures," 23 February 1991 (Administration); Army Regulation 40-10, "Medical Service Health Hazard Assessment Program In Support of the Army Matériel Acquisition Decision Process," 15 September 1983 (Administration and Safety Issues); Army Regulation 70-8: "Research, Development, and Acquisition, Personnel Performance and Training Program (PPTP), (Research Guidelines and Procedures; Training and Indoctrination)"; Army Regulation 70-8: "Research, Development, and Acquisition, Soldier-Oriented Research and Development in Personnel and Training," 31 July 1990 (Research Guidelines and Procedures Training and Indoctrination); Donald J. Atwood, Deputy Secretary of Defense, Department of Defense Directive Number 5000.1: "Defense Acquisition," 23 February 1991.

34. Oral communication by Colonel John Boslego, Deputy Director, WRAIR, to Shobita Parthasarathy (ACHRE Staff), 13 September 1995.

35. Lamiel, "Revised Chapter Drafts of the Advisory Committee."

36. Major Dale Vander Hamm, Chief, Human Use Review and Regulatory Affairs Division, Headquarters, U.S. Army Medical Research and Matériel Command, to Shobita Parthasarathy, ACHRE, 27 July 1995 ("Human Volunteers in U.S. Army Research in 1995").

37. Ibid., § ___.113.

38. Ibid., § ___.103(b) and (f), § ___.122 – ___.123.

39. Each agency has its own regulations regarding the oversight of compliance. For example, debarment procedures are specified for DHHS at 45 C.F.R. § 76. These procedures are summarized in a memorandum from Gary Ellis to OPRR staff, 5 February 1993 ("Compliance Oversight Procedures"), 3.

40. *California Health and Safety Code,* vol. 40B, § 24176 (1995).

41. Joseph V. Osterman, Environmental and Life Sciences, Office of the Director of Defense Research and Engineering, to Principal Deputy, Assistant to the Secretary of Defense (Atomic Energy), 27 February 1995 ("White House Advisory Committee on Human Radiation Experiments"). Larry Magnuson, M.D., CIA, in oral communication to Gary Stern, ACHRE.

42. Osterman to Assistant to the Secretary of Defense (Atomic Energy), 27 February 1995.

43. Notes of Gary Stern (ACHRE staff), regarding meeting with inspector general's staff, CIA, 7 March 1994 (8 March 1994).

44. Ibid.

45. Ellyn R. Weiss, Office of Human Radiation Experiments, DOE, to Daniel Guttman, ACHRE, 13 February 1995 ("This letter is in response to . . .").

46. Described in the DOE guidance, *Protecting Human Subjects at the Department of Energy, Human Subjects Handbook,* Office of Health and Environmental Research, 1992, "Review of Classified Research," unpaginated (ACHRE No. DOE-050694-A).

47. The FDA and DOD, memorandum of understanding of May 1987 ("Concerning Investigational Use of Drugs, Antibiotics, Biologics, and Medical Devices by the Department of Defense"), 4.

48. Veterans Administration, "Policy Manual MP-1" (21 November 1979), part 5, chapter 1, 5-8.

49. Janis Stoklosa, Office of Aerospace Medicine, NASA, in oral communication to Wilhelmine Miller, ACHRE, February 1995.

50. *Federal Policy for the Protection of Human Subjects,* § ___.116 (d).

51. Ibid., §___.101(i). This provision also allows for a statute or executive order to override the notification and publication requirements.

15

Research Proposal
Review Project

TWO of the biggest differences between research involving human subjects today and research involving human subjects as it was conducted in the 1940s, 1950s and 1960s, are the presence of applicable federal regulations and the articulation of rules of professional and research ethics. There is little question that these developments have had a significant effect on the protection of the rights and interests of human subjects. At the same time, however, there has been little systematic investigation of how much protection these developments have provided. As an Advisory Committee charged both with looking at the past and making recommendations about the future, we hoped to learn as much as we could about the state of contemporary human subjects research. We were particularly interested in exploring the extent to which the rights and interests of people currently involved as subjects of radiation research conducted or supported by the federal government appear to be adequately protected and whether the level of protection afforded these subjects was the same as that afforded the subjects of nonradiation research. The Advisory Committee's Research Proposal Review Project (RPRP) was designed to address these questions. By examining documents from a wide variety of research projects funded by

many agencies of the federal government, we hoped to offer insight into the general state of the protection of the rights and interests of human subjects.

During the course of the RPRP, the Committee reviewed documents from a random sample of research proposals involving human subjects and ionizing radiation that were approved and funded in fiscal years 1990 through 1993 by the Departments of Health and Human Services (DHHS), Defense (DOD), Energy (DOE), Veterans Affairs (VA), and the National Aeronautics and Space Administration (NASA); these are the only federal agencies that currently conduct human subjects research involving ionizing radiation.[1] We also reviewed a comparison sample of studies that did not involve ionizing radiation funded by the same agencies during the same period.

In this chapter, we first present the methodology and findings of the Research Proposal Review Project. We then report the results of an independent review of research proposals and documents conducted by one member of the Committee who also acted as a reviewer in the RPRP. The chapter closes with a discussion of our results in the context of current policies and practices in research involving human subjects.

METHODOLOGY OF THE RPRP

Obtaining Research Proposal Abstracts to Identify Studies of Interest

The RPRP involved the collection and review of documents related to recently funded, federally supported human radiation research. This included research supported or performed by the DOD, DOE, DHHS, NASA, and VA. Each agency funds intramural research conducted by agency staff members at various field sites and extramural research conducted outside the agency by contractors or grantees. The Advisory Committee requested and received abstracts or similar descriptions from these agencies for all intramural and extramural studies newly approved and funded between fiscal years 1990 and 1993 (that is, "new starts" in those fiscal years) that fell within two general categories: (1) studies involving the exposure of human subjects to research applications of ionizing radiation (or follow-up studies of such exposures); and (2) nonradiation research involving human subjects. These abstracts represented the "universe" of federally funded contemporary human research from which studies were then selected for review.

Selection of Studies Involving Ionizing Radiation

For purposes of the RPRP, a radiation experiment was defined as *any federally funded or performed investigation where the exposure of human subjects to ionizing radiation is an element of the research design*. In addition, we included follow-up or epidemiological studies of exposures of humans to ionizing radiation.[2] Any procedures involving radiation incidental to a subject's enrollment in a study (for example, a diagnostic x ray in research involving chemotherapy) were not considered experimental for purposes of the review.

To select studies to review from the many abstracts we received, nuclear medicine experts on the Advisory Committee staff first reviewed and stratified the study abstracts obtained according

to the biomedical categories that the Advisory Committee established for radiation research: tracer/biodistribution studies, studies involving potential therapeutics, studies involving potential diagnostics, and epidemiological/observational studies. These categories were intended to parallel roughly the various types of past radiation experiments identified by the Advisory Committee. We recognized that placing radiation experiments into discrete categories was a difficult task. The purpose of the categorization, however, was to sample proposals across the range of radiation research conducted on human subjects rather than to identify specific research as falling into strict categories.[3] Definitions of the biomedical categories used in the Research Proposal Review Project are listed in the accompanying box.

We then selected studies[4] to ensure that each funding agency and each biomedical category of human radiation research (tracer/biodistribution, therapeutic, diagnostic, and epidemiological/observational) was adequately represented in the random sample[5] of studies to be reviewed. Eighty-four radiation studies were selected from proposal abstracts provided by the agencies. These included 31 extramural proposals representing nonfederal research institutions,[6] primarily universities, and 53 intramural proposals[7] from the DHHS, DOE, DOD, NASA, and VA.

Selection of a Comparison Group of Nonradiation Studies

For purposes of selecting a comparison sample of nonradiation studies, the 84 radiation studies were reclassified according to the following categories: (1) federal funding agency, (2) extramural/intramural, and (3) cardiology/cancer/neither cardiology nor cancer.[8] Approximately half as many studies (41) were selected for the comparison sample and distributed in each of the three categories in comparable proportion to the distribution of radiation studies. We drew our sample of nonradiation studies from the same grantee institutions that were included in the radiation sample.

DEFINITIONS OF BIOMEDICAL CATEGORIES

*Tracer/biodistribution studies:*Studies involving the measurement of administered radioactive chemicals within the body (*in vivo*) using radiation detectors directed at the body from the outside, or in body fluids such as blood and urine in the test tube (*in vitro*).

Biodistribution studies are distinct from tracer studies in that their object of study is radioactive contaminants themselves, in order to understand their distribution and metabolism within the body. By contrast, tracer studies employ radio-labeled variants of ordinary biological chemicals to provide information on natural metabolic processes involving those chemicals. Tracer/biodistribution studies differ from research involving external sources of radiation (such as x rays), because tracer/biodistribution studies involve the administration of radioactive chemicals into a subject's body.[a]

Studies involving potential therapeutics: Studies that involve novel or nonvalidated uses of radiation for therapeutic purposes on sick individuals.

Studies involving potential diagnostics: Studies that involve experimental uses of radiological or nuclear medicine diagnostics (for imaging) that are experimental in that their efficacy has not been established. This includes research involving different types of radiation exposure as well as applications of established radiation imaging techniques (such as diagnostic x-rays or CAT scans), for new diagnostic purposes.

Epidemiological/observational: Studies of health effects in people who have experienced exposures to ionizing radiation. This research does not employ radiation, but attempts to understand health effects on humans exposed to ionizing radiation using follow-up studies, medical monitoring, and retrospective records reviews.

Data Sources

In total, the Advisory Committee identified for review 125 research proposals involving human subjects (84 involving ionizing radiation, and 41 not involving radiation) that were approved and funded by DHHS, DOE, DOD, NASA, or VA between fiscal year 1990 and fiscal year 1993.[9] Long-term epidemiological studies that were initiated before fiscal year 1990 and continued through this period were included in the review in cases where the methodology and/or consent procedures for such studies were found to have been updated in recent proposal renewals.

The Advisory Committee requested the fol-

lowing documents[10] for each of the 125 studies it identified for review:

1. Grant proposal submitted by investigator to federal agency;[11]
2. Institutional review board (IRB) application;[12]
3. Original consent form submitted to the IRB;
4. Consent form, as approved by the IRB;[13]
5. The IRB's final disposition letter;[14]
6. Documentation concerning any changes to the research design, methods, or consent form approved by the IRB after the IRB's initial approval of the study;[15]
7. If relevant, the application submitted to and the official letter of approval from the radioactive drug research committee (RDRC);[16]
8. If relevant, the application submitted to and the official letter of approval from any institutional human use committee other than the IRB or RDRC.

All of the relevant federal agencies and the 47 extramural grantee institutions to which the Advisory Committee submitted a request com-

---

a. Henry N. Wagner, Jr. and Linda E. Ketchum, *Living with Radiation—The Risk, The Promise* (Baltimore: The Johns Hopkins University Press, 1989), 77-78.

plied with this request. The willingness of institutions to voluntarily make available documents for review indicates their commitment to research ethics, which the Committee very much appreciates. The openness shown by the biomedical research community is important evidence of improvement in the ethics of human subject research over the fifty-year history reviewed by the Committee.

Review Process

Three basic elements were considered in developing a system to review the research materials supplied to the Advisory Committee: the procedures for obtaining informed consent, the balance of risks to potential benefits for the subject, and the selection and recruitment of subjects. An evaluation form was developed by a subcommittee of Committee members and staff to assist reviewers in organizing their assessments of the research documents (grant proposal, IRB application, RDRC application) and the consent form(s).

The documents for each proposal were reviewed by a team of two individuals, with at least one member of the Advisory Committee placed on each team, so that documents from every proposal were reviewed by at least one member of the Committee. Review teams consisted of either two Advisory Committee members or one Committee member and one staff member. One member of each team had expertise in research ethics, while the other had expertise in radiation science, radiation medicine, another branch of medicine, or epidemiology. Reviewers were never assigned documents from their own institution; they were also required to recuse themselves if they were well acquainted with the principal investigator of a proposal.

Documents were first reviewed independently by each reviewer and then by the reviewers together as a team. At the end of this process, each team completed a single evaluation form representing a joint assessment.

Limitations

The Research Proposal Review Project was designed to provide insight on an exploratory basis into the current practice of human subjects research conducted or supported by the U.S. government. The project was not undertaken with the expectation that our results would be generalizable to all research involving human subjects or to research sponsored by nongovernmental agencies. Of necessity, we reviewed documents from only a small sample of proposals for human subjects research funded in fiscal years 1990 through 1993. In a given year, DHHS supported 16,972 projects and subprojects involving human subjects research.[17] At the same time, however, our sample includes examples of both radiation and nonradiation research funded or sponsored by five different federal agencies across a variety of biomedical categories and medical specialties. Moreover, the proposals whose documents we received and reviewed were selected at random; there was no attempt to identify proposals that appeared from the outset to pose human subjects problems or high levels of risk and therefore no reason to suspect that the sample chosen was biased in favor of more problematic or higher-risk studies.

Within the Committee, reviewers rarely disagreed in their reviews. Although these reviews are based on interpretation and opinion in the context of Committee deliberation, it should be noted that so, too, are the evaluations of IRBs, on which the protection of human subjects now rests.

Perhaps the most significant limitation of this project is that the evaluation of each proposal was based only on the documents that were provided by the federal funding agency and grantee institution. The documentation we received was not always complete. Moreover, IRBs may have had access to sources of information not available to the Committee. Some IRBs invite principal investigators to make presentations at IRB meetings; others encourage reviewers to discuss proposals with principal investigators before IRB meetings. Thus, in some cases, IRBs may have reviewed the proposals evaluated by the Committee with a fuller and more accurate understanding of the project than was available to the Committee. It is therefore possible that some of the research projects that raised concerns for us based on the documents we reviewed, would, with the provision of additional information, be deemed unproblematic from a human subjects perspective. Conversely, it is possible that some of the research projects whose documents raised no concerns may nevertheless have inadequacies affecting the rights and interests of human subjects that we could not detect.

From the outset, the Committee neither desired nor thought it possible (because of our limited tenure and resources) to make judgments about the extent to which these 125 research projects were *in fact* being conducted in an ethically acceptable manner. This would have required a careful evaluation of far more than the documents that we received.

Neither IRB interactions with principal investigators nor documents speak to what actually happens between investigators, their assistants, and potential subjects. What investigators in fact say to potential subjects, the tone with which they say it, and the conditions under which the interaction takes place are pieces of information that were unavailable to the Advisory Committee in its review of the documents from contemporary human research proposals, just as they are generally unavailable to IRBs.

The Advisory Committee's review of research proposal documents thus was not intended to evaluate the performance of particular IRBs or the ethics of the conduct of particular investigators or specific insitutions. Rather, by examining documents from a wide variety of research projects funded by many agencies of the federal government, we hoped to offer insight into the general state of the protection of the rights and interests of human subjects.

FINDINGS OF THE RPRP[18]

In this section, we present the results of the RPRP. We begin with a general characterization of our overall assessment of the research documents. We also provide additional analysis of the impact of the level of risk and kind of experiment (nonradiation vs. radiation) on our evaluations. Next, we turn to a discussion of what the Committee found most troubling in these documents, organized around issues of understanding, voluntariness, and decisional capacity. Finally, we look at problems that were common in the sample as a whole, including the readability of consent forms and deficiencies in documentation.

Overall Assessment

Reviewer teams registered their overall assessment of each set of documents using a scale from 1 to 5, where 1 was taken to indicate no ethical concerns and 5 was taken to indicate serious ethical concerns. This scoring scale was used to assist reviewers in organizing their overall evaluations of the set of documents for each research proposal. These ratings were made in concert by the two reviewers after each had completed his or her own independent review. Ratings of 4 and 5 are grouped together in the discussion that follows because reviewers generally did not differentiate between the two; both ratings were used when documents raised serious ethical concerns for reviewers.[19]

For the total sample of documents from 125 radiation and nonradiation research proposals, two-thirds received ratings of either 1 (34%) or 2 (34%), while 18 percent received a rating of 3 and 14 percent received a rating of 4 or 5.

Level of Risk

Reviewers identified whether the research proposals as described in the documents involved minimal risk or greater than minimal risk of harm to subjects; 78 proposals were considered to involve greater than minimal risk (including 24 proposals that were evaluated as "maybe" greater than minimal risk[20]), while 47 proposals were considered to involve minimal risk.

There was a marked difference in the distribution of ratings between minimal-risk and greater-than-minimal-risk studies (Figure 1). Although a substantial number of greater-than-minimal-risk studies received ratings of 1 or 2, *all* of the studies that received 4s and 5s were considered greater than minimal risk.

Radiation versus Nonradiation Research

While about 70 percent of both radiation and nonradiation proposals received ratings of 1 or 2, a somewhat higher proportion of nonradiation studies than radiation studies received overall ratings of 4 or 5 (Figure 2). This difference could not be explained by differences in level of risk; the proportion of studies in the nonradiation subsample and the radiation subsample that involved greater than minimal risk was essentially the same. Perhaps the lower proportion of proposals in the radiation sample whose documents were rated as ethically problematic can be attributed to the second layer of scrutiny that is often afforded radiation studies during the initial review process. It must be noted, however, that because there were few studies that received ratings of 4 or 5, differences between radiation and nonradiation studies may not be significant.

Figure 1

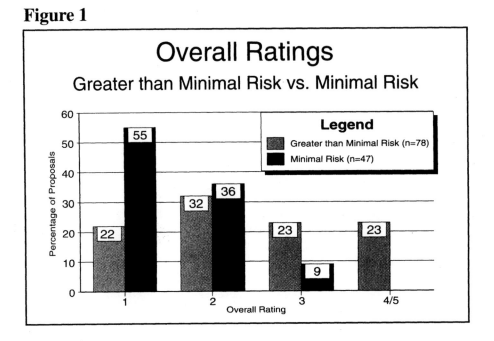

Figure 2

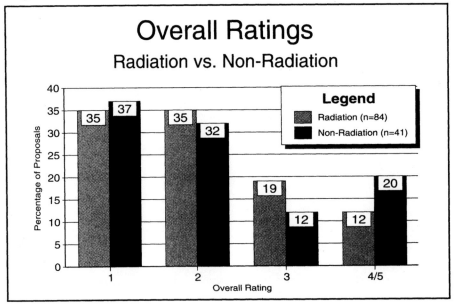

Issues Contributing
to the Overall Ratings

In this section we examine the kinds of problems that troubled reviewers in the documents from the 40 proposals that received ratings of 3, 4, or 5. These problems fell in to three categories: (1) factors likely to affect the adequacy of potential subjects' understanding of the research (other than questions of competence); (2) factors likely to affect the voluntariness of potential subjects' decisions about participation; and (3) approaches to the inclusion of subjects with limited or questionable decision-making capacity.

Factors Likely to Affect Understanding

Reviewers were likely to give a 3, 4, or 5 to proposals whose consent forms did a poor job of describing either what potential subjects stand to gain or what they stand to lose by participating in research. We looked carefully at how the consent forms presented the purpose of the study, its potential for direct benefits to the subject, the distinction between direct benefits and benefits to medical science, and alternatives to participation. How well consent forms communicated the realities of what it would be like to participate in the proposed research, including the likely impact on quality of life, also came under scrutiny. We were troubled, for example, by consent forms that, when compared with the information provided in the grant proposal or other research documents, appeared to overstate the therapeutic potential of research, either explicitly or indirectly. This issue was of particular concern to the Committee when the subjects being recruited were patients with poor prognoses. For example, one study, which was presented as primarily a toxicity study in the accompanying research documents, was cast differently in the consent form: "One objective is to find out how well patients respond to treatment. . . . If treatment works in your case, it may shrink your tumor or cause it to temporarily disappear, and/or prolong your life and/or improve the quality of your life. . . . Another objective of this study is to find out what kind of side effects this treatment causes and how often they occur."[21]

There also was significant concern about the use of the word *treatment* in consent forms for pharmacological studies. Phase I studies are designed to establish the maximum tolerated dose (MTD) for new chemotherapeutic agents and radiation regimens, which are then subjected to limited (Phase II) and then more extensive (Phase III) clinical trials to determine therapeutic effectiveness.[22] Although some Phase I studies contain elements of Phase II research and can appropriately be characterized as holding out at least a remote prospect of benefit to the subject, for some Phase I studies even the suggestion that subjects might benefit is inappropriate.

Reviewers were influenced in their overall assessments by inadequate descriptions of the physical risks of participating in the research. Reviewers were concerned, for example, when consent forms did not discuss the risks potential subjects faced in being removed from their standard treatments to be placed on an experimental protocol. In one study, patient-subjects were taken off cardiac medication in order to participate in a diagnostic study that offered no direct benefit to them. Any risks involved in the removal from this cardiac medication were not addressed in the consent form. The Advisory Committee also identified consent forms in which the possible lethality of drug treatments and radiation exposures was not adequately discussed. This occurred in contexts where patient-subjects generally faced far greater risks from their underlying illnesses, but, nevertheless, we felt that the consent forms should have been more forthcoming. A number of projects that involved combination drug treatments, for example, did not provide the potential subject with an estimate of the possibilities of death or major toxicities from a combination of drugs. One study involved a combination chemotherapy consisting of twelve different drugs but did not address the uncertainty of risk resulting from this new and investigational combination. Although the hazards and side effects for each drug were described individually, there was no discussion of overall risks and harms.

Even where consent forms described the risks of the research, there was often little mention of how participation would affect the subject's ability to function in daily life or how ill subjects might be made to feel during the course of the research. This omission was of particular concern to us when the implications for quality of life were markedly different depending upon whether a person decided to participate in the research or accept standard medical management, such as when standard management included only palliative care or watchful waiting. In one end-stage cancer study, for example, the consent form stated only that there may be "[o]ther general complications which may occur from combinations of chemotherapy drugs, including weight loss and loss of energy." The Advisory Committee was troubled that in such studies patient-subjects may not understand that although the research protocol might offer a chance to extend life, the time gained might be compromised by additional limitations in the quality of life resulting from participation in the study.

Reviewers also noted a number of problems in some consent forms for randomized clinical trials. For example, when some patient-subjects were randomized to receive the standard treatment while others would undergo an experimental procedure, reviewers commented that physical risks associated with the standard treatment or procedure were sometimes not adequately addressed in the consent forms. In one study of the effectiveness of a new compound for the decontamination of people who had ingested a radioisotope, although the grant proposal indicated that subjects would be randomized to receive a placebo, this information was not included in the consent form. In fact, the consent form only vaguely discussed the experimental procedures. "I [subject name] authorize [physician name]. . . to administer decorporation therapy utilizing the drug [name of drug]."

The Committee recognizes the difficulties facing investigators in communicating to potential research subjects a complex set of experimental procedures, side effects, long-term risks, trade-offs relative to alternatives, and other relevant information. This task is not impossible, however. We reviewed documents from several complex research proposals that at the same time had excellent consent forms.

For example, we reviewed documents from a proposal for a Phase I study of experimental antibody therapy that involved a number of possible risks; imposed a number of inconveniences including restrictions on sexual activities and a weeklong time commitment; and, as a Phase I study, offered little prospect of direct benefit to subjects. The consent form for this study addressed each of these issues in understandable language, briefly described how the monoclonal antibodies used in this research were derived, and explained that the U.S. Food and Drug Administration (FDA) permits experimental, new forms of therapy to be tested in a limited number of patient-subjects in Phase I studies. This consent form presented enough useful information to enable potential subjects to make an informed decision about whether to participate in the research, and it was not overly optimistic about the prospect of direct benefit to the patient-subject.

Another complex, greater-than-minimal-risk study with a good consent form involved an investigational radiation treatment, radiosurgery, for patient-subjects who had vascular disorders of the brain. The consent form for this study described the experimental procedures step-by-step with a very realistic picture of what participation would entail. Potential risks, possible benefits, and alternatives to participation for this experimental therapy were clearly presented. Furthermore, information that was likely to discourage some patients from enrolling—the possibility that participation in this study might limit the effectiveness of similar types of radiotherapy for the patient in the future—was disclosed in the consent form.

Factors Likely to Affect Voluntariness

As is discussed later in this chapter, the documents we reviewed often provided no basis on which to judge whether the participation of potential subjects was likely to be voluntary or not. In some cases, however, the information provided was sufficient to raise concerns. One was a neuroscience study that offered no prospect of medical benefit to potential subjects. Subjects were being recruited from among former cocaine addicts who were living in a residential treatment facility. Although compensation was not needed to reimburse subjects for travel expenses or loss of income, subjects were being offered $100 to participate. Reviewers were concerned that this cash payment might make it easier for those people struggling to break an addiction to get cocaine. Moreover, as part of the study, cocaine was injected into the body in order to measure brain uptake. Even if this procedure was not likely to have a physiologic effect upon the subjects, we were concerned that subjects may have been encouraged to participate because the research involved the injection of cocaine. We were also concerned about how their receiving cocaine as part of the research might affect the subjects' perceptions of themselves during the recovery process.

By contrast, the following text from a consent form for employee-subjects (colleagues of the investigators) who are smokers illustrates *exemplary* handling of the voluntariness issue in a minimal-risk study. The study, which involved no risk of physical harm to the subjects, was designed to measure environmental tobacco smoke.

Your participation in the experiments is entirely voluntary and you are free to refuse to take part. You may also stop taking part at any time. Because you are a colleague here at [research institution], we want to be especially clear on this point. We have approached you about the possibility of your volunteering for these experiments. Your refusal to participate or to continue will not be questioned by us, nor will it (or should it) be discussed further with anyone else.

Inclusion of Subjects With Limited Decisional Capacity

Several issues revolve around how certain factors that influence a subject's decisional capacity may affect his or her ability to understand the implications of participating in research. There is, for example, considerable controversy over how to conduct research ethically in emergency medicine when, because of the acute nature of the medical problem, the patient is temporarily incapacitated and no family members are available for consultation. The documents of one proposal raised some of these issues. In this example, 5 minutes were allotted to obtain consent from subjects who were recruited in the emergency room while their chest x-ray films were being processed. Under the stressful conditions of an emergency room and while experiencing chest pain, the decisional capacity of potential subjects was likely to be severely compromised. Reviewers expressed concern about the subjects' ability, in such a context, to comprehend the study adequately and then make a voluntary decision about whether or not to participate. In another study, women in preterm labor were recruited to participate in a study that involved collecting data about the infants born to these women. Although the proposal stipulates that "[n]o mother will be approached while under undo [*sic*] stress or in excessive pain," reviewers were nonetheless concerned about consent having been solicited during preterm labor.

The Advisory Committee also reviewed the documents of studies involving children and adults with questionable decision-making capacity, several of which raised serious ethical concerns.

Sixteen of the studies included in the Advisory Committee's review involved children as research subjects; 11 of these 16 studies, according to federal regulations, should have had assent forms as well as parental permission forms.[23] The documents we received on each of these proposals all included parental permission forms. We received assent forms for 8 of these 11 proposals. The 3 studies for which we did not receive assent forms all involved greater than minimal risk, 1 of which may not have offered any prospect of medical benefit to the children-subjects.

This last study illustrates a major issue in the ethics of research involving children. Current regulations permit the use of children as subjects in research that offers no prospect of direct medical benefit to them when the research poses no more than minimal risk. Nontherapeutic research on children posing more than minimal risk is permitted under special circumstances. A central, unresolved question is whether the administration of tracer amounts of radioactive materials to children can properly be classified as a minimal-risk intervention.

Eight studies in the project sample sought to recruit adult subjects with questionable decision-making capacity. 6 of the 8 appeared not to offer potential medical benefits to the subjects; two of the 6 were epidemiologic studies.

The Committee's concerns focused primarily on the remaining four studies, all of which involved diagnostic imaging with cognitively impaired persons, such as those with Alzheimer's disease. The imaging processes required that the subjects' movements be restricted, yet there was no discussion in the documents or consent form of the implications for the subjects of these potentially anxiety-provoking conditions. Nor was there discussion of the subjects' capacity to consent or evidence that appropriate surrogate decision makers had given permission for their participation. We were particularly troubled that two of these studies exposed subjects to greater than minimal risks. The question of whether or under what conditions adults with questionable decision-making capacity can be used as subjects of research that offers no prospect of benefit to them is unresolved in both research ethics and regulation. When such research puts potentially incompetent people at greater than minimal risk of harm, it is even more ethically problematic.

Common Problems With the Documents

We turn now to a discussion of issues that emerged often in the documents we reviewed, including documents that raised only minor concerns.

Consent Form Language

Although inappropriate reading level in a consent form was generally not sufficient in and of itself to result in ratings of 3, 4, or 5, it was sufficient for a rating of 2. A significant majority (nearly 80%) of the proposals receiving a 1 included consent forms that used a reading level appropriate for the study population. By contrast, the reading level was judged to be appropriate in no more than half of the remaining consent forms.

Reviewers raised a number of issues that they felt may have contributed to problematic reading levels in the consent forms. One such issue pertains to the complexity of the research being proposed. We were disturbed to find that in their attempts to convey complexities to the subject, investigators often drafted consent forms that

were too lengthy, highly technical, and generally unintelligible. Consider the following, for example: "The purpose of this study is to obtain a 'map' of brain cholinergic receptors. . . . This is done by administering, intravenously, small amounts of a radioactive substance that attaches to brain acetylcholine receptors and then producing a map of these receptors using Single Photon Emission Computed Tomography (SPECT)."

Still another consent form included language such as "[y]ou will then be positioned in a recumbent position," and "[a]nother possibility is poor regional function because of ongoing or intermittent ischemia at rest, resulting in anginal symptoms and global function that is worse than it can or should be."

A number of the consent forms included standard ("boilerplate") language that was often in a smaller type and distinct from the rest of the document. The presentation of information in this manner may have given subjects the impression that the information was less important and easily skipped. Sometimes these sections contained the only discussion of such critical topics as alternatives to participation, costs to the subject, confidentiality, potential benefits of participation, and voluntariness of participation.

The Advisory Committee found that intramural institutions often used a standard consent form that contained boilerplate language provided by their respective agencies. The following passage is an example of such language. It appeared in smaller type at the top of consent documents, clearly separated from the rest of the text:

We invite you (or your child) to take part in a research study at the [named institution]. It is important that you read and understand several general principles that apply to all who take part in our studies: (a) taking part in the study is entirely voluntary; (b) personal benefit may not result from taking part in the study, but knowledge may be gained that will benefit others; (c) you may withdraw from the study at any time without penalty or loss of any benefits to which you are otherwise entitled. The nature of the study, the risks, inconveniences, discomforts, and other pertinent information about the study are discussed below. You are urged to discuss any questions you have about this study with the staff members who explain it to you.

Reliance on Disclosures Not Subject to IRB Review

When patients are being approached to participate in research that has implications for the medical management of their illness, it is understandable and indeed desirable that patient-subjects discuss the proposed research with their treating physician. The Committee was disturbed, however, when consent forms indicated that the only presentation to potential subjects of key information about the research was to take place in such undocumented discussions. This suggests that it is difficult, if not impossible, for IRBs to judge whether potential subjects were being provided an adequate base of information on which to make an informed decision. There is no documentary record, either in the consent form or in other materials submitted to the IRB, of what potential subjects have or will be told about key aspects of research participation.

In some cases, consent forms indicated that subjects themselves were responsible for approaching physician-investigators for explanations of the choices available and guidance on how to compare the experimental protocol to standard treatment. Consider the following example:

Your (child's) doctor can provide detailed information about your (child's) disease and the benefits and risks of the various options available. You are (your child is) encouraged to discuss this with your (child's) doctor.

In this instance, it is unclear whether the phrase *your doctor* refers to the patient-subject's personal physician, a physician who is a member of the research team, or a physician who is both. This passage, as well as passages in several other consent forms, suggests that conversations between subjects and the doctors occurred *after* consent was given. Examples of this follow: "Severe and sometimes deadly side effects have occured when high doses of this drug have been given . . . You and your doctor will determine whether the benefits of such treatment outweigh the risk"; and "You will discuss the options with your physician and decide between . . . [surgical alternative] . . . or [medical alternative] . . ." Subjects in these studies may have received information critical to their decision making process only after giving their consent to partici-

pate in the research and without the IRB knowing the content of that information. This is particularly troublesome because these statements comprise the only discussions of side effects and alternatives, respectively, in these consent forms.

Other consent forms seemed to rely on disclosures that had already taken place by the time potential subjects were approached to give their consent, and so could not be afforded IRB review. One such consent form began, "The following is a summary of the information your doctors gave you when discussing this treatment with you. Please read it and ask any questions you may have." The summary that followed provided little specific detail. The Committee was left wondering whether the IRB was in a position to make a judgment about the adequacy of this prior disclosure.

Voluntariness

If an informed consent is to be a meaningful act of decisional autonomy, it is essential not only that the consent be based on adequate understanding but also that it be substantially free from coercive or manipulative influences. We found, however, that many proposal documents, including applications to IRBs, did not contain enough information to make a judgment about the likely voluntariness of subjects' consent decisions. For example, there was often insufficient or no information about who was soliciting a potential subject's consent and under what conditions.[24]

Often the only information in the documents reviewed that bore on issues of voluntariness was the inclusion in consent forms of boilerplate language to the effect that participation was voluntary. In most cases, the issue of voluntariness was simply ignored in proposal documents submitted to the IRBs and funding agencies, precluding us (and, presumably, IRBs) from making any judgments about the procedures employed to ensure voluntary decision making.

Scientific Merit

A controversy has long existed over whether the role of IRBs includes evaluation of the scientific merit of proposed research. Some argue that evaluation of scientific merit lies outside the scope of IRB review, while their opponents contend that it is impossible to do a proper assessment of the benefit-risk ratio without evaluating the potential contribution to science. Based on the documents we received, it was sometimes difficult to make judgments about scientific merit. In some cases, reviewers felt that they could not establish from the documents available to them whether there was sufficient scientific merit to warrant the exposure of human subjects to risk or inconvenience.

Psychosocial and Financial Risks

In research where psychosocial risks were clearly an issue, these risks were often inadequately addressed in proposal documents. A number of proposals that included neuropsychological batteries, for example, failed to discuss the potential anxieties that may result from participation in the study. The objective of one research project involved the inducement of sadness in the subject. Neither the consent form nor the research documents addressed the possibility that the sadness would not resolve itself quickly and that psychological counseling or other therapy might be necessary.

Four studies reviewed by the Advisory Committee involved DNA screening to determine the subjects' carrier status for a particular gene. None of the proposals for these studies addressed the potential psychosocial impact of learning about one's carrier status, including possible implications for other members of the subject's family or the potential for insurance discrimination. The availability of genetic counseling for these subjects was not mentioned in consent forms. Reviewers also were concerned that some proposals did not clearly explicate the types of tests that were included in what was referred to as "chronic disease screening" in the consent forms. This lack of specificity was particularly troubling for "chronic disease screens" that included human immunodeficiency virus (HIV) testing. Although the anxieties and social risks of HIV testing were likely to be addressed on a separate HIV-specific consent form, any study that requires HIV screening as part of its eligi-

bility criteria should make that clear to subjects so that those who do not wish to undergo HIV testing can decline participation.

Another area that was sometimes inadequately described in both consent forms and research documents was the financial cost to the subject of participating in the research. Costs were often briefly addressed in the boilerplate section of the consent form, but usually no project-specific information about actual expenses was offered to the subject. Reviewers were concerned that subjects might not appreciate the real costs and the possibility that insurance companies would be very reluctant to cover them. This omission was particularly troubling in studies involving seriously ill patient-subjects who may be at risk of spending much of their assets on research interventions at the end of life.

Justice in the Selection of Subjects

Most research documents did not include specific information about the subject populations that would be involved in the protocol. Unless IRBs are receiving more information on this topic than that provided in the documents reviewed by the Advisory Committee, they are clearly ill-equipped to address the social policy goal[25] of including women, minorities, and other groups in research. The racial and ethnic composition of the subject sample, for example, was specified in only one-quarter of the proposals whose documents were reviewed by the Committee.

The only frequently mentioned reason for excluding a person from participation in research was pregnancy. Pregnant women were explicitly excluded in 58 percent of the studies (73 of 125) and were explicitly included in only 5 percent (6 of 125) of the proposals. Pregnancy tests were often included in the eligibility screening procedures for women who were willing to participate in research. The RPRP sample also included 13 studies in which women who were not pregnant were expressly excluded from participation. There was no scientific reason to exclude women as subjects of research in any of these proposals. In two of these instances, women were excluded expressly because of the possibility that they might become pregnant.

The Committee's interpretation of the implications of these findings can be found in the "Discussion" section at the end of the chapter.

INDEPENDENT REVIEW
OF PROPOSALS

One member of the Advisory Committee, Jay Katz, served both as a reviewer for the RPRP and independently reviewed 93 proposals.[26] Katz's independent sample was drawn from the same pool of proposals from which the RPRP sample was drawn, included examples of both radiation and nonradiation research, and was based on the same sets of documents as the RPRP.[27] Although there is considerable overlap between the proposals included in Katz's review and those in the RPRP, the samples are not identical. Katz reviewed the first 93 proposals for which the Committee received documents, while the RPRP sample was drawn from the entire pool of proposals for which documents were received in order to achieve adequate representation by funding agency and type of research. In addition, a few of the studies reviewed by Katz were eliminated from eligibility in the RPRP because they did not fall within the biomedical categories established by the Committee.

Katz's review complements and strengthens the findings of the Research Proposal Review Project. Whereas the RPRP sought to investigate several basic issues regarding the conduct of human subjects research, including balance of risk to potential benefit, justice in the selection of subjects, the involvement of people with diminished decisional capacity, and the consent process, Katz focused exclusively on informed consent. In doing so, he asked himself two interrelated questions: (1) What can be learned about the contemporary informed consent process? and (2) How adequately does the process protect the rights and interests of research subjects? Although Katz appreciated that there was more to the IRB process than could be ascertained from the protocols and consent forms submitted to the IRB, he felt that consent forms constituted *written* documentation not only of what subjects ultimately agreed to but also what IRBs considered to be adequate written disclo-

sure for purposes of consent. With respect to these signed informed consent forms, he echoed a fellow Committee member's observation that, if such forms are not clearly written or are otherwise flawed in significant ways, it is likely that the oral interactions are similarly flawed.

Of the 93 proposals Katz reviewed, he identified 41 that posed greater than minimal risks to subjects and therefore that also raised significant and complex informed consent issues.[28] Of these 41 proposals, Katz found that 11 (26%) raised no or only minor ethical concerns and were analogous to those warranting a Committee rating of 1 or 2. Thirty protocols, however, raised ethical concerns about the informed consent process (analogous to a Committee rating of 3, 4, or 5). Of the 30 (74%) protocols that raised serious problems, Katz felt that 10 were "borderline" (analogous to a Committee rating of 3), and 20 raised serious ethical concerns of the sort analogous to those warranting a rating of 4 or 5 in the RPRP. Katz detailed the results of his review of these 20 problematic proposals for the Committee, and a summary of his findings specific to those proposals is presented here.

Physician-Investigators

In his review, Katz was struck by evidence of the dedication physician-investigators brought to their task. They were concerned, and so informed IRBs, about current treatments that were inadequate in eradicating disease or, at least, in prolonging life. Moreover, physician-investigators emphasized the importance of finding cures and not merely temporary or prolonged remissions.

Katz also noted that a number of the troublesome research proposals appeared to be part of an underlying "grand scientific design" to gain basic knowledge in such areas as cellular immunology or molecular biology, which might eventually lead to more clinical research about therapeutic effectiveness. The primary purpose of these studies was to advance knowledge for the sake of future patients, not to benefit present patients.

As investigators declared war on cancer and other ills, they often employed highly toxic agents to treat patients whose prognosis was grave. In their scientific protocols, the use of such agents was justified by arguing that only such aggressive approaches would ultimately lead to cure, although often only for future patients rather than present patient-subjects.

Katz, like the full Committee, was concerned that, at the same time, documents from these proposals were devoid of any discussion of the impact of the research on patient-subjects' quality of life, particularly in situations of terminal illness. He speculated that in their ultimate quest for finding cures, physician-investigators often paid more attention to increased longevity for present patient-subjects than to the quality of remaining life.

Patient-Subjects

To Katz, the ancient but questionable proposition that physicians and patients share an identity of interest in medical decision making becomes even more questionable in research settings where physician-investigators have dual allegiances: to their subject-patients and to their research objectives. As did those in the RPRP, Katz noted that consent forms for the troublesome proposals were often written in ways that made it difficult, if not impossible, for patient-subjects to come to a meaningful decision as to whether they wished to participate in research. Thus, patient-subjects seemed obliged to fall back on uninformed trust, based on a belief that physician-investigators will act only to ensure a patient-subject's therapeutic benefit.

Katz identified five specific problems with the informed consent process: (1) unclear purpose, (2) incomplete information regarding the consequences of participation in randomized studies, (3) confusing or incomplete discussion of risks, (4) exaggerated benefits, and (5) insufficiency of information in consent forms provided to IRBs. His concerns are elucidated below.

Specific Problems With the Informed Consent Process

Unclarity About Purpose

Katz found that the most striking element of the troublesome consent forms was the lack of a forthright and repeated acknowledgment that patient-subjects were *invited* to participate in human experimentation. All too quickly the language shifted to *treatment* and *therapy* when the latter was not the purpose and was only, at best, a by-product of the research. Like the other reviewers in the RPRP, Katz was particularly concerned with Phase I trials. As documented in some of the protocols in his examination, patient-subjects may suffer life-threatening toxicities that may, though rarely, kill them. Nevertheless, such studies are important for subsequent clinical trials and more widespread use in an attempt to save lives in the future. Katz's examination of consent forms revealed that investigators often did not take sufficient care to apprise patient-subjects of the purpose of Phase I studies. Although the dangers of the research are often mentioned, this information was often compromised when the "treatment" dimension of the research was emphasized. Katz concurred with a fellow Committee member who observed, through his participation in the RPRP, "Perhaps the consent form should not repeatedly emphasize that it is treatment, but I believe that it is the way it is perceived by the researchers themselves." Katz pointed out that the controversy over when, if ever, Phase I trials are to be regarded as *potentially* therapeutic has not been satisfactorily resolved with respect to the question: What must patient-subjects know? The President's Commission for the Study of Ethical Problems in Medicine and Biomedical and Behavioral Research,[29] when addressing Phase I trials, recommended that "patients not be misled about the likelihood (or remoteness) of any therapeutic benefit they might derive from such participation."[30] Katz's review of consent forms revealed that the Phase I purpose is often dismissed and the therapeutic benefits are highlighted. Thus, he was concerned that patient-subjects are likely to be confused about what is being asked of them.

The Consequences of Participation in Randomized Studies

A number of the troublesome proposals identified by Katz involved randomized clinical studies in which patient-subjects were assigned to two different experimental regimens to assess their comparative merits. These two procedures were generally described adequately in the consent forms. Patient-subjects, however, were generally not apprised of the already accumulated knowledge about possible therapeutic benefits to be derived from each regimen. Although protocols submitted to the IRB contained some, but often incomplete, information about the greater promise of one procedure over the other, patient-subjects rarely received such information.

In one protocol, for example, investigators clearly indicated that clinical experiences with the combined administration of chemotherapy and radiation had demonstrated its effectiveness against cancer. But since no scientific randomized clinical study had as yet been conducted, the investigators intended to submit half of the subjects to radiation alone. Consent forms provided no clues about what had already been learned from clinical experience and nonrandomized trials.

In another randomized trial, the research objective required that half of the patient-subjects submit to a mild treatment regimen, and the other half to a more intensive one. Katz noted that quality-of-life impairments imposed by random assignment to one research arm over another were not addressed in the consent forms. The consent forms also failed to address the fact that more intensive treatment regimens went counter to customary clinical practice of "watching and waiting," as the often slowly progressive nature of the cancers under investigation had led practitioners to recommend, in most cases, doing nothing or administering chemotherapy or radiation therapy only in low doses. Moreover, the risks inherent in both the mild and aggressive regimens were lumped together in the consent forms as if they were one and the same. The history of clinical experience with these particular cancers also was not discussed in the consent forms.

Discussion of Risks

The troublesome consent forms identified by Katz customarily listed an extremely detailed and separate discussion of all risks of the drugs, surgery, and/or radiation to be administered. Although he felt that federal regulations can be interpreted to require such detail, Katz, like the Advisory Committee as a whole, was concerned that such exhaustive treatment may serve only to overwhelm and numb patient-subjects. Only rarely were risks summarized or were risks of particular relevance to the research project highlighted. In almost none of the troublesome consent forms was there any comparative discussion of the impact on quality of life and toxic consequences of what investigators sometimes term *total therapy* (or of the physical and financial hardships imposed by countless research tests) on the one hand and of less toxic therapeutic alternatives that promise less but at least provide greater comfort for remaining life on the other.

For example, one study sought to explore the toxicity/efficacy of a new drug that may cause irreversible brain damage. That crucial piece of information, however, was not highlighted as a specific risk of the particular drug under investigation.

Another research project was designed to treat a cancer with a highly toxic drug, which had an expected mortality of up to 10 percent when used in a dosage greater than customary, as was contemplated in this "total therapy" research project. This fact, however, was not mentioned in the consent form. Although the patient-subjects had limited life expectancies, they probably would live longer than when a lethal drug toxicity would occur. Katz noted that another investigator simultaneously submitted the identical study to the same IRB (utilizing the same drug to combat the same disease), but with an exemplary protocol and consent form that discussed the expected 10 percent mortality rate without equivocation.

Presentation of Benefits

Like the RPRP reviewers, Katz found that benefits were often exaggerated in the troublesome consent forms. One consent form, for example, stated, "It is possible that the *treatment* [emphasis added] will cause the tumor to shrink or disappear or eliminate any symptoms and thus increas[e] life expectancy." Although this statement conveys a promise of benefit to the patient-subject, the protocol clearly indicates that any benefits would be fortuitous since they were neither an aspect of the research objective nor supported by evidence so far accumulated.

One consent form for a research project that was designed solely to establish the maximum tolerated dose of an intensive chemotherapy schedule stated, "It is not possible to predict whether or not any personal benefit will result from the use of the treatment program. A possible benefit could be the achievement of a remission." There was, however, no therapeutic intent in this proposal; physician-investigators were interested only in learning if it could be used safely in a subsequent randomized clinical trial. The subjects, however, could easily be led to believe that there was probable therapeutic benefit. Katz was particularly alarmed about the overstatement of benefits because patient-subjects so desperately long for such benefits.

Insufficiency of Information Provided to IRBs

In many cases, Katz found discrepancies between information provided in the protocol and that provided in the consent forms. This finding was not unlike that of the full Advisory Committee. Thus, an important question must be posed and eventually answered: Why was information that was available to IRBs not disclosed to patient-subjects?

According to the documents received, it seemed that even IRBs were often inadequately apprised of crucial information. In some cases, Katz noted that proposals were deficient in explicating the available knowledge about standard treatments, therapeutic effectiveness, and the impact of experimental procedures on quality of life. Although research is often a voyage into the unknown, investigators do possess preliminary guiding data that must be transmitted to IRBs. Only then can IRBs accurately evaluate consent forms and make certain that patient-subjects are provided with necessary information in order to make decisions about participation.

In one research project, for example, IRBs and, in turn, parents were insufficiently informed that the combination of radiation treatment and highly toxic chemotherapeutic agents used in the project exposed children to considerable risks that deserved careful scrutiny. The parents or guardians had two choices: to enroll their children in the study or to opt for standard treatments of either radiation or chemotherapy alone (depending also on the location of the cancer), with or without one of the chemotherapeutic agents that had considerable carcinogenic potential within five years. This example highlighted another, more general concern: that some patient-subjects may become part of inflexible research protocols when considerable clinical experience suggests that a patient-subject's medical condition may deserve an individualized treatment approach.

DISCUSSION

We turn now to a consideration of the implications of the results of the RPRP, as bolstered by Katz's review, for our understanding of the current status of human subjects protections. It should be reemphasized that these results were based solely on an evaluation of the documents available to the Committee. It is therefore possible that some of the research projects that raised concerns for us based on the documents we reviewed would, with the provision of additional information, be deemed unproblematic from a human subjects perspective. It is also possible that some of the research projects whose documents raised no concerns may nevertheless have inadequacies affecting the rights and interests of human subjects that we could not detect.

There is no evidence in this review that research in which human subjects are exposed to radiation is any more ethically problematic than other kinds of research involving human subjects; in fact, our results suggest that human subject protections may be more effective in radiation research then elsewhere, perhaps because some radiation research is reviewed by a radiation safety committee as well as an IRB. Because we failed to find any systematic differences between radiation research and non-radiation research in our review, our observations based on the RPRP results are directed at human subjects research generally, not solely at radiation research.

About 40 percent of the research whose documents we reviewed appeared to pose no greater than minimal risk to participants. Most of these studies raised no concerns about ethics, or only minor ones. Many studies that involved greater than minimal risks to subjects were similarly ethically unproblematic. Specifically, more than half of the greater-than-minimal-risk studies reviewed raised no or only minor concerns about ethics. There are important lessons to be learned from these studies. It is possible to conduct complex research that puts subjects at greater than minimal risk of harm in an ethically acceptable fashion. It is possible to develop good consent forms for this kind of research. Not only is it possible, but it appears that this happens frequently.

At the same time, our review suggests that there are significant deficiencies in some aspects of the current system for the protection of human subjects. We have evidence that the documents provided to IRBs often do not contain enough information about topics that are central to the ethics of research involving human subjects such as voluntariness of participation, fairness in the selection of subjects, and scientific merit. Although we have already noted that IRBs do not necessarily rely solely on documents in making their evaluations, clear, complete written documents are important. These documents form the core of the information upon which IRBs rely in protecting the rights and interests of human subjects; in some cases, they are the only source of information available. These documents also provide a written record of the research subject protection process for both administrative and historical purposes.

In some cases, the Committee found that it was difficult to assess the scientific merit of a protocol based on the documentation provided. This is particularly problematic for proposals in which the IRB provides the only opportunity for peer review, as is sometimes the case for research that is not funded by the federal government.

The Committee also found evidence suggesting that in some studies women are being excluded from participation in research, explicitly or presumably because of the possibility that they might become pregnant during the course of the study. This finding is disturbing in light of the fact that much of this research was undertaken after a national policy had been instituted, advocating the inclusion of women in research, and a general rejection of the mere possibility of pregnancy as a justifiable reason for not permit-

ting women to become research subjects.[31] The conditions under which pregnant women ought to be included as research subjects remain controversial. That pregnant women are frequently excluded from research was clearly evidenced in the RPRP; this occurred in more than half the studies in our review.

Some of the Committee's most serious concerns focus on informed consent. The results of the RPRP, as well as of Katz's review, suggest that some consent forms currently in use are flawed in morally significant respects, not merely because they are difficult to read but because they are uninformative or even misleading. These are consent forms that have been approved by an IRB, and still they are problematic, to the point where Committee reviewers viewed them as raising serious ethical issues. Most of these concerns centered on research involving patient-subjects with poor prognoses, people who are particularly vulnerable to confusion about the relationship of research to treatment. The consent forms to be used with such patient-subjects sometimes appeared to suggest a greater prospect of benefit than the research as described in the documents we reviewed warranted. In a few Phase I studies, any intimation that subjects would benefit appeared questionable. At the same time, the disadvantages of participation, particularly as they would affect quality of life, were sometimes inadequately described or not presented at all. The Committee recognizes that the consent form is only a document and is never to be confused with the entire process of soliciting informed consent, which includes far more than the form itself. It is possible that in some of these cases potential patient-subjects were provided more balanced and straightforward information in discussions with investigators or their own physicians. At the same time, however, the consent form as approved by the IRB is a powerful symbol of what the system considers an adequate disclosure. Moreover, this may convey to investigators that meeting ethical obligations to potential subjects requires the investigators to say nothing different and nothing more than what is approved on the consent form.

Our review also raises serious concerns about some research involving children and adults with questionable decision-making capacity. Although we looked at documents from only 125 proposals, we found examples of three controversial, unresolved issues in the ethics of research: research with patients in the midst of a potential medical emergency; research involving children that may offer them no prospect of direct benefit but that may put them at greater than minimal risk, depending on how minimal risk is understood; and research on adults with questionable decision-making capacity that offers them no benefit but that involves unpleasant procedures and exposes them to greater than minimal risk of harm.

All told, the documents of almost half the studies reviewed by the Committee that involved greater than minimal risk raised serious or moderate concerns. Katz, who focused exclusively on the informed consent process, had serious concerns about 50 percent of the greater-than-minimal-risk proposals he evaluated. These are findings that cannot be ignored. At the same time, our review provides evidence that research involving human subjects, even complex research, can and often is being conducted in an ethically responsible manner. The challenge is to identify what needs to be changed to ensure that all research involving human subjects is conducted in accord with the highest ethical standards.

NOTES

1. From 1988 to 1993, the CIA approved twelve proposals for human subjects research. However, none of these proposals involved ionizing radiation as an element of the research design.

2. Studies performed upon human tissue were included in this definition of radiation experiments only if a subject's exposure to ionizing radiation occurred prior to the collection of the tissue.

3. The matrix, shown in the chapter on the Research Proposal Review Project in a supplemental volume to this report, went through several adjustments to account for classification errors in which research abstracts were assigned to the incorrect biomedical category (owing largely to inadequate information in the abstracts).

4. The approach we used can be described as a modified quota sampling methodology. Quota sampling methodology is described briefly in Earl R. Babble, *Survey Research Methods* (Belmont, Calif.: Wadsworth Publishing Company, Inc., 1973), 107-108.

5. The final radiation study sample matrix can be found in the chapter on the Research Proposal Review

Project in a supplemental volume to this report. A matrix of these categories was constructed and a criterion was established that every cell of the matrix be filled with no fewer than 3 and no more than 5 studies from the contemporary period of research (fiscal years 1990 through 1993). Where there were more than 5 eligible studies per cell, a sample of 5 was randomly selected. The original target number of studies per cell was set at between 3 and 10; studies were identified for review according to this initial criterion. However, as the Advisory Committee approached the end of its tenure, the sampling of radiation studies was modified slightly. The Committee agreed that reviewing only 5 would enable the Research Proposal Review Project to be completed in enough time for the results to be included in this report without compromising the meaning of its findings. Accordingly, where 5 studies had not yet been reviewed in any given cell, the number of studies needed to reach 5 was randomly selected from those remaining.

6. Of the 225 research proposals originally chosen by the Advisory Committee, 91 were funded at 43 extramural institutions. As a number of these proposals were deleted during the review process, however, only 32 extramural institutions were represented in the final total sample. Although 31 extramural institutions were represented in the radiation sample, one institution was represented only in the nonradiation sample.

7. We originally believed that the ratio of the number of extramural to intramural studies would be much closer to 1:1; however, this was not the case with our radiation sample. The discrepancy can be attributed to the fact that VA studies are all intramural and that few extramural studies are funded by DOD.

8. The final nonradiation (and radiation) study sample matrix, broken down by funding agency, funding type, and disease, can be found in a supplemental volume.

9. The Advisory Committee received a total of 225 proposals. A number of these proposals were eliminated from the sample because they did not fall within the definitions (for radiation/nonradiation, biomedical category) established by the Advisory Committee. Additionally, approximately 40 proposals were deleted according to the modified sampling scheme that was conceived in order to pare down the number of studies reviewed to 125.

10. Grant proposals were requested directly from the funding agencies. All other materials for intramural studies were also requested from the respective agencies, while other materials for extramural studies were requested from the grantee institution performing the research.

11. Contains a detailed scientific research proposal including references to related work and relevant animal models. Descriptions of the informed consent process and subject selection are not usually well developed in this document. When the proposal was not available, a detailed protocol and summary of research

were used in lieu of the proposal. Intramural studies often did not include a full research proposal, but only the IRB application, consent form, and other supporting documents.

12. May sometimes include the grant proposal, but usually contains a more concise version of the research protocol with greater explanation of and specific attention to the use of human subjects.

13. The consent form provides a written record of what information is provided by the investigators to research subjects. The original and approved versions of the consent form reviewed together give some insight into the IRB review process for the particular research, the extent to which investigators understand the requirements for informed consent prior to submitting their study proposal for review by the IRB, and the IRB's required changes to the consent form and/or process prior to the approval. In some cases, only the consent form approved by the IRB was available.

14. Specific changes required for final approval of the proposal are often indicated in this document. When the disposition letter could not be obtained, a record of the IRB minutes for the meeting in which the project was approved served as a substitute.

15. Such documentation, which generally takes the form of correspondence or annual renewal forms, enabled the Committee to determine if, over time, there were improvements or further problems with the consent procedures, risks and benefits, and selection of subjects.

16. The RDRC application provides additional justification regarding the dosages of radioactive drugs administered to subjects, which helped the Committee assess risks to which subjects were exposed. The Advisory Committee sometimes received a radiation safety committee (RSC) application in lieu of or in addition to the RDRC application.

17. D. A. Henderson, University Distinguished Service Professor, The Johns Hopkins University School of Hygiene and Public Health, to Ruth Faden, Chair, ACHRE, 31 July 1995 ("Who would have believed . . .").

18. A more exhaustive report of the quantitative findings can be found in the Research Proposal Review Project chapter of a supplemental volume of this report. These findings include additional graphs of overall rating distributions according to federal agency, biomedical category, disease, and funding type.

19. All proposals receiving an overall rating of 4 or 5 were reviewed a second time by Advisory Committee and staff reviewers to identify the core ethical concerns involved in each such proposal.

20. In light of the analysis that showed that inclusion/exclusion of "maybe greater than minimal risk" studies with those studies that involved "greater than minimal risk" did not significantly affect the proportion of studies that received each overall rating, the Advisory Committee decided to evaluate these two groups of studies together.

21. In correspondence with RPRP extramural institutions, the Advisory Committee promised not to link findings in this report to any institution, investigator, or research proposal. Where individual RPRP proposals are discussed in the final report, no identifying information is provided.

22. The FDA approval process for the sale and use of drugs and medical devices proceeds through four phases of testing on humans, after being tested in animal models. In Phase I the goal is to establish the dose of a drug or treatment at which toxicity in humans results, in an effort to establish safe levels for human use. In Phase II the goal is to establish a therapeutically effective dose of a drug or treatment whose toxicity has been established. Phase III is testing of a drug or treatment whose therapeutic effectiveness has been shown, to determine its effectiveness as compared with existing drugs or treatments, in preparation for approval and marketing. Phase IV testing is postmarketing data collection to determine the longer term effects of the drug or treatment in a large group of patients, over time.

23. Of the 5 studies involving children that did not require assent forms, 3 involved sixteen- to eighteen-year-olds, and 2 were follow-up studies in which children were not actively recruited and were not likely to be subjects in the study.

24. It is important to note that the degree of professional expertise of the person soliciting consent is often considered important to the quality of the consent process. In some cases, consent is solicited by researchers or nurses working on the project, not by the principal investigators themselves. Although there is controversy over whether the influence of the investigator in the consent process is potentially coercive, many argue that it is the responsibility of the principal investigator to make sure that the subject's consent is informed and voluntary.

25. Institute of Medicine, Committee on the Ethical and Legal Issues Relating to the Inclusion of Women in Clinical Studies, *Women and Health Research: Ethical and Legal Issues of Including Women in Clinical Studies* (Washington, D.C.: National Academy Press, 1994).

26. Committee member Katz initiated this independent review of research proposals. Dr. Katz has been a scholar of the ethics of human experimentation for more than thirty years.

27. Most of these 100 studies reviewed by Professor Katz also appeared in the sample of studies "formally" evaluated by the review teams; however, several were cut from the sample over the course of the review process in the interests of managing the sample size. The studies included in Professor Katz's independent review sample that were eliminated from the "formal" reviews were otherwise relevant to the Research Proposal Review Project and provide additional depth to the scope of the Committee's review of research proposals.

28. Because Katz reviewed a convenient sample of proposals in the order received and did not apply the Committee's selection criteria, his sample included a number of minimal-risk studies that had been eliminated from the Committee's scope of review. It is, therefore, not surprising that there is a discrepancy between the proportion of minimal-risk studies assessed by Katz and those assessed by the full Committee.

29. The President's Commission for the Study of Ethical Problems in Medicine and Biomedical and Behavioral Research was chartered in 1979 to report biennially on the adequacy and uniformity of the federal rules and policies for the protection of human subjects in biomedical and behavioral research, as well as the adequacy and uniformity of their implementation.

30. President's Commission for the Study of Ethical Problems in Medicine and Biomedical and Behavioral Research, *Implementing Human Research Regulations: The Adequacy and Uniformity of Federal Rules and Their Implementation, for the Protection of Human Subjects Biennial Report No. 2* (Washington, D.C.: GPO, 1983), 2, 43.

31. The National Institutes of Health (NIH) Revitalization Act of June 1993 introduced new requirements to "ensure that women be included as subjects in each project of [clinical] research" and "conduct or support outreach programs for the recruitment of women and members of minority groups as subjects in projects of clinical research." A copy of this act can be found in Institute of Medicine, Committee on the Ethical and Legal Issues Relating to the Inclusion of Women in Clinical Studies *Women and Health Research*, appendix B. This report also stresses the need for the inclusion of *pregnant* women in clinical research and recommends "that NIH strongly encourage and facilitate clinical research to advance the medical management of preexisting medical conditions in women who become pregnant (e.g., lupus), medical conditions of pregnancy (e.g., gestational diabetes) and, conditions that threaten the successful course of pregnancy (e.g.,preterm labor)." Ibid., 16.

16

Subject Interview Study

IN reporting to the American people what we have learned about the current status of human subjects research, the Committee wanted to incorporate the voices and experiences of subjects themselves. What is it like to be a subject in biomedical research today? Why do people become research subjects, and what does participating in research mean to them?

To provide answers to these questions, the Advisory Committee conducted the Subject Interview Study (SIS), a descriptive study in which both patients who were research subjects and patients who were not research subjects were interviewed to determine whether they believed that they were participants in medical research, their general attitudes and beliefs about medical research, and if applicable, why they did or did not decide to participate in research. The Committee would have liked to have heard not only from patient-subjects but also from the many "healthy volunteers" who are critical to the success of much biomedical research. Unfortunately, time constraints made this impossible. Clinical research—research involving patients—does account for a large proportion of contemporary medical research involving human subjects, however, and it was toward this enterprise that the SIS was directed.

In this chapter, we report what patients and patient-subjects told us about research and what we learned about their experiences. We begin by describing the methodology of the SIS: how the patients were selected and how they were interviewed. Next, we report the results of these interviews, as well as the results of our review of the records of the patients to whom we talked. We close with a discussion of the limitations and implications of the SIS.

METHODOLOGY

The SIS included almost 1,900 patients at medical institutions across the country. To determine whether the experiences people had with radiation research were any different from those people had with nonradiation research, we interviewed patients in medical oncology, radiation oncology, and cardiology clinics. All of these patients participated in a Brief Survey (five to ten minutes). One hundred three of these patients, all of whom reported in the Brief Survey that they were research participants, also completed longer (roughly forty-five minutes) In-Depth Interviews, designed to give patients an opportunity to elaborate on their perceptions of research and their personal research experiences. Advisory Committee staff and consultants took primary responsibility for designing the SIS, recruiting institutions to participate in the study, conducting some of the interviews, and analyzing the data. Research Triangle Institute, a nonprofit organization, was hired to perform several tasks including conducting focus groups, piloting the interview instruments, conducting the majority of interviews, and performing most of the data entry.

Selection of Institutions

Five areas of the country were selected as sites for the SIS: Ann Arbor, Baltimore/Washington, Dallas/San Antonio, Raleigh/Durham, and Seattle/Tacoma. These sites were selected because they include institutions that receive some of the highest amounts of federal dollars for human subjects research and because we were trying to balance our sample with respect to geographic region, rural/urban settings, and expected ethnic mix. At each of these five sites, a university hospital, a VA hospital, and a community hospital were selected. If other federal

government or military hospitals were present at a site, the most highly funded of these institutions were included. A total of nineteen institutions were selected, as presented in table 1. Interviews were conducted at sixteen of the nineteen institutions selected. At the University of Washington Health Services Center and the Seattle Veterans Affairs Medical Center, the institutional review board (IRB) process could not be completed within the time constraints of the SIS. Baylor University Medical Center declined to participate in the study.

Recruitment of Patients

At each of the participating institutions, patients were recruited from the waiting rooms of three outpatient departments: medical oncology, radiation oncology, and cardiology. On the days that patients were seen in each of the departments, a

Table 1. Institutions Selected for the Subject Interview Study

Ann Arbor
> Ann Arbor Veterans Affairs Medical Center
> St. Joseph's Hospital
> University of Michigan Medical Center

Baltimore/Washington
> Baltimore Veterans Affairs Medical Center
> Clinical Center of the National Institutes of Health
> Greater Baltimore Medical Center
> The Johns Hopkins Hospital
> Walter Reed Army Medical Center

Dallas/San Antonio
> Baylor University Medical Center*
> Dallas Veterans Affairs Medical Center
> Parkland Memorial Hospital and the
> University of Texas Southwestern Medical Center at Dallas
> Wilford Hall Air Force Medical Center

Durham/Raleigh
> Duke University Medical Center
> Durham Veterans Affairs Medical Center
> Rex Hospital

Seattle/ Tacoma
> Madigan Army Medical Center
> Seattle Veterans Affairs Medical Center *
> Swedish Hospital
> University of Washington Health Services Center *

* Interviews were *not* conducted at these institutions.

clinic staff member informed patients arriving at the clinic that a study was being conducted to examine attitudes and beliefs about participation in medical research. The staff member also asked patients if they were willing to have a study interviewer approach them in the waiting room to see if they were willing to be interviewed. Interviewers approached a systematic sample of these patients and, following a brief description of the SIS, asked individuals to participate.

Each patient who agreed to participate in the Brief Survey completed a written consent form that authorized the SIS staff to consult one or more of the following sources to ascertain whether the patient was or had been a participant in a research project: doctors, investigators, research nurses, a research office, a research database, and their medical/research records. The survey, composed mostly of multiple-choice questions, was designed to take roughly five to ten minutes to administer. Patients completing the Brief Survey received $5.00 for their time and effort devoted to the study. All patients who indicated in the Brief Survey that they believed they were medical research participants were asked if they were willing to participate in an In-Depth Interview (roughly forty-five minutes). These interviews raised many of the topics from the Brief Survey. A sample of those who agreed to this further participation were contacted to arrange for an interview at a time and place convenient for them. Patients completing the In-Depth Interview received $25.00 to compensate them for their time and effort devoted to the study as well as to pay for any expenses related to participation in the study, such as transportation and parking.

A target of 150 Brief Surveys (50 each in medical oncology, radiation oncology, and cardiology) was set for each institution. A target of 100 total In-Depth Interviews was set for patients selected from all institutions. Both the Brief Survey instrument and the In-Depth Interview guide appear in a supplemental volume to this report. Electronic files containing the final data from the Brief Survey and transcripts from the In-Depth Survey are maintained along with other records of the Advisory Committee.

Data Collection and Analysis

Brief Survey

The Brief Survey instrument was refined based upon focus groups of patients conducted at two institutions not participating in the SIS: the University of North Carolina at Chapel Hill and Georgetown University. The instrument was then further refined based upon pilot testing at these same institutions. The instrument consisted predominantly of questions with multiple-choice answers addressing:

1. General attitudes toward medical research.
2. Any perceived differences in understanding among the following terms: *medical research, medical study, clinical trial, clinical investigation,* and *medical experiment.*
3. Beliefs about research participation.
4. Reasons for either participating in research or not participating in research (when applicable).
5. Demographic and other background information (such as race, sex, age, and employment/insurance status).

All survey forms were labeled with an identification number for each patient, rather than with patients' names. Data were entered into a computerized database and analyzed using standard statistical methods.

In-Depth Interview

An In-Depth Interview guide was developed based on the focus groups and pilot testing at the University of North Carolina at Chapel Hill and Georgetown University. The In-Depth Interview contained open-ended questions that allowed participants to speak more extensively about the issues addressed in the Brief Survey. For example, patients were asked to describe their attitudes about research generally, their own experience as research participants, how they arrived at their decision to participate, and the informed consent process for their particular project. All of the interviews were audiotaped and transcribed. All cassette tapes and transcripts were labeled with an identification number for each patient and never with patients' names. All transcripts were read in their entirety by Advisory Committee staff members, and then data were coded and analyzed using text analysis software.

Determination of Research Participation

To assess how well patients' reports of their participation in research matched their documented enrollment in research projects at the participating institutions, a mechanism for determining research participation was developed for each institution. In each instance, we sought documentation of participation in research from sources such as patients' medical or research records. This information was supplemented by information from investigators and research nurses.

A second level of review was conducted in those cases in which there was an apparent discrepancy between a patient's own description of having been, or not having been, a research subject and the documents or other sources of this information. A physician or a research nurse on the Advisory Committee staff reviewed the patient's interview, the patient's medical and research records, institutional databases, and other sources of information at the local institution for evidence that could either resolve or verify the discrepancy.

Expert Panel Assessments of the Research Projects

To identify some of the basic characteristics of the research projects in which the patients we interviewed were or had been participating, we convened an expert panel and asked this panel to make some preliminary judgments based on the information provided in the consent forms of these research projects. The panel consisted of eight physicians: specialists in oncology, radiation oncology, cardiology, nuclear medicine, and radiology, as well as general internists.

We attempted to secure a copy of an unsigned consent form for every research project in which a respondent in the SIS was a docu-

mented participant and that had been conducted at one of the study institutions. Although 336 consent forms were requested, only 236 were received in time to be reviewed by the expert panel.

Each consent form received was reviewed by the expert panel, which met for one day. After agreeing how the forms would be evaluated, the panel broke into four teams, each consisting of two physicians, one who had content area expertise in the project being reviewed and another who did not. If a team could not reach consensus on the evaluation of a particular consent form, it was brought to the larger group for review. If a consent form was received after this meeting of the panel it was sent to the panelists for review. The expert panel characterized the research projects on three dimensions: (1) type of research (therapeutic, diagnostic, or other); (2) degree of sickness of the population (expressed as a high burden for those with diseases such as AIDS, a medium burden for those with conditions such as hypertension, or a low burden for those who generally were healthy), and (3) incremental risk assumed by those who participated in the project compared with those who were not participating in the research project (measured as minimal or more than minimal incremental risk).

RESULTS

In this section, we present the results of the SIS. We begin with a description of the demographic characteristics of the patients we interviewed, as well as the basic characteristics of the research projects in which some of these patients were or had been participating. We then review what we learned about patients' general attitudes toward, and beliefs about, research and their understanding of some of the terms commonly used to describe research to potential subjects. This is followed by our results concerning patients' perceptions of whether they are, or are not, participants in research and the extent to which we were able to compare these perceptions with documents and other sources. We then discuss what patients said about the distinctions between research and treatment, and their reasons for deciding to participate. We also describe the characteristics of patients who reported that they declined to be research participants. Our discussion of results closes with what we learned from the SIS about the consent process and issues of voluntariness of participating in research.

Demographic Information

Brief Survey

A total of 1,882 patients completed the Brief Survey. The overall response rate was 95 percent. Patients predominantly were Caucasian (80%), more than sixty years old (53%), and male (59%). Other relevant demographic features are found in table 2.

In-Depth Interview

A total of 103 patients, representing fourteen of the sixteen institutions included in the overall study sample, were interviewed.[1] This sample also was predominantly Caucasian (74%) and male (54%) (see table 2). Due to technical or administrative difficulties with four interviews, only ninety-nine transcripts were available for analysis.[2]

Characteristics of the Research Projects

The characteristics of the projects in which patients participated are described in table 3. The expert panel categorized the disease burden associated with the projects reviewed as low (11%), medium (38%), and high (51%). Approximately half (48%) involved minimal incremental risk from research.

General Attitudes Toward and Beliefs About Biomedical Research

Brief Survey

In the Brief Survey, patients were asked a series of questions concerning their attitudes toward, and beliefs about, "medical research." Almost all the patients had positive impressions of medical research. Specifically, 52 percent reported a "very favorable" attitude toward research and 37 percent a "somewhat favorable" attitude. Only

5 percent of patients described themselves as having an unfavorable attitude. Controlling for multiple factors, the characteristics associated with more favorable general attitudes toward research included being older (age greater than sixty), being male, being a patient in radiation oncology rather than cardiology, and having reported currently being or having been a participant in research.[3]

More than two-thirds of the patients reported that they believed medical research usually or always advances science. More than 80 percent of the patients agreed that medical research does not involve unreasonable risks (86%). Nevertheless, some patients (9%) believed that research usually or always poses unreasonable risks to people. Controlling for multiple factors, the characteristics associated with holding this belief included being younger, being African-American, not having a college education, being in fair or poor general health, and not having any experience as a research participant.[4] Seven percent of patients believed that participants in medical research are usually or always pressured into participating. Patients more likely to believe that people are pressured into research were African-American and had an annual income of less than $25,000.[5]

Thirty-seven percent of patients believed that patients who participate in medical research are usually or always better off, medically, than similar patients who are not in medical research. Patients with a more positive view about research tended to be older, have incomes of less than $50,000 a year, and have had some experience as a research participant.[6]

In-Depth Interview

In the In-Depth Interviews, patients' general attitudes about research often seemed to be shaped by what their own research experiences had been, and patients generally had very positive things to say about their own experiences. Typically, they believed that the projects in which they were or had been participants had been explained thoroughly, that they had been treated kindly, and that they had received at least as much benefit as could have been expected. Moreover, the more experience people had with research, the

Table 2. Demographic Characteristics of SIS Patients

	Brief Survey (N = 1882) Percent	In-Depth Interview (N = 103) Percent
Sex		
Male	58.7	54.0
Female	41.3	46.0
Age Category		
Under 30	3.6	5.0
30 to 59	43.3	53.0
Over 59	53.1	42.0
Race/Ethnicity		
African-American	16.3	23.0
Caucasian	79.7	74.0
Latino	3.9	2.0
Other	2.9	1.0
Education		
Less than high school	21.5	9.0
High school graduate plus those with additional schooling	53.7	52.0
College graduate plus those with additional schooling	24.9	39.0
Annual Household Income		
Greater than $75,000	10.9	21.0
$50,000–$74,999	14.4	15.0
$25,000–$49,999	28.7	25.0
Less than $25,000	41.0	34.0
Not reported	5.1	5.0
Insurance*		
Private	87.2	65.0
Public	46.4	36.0
Veterans Administration	26.6	20.0
Not reported	1.4	12.0
Type of Institution		
Community	37.7	34.0
VA Medical	19.7	17.0
University	28.5	17.0
Government/Military	24.1	32.0

*A detailed breakdown, in schematic format, of the procedures and results reported in this section are found in a supplemental volume to this report.

more positive was their attitude. In addition, a few patients admitted that they had held a rather negative impression of research until they themselves had participated, at which time their impression changed. One respondent said, "I didn't know what to expect. In the beginning I was worried, you know. I was a little upset, a little frighten[ed] and everything. Once I got here, I found that the people were very nice, very professional, and they care about their patients . . . [Y]ou think that you are going to be a number, that they just may be cold and calculating, they['ll] just be thinking about just the data itself and you are just a number or something. But once I got here I found that . . . the doctors and nurses and everybody are very concerned about the indi-

Table 3. Characteristics of Research Projects

Type of Research	Patients' Report* N = 476	Expert Panel Report N = 236†
Therapeutic	65%	69%
Diagnostic	16%	18%
Other	14%	12%
Did not know	3%	N/A
Missing/Unreported	2%	1%

* Patients' Report refers to the patients' understanding of the type of research in which they were participating. Expert Panel Report refers to the expert panel's assessment of the type of research in which the patient was participating.

† Because of time constraints, the remaining consent forms were not reviewed.

vidual and you find that out because they take the time to know your name."[7] These findings are consistent with those from the Brief Survey, in which patients who currently were, or once had been, research participants had significantly more positive attitudes about research than those who had never participated.

When asked for their attitudes and beliefs about medical research generally (rather than about their own experiences), patients, again, had very positive things to say. Research was thought of as a promising endeavor, something that would advance knowledge and help other people: "[Research is] the only way advancement is made in the medical field particularly... [I]t's gotta be done at some point in time on human beings . . . and there are people who are alive today because of the people [who] did research projects."[8] Another respondent strongly endorsed research activities: "Overall I have to say clinical trials, medical experiments are the only way we're going to find any type of results . . . because you can . . . practice on guinea pigs, monkeys, or whatever, but the only way you're going to find out if any of these drugs are going to work is you're going to have to do it on a human being."[9] While patients articulated the necessity of conducting research, a few reiterated the importance of looking out for the interests of the human participants: "I think that . . . research is awfully important in all fields and . . .

the more it involves human life the more guarded one has to be about it."[10]

Terminology

Brief Survey

In the Brief Survey, patients were asked to compare the term *medical research* with one of four alternative terms: *clinical trial, clinical investigation, medical study,* or *medical experiment.* The term *medical experiment* evoked the most striking and negative associations. It was the only term to be evaluated as worse than the term *medical research* on all of the dimensions considered. Specifically, patients who were asked to compare medical experiments with medical research reported that patients in "medical experiments" were more likely to get unproven treatments and be at greater risk than patients in "medical research" and also that they were less likely to do better medically. By comparison, patients thought those in "medical research" were more likely than those in either "clinical investigations" or "clinical trials" to get unproven treatments and to be at greater risk, but they were more likely to do better medically. The term *medical study* got better ratings than the term *medical research* in every respect; medical studies were viewed as less risky, as less likely to involve unproven treatments, and as offering a greater chance at medical benefit.

In-Depth Interview

Distinctions in meaning among different terms for biomedical research also emerged from the In-Depth Interview. Elaborating on the findings of the Brief Survey, the terms *experiment* and *experimental*, for the vast majority of participants, meant that something was unproven, untested, or in the first stage of testing and was thereby riskier and perhaps scary. Some patients said they would become a participant in an "experiment" only if they were terminally ill. A few participants described quite explicit images of what experiments involved: "I envision all kinds of weird things done to the body and I assume that's not true, but also I envision a medical experiment maybe . . . done in a laboratory sealed up somewhere where no one even knows what [is] going on."[11] Another respondent said, "Medical experiment—almost sounds like Frankenstein to me."[12] When asked to explain the term *experiment,* patients often invoked the term *guinea pig* to convey the sentiment of being the "victim" of an experiment. For example, one respondent, when asked to define the term *medical experiment,* said, "That's where you get down to the human guinea pig . . . where they may be injecting medication or whatever they want to inject in someone and watching them for a reaction."[13]

In comparison with the term *experiment, clinical trial* and *clinical investigation* were not such evocative terms. Some patients gave hesitant or stumbling definitions or said they were not familiar with these terms. On the other hand, some patients did attach meaning to these terms, defining them as endeavors that were at an intermediate stage of inquiry, where researchers already know something about the topic and they are now trying the next step.

Patients were most likely to consider "study" a benign endeavor, akin to studying something in school: "*Study* brings to my mind more of using documentation for analysis. . . . With a study . . . you're looking at records. You look at past histories and so forth. . . . It is mostly paperwork, documents, or the books and things."[14]

Of the four terms offered, patients usually said they would prefer to be in a study. It was reported to be the least harmful because it was believed to be the least invasive. In comparison to experiments, which many patients believed involved "trying things out" on animals and/or humans, "studies," they felt, usually entailed gathering information and reviewing paperwork.

Personal Experience With Research

Brief Survey

Thirty percent (570) of the 1,882 patients interviewed reported that they were or had been participants in research (see table 4).[15] We were able to review records or consult other sources for 541 of these 570 cases. By these reviews, we were able to confirm research participation in 302 of 541 cases (56%). In another 203 of the 541 cases (38%), we were unable to find documentation to suggest whether or not the patient was participating or had participated in research. In the remaining 36 cases, the review by health professionals on the Advisory Committee staff concluded that these patients were probably in error and that they were not, indeed, research participants.[16] In summary, 16 percent (302 of 1,882) of the total sample, consistent with their reports, were former or current research participants. Also, assuming that most of the patients for whom research participation could not be verified but, consistent with their own reports were probably truly former or current participants (11%, or 203 of 1,882), then a total of 27 percent of the Brief Survey respondents were former or current research participants. By contrast, 2 percent of the total sample (36 of 1,882), were likely incorrect in their perception of themselves as being participants in research.

Sixty-five percent (1,223 of 1,882) of the patients interviewed reported that they were not and never had been participants in research. We were able to review records or consult other sources for 1,172 of these cases. In 23 of the cases, relevant records were unavailable to confirm participation. In our review of records and other sources, we did not find evidence of research participation for 1,080 of 1,149 patients. In 69 of these 1,149 cases, however, Advisory Committee health professional staff was able to confirm patients' participation in research. In 61 of these 69 cases, the

Table 4. Personal Experience With Participation in Research: Results of the Record Review*

Result of record review	Subjects who reported they were in research N=570	Subjects who said they were not in research N=1,223
In research	53% (302)	5% (69)
Couldn't tell	36% (203)	2% (23)
Probably not in research	6% (36)	88% (1,080)
No record review	5% (29)	4% (51)

*The numbers in this table should be interpreted in the context of the explanation and limitations presented in the text.

preliminary evidence for participation had included an informed consent form signed by the patient for enrollment in the research project. In summary, then, 60 percent of the total sample (1,080 of a total of 1,882) appear never to have been research participants—in the sense that there is no evidence to the contrary—and in another 1 percent of the sample (23 out of 1,882) it is unclear. By contrast, 4 percent (69 of 1,882) of the total sample were apparently incorrect in believing they never had been participants in research.

Although the Committee could not return to the 69 subjects to determine whether the apparent discrepancy was due to true lack of awareness or perhaps to other factors like confusion, misunderstanding of the question, or poor memory, we did attempt to take a closer look at these cases. These 69 patients came from all five geographic sites sampled in the SIS and were receiving care at every type of institution participating in the study (that is, university hospitals, government or military hospitals, Veterans Affairs medical centers, and community hospitals). These patients were interviewed in radiation oncology, medical oncology, and cardiology clinics. Their ages ranged from twenty-one to eighty-nine years of age; 30 were women and 39 were men; and the majority (53) were white (12 were African-American and 4 were of other ethnicities). Their educational background ranged from less than eighth grade to those with graduate or professional degrees.

The records of these 69 patients, who reported that they were not in research but for whom evidence of research participation was found, were subjected to extensive review and analysis by Advisory Committee health professional staff. According to this review, about half of these patients had been enrolled in research during the previous year.[17] The consent forms of 42 of the studies in which these patients were enrolled had been included in the sample of consent forms reviewed by our expert panel. According to the panel, the disease burden for those recruited for these 42 studies ranged from low (5 studies) to high (18 studies), with the remainder being medium (19 studies). Most of these studies involved the evaluation of treatment (23 studies), while some were diagnostic (13) or other types of studies (5).[18] Finally, of these 42 studies, 25 were determined by our experts as posing minimal incremental risk to subjects and 17 as posing more than minimal incremental risk.

In-Depth Interview

Patients completing the Brief Survey were recruited for the In-Depth Interview if they reported that they currently were or once had been participants in research. Through the review process described above, however, research participation could not be verified for 9 of the 99 In-Depth Interview patients, nor did the transcripts of these 9 patients suggest that they were research participants. Two of these 9 patients told stories about research participation that were confusing or unclear. Another 7 of the 9 seemed to believe that anything new or unknown, or, in a few instances, any tests, were research. One such respondent, with a rare medical condition without a known efficacious treatment, described the interventions she received and said, "Everything is experimental, they don't know how to cure it."[19] These 9 transcripts were excluded from further analysis.

Distinctions Between Research and Medical Care

While the Brief Survey did not address distinctions between medical treatment and research, this issue arose during the In-Depth Interview. Here, patients' descriptions of their research experiences often included descriptions of their physical conditions, their own health care providers, or the hospitals at which their research projects were conducted. Research experiences, particularly for those patients who reported being in research evaluating potential treatments, were inextricably interwoven with their medical care experiences. One respondent described her research experience "as a means of treating what I have."[20] Another respondent, when asked what she disliked about the project in which she was a participant, replied: "Nothing other than the fact that nobody likes to be sick and nobody likes to go to doctors."[21]

While patients, if asked, were quite able to identify which procedures, tests, and staff were associated with their research, they did not themselves readily make distinctions between research and medical treatment. Particularly for patients with serious medical diagnoses, research often was viewed as one of the treatment options for their medical conditions. Not surprisingly, then, some participants evaluated their research experience in terms of whether they believed it would provide them with clinical benefit. One respondent noted, "I see results that indicate that the chemotherapy that I'm taking is working, and therefore, that is adequate enough to satisfy me."[22]

Despite the tendency for some patients to fuse discussions of research and treatment, some clearly differentiated the two.[23] This was especially true for those who reported that they were in diagnostic, epidemiologic, or survey research.[24]

Deciding to Participate

Brief Survey

When asked whether specific factors contributed a lot, contributed a little, or did not contribute to their decision to participate in particular research projects, patients typically identified multiple motivations. Most patients reported that they had joined a research project to get better treatment (contributed a lot, 67%; a little, 11%) and because being in research gave them hope (contributed a lot, 61%; a little, 18%). Patients who cited the desire for better treatment as a reason for agreeing to be in research were more likely than other patients to be in a study that they viewed as "therapeutic," that related to the patient's medical condition, and that involved radiation.[25]

In addition to this emphasis on the possibility of better treatment and the bolstering of hope, 135 patients agreed with the statement that they "had little choice" but to participate and that this belief contributed a lot to their decision. While it is difficult to ascertain precisely what these patients understood this statement to mean, patients elaborated on this motivation in the In-Depth interviews, often saying that because of the serious nature of their medical condition and/or because other interventions had not been successful, they believed they had "little choice" but to try research. Patients reporting that they had little choice tended to categorize the projects in which they were subjects as treatment projects (compared with diagnostic or epidemiological), tended to report that the projects involved radiation, that they did not feel they had enough information, and that the research was related to their medical condition.[26]

Altruistic reasons also played a part in many patients' decisions to participate in research. Specifically, most patients reported that they looked at participation as a way to help others (contributed a lot, 76%; a little, 18%) and as a way to advance science (contributed a lot, 72%; a little, 21%). Patients also frequently said that they had joined research projects because it seemed like a good idea (contributed a lot, 48%; a little, 17%), the project sounded interesting (contributed a lot, 53%; a little, 24%), and they had no reason not to participate in medical research (contributed a lot, 56%; a little, 15%).

In-Depth Interview

In reporting how they had decided to participate in research, In-Depth Interview patients described many different processes, ranging from the very deliberate weighing of risks and benefits to the quicker decision of just taking action. Doctors (e.g., "my doctor," "the doctor," a particular doctor, or referring physician) were frequently identified as the key agent in the respondent's decision to participate in research.

Patients expressed a broad range of *reasons* they decided to participate in biomedical research. As in the Brief Survey, for people in therapeutic research, the primary reason for participating in research was to obtain benefits either through an experimental treatment they hoped would be better than standard treatment or through the closer medical attention they believed they would receive through research. One woman reported that she was participating in a treatment trial specifically to obtain an experimental drug that she believed looked promising. Furthermore, she wanted to receive it in a controlled environment where she could receive good follow-up and where researchers would document the drug's effects.[27] Another respondent commented that since doctors at the military hospital where he received his care were very busy, he could receive closer attention and obtain appointments more easily by enrolling in research.[28] Some patients who reported being in therapeutic research hoped that the research would give them more "time": "[A]ll I wanted at that point was five years to get my boys through high school";[29] "I want longevity . . . I don't see myself wanting to just pass away."[30] Some patients decided to be in research because they believed that newer therapies might inherently be better: "If there's something new on the market that might be better than the traditional program they've been using, why not try it?"[31]

Mirroring the Brief Survey finding that 31 percent of patients felt they had little choice in joining a research project, many In-Depth Interview patients who participated in therapeutic research remarked that they had joined because they believed they had "no choice," meaning they had no medical alternatives: "My doctor told me if I do not take the drug, in a couple of months I . . . [will] . . . die. So, I had no choice. Who wants to die? Nobody."[32] Another respondent said, "I had one more option as he [the doctor] put it."[33] Hope and desperation pervaded the remarks of many terminally ill patients. Patients said they wanted to "try anything" or that this was their "last resort." One man explained, "Well, what was driving me to say 'yes' was the hope that this drug would work. . . . When you reach that stage . . . and somebody offered that something that could probably save you, you sort of make a grab of it, and that's what I did."[34] This same patient noted that he had first declined what he had considered a very aggressive therapy, "because at that point everything was pretty okay and there was no need for me to do any wild things."[35] Later, when his condition worsened, he decided to participate in the research.

One of the most influential forces in patients' decisions to enroll was doctors' recommendations. One patient described the process of her enrollment: "He [the doctor] asked me if I wanted to go on it, and I said, 'If it's what you think I should do, yes, because you know more about it than I do.' . . . [H]e said, 'I think it would be a good idea to try it.'"[36]

Along these lines, a theme of trust overwhelmingly emerged. Patients trusted specific physicians, medical professionals more generally, or the overall research enterprise. Trust in specific physicians was straightforward: "Basically, y[ou] know, we trust Dr. [So-and-so] . . . [There] was no reason to, . . . get a second opinion from another doctor."[37] Another respondent exclaimed "Oh, I love that man. He has kept me alive and I obey him and I do what he tells me to do. . . ."[38] Some patients also communicated trust in the medical profession more generally: "I have this attitude. They know what they're doing. They wouldn't have you to do this if they didn't know what they were doing and . . . that's my attitude. . . ."[39] Finally, there were a few patients who expressed trust in the overall enterprise of medical research as well as its oversight. One respondent stated: "I do not feel like the drug would be on the market if it were going to harm me, and if it would help in any way . . . I'm very willing to participate in this and perhaps other studies."[40] Related were patients who said they decided to participate because of their trust in the institution where the research was being conducted. "I think I've got the best treatment down there [named hospital]. I don't think I could get any better."[41] Rare were the patients who had less "blind trust" and considered themselves to be more of a consumer: "I sort of take my own treatment in my head and tell them that I'm his client. It's not the other way around. . . ."[42]

Elaborating on responses to the Brief Survey, the majority of patients mentioned altruism as a reason to participate. This desire to help others took many forms, including helping others who had the same medical condition, advancing medical science more broadly, and contributing to society. Most frequently, those in therapeutic research seemed to voice a combined motivation of seeking benefit for themselves and hoping to achieve benefit for others. Very representative was the comment, "I was hoping, if not for me, at least for the next people coming along. . . ."[43]

For some patients who faced a life-threatening illness, participating in research seemed to offer them a greater sense of personal worth, a chance to contribute something of value to society. One woman said, "[I]f I can help find a cure for what seems to be so common [that is, cancer] these days, I would love to think I was part of finding that cure."[44] For a small number of patients, this notion of helping others went further, to be a duty or obligation: "[I thought], well, I don't have to do this, and then I thought, well, here I am benefiting from literally thousands and thousands of experiments that have gone before and that are helping to save my life and this one sounded [very] reasonable to me and I was happy to participate."[45] Similarly, one respondent replied, "I feel like that [participating in research and giving blood] is a moral obligation as a citizen. You put back into your community. . . . [O]pportunities to not only help yourself but other people are real important to me. . . ."[46]

Only three patients cited monetary reasons for participating in research.[47]

Deciding Not to Participate

It is also clear from the Brief Survey that not all patients approached to participate in a research project agree to do so. In fact, 191 (10%) of the 1,882 patients we spoke with told us that at some point they had made a decision not to participate in research. While 112 (59%) of these 191 patients had never decided to be in research, the remainder reported that at some time or other they had (39 were current research participants, and 40 were former research participants), suggesting that some patients discriminate between projects they are willing to participate in and those they are not. Patients who declined to participate in research ranged in age from twenty-one to eighty-three, with a median age of fifty-six. The patients were of both genders (53% male, 47% female), predominantly white (69%, with 27% African-American and the remainder being of other ethnicities), with wide educational backgrounds ranging from less than eighth grade to those with professional degrees.

We asked the 112 patients who had never been in research why, when they had been offered the chance, they had decided not to participate. The reasons that "contributed a lot" to their decision were that they wanted to know what treatment they were getting (64%); they wanted their medical decisions to be made by their doctors and themselves, not by researchers (56%); they believed that being in the medical research project was not the best way for them to get better (45%); and taking part in the medical research project would have been inconvenient (43%).

Consent and Voluntariness

Brief Survey

Overall, 83 percent of patients who told us they were current or former research participants remembered signing a consent form agreeing to take part in research. This was true for 88 percent of current research participants and 80 percent of former research participants. Most (90%) of the patients who believed that they were current or former participants in research reported that they felt they had enough information to make a good decision about whether to participate. This was the case for 95 percent of current research participants and 87 percent of former research participants.

Fewer than 2 percent of current or former research participants felt pressured by others in making a decision to participate. Six patients specifically said that they had been pressured by someone in the medical field (e.g., "my doctor"; "the hospital"); four patients reported having been pressured by someone in the military (e.g., "the military"; "Admiral on ship").

When patient-subjects were asked what they thought the policy was for dropping out of the study in which they were participating, 78 percent thought, correctly according to current research standards, that they could drop out at any time. A variety of other responses were also offered, ranging from not knowing the policy, to expressing that it was irrelevant (e.g., the entire project consisted of a single survey or blood test), to believing they had to stay in the research project.

In-Depth Interview

On the whole, patients who granted In-Depth Interviews recounted that the staff involved in conducting research explained research projects, gave participants time to read over the consent forms and confer with family and friends, and responded to participants' questions. One patient said explicitly, "It seemed to me that they were well prepared to answer any questions I would ask them."[48] Asked if research staff had provided her with as much information as she needed, one patient replied that they used "terminology . . . that I could relate to. They spoke in my language. That was a plus."[49]

The consent process, in general, and the consent form, in particular, held varying degrees of importance for patients. Most patients enrolled in survey or noninvasive projects did not attach a great deal of meaning to the consent form. One respondent, whose experimental procedure consisted of "just drawing some blood," thought, in fact, that his consent form went overboard.[50] For those patients who reported being in research evaluating potential treatments, the value of the consent form varied. For many, the decision to

participate seemed to have been made before the consent form was given to them, and they signed it almost as a formality. For a few, signing a consent form symbolized the first step on the path to getting better. Others, however, relied heavily upon the content of the form when deliberating about whether to participate. In addition, several patients noted that they held on to their consent forms, a few even offering them up for the interviewers' review.

The notion of trust also accompanied accounts of the consent process. For some participants, the consent form was the means by which patients could authorize trusted health professionals to do what they think is best. One respondent remarked, "[W]hatever the doctor was doing, well, that was all right. I consented to this and let the experts take over then."[51] This authorization for treatment meant abdicating attention to detail for some patients: "I'm the type of person, I don't read all this fine print and all this stuff and so forth. The lady said that we would like to experiment on your body to see what can be done . . . and it's to help me and so far, so good. . . ."[52]

While patients attached different levels of personal interest to the consent process, they were clear that the type of information typically conveyed in a consent process is exactly what they would need in order to decide about participating in research in the future. Patients overwhelmingly said that they would participate in a research project again if they had enough information and if the project were explained in sufficient detail by research staff: "I'd have to know the what fors, ifs, whys, what they're gonna do. . . ."[53] Or, "if somebody can't explain what they're going to do to me good enough, I wouldn't [do] it."[54] Furthermore, several patients stated that they would like to know why a particular study was being conducted and why certain procedures or techniques were necessary. "Communication," "information," and "honesty" were frequently identified by participants as essential in considering participation in any future research project.

For patients who described their own consent process, experiences generally were positive. A few patients reported problems, however. Three general problems were identified: (1) too much technical information that was difficult to read and understand,[55] (2) an overwhelming amount of information,[56] and (3) discussions occurring at stressful or inappropriate moments.[57] A few patients reported that during discussions with physicians or investigators they relied upon family members to help process the information conveyed.[58]

A few patients remarked upon the importance of contact among participants in research projects evaluating treatments. One respondent contrasted the type of information one research participant can provide to another versus that which a doctor can provide: "[It's] always nice to be able to . . . see somebody in the same boat or talk to [that person]. . . . because even though a doctor is very good in explaining thing[s]. . . . there are certain things that . . . only somebody who's going through the thing can really know what you're talking about."[59]

Consistent with findings from the Brief Survey in which 98 percent of patients reported that they were not pressured into participating in research, almost all the patients who gave In-Depth Interviews believed that the decision about whether to participate in research had been theirs to make and that they had not felt pressured into that decision. Indeed, many patients mentioned that they participated "voluntarily." One respondent said, "They wanted to know if I would be interested in this. Nobody was pushy. Nobody, they just said, 'Here it is, would you like to be involved in this program[?]'"[60] No one interviewed identified pressure from family members. More often, patients remarked that while they conferred with families and friends, the choice was ultimately their own: "My family. The people I work with . . . [E]verybody tells you you have to make up your own mind. . . . [N]obody's going to tell you what to do because it wouldn't work anyway. So nobody tried to influence me one way or the other. . . ."[61]

There were only a few patients who suggested that doctors tried to exert what was viewed as unwelcome or inappropriate influence. One respondent, who remarked in one portion of the transcript that she did not feel pressured, later reported, "[The doctor] sorta made a plug. He said, 'you know, if people like you refuse to get into this . . . we're never going to get any-

where.'"[62] Another respondent indicated that he felt pushed by one doctor to sign a consent form for a particular type of infusion treatment. "[T]hey say, well . . . go ahead and sign it . . . so we can . . . start you on the process, and I said, well, I want to read it. . . . And he said, have you signed it yet? And I said, 'I haven't read it yet. Oh, okay, well . . . we need you to sign it and then . . . make a copy and we'll just let you read it afterwards and I thought, what is going on? I mean, they had never ever kind of pushed it like that."[63]

Almost all patients reported that they had been told they could leave the research project at any time and that they *believed* that they could leave at any time. One respondent said, "[T]hey always told us all the way along, anytime you don't feel happy with this, we can quit . . . they said if you don't feel like you want to continue, you can quit anytime. There was no pressure on or nothing. . . ."[64] Similarly, "[I]t was made very clear up front[,] and then in the original package of material that they had, at any point in time for any reason in time any reason I wanted, you know, I didn't even have to have a reason, I could withdraw with no problem."[65] For some participants, the question of withdrawing seemed almost foreign because there was such trust in the research process. One respondent said, "[The thought of withdrawing] never entered my mind. I was going to let them make the decision because they were the ones that were watching the cancer. . . . I wasn't the expert. If they thought it was working, that was fine."[66] One respondent who was in the military believed that continued participation was required.[67] Another respondent, about to undergo a bone marrow transplant, reported being pressured both to enroll and to continue participation in a clinical trial. "They were really pushing this procedure [a drug to help raise white blood cell counts]. . . . It was very obvious to me that they wanted people to sign up for this bad, and I did not want to upset my doctor. . . . Y'know I'm totally helpless. I'm in his hands and so, part of it was, I wanted to keep him happy and, uh, there was some pressure."[68]

As described earlier, several patients in therapeutic research identified an intense desire to have some type of treatment. This not only influenced their decision to enroll, but also to remain in a research project. One respondent stated that participating in research "was through necessity. . . . [T]he thought never entered my mind that I would withdraw from this program."[69] Such sentiments also seemed to influence patients' desires to find research projects for which they might be eligible. "I said if something comes up that you think will benefit me, let me know . . . I wanted to be on that trial bad enough to where I gave [in to] the pressure."[70]

DISCUSSION

Limitations

Although we were able to involve different types of hospitals from five different areas across the country in this study, only sixteen hospitals were included in our sample. We have no way of knowing whether our findings would have been different if we had interviewed individuals at other hospitals. Similarly, we interviewed only medical oncology, radiation oncology, and cardiology patients who were not hospitalized but were receiving their care at outpatient clinics. Most of them were white, and many were more than sixty years of age. It is quite possible that other types of patients would have answered some questions differently from our patients and that healthy research subjects might have had different attitudes, beliefs, or motivations for participation than patients likely to have serious illnesses did. In the In-Depth Interview component of the study, only people who believed they were or had been research participants were included. The responses of people who had chosen *not* to participate in research, presumably, would be quite different. It should not be assumed, therefore, that our findings necessarily apply to the entire research enterprise.

An important research question in this project was the degree to which present-day patients know whether or not they are research participants. To answer this question, we interviewed patients and asked them whether they believed they were, or had been, participants in research, and then, with their permission, we checked their records for evidence of research participation. Although this approach provides an estimate of the degree to which present-day patients know or remember if they are research subjects, this estimate is likely to be very rough for two sets of reasons. First, interviewing patients in the way we did may not be the most accurate way of gauging their own understanding of participation in research. This is because they were often approached in a busy clinic setting by an interviewer they did not know. It is also likely that these patients were under stress at the time of the interview, either because of their upcoming appointment with their doctors or because of the very illnesses that brought them to the clinic. In addition, because of necessity the Brief Survey was designed to take only five to ten minutes to complete; we asked patients only about current or former research participation with single questions, rather than a series of questions designed to more completely capture those patients who had experience with research. Moreover, following our review of the medical and research records we did not go back to patients and ask them questions about research once we had an understanding of their medical history and documented research experience.

Second, despite significant attempts to gather information from multiple sources, the method of abstracting medical and research records we used may not have been comprehensive enough to locate all relevant evidence of research participation (e.g., records of research may not be retained at the same institution in which the Brief Survey was conducted, or research participation may have been in the distant past and records may no longer be available). A related problem is that some patients may have been enrolled in studies that purposely do not keep records of participation (e.g., studies where confidentiality is paramount). Finally, while trained abstractors examined records for all patients, health professionals on the Advisory Committee staff only reviewed records where patients' responses differed from the results of the initial records review conducted by the trained abstractors. Health professionals had only a one- to two-day window to perform this confirmatory search of documentation at each institution and thus were not always able to review relevant records because they were unavailable on short notice.

Because of these reasons, we do not know the degree to which our estimates are accurate regarding the proportion of patients whose responses about research participation differed from what we found in records.

Implications

A striking finding from this study is the frequency with which people with cancer and heart disease appear to come in contact with biomedical research in the course of their medical care. Notably, nearly 40 percent of the patients we talked with either believed they were or had been subjects in research, had records that showed that they were, or had reported that they had been offered the opportunity to be in research but had declined. Moreover, most patients thought that medical research was a good thing. They had favorable attitudes toward medical research generally, they believed that research did not involve unreasonable risks, and they believed that medical research usually or always advances science. Patients who are or had been participants in research had even more positive attitudes about research than those who had not.

There was evidence in this study that many patients feel free to refuse when physicians and researchers ask them to become research subjects. Nearly 200 patients told us that they had been offered an opportunity to participate, but had declined. Moreover, 40 percent of these patients had chosen at some other time to participate in research, indicating that at least some patients are discriminating in terms of the circumstances under which they are willing to participate in research. There also was little evidence that patients felt coerced or manipulated by health care providers or scientific investigators to participate in research. When we asked patients who were subjects if they had felt pressured by others into becoming research participants, these patient-subjects overwhelmingly said no. Not only did they give the impression that the initial decision to enter a research project was theirs, but many also informed us that they had been told frequently by the investigators that they could drop out of the study at any point, and the patients believed that this was so.

Although the vast majority of both African-American and white patients held favorable beliefs about research, such beliefs were held less often by patients who are African-American. Specifically, as compared with white patients, African-American patients were more likely to believe that people are pressured into research and more likely to believe that research poses unreasonable risks. These findings together suggest that for a small number of patients, distrust as a result of the troubled historical experience of African-Americans in research, as exemplified by the Tuskegee syphilis study, may persist.

We learned a great deal from this project about why patients choose to be in research. The overwhelming majority of the patients we interviewed who were participants in research were subjects in studies investigating medical treatments. Almost all of these patients said that they had enrolled in research because they thought it offered them their best chance of personal medical benefit. Moreover, for many of them, their doctors had recommended it. Often these patients had very serious illnesses and had tried many treatments unsuccessfully; the opportunity to be in research offered them hope that improvement might still be possible. Many of these patients specifically said that they had "no choice" but to participate. They had tried everything else to improve their condition, and nothing else had worked. These patients felt constrained to participate because of their medical situation, not by their providers or the research investigators.

Not surprisingly, then, when asked to describe the research project they were in, most of the patient-subjects we talked with described the project as part of their therapy. Although, when asked, these patients appeared to clearly understand which interventions were associated specifically with the research, they also conceived of the research as their medical treatment. And despite the recognition by most of these patients that the goal of the enterprise of medical research generally was to advance science, when asked about their own specific project, they often believed that the project would benefit them.

It is likely that in some, and perhaps in many, of these cases, it was indeed in the patient's medical best interest to be enrolled in a research project. As demonstrated by the recent push for access to investigational drugs on the part of people with HIV infection and other serious illnesses where there may seem to be no truly efficacious standard therapies, many patients believe that their best chance of extending life is to take treatments that are still experimental. In some cases, patient-subjects were participating in treatment studies involving agents available only through research because their illnesses may have had no known efficacious treatments. From the perspective that holds extending life to be the primary concern, it would be in the patients' best interests to be in the research.

It is a separate issue whether participation in research is in a patient's *overall* best interests. Investigational interventions for devastating, life-threatening illnesses may be a patient's best chance—however small—of extending life. However, this chance may be at the expense of the person's ability to function and enjoy life for the time affected by participation in the re-search. Furthermore, the history of experimentation demonstrates that such therapies might also shorten life rather than extend it. Unfortunately, we did not pursue whether these sorts of trade-offs were clearly understood by the patient-subjects we interviewed. In chapter 15, we report some data from the RPRP that bear on this question.

That patients viewed their participation as being in their best interests is consistent with patients' profound trust in their physicians, on whom they depend as their lifelines, and who they could not imagine offering something *not* in their best interests. We heard from several patients the belief that their doctors are the experts and that they know best what would be helpful. If a doctor recommended or even offered research, patients were certainly more inclined to decide to participate. The trust that patients placed in their physicians often was generalized to the medical and research community as a whole. Patient-subjects frequently expressed the belief that an intervention would not even be offered if it did not carry some promise of benefit; many certainly assumed that the intervention would not be offered if it posed significant risks.

It was largely because of this trust that most patient-subjects considered the consent process somewhat incidental to their decision to participate in research. When asked, almost all patients reported that they had been provided with information, their questions had been answered, and they had been satisfied with the consent process. Nevertheless, doctors' recommendations and patients' own beliefs that the research was their best chance or even their only hope made the research an obvious decision for many patients, and the consent process and consent form were viewed as somewhat of a formality.

This framing of research as therapy is consistent with the very language used to describe research projects. We learned that patients attach very different meanings to the different terms associated with medical research. *Experiments* are considered by patients to involve unproven treatments of greater risk, often invoking the image of human beings as "guinea pigs," while terms such as *clinical investigation* or *study* convey less uncertainty to patients and a greater chance of personal benefit.

The design of this study does not allow us to assess whether patients' expectations of benefit from their therapeutic trials were appropriate for the particular studies in which they were enrolled, or whether their expectations were exaggerated or unrealistic. Moreover, if patients' expectations were exaggerated in some way, we have no evidence to discern whether patients overestimated the expected benefit themselves or whether it was investigators who suggested that the research held more promise than was warranted. It is understandable that patients with poor prognoses may read hope into even the slimmest possibility of benefit. It also is understandable that some physicians, uncomfortable with having little to offer their seriously ill patients, might at such times inadvertently impart more hope than the clinical facts, strictly speaking, warrant.

Hope is a delicate and precious commodity for those with life-threatening illnesses. For clinicians, the balance between support of that hope and honesty is often difficult. At the same time, however, there is a world of moral difference between a physician emphasizing—even inappropriately—slim chances, in order to bolster waning hope, and a physician emphasizing slim chances in order to meet a recruitment goal for a clinical investigation. Feeding hope at the expense of candor is one thing; exploiting the desperation of those whose lives hang in the balance is another. Here again, our data are silent. We cannot know, insofar as physicians contributed to unrealistic expectations among these patient-subjects, how often this was the result of well-meaning reassurances or self-interested misrepresentations.

It seems very much related that we found that a small proportion of patients believed they were subjects in research when it appeared they were not, and other patients believed they were not research subjects when records suggested that they were. These confusions about whether a patient was in research occurred almost exclusively when patients were in (or thought they were in) research investigating potential therapeutic interventions. However, we found that these patients covered the full range in terms of education, income, sex, and race; they came from all three medical specialties studied and all types of hospitals.

At least three-quarters of the patients who apparently were mistaken when they reported they were not research subjects had actually signed consent forms authorizing participation in research. In addition to the limitations of our methods described earlier in the chapter, we can only speculate as to why the discrepancy exists between patients' perceptions and their records. Some patients may not have understood our question and may in fact have known they were research subjects all along. Other patients may not have understood what they were doing when they signed the consent form, perhaps believing that it was a consent for treatment. Still other patients may have had an adequate understanding that they were consenting to participate in research at the time they signed the form and then later forgot. This last explanation is not as troubling as the second, in that it suggests the possibility that in at least some cases valid consents were initially obtained, but it does raise questions about the meaningfulness of these patients' rights to withdraw from research. Such questions are obviously more meaningful in ongoing projects that involve continuing exposure to potential risk, in contrast to those studies where research participation is less burdensome, such as studies involving routine follow-up or only a minor change in a regular therapeutic regimen.

It is often the case in clinical research that the participation of ill people in research and the medical treatment they receive for their illnesses are identical. When this occurs, it is not surprising that some patients conflate their being in research with therapy to the point that they no longer understand or remember that they actually are in a research project. Ironically, it may be especially when patient-subjects feel well cared for that they are most likely to feel like a patient only, and not like a research subject. At the same time, many patient-subjects told us of being reminded by research staff that they could leave the project at any time for any reason. It seems doubtful that the patients we interviewed whose self-report of participation was not consistent with research records had such an experience.

Although most of the patients we interviewed listed a chance at medical benefit as a reason for participating in research, many patients also said that they had participated in research to help others. Some patients described the willingness to participate in research as a civic duty; others wanted to help members of their own families at risk for the same conditions, and still others saw being in research as a means of making a shortened life expectancy more meaningful. Participants in survey research and similar research projects were especially likely to say that they had joined in part because there was no reason not to do so, but also because they hoped they could help others or advance science by doing so. Several patients in therapeutic research who appreciated that there was only a slim chance that the research would provide them with personal benefit, offered that, as a result of their participation, they hoped at least that someone down the road would be better off, if not themselves. This willingness of patients to be altruistic should be tapped explicitly when recruiting participants for research, since it might help to underscore for patients that the primary objective of research is to create generalizable scientific knowledge rather than simply to offer them a chance for some medical benefit. In the end, it is only the benefit of furthering knowledge that can be honestly guaranteed to a potential research subject.

NOTES

1. Because of time constraints, no In-Depth Interviews were conducted with patients from the University of Michigan or the Baltimore VA Medical Center.

2. One audiotape of poor quality was never transcribed. Transcripts for three patients who stated clearly during the In-Depth Interview that they had never participated in research and who were inappropriately selected were also excluded.

3. All models were developed using multiple logistic regression, and the results are reported here as the baseline probability of a given response along with the approximate absolute difference due to a given factor. Each factor either adds to or subtracts from this baseline probability. Note that the baseline probability used in the models is not equal to the overall probability of corresponding response reported in the text. It is only a "baseline" within the context of these models.

4. Baseline probability of saying that medical research usually or always involves unreasonable risks was 19%; Age > 60, -6%; African-American, +11%; College degree, -6%; Good health, -8%; Research participant, -6%.

5. Baseline probability of feeling that potential subjects are usually or always pressured was 5%; African-American, +9%; Income=$25,000-$50,000, -2%; and Income > $50,000, -3%.

6. Baseline probability or saying those in research usually or always do better was 31%; Age > 60, +7%; Income > $50,000, -8%; Research experience, +9%.

7. Subject No. 335216-8, interview by Subject Interview Study staff (ACHRE), transcript of audio recording, 27 March 1995, lines 40-43, 170-175 (Research Project Series, Subject Interview Study).

8. Subject No. 551334-6, interview by Subject Interview Study staff (ACHRE), transcript of audio recording, 14 March 1995, lines 706-714 (Research Project Series, Subject Interview Study).

9. Subject No. 335213-5, interview by Subject Interview Study staff (ACHRE), transcript of audio recording, 28 March 1995, lines 1663-1668 (Research Project Series, Subject Interview Study).

10. Subject No. 442107-9, interview by Subject Interview Study staff (ACHRE), transcript of audio recording, 29 March 1995, lines 458-460 (Research Project Series, Subject Interview Study).

11. Subject No. 552106-7, interview by Subject Interview Study staff (ACHRE), transcript of audio recording, 14 March 1995, lines 311-315 (Research Project Series, Subject Interview Study).

12. Subject No. 442107-9, interview by Subject Interview Study staff (ACHRE), transcript of audio recording, 29 March 1995, line 432 (Research Project Series, Subject Interview Study).

13. Subject No. 333208-7, interview by Subject Interview Study staff (ACHRE), transcript of audio

recording, 6 March 1995, lines 745-748 (Research Project Series, Subject Interview Study).

14. Subject No. 333256-6, interview by Subject Interview Study staff (ACHRE), transcript of audio recording, 16 March 1995, lines 361-366 (Research Project Series, Subject Interview Study).

15. A detailed breakdown, in schematic format, of the procedures and results reported in this section is found in a supplemental volume to this report.

16. A variety of reasons suggested that although patients reported that they were research participants, review of their medical records suggested that they were not. For instance, in comparing patients' self-reports with their records, what they had called "research" actually was standard clinical care that they were receiving.

17. Data for this analysis were available for 54 of 69 individuals with discordant responses: 54% had enrolled after 1 January 1994; 65% after 1 January 1993; and 72% after 1 January 1991.

18. Data regarding type classification are missing for one study in this group.

19. Subject No. 552212-3, interview by Subject Interview Study staff (ACHRE), transcript of audio recording, 13 March 1995, lines 34-35 (Research Project Series, Subject Interview Study).

20. Subject No. 335227-5, interview by Subject Interview Study staff (ACHRE), transcript of audio recording, 24 March 1995, lines 226-227 (Research Project Series, Subject Interview Study).

21. Subject No. 552126-5, interview by Subject Interview Study staff (ACHRE), transcript of audio recording, 13 March 1995, lines 186-187 (Research Project Series, Subject Interview Study).

22. Subject No. 221202-5, interview by Subject Interview Study staff (ACHRE), transcript of audio recording, 14 March 1995, lines 403-405 (Research Project Series, Subject Interview Study).

23. One respondent remarked, "Wouldn't it be great" if this particular protocol "that treats you for [a] shorter period of time [with] high doses of chemotherapy" proved as beneficial as treatment over "years and years" (Subject No. 335215-0, interview by Subject Interview Study staff [ACHRE], transcript of audio recording, 24 March 1995, lines 536-539 [Research Project Series, Subject Interview Study]). It was clear that this respondent conceptualized research as an endeavor aimed at increasing knowledge about unproven interventions, rather than understanding research as a form of medical care.

24. Research experience for these patients tended to be described more dispassionately as a one-time event that stood apart from their therapeutic needs. Indeed, the patients tended to minimize the effect that the research experience had for them personally: deciding to join required little deliberation, and participating required little effort. One respondent made it clear repeatedly that her research experience was "just" an interview (Subject No. 334148-4, interview by Sub-

ject Interview Study staff [ACHRE], transcript of audio recording, 3 March 1995, lines 73, 105, 143-144 [Research Project Series, Subject Interview Study]).

A respondent who participated in survey research agreed out of a willingness to help: "If I can help on anything, I want to be able to do it . . . . 'course my wife thinks if we can help in any research, we're both willing to do it" (Subject No. 443321-5, interview by Subject Interview Study staff [ACHRE], transcript of audio recording, lines 108-109, 125-126 [Research Project Series, Subject Interview Study]). Finally, in notes kept by interviewers and in debriefing sessions with interviewers, interviewers reported that most patients who had participated in survey research simply did not have a lot to say compared with other patients.

25. Groups for which there was marginal statistical evidence for increased frequency for this belief were African-Americans (versus Caucasians) and those who were retired or unemployed. In this model, the baseline probability of contributing a lot to the decision was 13%; Treatment study +27%; Involved radiation 13%; African-American +8%, Employed -5%; and Research related to condition +20%.

26. Baseline probability of contributing a lot to their decision was 11%; Treatment study +10%; Radiation +6%; Had enough information 7%; Research related to condition +25%.

27. Subject No. 335215-0, interview by Subject Interview Study staff (ACHRE), transcript of audio recording, 24 March 1995, lines 11-72 (Research Project Series, Subject Interview Study).

28. Subject No. 553109-0, interview by Subject Interview Study staff (ACHRE), transcript of audio recording, 10 March 1995, lines 252-269 (Research Project Series, Subject Interview Study).

29. Subject No. 441227-6, interview by Subject Interview Study staff (ACHRE), transcript of audio recording, 10 March 1995, lines 542-544 (Research Project Series, Subject Interview Study).

30. Subject No. 221202-5, interview by Subject Interview Study staff (ACHRE), transcript of audio recording, 14 March 1995, lines 350-352 (Research Project Series, Subject Interview Study).

31. Subject No. 333208-7, interview by Subject Interview Study staff (ACHRE), transcript of audio recording, 6 March 1995, lines 30-32 (Research Project Series, Subject Interview Study).

32. Subject No. 333215-2, interview by Subject Interview Study staff (ACHRE), transcript of audio recording, 3 March 1995, lines 194-195 (Research Project Series, Subject Interview Study).

33. Subject No. 114229-8, interview by Subject Interview Study staff (ACHRE), transcript of audio recording, 14 March 1995, line 119 (Research Project Series, Subject Interview Study).

34. Subject No. 332250-0, interview by Subject Interview Study staff (ACHRE), transcript of audio recording, 28 March 1995, lines 105-109 (Research Project Series, Subject Interview Study).

35. Subject No. 332250-0, interview by Subject Interview Study staff (ACHRE), transcript of audio recording, 28 March 1995, lines 208-210, 188-193 (Research Project Series, Subject Interview Study).

36. Subject No. 552264-4, interview by Subject Interview Study staff (ACHRE), transcript of audio recording, March 1995, lines 432-434 (Research Project Series, Subject Interview Study).

37. Subject No. 114229-8, interview by Subject Interview Study staff (ACHRE), transcript of audio recording, 14 March 1995, lines 194-198 (Research Project Series, Subject Interview Study).

38. Subject No. 114217-3, interview by Subject Interview Study staff (ACHRE), transcript of audio recording, 10 March 1995, lines 50-51 (Research Project Series, Subject Interview Study). A powerful instance of the finding of trust in physicians' recommendations and requests was the participant who said, "He [a physician] asked me would I do it and I told him, 'Yeah.' I didn't think that he would harm me [in] any kind of way, hurt me in any kind of way, so I told him, 'Yeah.' He couldn't get I don't believe . . . anybody else to do it" (Subject No. 441311-8, interview by Subject Interview Study staff [ACHRE], transcript of audio recording, 28 March 1995, lines 155-158 [Research Project Series, Subject Interview Study]).

39. Subject No. 332324-3, interview by Subject Interview Study staff (ACHRE), transcript of audio recording, 24 March 1995, lines 309-311 (Research Project Series, Subject Interview Study). Another reported, "There's not a lot that you can control when you're sick so you have to rely on your doctor . . . if he suggests that you should go into a research project, I think you should really take his advice or her advice, whatever it may be . . .[B]ecause if you take the time to get yourself a good doctor and they're involved in research, they would never steer you wrong" (Subject No. 552244-6, interview by Subject Interview Study staff [ACHRE], transcript of audio recording, 5 March 1995, lines 617-675 [Research Project Series, Subject Interview Study]).

40. Subject No. 443241-5, interview by Subject Interview Study staff (ACHRE), transcript of audio recording, 2 March 1995, lines 67-70 (Research Project Series, Subject Interview Study).

41. Subject No. 333208-7, interview by Subject Interview Study staff (ACHRE), transcript of audio recording, 6 March 1995, lines 381-383 (Research Project Series, Subject Interview Study).

42. Subject No. 552143-0, interview by Subject Interview Study staff (ACHRE), transcript of audio recording, 15 March 1995, lines 327-329 (Research Project Series, Subject Interview Study).

43. Subject No. 223212-2, interview by Subject Interview Study staff (ACHRE), transcript of audio recording, 16 March 1995, lines 233-234 (Research Project Series, Subject Interview Study). A few patients had very specific others in mind. One respondent, for instance, enrolled in a genetic study of co-

lon cancer, said, "Because if it's hereditary and it sure seems [so] in my situation . . . I'm concerned about my daughter. I'm concerned about her kids, and [it] goes on and on and on . . . " (Subject No. 221240-5, interview by Subject Interview Study staff [ACHRE], transcript of audio recording, 15 March 1995, lines 334-337 [Research Project Series, Subject Interview Study]). As patients became sicker, altruism played a larger note for some. For example, one respondent explained that over the course of his disease and enrollment in numerous research projects, his reasons for participating in research had become more altruistic: "[I]t will never cure me . . . I'll be dead in the next couple of years . . . but if they can find something that can save someone else [I'll be happy] . . . [W]hen you first go in . . . you're kind of dealing with whatever . . . disease you're dealing with. . . . There's this hope factor that's there, that you think, 'Well, maybe this is going to work. Maybe I'm going to— it's going to help me. . . .' [But now] I don't have the expectations that . . . I did . . . seven or eight years ago . . . I'm realistic. It might help. It might not. But, you know, they're going to find out something that's going . . . to help somebody else and you have to think of it that way" (Subject No. 335213-5, interview by Subject Interview Study staff [ACHRE], transcript of audio recording, 28 March 1995, lines 598-600, 1234-1238, 1294-1299 [Research Project Series, Subject Interview Study]).

44. Subject No. 443252-2, interview by Subject Interview Study staff (ACHRE), transcript of audio recording, 1 March 1995, lines 198-200 (Research Project Series, Subject Interview Study).

45. Subject No. 442107-9, interview by Subject Interview Study staff (ACHRE), transcript of audio recording, 29 March 1995, lines 120-124 (Research Project Series, Subject Interview Study).

46. Subject No. 443218-3, interview by Subject Interview Study staff (ACHRE), transcript of audio recording, 24 February 1995, lines 545-553 (Research Project Series, Subject Interview Study).

47. Only one respondent noted that he participated simply because he wanted the money that was being paid to participants in his research project (Subject No. 551145-6, interview by Subject Interview Study staff [ACHRE], transcript of audio recording, 11 March 1995, line 60 [Research Project Series, Subject Interview Study]). Another respondent with breast cancer stated plainly that she, as someone without health insurance, had enrolled in research to get treatment and "didn't have to worry about trying to pay something back later on" (Subject No. 335216-8, interview by Subject Interview Study staff [ACHRE], transcript of audio recording, 27 March 1995, lines 30-33 [Research Project Series, Subject Interview Study]).

48. Subject No. 335227-5, interview by Subject Interview Study staff (ACHRE), transcript of audio recording, 24 March 1995, lines 47-48 (Research Project Series, Subject Interview Study).

49. Subject No. 333256-6, interview by Subject Interview Study staff (ACHRE), transcript of audio recording, 16 March 1995, lines 332-333 (Research Project Series, Subject Interview Study).

50. Subject No. 221240-5, interview by Subject Interview Study staff (ACHRE), transcript of audio recording, 15 March 1995, lines 359-370 (Research Project Series, Subject Interview Study).

51. Subject No. 333208-7, interview by Subject Interview Study staff (ACHRE), transcript of audio recording, 6 March 1995, lines 407-409 (Research Project Series, Subject Interview Study). Another respondent had a similar response: "[T]o me they are the doctors and once I had gotten those doctors and I trusted them. It was pretty much up them. I wanted to know what I was going to be going through as far as what to expect . . . physically . . . [b]ut a lot of the little nitty-gritty detail, I did not even want to know" (Subject No. 114250-4, interview by Subject Interview Study staff [ACHRE], transcript of audio recording, 10 March 1995, lines 274-283 [Research Project Series, Subject Interview Study]).

52. Subject No. 332324-3, interview by Subject Interview Study staff (ACHRE), transcript of audio recording, 24 March 1995, lines 156-161 (Research Project Series, Subject Interview Study).

53. Subject No. 441204-5, interview by Subject Interview Study staff (ACHRE), transcript of audio recording, 8 March 1995, lines 171-172 (Research Project Series, Subject Interview Study).

54. Subject No. 552365-9, interview by Subject Interview Study staff (ACHRE), transcript of audio recording, 14 March 1995, lines 399-401 (Research Project Series, Subject Interview Study).

55. There were a number of people who said that the written material was difficult to read: "I read some of the literature and it didn't mean a hill of beans to me because I didn't know anything about medical science . . . but . . . like I say, if it's to help me, I'll go in . . ." (Subject No. 332324-3, interview by Subject Interview Study staff [ACHRE], transcript of audio recording, 24 March 1995, lines 189-192 [Research Project Series, Subject Interview Study]). One respondent drew attention to the overly technical language used in forms: "You kind of think, 'Hmmmm. What do these things really mean?'. . . [You hear that] your follicles might fall. . . . you're thinking, 'follicles fall?' My hair . . . [T]hey're slick at . . . [how] they present stuff . . . " (Subject No. 335213-5, interview by Subject Interview Study staff [ACHRE], transcript of audio recording, 28 March 1995, lines 742-748 [Research Project Series, Subject Interview Study]). A few patients said the forms were unnerving: "[I]t'd be more reassuring for the person . . . that's going to be involved in the research to receive some positive, more positive language in the protocol itself. . . . [C]ertainly in a way it's good that they let you know these things. . . . on the other hand, it just scares people sometimes" (Subject No. 335227-5, interview by Subject Interview Study staff [ACHRE], transcript of audio recording,

24 March 1995, lines 89-92, 114-116 [Research Project Series, Subject Interview Study]).

56. One respondent noted that cancer patients such as herself receive a deluge of technical information to digest: "We were sorta bombarded with information and I just made my mind up at that one appointment to go with the study" (Subject No. 443311-6, interview by Subject Interview Study staff [ACHRE], transcript of audio recording, 28 February 1995, lines 212-214 [Research Project Series, Subject Interview Study]). Another replied, "[T]hey do give you all the available information, almost too much, because you can't absorb it all at once, and I brought home all these little books and the books are good and you just get sick of it . . ." (Subject No. 333250-9, interview by Subject Interview Study staff [ACHRE], transcript of audio recording, 8 March 1995, lines 84-87 [Research Project Series, Subject Interview Study]).

57. One respondent, approached in his hospital room the night before scheduled brain surgery to consider enrolling in a clinical trial for anesthesia, felt that the timing of consent was poor: ". . . I felt the timing could have been a little better, because I was concerned about sleeping and being rested. . . . [I]t might have been better a day earl[ier] . . ." (Subject No. 442107-9, interview by Subject Interview Study staff [ACHRE], transcript of audio recording, 29 March 1995, lines 155-157, 174-175 [Research Project Series, Subject Interview Study]).

58. One respondent who spoke only broken English reported that she relied upon her husband to gather and make sense of information that staff relayed about her therapeutic research project.

59. Subject No. 114229-8, interview by Subject Interview Study staff (ACHRE), transcript of audio recording, 14 March 1995, lines 481-486. Similarly, another respondent argued that "it's important that the people who are on the protocol talk," particularly since in this forum participants can more quickly exchange information about what "is going to happen. . . ." Talking can convey the information "quicker than if it's on a piece [of paper] . . ." (Subject No. 335213-5, interview by Subject Interview Study staff [ACHRE], transcript of audio recording, 28 March 1995, lines 1403-1404, 1417-1420 [Research Project Series, Subject Interview Study]).

60. Subject No. 333208-7, interview by Subject Interview Study staff (ACHRE), transcript of audio recording, 6 March 1995, lines 27-29. Even when some patients noted that a doctor's recommendation had influenced them, they did not construe this as "pressure": "[D]on't misunderstand me, [my doctor] didn't influence me in [any] way . . . [but] he thought it would be a good program for my type of cancer" (Subject No. 552126-5, interview by Subject Interview Study staff [ACHRE], transcript of audio recording, 13 March 1995, lines 125-127 [Research Project Series, Subject Interview Study]). Another respondent

noted that she thought that the staff wanted her to enroll, "but they [were] not pushing anything" (Subject No. 113122-6, interview by Subject Interview Study staff [ACHRE], transcript of audio recording, 11 March 1995, line 462 [Research Project Series, Subject Interview Study]).

61. Subject No. 552126-5, interview by Subject Interview Study staff (ACHRE), transcript of audio recording, 13 March 1995, lines 102-105 (Research Project Series, Subject Interview Study).

62. Subject No. 223201-5, interview by Subject Interview Study staff (ACHRE), transcript of audio recording, 13 March 1995, lines 136-138 (Research Project Series, Subject Interview Study).

63. Subject No. 335213-5, interview by Subject Interview Study staff (ACHRE), transcript of audio recording, 28 March 1995, lines 687-700 (Research Project Series, Subject Interview Study). Another respondent, who did not remember signing a consent form reported, the doctor "just recommended me to go [on the drug]" (Subject No. 441311-8, interview by Subject Interview Study staff [ACHRE], transcript of audio recording, 28 March 1995, lines 63 [Research Project Series, Subject Interview Study]). Another respondent reported that although she did not recall signing a consent form, she later discovered that she had. One procedure to which she had consented in written form was not something she wanted to go through, however. This respondent explained the confusion in part to the fact that she . . . "never thought about the study" . . . because she was worried about . . . "[having] to be cut again . . ." (Subject No. 443226-6, interview by Subject Interview Study staff [ACHRE], transcript of audio recording, 16 March 1995, lines 246-255, 348-361, 446-448 [Research Project Series, Subject Interview Study]).

64. Subject No. 223212-2, interview by Subject Interview Study staff (ACHRE), transcript of audio recording, 16 March 1995, lines 195-200 (Research Project Series, Subject Interview Study).

65. Subject No. 552143-0, interview by Subject Interview Study staff (ACHRE), transcript of audio recording, 15 March 1995, lines 205-208 (Research Project Series, Subject Interview Study).

66. Subject No. 333208-7, interview by Subject Interview Study staff (ACHRE), transcript of audio recording, 6 March 1995, lines 250-252, 256-257, 260-261 (Research Project Series, Subject Interview Study).

67. This respondent described that others in the study had been "on orders to leave here" (i.e., to go to another military base), . . . and were "not allowed to [leave]. The doctor told them that they could not . . . [and] . . . had the orders changed because they were enrolled in an intense medical program research program" (Subject No. 333301-0, interview by Subject Interview Study staff [ACHRE], transcript of audio recording, 9 March 1995, lines 779-783 [Research Project Series, Subject Interview Study]).

68. The respondent went on to identify sources of overt and covert pressure for him to remain in the trial: "[T]he response was kind of like trying to convince me to just finish it up and that was always the response, anytime I had reservations there was somebody there to . . . talk about those reservations, but in the course of doing it really trying to convince me that it's ok[ay]. . . . " The respondent also said, "When they asked for a bone marrow biopsy I said I'm not gonna do it, so I just dropped out at that point, and she says, you know if we don't do that then I mean, it's not valid . . . it defeats all those days . . ." (Subject No. 551334-6, interview by Subject Interview Study staff [ACHRE], transcript of audio recording, 14 March 1995, lines 303-305, 326-328, 330-332, 562-566, 652-656, [Research Project Series, Subject Interview Study]).

69. Subject No. 553215-5, interview by Subject Interview Study staff (ACHRE), transcript of audio recording, 10 March 1995, lines 96-99, 205-206. Another respondent explained, ". . . I knew that I could stop at any point and time. I was aware of that . . . I knew that I could, but I didn't have anything else . . . I didn't know what stopping was going to do either . . . then I thought well, if I stop, what do I do[?]" (Subject No. 334110-4, interview by Subject Interview Study staff [ACHRE], transcript of audio recording, 28 March 1995, lines 483-488 [Research Project Series, Subject Interview Study]).

70. Subject No. 552143-0, interview by Subject Interview Study staff (ACHRE), transcript of audio recording, 15 March 1995, lines 37-38, 217-218 (Research Project Series, Subject Interview Study).

Discussion of Part III

THE Committee undertook the efforts described in part III of this report in order to gain insight into the current status of protections in human radiation research and research with human subjects generally. An important finding of part III is that with respect to the rights and interests of human subjects there appear to be no differences between radiation research and other research.

Compared with what we have learned about the 1940s, 1950s, and 1960s, there have been many changes in the climate and conduct of human subject research. The most obvious change is the regulatory apparatus described in chapter 14, which was not in place in that earlier time. The rules of research ethics are also more articulated today than they were then, as exemplified by the evolution of the concept of informed consent. Although the basic moral principles that serve as the underpinning for research ethics are the same now as they were then, some of the issues of greatest concern to us today are different, or have taken on a different cast, from those of earlier decades.

In our historical inquiry, for example, we concentrated on cases that offered subjects *no* prospect of medical benefit; they were instances of "nontherapeutic research" in the strictest sense. That is, these were experiments in which there was *never* any basis or expectation that subjects could benefit medically—both the design and the objectives precluded such a possibility. Most of the human radiation experiments that were public controversies when the Advisory Committee was appointed were of this type. The basic moral concern they raised was whether people had been used as mere means to the ends of scientists and the government; this would have occurred if the subjects could not possibly have benefited medically from being in the research and they had not consented to this use of their person.

As we noted in chapter 4, the ethical issues raised by research that is nontherapeutic in this strict sense are stark and straightforward. Because risks to subjects cannot be offset by the possibility that they might benefit medically, there is rarely justification for nontherapeutic research that puts subjects at significant risk. Participation in such research is always a burden and never a benefit to the individual subject, making questions of justice straightforward as well. And, at least theoretically, there are no subtleties involved in disclosing to potential subjects that they cannot possibly benefit medically from participating in the research, although problems do emerge concerning what incentives are appropriate to induce people to become research subjects when, considerations of altruism aside, it is otherwise not in their interests to do so.

Today, we still conduct nontherapeutic medical research on human subjects. Much research in physiology offers subjects no prospect of medical benefit, as does every protocol that calls for "normal controls." Although nontherapeutic research frequently involves subjects who are healthy, it also often involves patient-subjects as well. It is, of course, still appropriate to be concerned about the ethics of such research, as it is with all research. For example, we were particularly troubled in our Research Proposal Review Project by documents that suggested that adults of questionable competence were being used as subjects of research from which it appeared they could not benefit medically and where the authorization for this use was unclear.

Much research involving human subjects does not, however, fit this nontherapeutic paradigm. Many of our most pressing ethical questions concern research that raises at least the specter of potential medical benefit to the patient-subject. For example, unlike the plutonium experiments with hospitalized adults or the iodine 131 experiments with hospitalized children, in which there was no possibility that the patient-subjects could have benefited medically, in the modern Phase I trial, which is conducted to establish toxicity, there is at least the possibility of therapeutic benefit, however slim. Thus, although Phase I trials often impose significant burden and risk on subjects, they are not nontherapeutic in the strict sense. And, in contrast with Phase I trials, in much research involving patient-subjects there is a real prospect that subjects will benefit medically from their participation. In many of these cases, being a research subject is clearly in the medical best interests of the patient.

As Otto Guttentag observed in the 1950s (see chapter 1), it is the possibility of medical benefit that creates much of the moral tension in human subject research. Physician researchers are often torn between the demands of a research project and the needs of particular patients. Today this tension has taken on special significance, with the immense growth of research at the bedside and the frequency with which the medical care of seriously ill patients is intertwined with clinical research. In our Subject Interview Study, for example, at least a third of the patients interviewed had some contact with medical research.

It is these considerations that led us to focus the efforts reported on in part III, and particularly the SIS, on research involving patient-subjects. The Committee regrets that we did not have the resources to conduct a similar study with subjects who are not also patients. It would have been particularly useful to have conducted such a study with subjects who are military personnel not currently in medical care. This would have allowed us to investigate other important sources of tension in the ethics of research, including the tension between giving orders and soliciting consent and between occupational monitoring and research.

Although the SIS and the RPRP employed radically different methodologies and were directed toward different research questions, both projects speak to the ethical issues raised by the conduct of human research in a medical context, a context dominated by the human needs to be healed and to heal.

The findings of the SIS underscore what other, smaller studies also have identified—that patient-subjects generally decide to participate in medical research because they believe that being in research is the best way to improve their medical condition.[1] In the SIS, we could not determine whether the patients had unrealistic expectations about how likely it was that they might benefit from being in research, or in what form that benefit might take. Other empirical studies suggest that some subjects do have an inadequate, sometimes exaggerated understanding of the potential benefits of the research in which they are participating.[2] In the RPRP, we reviewed consent forms that appeared to over-promise what research could likely offer the ill patient and underplay the effect of the research on the patient's quality of life. These were the kinds of disclosures that could easily be interpreted by a patient desperate for hope as offering much more than realistically could be expected. Not surprisingly, this problem was the most acute in certain Phase I trials that, while not being nontherapeutic in the strict sense, appeared to offer only a remote possibility of benefit to the patient-subject. In oncology, for example, it is estimated that in only about 5 percent of subjects enrolled in Phase I chemotherapy studies does the tumor respond to the drug,[3] and it is often unclear even then what the tumor response means from a patient's point of view.[4] To say that there is *no* prospect that the patients might benefit medically is questionable; there are enough cases of patients being helped in Phase I trials to make such a stark claim problematic. Beneficial effects of Phase I trials have a very low probability, but do occur.[5] At the same time, however, any suggestion of the possibility of benefit has the potential to be magnified many times over by patients with no good medical alternatives. It is understandable that physicians, faced with the prospect of little or nothing to offer seriously ill patients, may sometimes impart more hope than the clinical facts warrant. At the same time, however, desperate hopes are easily manipulated.

Consider, for example, a recent report of a small study of patient-subjects participating in Phase I clinical oncology trials.[6] Despite the predictably low likelihood of medical benefit for subjects in Phase I trials, all of the patient-subjects surveyed about their reasons for participating said their decision was motivated in large part by the possibility of therapeutic benefit, and nearly three-quarters cited trust in their physician as motivating their decision to participate. Only one-third listed altruism as a major motivating factor. These results support what we found in the SIS—that patient-subjects view research participation as a way of obtaining the best medical care, even when participating in research holds out very little prospect of direct benefit. This phenomenon, which is especially relevant when some subjects receive a placebo as a part of the research, has been dubbed the therapeutic misconception.[7] This phenomenon is not confined to patient-subjects' perceptions of benefit from research; at least one study has shown that physician-investigators also overestimate the potential benefits to subjects participating in Phase I oncology trials.[8]

One of the most powerful themes to emerge from the SIS is the role of trust in patients' decisions to participate in research, a finding that has been observed in other studies as well.[9] It was common for patients in the SIS to say that they had joined a research project at the suggestion of their physician and that they trusted that their physician would never endorse an option

that was not in their best interest. This trust underscores the much-discussed tension in the role of physician-investigator,[10] whose duties as a healer and as a scientist inherently conflict. This trust that patients place in their physicians often is generalized to the medical and research community as a whole. Some patients expressed faith not only in their doctors but also in the institutions where they were receiving medical care. These patients believed that hospitals would never permit research to be conducted that was not good for the patient-subjects. The trust that patients have in physicians and hospitals underscores the importance of the Committee's concern, based on our review of the documents in the RPRP, that IRBs may not always be properly structured to ensure that the medical interests of ill patients are adequately protected. In some cases, the scientific information to make such judgments was not included in the documents we received. Even with adequate information, IRBs may lack sufficient expertise to evaluate the science or implications for medical care of particular proposals. As we heard from some IRB chairs, they may also lack the staff or the respect and authority within their institutions to function adequately to protect subjects.

The theme of trust discerned in the SIS also has implications for how properly to view the role of informed consent in protecting the rights and interests of human subjects. For many of the patients who based their decision to be in research on their trust in their physicians, the informed consent process and the informed consent form were of little importance. IRBs can serve the interests of these patients best by being vigilant in their review of risks and benefits and attending to questions of fairness in the selection of subjects. On the other hand, we also found in the SIS that sizable numbers of patients had refused offers to participate in research and that some patients who had consented to be research subjects had made efforts to learn what they could about the research opportunity. For these patients, the informed consent process likely served an important moral function.

From these seemingly conflicting results we can conclude both that the informed consent requirement is crucial to protect the autonomy rights of those potential subjects who choose to exercise them, but that it is naive to think that informed consent can be relied upon as the major mechanism to protect the rights and interests of patient-subjects. Taken together, the results of our two projects suggest that it *is* important to correct the deficiencies identified in the RPRP with respect to informed consent. Our results also underscore, however, the importance of an IRB review that focuses on whether the proposed research is a reasonable, ethically acceptable option to offer the patient, in light of available alternatives and the risks and potential benefits of the proposed research for the subject, including impact on quality of life. An alternative, the practice of adding detail to consent forms as a way of further informing potential subjects who often have a difficult time understanding risks, benefits, and purposes of research,[11] is unacceptable; by confusing subjects, it offers less, rather than greater protection. For the many patients who continue to rely on the expertise and good will of physicians and hospitals in deciding whether to participate in research, rigorous review on the part of IRBs and rigorous commitment on the part of physicians to honor the faith entrusted to them are the important protections.

The SIS and the RPRP also both speak to the current confusion between research and "standard care" in medical practice.[12] The same therapy that is part of a research protocol, and therefore must receive IRB approval, can proceed outside of the research setting and not be subject to IRB oversight. This leads to understandable confusion on the part of subjects as to whether they are participating in research, receiving standard care, or some combination. It is thus perhaps not surprising that research subjects occasionally seem unaware of their participation in research, even when there is evidence they have signed consent forms.[13] This finding was observed in the SIS, though the methodology of the study did not allow us to probe the reasons some subjects appeared unaware of their participation.

The confusion between research and alternative medical interventions is mirrored in the language used to communicate to patients in the informed consent process and in the language of patients themselves. In the SIS, the patients surveyed viewed *experiments* as involving unproven treatment of greater risk, while *clinical investigation* or *study* conveyed less uncertainty and were perceived as offering a greater chance of personal benefit. None of the consent forms we reviewed in the RPRP used the term *experiment*.

CONCLUSION

In addition to the role they played in helping the Advisory Committee come to our conclusions, the RPRP and SIS should be understood as adding to the body of research undertaken to try to understand the strengths and weaknesses of the system to protect the rights and interests of human subjects.

In the end, patients' reasons for participating in research must more accurately reflect the benefits they may reasonably expect. Altruistic motivation can be more fruitfully tapped, both for the benefit it provides to the advancement of science and to underscore for patients that the primary objective of some research is to create generalizable scientific knowledge rather than to offer personal benefit to them. Subjects are much more likely to have a positive view of biomedical research if they feel they understand what prospects research holds for them. The good news in the endeavor of human subject research is that subjects are willing to participate, and in the process entrust their care to researchers; however, that trust cannot be taken for granted as it sometimes has been in our history.

Increasingly it is being argued that it is generally advantageous for patients to participate in research; the distinction between standard care and research, if it was ever clear, is viewed as growing dimmer all the time.[14] As a consequence, the debate over subject selection has changed entirely. As we discussed in parts I and II, in the past a central concern was that certain populations, considered vulnerable to exploitation because of their relative powerlessness, were inequitably bearing the burdens of the risks of research. Today, the concern is that the same populations may have inequitable access to research and therefore individuals and the communities of which they are a part may be denied a fair share of the benefits of research participation. While this is a valid moral concern, the results of the SIS and the RPRP suggest that it remains important to be attuned to issues of vulnerability. While patients with serious illnesses may stand to gain the most from participating in medical research, they also are among the most vulnerable to its risks.

It also is important to underscore the finding in the RPRP that in both studies involving minimal risk and those involving greater risk, research with ill patient-subjects can proceed ethically and consent can be properly obtained. The research enterprise is too important to jeopardize by inadequate protections for subjects. Tensions and potential conflicts exist throughout the research process, and so we must be sure to acknowledge and address them squarely. This is the goal of the next and final part of the Advisory Committee's report.

NOTES

1. Barrie R. Cassileth, Edward J. Lusk, David S. Miller, and Shelley Hurwitz, "Attitudes Toward Clinical Trials Among Patients and the Public," *Journal of the American Medical Association* 248, no. 8 (1982): 968–970; and Christopher Daugherty, Mark J. Ratain, Eugene Grocowski et al., "Perceptions of Cancer Patients and Their Physicians Involved in Phase I Trials," *Journal of Clinical Oncology* 13, no. 5 (1995): 1062–1072.

2. Ira S. Ockene et al., "The Consent Process in the Thrombolysis in Myocardial Infarction (TIMI—Phase I) Trial," *Clinical Research* 39 no. 1 (1991): 13–17; Roberta M. Tanakanow, Burgunda V. Sweet, and Jill A. Weiskopf, "Patients' Perceived Understanding of Informed Consent in Investigational Drug Studies," *American Journal of Hospital Pharmacy* 49 (1992): 633–635; Doris T. Penman et al., "Informed Consent for Investigational Chemotherapy: Patients' and Physicians' Perceptions," *Journal of Clinical Oncology* 2 no.7 (1984): 849–855; Henry W. Riecken and Ruth Ravich, "Informed Consent to Biomedical Research in Veterans Administration Hospitals," *Journal of the American Medical Association* 248 no. 3 (1982): 344–348; Gail A. Bujorian, "Clinical Trials: Patient Issues in the Decision-Making Process," *Oncology Nursing Forum* 15, no. 6 (1988): 779–783; Niels Lynoe et al., "Informed Consent: Study of Quality of Information Given to Participants in a Clinical Study," *British Medical Journal* 303 (1991): 610–613; and Daugherty et al., "Perceptions of Cancer Patients and Their Physicians Involved in Phase I Trials."

3. D. D. Von Hoff and J. Turner, "Response Rates, Duration of Response, and Dose Response Effects in Phase I Studies of Antineoplastics," *Investigational New Drugs* 9 (1991): 115–122; E. Estey et al., "Therapeutic Response in Phase I Trials of Antineoplastic Agents," *Cancer Treatment Report* 70 (1986): 1105–1155; and G. Decoster, G. Stein, and E. E. Holdener, "Responses and Toxic Deaths in Phase I Clinical Trials," *Annals of Oncology* 2 (1990): 175–181.

4. Decoster, Stein, and Holdener, "Responses and Toxic Deaths in Phase I Clinical Trials," 175–181.

5. There have been very few occasions reported in the literature of Phase I trials having medical benefit for patient-subjects. While very rare, that such benefit occurs at all further complicates the difficulty over what to say to patients at the end of the medical road who are considering enrolling in a Phase I trial. See, for example, M. Kaminski, "Radioimmunotherapy of Bcell Lymphoma with I-131 Anti-B1," *New England Journal of Medicine* 329, no. 7 (1993): 459–465.

6. Daugherty et al., "Perceptions of Cancer Patients and Their Physicians Involved in Phase I Trials."

7. Paul S. Appelbaum, Loren H. Roth, and Charles Lidz, "The Therapeutic Misconception: Informed Consent in Psychiatric Research," *International Journal of Law and Psychiatry* 5 (1982): 319–329; Paul S. Appelbaum et al., "False Hopes and Best Data: Consent to Research and the Therapeutic Misconception," *Hastings Center Report*, April 1987, 20–24.

8. Decoster, Stein, and Holdener, "Responses and Toxic Deaths in Phase I Clinical Trials."

9. Cassileth et al., "Attitudes toward Clinical Trials among Patients and the Public"; Susan M. Newburg, Anne E. Holland, and Lesly A. Pearce, "Motivation of Subjects to Participate in a Research Trial," *Applied Nursing Research* 5, no. 2 (1992): 89–104; Penman et al., "Informed Consent for Investigational Chemotherapy: Patients' and Physicians' Perceptions"; and Daugherty et al., "Perceptions of Cancer Patients and their Physicians Involved in Phase I Trials."

10. Nancy M. P. King, "Experimental Treatment: Oxymoron or Aspiration?" *Hastings Center Report* 25, no. 4 (1995): 6–15; Jay Katz, "The Regulation of Human Experimentation in the United States—A Personal Odyssey," *IRB* 9, no. 1 (1987): 1, 5–6; and J. R. Maltby, and C. J. Eagle, "Patient Recruitment for Clinical Research [letter, comment]," *Canadian Journal of Anaesthesia* 40, no. 9 (1993): 897–898.

11. Ockene et al., "The Consent Process in the Thrombolysis in Myocardial Infarction (TIMI—Phase I) Trial"; Tanakanow et al., "Patients' Perceived Understanding of Informed Consent in Investigational Drug Studies"; Penman et al., "Informed Consent for Investigational Chemotherapy: Patients' and Physicians' Perceptions"; Riecken et al., "Informed Consent to Biomedical Research in Veterans Administration Hospitals"; Bujorian et al., "Clinical Trials:

Patient Issues in the Decision-Making Process"; Lynoe et al., "Informed Consent: Study of Quality of Information Given to Participants in a Clinical Study"; and Daugherty et al., "Perceptions of Cancer Patients and their Physicians Involved in Phase I Trials."

12. King, "Experimental Treatment: Oxymoron or Aspiration?"

13. Riecken et al., "Informed Consent to Biomedical Research in Veterans Administration Hospitals."

14. King, "Experimental Treatment: Oxymoron or Aspiration?"

Coming to Terms with the Past, Looking Ahead to the Future

PART IV
OVERVIEW

IN part IV we present the overall findings of the Advisory Committee's inquiry and deliberations and the recommendations that follow from these findings.

In chapter 17, findings are presented in two parts, first for the period 1944 through 1974 and then for the contemporary period. These parts, in turn, are divided into findings regarding biomedical experiments and those regarding population exposures.

We begin our presentation of findings for the period 1944 through 1974 with a summation of what we have learned about human radiation experiments: their number and purpose, the likelihood that they produced harm, and how human radiation experimentation contributed to advances in medicine. We then summarize what we have found concerning the nature of federal rules and policies governing research involving human subjects during this period, and the implementation of these rules in the conduct of human radiation experiments. Findings about the nature and implementation of federal rules cover issues of consent, risk, the selection of subjects, and the role of national security considerations.

Our findings about government rules are followed by a finding on the norms and practices of physicians and other biomedical scientists for the use of human subjects. We then turn to the Committee's finding on the evaluation of past experiments, in which we summarize the moral framework adopted by the Committee for this purpose. Next, we present our findings for experiments conducted in conjunction with atmospheric atomic testing, intentional releases, and other population exposures. The remaining findings for the historical period address issues of government secrecy and record keeping.

There is an asymmetry in our findings on human radiation experiments and intentional releases. In both cases, we discuss their number and purpose, the likelihood that they produced harm, and what is known about applicable government rules and policies. In the case of human radiation experiments, we also have a finding on the benefits to medicine—and thus to all of us—

that human radiation research during this period produced. We do not, however, have a corresponding finding on the benefits of the intentional releases of the period, benefits that would presumably have been to the national defense and, thus again, to all of us. Although the members of the Committee are positioned to comment on contributions to medicine and medical science, we do not have the expertise to evaluate contributions to the national defense and thus could not speak to this issue.

Our findings for the contemporary period summarize what we have learned about the rules and practices that currently govern the conduct of radiation research involving human subjects, as well as human research generally, and about the status of government regulations regarding intentional releases.

Chapter 18 presents the Committee's recommendations to the Human Radiation Interagency Working Group and to the American people. The Committee's inquiry focused on research conducted by the government to serve the public good—the promotion and protection of national security and the advancement of science and medicine. The pursuit of these ends—today, as well as yesterday—inevitably means that some individuals are put at risk for the benefit of the greater good. The past shows us that research can bear fruits of incalculable value. Unfortunately, however, the government's conduct with respect to some research performed in the past has left a legacy of distrust. Actions must be taken to ensure that, in the future, the ends of national security and the advancement of medicine will proceed only through means that safeguard the dignity, health, and safety of the individuals and groups who may be put at risk in the process.

The needed actions are in four dimensions:

First, the nation must provide for appropriate remedies as it comes to grips with the past.

Second, the nation must provide improved means to better ensure that those who conduct research involving human subjects act in a manner consistent with the interests and rights of those who may be put at risk and consistent with the highest ethical standards of the practice of medicine and the conduct of science.

Third, the nation must ensure that special care is taken to prevent abuses in the conduct of human subject research and environmental releases in a context where these activities must occur in secret.

Fourth, the nation must ensure that records are kept so that a proper accounting can be made to those who are asked to bear risks, particularly when any or all of the risk taking involves secrecy. Moreover, these records should be made available to the public at large on a timely basis consistent with legitimate national security requirements.

The Committee's recommendations address these four areas—remedies for the past, practices to govern the future of biomedical experimentation, practices to govern the future exposure of citizens to biomedical research or environmental releases from secret activities, and provisions for record keeping and public access to records.

We wish to note here the limits of our framework for remedies for past harms or wrongs for subjects of human radiation experiments.* First, we are addressing questions of remedies from the perspective of what, ethically, ought to be done. We recognize that some of the remedies we propose, including financial compensation, may not be available under current federal law. To the extent that such remedies are not available under current law, we encourage the administration to work with Congress to develop such remedies through legislation or other appropriate means.

Second, the Committee has focused on past experiments in which there was no possibility that subjects could derive medical benefit from being in the research or in which the potential for this benefit is in dispute. These were the

experiments that raised the greatest public concern. They were also the experiments that raised the greatest concern for most members of the Committee when we considered the 1944–1974 period. This was a time, as noted throughout this report, when it was common for physicians to use patients as research subjects without the patients' knowledge or consent. It was also a time, however, when physicians were ceded considerable moral authority both by patients and by society to decide for patients what medical treatments they should receive. This authority extended, as well, to deciding whether a patient should be a subject in therapeutic medical research, provided that this decision was based on a good faith judgment by the physician that it was in the patient's medical best interest to be a subject in the research and thus that any risks of the research were acceptable in light of the possibilities for medical benefit. Even at the time, however, physicians did not have the moral authority to use patients, without their knowledge or consent, as subjects in research in which there was *no* expectation that they could benefit medically.

The Committee appreciates that simply because the moral context of the doctor-patient relationship during the 1944–1974 period was different from today's, this does not mean that all therapeutic research was always or even often conducted in an ethical fashion. We also appreciate that the risks of therapeutic research were often considerable and that it is likely that some patient-subjects were harmed unnecessarily as a consequence. However, the moral problems presented when people in the 1944–1974 period were used as subjects of research from which they *could not benefit* medically are both more straightforward and more compelling. We therefore felt obligated to expend our limited resources on historical and moral analysis of these kinds of experiments. We do not address whether or under what conditions remedies should be provided for injuries or offenses related to research that offered a plausible prospect of medical benefit to subjects and we leave that work to others.

Third, even in those experiments where there was no prospect of medical benefit, limited Committee resources, and the overall Committee mandate, precluded the type of fact-inten-

---

* In accordance with our charter, these recommendations apply to human radiation experiments conducted from 1944 to 1974 that were supported by the government, whether the support was in the form of funding (including funding for data gathering in conjunction with exposure of patient-subjects to radiation) or other means, such as the provision of equipment or radioisotopes, and regardless of whether the research was performed by federal employees or nonfederal investigators. Although we focus here on human research involving exposure to ionizing radiation, the moral justification for these recommendations is not specific only to experiments involving radiation.

sive individual investigation that would give rise to a recommendation of compensation in individual cases. The Committee did not have the ability to locate and evaluate the research and medical records of countless individual subjects. As a consequence, for example, we were not able to make judgments about whether, in individual cases, subjects had suffered physical harm attributable to their involvement in research.

Fourth, we note that the Committee was not unanimous in its decision to make a recommendation for remedies for people who were subjects in experiments that offered them no prospect of medical benefit but who were not physically harmed as a consequence (recommendation 3). Three Committee members elected not to support this recommendation.

The entire Committee believes that people who were used as research subjects without their consent were wronged even if they were not harmed. Although it is surely worse, from an ethical standpoint, to have been both harmed and wronged than to have been used as an unwitting subject of experiments and suffered no harm, it is still a moral wrong to use people as a mere means. Although what we know about the practices of the time suggests it is likely that many people who were subjects in nontherapeutic research were used without their consent or with what today we would consider inadequate consent, in most of these cases, we have almost no information about whether or how consent was obtained. Moreover, in most of these cases, the identities of the subjects are not currently known; even if considerable resources were expended, it is likely that most of their identities would remain unknown. The Committee is not persuaded that, even where the facts are clear and the identities of subjects known, financial compensation is necessarily a fitting remedy when people have been used as subjects without their knowledge or consent but suffered no material harm as a consequence; the remedy that emerged as most fitting was an apology from the government.

The Committee struggled with and ultimately was divided on the issue of whether to recommend that the government extend an apology under the circumstances just described. While all members agreed that a goal of all the Com-

mittee's recommendations is, in the words of one member, to "bind the nation's wounds," we disagreed about how best to accomplish that end when debating whether we should recommend such an apology. Our deliberations were complicated by what we all agreed was a murky historical record. In the case of some experiments, there was evidence of some disclosure or some attempt to obtain consent, and the issue emerged as to how poor these attempts must be for an apology still to be in order. In other cases, there was simply too little documentary evidence to draw any conclusions about disclosure or consent. In most cases, as noted above, the identities of subjects are unknown and are unlikely to be uncovered even with an enormous expenditure of resources.

The Committee members who concluded that it was not appropriate to recommend that a government apology be extended did not all reach this conclusion for the same reasons. Among the reasons put forward were that it would be impossible to craft a recommendation for an apology in such a way as to avoid the divisiveness that could result from apologizing to some but not all of those who view themselves as victims of this kind of human radiation experiment. There was concern that if the criteria for who should receive an apology were too narrow, some people would resent not qualifying for an apology; conversely, if the criteria were too broad and included large numbers of people, the generality of the apology would diminish its meaningfulness. It was also argued that a recommendation for an apology should not be made because of the difficulties in crafting the criteria for eligibility in the face of an incomplete historical record. Another reason for not recommending an apology was that during the 1944–1974 period many people were used as subjects of research that did not involve radiation, for which there was no prospect of medical benefit and consent was not obtained from them, and these people would not be included in a recommendation from us for an apology.

The Committee members who favored an apology took the position that justice requires that an apology from the government is due in research that it sponsored, where it can be determined that an apology is deserved and the identities of subjects who were wronged can be known. They do not believe that the recommendation to apologize rests on the likelihood that it will lead to more healing than divisiveness. Rather, these Committee members hold that an apology is a just remedy for those who were wronged and that it should not be withheld only because there are other cases that are likely to have been morally similar but for which a recommendation of an apology could not be made because the evidence was unclear or unavailable. Making a specific apology in those cases where the facts are clear today would not for these Committee members preclude apologies being extended to other subjects in the future, should new information come to light.

All Committee members agreed that it was appropriate that the subjects of the experiments at the Fernald State School in Massachusetts receive an apology, but divisions within the Committee arose when we tried to determine how to differentiate them from the subjects of studies similar to those conducted at Fernald about which less is known in relation to disclosure and consent.

Fifth, the Committee notes that our recommendations for remedies are directed solely to the executive and legislative branches of the federal government; they are not recommendations for exclusive remedies intended to bar the opportunity to seek redress from other parties or the courts. Those who believe they or their family members have been wronged or injured should be free also to seek relief from appropriate institutions or from individuals; the Committee does not intend to suggest the limiting of any rights to do so.

Finally, the framework for remedies for former subjects of human radiation experiments that the Committee proposes in our recommendations limits the availability of compensation from the federal government to what is likely to be a small number of people. In developing the framework we were concerned about the impact of recommending criteria that would result in compensation in some cases but not in others. The Committee sought and heard testimony from hundreds of witnesses, over months of deliberation, many of whom were emotional and heart-rending in sharing their experiences. Often

these witnesses expressed considerable anguish over the pain that they and their families suffer because of their belief that they have been or might yet be harmed, and some advanced the view that compensation is appropriate. It was very painful for the Committee to recognize that often we had neither the resources nor the mandate to investigate all these compelling stories. The Committee concluded that an appropriate service we could render was to shed light on this dark period in our history by articulating the historical record to the best of our ability. But it is equally important that, the historical record having been spelled out, we as a nation move forward. The most fitting way to acknowledge the wrongs and harms that were done to others in the past, and to honor their contributions to the nation, is for the government to take steps to ensure that what they experienced will not happen again.

Thus, many of our recommendations are directed not to the past but toward the future. The Committee calls for changes in the current federal system for the protection of the rights and interests of human subjects. These include changes in institutional review boards; in the interpretation of ethics rules and policies; in the conduct of research involving military personnel as subjects; in oversight, accountability, and sanctions for ethics violations; and in compensation for research injuries. Unlike the 1944–1974 period, in which the Committee focused primarily on research that offered subjects no prospect of medical benefit, our recommendations for the future emphasize protections for patients who are subjects of therapeutic research, as many of the contemporary issues involving research with human subjects occur in this setting. We also call for the adoption of special protections for the conduct of human research or environmental releases in secret, protections that are not currently in place.

We realize, however, that regulations and policies are no guarantee of ethical conduct. If the events of the past are not to be repeated, it is essential that the research community come to increasingly value the ethics of research involving human subjects as central to the scientific enterprise. We harbor no illusions about the Pollyanna-ish quality of a recommendation for professional education in research ethics; we call for much more. We ask that the biomedical research community, together with the government, cause a transformation in commitment to the ethics of human research. We recognize and celebrate the progress that has occurred in the past fifty years. We recognize and honor the commitment to research ethics that currently exists among many biomedical scientists and many institutional review boards. But more needs to be done. The scientists of the future must have a clear understanding of their duties to human subjects and a clear expectation that the leaders of their fields value good ethics as much as they do good science. At stake is not only the well-being of future subjects, but also, at least in part, the future of biomedical science. To the extent that that future depends on public support, it requires the public's trust. There can be no better guarantor of that trust than the ethics of the research community.

Finally, our examination of the history of the past half century has helped us understand that the revision of regulations that govern human research, the creation of new oversight mechanisms, and even a scrupulous professional ethics are necessary, but are not sufficient, means to needed reform. Of at least equal import is the development of a more common understanding *among the public* of research involving human subjects, its purposes, and its limitations. Furthermore, if the conduct of the government and of the professional community is to be improved, that conduct must be available for scrutiny by the American people so that they can make more informed decisions about the protection and promotion of their own health and that of the members of their family. It is toward that end that we close our report with recommendations for continued openness in government and in biomedical research. It is also toward that end that this report is dedicated. Some of what is regrettable about the past happened, at least in part, because we as citizens let it happen. Let the lessons of history remind us all that the best safeguard for the future is an informed and active citizenry.

17

Findings

Findings for the Period 1944–1974

BIOMEDICAL EXPERIMENTS

FINDING 1

The Advisory Committee finds that from 1944 through 1974 the government sponsored (by providing funding, equipment, or radioisotopes) several thousand human radiation experiments. In the great majority of cases, the experiments were conducted to advance biomedical science; some experiments were conducted to advance national interests in defense or space exploration; and some experiments served both biomedical and defense or space exploration purposes.

These experiments were conducted by researchers affiliated with government agencies, universities, hospitals, and other research institutions. Only fragmentary information survives about most experiments.

FINDING 2

The Advisory Committee finds that the majority of human radiation experiments in our database in-volved radioactive tracers administered in amounts that are likely to be similar to those used in research today. Most of these tracer studies involved adult subjects and are unlikely to have caused physical harm. However, in some nontherapeutic tracer studies involving children, radioisotope exposures were associated with increases in the potential life-time risk for developing cancer that would be considered unacceptable today. The Advisory Committee also identified several studies in which patients died soon after receiving external radiation or radioisotope doses in the therapeutic range that were associated with acute radiation effects.

Review of available information indicates that the majority of the approximately 4,000 human radiation experiments in the Advisory Committee database involved the use of radioisotopes as tracers in research designed to measure physiological processes in either normal or diseased states. These experiments were not typically aimed at measuring the biological effects of radiation itself. However, information on the majority of experiments in our database was fragmentary and thus did not allow for detailed estimates of dosimetry or examination of issues of experimental design and subject selection.

To supplement the information in our database and provide context to our analysis, we independently reviewed archival documents from AEC-mandated institutional local isotope committees. These local use committees were part of a larger AEC program that facilitated the distribution of radioisotopes for use in government-sponsored human subjects research in the 1947–1974 period and involved the review of experimental risk on an individual basis to ensure that human uses of isotopes were within accepted risk standards of the day. We thus used these materials as an indicator of isotope use and regulatory practices at that time.

While we recognize the limitations of the data available to us, our evaluation suggests that most tracer studies conducted during the period 1944–1974 likely involved low doses that did not cause any acute or long-term effects. The Advisory Committee cannot rule out, however, the possibility that some people were or will be harmed as a consequence of their involvement in these experiments.

The Committee did identify some nontherapeutic tracer experiments involving the administration of iodine 131 to children that may have raised the subsequent risk of developing thyroid cancer to levels that would be considered unacceptable today. Based on the average risk estimate for each experiment, approximatedly 500 individuals were exposed to greater than minimal risk. (The Committee used a threshold of greater than or equal to one excess case of cancer per 1,000 subjects for categorizing experiments as greater than minimal risk.) Combining the average risk estimates for each experiment, this translates into an expected excess of 1.3 incident cases of thyroid cancer for the entire group. Fortunately, unlike many other cancers, thyroid cancer is curable in more than 90 percent of cases; therefore, it is unlikely that, even if cancers developed, these exposures caused any premature deaths. Furthermore, although there is strong scientific evidence that radiation doses delivered over a short period of time from external sources can result in increases in cancer incidence at specific sites, comparable data suggest that the carcinogenic effects of isotope exposures are less than those of external irradiation. The difference in carcinogenic effect is thought to be due to the relatively low dose rate of the isotope exposure, which allows for effective repair of radiation damage.

One additional isotope study involving the administration of radioiron to pregnant women has been linked to a possible increase in cancers in children who were exposed in utero. However, the small number of observed cancers as well as considerable uncertainties in the amount of radioisotope administered have made the determination of causality difficult. Finally, the Committee found some experiments where radioisotope exposures were associated with either acute or chronic physiologic changes of uncertain clinical significance, pathologic evidence of kidney damage secondary to chemical and radiation toxicity in some patients injected with uranium, and radiographic evidence of minimal bone changes in some long-term survivors of plutonium injections.

Studies that involved radiation doses in the therapeutic range were for the most part performed on patient-subjects where there was, at least arguably, a prospect that the subjects might benefit medically from the exposure. However, the TBI and experimental gallium treatments, in which patients suffered symptoms of acute radiation sickness and died soon after treatment, raise the question of whether their deaths were hastened by the radiation treatments. Resolution of this issue requires review of individual medical histories, which the Advisory Committee could not undertake.

FINDING 3

The Advisory Committee finds that human radiation experimentation during the period 1944 through 1974 contributed significantly to advances in medicine and thus to the health of the public.

Human radiation research was essential to the development of new therapies such as the use of radioactive iodine to treat thyroid cancer; the use of phosphorus to treat blood diseases such as polycythemia vera; and the use of radioactive strontium as a palliative in prostate and other cancers metastasized to the bone. Diagnostic uses of radionuclides developed during this period include scanning techniques to identify tumors and radiolabeling techniques that help evaluate a variety of cardiac diseases. The quality of images produced by external sources of radiation also improved dramatically between 1944 and 1974, making possible, for example, techniques such as balloon angioplasty to open occluded arteries.

FINDING 4

The Advisory Committee finds that some govern-ment agencies required the consent of some research subjects well before 1944. These requirements gen-erally did not stipulate what was meant by consent, however, nor did they generally indicate whether investigators were obligated to disclose specific in-formation to potential subjects. The government did not have comprehensive policies requiring the con-sent of all subjects of research, including both healthy subjects and patient-subjects, until 1974.

4a. *Research Involving Healthy Subjects:* In the 1920s, the Army promulgated a regulation con-cerning the use of "volunteers" for medical research. In 1932, the secretary of the Navy required that subjects of a proposed experiment be "informed volunteers." In 1942 the require-ment that healthy subjects be informed volun-teers was also articulated by the Committee on Medical Research, which oversaw war-related research for the Executive Office of the Presi-dent. In 1953, the principle of consent articu-lated in the Nuremberg Code was adopted by the Department of Defense in a Top Secret memo-randum from Secretary of Defense Charles Wil-son regarding human research related to atomic, biological, and chemical warfare (this document is known as the Wilson memorandum); in 1954, this application of the Nuremberg Code was expanded by the Army Office of the Surgeon General as an unclassified policy for all research with "human volunteers." A policy of requiring researchers to obtain consent was adopted by the Clinical Center, the research hospital of the National Institutes of Health, in 1953; by the Atomic Energy Commission in 1956; and by the Air Force in 1958. In the 1960s, all branches of the Department of Defense promul-gated regulations requiring the consent of healthy subjects, and the Isotopes Distribution Division of the AEC included in its guide for researchers a requirement of consent from all subjects. In 1966, the surgeon general of the Public Health Service issued a policy requiring the consent of all subjects of research conducted or funded by PHS; also in the late 1960s, the Veterans Administration codified in its oper-ating manual a requirement of consent from all research subjects. In 1972, the National Aero-nautics and Space Administration adopted

similar consent requirements, although excep-tions were made for certain subject popula-tions, such as astronauts. In 1974, the Public Health Service policy was promulgated as a regulation for all contracts and grants of the Department of Health, Education, and Wel-fare. The CIA did not formally adopt consent requirements until 1976, when an executive order mandated that it follow the 1974 regu-lations of DHEW concerning research involv-ing human subjects.

4b. *Research Involving Patient-Subjects:* In an April 1947 letter, the AEC general manager stated the AEC's understanding that AEC contract re-searchers would inform patient-subjects of the risks associated with a research intervention and that patient-subjects express a willingness to re-ceive the intervention. In a second letter, written in November 1947, the general manager specifi-cally stipulated that the AEC require researchers to obtain "informed consent in writing" from patient-subjects where "substances known to be or suspected of being poisonous or harmful" were given to human beings. In 1948, the AEC per-mitted the administration of "larger doses [of radioisotopes] for investigative purposes," but only with the patient-subject's consent. In 1953, the NIH Clinical Center required consent from all patient-subjects and specified that written con-sent was to be obtained from patient-subjects in-volved in high- or uncertain-risk experiments. In the early 1960s, several government agencies adopted consent provisions for investigational drugs; these requirements applied to some radio-isotope experiments with patients. In 1965, the AEC required that consent be obtained from all subjects, including patient-subjects, who were administered radioisotopes for experimental purposes, except when it appeared "not feasible" or not in the patient's "best interest." By 1967, the VA required the consent of all patient-sub-jects. As noted in Finding 4a above, in 1965 the AEC required that consent be obtained from all subjects administered radioisotopes for experi-mental or nonroutine uses. In 1966 the surgeon general of the Public Health Service issued a policy requiring the consent of all subjects of research conducted or funded by PHS, includ-ing patient-subjects. Exceptions to this require-ment were permitted for only certain kinds of social science research posing minimal risk. A

1972 NASA policy applied to all subjects of research, presumably including patient-subjects. By 1973, all the branches of the military had promulgated regulations requiring the consent of patient-subjects. In 1974, the PHS policy was promulgated as a regulation for all contracts and grants of DHEW.

FINDING 5

The Advisory Committee finds that government agencies did not generally take effective measures to implement their requirements and policies on consent to human radiation research.

Evidence of the implementation of the AEC's consent requirements stated in April and November 1947 letters from the general manager is slim. A document suggests that the April 1947 requirement for a signed statement from two physicians testifying to consent was satisfied in at least one case. However, the Advisory Committee did not find evidence that this or other requirements stated in the 1947 letters were embodied as a provision of AEC contracts involving human subject research or otherwise routinely communicated to contract researchers. Further, there was no reference to the requirements stated in these letters or to the letters themselves in the written material disseminated to researchers by the AEC's program for distributing radioisotopes for "human uses." Moreover, requests for guidance concerning human use policies from investigators at AEC-operated research facilities suggest that the 1947 requirements were not routinely disseminated. Subsequent requirements that healthy subjects be informed volunteers and that consent be obtained from seriously ill patients receiving higher doses of radioisotopes were more widely communicated; we have not been able to determine the extent to which they were actually implemented.

Secretary of Defense Wilson's February 1953 Top Secret memorandum detailing requirements for research with human subjects was rewritten as an unclassified June 1953 directive from the secretary of the Army. It is difficult to determine why these requirements were applied to some activities and not to others. For example, elements of some of these requirements appear to have been implemented in some experiments conducted in conjunction with atomic bomb tests and not in others. In 1954, these requirements were adopted by the Army surgeon general as applicable to all research involving "human volunteers." This 1954 statement was transmitted to contractors as a "nonmandatory guide." However, there is some evidence that the Army sought to include this statement as a condition in at least some contracts.

Evidence of implementation of the NIH Clinical Center's 1953 policy requiring that information be provided to and consent obtained from all subjects is difficult to find; in most cases involving patient-subjects, documentation would not have been required in writing. By contrast, the use of healthy subjects in the Clinical Center required written consent and the "normal volunteer program" appears to have involved greater supervision to ensure that consent was obtained from these subjects.

FINDING 6

The Advisory Committee finds that from at least 1946 some government agencies had in place procedural mechanisms for reviewing the acceptability of risks to human subjects from exposure to radioisotopes. By 1974, the government had policies requiring review of the acceptability of risks to human subjects in other forms of research, including research involving exposure to external radiation.

Beginning in 1946 the Manhattan Project, and from 1947 onward the AEC, required some investigators seeking to conduct experiments using radioisotopes supplied by the government to have the risks to subjects reviewed by a committee at the institution where the work was to be conducted and in some cases by the AEC's Subcommittee on Human Applications as well. The AEC required that local committees be composed of at least three physicians or researchers with relevant expertise regarding radiation safety and medical applications. By 1949, it was clear that this policy applied to all investigators using radioisotopes supplied by the AEC.

In 1953 prior group review for risk was also begun at the NIH Clinical Center for proposed human research that involved unusual hazard.

No such requirement applied to research funded by NIH but conducted at universities and other nongovernmental research facilities until 1966, when the PHS required that all institutions establish a local peer review committee to evaluate the adequacy of the protection provided to human subjects in each proposal. This requirement was promulgated as an institutional policy by the DHEW in 1971.

In 1953, by adopting the Nuremberg Code, the secretary of defense and the Department of the Army endorsed several principles intended to minimize risk in research with human subjects, at least in regard to the atomic, biological, and chemical warfare experiments that were subject to this policy. In the DOD, both the purpose of proposed research and the level of risk were subjected to prior review through the military chain of command. This was previously required by the Navy at least from 1943, and the Air Force from 1952. However, the extent to which these requirements covered particular research activities (such as healthy subjects vs. patients; radioisotopes vs. external radiation) and particular institutions (such as contractors vs. in-house research) differs and is difficult to reconstruct. Also difficult to reconstruct is the extent to which the risk protection principles of the Nuremberg Code were implemented. In the mid-1960s, concurrent with the adoption of regulations related to investigational drug testing, the DOD and each military service adopted provisions requiring the establishment of a "review board" or committee to oversee proposed research projects involving new drugs. In some cases, such as with the Air Force beginning in 1965, this committee also served to evaluate all other proposals involving human subjects. During this period, the VA also established a review board mechanism for research involving new drugs and investigational procedures.

However, there is no evidence that a parallel mechanism for reviewing the risks of research involving external radiation was in place.

From its 1947 birth, the AEC, as part of its policy to promote the peaceful use of radioisotopes, required private institutions that wished to obtain radioisotopes for "human uses" (including human experimentation as well as patient treatment) to establish local review committees. These committees reviewed proposed human uses under guidelines provided by the AEC's own Subcommittee on Human Applications of the Advisory Committee on Iso-tope Distribution Policy. This AEC subcommittee reviewed these applications, providing a second level of oversight of risk. By 1949, the AEC's own labs were required to establish local committees and to have human use applications, reviewed by the same AEC Subcommittee on Human Applications. The control of risk, and the assurance of safety to all those involved (including doctors and other health care workers), was a primary purpose of the reviews. The Advisory Committee lacked sufficient evidence to determine whether the system was implemented in all of the many institutions that used government-supplied radioisotopes for human subjects research or whether the system was always adhered to in any particular institution.

In addition to providing for the review of research proposals, the AEC dramatically increased the number of qualified personnel by offering training courses in the safe handling and use of radioisotopes. As individual procedures became routine, the degree of review was lessened; as specific institutions became more experienced, more reviewing authority was delegated to them.

The primary function of the system was to reduce the physical hazards of using radioisotopes, not to enforce any policies regarding consent of subjects. (See chapter 6.)

FINDING 7

The Advisory Committee finds that the government program of distributing radioisotopes for use in human subject research included procedures for the review of risk. These were widely implemented by researchers and institutions that used isotopes obtained from the AEC for human experimentation.

FINDING 8

The Committee finds that for the period 1944 to 1974 there is no evidence that any government statement or policy on research involving human subjects contained a provision permitting a waiver of consent requirements for national security reasons.

Neither the AEC nor the DOD included national security exceptions in their written rules on human subjects research. For example, the 1953 Wilson memorandum adopting the Nuremberg Code was expressly applicable to human experimentation related to atomic, biological, and chemical warfare and did not provide for any "national security" exception.

The Committee notes that much documentation related to the CIA's program of secret experimentation, including MKULTRA, has long since been destroyed, and, therefore, we cannot state with certainty what policy(ies) underlay human experiments in these programs or whether such policies included national security exceptions.

FINDING 9

9a. The Advisory Committee finds that government agencies had no requirements or policies to ensure equity in the selection of subjects for research conducted or funded by the federal government during the period 1944 through 1974.

The only reference during this period to issues of equity in the selection of subjects in agency documents reviewed by the Advisory Committee is in an influential DHEW guide to recipients of federal research funds published in 1971, popularly known as the Yellow Book. The Yellow Book notes a "particular concern" about research involving "groups with limited civil freedom."

9b. Because of the limited data available on the universe of experiments identified by the Committee, the Committee was unable to determine whether during the period 1944 through 1974 people who were socially disadvantaged were more likely than more socially advantaged people to be used as subjects in human radiation experiments generally or in those experiments that offered no prospect of medical benefit or posed greater risks. The Advisory Committee finds, however, that some of the biomedical experiments reviewed by the Committee that were ethically troubling were conducted on institutionalized children, seriously ill and sometimes comatose patients, African-Americans, and prisoners.

The Committee was troubled by the selection of subjects in many of the experiments we reviewed. These subjects often were drawn from relatively powerless, easily exploited groups, and many of them were hospitalized patients. As noted in Finding 9a, there were during this period no federal rules or policies directed at fairness in the selection of research subjects, and no norms or practices within the biomedical research community specifically addressing considerations of fairness. This silence on questions of justice in the conduct of human research was characteristic not only of radiation research but also of the entire research enterprise. While we note here cases that provoked concern, we were unable to determine the extent to which there were systemic injustices in the selection of research subjects in human radiation research because in most cases we were unable to determine any of the characteristics of the subjects involved in the experiments we catalogued.

FINDING 10

The Advisory Committee finds that even as early as 1944 it was conventional for physicians and other biomedical scientists to obtain consent from healthy subjects of research. By contrast, during the 1944–1974 period but especially through the early 1960s, physicians engaged in clinical research generally did not obtain consent from patient-subjects for whom the research was intended to offer a prospect of medical benefit. Even where there was no such prospect, it was common for physicians to conduct research on patients without their consent. It also was common, however, for physicians to be concerned about risk in conducting research on patient-subjects and, in the absence of a prospect of offsetting medical benefit, to restrict research uses of patients to what were considered low- or minimal-risk interventions.

Perhaps the best-known example of the use of informed volunteers in research conducted at the turn of the century is the yellow fever research by military scientist Walter Reed. In the Advisory Committee's Ethics Oral History Project, several of Reed's military successors who were active in the 1940s and 1950s gave similar examples of voluntary consent from healthy subjects in the context of work on typhus and malaria. In 1946, the

American Medical Association (AMA) articulated the principle that human subjects must give "voluntary consent." In 1947, the prosecution's expert witness at the Nuremberg Medical Trial, Dr. Andrew Ivy, who had helped shape the AMA principle in conjunction with his role at Nuremberg, asserted that this was a standard by which physicians were guided in the use of human beings in medical experiments and that this standard was in "common practice" prior to its articulation by the AMA in 1946. Precisely what Ivy meant by this claim is unclear. Although there are doubtless instances in which this standard of voluntary consent was not followed, it does seem to have been widely recognized and adhered to among investigators whose research involved healthy subjects.

By contrast, various sources confirm that it was not conventional to obtain consent from patient-subjects. These sources include documents from the period in which the conduct of clinical research was discussed as well as the Committee's Ethics Oral History Project, in which physicians active in research in the 1940s and 1950s agreed that consent played little or no role in research with patient-subjects, even where there was no expectation that the patient would benefit medically from the research. At the same time, however, there was also agreement that, where patients were used as subjects in nontherapeutic research, the research usually posed little or no risk to the patients.

FINDING 11

11a. The Advisory Committee finds that the government and government officials are morally responsible in cases in which they did not take effective measures to implement the government's policies and requirements, and the medical profession and biomedical scientists are morally responsible for instances in which they failed to adhere to the professional norms and practices of the time.

The Advisory Committee was concerned that our conclusions about actions taken in the past be rendered fairly. Clearly, if government agencies had rules or requirements for the use of human subjects at the time, and if these requirements were sound from our point of view and

consistent with basic moral principles, then agencies and agency officials had just as much moral responsibility to implement those requirements as those in analogous positions would have today, or in any day, with respect to current sound government requirements.[1] We have found that some government agencies did in fact have such requirements (see Findings 4 and 6).

Similarly, if the medical profession and the research community generally had recognized norms and practices for the conduct of research with human subjects, and if these norms and practices were sound, then physician-investigators and other scientists operating in the past had just as much responsibility to adhere to those norms and practices as those in analogous positions would have today with respect to current norms and practices that are morally sound. The Committee found evidence that the medical profession had such norms with respect to obtaining consent from healthy subjects, although physicians engaged in clinical research did not generally seek consent from patient-subjects. The Committee also found evidence of professional norms concerning acceptability of risk to subjects (see Finding 10).

11b. The Advisory Committee finds that by today's standards we consider it wrong that our government did not take effective measures to adequately protect the rights and interests of all human subjects of research. We also find that by today's standards we consider it wrong that medical and other professions engaged in human research did not have norms and practices of consent for all subjects of research.

There is today a well-established consensus about the basic principles that should govern the use of human subjects of research. There is also wide agreement that the government has an obligation to protect the rights and interests of all human research subjects and that the medical and other professions engaged in research are obligated to have norms and practices of consent for all human subjects of research. The failure to have such conditions in place would today be considered wrong.

11c. The Advisory Committee finds that government officials and investigators are blameworthy for not having had policies and practices in place to protect the rights and interests of human subjects who were used in research from which the subjects could not possibly derive medical benefits (nontherapeutic research in the strict sense). By contrast, to the extent that there was reason to believe that research might provide a direct medical benefit to subjects, government officials and biomedical professionals are less blameworthy for not having had such protections and practices.

We also find that, to the extent that research was thought to pose significant risk, government officials and investigators are more blameworthy for not having had such protections and practices in place. By contrast, to the extent that research was thought to pose little or no risk, government officials and biomedical professionals are less blameworthy for not having had such protections and practices.

Today we consider policies and practices to protect the rights and interests of human subjects to be *as important* in research that offers participants a prospect of medical benefit as in research that does not. Government regulations and the rules of professional and research ethics apply equally to both kinds of research. In the 1940s, 1950s, and 1960s, however, patients and society generally accorded doctors more authority to make decisions for their patients than they do today. It was both commonplace and considered appropriate for a physician to determine what treatments a patient should receive without necessarily consulting the patient, provided that the decision was based on the physician's judgment about what would be in the patient's best interest. This authority generally extended to decisions about whether a patient's interest would be served by being a subject in medical research. Judgments about the blameworthiness of officials and physician-investigators for not having had policies and practices to protect the rights of human subjects in research that offered a prospect of medical benefit, such as requirements of consent, are mitigated by this historical context.

However, even at the time, government officials and biomedical professionals should have recognized that when research offers *no prospect* of medical benefit, whether subjects are healthy or sick, research should not proceed without the person's consent. It should have been recognized

that despite the significant decision-making authority ceded to the physician within the doctor-patient relationship, this authority did not extend to procedures conducted solely to advance science without a prospect of offsetting benefit to the person. This finding is supported by the moral principle, deeply embedded in the American experience, that individuals may not be used as mere means toward the ends of others. We also note that at its 1947 beginning, officials of and biomedical advisers to the AEC were clearly aware of the issues raised when patients, as well as healthy people, were used as subjects in nontherapeutic research without their consent.

The Advisory Committee has also determined that government officials and scientific investigators at the time recognized that research could put subjects at risk of harm, that they had an obligation to determine that the risks imposed were reasonable, and that research that posed greater or more uncertain risks was more problematic than research whose risks were lower. Sometimes government officials and investigators took steps to protect subjects from unnecessary or unacceptable risks. These steps included in some cases a requirement of group review of research proposals and the obtaining of consent of the subjects, particularly where risks were considered worrisome. But these steps were not consistently or uniformly taken.

POPULATION EXPOSURES

FINDING 12

The Advisory Committee finds that some service personnel were used in human experiments in connection with tests of atomic bombs. The Committee finds that such personnel were typically exposed to no greater risks than the far greater number of service personnel engaged in similar activities for training or other purposes. The Committee further finds that there is little evidence that the 1953 secretary of defense Nuremberg Code memorandum was transmitted to those involved with human experiments conducted in conjunction with atomic testing. However, some of the requirements con-

tained in the memorandum were implemented in the case of a few experiments, apparently independently of the memorandum. The Committee also finds that the government did not create or maintain adequate records for both experimental and nonexperimental participants.

More than 200,000 service personnel participated in nuclear weapons tests from 1946 to the early 1960s. The vast majority of those who participated were engaged in management of the tests, training maneuvers, or data-gathering activities. In the range of 2,000 to 3,000 of these participants were research subjects. In many cases these research subjects engaged in activities, and were subjected to risks, essentially identical to those engaged in by many more people who were not research subjects. The purpose of this human subject research was not to measure the biological effects of radiation. Rather, for example, researchers sought to measure the psychological and physiological effects of participation in bomb tests, the levels of radiation to which individuals who flew in and around atomic clouds were exposed, and the effects of intense light from the bomb blast on the eyes.

The Advisory Committee found little evidence that the 1953 Wilson memorandum on human experimentation in connection with atomic, biological, and chemical warfare (or an Army implementing document) was transmitted to those involved in bomb-test-related experimentation. In interviews with Committee staff, some of those involved in the experimentation stated that they were unaware of the memorandum. However, there is evidence that in some of the experiments consent was provided for, but this was likely independent of the 1953 policy.

The military took successful precautions against exposure to radiation levels that were likely to produce acute effects. However, bomb-test participants were exposed to lower levels of radiation, which might conceivably have effects on some participants over the longer term. The evidence shows that those who managed the tests were aware of the potential, however small, that injury might result years later from such exposures. In recent years, as the government and vet-

erans have sought to reconstruct the extent of exposure and resulting injury, it has become apparent that the government did not uniformly create records that would permit all individuals to efficiently and confidently know the extent of their exposure, did not create records that would permit reconstruction of the identity and location of all those who participated at the tests, did not adequately undertake to link medical and exposure records, and did not adequately maintain those records that were initially created.

FINDING 13

The Advisory Committee finds that during the 1944–1974 period the government intentionally released radiation into the environment for research purposes on several hundred occasions. In only a very few of these cases was radiation released for the purpose of studying its effect on humans.

The Advisory Committee's charter identified thirteen releases: one related to the testing of intelligence equipment (the "Green Run"), eight radiological warfare tests, and four releases of radioactive lanthanum ("RaLa") to test the mechanism of the atomic bomb. The Advisory Committee received information on more than sixty radiological warfare releases that took place in the period 1949–1952 and on the nearly 250 RaLa releases that took place in the period 1944–1961. We identified further intentional releases of a kind that were not described in the charter. These included the release of radiation to study its environmental pathways and the release of radiation in connection with outdoor safety tests and tests related to the development of nuclear reactors, as well as to the development of nuclear-powered rockets and airplanes.

Most releases took place in and around the sites that constitute the nation's nuclear weapons complex, notably Oak Ridge, Tennessee; Hanford, Washington; Los Alamos, New Mexico; the Nevada nuclear weapons test site; and the Idaho National Engineering Laboratory. Releases related to radiological warfare tests took place primarily at the Dugway Proving Ground in Utah. Radioactive material was also released into the environment for research purposes at other locations; for example, fallout from the Nevada Test Site was inserted into the tundra of Alaska.

FINDING 14

The Advisory Committee finds that for both the Green Run (at Hanford) and the RaLa tests (at Los Alamos), where dose reconstructions have been undertaken, it is unlikely that members of the public were directly harmed solely as a consequence of these tests.

It is impossible to distinguish any harm due to these releases from other sources of exposure, particularly at Hanford, where the amount of radioactivity intentionally released by the Green Run was 1 percent of the amounts released by routine operations of the Hanford facility in the 1945–1947 period. The risks of thyroid disease from all past operations of the Hanford plant are currently under study; however, the Advisory Committee estimates that the contribution of the Green Run to any such risks amounts to substantially less than one case. No dose reconstruction has been undertaken for the radiological warfare field tests at the Dugway Proving Grounds. Most of the intentional releases the Advisory Committee has studied, including all those identified in our charter, involved radioactive materials with short-enough half-lives that they quickly decayed and therefore pose no risk to health from continuing exposure.

FINDING 15

The Advisory Committee finds that during the period from 1944 to about 1970 there was no system of environmental laws and regulations governing the conduct of intentional releases analogous to that currently in place. However, those responsible for intentional releases during this period recognized the possible health risks from environmental releases and that risks had to be considered in making policy decisions about such releases.

In the case of the Green Run, guidelines existed for routine (or normal operating) environmental releases of radioactive iodine but were exceeded; in the case of radiological warfare tests,

a safety panel was created. These and other releases specified in the Advisory Committee's charter were conducted in secret because of a combination of concerns about national security and public reaction. The Atomic Energy Act of 1954 began the formal public system of safety regulation of environmental releases of radiation. It was not until the National Environmental Policy Act of 1969 that public review of federal actions likely to have a significant impact on the environment was institutionalized.

FINDING 16

The Committee finds that, as a consequence of exposure to radon and its daughter products in underground uranium mines, at least several hundred miners died of lung cancer and surviving miners remain at elevated risk. As a consequence of a U.S. hydrogen bomb test conducted in 1954, several hundred residents of the Marshall Islands and the crew of a Japanese fishing boat developed acute radiation effects. Some of the Marshall Islanders subsequently developed benign thyroid disorders and thyroid cancer as a result of the radiation exposure. Surviving Marshallese also may remain at elevated risk of thyroid abnormalities.

The miners, who were the subject of government study as they mined uranium for use in weapons manufacturing, were subject to radon exposures well in excess of levels known to be hazardous. The government failed to act to require the reduction of the hazard by ventilating the mines, and it failed to adequately warn the miners of the hazard to which they were being exposed, even though such actions would likely have posed no threat to the national security.

Some Marshallese exposed during the 1954 bomb test received radiation doses substantially in excess of those considered safe, both at the time and today. One Marshallese exposed as a baby died of leukemia in 1972, which may have been as a consequence of exposure during the test. In 1954, twenty-eight U.S. servicemen manning a weather station on Rongerik Atoll also received doses of radiation substantially in excess of those considered safe at the time and today. The Advisory Committee does not know whether any of the servicemen suffered long-term harm as a result of their exposure. Twenty-three Japanese fishermen were irradiated as a result of the fallout from the 1954 bomb test. The exposed Marshallese population received additional doses of radiation from later bomb tests and residual radiation in the food chain, which continues to this day. The U.S. government—initially the Navy and then the AEC and its successor agencies—has provided care to the Marshallese ever since for radiation-related illnesses while conducting research on this population to determine radiation effects. For many years the distinction between research and clinical care was not adequately explained to the Marshallese.

FINDING 17

The Committee finds that since the end of the Manhattan Project in 1946 human radiation experiments (even where expressly conducted for military purposes) have typically not been classified as secret by the government. Nonetheless, important discussions of human experimentation took place in secret, and information was kept secret out of concern for embarrassment to the government, potential legal liability, and concern that public misunderstanding would jeopardize government programs. In some cases, deception was employed. In the case of the plutonium injection experiments, government officials and government-sponsored researchers continued to keep information secret from the subjects of several human radiation experiments and their families, including the fact that they had been used as subjects of such research. Some information about the plutonium injections, including documentation showing that data on these and related human experiments were kept secret out of concern for embarrassment and legal liability, was declassified and made public only during the life of the Advisory Committee.

Human experimentation conducted during the Manhattan Project was carried out in secret. Since 1947 (when the Atomic Energy Commission began operations and the military services were unified under a secretary of defense) human

radiation experiments have rarely been protected as classified secrets. However:

In 1947 AEC biomedical advisers publicly urged that biomedical research be kept secret only where required by national security. At the same time, AEC officials and advisers secretly determined that reports on human radiation experiments should not be declassified where they contained information that was potentially embarrassing or a cause of legal liability. Upon requests for declassification, research reports involving human radiation experiments and other human radiation exposures were reviewed for their effects on public relations, labor relations, and potential legal claims.

In 1947 AEC officials and advisers conducted discussions about human subject research policy; some of these discussions were conducted in secret meetings, and the statements of requirements that were articulated, while not secret, evidently were little disseminated. Similarly, 1949–1950 AEC/DOD discussions of the terms on which human radiation experiments could be conducted were either secret or the substance of the discussions was given limited public distribution. In 1952, Department of Defense biomedical advisory groups also engaged in secret or restricted discussions of policy, which led to the 1953 issuance of the Wilson memorandum, which was itself issued in Top Secret.

Government officials and experts did not squarely and publicly address the existence and scope of government-supported human radiation experimentation. For example, in the late 1940s and early 1950s the AEC denied to the press and citizens that it engaged in human experimentation, even though the AEC's highly visible radioisotope distribution program had been created to provide the means for, among other things, human experimentation.

Project Sunshine, a worldwide program of data gathering, including human data gathering to measure the effects of fallout, was kept secret from its 1953 inception until 1956, and AEC officials and researchers employed deception in the solicitation of bones of deceased babies from intermediaries with access to human remains. It appears that concern for public relations played a key role in keeping the human data gathering, and the very existence of Project Sunshine, secret.

FINDING 18

All the intentional releases identified in the Advisory Committee's charter, as well as the several hundred other releases that were essentially of the same types, were conducted in secret and remained secret for many years thereafter. All involved some stated national security purpose, which may have justified some degree of secrecy. Despite continued requests from the public that stretch back well over a decade, however, some information about intentional releases was declassified and made publicly available only during the life of the Advisory Committee.

The Committee's review indicates that internal proposals that the public be informed about the existence of the radiological warfare program were rebuffed on grounds that public misunderstanding might jeopardize the program.

Citizens learned of the 1949 Green Run in 1986, and then only following close review of documents requested from the government by members of the public. Portions of a key surviving report on the Green Run were not declassified until 1994. Similarly, although 250 intentional releases near the land of the Pueblo Indians in New Mexico took place between 1944 and 1961, the Pueblo do not appear to have been informed of the full scope of the program until 1994. Documentation on these midcentury tests is only now being declassified.

FINDING 19

The Advisory Committee finds that the government did not routinely undertake to create records needed to ensure that secret programs could be understood and accounted for in later years and that it did not adequately maintain such records where they were created. The Committee further finds that many important record collections (including records that were not initially classified) have been maintained in a manner that renders them practically inaccessible to those who need them, thereby limiting the utility of the records to the government itself, as well

as the public's rights under the Freedom of Information Act.

Where citizens are exposed to potential hazards for collective benefit, the government bears a burden of collecting data needed to measure risk, of maintaining records, and of providing the information to affected citizens and the public on a timely basis. The need to provide for ultimate public accounting, as was recognized by early AEC leadership, is particularly great where risk taking occurs in agencies that do much of their work in secret. The government did not routinely or adequately create and maintain such records for relevant human radiation experiments, intentional releases, and service personnel exposed in conjunction with atomic bomb tests.

Where records were initially created, important collections have been lost or destroyed over the years. These include the classified records of the Atomic Energy Commission's Intelligence Division; secret records that were kept in anticipation of potential liability claims from service personnel exposed to radiation;[2] records relating to the secret program of experimentation conducted by the CIA (MKULTRA); nonclassified records of VA hospitals regarding the thousands of experiments that, the VA told the Advisory Committee, were conducted there; and nonclassified files of the AEC's Isotope Distribution Program relating to the many licenses for "human use" it granted in the period 1947 to 1955. The Committee notes that laws governing government records provide for routine destruction of older records; however, we also found that some records documenting the destruction of records had been lost or destroyed.

Public witnesses and others repeatedly expressed doubt to the Advisory Committee about the credibility of the government's efforts to respond to requests for documents. The Advisory Committee's experience indicates that shortcomings in government response to Freedom of Information Act requests, which may be interpreted by citizens as deliberate nondisclosure, may often occur because the agencies themselves lack adequate road maps to the records that still exist and lack records needed to determine whether collections of importance to the public have been lost or destroyed. In the absence of the efforts put forth by the Human

Radiation Interagency Working Group, thousands of documents that have now been made public would not have been located.

Findings for the Contemporary Period

BIOMEDICAL EXPERIMENTS

FINDING 20

The Advisory Committee finds that human research involving radioisotopes is currently subjected to more safeguards and levels of review than most other areas of research involving human subjects. The Advisory Committee further finds that there are no apparent differences between the treatment of human subjects of radiation research and human subjects of other biomedical research.

Today, research involving either external radiation or radioactive drugs usually undergoes an additional layer of review for safety and risk. Most medical institutions have a radiation safety committee (RSC) responsible for evaluating the risk of radiation research and other medical activities and limitation of radiation exposure of both employees and subjects. Research and medical institutions that perform basic research involving human subjects and radioactive drugs must also have studies reviewed and approved by a radioactive drug research committee (RDRC), a local institutional committee approved by the Food and Drug Administration, to ensure that safeguards in the use of such drugs are met. These steps are in addition to the review of risks and benefits undertaken for all research, whether radiation or nonradiation, by local institutional review boards.

In the Advisory Committee's two empirical projects examining current practices in human subject research, we found no meaningful differences between radiation research and human research in other fields.

FINDING 21

The Advisory Committee finds that today research involving human subjects sponsored by the government may be classified and conducted in secret, but it must comply with the provisions of the Common Rule.

It is permissible today to perform classified research on human subjects, although it is our understanding that classified research occurs relatively rarely. Like unclassified research, such research is covered by the protections enunciated in the Common Rule. There may be significant problems in the application of the Common Rule to classified research, however. One problem concerns the possible need for security clearances if institutional review boards are to appropriately protect the interests of human subjects. Written guidance on this question differs among the agencies. Of particular concern is whether only those members of the IRB who are employees of the agency will possess security clearances and thus be able to participate in reviewing classified projects.

Another issue of concern is that for classified research involving no more than minimal risk, as with any such research, the Common Rule allows IRBs to waive any or all elements of informed consent if, among other things, it is not practicable for the research to be carried out without such a waiver.* The Committee believes, however, that research conducted in secret should never be permitted on human subjects without the subjects' informed consent. The question of what must be disclosed to potential subjects in order for them to make an informed decision about participating in classified research, including whether an adequate disclosure can be made to people who do not have security clearances, is an important issue not addressed in the Common Rule.

---

*Common Rule, __.116(d). Under the Common Rule, four requirements must be met for an IRB to waive the rule's informed consent requirements: "(1) the research involves no more than minimal risk; (2) the waiver or alteration will not adversely affect the rights and welfare of the subjects; (3) the research could not practicably be carried out without the waiver or alteration; and (4) whenever appropriate, the subjects will be provided with additional pertinent information after participation."

FINDING 22

The Advisory Committee finds that, in comparison with the practices and policies of the 1940s and 1950s, there have been significant advances in the protection of the rights and interests of human subjects of biomedical research. However, we also find that there is evidence of serious deficiencies in some parts of the current system for the protection of the rights and interests of human subjects.

Based on the Advisory Committee's review, it appears that about 40 to 50 percent of human subjects research poses no more than minimal risk of harm to subjects. In our review of research documents that bear on human subjects issues, we found no problems or only minor problems in most of the minimal-risk studies we examined. In our review of documents we also found examples of complicated, higher-risk studies in which human subjects issues were carefully and adequately addressed and that included excellent consent forms. In our interview project, there was little evidence that patient-subjects felt coerced or pressured by investigators to participate in research. We interviewed patients who had declined offers to become research subjects, reinforcing the impression that there are often contexts in which potential research subjects have a genuine choice.

At the same time, however, we also found in our review of documents examples in which human subjects issues were carelessly and inadequately addressed. These disparities suggest that there is substantial variation in the performance of institutional review boards.

We found serious deficiencies in our review of research proposal documents in several areas central to the ethics of research involving human subjects. Specifically, these documents often failed to provide sufficient information with which judgments could be made about the likely voluntariness of participation and about the characteristics of and justification for the subjects selected for study. It also was often difficult to assess, again because of insufficient information, whether the likely merits of the research warranted the imposition of risk or inconvenience on human subjects. We also found serious deficiencies in many of the consent forms we reviewed, including the consent forms of some minimal-risk studies.

Most of the Advisory Committee's concerns focus, however, on research that exposes subjects to greater than minimal risk. We found evidence of confusion over the distinction between research and therapy in interviews with patients, in the research documents reviewed, and in public testimony. This confusion appears to be borne out of a combination of trust in physicians and an inadequate understanding of the differences among innovative practice, therapeutic research, and accepted modes of therapy. The Advisory Committee's empirical studies suggest that there is reason to worry that patient-subjects who have serious illnesses may have unrealistic expectations both about the possibility that they will personally benefit by being a research subject and about the discomforts and hardships that sometimes accompany research.

The Advisory Committee is also concerned about research we reviewed involving adult subjects of questionable capacity. In the documents made available to the Advisory Committee, there was little discussion of the implications of diminished capacity for the process of consent and authorization to participate in research, even in studies that appeared to offer no prospect of medical benefit to subjects. In addition, the Advisory Committee is concerned about the failure of federal regulations to address the conduct of research involving institutionalized children.

POPULATION EXPOSURES

FINDING 23

The Advisory Committee finds that events that raise the same concerns as the intentional releases in the Advisory Committee's charter could still take place in secret under current environmental laws and regulations.

Today the law provides that environmental reviews may be conducted in part or even in whole in secret, thereby eliminating provision for public notice and comment. In classified programs, the government must still comply with environmental standards, and the Environmental Protection Agency must oversee and review environmental compliance. However, the EPA has not maintained records of environmental releases where the reviews were conducted in whole or in part in secret. Environmental laws and regulations that limit quality or quantity of a release also contain provisions allowing exemptions for national security. In principle, the President or the secretary of energy (in the case of the Atomic Energy Act) could invoke these exemptions to permit releases that would otherwise exceed risk standards.

NOTES

1. The qualification of agency officials' responsibility to implement "sound" requirements refers to the moral quality of the requirements. There may be other reasons for a society to hold its government officials responsible for implementing duly authorized rules, such as prudential needs for orderliness and predictability. But we would not hold an official morally blameworthy for failing to implement a requirement that is morally unsound. In that case his or her role-related responsibilities are superseded by basic ethical principles.

2. As discussed in chapter 10, the precise nature of all the records that were kept remains to be determined.

Recommendations

Recommendations for Remedies Pertaining to Experiments and Exposures During the Period 1944–1974[*]

BIOMEDICAL EXPERIMENTS

RECOMMENDATION 1

The Advisory Committee recommends to the Human Radiation Interagency Working Group that the government deliver a personal, individualized apology and provide financial compensation to the subjects (or their next of kin) of human radiation experiments in which efforts were made by the government to keep information secret from these individuals or their families, or from the public, for the purpose of avoiding embarrassment or potential legal liability, or both, and where this secrecy had the effect of denying individuals the opportunity to pursue potential grievances.

The Advisory Committee has found three cases to which the above applies. These are the surviving family members of:

1. The eighteen subjects of the plutonium injection experiments;
2. The subject of a zirconium injection experiment, known only as Cal-Z; and
3. Several subjects of total-body irradiation experiments conducted during World War II.[1]

Deliberate attempts by public officials in trusted and often sensitive government positions to conceal the fact of participation from subjects or their families, particularly in the absence of sufficient national security justification and for the declared purpose of avoiding potential liability and public embarrassment, are assaults upon the foundations of individual privacy and self-determination. Such actions violate an individual's right to information about him or herself and must be taken with the utmost seriousness.

In the cases listed above, this secrecy served to prevent people who may have been wronged from seeking redress within their lifetimes.

_____

[*]In preparing these recommendations, the Advisory Committee addressed only the question of whether the federal government owes remedies to subjects or their surviving immediate family members. The remedies identified below are not intended to preclude any remedies that subjects or their family members may otherwise be entitled to from nonfederal institutions or from individuals.

Secrecy regarding the participation of particular subjects was maintained until as late as 1974. Documents showing that the government kept information secret about particular 1940s experiments on grounds of potential liability and embarrassment remained secret until retrieved by the committee in 1994. Even though at the time justice might not have required financial compensation for the failure to disclose information in the absence of direct physical harm, the fact that the government's actions limited the opportunity of these subjects to seek justice is undeniable. Because of the offensiveness of the government's actions, justice today warrants a remedy of financial compensation.

Moreover, efforts to cover up governmental wrongdoing are assaults upon the polity itself, and not just upon the directly affected individual, because such efforts undermine the ability of a civil society to ensure that the government and its agents act within the rule of law. Such a situation warrants the extension of compensation to the next generation.

Implementation

Congress may need to consider legislation to provide compensation for the immediate families of the subjects in the plutonium injection experiments whose identities are known. The identities of the subject known as Cal-Z, as well as the subjects in the wartime total-body irradiation experiments, are not now known. Should their identities come to light, they or their families also should be compensated. In addition, should additional cases be identified that satisfy the criteria outlined above, further legislation should be enacted or other steps taken to provide those individuals or their family members with similar compensation.

RECOMMENDATION 2

The Advisory Committee recommends to the Human Radiation Interagency Working Group that for subjects of human radiation experiments that did not involve a prospect of direct medical benefit to the subjects, or in which interventions considered to be controversial at the time were presented *as conventional or standard practice, and physical injury attributable to the experiment resulted, the government should deliver a personal, individualized apology and provide financial compensation to cover relevant medical expenses and associated harms (pain, suffering, loss of income, disability) to the subjects or their surviving immediate family members.* *

The Advisory Committee has identified several experiments that are candidates for remedies to former subjects under this recommendation; these are described below in the section on implementation.

When the government puts an individual at risk in order to serve some collective national interest, it must take steps to ensure that the rights and interests of the individual are adequately protected. The Advisory Committee presumes, however, based on our understanding of the historical context, that such steps were not uniformly undertaken. As a consequence, it is possible that a citizen who was harmed as a result of participation in nontherapeutic research did not adequately consent to this use of his or her person. That the government did not have a system in place to ensure that individuals were not wronged by their use as research subjects in nontherapeutic research without their adequate consent, when that use resulted in harm, warrants a personal, individualized apology and financial compensation to subjects or to their surviving immediate family members.

Analogous cases exist to support this recommendation. In awarding substantial compensation to victims (or their families) of the CIA's MKULTRA experiments who were killed or suffered other serious harm, Congress and the

---

* The Advisory Committee was convened in response to concerns about human radiation experiments that offered no prospect of medical benefit to human subjects. In our historical analysis, the experiments we investigated either offered no prospect of medical benefit or they involved interventions alleged to be controversial at the time (see Overview to Part II). As a consequence, the Advisory Committee focused its consideration of remedies for subjects of human radiation experiments only on those experiments that fit these descriptions. The Committee makes no recommendations about whether, or under what conditions, remedies are appropriate for subjects of human radiation experiments that were considered at the time to offer a plausible prospect of medical benefit to subjects.

courts recognized that individuals used for government purpose without direct benefit to the experimental subject and without their consent deserved substantial awards.[2]

Nothing in this recommendation should be taken as having implications for how future policies governing compensation for research injuries should be constructed.

Implementation

Of the experiments that the Advisory Committee studied in detail, we have identified several that are candidates for remedies under this recommendation. These are as follows: the total-body irradiation (TBI) experiments (should it be determined that TBI was considered at the time to be a controversial treatment for patients with "radioresistant" tumors, and it was not presented as such to potential subjects, and should a determination of harm attributable to the experiments be made); the testicular irradiation experiments using prisoners as subjects (should a determination of harm attributable to the experiments be made); the uranium injection experiments at Rochester and Boston (should a determination of harm attributable to the experiments be made); and some of the iodine 131 experiments involving children (should a determination of harm attributable to the experiments be made). Because of the scope of the Advisory Committee's charge and our limited tenure, we were not in a position to undertake the individualized and detailed fact-finding required to resolve the uncertainties in each of these cases, including the evaluation of medical and research records of all the patients or subjects involved.

In addition, two experiments that the Committee did not study in detail, the iodine 131 experiment in Alaska and the Vanderbilt radioiron nutrition experiments, are currently in legal proceedings in which claims of harm have been made.

If an appropriate forum such as the courts or a properly constituted review committee determines that subjects were harmed as a consequence of nontherapeutic research, or as a consequence of research in which controversial treatments were presented to patients as con-

ventional or standard therapy, it is the Advisory Committee's view that the government should take steps to ensure that the remedies of apology and financial compensation are awarded.

The question of causation is key to any such determination. The Advisory Committee has heard from many public witnesses regarding the standards of proof and presumptions involved in the administration of existing radiation compensation statutes, which cover atomic bomb testing and uranium mining. In those cases the nature of the exposure for all applicants is relatively uniform and well defined, and the exposures have been the subject of a relatively large amount of study; by contrast, in the case of human radiation experiments, each experiment may present a different set of circumstances. In some cases, as in the administration of iodine 131, there is considerable knowledge of the relation between exposure and subsequent injury. In many other situations, less is known.

A decision should be made about how strict a causal association ought to be required, with a more strict standard making financial compensation available to fewer individuals. Whether the standard for presuming "causation" should be strict or loose is a policy decision that depends on values, not science. The standards/values problem speaks both to what should be done about whether the illness should be treated as experiment-related for purposes of compensation if (1) it is impossible to determine the likely range of association between the exposure and the illness (because the facts about dose or method of exposure are not available); and (2) the likely range of association is broad or the probability of association between the exposure and the illness is low.

To determine reasonable medical expenses, a schedule of projected medical costs appropriate for reimbursement could be created for specific diagnoses, rather than compensating for actual costs incurred. This approach would relieve the burden on the subject or immediate family members to prove actual costs, streamline the process for determining level of compensation, and allow for compensation for costs not yet incurred.

RECOMMENDATION 3

The Advisory Committee recommends to the Human Radiation Interagency Working Group that for subjects who were used in experiments for which there was no prospect of medical benefit to them and there is evidence specific to the experiment in which the subjects were involved that (1) no consent, or inadequate consent, was obtained, or (2) their selection as subjects constituted an injustice, or both, the government should offer a personal, individualized apology to each subject.*

The Committee believes that people who were used as research subjects without their consent were wronged even if they were not harmed. Although it is surely worse, from an ethical standpoint, to have been both harmed and wronged than to have been used as an unwitting subject of experiments and suffered no harm, it is still a moral wrong to use people as a mere means without their consent. Although what we know about the practices of the time suggests it is likely that many people who were subjects in nontherapeutic research were used without their consent or with what today we would consider inadequate consent, in most of these cases we have almost no information about whether or how consent was obtained. Moreover, in most of these cases, the identities of the subjects are not currently known; even if considerable resources were expended, it is likely that most of their identities would remain unknown.

The Committee is not persuaded that, even where the facts are clear and the identities of subjects known, financial compensation is necessarily a fitting remedy when people have been used as subjects without their knowledge or consent but suffered no material harm as a consequence; the remedy that emerged as most fitting was an apology from the government.

The Committee struggled with the issue of whether to recommend that the government extend such an apology. Our deliberations were complicated by what we all agreed was a murky historical record. In the case of some experiments, there was evidence of some disclosure or some attempt to obtain consent, and the issue emerged as to how poor these attempts must be for an apology still to be in order. In the great majority of cases, there was simply too little documentary evidence to draw any conclusions about disclosure or consent. In most cases, as noted above, the identities of subjects are unknown and are unlikely to be uncovered even with a substantial expenditure of resources.

What kind of evidence is necessary to determine that an apology is warranted? In the preceding recommendation, the remedy is linked to evidence of harm to particular individuals. While requiring evidence of harm specific to individuals, we did not require such specific evidence of lack of consent. Rather, in that recommendation, we presumed that the government did not uniformly undertake steps to ensure that the rights and interests of individual subjects were adequately protected, and thus that it is possible that people who were harmed as a result of participation in research did not adequately consent to this use of their person. In this recommendation, by contrast, a remedy is linked to a showing that people were *wronged*, not harmed. Thus the Committee believes that an apology should be offered only where there is evidence specific to an experiment or subject that no consent, or inadequate consent, was obtained, or the subject's selection constituted an injustice, or both.

The Committee believes that, among those experiments we have had the opportunity to review in depth, there is sufficient evidence that wrongs were committed against the children who participated in the experiments at the Fernald School. This case is discussed in detail in chapter 7.*

In recommending an apology to individuals who were subjects of these experiments, the Committee wishes to emphasize that there are likely many other instances in which an apology is warranted but for which experiment-specific factual support is not currently available.[3]

---

*For a discussion of the Committee's deliberations about this recommendation, see "Overview to Part IV."

*Several other experiments studied by the Committee are candidates for remedies under Recommendation 2. Where it is determined that subjects in these experiments were not harmed, they may be due an apology under this recommendation if it is determined that they were wronged.

RECOMMENDATION 4

In the research that we reviewed for this recommendation, the Advisory Committee has found no subjects of biomedical experiments for whom there is a need to provide notification and medical follow-up for the purpose of protecting their health. In the event that other experiments of concern come to light in the future, we recommend to the Human Radiation Interagency Working Group that subsequent decisions for notification be based on evaluation of both the level of risk from radiation exposure and the potential medical benefit from medical follow-up in exposed individuals.

Additionally, the Advisory Committee has found no evidence to indicate that the subjects of human radiation experiments we reviewed would have had greater likelihood of incurring heritable (genetic) effects than the general population and thus does not recommend notification or medical follow-up for descendants of subjects of human radiation experiments.

In formulating this recommendation, the Advisory Committee considered those subjects for whom there is a significant risk of developing a radiation-related disease that has not yet occurred, or has occurred but may still be undetected or untreated, and in whom there might be an opportunity to prevent or minimize potential health risks through detection and treatment. In considering notification, we focused only on biomedical experiments, as stated in our charter.

The Advisory Committee based its present recommendation on the specific guidelines stated below and recommends that future decisions for medical notification and follow-up of subjects of government-sponsored human radiation experiments not examined by the Committee, or that have not yet come to light, be based on these same guidelines, as follows:

1. The subject was placed at increased lifetime risk for development of a fatal radiation-induced malignancy. The level of increased risk was set by the Advisory Committee at 1/1,000 remaining lifetime risk and an excess relative risk of greater than 10 percent (organ specific). This level of risk was arbitrarily chosen by the Advisory Committee. When compared with the normal risk of dying of cancer (220 out of 1,000), this level of risk is small. The Advisory Committee chose this small remaining lifetime risk as a reasonable initial criterion to decide if an analysis of the utility of screening and intervention (criterion 2 below) was needed.

2. There is a recognized medical benefit from early detection and treatment of the cancer, which outweighs whatever medical risks are associated with detection and treatment interventions. In addition, the government should consider the public health and financial costs as well as the potential benefits before making a decision to offer such a notification and screening program.

Eligible subjects for whom medical follow-up to protect health is recommended should be notified of their participation in a human radiation experiment, and voluntary screening programs offered to them. Such a program should include adequate disclosure of both the nature of the potential benefits as well as the potential risks of medical follow-up, which might include some of the following aspects:

- medical harm, discomfort, inconvenience, or anxiety from the screening test itself or subsequent follow-up exams;
- the possibility of incorrect test results, either false positive or false negative;
- the possibility of stigmatization by friends, family, employers, or life/health insurance carriers;
- the costs to themselves of the screening program (if any) and subsequent medical tests and treatments.

Thus the Advisory Committee's recommendations for notification and medical follow-up of individuals who were subjects of a human radiation experiment depend equally on risk estimates and the medical utility of early detection and treatment for changing the course of disease or the quality or length of life in such an exposed individual, as shown in the accompanying table.

The Advisory Committee database includes articles and other documents describing approxi-

DETERMINATION OF THE NEED FOR NOTIFICATION AND MEDICAL FOLLOW-UP

		Risk Analysis (For Development of Fatal Cancer)	
		Remaining Lifetime Risk ≥ 1/1,000 AND RR ≥ 10%	Remaining Lifetime Risk < 1/1,000 OR RR < 10%
Medical Benefit from Early Detection and Treatment	Yes	RECOMMEND NOTIFICATION AND MEDICAL FOLLOW-UP	NO NOTIFICATION
	No	NO NOTIFICATION	NO NOTIFICATION

mately 4,000 government-sponsored human radiation experiments. Because of the limited data available on most of these, and the Advisory Committee's limited resources, it has not been feasible for the Advisory Committee to systematically apply the two criteria described above to the majority of experiments identified within its database. The Advisory Committee therefore selected for review types of experiments that seemed most likely to include subjects who might still be alive and meet the risk criteria chosen by the Committee and who might medically benefit from notification and medical follow-up.

Specifically, the Advisory Committee has reviewed twenty one studies involving three types of experiments:

1. Children who received iodine 131;
2. Prisoners subjected to testicular irradiation; and
3. Children and military personnel exposed to nasopharyngeal radium treatments.

Following this detailed analysis, the Advisory Committee concluded that none of the experiments examined satisfied both of the guidelines identified in this recommendation. If in the future new methods of screening are developed or new information about increased risk is discovered, then these experiments should be reevaluated to assess whether they meet the criteria. (For a full discussion, see the addendum on medical notification and follow-up at the end of this chapter.)

Though it was beyond the scope of the Advisory Committee to evaluate individually all the experiments in our database, the results of our review of these carefully selected studies suggest that the remaining experiments would be unlikely to meet the proposed criteria for notification and medical follow-up. However, another important group of studies not considered in detail by the Advisory Committee were tracer studies in pregnant and nursing women.

It is possible that experiments that would satisfy the Committee's criteria for notification and medical follow-up will be identified. Implementation of a notification and medical follow-up program would have to be done carefully if a follow-up program is to provide former research subjects with greater health benefit than harm. Considerable effort would be needed to educate both subjects and physicians about the realistic benefits and the possible harms of medical follow-up, as well as the specific screening modalities and follow-up care that would be indicated. It is particularly important to distinguish follow-up that is intended to benefit medical science from follow-up that is intended to medically benefit patients. An additional concern is that, for most experiments, no list of subjects exists. Performing screening tests in people who are incorrectly identified as having an increased risk is unlikely to result in any benefit and may result in harm.

The Advisory Committee also recognizes that individuals who have received therapeutic radiation treatments, either in a purely clinical setting or research setting, may have been exposed to substantially higher doses of radiation and should seek medical follow-up pursuant to the advice of their treating physician.

With regard to the need to notify descendants of subjects of human radiation experiments of potential genetic effects, it is likely that the risk of radiation-induced mutations is small in relation to natural rates. Thus it would be impossible to distinguish whether the condition was caused by the parent's radiation exposure or by other factors. Based on these considerations, the Advisory Committee does not recommend notification and medical follow-up for descendants of subjects of radiation experiments.

In the event that specific genetic effects attributable to radiation exposure could be identified in a particular population of descendants at some future time, the guidelines would be the same as those previously outlined for subject populations—there would need to be evidence to indicate that early intervention would change the course of a particular disease before notification and follow-up would be recommended.

POPULATION EXPOSURES

In recent years Congress has enacted a body of laws to provide relief to service personnel exposed to radiation in connection with atmospheric nuclear tests, citizens who lived downwind from the tests, and workers who mined uranium to be used by the government in nuclear weapons production. These include the Veterans Dioxin and Radiation Exposure Compensation Standards Act of 1984, the Radiation-Exposed Veterans Compensation Act of 1988, and the Radiation Exposure Compensation Act of 1990.

In the Committee's view, these existing laws provide the framework on which to base continued provision for relief. In the interim since these laws were passed, experience with the laws and more current scientific knowledge strongly suggest the need for revisiting the laws and their administration and for extending their coverage to similarly situated groups—such as those exposed to intentional releases—who are not now covered.

The following recommendations address the circumstances of groups exposed to intentional releases, service personnel who were exposed in connection with nuclear weapons tests, and workers who mined uranium for use in government programs. We also address the circumstance of the citizens of the Republic of the Marshall Islands, for whom a different framework of remedies has been fashioned.

RECOMMENDATION 5

The Advisory Committee recommends to the Human Radiation Interagency Working Group that it, together with Congress, give serious consideration to amending the provisions of the Radiation Exposure Compensation Act of 1990 to encompass other populations environmentally exposed to radiation from government operations in support of the nuclear weapons program, should information become available that shows that areas not covered by the legislation were sufficiently exposed that a cancer burden comparable to that found in populations currently covered by the law may have resulted.

The Advisory Committee did not have the time or resources to undertake our own epidemiologic studies of the cancer burden surrounding the Hanford facility in Washington state, where the Green Run took place. The preliminary radioiodine dose estimates now available raise the issue of whether the releases from Hanford may have caused cancers. The Advisory Committee found that the Green Run itself contributed only a very small portion of that cancer burden, so small that it would be impossible to attribute any cancers to the Green Run as opposed to other sources (including routine Hanford releases). The Advisory Committee believes that in addressing the Green Run intentional release, the appropriate response is to redress injury without regard to whether exposures were in the course of routine or research activities. There would be no practical way to make this distinction, if it were desired. We also note that the Radiation Exposure Compensation Act provides relief for downwinders and uranium miners without regard for whether they were subjects of research (and in many cases they were not).

RECOMMENDATION 6

The Advisory Committee recommends to the Human Radiation Interagency Working Group that it, together with Congress, give serious consideration to reviewing and updating epidemiological tables that are relied upon to determine whether relief is appropriate for veterans who participated in atomic testing so that all cancers or other diseases for which there is a reasonable probability of causation by radiation exposure during active military service are clearly and unequivocally covered by the statutes.

Congress has provided for compensation for veterans who participated in atmospheric atomic tests or the American occupation of Hiroshima or Nagasaki, Japan. The provision of compensation depends on evidence that the veteran has sustained disability from a disease that may be related to radiation exposure.

The Veterans Dioxin and Radiation Exposure Compensation Standards Act of 1984 required the Veterans Administration to write a rule governing entitlement to compensation for radiation-related disabilities. The resulting regulation contains criteria for adjudicating radiation claims, including consideration of a radiation-dose estimate and a determination as to whether it is at least as likely as not that the claimed disease resulted from radiation exposure. The Radiation-Exposed Veterans Compensation Act of 1988 provides that a veteran who was present at a designated event and subsequently develops a designated radiogenic disease may be entitled to benefits without having to prove causation.[4]

The Committee recommends that the radio-epidemiological tables prepared by the National Institutes of Health in 1985, which identify diseases that may be causally connected to radiation exposures, be updated. The Committee understands that the Department of Veterans Affairs agrees with this recommendation.

The Advisory Committee further recommends to the Human Radiation Interagency Working Group that it review whether existing laws governing the compensation of atomic veterans are now administered in ways that best balance allocation of resources between financial compensation to eligible atomic veterans and administrative costs, including the costs and scientific credibility of dose reconstruction.

While the Committee's inquiry focused on participants at atmospheric testing who were subjects of experimentation, the Committee found that the risks to which experimental subjects were exposed were typically similar to those to which many other test participants were subjected. Those service members who were participants in the experiments reviewed by the Advisory Committee would, as veterans of atmospheric atomic tests, be eligible for relief under the laws enacted in 1984 and 1988, as amended, concerning radiation-exposed veterans.

The Committee found that the government did not create or maintain adequate records regarding the exposures of all participants, the identity and test locale of all participants, and the follow-up, to the extent it took place, of test participants. Witnesses before the Advisory Committee, and others who communicated with us by mail, telephone, and personal visit, expressed strong concerns about the adequacy and operation of the current laws, including, specifically, record-keeping practices. Although the Committee did not have the time or resources to pursue these concerns to the degree they merit, we believe that the concerns expressed by veterans and their family members deserve attention, and we urge the Human Radiation Interagency Working Group in conjunction with Congress to address these concerns promptly. The concerns reported to us include the following:

1. The listing of diseases for which relief is automatically provided—the "presumptive" diseases provided for in the 1988 law—is incomplete and inadequate.
2. The standard of proof for those without a presumptive disease is impossible to meet and, given the questionable condition of the exposure records retained by the government, inappropriate.
3. The statutes are limited and inequitable in their coverage; for example, the inclusion of those exposed at atmospheric tests does not protect those who were exposed to equal amounts of radiation in activities such as cleanup at Enewetak atoll.
4. The time and expense needed to prosecute a claim is too great. For example, veterans whose claims are initially denied at the VA regional offices and are seeking appeal of the initial decision receive a form letter stating that it will take at least twenty-four months to process their appeal.

5. Time and money spent on contractors and consultants in administering the program would be better spent on directly aiding veterans and their survivors.

RECOMMENDATION 7

The Advisory Committee recommends to the Human Radiation Interagency Working Group that it, together with Congress, give serious consideration to amending the provisions of the Radiation Exposure Compensation Act of 1990 relating to uranium miners in order to provide compensation to all miners who develop lung cancer after some minimal duration of employment underground (such as one year), without requiring a specific level of exposure. The act should also be reviewed to determine whether the documentation standards for compensation should be liberalized.

The uranium miners were exposed to extremely high levels of radon daughters, which were recognized at the time to be hazardous yet were not controlled by the government, despite the availability of feasible means to ventilate the mines. Furthermore, the government studied the miners without disclosing the purposes of the examinations or warning them of the hazards to which they were exposed. As a result of their continued exposure, hundreds of miners developed lung cancer or nonmalignant respiratory diseases that could have been prevented, and many of them have died.

In recognition of this tragedy, Congress included provisions for compensating certain uranium miners in the Radiation Exposure Compensation Act of 1990 (RECA). However, the criteria for compensation set in this act were far more stringent than for the two other groups (atomic veterans and downwinders of the Nevada Test Site) for which compensation was provided, despite the fact that the risks were far higher for the uranium miners.

Since 1990, additional scientific information has become available to support the view that radon exposure is responsible for a much higher proportion of the lung cancer cases among the miners than had been previously thought. In particular, the act's current requirement of a minimum of 200 WLM (working level months)

exposure for nonsmokers or 300 to 500 WLM (depending on age) for smokers translates to quite large probabilities of causation, according to a recent report by the National Cancer Institute.[5] That analysis finds little evidence to support a distinction between smokers and nonsmokers and suggests that a majority of lung cancer deaths among Colorado white miners and New Mexico Navajo miners are attributable to radon exposure. Furthermore, it finds that the lung cancer risk is strongly modified by a number of factors and uncertainties that are not accounted for in the total dose; thus, for many miners, the level of exposure that would merit compensation on the basis of the principle of "balance of probabilities" might be far lower than the present criteria. In particular, no exposure measurements are available for 90 percent of the years in most mines, so that any requirement to reconstruct exposure histories is likely to require some degree of extrapolation or estimation and be quite uncertain. Furthermore, many mines have since gone out of business, so that records needed to establish an exposure history are simply unavailable.

Also since 1990, there has been considerable experience with the administration of the act, and apparently much of it has been negative. The Advisory Committee took extensive testimony regarding the difficulties faced by miners in meeting the documentation requirements, particularly those related to the requirement to provide a reconstruction of their radon dose. For these practical reasons, and in light of the additional information, we suggest that the requirement that a miner demonstrate that he had been exposed to a certain minimum cumulative dose be replaced by a simple requirement that he worked underground for a certain minimum length of time. Since more than half the lung cancer deaths in the cohort who worked at least one month underground appear to be attributable to radon, we suggest that minimum length of service be set quite low, preferably not more than a year. At most this should then lead to compensation being awarded to twice as many miners as would be entitled to it under the balance of probabilities principle, while not denying it to any who are entitled to it.

The grave injustice that the government did to the uranium miners, by failing to take action to control the hazard and by failing to warn the miners of the hazard, should not be compounded by unreasonable barriers to receiving the compensation the miners deserve for the wrongs and harms inflicted upon them as they served their country.

RECOMMENDATION 8

The Advisory Committee supports the Department of Energy's program of medical monitoring and treatment for the exposed inhabitants of the Marshall Islands atolls of Rongelap and Utirik and recommends that this program be continued as long as any member of the exposed population remains alive. Furthermore, the Advisory Committee recommends that the program be reviewed to determine if it is appropriate to add to the program the populations of other atolls to the south and east of the blast whose inhabitants may have received exposures sufficient to cause excess thyroid abnormalities. The Advisory Committee also recommends that consideration be given to the involvement of the Marshall Islanders in the design of any further medical research to be conducted upon them and the Advisory Committee recommends that the Human Radiation Interagency Working Group consider establishing an independent panel to review the status and adequacy of the current program of medical monitoring and medical care provided by the United States to the exposed population of the Marshall Islands.

The 1954 Bravo hydrogen bomb test caused the populations of several Marshall Islands atolls to be exposed to hazardous levels of radiation. The United States has provided a medical follow-up program that combines research on radiation effects with treatment for radiation-related illnesses. It is noteworthy that as a result of the ongoing program to study radiation effects, many cases of thyroid disease were detected and treated, but not all exposed Marshallese received the benefits of the program. The people of Ailuk, for example, who according to early reports received about the same exposure as the people of Utirik, were never evacuated from their atoll and were not followed up medically, even though they received a radiation dose of more than six roentgens. Moreover, an epidemiological study reported in the *Journal of the American Medical Association* in 1987 demonstrated that inhabitants of several atolls to the east and south of Bikini had elevated levels of thyroid disease and that there was a "strong inverse linear relationship" between incidence of thyroid nodules and distance from the blast. It should also be noted that the exposed populations received additional doses of radiation over the years from later bomb tests and residual radiation on the atolls. The medical program is ongoing, but Congress has the authority to reduce or eliminate funding.

Available evidence indicates that many Marshallese—it is impossible to identify specific individuals—were not adequately informed about the purposes of the medical tests to which they were subjected. There is also evidence in the documentary record that the Marshallese often did not understand the relationship between the research and medical care components of the medical follow-up program. For example, Dr. Robert A. Conard headed the program, and according to his report on twenty years of medical treatment and monitoring, "the people did not always understand the need for the examinations, or their results." Although this situation has improved in recent years, it would nevertheless be appropriate to consult with the Marshallese in the design and implementation of further medical research so as to minimize any possibility of misunderstanding and to ensure that the priorities of the Marshallese are a consideration in the planning of such research.

The Advisory Committee supports the continuation of the Department of Energy's program of medical monitoring and medical care for the exposed inhabitants of the Marshall Islands. Questions have been raised during the course of our deliberations as to whether this program is running as well as it should, both with respect to the research and monitoring activities conducted by Brookhaven National Laboratory (BNL) and with respect to the medical care provided. In particular, the issue has emerged whether the medical care ought to be expanded to include treatment for conditions that are not radiogenic as a further remedy to Marshallese people who were exposed, however inadvertently, as a result of weapons tests. The Advisory Committee did not have the resources to pursue these issues, but we believe that they deserve serious consideration. One mechanism through which this could be accomplished is the establishment of an independent panel to review the program with input from the Marshallese as to the panel's composition.

Recommendations for the Protection of the Rights and Interests of Human Subjects in the Future

While we were constituted to consider issues related to human radiation experiments, in critical (but not all) respects, the government regulations that apply to human radiation research do not differ from those that govern other kinds of research. In comparison with the practices and policies of the 1940s and 1950s, there have been significant advances in the protection of the rights and interests of human subjects. These advances, initiated primarily in the 1970s and 1980s, culminated in the adoption of the Common Rule throughout the federal government in 1991. Although the Common Rule now affords all human subjects of research funded or conducted by the federal government the same basic regulatory protections, the work of the Advisory Committee suggests that there are serious deficiencies in some parts of the current system. These deficiencies are of a magnitude warranting immediate attention.

The Committee was not able to address the extent to which these deficiencies are a function of inadequacies in the Common Rule, inadequacies in the implementation and oversight of the Common Rule, or inadequacies in the awareness of and commitment to the ethics of human subject research on the part of physician-investigators and other scientists. We urge that in formulating responses to the recommendations that follow, the Human Radiation Interagency Working Group consider each of these factors and subject them to careful review.

RECOMMENDATION 9

The Advisory Committee recommends to the Human Radiation Interagency Working Group that efforts be undertaken on a national scale to ensure the centrality of ethics in the conduct of scientists whose research involves human subjects.

A national understanding of the ethical principles underlying research and agreement about their importance is essential to the research enterprise and the advancement of the health of the nation. The historical record makes clear that the rights and interests of research subjects cannot be protected if researchers fail to appreciate sufficiently the moral aspects of human subject research and the value of institutional oversight.

It is not clear to the Advisory Committee that scientists whose research involves human subjects are any more familiar with the *Belmont Report*[6] today than their colleagues were with the Nuremberg Code forty years ago. The historical record and the results of our contemporary projects indicate that the distinction between the ethics of research and the ethics of clinical medicine was, and is, unclear. It is possible that many of the problems of the past and some of the issues identified in the present stem from this failure to distinguish between the two.

The necessary changes are unlikely to occur solely through the strengthening of federal rules and regulations or the development of harsher penalties. The experience of the Advisory Committee illustrates that rules and regulations are no guarantee of ethical conduct. The Advisory Committee has also learned, in responses to our query of institutional review board (IRB) chairs, that many of them perceive researchers and administrators as having an insufficient appreciation for the ethical dimensions of research involving human subjects and the importance of the work of IRBs. The federal government must work in concert with the biomedical research community to exert leadership that alters the way in which research with human subjects is conceived and conducted so that no one in the scientific community should be able to say "I didn't know" or "nobody told me" about the substance or importance of research ethics.

The Advisory Committee recommends that the Human Radiation Interagency Working Group institute, in conjunction with the biomedical community, a commitment to the centrality of ethics in the conduct of research involving human subjects. We urge that careful consideration be given to the development of effective strategies for achieving this change in the culture of human subjects research, including, specifically, how best to balance policies that mandate the teaching of research ethics with policies that encourage and support private sector initiatives. It may be useful to commission a study or convene an advisory panel charged with developing and perhaps implementing recommendations on how best to approach this challenge for the research community.[7]

The Committee suggests that such an examination include consideration of the following:

- Extending to all federal grant recipient institutions and all students and trainees involved or likely to be involved in human subject research the current federal requirement that institutions receiving NIH National Research Service Award training grants offer programs in the responsible conduct of research.
- The role of accrediting bodies such as the Joint Commission on Accreditation of Healthcare Organizations (JCAHO).
- Establishing competency in research ethics as a condition of receipt of federal research grants, both for institutions and individual investigators.
- Incorporating of research ethics, and the *differences* between the ethics of research involving human subjects and the ethics of clinical medical care, into curricula for medical students, house staff, and fellows.

- Encouraging the nation's leaders in biomedical research to spearhead efforts to elevate the importance of research ethics in science.

RECOMMENDATION 10

The Advisory Committee recommends to the Human Radiation Interagency Working Group that the IRB component of the federal system for the protection of human subjects be changed in at least the five critical areas described below.

1. Mechanisms for ensuring that IRBs appropriately allocate their time so they can adequately review studies that pose more than minimal risk to human subjects. This may include the creation of alternative mechanisms for review and approval of minimal-risk studies.

The majority of the Advisory Committee's concerns in its Research Proposal Review Project centered on research that exposed subjects to greater than minimal risk of harm. If human subjects are to be adequately protected, such research must be carefully scrutinized. However, higher risk research is often complex, and careful review is time-consuming and difficult. The Advisory Committee heard from several chairs of IRBs who underscored the difficulties their committees experience in finding the time to adequately review such research. Members of IRBs have only so many hours they can devote to review of proposals. This problem of inadequate time appears to have worsened in recent years. Institutional review boards are required to review research proposals prior to their review for funding by the National Institutes of Health. As the probability that a proposal will be approved for funding has decreased over time, due to increasing competition for limited research monies, the number of proposals being submitted to NIH from many institutions has significantly increased. This has resulted in a substantial increase in the workload of some IRBs, whose members are spending considerable time reviewing proposals that are never implemented. Without guidance from the federal government, and perhaps regulatory relief, IRBs may not have the flexibility necessary to concentrate their efforts where subjects are in greatest need of protection—on the proposals that pose the greatest risks to subjects and that are actually implemented.

2. Mechanisms for ensuring that the information provided to potential subjects (1) clearly distinguishes research from treatment, (2) realistically portrays the likelihood that subjects may benefit medically from their participation and the nature of the potential benefit, and (3) clearly explains the potential for discomfort and pain that may accompany participation in the research.

The Advisory Committee's empirical studies and public testimony suggest that there may be considerable confusion in the minds of many members of the public concerning what is "research" or "experimentation," and what is simply an application of a new technology or even standard medical care. There is reason to worry that participants in research may have unrealistic expectations both about the possibility that they will personally benefit from participation and about the discomfort, pain, and suffering that sometimes accompany some research. This seemed particularly to be the case in Phase I and Phase II drug trials. It is important that in the informed consent process it is clearly communicated to the potential subject, particularly the potential patient-subject, that the primary intent of "research" is to advance medical knowledge and not to advance the welfare of particular subjects. Inadequate and potentially misleading information about potential benefits and harms, and about the trade-offs between enrollment in research and standard or conventional treatment, was one of the major problems identified by the Advisory Committee in our Research Proposal Review Project.

3. Mechanisms for ensuring that the information provided to potential subjects clearly identifies the federal agency or agencies sponsoring or supporting the research project in whole or in part and all purposes for which the research is being conducted or supported.

A morally complicating factor in several of the human radiation experiments the Advisory Committee has studied is the tendency to disclose to subjects only the medical purpose of the research (if that) and not those purposes of the research that advance interests other than medical science or the sponsorship of agencies other than DHEW/DHHS. For example, in the case of the total-body irradiation experiments, the data gathered from the research had a military purpose quite distinct from questions of cancer therapy. The purpose and funding source may be relevant to a person's decision to participate in human subject research and should be disclosed.

4. Mechanisms for ensuring that the information provided to potential subjects clearly identifies the financial implications of deciding to consent to or refuse participation in research.

Many of the consent forms that the Committee reviewed as part of the Research Proposal Review Project were silent on the subject of financial costs. However, knowing whether being in research costs or saves them money may be necessary for potential subjects to make an informed decision about whether to participate. Potential subjects need to know whether the interventions that are part of the research are free or must be paid for and—if there are any financial costs—what they are, the likelihood that third-party payers will pay for these research-related medical services, and the extent to which the research institution will assist patient-subjects in securing third-party payment or reimbursement.

5. Recognition that if IRBs are to adequately protect the interests of human subjects, they must have the responsibility to determine that the science is of a quality to warrant the imposition of risk or inconvenience on human subjects and, in the case of research that purports to offer a prospect of medical benefit to subjects, to determine that participating in the research affords patient-subjects at least as good an opportunity of securing this medical benefit as would be available to them without participating in research.

In research involving human subjects, good ethics begins with good science. In our Research Proposal Review Project, the Advisory Committee was unable to evaluate the scientific merit of a significant number of proposals based on the documents provided by institutions. We suspect that this occurred in part because there is ambiguity about the role that IRBs should play with respect to evaluation of scientific merit and, thus, that documents submitted to IRBs may be inadequate in this area. The Advisory Committee also heard dissatisfaction with this ambiguity in our interviews and oral histories of researchers and from chairs of IRBs. If the science is poor, it is unethical to impose even minimal risk or inconvenience on human subjects. Although the fine points of the relative merit of research proposals are best left to study sections and other review mechanisms specially constituted to make such judgments, IRBs must be situated to assure themselves that the science they approve to go forward with human subjects satisfies some minimal threshold of scientific merit. In some cases, the IRB may be the only opportunity for this kind of scientific review.

In our Subject Interview Study interviews with patient-subjects, we confirmed that patient-subjects often base their decisions to participate in research on the belief that physicians, and research institutions generally, would not ask them to enter research projects if becoming a research subject was not in their medical best interests. For these patients, even the most candid, clearly written consent form affords little protection, for both the consent form and the consent process are of little interest to them. For patient-subjects whose decisions to participate in research are based on trust, and not on an assessment of disclosed information, the IRB review is of special importance. It is the only source of protection in the *federal* system for regulating human research positioned to ensure that their participation in research does not compromise their medical interests. Such a determination, however, often requires more specialized clinical expertise than any one IRB can possess. Federal policy must make it clear that IRBs have the responsibility to make this determination, but it must also allow mechanisms to be devised at the local level that permit this responsibility to be satisfied in an efficient and effective manner.

RECOMMENDATION 11

The Advisory Committee recommends to the Human Radiation Interagency Working Group that a mechanism be established to provide for the continuing interpretation and application of ethics rules and principles for the conduct of human subject research in an open and public forum. This mechanism is not provided for in the Common Rule.

Issues in research ethics are no more static than issues in science. Advances in biomedical research bring new twists to old questions in ethics and sometimes raise new questions altogether. No structure is currently in place for interpreting and elaborating the rules of research ethics, a process that is essential if research involving human subjects is to have an ethical framework responsive to changing times. Also, for this framework to be effective, any changes or refinements to it must be debated and adopted in public; otherwise, the framework will fail to have the respect and support of the scientific community and the American people, so necessary to its success.

Three examples of outstanding policy issues in need of public resolution that the Advisory Committee confronted in our work are presented below:

1. Clarification of the meaning of minimal risk in research with healthy children, including, but not limited to, exposure to radiation.
2. Regulations to cover the conduct of research with institutionalized children.
3. Guidelines for research with adults of questionable competence. Of particular concern is more-than-minimal-risk research that offers adults of questionable competence no prospect of offsetting medical benefit.

Current regulations permit the involvement of children as subjects in research that offers no prospect of medical benefit to participants when the research poses no more than minimal risk. An important question that has come to the Advisory Committee's attention, both in the literature and in our Research Proposal Review Project, is whether research proposing to expose healthy children to tracer doses of radiation constitutes minimal risk. The uncertainty surrounding this issue calls into question the adequacy of the federal regulations, as currently formulated, in providing guidance for this category of research. This is a policy question that ought to be discussed and resolved in a public forum at the national level, not left to the deliberations of individual IRBs.

Current regulations do not provide any special protections for children who are institutionalized unless they are also wards of the state. Thus, researchers and IRBs have no more guidance from the federal government on the ethics of conducting such research than was available at the time of the Fernald and Wrentham experiments, decades ago.

The Advisory Committee also confronted in its Research Proposal Review Project another issue of research policy deserving public debate and resolution in a public forum. This is the issue of whether and under what conditions adults of questionable capacity can be used as subjects in research that puts them at more than minimal risk of harm and from which they cannot realize direct medical benefit. It is important that the nation decide together whether or under what conditions it is ever permissible to use a person toward a valued social end in an activity that puts him or her at risk but from which the person cannot possibly benefit medically.

RECOMMENDATION 12

The Advisory Committee recommends to the Human Radiation Interagency Working Group that at least the following four steps be taken to improve existing protections of the rights and interests of military personnel with respect to human subject research.

1. Review of policies and procedures: Policies and procedures governing research involving human subjects should be reviewed to ensure that they (1) clearly state that participation as research subjects by members of the armed services is voluntary and without repercussions for those who choose not to participate; and (2) clearly distinguish those activities that are research and therefore discretionary on the part of members of the armed services from other activities that are obligatory, such as training maneuvers and medical interventions intended to protect the troops.

2. Appreciation of regulations: Education in applicable human subjects regulations should be a component of the training of all officers and investigators who may be involved in decisions regarding research on human subjects. Mechanisms are needed to ensure that officers expected to have command responsibilities and all officers engaged in research, development, testing, and evaluation have an adequate appreciation of the regulations (including DOD regulations and directives, and service regulations) that bear on the conduct of research involving human subjects, including an appreciation of the conditions under which such regulations apply, the role of officers in interpreting such regulations, and how such regulations are to be implemented.

3. Maximizing voluntariness: The service secretaries should consider the situations under which it would be appropriate to make obligatory two practices for maximizing voluntariness that have been employed on an ad hoc basis in some military research: first, that unit officers and senior noncommissioned officers (NCOs) who are not essential as volunteers in the research be excluded from recruitment sessions in which members of units are informed of the opportunity and asked to participate in research by investigators; and second, that an ombudsman not connected in any way with the proposed research be present at all such recruitment sessions to monitor that the voluntariness of participation is adequately stressed and that the information provided about the research is adequate and accurate.

The Advisory Committee recommends consideration of steps 1 through 3 above in light of our examination of history that makes plain how difficult it often is in a military context to distinguish an order from a request for voluntary participation and to distinguish research from training. (These tensions are similar in many respects to tensions in the clinical context between research and treatment.) Although the military has a long tradition of commitment to the use of volunteers in research and has introduced significant advances in the military's system of protection for human subjects since the 1940s and 1950s, without constant attention to these inherent tensions, the potential for confusion and inappropriate practice continues.

The military setting, with its strict hierarchical authority structure and pervasive presence in the lives of its members, poses special problems for ensuring the voluntariness of participation in research activities. Thus, although the DOD has adopted and implemented the consent requirements of the Common Rule, additional procedural safeguards and educational activities for officers may be warranted to counteract the generalized deference to authority inherent in military culture. Also, because the opportunity to serve the nation as subjects in defense-oriented research projects is closely akin to the demands placed on members of the military in their routine duties, it is desirable to emphasize the distinction between research and course-of-duty risks both in consent procedures and in officer training programs.

The Advisory Committee recognizes that additional procedural requirements in soliciting research volunteers and augmenting already demanding training curricula would have administrative costs and, to a limited extent, would shift organizational priorities. It is the Advisory Committee's understanding that the DOD is preparing to revise its directive implementing the Common Rule and that the Advisory Committee's recommendations with respect to steps 1 through 3 above are a timely contribution to the department's deliberations.

Military personnel are exposed to both short- and long-term risks in the course of training and regular duty activities as well as when they participate in biomedical or behavioral experiments. The demarcation of those activities that are research in contrast with those that constitute routine duty assignments and medical care in the military context is not always easy to discern from the standpoint of the potential subject-member of the military. Indeed, except in medical settings where research studies are regularly performed and military testing sites that conduct weapons, matériel, and performance trials routinely, officers as well as their troops may be uncertain as to whether the status of particular exercises is research or training. Greater clarity in communications to potential subjects about the genuinely voluntary nature of participation in research projects and procedural safeguards in recruiting volunteers could improve their understanding of what they are being asked (rather than required) to do. Likewise, educating officers throughout the military services who may be in a position to solicit volunteers for research studies as to the distinctive rights of research subjects and the particular duties to protect subjects of research from both harm and violations

of rights would make the Common Rule protections of subjects more effective.

4. Maintenance of a registry: The secretaries of the Navy and the Air Force should be directed to adopt the policy of the Army, as detailed in Army Regulation 70–25, to maintain a registry of all volunteers in human studies and experiments conducted under research and development programs. Such registries make it easier to confirm participation in research by subjects and facilitates their long-term follow-up.

In analyzing the record of atomic bomb testing, the Advisory Committee has found that military personnel were exposed to radiation and nonradiation risks as participants in experiments that were conducted in conjunction with the tests, and as participants in other activities connected to the testing. While these activities were not intended to measure biological effects of ionizing radiation, the exposure to radiation risk was incurred without adequate provision for the maintenance of records to document exposures or in order to allow for monitoring and follow-up of those who were exposed. Army regulations now provide for a registry of participants in experiments conducted under the authority of the Army's research and development program. This tool for long-term monitoring and follow-up in the case of exposures to risks unknown at the time of participation should be employed by the other services as well.

RECOMMENDATION 13

The Advisory Committee recommends that the Human Radiation Interagency Working Group take steps to improve three elements of the current federal system for the protection of the rights and interests of human subjects—oversight, sanctions, and scope.

1. Oversight mechanisms to examine outcomes and performance. In most federal agencies, current mechanisms of oversight of research involving human subjects are limited to audits for cause and a review of paperwork requirements. These strategies do not provide a sufficient basis for ensuring that the current system is working properly. The adequate protection of human subjects requires that the system be subjected to regular, periodic evaluations that are based on an examination of outcomes and performance and that include the perspective and experiences of subjects of research as well as the research community. The Committee recommends that the Human Radiation Interagency Working Group consider new methods of oversight that focus on outcomes and performance of the system of protection of human subjects. The Committee's Subject Interview Study and Research Proposal Review Project, for example, yielded important and heretofore unavailable information about the current status of human subjects protections that could never be obtained from either an oversight policy that audits only "for cause" or a review that determines only whether paperwork requirements have been satisfied.

We realize that resources available for oversight are limited and that there may be real constraints on what, practically, can be achieved. At the very least, we urge that in the setting of priorities for limited oversight dollars, a premium be placed on methods that permit an examination of what the system is actually producing with respect to the outcome of human subjects protections, in contrast to methods that focus on process.

2. Appropriateness of sanctions for violations of human subjects protections. The Committee recommends that the Human Radiation Interagency Working Group review and evaluate the options available to the government when it is determined that there has been a violation of the Common Rule in the conduct of federally sponsored research involving human subjects. The object of this review is to determine whether the current structure of sanctions that can be imposed on investigators and grantee institutions is appropriate to the seriousness with which the nation takes violations of the rights and interests of human subjects. This structure includes mechanisms for detecting violations (including issues of oversight discussed above), severity of sanctions, and dissemination of policies on sanctions to investigators and institutions. We are particularly concerned that, even in the absence of research-related injury, there be clear and severe penalties for investigators who use human subjects without their consent. Although at least one state authorizes civil and criminal penalties for failure to obtain a subject's

consent,[8] in most jurisdictions civil litigation is unlikely to result in penalties to investigators for failing to obtain consent from subjects if the subjects have not been physically injured. The Committee is aware that the Common Rule provides for sanctions of violations of its provisions, including the withdrawal of multiple project assurances and, with that action, research funding. It is not clear, however, that this system of sanctions functions well; nor is it clear that it adequately addresses the public's concerns that those who abuse the trust of research subjects be dealt with accordingly.

3. Extension of human subjects protections to nonfederally funded research. While some nonfederally funded research is performed voluntarily in accordance with the Common Rule, there is a need to assess the level of research performed outside its requirements and to consider action to ensure that all subjects are afforded the protections it offers. The Committee was charged with reviewing only federally funded research, and we limited our inquiries accordingly. However, we are aware that important areas of research are conducted largely independently of federal funding—for example, some research on reproductive technologies. We recommend that the Human Radiation Interagency Working Group take steps to ensure that all human subjects are adequately protected.

RECOMMENDATION 14

The Advisory Committee recommends that the Human Radiation Interagency Working Group review the area of compensation for research injuries of future subjects of federally funded research, particularly reimbursement for medical costs incurred as a result of injuries attributable to a subject's participation in such research, and create a mechanism for the satisfactory resolution of this long-standing social issue.

A system of compensation for research injuries has been contemplated since at least the late 1940s, when the Army debated, but ultimately rejected, suggestions to establish a "uniform" program for compensating prisoner volunteers who were injured during experiments involving malaria and hepatitis. Beginning in the 1970s, a number of government-sponsored ethics panels endorsed the provision of compensation for research injuries, culminating with the President's Commission for the Study of Ethical Problems in Medicine and Biomedical and Behavioral Research (President's Commission) in 1982. Since then, experts and commentators have continued to support this position.[9]

In our deliberations concerning retrospective remedies for injured research subjects, the Advisory Committee was unable to reference a federal policy or guide for a fair system of compensation of research subjects, as no policy exists even today. So that years from now others do not have to revisit and struggle with this issue, the federal government must take steps now to address the issue of compensation for injured research subjects. These steps should include consideration of the approach recommended by the President's Commission in its report, *Compensating for Research Injuries: The Ethical and Legal Implications of Programs to Redress Injured Subjects.*[10]

The President's Commission summarized the basic argument for compensation as follows:

Medical and scientific experimentation, even if carefully and cautiously conducted, carries certain inherent dangers. Experimentation has its victims, people who would not have suffered injury and disability were it not for society's desire for the fruits of research. Society does not have the privilege of asking whether this price should be paid; it is being paid. In the absence of a program of compensation of subjects, those who are injured bear both the physical burdens and the associated financial costs. The question of justice is why it should be these persons, rather than others, who are to be expected to absorb the financial, as well as the unavoidable human costs of the societal research enterprise which benefits everyone.[11]

The Advisory Committee urges not only consideration of a compensation policy for physical injuries attributable to research but also that consideration be given to appropriate remedies for subjects who have suffered dignitary harms, even in the absence of physical injury. Subjects so wronged have little recourse in the current system; litigation in the absence of physical injury is unlikely to provide relief to people who have been used as subjects without their adequate consent. If it is determined that financial compensation is not generally an appropri-ate remedy in the absence of physical injury, consideration should be given to other remedies that would be fitting.

Recommendations for Balancing National Security Interests and the Rights of the Public

RECOMMENDATION 15

15a: The Advisory Committee recommends to the Human Radiation Interagency Working Group the adoption of a federal policy requiring the informed consent of all human subjects of classified research and that this requirement not be subject to exemption or waiver. In all cases, potential subjects should be informed of the identity of the sponsoring federal agency and that the project involves classified information.

15b: The Advisory Committee recommends to the Human Radiation Interagency Working Group the adoption of a federal policy requiring that classified research involving human subjects be permitted only after the review and approval of an independent panel of appropriate nongovernmental experts and citizen representatives, all with the necessary security clearances. This panel should be charged with determining (1) that the proposed experiment has scientific merit; (2) that risks to subjects are acceptable and that the balance of risk and potential benefit is appropriate; (3) that the disclosure to prospective subjects is sufficiently informational and that the consent solicited from subjects is sufficiently voluntary; and (4) whether potential subjects must have security clearances in order to be sufficiently informed to make a valid consent decision, and if so, how this can be achieved without compromising the privacy and voluntariness of potential subjects. Complete documentation of the panel's deliberations and of the informed consent documents and process should be maintained permanently. These records should be made public as soon as the national security concern justifying secrecy no longer applies.

Although the Advisory Committee believes that the interests of both science and potential subjects are best served when research involving human subjects is conducted in the open, a public policy prohibiting the conduct of human subject research in secret is unwise. Important national security goals may suffer if human subjects research projects making unique and irreplaceable contributions were foreclosed. More citizens may suffer harms for lack of such information than would be harmed if adequately safeguarded human subjects research was conducted in secret.

It also is possible that a prohibition on classified human subjects research would be circumvented through redefinition of activities or disregarded outright. If this were to occur, the participants in such activities could end up less well protected than if they were bona fide research subjects.

The Advisory Committee believes, however, that the classification of human subject research ought properly to be a rare event and that the subjects of such research, as well as the interests of the public in openness in science and in government, deserve special protections. The Advisory Committee does not believe that continuing with the current federal policy governing the protection of human subjects, which does not provide any special safeguards or procedures for classified research, is adequate.

In the current political context, classified human subjects research occurs relatively rarely. Existing policy may prove an inadequate safeguard of individual rights and welfare, however, if in the future national security crises occur that generate a perceived need for classified research. The history of human experimentation conducted in the interests of strengthening and protecting national security that the Advisory Committee has examined demonstrates how the rights and interests of citizens can be violated in secret research. The convergence of elements of secrecy, urgent national purposes, and the essential vulnerability of research subjects, owing to differentials in information and power between those conducting research and those serving as subjects, could again lead to abuses of individual rights and, upon subsequent revelation, the erosion of public distrust in government.

The Advisory Committee is particularly concerned about two aspects of current policy—exceptions to informed consent requirements and the absence of any special review and approval process for human research that is to be classified. The current requirement for the informed consent of research participants is not absolute, leaving open the possibility that subjects may serve as mere tools of the state in the interests of national security if consent is waived. A strengthened requirement for the informed consent of research subjects in classified research should safeguard against the merely instrumental use of individual people to serve national purposes.

Institutional review boards of government agencies are not sufficiently independent of the interests of the organizations of which they are a part to set aside considerations of organizational mission when considering research construed as having the greatest national priority. Thus, determination by an agency IRB that a waiver of informed consent is warranted, or that sufficient information about a study remains in a censored protocol description for a potential subject's review, inadequately protects subjects' interests and rights and does not adequately safeguard the public's trust. By contrast, an independent panel should be less subject to unintended bias than that of an IRB of a federal agency whose mission is to protect and promote national security.

Although the Advisory Committee acknowledges that both the formation of an independent review panel and an absolute informed consent requirement create opportunities for information leaks or security breaches and delays in the progress of urgent research, these disadvantages are surmountable and are more than balanced by the increased vigilance afforded the rights and interests of citizens and the safeguarding of the public's trust in government.

RECOMMENDATION 16

The Advisory Committee recommends to the Human Radiation Interagency Working Group that improvements be made in the protections of the public's rights and interests with respect to intentional releases.

16a. The Advisory Committee recommends to the Human Radiation Interagency Working Group that an independent review panel review any planned or intended environmental releases of substances in cases where the release is proposed to take place in secret or in circumstances where any aspect of the environmental review process required by law is conducted in secret.

In conducting its review, the independent panel should ensure that (1) secrecy is limited to that required for reasons of national security; (2) records will be kept on the nature and purpose of the release, the rationale for not informing the public (including workers and service personnel, as well as affected citizens), and alternative means of gathering data that were considered; (3) actions to mitigate risk were considered and will be taken; and (4) actions will be taken to measure the actual effect of the release on the environment and human health and safety, to the extent that measurements are deemed needed and feasible. The panel should also review the conditions on which any information kept secret should be made public, with a view toward ensuring the release of information as soon as practicable, consistent with any legitimate national security restrictions. The panel should report to Congress periodically on the number and nature of releases it has reviewed.

The Advisory Committee does not conclude that intentional releases can never be conducted in secret. It does conclude that, to the extent that the government proposes to conduct an intentional release that involves elements of secrecy, there must be independent review to ensure that the action is needed, that risk is minimized, and that records will be kept to make sure a proper accounting is made to the public at the earliest date consistent with legitimate national security concerns.

The Advisory Committee found that the government has sponsored numerous intentional environmental releases of radiation for research purposes. In many cases these releases were conducted in secret, without warning to the surrounding populations. While the risks posed by these releases appear to have been relatively small, in many cases little data remain on the precise measure of these risks or on actions taken to minimize risk and to ensure that unknowing

citizens did not inadvertently expose themselves to greater risks than necessary. In addition, the Committee found that the risks and concerns posed by intentional releases for research purposes—in terms of both the magnitude of radiation exposure and the consequences of secret keeping—sometimes did not differ qualitatively from those posed by "routine" operational releases of radiation. Most notably, the radiation risk posed by the Green Run, a relatively large intentional release, was a fraction of that posed by radiation released in the normal course of operation of Hanford in the mid-1940s.

This recommendation is intended to apply to all secret releases of substances into the environment, not merely to substances determined to be hazardous. The Committee believes that the operative concern is secrecy; even if the substance released is entirely harmless, the backdrop of secrecy is sufficient to create a climate of distrust. The Committee did not have the expertise, however, to determine whether so broad a sweep was feasible. At minimum, the Committee recommends that any secret release of a substance that would necessitate an environmental impact statement be required to have a review by an independent panel.

Today, federal environmental laws and rules provide for environmental impact statements, which are subject to review, in instances in which the federal government proposes actions with a substantial effect on the environment. However, the rules also provide that part—or even all—of such reviews may be conducted in secret. In fact, reviews that are secret in whole or part do take place.

The Environmental Protection Agency has the authority and responsibility to oversee all environmental impact reviews, including those conducted in secret. However, the Advisory Committee's inquiries indicate that EPA's role in the review of secret impact statements has been limited. Moreover, the decades of secret keeping regarding intentional releases have created a basis for distrust, particularly among those living in potentially affected communities. Even today, there is little practical means by which the public can know the full extent (whether or not great) of environmental decision making and action that is being kept secret. The location of responsibility for review of these activities in a single panel that is itself accountable and that is independent of agencies that conduct releases should be a means to restoring lost trust.

16b. The Advisory Committee recommends to the Human Radiation Interagency Working Group that an appropriate government agency, currently the Environmental Protection Agency, maintain a program directed at the oversight of classified programs, with suitably cleared personnel. This program should maintain critical records, such as environmental impact statements and environmental permits, permanently. The agencies subject to regulation should ensure the timely consideration of environmental impacts and oversight and the timely provision of all necessary clearances. EPA should provide regular unclassified reports to Congress describing the extent of its activities as well as any significant problems.

The requirements of environmental law apply to activities of the federal government, regardless of whether those activities are classified. However, classification complicates the process of regulatory oversight by the EPA or any other regulatory agency and limits the ability to report to the public and for the public to express its own concerns. Furthermore, secrecy has been used to shield activities that raise public health concerns.

For these reasons, the responsibility for environmental oversight is magnified for secret programs. There is no fundamental barrier to effective oversight—at least some regulators can be given the necessary clearances. However, ensuring timely and effective oversight requires cooperation between the regulated agency and the regulatory agency to establish the necessary oversight procedures. These mechanisms are not fully in place. For example, the EPA office with the statutory responsibility to review environmental impact statements maintains no records of classified environmental impact statements and has not historically had individuals cleared to review the most highly classified defense programs. The EPA office responsible for overseeing federal compliance with environmental regulations has just begun to establish mechanisms for overseeing secret programs.

Recommendations on Openness

RECOMMENDATION 17

The Advisory Committee recommends that the Human Radiation Interagency Working Group take steps to ensure the continued application of the lessons learned from the Human Radiation Interagency Working Group's efforts to organize and make accessible to the public, and the government itself, the nation's historical records.

The Committee's experience confirms that with presidential directive and the strong and continued support of a multiagency records search team, substantial amounts of the nation's documentary heritage can be located and retrieved. Through the research process, important lessons were learned about ways in which to improve the accessibility and usefulness of this documentary record to both the public and the government.

We are aware that government resources are stretched thin and may well be diminishing. However, the nation's records are a precious asset that the government created, and holds in trust, for its citizens. This asset, and the commitment made to the public through the enactment of the Freedom of Information Act, is of limited value if the government itself cannot access its records as citizens rightfully expect it should. The Committee's experience confirms that there is an intense public interest in using these records, a public willingness to volunteer time and intelligence needed to help organize and research them, and great opportunity to make them available in ways that will permit citizens to do so.

The Committee recommends that the Human Radiation Interagency Working Group effect the following five steps to increase both government and citizen access to information about the past. The implementation of these steps might best be accomplished by the designation of an individual or entity with responsibility and appropriate authority for their effectuation.[12]

1. The most important historical collections

should be entrusted to the National Archives. The agencies and the National Archives should review the extent to which this is now being done and develop policies to hasten the transfer of agency records to the National Archives.

Federal law basically requires that permanent records be transferred to the National Archives when (1) they are more than thirty years old; or (2) earlier if the originating agency no longer needs to use the records for the purpose for which they were created or in its regular current business, or if agency needs will be satisfied by use of the records at the National Archives.

Nonetheless, many portions of older collections have been appraised as permanently valuable but are not at the National Archives. For example, the Committee found that a great number of AEC headquarters records of substantial interest to the Committee and the public are still held by DOE either at its headquarters or at the Washington National Records Center (these include the only collection of general manager files, the post-1958 Executive Secretariat files, virtually all the Division of Military Application files, and most of the files of the Division of Biology and Medicine). In the case of the Department of Defense, the records of the Office of the Secretary of Defense largely remain at the Washington National Records Center or with the Office of the Secretary of Defense.[13]

The public's ability to access records held by agencies is limited because (1) most agencies do not know in detail what records they still hold, and even if folder listings exist, they are not publicly available for the most part; (2) there has generally been little declassification review of these records; (3) there is no requirement that agencies permit access to even completely unclassified or declassified collections; and (4) most agencies have very limited facilities to accommodate researchers. The public's ability to gain access to documents in federal records centers is also limited because (1) the task of examining the basic inventory forms (SF-135s)[14] to determine what is in a record group is time-consuming, and in many cases, the SF-135s do not adequately describe the records; (2) there has generally been very little declassification review of these records; and (3) permission must be obtained from the appropriate agencies to review

even completely unclassified or declassified collections; this permission process can be time-consuming and agencies can impose restrictions, such as permitting review but not copying.

Locating records at the National Archives has the following advantages: (1) there is generally at least some type of finding aid and, in some cases, folder listings prepared by the National Archives or the agencies when the records were sent; (2) archivists are available to assist researchers; (3) there is complete access to unclassified and declassified collections (unless Privacy Act or similar restrictions apply); and (4) many classified records at the National Archives (among the exceptions are Restricted Data records and records dealing with intelligence) are properly the subject of an informal and usually very quick in-house declassification review process called Special Declassification Review. Under Special Declassification Review, records are often reviewed within months, versus the years it takes under the Freedom of Information Act or Mandatory Declassification Review.

2. Agencies should make readily available all existing inventories, indices, folder listings, and other finding aids to record collections now under agency control. Classified finding aids should undergo declassification review, and declassified versions of these finding aids should also be made available.

Finding aids or indices to federal government records holdings are an invaluable tool, without which it would be practically impossible to locate documents of interest from among the hundreds of thousands of boxes of records maintained by the government.

Many collections of records still held by agencies have finding aids or indices that have been inaccessible to the public, either because they simply have never been made available or because they are classified. Finding aids should be made available to the public in a headquarters office, regional offices (including all field site reading rooms), and ultimately, on the Internet. (This recommendation does not call for the creation of indices where they do not currently exist.)

For example, folder listings (which provide the titles of records files) exist for many of the AEC headquarters record collections that are still at DOE or at the Washington National Records Center. These include, among others, the only known collection of general manager's files from 1947 through 1974, all of the Division of Military Applications files from 1947 through 1974, all of the Executive Secretariat files from 1959 through 1974, and most of the Division of Biology and Medicine files from 1947 through 1974. Without the folder listings it would have been difficult for the Advisory Committee to locate particular collections of interest and, even if located, to determine the documents to be reviewed. The folder listings, however, have not been generally available to the public.

Similarly, the DOE's Oak Ridge Operations Office vault contains more than 7,000 cubic feet of classified records. The Committee found that the Records Holding Task Group (RHTG) collection in this vault (about 300 cubic feet) contained many documents of interest to the Committee, which were typically readily declassifiable. This collection has an index; however, the index is classified.

In the case of the National Archives, finding aids are generally available. However, there are fifteen National Archives facilities around the country. Currently, the only means of determining exactly what records are at a particular branch is to contact that branch directly. This is a time-consuming process, and there are understandable limits on the number of pages of finding aids archivists can copy and send to any person (a single finding aid can total hundreds of pages). It would be much simpler and easier for the public to be able to review the finding aids from all fifteen branches at any one of them.

3. The Human Radiation Interagency Working Group should ensure the development of policies to improve public access to records held by agencies or deposited in federal records centers.

In the case of a vast amount of records, particularly those not yet transferred to the National Archives, the available descriptions are often too broad or incomplete to provide meaningful clues to the contents of boxes. Thus, a Freedom of Information Act request that seeks all information on a given topic may well receive a response that ignores information located in boxes or files that are not clearly labeled or indexed. Under

these circumstances, searches may be more fruitfully conducted by citizens with an interest in, and understanding of, the subject of the search. However, because so many of the nation's records collections are off-limits to the public, even citizens who are willing to help are often precluded from lending a hand.

Many collections of interest to citizens contain no classified documents and can be made directly accessible to them. However, the Committee reviewed collections, particularly those containing decades-old records, where the entire collection was classified because it housed a small number of classified documents. For example, Record Group 326 at the College Park National Archives has approximately 160 feet of Metallurgical Laboratory/Argonne National Laboratory documentation that should be of significant historical interest. The collection itself is classified and currently inaccessible to citizens. The Committee's examination of large portions of the collection found very few classified documents, and when found, these documents were immediately declassified.

Executive Order 12958, issued by President Clinton on April 17, 1995 ("Classified National Security Information"), provides broadly for the automatic declassification (with specific exceptions) of all records that are more than twenty-five years old. In implementing the order, agencies should target collections that can be relatively quickly reviewed and made available to the public in their entirety.

4. Agencies should maintain complete records, available to the public, of document destruction.

Government records management rules provide for the destruction at varying dates in the future of all records that are appraised as temporary (that is, nonpermanent). They also provide that records be kept where certain collections, including classified records, are destroyed. But the Committee found that records of destruction are themselves routinely destroyed.

For example, upon Committee inquiry, DOE investigation revealed that the files of the AEC's Intelligence Division had been substantially destroyed during the 1970s and as late as 1989. (These files may have contained data on intentional releases, experimentation performed by the AEC for other agencies, and on the rules and practices of secret keeping regarding human data gathering). The DOE's inquiry found individuals who stated that they destroyed substantial records and that records of destruction were made. However, in accordance with DOE rules, the "certificates of destruction" were themselves later destroyed.[15] As another example, documents provided by the Department of Veterans Affairs and the Department of Defense indicate that, in 1947, the government contemplated the keeping of secret records in anticipation of potential liability claims from service personnel exposed to radiation and that some such records were kept. However, despite substantial search efforts by the DOD and the VA, the specific identity of the records referred to has not yet been determined.[16]

The Committee presumes that the vast majority of these records were destroyed in the routine course of business. Nonetheless, where records recording the destruction of important collections of records are themselves destroyed, the public cannot know whether important records have been destroyed (or merely are lost) and cannot be easily assured that destruction was in the routine course of business.

5. The Human Radiation Interagency Working Group should review and develop policies concerning public access to records generated or held by private contractors and institutions receiving federal funding.

Since World War II, the government has relied on contractors and grantees to perform an increasing number of governmental activities, including government-sponsored biomedical research. When the Advisory Committee undertook to locate information on particular government-sponsored radiation experiments, it was often told by federal agencies that, if such information was created, it would have been maintained only by nonfederal entities or investigators and not the government itself.

Where an activity is conducted by government employees (for example, researchers working in the facilities of the National Institutes of Health's Clinical Center), citizens have a right to seek access to information relating to that activity under the Freedom of Information Act. A similar right of access often does not apply, however, where a similar or even identical activity is conducted, also on federal funds, at nonfederal facilities.[17]

From the citizen's vantage point, the right to know about a government-funded activity should not depend on whether that activity is conducted directly by the government or by a government-funded private institution. At the same time, nonfederal institutions are not governmental agencies, and there may be good reasons they should not be burdened with identical obligations to retain records and to provide information to the public.

Rules are needed that accommodate both the citizen's right to know about the conduct of the government and the relevant differences between nonfederal and federal institutions with respect to duties to create and maintain publicly accessible records.[18] To ensure consistent and informed governmentwide treatment of the question, the Human Radiation Interagency Working Group may wish to call on the Office of Management and Budget (OMB) and the Office of Federal Procurement Policy (OFPP) to review the current right of members of the public to gain access to the records of government grantees and contractors.

RECOMMENDATION 18

18a: The Advisory Committee recommends to the Human Radiation Interagency Working Group that the CIA's record-keeping system be reviewed to ensure that records maintained by that agency are accessible upon legitimate request from the public or governmental sources. This review could be performed by the CIA inspector general or an oversight panel.

18b: The Advisory Committee recommends that all records of the CIA bearing on programs of secret human research, such as MKULTRA and the related CIA human behavior projects from the late 1940s through the early 1970s, including Bluebird, Artichoke, MKSEARCH, MKDELTA, Naomi, Chance, Often, and Chickwit, become a top priority for declassification review with the expectation that most, if not all, of these documents can be declassified and made available to the public.

These recommendations are intended to ensure that the public and the government have practical access to historical records of the CIA (where access is otherwise appropriate) and to address long-standing public interest and concerns regarding secret human experiments conducted or sponsored by the CIA.

The framework of the records collections of all the Human Radiation Interagency Working Group agencies, save the CIA, is visible to the public. This is the case even in agencies, such as the Defense Nuclear Agency, where historical research records are largely classified.

While documents showing CIA participation in midcentury DOD-sponsored discussions of human experimentation were obtained from DOD, DOE, and the public National Archives, the CIA was not able to locate such documents in its own files and states that the CIA's role in these discussions was sufficiently minor that such records would not have been kept. The Advisory Committee also notes the recent report to the attorney general of the BNL Task Force, which was investigating a bank-related scandal: "While we benefited from extensive cooperation and assistance from the CIA's Office of General Counsel, the CIA's ability to retrieve information is limited. Records are 'compartmentalized' to prevent unauthorized disclosure; only some of those records are retrievable through computer databases; no database encompasses all records; and not all information is recorded. In the course of our work, we learned of 'sensitive' components of information not normally retrievable and of specialized offices that previously were unknown to the CIA personnel assisting us."[19]

In addition, while the Advisory Committee has found no evidence to show that the CIA conducted or sponsored human radiation experiments, numerous documents, some of which remain partially classified, make reference to possible CIA interest in this area. Although Advisory Committee staff has reviewed all of the available classified information concerning human radiation experiments and requested that it be declassified, the public does not as yet have the benefit of such access.

Twenty years after they were first revealed to the public, there continues to be a strong public interest in the CIA's "mind control" programs.

The Advisory Committee received numerous queries about MKULTRA and the other related programs from scholars, journalists, and citizens who have been unable to review the complete record. Although these CIA projects were the subject of significant governmental inquiry in the mid to late 1970s—by the Senate and House committees and by the presidentially appointed Rockefeller Commission—and a substantial portion of the records have been declassified and released to the public, a number of documents remain classified, and many of the documents that have been released contain numerous redactions. This has made it extremely difficult to understand the full context of the activities or to clarify discrepancies or uncertainties in the record.

A number of the declassified documents make reference to radiation experiments. However, because of the redactions, it is impossible for the public to determine from these documents whether there is additional, secret information about radiation activities. (Advisory Committee staff have reviewed the full text of these documents.) For example, the 1963 CIA inspector general report on the inspection of MKULTRA, which was declassified in redacted form in 1975, stated that "radiation" was one of the avenues explored under MKULTRA. But because so much of that document was redacted, the public reader might reasonably suspect that there is more information about radiation in the report. At the request of the Advisory Committee, the CIA re-released this document, and a handful of others, with minimal redactions.

However, few other such documents have been re-reviewed for declassification in almost twenty years. Since most of the classified CIA documents concerning MKULTRA and related programs that Advisory Committee staff reviewed were declassified upon request, the Advisory Committee believes that if the rest of these records were reviewed for historical declassification, most, if not all, of the records could be declassified without harming the national security.

So long as documents about secret human experiments are withheld from the public, it will be impossible to put to rest distrust with the conduct of government. The rapid, public release of the remaining documents about

MKULTRA and other secret programs would be a fitting close to an unhappy chapter in the nation's history.

Addendum to Recommendation 4: Medical Notification and Follow-up

The Advisory Committee's charter requires that we consider the issue of notice to experimental subjects of potential health risk and the need for medical follow-up:

If required to protect the health of individuals who were subjects of a human radiation experiment, or their descendants, the Advisory Committee may recommend to the Human Radiation Interagency Working Group that an agency notify particular subjects of an experiment, or their descendants, of any potential health risk or the need for medical follow-up [Sec. 4.c.].

The basic intent of this provision is not directed at subjects who have already died, or at subjects who have already become ill and been treated. It is primarily aimed at asymptomatic subjects who remain at significant risk for the development of radiation-induced cancers. Because at least two and as many as five decades have passed since the experiments took place, most of those who may eventually develop cancer as a result of the experiment will already have developed symptoms and sought treatment. However, some subjects may still be at risk and thus arguably might benefit from medical follow-up.

The initial consideration in deciding whether to implement a program of active notification and medical follow-up is the identification of populations of subjects who have been put at significant risk for the development of radiogenic cancers. The magnitude and focus of these risk estimates are driven by the specific organs placed at highest risk from the particular radiation exposure (for example, thyroid being the organ at greatest risk in the iodine 131 experiments, testes in the Oregon and Washington prisoner experiments, and the brain for the nasopharyngeal radium experiments). Risk estimates are calculated for each target organ according to a number of assumptions that may include adjustments for variables such as age at exposure, sex, or type of radiation (isotope vs. external beam) and are generally expressed in terms of excess cancer incidence/mortality for a given population over a specified period at a specified dose.

The Advisory Committee adopted an excess site-specific cancer mortality (death) greater than 1 case in 1,000 (lifetime) as a criterion for determining that a subject had been placed at increased risk. However, because of the substantial passage of time since the initial exposure, the criteria for consideration of active notification were set at 1/1,000 future or remaining lifetime risk and an excess relative risk of greater than 10 percent (organ specific). This level of risk was arbitrarily chosen by the Advisory Committee. When compared with the normal risk of developing cancer (220 out of 1,000), this level of risk is small. The Advisory Committee chose this small remaining lifetime risk as a reasonable initial criterion to decide if a more in-depth analysis of the effectiveness of screening and intervention was needed.

Once a population has been determined to have an increased remaining lifetime risk for radiogenic cancer mortality, a second criterion must be satisfied before a government-funded medical follow-up program is recommended, namely whether the exposed individuals would likely benefit from a program of early detection or early treatment of the malignancy. Effective screening procedures for the detection of an early-stage cancer exist only for a limited number of cancer sites. Moreover, the lack of specificity of all diagnostic screening tests results in a significant number of "false positives" (a positive test result in an individual who in truth is not affected), resulting in unnecessary and potentially hazardous medical procedures that may cause health problems in and of themselves. On the other hand, most diagnostic tests are also imperfectly sensitive, meaning that some individuals who actually have the disease will be falsely reassured that they are cancer free and may thereby delay seeking attention when it becomes symptomatic. To this end the Advisory Committee has adopted the following criteria for assessing the value of screening, preventive, or therapeutic measures for exposed subjects of biomedical experiments:[20]

1. The condition must have a significant effect on the quality or length of life.

2. The condition must have an asymptomatic period during which it can be detected by available screening methods.

3. These screening methods must have high sensitivity and specificity.

4. Treatment in the asymptomatic phase must yield a therapeutic result superior to that obtained by delaying treatment until symptoms appear.

5. The medical benefits of screening and early treatment must outweigh any detrimental medical effects or risks.

These criteria were applied to each exposed population at significant risk for development of a malignancy and evaluated according to the organ(s) at risk from radiation exposure. In each case, the conditions enumerated above must be satisfied before specific medical follow-up would be recommended.

Details of the Advisory Committee's risk calculations can be found in chapters 7 and 9. To summarize, the Advisory Committee found no experiments involving iodine 131 administration to children that met our 1/1,000 criterion for remaining lifetime risk of dying of cancer; even in the most highly exposed individuals, risks were estimated to be 1/2,000 (remaining lifetime risk). In addition, the U.S. Preventive Services (USPS) Task Force concluded that "routine screening for thyroid disorders is otherwise not warranted in asymptomatic adults or children." Although it has been suggested that people placed at risk for development of thyroid carcinoma following high-dose external irradiation to the upper body may benefit from regular physical examination of the thyroid, there are no data to support a similar risk or benefit for those who have been exposed to diagnostic or therapeutic doses of iodine 131.[21]

The Advisory Committee recognizes that in addition to the very small risk of a fatal thyroid cancer, individuals exposed as children to iodine 131 also have a larger risk of a nonfatal thyroid cancer or benign tumor, a lifetime risk that in many of the experiments we considered exceeded 1/1,000 and in a few individuals exceeded 1/100. We recognize that such conditions may require medical treatment and may be associated with considerable anxiety and discomfort. After considerable discussion, however, the Committee concluded that notification was not warranted for the purpose of detecting such conditions early, on several grounds. First, the prognosis for such conditions under standard clinical care is excellent, and there is no evidence that early detection improves the outcome. Second, even among the subgroup of about 200 children exposed to this level of risk, the number of excess cancers expected is less than one, whereas the normal prevalence in an unexposed population is about 20 to 30 percent. Third, many thyroid cancers that are detectable by screening may have no clinical significance. Finally, the most effective means of screening for thyroid cancer remains palpation, which has low sensitivity and low specificity.

For the prisoners subjected to testicular irradiation, the Advisory Committee estimates that even the most heavily exposed individual (600 rad to the testicles) would have a risk of only 0.4/1,000[22] of developing a fatal cancer, which does not attain our stated criterion. Furthermore, the USPS Task Force has concluded that "there is insufficient evidence of clinical benefit or harm to recommend for or against routine screening of asymptomatic men [other than those with a history of cryptorchidism, orchiopexy, or testicular atrophy] for testicular cancer."[23] These considerations lead the Advisory Committee to recommend against any program of active notification of these subjects. However, subjects who voluntarily request medical check-up or counseling should have such provided in a standard clinical setting.

For the children who received nasopharyngeal radium treatments, the Advisory Committee has estimated that the lifetime risk of tumors to the central nervous system (brain), head, and neck regions is approximately 4.35/1,000 and the excess relative risk is about 62 percent, both with considerable uncertainties.[24] Although these experiments were conducted in the 1940s and much of the risk has probably already been expressed, it is still possible that the future risk is greater than or equal to our arbitrary 1/1,000 risk criterion. However, at greatest risk are the brain, and head and neck tissues, for which there is neither an accepted nor recommended screening procedure.[25] Thus, while the subjects in these experiments meet the Advisory Committee's arbitrary 1/1,000 criterion for consideration for notification and medical follow-up (criterion 1 in Recommendation 4, above), the utility of such a program has not been demonstrated, so criterion 2 of Recommendation 4 is not satisfied. Adult military personnel who participated in trials of this procedure received significantly lower radiation exposures, did not attain our arbitrary 1/1,000 criterion for risk, and would similarly fail to meet the criteria in guideline 2. Therefore, the Advisory Committee does not recommend notification and medical follow-up of children or adults in this group of experiments.

The Advisory Committee's charter also requires that we consider the need for notification of descendants of experimental subjects for purposes of health protection. The rationale for considering notification in this instance derived from the assumption that the offspring of former subjects might be at risk for disease or disability as a consequence of inherited mutations resulting from their parent's previous radiation exposure. The weight of evidence suggests that the risk of heritable genetic effects from the radiation exposures in the experiments we reviewed is very small, although it is possible that some offspring of exposed individuals might carry mutations that were caused by radiation.[26] Moreover, in most medical experiments involving external sources of radiation, efforts are made to shield the gonads (ovaries/testes) as much as possible. With the exception of the testicular irradiation experiments, where subjects agreed to undergo vasectomy to prevent transmission of any mutations that might have occurred, experiments involving external irradiation are likely to have produced relatively small gonadal doses, as would those experiments involving tracers. Even

therapeutic studies involving internal radionuclides would generally involve only modest gonadal doses. Thus, in the vast majority of experiments, it is likely that the risk of radiation-induced mutations is small in relation to natural rates.

In addition to cancer and genetic effects, there are only a small number of well-established effects of radiation, including severe mental retardation among those exposed in utero (particularly between eight and fifteen weeks of gestation), sterility, cataracts, and hypothyroidism. Unlike cancer and genetic effects, however, these other endpoints appear to be "deterministic" effects that appear only after high doses that are unlikely to have been received by subjects in the experiments under consideration for notification. The Advisory Committee heard extensive public testimony about a range of other conditions that those testifying thought might be related to radiation exposures. However, the Advisory Committee believes that a program of active notification must be grounded on currently accepted scientific evidence concerning the conditions that are likely to be caused by radiation.

NOTES

1. AEC documents reveal that in order for one researcher to publish a report on his TBI research, he had to respond to the AEC's concerns about potential public relations and legal liability consequences and did so by deleting information that might permit identification of patients. See chapter 8.

2. These awards included $750,000 in 1976 by Congress to the Olson family, $703,000 in 1987 by court order to the Blauer family, and $750,000 in 1988 by court order to nine Canadians for nonfatal brainwashing experiments. See chapter 3.

3. For example, based on facts available to the Committee, those Alaskans who were subjects of Air Force-sponsored radioisotope research (see chapter 12) and the pregnant women who were subjects of radioisotope research at Vanderbilt University (see chapter 7) may also be owed an apology. However, the Committee conducted only limited inquiry into these cases. The Advisory Committee did not attempt a full factual inquiry into the Alaskan research, which is the subject of an inquiry by a committee of the Institute of Medicine and the National Research Council, whose report is pending. The Vanderbilt research is currently the subject of litigation that may provide for fuller development of the facts.

4. Veterans who participated in weapons tests are also eligible for relief under the Radiation Exposure Compensation Act of 1990, which, however, requires claimants to elect the monetary remedy to the exclusion of other benefits to which a veteran may be eligible. We also note the Veterans Exposure Amendments of 1992.

5. National Cancer Institute, National Institutes of Health, *Radon and Lung Cancer Risk: A Joint Analysis of 11 Underground Miner Studies* (Washington, D.C.: National Institutes of Health Publication No. 94-3644, January 1994).

6. *The Belmont Report: Ethical Principles and Guidelines for the Protection of Human Subjects of Research,. Report of the National Commission for the Protection of Human Subjects of Biomedical and Behavioral Research* (Washington, D.C.: GPO, 1979).

7. The convening of a national panel could assist as well with the implementation of Recommendations 10 and 11.

8. *California Health and Safety Code*, vol. 40B, sec. 24176 (1995).

9. For example, in 1994, the Institute of Medicine's Committee on the Ethical and Legal Issues Relating to the Inclusion of Women in Clinical Studies recommended that the National Institutes of Health review the area of compensation for research injury. See *Women and Health Research* (Washington, D.C.: National Academy Press, 1994), 169 and appendix D to that volume titled "Compensation for Research Injuries."

10. President's Commission for the Study of Ethical Problems in Medicine and Biomedical and Behavioral Research, *Compensating for Research Injuries: The Ethical and Legal Implications of Programs to Redress Injured Subjects, Vol. 1, Report* (Washington, D.C.: GPO, June 1982).

11. Ibid., 50.

12. While lessons such as those identified above have been learned, by the same token, it seems unlikely that they will be fully taken advantage of unless some individual or entity is designated with responsibility to ensure that this takes place.

13. The post-World War II records of the Army Office of the Surgeon General are also located primarily either at the Washington National Records Center or with the Office of the Surgeon General. Similarly, very few of the post-World II records of the Chemical Corps and its successors are located at the National Archives but are mostly found at the Washington National Records Center or the successors.

14. Standard Form 135 (SF-135) is the transmittal form agencies use when shipping records to a federal records center. A folder listing is supposed to accompany all shipments of records, with the exception of the relatively rare classified SF-135, the forms are available for examination by the public.

15. "Destruction of the U.S. Atomic Energy Commission Division of Intelligence Files," report by the Office of Human Radiation Experiments, 26 August 1994.

16. As noted in chapter 10, an investigation by the VA concluded that the "confidential Atomic Medicine Division" evidently contemplated was not activated; nonetheless, remaining documents indicate that certain records were kept in anticipation of potential liability claims. As noted further in chapter 10, the precise nature of all records at issue cannot be conclusively determined.

17. Government contractor records have been found to be beyond the reach of the Freedom of Information Act because contractors are not "agencies" who maintain "agency records," a condition required by the act. However, regulations that govern contractors may bring records that contractors maintain under the act. For example, a recent Department of Energy regulation (10 C.F.R. § 1004.3[e], 59 Fed. Reg. 63883 [12 December 1994]), provides that even if a contractor-held document fails to qualify as an "agency record" it may be subject to the act if the contract provides that the document in question is the property of DOE. For a discussion of the application of this rule, see *Cowles Publishing Company*, Decision and Order of the Department of Energy, Case No. VFA-0018, 28 February 1995.

18. In making this recommendation, the Advisory Committee emphasizes that we do not intend to alter privacy restrictions that currently limit access to records related to biomedical research (such as personal medical records).

19. 21 October 1994 Addendum to the BNL Task Force-Final Report from John Hogan, Acting Assistant U.S. Attorney, Northern District of Georgia and Counselor to the Attorney General to the Attorney General (ACHRE No. CORP-060595-A), 2-3.

20. Adapted from U.S. Preventive Services Task Force, *Guide to Clinical Preventive Services: An Assessment of the Effectiveness of 169 Interventions* (Baltimore: Williams & Wilkins, 1989), xxix-xxxii; and P. S. Frame, "A Critical Review of Adult Health Maintenance," *Journal of Family Practice* 22 (1986): 341, 417, 511.

21. National Research Council, Board on Radiation Effects Research, Committee on the Biological Effects of Ionizing Radiations, *Health Effects of Exposure to Low Levels of Ionizing Radiation: BEIR V* (Washington, D.C.: National Academy Press, 1990), 5, 287-294.

22. See footnote on testicular risk analysis in chapter 9.

23. U.S. Preventive Services Task Force, *Guide to Clinical Preventive Services*, 77.

24. See footnote on children's risk analysis in chapter 7.

25. U.S. Preventive Services Task Force, *Guide to Clinical Preventive Services*.

26. See "The Basics of Radiation Science" section of the Introduction.

Statement by Individual
Committee Member

STATEMENT BY COMMITTEE MEMBER JAY KATZ

W E were assigned two tasks: to examine the past and to examine the present. Telling the full story of government sponsored Cold War human radiation experiments serves many important purposes—remembrance, warning, healing. Ultimately, however, the value of knowing the past resides in the lessons it can teach us for the present and future. Thus, the central question is this: Do current regulations of human experimentation adequately protect patient-subjects? Here I have the most serious reservations about our Report.

In summary, my conclusions are these: (1) In the quest to advance medical science, too many citizen-patients continue to serve, as they did during the Cold War period, as means for the sake of others. (2) The length to which physician-investigators must go to seek "informed consent" remains sufficiently ambiguous so that patient-subjects' understanding of the consequences of their participation in research is all too often compromised. (3) The resolution of the tensions inherent in the conduct of research—*i.e.*, respect for citizen-patients' rights to, and interest in, self-determination on the one hand and the imperative to advance medical science, on the other—confronts government officials with policy choices that they were unwilling to address in any depth during the Cold War or for that matter in today's world. (4) Our Recommendations only touch on these problems and at times make too much of the safeguards that have been introduced since 1974. The present regulatory process is flawed. It invites in subtle, but real, ways repetitions of the dignitary insults which unconsenting citizen-patients suffered during the Cold War.

Medical research is a vital part of American life. The Federal government allocates billions of dollars to human research, and the pharma-

ceutical industry spends many more billions to develop new drugs and medical devices. And research is by and large conducted with patients. Since all of us at one time or another will be patients, we are readily available subjects for research. Thus, the protection of the rights and interests of citizen-research subjects in a democratic society is a major societal concern.

Let me introduce my Reservations by offering some preliminary remarks about the current regulatory scheme and the history of consent. The contemporary regulatory scheme provides insufficient guidance for addressing one basic question: When, if ever, should conflicts between advancing medical knowledge for our benefit and protecting the inviolability of citizen-subjects of research be resolved in favor of the former? Inviolability, unless patient-subjects agree to invasions of mind and body, requires punctilious attention to disclosure and consent and, in turn, imposes considerable burdens on physician-investigators—be it taking the necessary time to converse with patient-subjects or, if necessary, making discomforting disclosures. Moreover, taking informed consent seriously may slow the rate of medical progress with painful consequences to investigators' work and to society. These dilemmas must be resolved forthrightly, instead of allowing them to be "resolved" by discretionary subterfuge.

Neither the drafters of the 1974 Federal Regulations nor the members of the research community were willing to respond to the reality that taking informed consent seriously in this new age of informed consent confronted them with problems that required sustained and thoughtful exploration. Implementation would also turn out to be a most formidable task because of physicians' low regard for patient consent throughout medical history. The Committee's analysis of the informed consent requirements in existence during the Cold War and earlier in the 20th century acknowledges, but not sufficiently so, that the millennia-long history of medical custom casts a dark shadow over what transpired during the Cold War.

Patient consent, until most recently, has not been enshrined in the ethos of Hippocratic medicine. As I once put it, the idea of patient autonomy is not to be found in the lexicon of medicine. It is important to be aware of this history; for it explains why our Findings on contemporary research practices, which time constraints prevented us from probing in sufficient depth, revealed deficiencies in the informed consent process, both at the levels of physician-investigator interactions with their patient-subjects and of IRB review. This is not surprising; for not only does it take time to change historical practices, it also requires more thoughtful rules and procedures than currently exist.

My reading of the Cold War record suggests that governmental officials in concert with their medical advisers at best paid lip service to consent. Whenever they considered it, they worried mostly about legal liability and embarrassment. They were not worried or embarrassed about their willingness to conscript unconsenting patient-subjects to serve as means in plutonium and whole body radiation experiments. All this is a frightening example of how thoughtlessly human beings, including physicians, can treat human beings for "noble" purposes. Most references to consent (with rare exceptions) that we uncovered in governmental documents or in exchanges between officials and their medical consultants were meaningless words, which conveyed no appreciation of the nature and quality of disclosure that must be provided if patient-subjects were truly to be given a choice to accept or decline participation in research. Form, not substance, punctuated most of the policies on consent during the Cold War period. The drafters of the Federal Regulations would eventually build their rules on this shaky historical foundation, disregarding in the process that the imprecision of their policies invited physician-investigators not to alter decisively customary Hippocratic practices.

The long established tradition of obtaining consent from healthy subjects is a separate story; for this tradition did not extend to patients or patient-subjects. Put another way, the latter were quarantined from disclosure and consent. In our Finding 10, this was clearly stated: "[D]uring the 1944–1974 period . . . physicians engaged in clinical research generally did not obtain consent from patient-subjects for whom the research was intended to offer a prospect of medical benefit." Therefore, it should come as no surprise, as noted in our Report, that when a decision was reached in 1951 not to pursue radiation research with prisoners or healthy subjects in connection with an important defense project, "the military immediately contracted with a private hospital to study patients being irradiated for cancer treatment." Patients have always been the most vulnerable group for purposes of research.

From the perspective of history no significant conclusions can be drawn about ethical consent standards that "should" have existed for research with *patients* by drawing attention to consent requirements that existed for *healthy volunteers*. When persons became patients, the rules of consent changed. This observation also has relevance for the impact of the Nuremberg Code on the conduct of research. The Code emerged from contexts not only of research with non-patients but also of sadistic and brutal disregard for the sanctity of human life, unparalleled in the annals of Western research. American physician-investigators, therefore, found it doubly easy to consider the pronouncements of the Allied Military Tribunal irrelevant to their practices.

Let me interject here a few brief remarks about risks: Taking risks is inevitable in research. After all, research is by its nature a voyage into the unknown. To pierce uncertainty, to gain scientific knowledge requires risk taking. And, as our Report makes clear, physician-investigators and government officials as well have generally been attentive, whenever physical risks needed to be taken, to minimize them. But such care notwithstanding, research requires taking risks; for example, research with highly toxic agents affects the quality and extent of remaining life. In our review of contemporary research we identified many instances where patient-subjects were unknowingly exposed to such risks, which have both physical and emotional dimensions.

Scientific studies in today's world often involve patient-subjects whose prognosis is dire— the most vulnerable of all disadvantaged groups—and for whom no effective or curative treatments exist. In these situations hope can readily be exploited by intimating that research interventions *may* also benefit patient-subjects, even though the experiment's objectives are in the service of gaining scientific knowledge. Embarking on this slippery slope begins with investigators' rationalizations which justify experimental interventions on grounds of "possible" therapeutic benefits; it continues with apprising patient-subjects insufficiently of the slings and arrows of the experimental component; and it ends with feeding into patient-subjects' own dispositions to deny the truth. In sum, by obliterating vital distinctions between therapy and research, investigators invite subjects to collude with them in the hazy promise of therapeutic benefits. Put another way, the "therapeutic illusion," as one commentator felicitously called it, can lead physician-investigators to emphasize the possible (though unproven) therapeutic benefits of the intervention and, in turn, to minimize its risks, particularly to the quality of (remaining) life. Such considerations played a role in the total body radiation experiments discussed in our Report.

In my Reservations I want to emphasize, however, the centrality of dignitary, not physical, injuries in any appraisal of the ethics of research. This is the uncompromising message of the Nuremberg Code's first principle on voluntary consent, a message which during the Cold War period physician-investigators found impossible to accept. But the problem goes deeper than that. The Code, without extensive exegesis, could not serve as a viable guide for the conduct of medical research. This made its disregard easy and in the process, the central message which the judges tried to convey in their majestic first principle was also lost. Thus too much can be made, as our Report does, of Secretary of Defense Wilson's memorandum endorsing the Nuremberg Code. To hold him culpable for not *implementing* the Code makes little sense. If he is culpable of anything, it

is for *promulgating* it without first having sought thoughtful advice about what needed to be explicated to make it a viable statement for research practices. Merely embracing the Code invited, indeed guaranteed, neglect.

Finally, from the perspective of history I want to note that only since the early 1960's was the importance of consent given greater attention. Among the social forces that contributed to this development two stand out: Judges' promulgation of a new legal doctrine of informed consent, based on the Anglo-American premise of "thoroughgoing self-determination." And the explorations by a new breed of bioethicists, recruited from philosophy and theology, of the relevance of such principles as autonomy, self-determination, beneficence, and justice to medical decision-making. Their novel and powerful arguments, so alien to the medical mind, disturbed the sleep of the medical community. Physicians had a particularly hard time in coming to terms with the idea of patient autonomy. To this day, I believe, this principle has only gained a foothold in the ethos of medical practice and research.

In our Report we emphasize the primacy of patient-subject autonomy in research. It led us to conclude in our Interim Report that "[a] cornerstone of modern research ethics [is] informed consent." I agree with this statement of principle. From the 1963 beginnings of my work in human experimentation, I have championed the idea of respect for autonomy and self-determination in all interactions between physician-investigators and patient-subjects. But I introduced one major qualification when I wrote that only when the Nuremberg Code's first principle on voluntary consent

is firmly put into practice can one address the claims of science and society to benefit from science. Only then can one avoid the dangers that accompany a balancing of one principle against the other that assigns equal weight to both; for only if one gives primacy to consent can one exercise the requisite caution in situations where one may wish to make an exception to this principle for *clear and sufficient* reasons.

I mention this here because the final and most far-reaching recommendation for change that I shall soon propose is based on two premises: (1)

that any exception to the principle of individual autonomy, since it tampers with fundamental democratic values, must be rigorously justified by clear and sufficient reasons; and (2) that such exception cannot be made by investigators or IRBs but only by an authoritative and highly visible body.

I now turn to our Research Proposal Review Project. The Committee's review of contemporary research reveals that of the greater-than-minimal-risk studies (which are the ones that raise complex informed consent issues) 23% were ethically unacceptable and 23% raise ethical concerns. My own independent review tells a grimmer story: 50% raise serious ethical concerns and an additional 24% raise ethical concerns that cannot be taken lightly. Since I focused exclusively on the informed consent process, the differences in our Findings can perhaps in part be explained on that basis. My data, like the Committee's, were the protocols submitted to IRBs and the informed consent forms signed by patient-subjects. I appreciate that the evidence available to us does not reflect what patient-subjects might have been told during oral communications. But if the protocols and patient-subject consent forms are flawed in significant ways, it is likely that the oral interactions are similarly flawed. Moreover, since IRBs are charged to pay particular attention to the informed consent process, I contend that IRBs should not have approved the problematic consent forms in the form they were submitted. The forms often seem to "sell" research rather than to convey a sense of caution that invites reflective thought.

I had expected to discover problems, but I was stunned by their extent. Consider what we observed in Chapter 15 and what is described there in greater detail: The obfuscation of treatment and research, illustrated most strikingly in Phase I studies, but by no means limited to them; the lack of disclosure in randomized clinical trials about the different consequences to patient-subjects' well being if assigned to one research arm or the other; the administration of highly toxic agents, in the "scientific" belief that only the knowledge gained from "total therapy" will *eventually* lead to cures, but without disclosure of the impact of such radical interventions

on quality of life or longevity. I do not wish to minimize the impact of making total disclosure on patient-subjects' and physician-investigators' hopes and fears. Yet, nagging questions remain: What are "clear and sufficient reasons" which permit tampering with disclosure and consent; and, if permissible, who decides?

Our Recommendations do not go far enough in remedying the flawed nature of our current regulations which appear to rely so heavily on informed consent, but which in practice I contend, bypass true informed consent. Here I can only make a few comments about the changes required if we wish to protect adequately the rights and interests of subjects of research:

(1) *Informed consent* is central to such protections. The drafters of the Federal regulations have acknowledged that fact. They have failed, however, to take responsibility for making these requirements meaningful ones. Thus, patient-subjects now all too often give a spurious consent; a "consent" that can readily mislead physician-investigators into believing that they have received the authority to proceed when in fact they have not.

(a) The Federal regulations imply that the principle of respect for patient-subjects' autonomy is central to the regulatory scheme. Leaving it at that is not enough; for the principle requires commentary so that physician-investigators will have a more thoroughgoing appreciation of the moral issues at stake whenever they ask human beings to serve as means for the ends of others. Only then will they learn, for example, that to take informed consent seriously requires them to spend considerable time with prospective patient-subjects and to engage them in searching conversations. In these conversations they must disclose (a) that their subjects are not patients or, to the extent they are patients, that their therapeutic interests will be subordinated in specified ways to scientific interests; (b) that it is problematic (and in what ways) whether their welfare will be better served by placing their medical fate in the hands of a practitioner rather than a physician-investigator; (c) that in opting for the care of a physician they may be better or worse off and for such and such reasons; (d) that research is governed by a research protocol and a research question and therefore patient-subjects' interests and needs have to yield (and to what extent) to the claims of science; etc.

Such disclosure obligations are formidable ones. They need to be fulfilled in a manner that will give patient-subjects a clear appreciation of the difference between research and therapy, and in the spirit that disabuses them of the belief, so widely held—as our Subject Interview Study demonstrates—that everything the investigator proposes serves their best therapeutic interests.

The Cold War experiments teach us that misplaced trust can deceive; that trust must be *earned* by prior disclosures of what research participation entails. I agree, as our Recommendation 9 proposes, that scientists should be educated "to ensure the centrality of ethics in [their] conduct." To accomplish that educational task, however, requires policies that more clearly delineate the ambit of discretion which investigators can exercise in the conduct of research.

(b) Current criteria for informed consent encourage, perhaps even mandate, overwhelming patient-subjects with information on every conceivable risk and benefit as well as on the scientific purpose of the study. Adherence to these mandates has led, and justifiably so, to concerns about the incomprehensibility of the informed consent forms that patient-subjects must sign. Much thought, and then guidance, has to be given to IRBs and investigators as to the essential information they most provide; *e.g.*, alternatives, uncertainties, essential risks, realistic benefits as well as the impact of participation—known and conjectured—on the quality of future (or remaining) lives. Many of the informed consent forms I have examined fail to emphasize the risks germane to the research protocol; instead they go into numbing detail on risks that can be summarized. To put it bluntly: Informed consent criteria in today's world, at least in the ways they are communicated to patient-subjects, often serve purposes of obscuring rather than clarifying what participation in research entails.

(2) Though *IRBs* serve important functions, they do not have the capacity, if only by virtue of composition and lack of time, either to modify consent standards (including the ones I have just proposed) or, more generally, to make any other decisions that could affect the fundamental constitutional rights and personal interests of subjects of research. IRBs should not have the authority to decide how to balance competing principles in situations where the competence of subjects' consent is in question, or where consent cannot be obtained because patient-subjects suffer from a life-threatening condition, or where other complex issues need to be resolved, as illustrated in our Chapter on the total body radiation experiments. Such fateful decisions are beyond their competence.

Moreover, IRBs work in a climate of low visibility, another species of secrecy about which we expressed so much concern in Chapter 13. These and other complex ethical problems should only be resolved by an *accountable* and highly *visible* national Body. That Body then can provide IRBs with guidelines that will better inform their deliberations. I would like to note here, but only in passing, that the Body I envision will lighten IRBs' tasks; for example, by fashioning policies for cursory review of the many minimal/no risk studies, or by being available for advisory opinions whenever IRBs are confronted with new ethical problems. (IRBs now spend an inordinate amount of time on such problems which they should not resolve in the first place.) The national Body should not review individual research projects except when investigators and IRBs disagree. Finally, a national Body is needed for another reason as well: The considerable pressure for approval of protocols to which IRBs are subjected by the scientists at their institutions.

(3) Already in 1973, when I served on HEW's Tuskeegee Syphilis Study Ad Hoc Advisory Panel, we proposed in our Final Report that Congress establish a permanent body—we called it the *National Human Investigation Board*—with the authority to regulate at least all Federally supported research involving human subjects. We recommended that this Board should not only *promulgate* research policies but also *administer* and *review* the human experimentation process. Constant interpretation and review by a Body whose decisions count by virtue of the authority invested in them can protect both the claims of science and society's commitment to the inviolability of subjects of research.

A most important task which such a Board would face in formulating research policies is to delineate exceptions to the informed consent requirement when competing principles require it. For example, when might it be permissible for IRBs to "defer consent" (or more correctly, to allow physician-investigators to proceed without consent) with patient-subjects suffering from acute head trauma? Conscripting citizen-patients to anything they have not consented to is deeply offensive to democratic values and, if necessary, requires public approval. Greater public participation in the formulation of research policies is vital, and the Board must therefore establish procedures for the publication of all its major policy and advisory decisions, particularly those where compromises seem warranted between the advancement of science and the protection of subjects of research. Publication of such decisions would not only permit their intensive study both *inside* and *outside* the medical profession but would also be an important step toward the case-by-case development of policies governing human experimentation. If we are truly concerned about the baneful effects of secrecy on public trust, what I propose here could restore trust.

There is, of course, much more to consider, and I have written about it elsewhere. I hope, however, that I have said enough to suggest that the problems inherent in research with human subjects—advancing science and protecting subjects of research—are complex. Society can no longer afford to leave the balancing of individual rights against scientific progress to the low-visibility decision-making of IRBs with regulations that are porous and invite abuse. The important work that our Committee has done in its evaluation of the radiation experiments conducted by governmental agencies and the medical profession during the Cold War once again confronts us with the human and societal costs of too relentless a pursuit of knowledge. If this is a price worth paying, society should be forced to make these difficult moral choices in bright sunlight and through a regulatory process that constantly strives to articulate, confront, and delimit those costs.

We have judged the past and judgments of the past become most relevant when they teach us lessons for the present and future. Yet, we did not judge the present with sufficient care. If the problem was time, I wanted to take the time to offer my judgments. I also took the time and "took [the road] less traveled by" because much is at stake in the quest for advancing medical science that speaks not only to progress in the conquest of disease but to other moral values as well.

Official Documents

Presidential Documents

Executive Order 12891 of January 15, 1994

Advisory Committee on Human Radiation Experiments

By the authority vested in me as President by the Constitution and the laws of the United States of America, it is hereby ordered as follows:

Section 1. *Establishment.* (a) There shall be established an Advisory Committee on Human Radiation Experiments (the "Advisory Committee" or "Committee"). The Advisory Committee shall be composed of not more than 15 members to be appointed or designated by the President. The Advisory Committee shall comply with the Federal Advisory Committee Act, as amended, 5 U.S.C. App. 2.

(b) The President shall designate a Chairperson from among the members of the Advisory Committee.

Sec. 2. *Functions.* (a) There has been established a Human Radiation Interagency Working Group, the members of which include the·Secretary of Energy, the Secretary of Defense, the Secretary of Health and Human Services, the Secretary of Veterans Affairs, the Attorney General, the Administrator of the National Aeronautics and Space Administration, the Director of Central Intelligence, and the Director of the Office of Management and Budget. As set forth in paragraph (b) of this section, the Advisory Committee shall provide to the Human Radiation Interagency Working Group advice and recommendations on the ethical and scientific standards applicable to human radiation experiments carried out or sponsored by the United States Government. As used herein, "human radiation experiments" means:

(1) experiments on individuals involving intentional exposure to ionizing radiation. This category does not include common and routine clinical practices, such as established diagnosis and treatment methods, involving incidental exposures to ionizing radiation;

(2) experiments involving intentional environmental releases of radiation that (A) were designed to test human health effects of ionizing radiation; or (B) were designed to test the extent of human exposure to ionizing radiation.

Consistent with the provisions set forth in paragraph (b) of this section, the Advisory Committee shall also provide advice, information, and recommendations on the following experiments:

(1) the experiment into the atmospheric diffusion of radioactive gases and test of detectability, commonly referred to as "the Green Run test," by the former Atomic Energy Commission (AEC) and the Air Force in December 1949 at the Hanford Reservation in Richland, Washington;

(2) two radiation warfare field experiments conducted at the AEC's Oak Ridge office in 1948 involving gamma radiation released from non-bomb point sources at or near ground level;

(3) six tests conducted during 1949–1952 of radiation warfare ballistic dispersal devices containing radioactive agents at the U.S. Army's Dugway, Utah, site;

(4) four atmospheric radiation-tracking tests in 1950 at Los Alamos, New Mexico; and

(5) any other similar experiment that may later be identified by the Human Radiation Interagency Working Group.

The Advisory Committee shall review experiments conducted from 1944 to May 30, 1974. Human radiation experiments undertaken after May 30 1974 the date of issuance of the Department of Health, Education, and Welfare ("DHEW") Regulations for the Protection of Human Subjects (45 C.F.R. 46), may be sampled to determine whether further inquiry into experiments is warranted. Further inquiry into experiments conducted after May 30, 1974, may be pursued if the Advisory Committee determines, with the concurrence of the Human Radiation Interagency Working Group, that such inquiry is warranted.

(b)(1) The Advisory Committee shall determine the ethical and scientific standards and criteria by which it shall evaluate human radiation experiments, as set forth in paragraph (a) of this section. The Advisory Committee shall consider whether (A) there was a clear medical or scientific purpose for the experiments; (B) appropriate medical follow-up was conducted; and (C) the experiments' design and administration adequately met the ethical and scientific standards, including standards of informed consent, that prevailed at the time of the experiments and that exist today.

(2) The Advisory Committee shall evaluate the extent to which human radiation experiments were consistent with applicable ethical and scientific standards as determined by the Committee pursuant to paragraph (b)(1) of this section. If deemed necessary for such an assessment, the Committee may carry out a detailed review of experiments and associated records to the extent permitted by law.

(3) If required to protect the health of individuals who were subjects of a human radiation experiment, or their descendants, the Advisory Committee may recommend to the Human Radiation Interagency Working Group that an agency notify particular subjects of an experiment, or their descendants, of any potential health risk or the need for medical follow-up.

(4) The Advisory Committee may recommend further policies, as needed, to ensure compliance with recommended ethical and scientific standards for human radiation experiments.

(5) The Advisory Committee may carry out such additional functions as the Human Radiation Interagency Working Group may from time to time request.

Sec. 3. *Administration.* (a) The heads of executive departments and agencies shall, to the extent permitted by law, provide the Advisory Committee with such information as it may require for purposes of carrying out its functions.

(b) Members of the Advisory Committee shall be compensated in accordance with Federal law. Committee members may be allowed travel expenses, including per diem in lieu of subsistence, to the extent permitted by law for persons serving intermittently in the government service (5 U.S.C. 5701–5707).

(c) To the extent permitted by law, and subject to the availability of appropriations, the Department of Energy shall provide the Advisory Committee with such funds as may be necessary for the performance of its functions.

Sec. 4. *General Provisions.* (a) Notwithstanding the provisions of any other Executive order, the functions of the President under the Federal Advisory Committee Act that are applicable to the Advisory Committee, except that of reporting annually to the Congress, shall be performed by the Human Radiation Interagency Working Group, in accordance with the guidelines and procedures established by the Administrator of General Services.

(b) The Advisory Committee shall terminate 30 days after submitting its final report to the Human Radiation Interagency Working Group.

(c) This order is intended only to improve the internal management of the executive branch and it is not intended to create any right, benefit, trust, or responsibility, substantive or procedural, enforceable at law or equity by a party against the United States, its agencies, its officers, or any person.

THE WHITE HOUSE,
January 15, 1994.

[FR Doc. 94–1531
Filed 1–18–94: 4:37 pm]
Billing code 3195–01–P

CHARTER

ADVISORY COMMITTEE ON HUMAN RADIATION EXPERIMENTS

1.   **Committee's Official Designation**

Advisory Committee on Human Radiation Experiments (the
"Advisory Committee" or "Committee").

2.   **Authority**

Executive Order No. 12891.

3.   **Objectives and Scope of Activities**

There has been established a Human Radiation Interagency
Working Group (the "Interagency Working Group"), the members
of which include the Secretary of Energy, the Secretary of
Defense, the Secretary of Health and Human Services, the
Secretary of Veterans Affairs, the Attorney General, the
Administrator of the National Aeronautics and Space
Administration, the Director of Central Intelligence, and
the Director of the Office of Management and Budget. As set
forth in section 4 of this Charter, the Advisory Committee
shall provide to the Interagency Working Group advice and
recommendations on the ethical and scientific standards
applicable to human radiation experiments carried out or
sponsored by the United States Government. As used herein,
"human radiation experiments" means:

(1)   Experiments on individuals involving intentional
      exposure to ionizing radiation. This category does not
      include common and routine clinical practices, such as
      established diagnosis and treatment methods, involving
      incidental exposures to ionizing radiation.

(2)   Experiments involving intentional environmental
      releases of radiation that (A) were designed to test
      human health effects of ionizing radiation; or (B) were
      designed to test the extent of human exposure to
      ionizing radiation.

Consistent with the provisions set forth in section 4 of
this Charter, the Advisory Committee also shall provide
advice, information and recommendations on the following
experiments:

1

(1) The experiment into the atmospheric diffusion of radioactive gases and test of detectability, commonly referred to as "the Green Run test," by the former Atomic Energy Commission (AEC) and the Air Force in December 1949 in Hanford, Washington;

(2) Two radiation warfare field experiments conducted at the AEC's Oak Ridge office in 1948 involving gamma radiation released from non-bomb point sources at or near ground level;

(3) Six tests conducted during 1949-1952 of radiation warfare ballistic dispersal devices containing radioactive agents at the U.S. Army's Dugway, Utah site;

(4) Four atmospheric radiation-tracking tests in 1950 at Los Alamos, New Mexico; and

(5) Any other similar experiments which may later be identified by the Interagency Working Group.

The Advisory Committee shall review experiments conducted from 1944 to May 30, 1974. Human radiation experiments undertaken after May 30, 1974, the date of issuance of the Department of Health, Education and Welfare Regulations for the Protection of Human Subjects (45 C.F.R. 46), may be sampled to determine whether further inquiry into experiments is warranted. Further inquiry into experiments conducted after May 30, 1974, may be pursued if the Advisory Committee determines, with the concurrence of the Interagency Working Group, that such inquiry is warranted.

4. **Description of Duties for Which Committee is Responsible**

The duties of the Advisory Committee are solely advisory and shall be:

a. The Advisory Committee shall determine the ethical and scientific standards and criteria by which it shall evaluate human radiation experiments, as set forth in section 3 of this Charter. The Advisory Committee shall consider whether (A) there was a clear medical or scientific purpose for the experiments; (B) appropriate medical follow-up was conducted; and (C) the experiments' design and administration adequately met the ethical and scientific standards, including standards of informed consent, that prevailed at the time of the experiments and that exist today.

2

b.   The Advisory Committee shall evaluate the extent to which human radiation experiments were consistent with applicable ethical and scientific standards as determined by the Committee pursuant to paragraph (a) of this section. If deemed necessary for such an assessment, the Advisory Committee may carry out a detailed review of experiments and associated records to the extent permitted by law.

c.   If required to protect the health of individuals who were subjects of a human radiation experiment, or their descendants, the Advisory Committee may recommend to the Interagency Working Group that an agency notify particular subjects of an experiment, or their descendants, of any potential health risk or the need for medical follow-up.

d.   The Advisory Committee may recommend further policies, as needed, to ensure compliance with recommended ethical and scientific standards for human radiation experiments.

e.   The Advisory Committee may carry out such additional functions as the Interagency Working Group may from time to time request.

5.   **To Whom the Advisory Committee Reports**

The Advisory Committee shall report to the Interagency Working Group.

The Advisory Committee shall submit its final report to the Interagency Working Group within one year of the date of the first meeting of the Advisory Committee, unless such period is extended by the Interagency Working Group. The Advisory Committee shall issue an interim report not more than six months after the date of the first meeting of the Advisory Committee. That interim report shall advise the Interagency Working Group on the status of the Advisory Committee's proceedings and the likelihood that the Committee will be able to complete its duties within one year of the date of the first meeting of the Advisory Committee.

6.   **Duration and Termination Date**

The Advisory Committee shall terminate thirty days after submission of its final report to the Interagency Working Group. This Charter shall expire one year plus thirty days after the first meeting of the Advisory Committee, subject to renewal and extension by the President.

3

7. **Agency responsible for providing financial and <u>administrative support to the Advisory Committee</u>**

Financial and administrative support shall be provided by the Department of Energy.

8. **Estimated Annual Operating Costs**

$3 million.

9. **Estimated Number and Frequency of Meetings**

The Advisory Committee shall meet as it deems necessary to complete its functions.

10. **Subcommittee(s)**

To facilitate functioning of the Advisory Committee, subcommittee(s) may be formed. The objectives of the subcommittee(s) are to make recommendations to the Advisory Committee with respect to matters related to the responsibilities of the Advisory Committee. Subcommittees shall meet as the Advisory Committee deems appropriate.

11. **Members**

Up to a maximum of fifteen Advisory Committee members shall be appointed by the President for a term of one year, which may be extended by the President. Committee members shall be compensated in accordance with federal law. Committee members may be allowed travel expenses, including per diem in lieu of subsistence, to the extent permitted by law for persons serving intermittently in the government service (5 U.S.C. §§ 5701-5707).

12. **Chairperson**

The President shall designate a Chairperson from among the members of the Advisory Committee.

4

Appendices

Acronyms and Abbreviations

ACBM — Advisory Committee for Biology and Medicine (a civilian advisory panel established in late 1947 to advise AEC's DBM on various aspects of biomedical research; dissolved in 1974)

ACR — American College of Radiology (professional society)

AEB — Army Epidemiological Board (established in 1942; through a series of various commissions, whose members were civilian health professionals, sponsored studies of infectious diseases of interest to military; succeeded by Armed Forces Epidemiological Board in 1949)

AEC — Atomic Energy Commission (established by the Atomic Energy Act of 1946 and inherited most functions of the MED; succeeded in 1974 by ERDA and NRC)

AFEB — Armed Forces Epidemiological Board (1949 successor to AEB)

AFMPC — Armed Forces Medical Policy Council (established by the secretary of defense in January 1951; formerly the Office of Medical Services [OMS]; members included a civilian physician as chairman, other civilians from medicine or related fields, and the surgeons general of the three services; developed basic medical and health policies for DOD and reviewed the medical and health aspects of the policies, plans, and programs of other DOD agencies; succeeded by the ASD [H&M] in late 1953)

AFPC — Armed Forces Policy Council (established under National Security Act of 1947, this panel advised the secretary of defense on broad policy matters and specific issues as requested; its initial members included the secretary and deputy secretary of defense; the secretaries of the Air Force, Army, and Navy; the chairman of the JCS; chiefs of staff of the Air Force and the Army; and chief of naval operations)

AFSWP — Armed Forces Special Weapons Project (established by the secretaries of war and the Navy in January of 1947; inherited certain functions of the MED in the areas of nuclear weapons development, testing, storage, and training of personnel; succeeded by DASA in 1958)

AMA — American Medical Association (professional society)

ANL — Argonne National Laboratory (established in 1946 and operated by the University of Chicago; inherited many of the facilities and functions of Met Lab; one of the three original national laboratories, the others are BNL and ORNL, established in 1946 and 1947, respectively)

AR   Army regulation (policy directive)

ASD (H&M)   assistant secretary of defense (health and medicine) (succeeded the AFMPC in 1953; provided advice and assistance on health and medical aspects of DOD policies, plans, and programs and collaborated with ASD [R&D] in the development of policies and the review of requirements for biomedical research by DOD)

ASD (R&D)   assistant secretary of defense (research and development) (replaced the RDB in 1953; provided advice and assistance to the secretary of defense on R&D policies, plans, and programs, developed an integrated DOD R&D program, assigned specific responsibilities for R&D programs where unnecessary duplication would be eliminated by such action, examined the interaction of R&D and strategy and advised the JCS, and reviewed proposed R&D budgets and made recommendations thereon; succeeded by ASD [R&E] in 1957)

ASD (R&E)   assistant secretary of defense (research and engineering) (combined the offices of ASD [R&D] and the assistant secretary of defense [engineering]; succeeded by the director of defense research and engineering [DDR&E] in 1958)

BNL   Brookhaven National Laboratory (established by the MED in 1946 and operated by the Associated Universities; created to facilitate cooperation between universities and the federal government in performing research in physics and nuclear science)

BuMed   Bureau of Medicine and Surgery (operates Navy's hospitals and medical research centers, as well as sponsoring most of its outside biomedical research)

CDC   Centers for Disease Control and Prevention

CEQ   Council on Environmental Quality (three-member panel within EOP, established by National Environmental Policy Act; has environmental oversight responsibilities)

C.F.R.   Code of Federal Regulations (compilation of federal regulations available from the Government Printing Office and in many public and private libraries)

CHR   Center for Human Radiobiology (created within Argonne National Laboratory in the late 1960s)

CMR   Committee on Medical Research (established in 1942 under OSRD to sponsor nonradiation-related biomedical research of interest to the military; disestablished in late 1946)

CMS   Committee on Medical Sciences (RDB committee in existence from 1948 to late 1953 that reviewed, evaluated, and made recommendations on all biomedical research conducted by or for DOD entities; members included both civilian and military health professionals; from late 1953 to 1957, an advisory group to ASD [R&D] and ASD [R&E], functions transferred to the Committee on Science in 1957)

DASA — Defense Atomic Support Agency (1958 AFSWP successor)

DBM — Division of Biology and Medicine (established in early 1948 to direct and coordinate all AEC biomedical research activities; became the Biological and Environmental Research Division with the creation of ERDA in 1974)

DDR&E — director of defense research and engineering (succeeded ASD [R&E] in 1958, reviewing, evaluating, and directing all R&D conducted by or for DOD)

DHEW — Department of Health, Education, and Welfare (DHHS predecessor, established in 1953)

DHHS — Department of Health and Human Services (1980 DHEW successor; the principal federal agency charged with advancing the health of Americans and providing essential human services)

DNA — Defense Nuclear Agency (1971 successor to DASA)

DOD — Department of Defense (new name established in 1949 for the National Military Establishment, which had been created under the National Security Act of 1947 to replace the War and Navy Departments)

DOE — Department of Energy (1977 successor to ERDA)

EOP — Executive Office of the President

EPA — Environmental Protection Agency (federal agency charged with monitoring the quality of the environment)

ERDA — Energy Research and Development Administration (succeeded AEC in 1974, with responsibilities for civilian nuclear power and isotope licensing and distribution transferred to the newly created Nuclear Regulatory Commission; succeeded by DOE in 1977)

FDA — Food and Drug Administration (established as part of the Department of Agriculture in 1862; became a regulatory agency in 1906; transferred to Federal Security Agency in 1940, which became HEW in 1953; became part of PHS in 1968; enforces laws to ensure the safety and efficacy of foods, food additives, drugs, biologics, cosmetics, and medical devices)

HEDR — Hanford Environmental Dose Reconstruction (established by DOE, later transferred to Centers for Disease Control, this project assesses human exposures to ionizing radiation due to radioactive emissions from the Hanford, Washington, plutonium-production plant)

HEW — See *DHEW*

HHS — See *DHHS*

HURB — Human Use Review Board (within Army surgeon general's office, reviews proposed research involving greater than minimal risk)

ICRP — International Commission on Radiological Protection (international body of scientific experts, created in 1928, which functions on an international basis as the NCRP does within the United States)

IG — inspector general (office in federal departments and agencies that conducts and supervises audits, investigations, and inspections of department and agency operations)

INEL — Idaho National Engineering Laboratory (originally named the National Reactor Testing Station, INEL was established in 1949 as a remote site to work with experimental civilian and military reactors)

IRB   institutional review board (See Glossary)

JCAE   Joint Committee on Atomic Energy (congressional committee established under the Atomic Energy Act of 1946 to oversee AEC; disestablished in 1974).

JCS   Joint Chiefs of Staff

LANL   Los Alamos National Laboratory (established as Los Alamos Scientific Laboratory by the MED in 1943; operated by the University of California since it was established; originally created to design and build a fission bomb; designated a national laboratory in 1977)

LBL   Lawrence Berkeley Laboratory (1971 successor to UCRL)

LLNL   Lawrence Livermore National Laboratory (successor to the Livermore weapons lab which had been established in 1952 as the second weapons lab and had been operated by UCRL)

MED   Manhattan Engineer District, also popularly known as the Manhattan Project (established in 1942 within the U.S. Army to build the atomic bomb; functions transferred to AEC and AFSWP in 1947)

MetLab   Metallurgical Laboratory (University of Chicago-based MED laboratory established in 1942; most functions transferred to ANL in 1946)

MKULTRA   A domestic CIA program in the 1950s and 1960s involving human experimentation to investigate control of human behavior through the use of chemical, biological, and other means (including psychoactive drugs, psychology, and possibly radiation)

MLC   Military Liaison Committee (established under the Atomic Energy Act of 1946; chaired by a civilian, its other members included two senior officers from each of the three services; advised the secretary of defense and Joint Chiefs of Staff on priorities for DOD atomic energy R&D, which component should conduct it, and liaisoned with the AEC on DOD activities)

MPA   multiple-project assurance (research institution's assurance, covering a number of different research projects, to OPRR or the funding agency that the institution will comply with federal human subjects protection policy)

MPBB   maximum permissible body burden (amount of radioactivity that, if deposited in the body, is estimated to deliver the highest allowable dose rate to the most critical organ over a defined period of time)

NASA   National Aeronautics and Space Administration (established in 1958; agency responsible for the development of space aviation, technology, and exploration)

NCI   National Cancer Institute (established in 1937, part of NIH)

NCRH   National Center for Radiological Health (1967 successor to PHS's radiological health and safety program; conducted biological and epidemiological research on radiation effects)

NCRP   National Committee on Radiological Protection and

Measurements (1946 successor to Advisory Committee on X-ray and Radium Protection, known after 1964 as National Council on Radiation Protection and Measurements; an independent body of scientific experts, it recommends limits for occupational exposure that are widely followed and periodically issues reports on special topics)

NEPA (1) Nuclear Energy for the Propulsion of Airplanes (1946–1961 Air Force program for developing nuclear-powered bomber) (2) National Environmental Policy Act of 1969 (Federal statute requiring that the U.S. government consider and publicize the environmental impact of its actions)

NIH National Institutes of Health (part of PHS; begun as a one-room Laboratory of Hygiene in 1887, now the world's largest biomedical research facility; based in Bethesda, Maryland; conducts and sponsors research dedicated to health promotion and the discovery of causes, prevention, and cure of diseases)

NIOSH National Institute for Occupational Safety and Health (part of CDC)

NRC Nuclear Regulatory Commission (established in 1974 as a successor to AEC to run civilian nuclear power program and radioisotope licensing and distribution program)

NTPR Nuclear Test Personnel Review (DNA program established in 1978 to, among other things, compile unclassified histories of atmospheric nuclear weapons tests, determine which DOD civilian and military personnel were present at these tests, and establish their exposure levels at the tests)

NYOO New York Operations Office (AEC regional office)

OPRR Office for Protection from Research Risks (established within NIH in 1966 to educate investigators and others about research ethics and to implement regulations for the protection of human and animal subjects)

ORAU Oak Ridge Associated Universities (1966 successor to ORINS)

ORINS Oak Ridge Institute of Nuclear Studies (established in 1946, and operated initially by a consortium of fourteen Southeastern universities under AEC contract beginning in 1947; a research and training site for users of radioisotopes in medicine and site of biomedical research)

ORISE Oak Ridge Institute for Science and Education (1991 successor to ORAU)

ORO Oak Ridge Operations Office (AEC/ERDA/DOE regional office)

ORNL Oak Ridge National Laboratory (established in 1947, succeeding Clinton Labs; has conducted a wide range of research for AEC, ERDA, and DOE)

OSG Army Office of the Surgeon General (operates Army's hospitals and medical research centers, as well as sponsoring most of its outside biomedical research)

OSRD Office of Scientific Research and Development (through numerous committees, coordinated and directed all nonatomic energy R&D of the War and Navy Departments from 1942 to 1946; succeeded by the

Joint Research and Development Board)

PBI partial-body irradiation

PHS Public Health Service (the federal government's principal health agency, restructured three times since World War II, now one of five operating divisions of DHHS; functions to improve public health through the promotion of physical and mental health and the prevention of disease, injury, and disability)

R&D research and development

RDB Research and Development Board (reviewed, evaluated, and directed all research and development conducted by or for DOD; functions transferred to ASD [R&D] and ASD [R&E] in late 1953)

RDRC radioactive drug research committee (reviews proposed use of radioactive drugs within an institution)

RSC radiation safety committee (monitors radiation safety within an institution)

RW radiological warfare

SAM School of Aviation Medicine

(Air Force component; conducted radiobiology research beginning in the late 1940s; coordinated efforts with other government agencies)

TBI total-body irradiation

UCRL University of California Radiation Laboratory (lab established in 1936 by Ernest Lawrence on the Berkeley campus; conducted a wide range of research for the MED and AEC; operated the Livermore weapons lab from its establishment in 1952; redesignated the Lawrence Berkeley Laboratory in 1971)

UCSF University of California at San Francisco (biomedical research site)

U.S.C. United States Code (compilation of congressionally enacted laws available in many public and private libraries)

VA Department of Veterans Affairs (successor to 1930–1989 Veterans Administration)

WMA World Medical Association (professional organization; issued Helsinki Declaration in 1964)

Glossary

Terms in *italics* appear in the Glossary as separate entries.

Alpha radiation See *Ionizing radiation.*

Association In statistics, the correlation or relationship between one factor and one or more other pertinent factors as demonstrated by experimental data.

Atomic bomb An explosive device in which a large amount of energy is released through the nuclear *fission* of uranium or plutonium. The first atomic bomb test, known as the Trinity Shot, took place in the desert north of Alamogordo, New Mexico, on July 16, 1945. Several weeks later, an atomic bomb was used for the first time as an instrument of war, detonating over the Japanese cities of Hiroshima (August 6) and Nagasaki (August 9).

Atomic pile See *Nuclear reactor.*

Becquerel See *Units of radioactivity.*

Beta radiation See *Ionizing radiation.*

Biodistribution The pattern and process of a chemical substance's distribution through the body.

Biological dosimeter See *Dosimeter.*

Biopsy The removal and/or examination of tissues, cells, or fluids from a living body for the purposes of diagnosis or experimental tests.

Biophysics The application of physical principles and methods to the study of the structures of living organisms and the mechanics of life processes.

Body burden The amount of a radioactive material present in a body over a long time period. It is calculated by considering the amount of material initially present and the reduction in that amount due to elimination and radioactive decay.

It is commonly used in reference to *radionuclides* having a long biological *half-life.* A body burden that subjects the body's most sensitive organs to the highest dose of a particular *radionuclide* that regulators allow is known as a *maximum permissible body burden (MPBB).*

Bone marrow infusion The injection of bone marrow (an essential tissue producing red and white blood cells and platelets) into the body, used primarily to replace bone marrow destroyed by disease or in the course of *radiation* and other therapies for certain types of cancer.

Carcinogen A material that can initiate or promote the development of cancer. Well-known carcinogens include saccharine, nitrosamines found in cured meat, certain pesticides, and *ionizing radiation.*

Chain reaction The process by which the *fission* of a nucleus releases neutrons, causing other nuclei to undergo fission in turn. Both the *atomic bomb* and the *nuclear reactor* use a chain reaction to generate energy.

Clinical trial A research study involving human subjects, designed to evaluate the safety and effectiveness of new therapeutic and diagnostic treatments.

Common Rule The 1991 federal regulation that provides the basic procedures and principles that are to be followed in the conduct of human subject research sponsored by federal agencies.

Critical mass The amount of fissionable material (uranium 235 or plutonium 239) sufficient to sustain a nuclear *chain reaction.*

Curie See *Units of radioactivity.*

567

Cyclotron A device that uses alternating electric fields to accelerate subatomic particles (a particle smaller than an atom, such as an alpha particle or a proton). When these particles strike ordinary nuclei, *radioisotopes* are formed. For his work in developing the cyclotron in the early 1930s, Ernest Lawrence of the University of California received the 1939 Nobel Prize in Physics.

Deterministic effect An effect, such as kidney damage, whose severity increases with increasing dose of radiation or other agent.

Diagnostic procedure A method used to identify a disease in a living person.

Dosage The prescribed amount of medicine or other therapeutic agent administered to treat a given illness.

Dose In radiology, a measure of energy absorbed in the body from *ionizing radiation,* measured in *rad.*

Dose reconstruction The process of using information about an individual's past exposures to *ionizing radiation* as well as general knowledge about the behavior of radioactive materials in the human body and in the environment to estimate the *dose* of *radiation* that someone has received.

Dosimeter An instrument that measures the dose of *ionizing radiation. A biological dosimeter* is a biological or biochemical indicator of the effects of exposure, such as a change in blood chemistry or in blood count. A highly accurate *biological dosimeter* has yet to be found.

Dosimetry The measurement and calculation of radiation doses.

Endocrinology The study of the body's hormone-producing glands, such as the thyroid, pituitary, and adrenal glands, and the functions of the hormones they synthesize and secrete.

Epidemiology The study of the determinants (risk factors) and distribution of disease among populations.

Fallout *Radioactive* debris that falls to earth after a nuclear explosion.

Fission The division of an atomic nucleus into parts of comparable mass. Generally speaking, fission may occur only in heavier nuclei, such as *isotopes* of uranium and plutonium. *Atomic bombs* derive energy from the *fission* of uranium or plutonium.

Fission product An atom or nucleus that results from the *fission* of a larger nucleus.

Fusion The combining of two light atomic nuclei to form a single heavier nucleus, releasing energy. *Hydrogen bombs* derive a large portion of their energy from the fusion of hydrogen *isotopes.*

Gamma radiation See *Ionizing radiation.*

Genetic effects Changes in a person's germ calls (sperm or ova) that are transmissible to future generations. Such changes result from mutations in genes within the germ cells.

Gray see *Units of radioactivity.*

Half-life The average time required for one-half of the amount of radioactivity of a *radionuclide* to undergo *radioactive* decay. For material with a half-life of one week, half of the original amount of activity will remain after one week; half of that (one-quarter of the original amount) will remain after two weeks; and so on.

Health physics A branch of physics specializing in accurate measurement of agents, such as *ionizing radiation,* which can have effects on human health.

Hydrogen bomb (also known as a thermonuclear weapon) An explosive weapon that uses nuclear *fusion* to release energy stored in the nuclei of hydrogen *isotopes.* The high temperatures essential to *fusion* are attained by detonating an *atomic bomb* placed at the H-bomb's structural center. The United States tested the first hydrogen bomb in 1954 at the Pacific Test Site.

Institutional review board (IRB) Under the Common Rule, a local review board convened by any institution conducting federally sponsored human subject research, vested with the responsibility to review research proposals to ensure compliance with federal research regulations.

Internal emitter A radioisotope incorporated into a tissue in the body that decays in place and continuously exposes that tissue to ionizing radiation.

Ionization The process by which a neutral atom or molecule loses or gains electrons, thereby acquiring a net electrical charge. When charged, it is known as an ion.

Ionizing radiation Any of the various forms of radiant energy that causes *ionization* when it interacts with matter. The most common types are *alpha radiation,* made up of helium nuclei; *beta radiation,* made up of electrons; and *gamma* and *x radiation,* consisting of high-energy particles of light (photons).

Irradiation Exposure to radiation of any kind, especially *ionizing radiation.*

Isotope A species of nucleus with a fixed number of protons and neutrons. The term *isotope* is usually used to distinguish nuclear species of the same chemical element (i.e., those having the same number of protons, but different numbers of neutrons), such as iodine 127 and iodine 131.

Latency period The time between when an exposure occurs and when its effects are detectable as an injury or illness.

Maximum Permissible Body Burden (MPBB) see *Body burden*

Metabolism The manner in which a substance is acted upon (taken up, converted to other substances, and excreted) by various organs of the body.

Natural background radiation *Ionizing radiation* that occurs naturally. Its principal sources are cosmic rays from outer space, *radionuclides* in the human body, and radon gas (a decay product of natural uranium in the earth's crust).

Nuclear medicine A branch of medicine specializing in the use of *radionuclides* for diagnostic and therapeutic purposes.

Nuclear reactor A device containing *fissionable* material in sufficient quantity and suitable arrangement to maintain a controlled, self-sustaining nuclear *chain reaction.*

Nuclide A type of nucleus with a fixed number of protons and neutrons. The term *nuclide* is usually used to distinguish nuclear species of different chemical elements (i.e., those having different numbers of protons and neutrons), such as iodine 127 and uranium 235.

Partial-Body Irradiation (PBI) Exposure of part of the body to external *radiation.*

Permissible dose In the judgment of a regulatory or advisory body, such as the National Committee on Radiation Protection, the amount of *radiation* that may be received by an individual within a specified period.

Principal investigator The scientist or scholar with primary responsibility for the design and conduct of a research project.

Protocol The formal design or plan of an experiment or research activity; specifically, the plan submitted to an *institutional review board* for review and to a government agency for research support. Protocols include a description of the research design or methodology to be employed, the eligibility requirements for prospective subjects and controls, the treatment regimen(s), and the methods of analysis to be performed on the collected data.

Rad See *Units of radiation*

Radiation The emission of waves transmitting energy through space or a material medium, such as water. Light, radio waves, and *x rays* are all forms of *radiation.*

Radiation biology See *radiobiology.*

Radiation oncology A branch of medicine specializing in the treatment of cancer with *radiation.* Radiation therapy and *radiotherapy* are equivalent terms.

Radiation sickness Acute physical illness caused by exposure to *doses* of *ionizing radiation* large enough to cause toxic reactions. This can include symptoms such as nausea, diarrhea, headache, lethargy, and fever.

Radioactive decay The process by which the nucleus of a radioactive isotope decomposes and releases *radioactivity.* For example, carbon 14 (a *radioisotope* of carbon) decays by losing a beta particle, thereby becoming nitrogen 14, which is unstable.

Radioactivity The decay of unstable nuclei through the emission of *ionizing radiation.* The resulting nucleus may itself be unstable and undergo *radioactive decay.* The process stops only when the decay product is stable.

Radiobiology Branch of biology specializing in the study of the effects of *radiation* on biological molecules, cells, tissues, and whole organisms, including humans. *Radiobiology* seeks to discover the molecular changes responsible for radiation effects such as cancer induction, genetic changes, and cell death.

Radiogenic A term used to identify conditions observed to be caused by exposure to *ionizing radiation*, such as certain kinds of cancer.

Radioisotope A radioactive isotope. Radioisotopes are used in medical research as *tracers*. See also *isotope, nuclide*, and *radionuclide*.

Radiological weapons Weapons that use *radioactive* materials to cause *radiation* injury.

Radionuclide A *radioactive nuclide*. Often used to distinguish *radioisotopes* of different chemical elements, such as iodine 131 and uranium 239.

Radiopharmaceuticals Drugs (compounds or materials) that may be labeled or tagged with a *radioisotope*. In many cases, these materials function much like materials found in the body and do not produce special pharmacological effects. The principal risk associated with these materials is the consequent exposure of the body or certain tissues to *radiation*.

Radioresistance The degree of resistance of organisms or tissues to the harmful effects of *ionizing radiation*.

Radiosensitivity The degree of sensitivity of organisms or tissues to the harmful effects of *ionizing radiation*.

Radiotherapy See *radiation oncology*.

Rem See *Units of radiation*.

Rep See *Units of radiation*.

Roentgen See *Units of radiation*.

Tolerance dose See *Permissible dose*.

Total-Body Irradiation (TBI) Exposure of the entire body to external *radiation*.

Tracer A distinguishable substance, usually *radioactive*, administered to determine the distribution and/or *metabolism* of materials in the body. In 1923, George Hevesy was the first investigator to use an *isotope* (radioactive thorium) in metabolic studies, exploring lead transport in the bean plant. Metabolic studies proliferated after World War II, when with the development of the *cyclotron*, radioisotopes of various atoms became more widely avail-able. *Isotopes* commonly used as *tracers* today in-clude carbon 14, iodine 131 and phosphorus 32.

Transuranic elements Radioactive elements with atomic numbers (i.e., the number of protons in the nucleus) greater than 92. Only two of these elements (plutonium in minute amounts and neptunium) occur in nature; the others are produced in minute amounts through the radioactive decay of uranium. The first transuranic elements were discovered as synthetic *radioisotopes* at the University of California at Berkeley and the Argonne National Laboratory in the 1930s and 1940s.

Units of radiation The basic unit of *radiation* exposure is the *roentgen*, named after Wilhelm Roentgen (discoverer of x rays). It is a measure of *ionization* in air, technically equal to one ESU (electrostatic unit) per cubic centimeter, due to radiation. A *rep (roentgen* equivalent physical) is an archaic measure of skin exposure to a dose of *beta radiation* having an effect equivalent to 1 *roentgen* of x rays. The basic unit of *radiation* absorbed by the body is the *rad*, technically equal to 100 ergs (energy unit) per gram of exposed tissue. One *roentgen* corresponds to roughly 0.95 *rad*. The *rem (roentgen* equivalent in man) is a unit of effective dose, a dose corrected for the varying biological effectiveness of various types of *ionizing radiation*. The currently accepted unit of radiation is the *gray (Gy)*, the International System unit of absorbed dose, equal to the energy imparted by ionizing radiation to a mass of matter corresponding to one joule per kilogram.

Units of Radioactivity The *becquerel (Bq)*, named after the physicist Henri Becquerel (the discoverer of *radioactivity*), is a measure of radioactivity equal to one atomic disintegration per second. The *curie (Ci)*, whose name honors the French scientists Marie and Pierre Curie (the discoverers of radium), is a standard based on the radioactivity of 1 gram of radium. It is equal to 3.7×10^{10} *becquerels*.

X rays Invisible, highly penetrating electromagnetic *radiation* of a much shorter wavelength than visible light, discovered in 1895 by Wilhelm C. Roentgen. Most applications of X rays are based on their ability to pass through matter. They are dangerous in that they can destroy living tissue, causing severe skin burns on human flesh exposed for too long a time. This property is applied in x-ray therapy to destroy diseased cells. See *Ionizing radiation*.

Selected Bibliography

The books and articles listed below provide additional reading in scientific, ethical, and historical literature. A comprehensive bibliography of primary and secondary sources used in this report appears in the supplemental volume *Sources and Documentation*.

RESEARCH ETHICS

Annas, George J., and Michael A. Grodin, eds., *The Nazi Doctors and the Nuremberg Code: Human Rights in Human Experimentation*. New York: Oxford University Press, 1992.

Beecher, Henry K. "Ethics and Clinical Research." *New England Journal of Medicine* 274 (1966): 1354–1360.

Bok, Sissela. *Secrets: On the Ethics of Concealment and Revelation*. New York: Vintage Books, 1989.

Couglin, Steven S., and Tom L. Beauchamp. *Ethics in Epidemiology*. New York: Oxford University Press, forthcoming.

Faden, Ruth, and Tom L. Beauchamp. *A History and Theory of Informed Consent*. New York: Oxford University Press, 1986.

Gray, Bradford H. *Human Subjects in Medical Experimentation: A Sociological Study of the Conduct and Regulation of Clinical Research*. New York: John Wiley and Sons, 1975.

Grodin, Michael A., and Leonard H. Glantz, eds., *Children as Research Subjects: Science, Ethics, and Law*. New York: Oxford University Press, 1994.

Institute of Medicine, Committee on the Ethical and Legal Issues Relating to the Inclusion of Women in Clinical Studies. *Women and Health Research: Ethical and Legal Issues of Including Women in Clinical Studies*. Vol. 1. Washington, D.C.: National Academy Press, 1994.

Katz, Jay. *Experimentation with Human Beings: The Authority of the Investigator, Subject, Professions, and State in the Human Experimentation Process*. New York: Russell Sage Foundation, 1972.

Katz, Jay. *The Silent World of Doctor and Patient*. New York: Free Press, 1984.

Lederer, Susan. *Subjected to Science: Human Experimentation in America after the Second World War*. Baltimore: Johns Hopkins University Press, 1995.

Levine, Robert J. *Ethics and Regulation of Clinical Research*. 2d ed. Baltimore: Urban and Schwarzenberg, 1986.

Lifton, Robert Jay. *Nazi Doctors: Medical Killing and the Psychology of Genocide*. New York: Basic Books, 1986.

Orlans, F. Barbara. *In the Name of Science: Issues in Responsible Animal Experimentation*. New York: Oxford University Press, 1993.

Rothman, David J. *Strangers at the Bedside: A History of How Law and Bioethics Transformed Medical Decision Making*. New York: Basic Books, 1991.

Veatch, Robert M. *The Patient as Partner: A Theory of Human-Experimentation Ethics*. Bloomington: Indiana University Press, 1987.

RADIATION-RELATED SCIENCE

Brucer, Marshall. *A Chronology of Nuclear Medicine*. St. Louis, Mo.: Heritage Publications, 1990.

Committee on the Biological Effects of Ionizing Radiation. Board on Radiation Effects Research. Commission on Life Sciences. National Research Council. *Health Risks of Radon and other Internally Deposited Alpha-Emitters: BEIR IV*. Washington, D.C.: National Academy Press, 1988.

Committee on Biological Effects of Ionizing Radiation, Board on Radiation Effects Research, National Research Council. *Health Effects of Exposure to Low Levels of Ionizing Radiation: BEIR V*. Washington, D.C.: National Academy Press, 1990.

Conklin, James J., and Richard I. Walker, eds., *Military Radiobiology*. Orlando, Fla.: Academic Press, 1987.

Eisenberg, Ronald L. *Radiology: An Illustrated History*. St. Louis: Mosby-Year Book, 1992.

Gofman, John W. *Radiation and Human Health.* San Francisco: Sierra Club Books, 1981.

Hennekens, Charles H., and Julie E. Buring. *Epidemiology in Medicine.* Edited by Sherry L. Mayrent. Boston: Little, Brown and Company, 1987.

Martin, Alan, and Samuel A. Harbison, eds., *An Introduction to Radiation Protection.* 3d ed. New York: Chapman and Hall, 1986.

McAfee, J. G., R. T. Kopecky, and P. A. Frymoyer. "Nuclear Medicine Comes of Age: Its Present and Future Roles in Diagnosis." *Radiology* (1990): 609–620.

Mettler, Fred A., Jr., and Arthur C. Upton. *Medical Effects of Ionizing Radiation.* 2d ed. Philadelphia: W. B. Saunders, 1995.

Schapiro, Jacob. *Radiation Protection: A Guide for Scientists and Physicians.* 3d ed.; Cambridge, Mass.: Harvard University Press, 1990.

United Nations Scientific Committee on the Effects of Atomic Radiation. "Sources and Effects of Ionizing Radiation." *UNSCEAR 1993 Report to the General Assembly, with Scientific Annexes.* New York: United Nations, 1993.

Upton, Arthur C. "The Biological Effects of Low-Level Ionizing Radiation." *ScientificAmerican* (February 1982): 41–49.

HISTORY AND BIOGRAPHY

Bradley, David. *No Place to Hide.* Boston: Little, Brown and Company, 1948.

Bush, Vannevar. *Pieces of the Action.* New York: William Morrow, 1970.

Conard, Robert A. *Fallout: The Experiences of a Medical Team in the Care of a Marshallese Population Accidently Exposed to Fallout Radiation.* New York: Brookhaven National Laboratory, 1992. Available from National Technical Information Service, U.S. Department of Commerce, 5285 Port Royal Road, Springfield, VA 22161.

D'Antonio, Michael. *Atomic Harvest: Hanford and the Lethal Toll of America's Nuclear Arsenal.* New York: Crown Publishing, 1993.

Divine, Robert A. *Blowing on the Wind: The Nuclear Test Ban Debate 1954–1960.* New York: Oxford University Press, 1978.

Eichstaedt, Peter H. *If You Poison Us: Uranium and Native Americans.* Santa Fe, N.M.: Red Crane Books, 1994.

Eisenbud, Merril. *An Environmental Odyssey: People, Pollution and Politics in the Life of a Practical Scientist.* Seattle: University of Washington Press, 1990.

Fradkin, Phillip L. *Fallout: An American Nuclear Tragedy.* Tucson: University of Arizona Press, 1989.

Gallagher, Carole. *American Ground Zero: The Secret Nuclear War.* Boston: MIT, 1993.

Gerber, Michele Stenehjem. *On the Home Front: The Cold War Legacy of the Hanford Nuclear Site.* Lincoln: University of Nebraska Press, 1992.

Glasser, Otto. *Wilhelm Conrad Roentgen and the Early History of the Roentgen Rays.* San Francisco: Norman Publishing, 1993.

Hacker, Barton C. *The Dragon's Tail: Radiation Safety in the Manhattan Project, 1942–1946.* Berkeley: University of California Press, 1987. A history of radiation safety in the Manhattan Project, 1942–1946.

———. *Elements of Controversy: The Atomic Energy Commission and Radiation Safety in Atomic Weapons Testing, 1947–1974.* Berkeley: University of California Press, 1994.

Hershberg, James. *Harvard to Hiroshima and the Making of the Nuclear Age.* New York: Alfred A. Knopf Publications, 1993.

Hewlett, Richard G., and Oscar E. Anderson, Jr. *The New World: A History of the United States Atomic Energy Commission, Volume 1: 1939–1946.* University Park: Pennsylvania State University Press, 1962.

———and Francis Duncan. *Atomic Shield: A History of the Unites States Atomic Energy Commission, Volume II: 1947–1952.* University Park: Pennsylvania State University Press, 1969.

———and Jack M. Moll. *Atoms for Peace and War: Eisenhower and the Atomic Energy Commission: 1953–1961.* Berkeley: University of California Press, 1989.

Johnson, Charles W., and Charles O. Johnson. *City Behind a Fence: Oak Ridge, Tennessee, 1942–1946.* Knoxville: University of Tennessee Press, 1981.

Kathren, Ronald L., Jerry B. Gough, and Gary T. Benefiel, eds. *The Plutonium Story: The Journals of Professor Glenn T. Seaborg, 1939–1946.* Columbus, Ohio: Battelle Press, 1994.

Kevles, Daniel J. *The Physicists: The History of a Scientific Community in Modern America.* New York: Vintage Books, 1979.

Lindee, Susan. *Suffering Made Real: American Science and the Survivors of Hiroshima.* Chicago: University of Chicago Press, 1994.

Mazuzan, George T., and J. Samuel Walker. *Controlling the Atom: The Beginnings of Nuclear Regulation, 1946–1962.* Berkeley: University of California Press, 1992.

Price, Don K. *The Scientific Estate.* New York: Oxford University Press, 1965.

Quinn, Susan. *Marie Curie: A Life.* New York: Simon & Schuster, 1995.

Rhodes, Richard. *The Making of the Atomic Bomb.* New York: Simon & Schuster, 1986.

———. *Dark Sun: The Making of the Hydrogen Bomb.* New York: Simon & Schuster, 1995.

Rosenberg, Howard L. *Atomic Soldiers: American Victims of Nuclear Experiments.* Boston: Beacon Press, 1980.

Smyth, Henry DeWolf. *Atomic Energy for Military Purposes: The Official Report on the Development of the Atomic Bomb under the Auspices of the United States Government: 1940–1945.* Stanford, Calif.: Stanford University Press, 1989.

Stannard, J. Newell. *Radioactivity and Health: A History.* Oak Ridge, Tenn.: Office of Scientific and Technical Information, 1988.

Starr, Paul. *The Transformation of American Medicine.* New York: Basic Books, 1982.

Taylor, Telford. *The Anatomy of the Nuremberg Trials: A Personal Memoir.* New York: Alfred A. Knopf, 1992.

Udall, Stewart L. *The Myths of August: A Personal Exploration of Our Tragic Cold War Affair with the Atom.* New York: Pantheon Books, 1994.

Walker, J. Samuel. *Containing the Atom: Nuclear Regulation in a Changing Environment, 1962–1971.* Berkeley: University of Califonia Press, 1992.

Weart, Spencer R. *Nuclear Fear: A History of Images.* Cambridge, Mass.: Harvard University Press, 1988.

Weisgall, Jonathan. *Operation Crossroads.* Annapolis, Md.: Naval Institute Press, 1994.

Public Comment Participants

Unless otherwise noted, full committee meetings, with opportunity for public comment, took place in Washington, D.C. There was also one full committee meeting in San Francisco. In addition, the Committee convened panels of its members to take testimony in Cincinnati, Spokane, Sante Fe, and Knoxville on the dates listed. Where known, cities and affiliations are noted.

April 21–22, 1994
Gwendon Plair, Concerned Relatives of Cancer Study Patients

May 18–19, 1994
E. Cooper Brown, Executive Commission, Task Force on Radiation and Human Rights
H. W. Cummins, Human Radiation Experiments Litigation Project
Fred Allingham, National Association of Radiation Survivors
Daryl Kimball, Physicians for Social Responsibility

June 13–14, 1994
Tod Ensign, Citizen Soldier, NY
John McCarthy, Sacramento Radiation Survivors Group
Thomas Smith, National Association of Radiation Survivors
Pat Broudy, National Association of Atomic Veterans

July 5–6, 1994
Wilfred Kendall, Representative of the Embassy of the Republic of the Marshall Islands
Tony deBrum, Representative of the Embassy of the Republic of the Marshall Islands
Jonathan Weisgall, Attorney representing Bikini Islands

E. Cooper Brown, Executive Commission, Task Force on Radiation and HumanRights*

July 25–26, 1994
Stewart Udall, former U.S. Secretary of the Interior
Eugene Trani, Virginia Commonwealth University
Hermes Kontos, Virginia Commonwealth University
John Jones, Virginia Commonwealth University
Chris Zucker, Disability Advocates of New York, Inc., Albany, NY
Pat Broudy, National Association of Atomic Veterans*
Dr. Oscar Rosen, National Association of Atomic Veterans
Catherine Variano, South Bend, ID
Janet Gordon, Citizen's Call, UT

September 12–13, 1994
Ruth Blaz, Hollywood, FL
Cliff Honicker, Environmental Health Studies Project, Knoxville, TN
Francis Brown, Southwick, MA
Tod Ensign, Citizen Soldier, NY*
Pat Broudy, National Association of Atomic Veterans*

San Francisco, October 11–13, 1994
Nancy Lynch, Santa Barbara, CA
Jackie Maxwell, Menlo Park, CA
Vernon Sousa, San Francisco, CA
Gwynne Borroughs, Chico, CA

*Indicates that the participant spoke at a previous meeting

Israel Torres, Niporno, CA
Audrey Hack, Union City, CA
Richard Harley, Bakersfield, CA
Donald Arbitlit, San Francisco, CA
Geoffrey Sea, International Radiation
  Survivors, Oakland, CA
Harold Bibeau, Portland, OR
Cheri Anderson, Placerville, CA
Tom Wilson, Placerville, CA
Michael Yesley, Los Alamos National
  Laboratory
Lynn Stembridge, Hanford Education
  Action League, Spokane, WA
Trisha Pritikin, Berkeley, CA
Lois Camp, Hanford Downwinder Health
  Effects Group
Darcy Thrall, Richland, WA
Dr. Bernard Lo, San Francisco, CA
Jackie Cabasso, Oakland, CA
Marylia Kelley, Livermore, CA

Cincinnati, October 21, 1994
U.S. Representative Rob Portman (OH)
Gwendon Plair, Concerned Relatives of
  Cancer Study Patients*
Doris Baker, Cincinnati, OH
Gloria Nelson
Richard Casey
Lisa Crawford, Fernald Residents for
  Environmental Safety and Health
Herbert Varin
Leslie Lynch
Professor Martha Stephens, University of
  Cincinnati
Bob Phillips
Lillian Pagano
Sherry Brabant
Otisteen Goodwin
Clifford Tidwell
Owen Thompson
Dr. Joseph Steger, President, University of
  Cincinnati
Stan Chesley, Former Chairman of the
  Board, University of Cincinnati
David Thompson, Attorney, Cincinnati
  lawsuits
Kenneth Kendall
Tom Wilkenson
Tom Row, Oak Ridge National Laboratory
Joe Larkins

Monica Ray
Gene Branham
Dorothy Sweety
Pat Wheeler
Katherine Hager
Robert Hager
H. W. Cummins, Radiation Health
  Effects Public Law Group*
Ruth Blaz, Hollywood, FL*
Manuel Blaz, Hollywood, FL
Jackie Kitrell, American Environmental
  Health Studies Project
Ann Hopkins
Mary Mueller
Daryl Kimball, Physicians for Social
  Responsibility*
Vina Colley, Portsmouth/Piketon Residents
  or Environmental Safety and Security
Diana Salisbury, Portsmouth/Piketon
  Residents for Environmental Safety
  and Security
Geoffrey Sea, International Radiation
  Survivors, Oakland, CA*

November 14–15, 1994
Marcia Haggard, Silver Spring, MD
Dr. Kathy Platoni, Beaver Creek, OH
Dr. Dennis Nelson, Kensington, MD
Mayor George Ahmaogak, North Slope
  Borough Assembly, AK
Rossman Peetok, North Slope Borrough
  Assembly, AK

Spokane, November 21, 1994
Leonard Schroeter, Seattle, WA
Gertie Hanson, Citizens Against Nuclear
  Weapons and Exterminations
Al Conklin, Department of Health, WA
Harold Bibeau, Portland, OR*
Catherine Knox, Department of
  Corrections, OR
Jim Thomas, Seattle, WA
Trisha Pritikin, Berkeley, CA*
Fred Larson, Ocean Park, WA
Brenda Weaver, Spokane, WA
Tom Bailie, Mesa, WA
Lynn Grover, Mesa, WA
Michelle Grover, Mesa, WA
Geoffrey Sea, International Radiation
  Survivors, Oakland, WA*
Kathy Jacobovitch, Vashon Island, WA

JoAnne Watts, Grants Pass, OR
Theresa Potts, Couer d'Alene, ID
Tom Cooper, Couer d'Alene, ID
Kay Sutherland, Walla Walla, WA
Beverly Aleck, Anchorage, AK
Sherri Lozon, Nez Pierce Tribe
Jeanne Haycraft, Enterprise, OR
Darcy Thrall, Benton City, WA*
Lynne Stembridge, Hanford Education
    Action League*
Lois Camp, LaCrosse, WA*
Lynn Horn, Spokane, WA
Charlie Miller, Spokane, WA
Charles Lombard, Spokane, WA
Curt Leslie, Wallua, WA
Rex Harter, Mesa, WA
David Vanderbilt, Ione, WA
Wendell Ogg, Knoxville, TN
Iris Hedman Othello, WA

December 15–16, 1994
Doris Baker, Cincinnati, OH*
Vina Colley, Portsmouth/Piketon Residents
    for Environmental Safety and Security*
Diana Salisbury, Sardinia, OH*
Lenore Fenn, Lexington, MA
Peter Lewis, Uniontown, PA
Professor Robert Proctor, Pennsylvania
    State University
William Jackling, Honeye Falls, NY
Fred Boyce, Norwell, MA
Pat Broudy, National Association of
    Atomic Veterans*

Santa Fe, January 30, 1995
Stewart Udall, Former U.S. Secretary of
    the Interior*
Ray Michael, Truth or Consequences, NM
Darcy Thrall, Benton City, WA*
Tyler Mercier, Santa Fe, NM
DH Bob Hofmann, Mountain Home, AR
Theodore Garcia, Las Cruces, NM
Bill Holmes, Fulsom, CA
Manuel Pino, Mesa, AZ
Alvino Wacanda, Laguna-Acoma Delegation,
    Paguato, NM
Curtis Francisco, Laguna-Acoma Delegation,
    Pueblo, NM
Dorothy Purley, Laguna-Acoma Delegation,
    Paguato, NM
Harry Lester, Albuquerque, NM

Milton Stadt, Victor, NY
Clyde Gardner, Edgewood, NM
Stanley Paytioma, Pueblo of Acoma,
    Acoma, NM
John Taschner, Los Alamos National
    Laboratory
Don Petersen, Los Alamos National
    Laboratory
George Voelz, Los Alamos National Laboratory
Joe Nardella
Timothy Benally, Shiprock, NM
Carlos Pacheco, Santa Fe, NM
Rosalie Jones, West Jordan, UT
Bernice Brogan, West Valley City, UT
Barney Bailey, Lovington, NM
Robert Stapleton, Ventura, CA
Linda Terry, Albuquerque, NM
Sue Dayton, Tijares, NM
Ernest Garcia, Chair, National
    Contaminated Veterans of America,
    Albuquerque, NM
Dale Howard, Las Lunas, NM
Coy Overstreet, Dickens, TX
Denise Nichols, USAF Major, Retired
Langdon Harrison, Albuquerque, NM
Ray Koonuk, Mayor, Point Hope, AK
Jack Schaefer, Point Hope, AK
Caroline Cannon, Point Hope, AK
Dr. Chellis Glendinning, Chimayo, NM
John Sheahan, Albuquerque, NM
Damacio Lopez, Bernalillo, NM
Phil Harrison, Uranium Radiation Victims
    Committee, Shiprock, NM
John Fowler
Renda Fowler
Bill Tsosie
Glenn Stuckey, Albuquerque, NM
Robert McConaghy

January 19–20, 1995
Joan McCarthy
Charles McKay, Severna Park, MD
Pat Broudy, National Association of
    Atomic Veterans*

February 15–16, 1995
Alex Reinhart, Braintree, MA
Wilfred Kendall, Embassy of the Republic of
    the Marshall Islands*
Senator Henchi Balos, Republic of the
    Marshall Islands

Holly Barker, Embassy of the Republic of
the Marshall Islands
E. Cooper Brown, Executive Commission Task
Force on Radiation and Human Rights*
Cliff Honicker, Environmental Safety
Studies Project, Knoxville, TN*
Pat Broudy, National Association of
Atomic Veterans*
Jonathan Weisgall, Attorney representing
the Bikini Islands*

Knoxville, March 2, 1995
Paul White, Oak Ridge, TN
Dorothea Gay Brown, Knoxville, TN
Betty Freels, Clinton, TN
Mary Bunch, Clinton, TN
Margaret Jacobs, Harriman, TN
Dorothy McRight, Nashville, TN
David Lee, Knoxville, TN
Gary Litton, Oak Ridge, TN
Dr. Karl Morgan, Oak Ridge, TN
Dr. Gary Madsen, Utah State University
Richard Sheldon, Knoxville, TN
Janice Stokes, Clinton, TN
Shirley Rippletoe, Old Hickory, TN
Bill Clark, Knoxville, TN
Dr. Helen Vodopick, Oak Ridge, TN
Claudia Soulyarette, Oak Ridge, TN
Dick Smyser, *The Oak Ridger*
Gertrude Copeland, Brentwood, TN
Dr. William Burr, Oak Ridge, TN
Dr. Bill Bibb, Oak Ridge, TN
Dr. Shirley Fry, Oak Ridge, TN
Acie Byrd, Washington, D.C.
Reba Neal, Coalfield, TN
Emma Craft, White Bluff, TN
Mary Hamm, Goodletsville, TN
Ron Hamm, Goodletsville, TN
Venia Lazenby, Mt. Juliet, TN
Dot McLeod, Lake Park, GA
Mary Lynn Stanley, Wrightsville, GA
Richard Vaughn, Franklin, TN
Dr. Frank Comas, Knoxville, TN
Ann Sipe, Oak Ridge, TN
Freda Jo Burchfield, Morristown, TN
Barbara Humphreys, Louisville, TN
Wilton McClure, Tony, AL
Earl McClure, Nashville, TN
Irene Sartain, Nashville, TN
Bruce Lawson, Oak Ridge, TN

Jeff Hill, Oak Ridge, TN
Patricia Jedlica, Spring City, TN
Carolyn Szetela, Nashville, TN
Doris Baker, Cincinnati, OH*
Gloria Nelson, Cincinnati, OH
Ann Marie Harrod, Nashville, TN

March 15–17, 1995
U. S. Senator Paul Wellstone (MN)
Ernest Sternglass, University of Pittsburgh
Elmerine Whitfield Bell, Dallas, TX
E. Cooper Brown, Executive Commission,Task
Force on Radiation and Human Rights*
Dr. Oscar Rosen, National Association for
Atomic Veterans*
Glenn Alcalay, New York, NY
Denise Nelson, Bethesda, MD
Chris DeNicola, New Orleans, LA
Valerie Wolfe, New Orleans, LA
Claudia Mullen, New Orleans, LA
Suzanne Starr, Chimayo, NM
Steven Schwartz, Washington, D.C.

April 10–12, 1995
Gwendon Plair, Concerned Relatives
of Cancer Study Patients,
Cincinnati, OH*
James Tidwell, Cincinnati, OH
Barbara Tatterson, Cincinnati, OH
Joseph Peterson, Carson City, NV
Banny deBrum, Acting Ambassador,
Embassy of the Republic of the Marshall
Islands
Rebecca Harrod Stringer, St. Augustine, FL

May 8–10, 1995
Doris Baker, Cincinnati, OH*
Barbara Tatterson, Cincinnati, OH*
Herbert Varin, Cincinnati, OH*
Clifford Tidwell, Cincinnati, OH*
Beatrice Tidwell, Cincinnati, OH
Zettie Smith, Cincinnati, OH
Pat Broudy, National Association of
Atomic Veterans*

June 21–23, 1995
Anthony Roisman, National Committee
of Radiation Victims, Washington, DC
Geoffrey Sea, Task Force on Radiation
and Human Rights, Oakland, CA*
Phil Harrison, Uranium Radiation Victims
Committee, Albuquerque, NM*

Rachel Greene, Hyattsville, MD
Zina Greene, Washington, DC
Julie Boddy, Tacoma Park, MD

July 17–19
Senator Tony deBrum,(w/Ambassador Wilfred
    Kendall, Phillip Muller), Marshall Islands*
Dr. Bernard Aron, Cincinnati, OH
Dr. David Egilman, Braintree, MA*

Dr. Oscar Rosen, National Association of
    Atomic Veterans*
Dr. Dennis Nelson, Bethesda, MD
Ms. Mary Mueller, Task Force on Radiation
    and Human Rights
Mr. Acie Byrd, Task Force on Radiation
    and Human Rights*

A Citizen's Guide to the Nation's Archives

WHERE THE RECORDS ARE AND HOW TO FIND THEM

SOME INITIAL QUESTIONS AND ANSWERS

How can I find out if I or my relative was in a radiation experiment?

This was one of the most commonly asked questions from the hundreds of individuals who contacted the Committee. There is no simple answer. Medical records are the place to start. They should provide information on what condition you or your relative was treated for, what treatment was actually given, and who administered this treatment. See part III.A for further details.

How can I obtain medical records? What should I do with them once I have them?

You have a legal right to your own medical records and, with the proper authorization, to a relative's medical records. By contacting the facility where the treatment occurred, you should be able to request and obtain the records. The next step is to have a qualified medical professional review the records to ascertain whether the treatment administered was acceptable for the patient's condition. See part III.B for further details.

What ACHRE materials are available to the public? Where are they stored and who can look at them?

All documents obtained by and produced by the Advisory Committee are public informa-

tion, available to anyone. A large portion of Committee materials is available through the Internet. Hard copies of all materials will be stored at the National Archives and Records Administration in Washington, D.C. See part IV.A for further details.

Whom should an individual call to request an investigation into his or her particular case?

This was another very common question. No office currently exists that is specifically chartered to investigate individual cases with respect to human radiation experiments. That is one reason for this guide: to provide individual citizens with enough guidance to begin their own investigations. See part II for details.

Where should an individual researcher turn to learn more about radiation experiments with government involvement?

Researchers can use a number of resources, including the ACHRE collection. If more information is desired, the federal agencies have reported to the Advisory Committee that the public may contact their designated offices. See part II for details.

Whom should the public work through after the Advisory Committee is disbanded?

No extant government body is chartered to provide such guidance. It is the purpose of this appendix to provide individuals with enough direction to begin their own investigations.

579

CONTENTS

... I have been to Oak Ridge, Tennessee, and Washington, D.C. I have seen a lot of documents. I have learned some of the codes, so please don't try to shaft me. I know a lot. The records are not here in Cincinnati, all of them on my grandmother. And I have been trying to find them. And I just would like to know where the rest of them are. So please, will you help me find them?

—Citizen at the ACHRE public forum in Cincinnati, 21 October 1994

As the Advisory Committee traveled across the country taking public testimony, it heard citizens describe many of the same experiences over and over. One common thread that struck a particularly responsive chord with the Committee was the sheer frustration felt by many, even experienced researchers, who had tried to find their own records or to find out the details of government programs. The difficulty we have all faced in doing this research yields an important lesson: The government must be honest about the nature and purposes of the studies it sponsors and conducts; in sponsoring human experimentation, it has an even higher obligation to keep a fair record and provide those involved with meaningful access. The Advisory Committee has done what it can to open the door to our nation's archives. We all must see that it remains open.

This appendix is intended to help. For those who want to know whether a relative was involved in an experiment, and for historians, journalists, and others with a more general interest in human radiation experiments (HREs) and the general topic of government-sponsored research, the following pages discuss what to ask for, whom to write to, and where to go.

The Advisory Committee's records are one important place to turn (see part IV). It should be understood, though, that the Advisory Committee did not find everything there is to find about human radiation experiments, nor could we review what we did find in the detail we would have preferred. Moreover, neither the Advisory Committee nor the agencies, generally speaking, sought the medical records of individuals. But there is much information that we did recover, and the efforts of the Advisory Committee and the agencies have increased the likelihood that citizens will be able to find the personal documents they need.

This *Guide* has four parts: part I is an introduction to finding and using federal records; part II covers agency facilities and services, including what information is available at which agencies, and where to go and how to get it; part III focuses on finding medical records. And part IV is an introduction to the records collected and created by the Advisory Committee.

PART I: FINDING FEDERAL RECORDS

Finding the most general information about the activities of the federal government can be as easy as picking up the telephone or looking in a reference book, but those approaches do not provide the detail necessary to understanding how a program operates or why it does what it does. Finding information like this requires research, and research in government documents may require time and effort. The government's records are stored in a sprawling, decentralized, and sometimes haphazard system, and particular records are often hard to locate. It may be difficult simply to determine whether the records still exist. Federal records laws and rules provide for the periodic review and destruction of certain categories of records. However, the Committee found that the documents that recorded the destructions of other documents were themselves often later destroyed. Thus, it is often difficult to know for certain whether particular documents have been destroyed or are simply hard to locate.

This part of the *Citizen's Guide* provides information that will allow the researcher to focus more quickly on where the desired information may be or, that being determined, how to go about retrieving it.

Types and Sources of Federal Information

Although there are many ways to categorize the types of information citizens seek, the one that will have the most profound effect on what to look for and where to look for it is whether the citizen is interested in *records of individual experience* or in *program records*. Records of individual experience are those that document the history of a particular person—medical records, personnel records, tax returns, memberships, and so forth—and are usually kept for the private use of that person and the institution whose relationship they record. Such records will only rarely include information about a program in which the individual participates. For example, an individual's medical records will not likely contain information on the government program that funded the medical research or the ethical guidelines applicable to the use of human subjects in the program.

Program records, on the other hand, document the purposes, organization, staffing, and funding of an activity—minutes, proceedings, memorandums, proposals, contracts, and so forth—and are likely to be available to the public in some form. Such records will only rarely contain information about individuals. For example, agency records on a biomedical research program will not contain the names of the patients involved in it or their medical histories.

As is obvious from these descriptions, records of individual experiences and program records hold very different types of information.[1] The significance for the researcher is that the two types of records are kept in different places, and his or her approach to finding the information must reflect this fact. For example, if information about the physical condition and treatment of an individual is what is wanted—that is, medical facts—a search for medical records is likely to be more useful than a search for records of experiments. Medical information about the condition and treatment of experimental subjects is generally contained in medical records and not in the scientific records of experiments.[2] On the other hand, information about a study in which citizens participated is unlikely to be found in their medical records, but in the investigator's records and those of his institution and the study's sponsors.

Further information on finding program records, which are generally publicly available in large repositories, may be found in the remainder of part I and in parts II and IV. Those sections also provide information on government-held records of personal experience. For information on finding medical records, see part III.

Aids for Focusing Research

Because the federal government is vast, it is vitally important to identify as quickly as possible the government components whose records may contain the needed information; as will be discussed in the section below on the National Archives, that understanding is also important to using the records once they are found. Unfortunately, one of the things that the Advisory Committee learned in our research is that many government agencies do not have complete information on all the programs they sponsored through the years or on the records that were created or preserved or where they are located. And, furthermore, there is no central, comprehensive source of information for the history of the federal government: Even the collections of the National Archives do not reflect the full and complete history of the government and its programs. In some cases extensive research was required to discover or to understand the histories of certain parts of the agencies in order to identify the organizational components whose work was potentially relevant to the Advisory Committee's research. Only then could the search for records begin.

Fortunately, however, much unearthing of the histories of government organizations and locating of pertinent records has been done by agency personnel and Advisory Committee staff. The fruits of these efforts are available in three resources that may assist the citizen researcher in finding agency information. First, the relevant organizational components; the location, classification, and review of their records; and what records were never located are all described in great detail on an agency-by-agency basis in the supplemental volume, *Sources and Documentation*. This volume serves as an excellent guide for those doing their own research.[3] The second is the ACHRE collection itself; as explained in more detail in part IV below, most records in the ACHRE collection can be traced to the agency collection and repository from which they came. The third is the February 1995 Department of Energy publication, *Human Radiation Experiments: The Department of Energy Roadmap to the Story and the Records* and its July 1995 supplements (see part II, below).[4] This work describes in considerable detail many relevant DOE record collections that are located at various repositories in the Washington, D.C., area and the national laboratories around the nation (see part II, below, for further information about the laboratories). We note that, in addition to resources created during the life of the Committee, agencies may have created other guides to agency history and records collections. See, for example, "A Guide To Resources on the History of the Food and Drug Administration," Food and Drug Administration, History Office.

Where Federal Records Are

Unless they have been lost or destroyed, almost all federal records[5] created since the founding of the Republic are in agency files, stored at a federal records center, or preserved in the National Archives.[6] Generally, agencies are required to transfer to the National Archives records that are of sufficient historical or other value to warrant preservation. Documents are transferred when they are thirty years old or, regardless of age, when the originating agency no longer needs them for its regular business and will be satisfied accessing them through the National Archives.

In actual practice, few, if any, agencies have fully complied with these requirements. Most records are still under the control of the agencies that created them, though some are stored with the National Archives and Records Administration (NARA). Even for quite old records, therefore, the citizen will often find it necessary to look beyond the National Archives into the federal record centers and the agencies. The use of these three repositories is described below; further information on the agencies is contained in part II.

National Archives

Collections. NARA does not refile the records it receives according to some grand theoretical scheme but, rather, preserves them in as close to their original order as is practical, arranging them according to *provenance*.[7] This means that the structure and organization of records in the

National Archives reflects the structure and organization of the office that created them, using the same divisions and titles that were used by the office originally. For this reason, all records of an individual agency—or in the cases of very large agencies such as the military services, the records of various commands, headquarters, and other major organizational units—are placed by the National Archives in a separate *record group* with a distinctive title and number. The approximately 475 record groups at the National Archives vary in size from less than 100 cubic feet to tens of thousands of feet. Record groups are divided into subdivisions called *entries* that often hold the records of a single division, department, bureau, or office. The access tool generally used to find basic information in a record group (e.g., brief descriptions of individual entries) is the *finding aid* created by the National Archives. Not all record groups have finding aids, however, and some older ones have not been kept up to date. The archivists who work with the record groups are often an invaluable source of information as well.

Services. The National Archives is the one repository holding agency records specifically charged with accommodating the public. In addition to a staff of professional archivists, the Archives provide large research rooms, copiers, and complete access to unclassified and declassified collections.

The National Archives has two major public facilities in the Washington area: the National Archives, Pennsylvania Avenue between 7th and 8th Streets, N.W., Washington, D.C., and the National Archives at College Park ("Archives II"), 8601 Adelphi Road, College Park, Maryland 20740-6001. (Telephone 202-501-5400 to request reference help, or write Reference Services Branch, National Archives and Records Administration, Washington, D.C. 20408.) Research hours at both the downtown Washington and College Park facilities are 8:45 a.m. to 9:00 p.m., Tuesday, Thursday, and Friday; 8:45 a.m. to 5:00 p.m., Monday and Wednesday; and 8:45 a.m. to 4:45 p.m. on Saturday, except federal holidays.

Records that are generated by regional offices are maintained in regional archives:

Anchorage, Alaska: 654 W. 3rd Avenue, 99501; 907-271-2441

Chicago, Illinois: 7358 S. Pulaski Road, 60629; 312-581-7816

Denver, Colorado: Building 48, Denver Federal Center, 80225; 303-236-0817

East Point, Georgia: 1557 St. Joseph Avenue, 30344; 404-763-7477

Fort Worth, Texas: 501 W. Felix Street, 76115; 817-334-5525

Kansas City, Missouri: 2312 E. Bannister Road, 64131; 816-926-6272

Laguna Niguel, California: 24000 Avila Road, 92677; 714-643-4241

New York, New York: 201 Varick Street, 10014; 212-337-1300

Philadelphia, Pennsylvania: 9th and Market Streets, 19107; 215-597-3000

San Bruno, California: 1000 Commodore Drive, 94066; 415-876-9018

Seattle, Washington: 6125 Sand Point Way N.E., 98115; 206-526-6507

Waltham, Massachusetts: 380 Trapelo Road, 02154; 617-647-8100

For Freedom of Information Act (FOIA) and Privacy Act requests, speak with the archivists who work with the record group concerned, or write: Office of the National Archives, National Archives and Records Administration, Washington, D.C. 20408; telephone 202-501-5300. For further information see the section on Rights and Restrictions on Access to Information, below.

Federal Records Centers

When an agency determines that it no longer needs to house a group of records it can transfer them to a federal records center in its geographical area. Federal records centers have been established solely to assist the agencies in the storage and processing of their records. There is no requirement that any agency transfer its records to a records center. Although the records centers are managed by NARA, the agencies retain legal custody and control of the records.

Collections. Records held in federal records centers are also organized into record groups (using the same titles and numbers as at the National

Archives), but are not further broken down into entries. Instead, a record group at a records center consists simply of a series of *accessions*, the shipments of records added to it. Record groups may contain from a few to thousands of accessions, and an individual accession may hold one to many hundreds of boxes of records. Unfortunately, there are no archivists or finding aids at federal records centers to assist the public. The only means of determining what is in a record group is by examining the *Standard Form 135 (SF-135)* prepared by the agency for each individual shipment. These forms contain a great deal of information, including the accession number, name and address of the office shipping the records, point of contact, security classification of the records, quantity of records in cubic feet, and a description of the records that often includes a folder listing.[8] The examination of SF-135s can be a very tedious process, for they may total many thousands of pages.

Services. The public does not have free access to records at a federal records center, not even to completely unclassified or declassified accessions. Permission first must be obtained from the agency that owns the records, and this can be a time-consuming process. Personnel at the federal records centers will provide information on who should be contacted at an agency about obtaining such permission.

The one federal records center in the Washington, D.C., area is the Washington National Records Center, 4205 Suitland Road, Suitland, Maryland 20409; telephone 301-763-7000. The hours are 8:00 a.m. to 4:30 p.m., Monday through Friday except federal holidays. There are thirteen regional federal records centers, which hold records generated by federal offices in that particular geographical region of the nation. Many, but not all, are located in the same place as the regional National Archives:

Bayonne, New Jersey: Building 22, Military Ocean Terminal, 07002; 201-823-7161

Chicago, Illinois: 7358 S. Pulaski Road, 60629; 312-352-0164

Dayton, Ohio: 3150 Springboro Road, 45439; 513-225-2878

Denver, Colorado: Building 48, Denver Federal Center, 80225; 303-236-0804

East Point, Georgia: 1557 St. Joseph Avenue, 30344; 404-763-7476

Fort Worth, Texas: Building 1, Fort Worth Federal Center, 76115; 817-334-5515

Kansas City, Missouri: 2312 E. Bannister Road, 64131; 816-926-7271

Laguna Niguel, California: 24000 Avila Road, 92677; 714-643-4420

Philadelphia, Pennsylvania: 5000 Wissahickon Avenue, 19144; 215-951-5588

San Bruno, California: 1000 Commodore Drive, 94600; 415-876-9015

Seattle, Washington: 6125 Sand Point Way N.E., 98115; 206-526-6501

St. Louis, Missouri: National Personnel Records Center, 9700 Page Boulevard, 63132; 314–263–7201

Waltham, Massachusetts: 380 Trapelo Road, 02154; 617–647–8745

FOIA requests for records in the custody of the federal records centers must be submitted to the federal agency that transferred the records to the federal records center. Records center personnel will provide addresses and contacts. For further information, see the section "Access to Information: Rights and Restrictions," below.

Records Still Held by Agencies

Several agencies retain great volumes of records that have never been sent to the National Archives or a federal records center. Such records may be stored at any number of places, including internal record storage facilities and history offices. With a few exceptions, these collections are generally less well organized and described than those at the National Archives or federal records centers. Furthermore, most agencies have only a limited ability to accommodate researchers. The names, addresses, and telephone numbers of the locations where the agencies store records are available in part II, below.

Access to Information: Rights and Restrictions

This section addresses some government policies that control access to information—on privacy, freedom of information, and national security classification—and some of a citizen's rights to information and how to exercise them.

Privacy and Freedom of Information

The Privacy Act and the Freedom of Information Act (FOIA)[9] are the most critical components of the legal framework that supports public access to federal records. The Privacy Act defines certain types of information as privileged to the individual, and during his or her lifetime it prevents their public dissemination or their use for purposes other than those originally authorizing their collection. This means, for example, that one agency may not share personal information about citizens with another government agency, and it means that one person may not have access to such information about any other person without authorization. This protection of privacy extends to records in the National Archives as well. The Freedom of Information Act guarantees, with some categories of exceptions, that all records created by the executive branch of the federal government are available to citizens. Among those exemptions are a privacy clause that broadens the scope of the Privacy Act by extending protection to personnel and medical files by category rather than limiting protection to the lifetime of individuals, and a national defense and foreign policy clause that precludes one from obtaining certain classified information under FOIA.

The next two sections discuss the effect of these laws on obtaining information based on the names of individuals, and the procedures and requirements for making Freedom of Information Act Requests.

Name Searches

Access by citizens to federal records that are retrievable by the names of individuals or other personal information are controlled by the Privacy Act and by the privacy clause of the Freedom of Information Act. The Privacy Act restricts access to information contained in what are called *Privacy Act systems of records*, records arranged by the names of individuals or other personal information. In general, during an individual's lifetime, records retrieved by the use of personal information are available only to that person or with his or her authority,[10] although redacted copies of such documents—that is, copies from which private information has been removed—may be available if the records are retrievable in some other way.[11] If, therefore, a citizen is interested in obtaining records that concern him or her or, with the appropriate authority, those that concern a close relative, there should be no legal restrictions on access; to the extent, however, that a citizen wishes more information about other individuals who are mentioned in those records, there may be considerable difficulty. In such cases one would probably have greater success identifying the program in which he or she participated, determining where the records of that program are housed, and extracting information from those records.

FOIA Requests

In general, the Freedom of Information Act requires that the individual[12] make inquiry in writing[13] directly to the appropriate agency, in conformity with the established procedures of the agency, and that agreement on the payment or inapplicability of fees is reached between the requester and the agency. The first requirement usually is understood to include identification of the records in which the information is to be found. Agencies are not required to do research for the citizen but only to conduct "reasonable searches" of their records in an attempt to meet the request.[14] The second requirement recognizes that different agencies may have different procedures for handling public inquiry.[15] The third requirement permits the agency to determine before accepting the request that the requestor will pay all the applicable fees or, in the alternative, that there are valid grounds for waiving the fees.[16]

Once an agency has accepted a FOIA request, the law establishes very short periods of time for the agency to respond. If the request is accepted, the agency is obligated to decide within ten working days of acceptance whether or not it will provide the information within a reasonable length of time,[17] and if the request is denied and an appeal is made, it must provide a response within twenty working days. In actual practice, however, agencies rarely meet these time limits. Depending on the backlog of requests, the number of other agencies that must be contacted, and other factors, a FOIA can take one to five years to process.

Agencies are most likely to reproduce and mail copies of records to requesters, but they are not required to do so and are permitted to provide access to the records at a central location (see also the information on the FOIA reading rooms and offices at the agencies in Part II).

If an agency denies a request in whole or in part, the requester then has the right to make one administrative appeal. If after these the requester is still not satisfied, the only recourse is federal court.

Classified Records

There is a vast number of records at the National Archives, in the federal records centers, and at the agencies that are still classified and therefore unavailable to the public. The government is obliged by executive order to review its records periodically for declassification, but citizens may request a review on their own initiative. Submitting a request, of course, does not guarantee that the records will be declassified either in whole or in part. The government authorities conducting the review may conclude that the documents should remain classified.

There are three methods under which the public can request that documents at the National Archives be reviewed for declassification. The first is under FOIA, and the second is under the Mandatory Declassification Review (MDR) provisions of Executive Order 12958 of April 17, 1995.[18] Under both methods, a request is submitted to the National Archives (rather than to the agency that generated the records), whose archivists will provide information on how the request should be handled further. The third method for requesting declassification is under the Special Declassification Review procedure. This informal procedure, which is only applicable to records at the National Archives, is much quicker than either FOIA or MDR, but there are some records—intelligence records, for example—that cannot be reviewed in this way. The archivists working with the records should be consulted to determine whether a Special Declassification Review may be used.

To access classified collections at federal records centers or agencies, either a FOIA or an MDR request must be submitted to the agency.

Classified records that turn up in the course of a document search are sent through declassification review. There is no Special Declassification Review procedure at federal records centers or agencies.

PART II: AGENCY INFORMATION AND SERVICES

As part of the Advisory Committee's effort to improve citizen access to information, we asked the agencies providing information to the Committee—chiefly the members of the Interagency Working Group and the Nuclear Regulatory Commission—to respond to a series of questions concerning the handling of private information requests. We asked how citizens should make requests, what services the agencies would provide, what information resources were available, and how agencies would handle requests for information held by agency contractors and grantees. Each agency's response is summarized in its section, below. Those sections also include general information obtained from the *U.S. Government Manual*,[19] including the location of FOIA reading rooms and offices.[20]

Department of Energy

General

DOE maintains a Freedom of Information Act Reading Room at its headquarters in Washington. The address is FOIA Reading Room, Forrestal Building, Room 1E-190, Department of Energy, 1000 Independence Avenue, S.W., Washington, D.C. 20585; telephone 202-586-6020. The reading room is open 9:00 a.m. to 4:00 p.m., Monday through Friday, except federal holidays. General information on filing FOIA requests may be obtained from the FOIA office, 202-586-5955.

As described both in *Sources and Documentation* and in *Human Radiation Experiments: The Department of Energy Roadmap to the Story and the Records*, the History Division at DOE headquarters has custody of many collections of records. The relatively few unclassified and declassified collections that the division maintains can be examined at its office in DOE's Germantown facility: U.S. Department of Energy, History Division, HR-76, Room F031, 19901 Germantown Road, Germantown, Maryland 20874-1290; telephone 301-903-5431. The hours are from 8:00 a.m. to 4:00 p.m., Monday through Friday, except federal holidays. An appointment must be made, as there is limited space to accommodate the public.

In addition, the national laboratories around the nation hold a huge volume of records. Information at those locations is available as follows:

Argonne National Laboratory: There is no reading room at Argonne, but citizens may write: Argonne National Laboratory, Office of Public Affairs, 9700 South Cass Avenue, Argonne, Illinois 60439; 708-252-5575.

Brookhaven National Laboratory: There is no reading room at Brookhaven, but citizens may write: Brookhaven National Laboratory, Office of Public Affairs, Building 134, P.O. Box 5000, Upton, New York 22973; 516-282-2345.

Hanford: DOE Public Reading Room, P. O. Box 999—Mail Stop H2–53, Richland, Washington 99352; 509-376-8583. This facility is in the library at Washington State University—Tri-Cities Campus, 100 Sprout Road, Richland, Washington. The hours are 8:00 a.m.-noon and 1:00–4:30 p.m., Monday through Friday.

Los Alamos National Laboratory: Public Reading Room, 1350 Central Avenue—Suite 101, Los Alamos, New Mexico 87544; 505-665-2127 or 800-343-2342. The reading room is open 9:00 a.m.–5:00 p.m., Monday through Friday, except federal holidays.

Idaho National Engineering Laboratory: DOE Idaho Operations Public Reading Room, 1776 Science Center Drive, Idaho Falls, Idaho 83415-2300; 208-526-9162. The hours are 8:00 a.m.–5:00 p.m., Monday through Friday, except federal holidays.

Lawrence Livermore National Laboratory: There is no reading room at Lawrence Livermore, but citizens may write: Area Relations—Mail Stop L404, Lawrence Livermore National Laboratory, P.O. Box 808, Livermore, California 94550.

Oak Ridge National Laboratory: Oak Ridge Operations (ORO) Public Reading Room, 55 Jefferson Circle, Oak Ridge, Tennessee 37831; 615-241-4780. The hours are 8:00–11:30 a.m. and 12:30–5:00 p.m., Monday through Friday, except federal holidays. In addition to the laboratory itself (ORNL), the Oak Ridge complex also encompasses the regional DOE office (ORO) and an independent research institute (ORISE) operated by a consortium of universities. The regional office may be contacted by writing: Oak Ridge Operations Office (ORO), P.O. Box 2001, Oak Ridge, Tennessee 37831. The research institute may be contacted by writing: Oak Ridge Institute for Science and Education (ORISE), P.O. Box 117, Oak Ridge, Tennessee 37831–0117.

Information on Human Radiation Experiments

In response to the Advisory Committee's request, the Department of Energy has provided the following information about its resources and services for citizens inquiring about human radiation experiments.

General Department of Energy information about human radiation experiments sponsored by DOE and its predecessors, and referrals, may be requested through the Radiation Research Helpline (1-800-493-2998) or by writing to the Department of Energy, Office of Human Radiation Experiments (OHRE), EH-8, 1000 Independence Avenue, S.W., Washington, D.C. 20585.

The largest body of pertinent records is maintained by the Coordination and Information Center (CIC).[21] All CIC material is declassified, screened, and redacted for public dissemination. The CIC may be contacted by writing to the Coordination and Information Center, 3084 South Highland Street, Las Vegas, Nevada 89109, or by calling 702-295-0731. Although generally equivalent for DOE-related human radiation experiment records, the ACHRE and CIC collections are not identical: The ACHRE collec-

tion contains most but not all of CIC's Human Radiation Experiments records series and has some DOE records not represented in CIC collections. For further information on CIC documentation, see "How to Go From the ACHRE Collection to Agency Records" in part IV, below.

Medical records should be requested from the facility where the medical services were performed. Current or former DOE employees may obtain their medical records from the site where they worked or from the National Personnel Record Center in St. Louis, Missouri, which may be contacted directly (314-538-3882).[22] Dosimetry records documenting occupational radiation exposures are maintained for both government and contractor personnel; they should be requested from the DOE manager at the site where exposure may have occurred. DOE also maintains a consolidated collection of dosimetry records related to weapons testing, including both civilian and military information. Information may be requested by writing to the Dosimetry Research Program (DRP), P.O. Box 98521, Las Vegas, Nevada 89193–8521, or by calling 702-295-0731. DOE will also help to identify and locate records that are not in the custody of the department, although citizens must contact those institutions or individuals themselves.

Several DOE departments have created finding aids that may be useful in finding HRE records: (1) As mentioned above, the report *Human Radiation Experiments: The Department of Energy Roadmap to the Story and the Records*, prepared by the Office on Human Radiation Experiments, provides summaries of that office's findings and descriptions of some relevant record collections. (2) An electronic index to pertinent CIC holdings is available at the CIC and OHRE offices and at DOE's reading rooms. Citizens may request searches or do their own at those locations. (3) For those with Internet access, recently declassified documents are available from DOE's Office of Scientific and Technical Information through its World Wide Web[23] presence, Opennet (http://www.doe.gov/html/osti opennet/opennet1.html). And another group of databases on the Internet, created by OHRE, provide full access to the documents in the CIC collection. (Further information about OHRE and this complex of databases [called HREX] may be obtained from its World Wide Web site, http://www.ohre.doe.gov.) Finally, OHRE issued a supplement to its February 1995 report in July 1995 entitled *Human Radiation Experiments Associated with the U.S. Department of Energy and Its Predecessors*.[24] This volume adds to the information reported in the February 1995 volume, and also includes summaries of the nearly 150 HREs reported by DOE.

Department of Defense

General

The Department of Defense's Freedom of Information Act offices may be contacted as follows: DOD, 703-697-1180; Army, 703-607-3452; Navy, 703-697-1459; Air Force, 703-697-3467.

Information on Human Radiation Experiments

In response to the Advisory Committee's request, the Department of Defense has provided the following information about its resources and services for citizens inquiring about human radiation experiments.

Information concerning human radiation experiments sponsored or conducted by the Department of Defense is available chiefly through the Radiation Experiments Command Center (RECC), the DOD equivalent of DOE's Office of Human Radiation Experiments. RECC is operated under contract by Science Applications International Corporation (SAIC). The primary method of contacting RECC is by referral from the DOE Radiation Research Helpline (1-800-493-2998)—RECC does not provide direct telephone assistance. Citizens may also write directly to RECC: Radiation Experiments Command Center, 6801 Telegraph Road, Alexandria, Virginia 22310-3398. Individuals contacting RECC will be requested to fill out a survey form to facilitate the search for records responsive to their requests. The RECC collection and the ACHRE collection of DOD materials are generally equivalent. For further information on RECC documentation, see "How to Go From the ACHRE Collection to Agency Records" in part IV, below.

RECC does not keep medical records but will assist those who request them by contacting the appropriate facility and referring the individual there. Active duty military personnel will find their complete medical records at their current duty stations; upon retirement or discharge, their files are transferred to the National Personnel Records Center in St. Louis. Former military personnel may contact the center directly (314-538-3882).[25]

RECC maintains a database of information on human radiation experiment documents identified during DOD's search and a database of secondary information concerning the history and policy behind the activities. Case files on individuals exposed to radiation are being created and categorized by exposure. RECC will also help citizens contact private institutions involved in DOD-sponsored programs, within the limits of the Privacy Act.

Another DOD resource is the Nuclear Test Personnel Review Program (NTPRP) operated by the Defense Nuclear Agency (DNA), which has obtained a considerable volume of records and information related to military and civilian participants in atmospheric nuclear tests between 1945 and 1962. Unclassified and declassified records that do not contain privacy information can be reviewed by the public at a special library at DNA headquarters. The program also provides certain informational and referral services to participants. The address is Defense Nuclear Agency, Nuclear Test Personnel Review Program, 6801 Telegraph Road, Alexandria, Virginia 22310; telephone 1-800-462-3683. Additional services may be available through the VA's Ionizing Radiation Registry Examination Program (see VA section, below).

Department of Health and Human Services

General

There is no general reading room for the Department of Health and Human Services, nor for its research divisions, the Public Health Service and the National Institutes of Health.[26] Each institute of NIH[27] maintains its own information facilities, including its own office of public affairs. For help in identifying the sort of information needed and how to obtain it, a good place to start is the National Library of Medicine, 8600 Rockville Pike, Bethesda, Maryland. The general information line for NIH is 301-496-4000.

Information on Human Radiation Experiments

In response to the Advisory Committee's request, the Department of Health and Human Services has provided the following information about its resources and services for citizens inquiring about human radiation experiments.

DHHS sponsors two types of research—intramural ("within the walls"), research conducted by DHHS staff members, and extramural ("outside the walls"), research conducted outside DHHS by contractors or grantees. DHHS keeps medical records only for individuals who participated in intramural research. Inquiries concerning such records should be directed in writing to the Deputy Assistant Secretary for Health/Communications, Department of Health and Human Services, Hubert Humphrey Building—Room 701H, 200 Independence Avenue, S.W., Washington, D.C. 20201.

There are four DHHS databases that may help identify potential human radiation experiments. The first is the Clinical Center intramural protocol database (also called the Protocols by Institute database), which was created at the Advisory Committee's request to index information about NIH intramural research. This database was completed in February 1995 and contains more than 5,000 entries for the period 1953 through November 1994. More recent information on extramural research is included in the CRISP (Computer Retrieval of Information on Scientific Projects) database, which contains records for all PHS extramural projects and for NIH and Food and Drug Administration (FDA) intramural projects. The most comprehensive database is called IMPAC and includes information on awards as far back as 1944, although not all programs are included for their entire tenure and the information on early awards is limited. Finally, the National Library

of Medicine (NLM) is creating a database with entries for all articles written by investigators whose human radiation experiments were supported by NIH. (Thus the database will contain citations for both radiation and nonradiation research.) NLM expects the database will eventually contain approximately 100,000 entries.

DHHS has a contractual relationship with its contractors and grantees that limits its access to the records they create to those occasions required by agency functions. Consequently, although DHHS will help citizens identify the independent researchers and institutions that hold their medical records, it asks that the initial contact be made by the citizen. If that approach is unsuccessful, DHHS will attempt to obtain the records. Citizens are encouraged to contact DHHS to make a precise determination of whom to contact and what information to include in their inquiries.

Department of Veterans Affairs

General

The VA maintains a reading room at its central office in Washington, D.C., where citizens may inspect or copy VA records available to the public. The address is Room 170, 810 Vermont Avenue, N.W., Washington, D.C. 20420; telephone 202-233-2356. For further information, contact the Office of Public Affairs, Department of Veterans Affairs, 810 Vermont Avenue, N.W., Washington, D.C. 20420; telephone 202-273-5700.

Information on Human Radiation Experiments

In response to the Advisory Committee's request, the Department of Veterans Affairs has provided the following information about its resources and services for citizens inquiring about human radiation experiments.

The VA is continuing to look for information on human radiation experiments in its own records and will assist citizens in identifying nongovernment records related to their case histories. It has also published a fact sheet, "Information for Veterans Exposed to Radiation"

(November 1994). Requests for information about participation in experiments may be made directly to the director of the appropriate VA medical center or to the director of the regional VA office (toll-free 1-800-827-1000). The VA maintains an Ionizing Radiation Registry Examination Program for veterans who may have been exposed to the ionizing radiation while on active duty in the period 1945–1962. Information about the program may be requested in writing from: Director, Environmental Epidemiology Service (103E), Department of Veterans Affairs, 1120 20th Street, Suite 950, Washington, D.C. 20036–3406, telephone 202-606-5420. Additional information may be requested from DOD's Nuclear Test Personnel Review Program (see DOD section, above).

National Aeronautics and Space Administration

General

The NASA Headquarters Information Center is in Room 1H23, 300 E Street, S.W., Washington, D.C. 20546, and is open 8:00 a.m. to 4:30 p.m., Monday through Friday, except federal holidays. For information about holdings, telephone 202-358-1000.

Information on Human Radiation Experiments

In response to the Advisory Committee's request, the National Aeronautics and Space Administration has provided the following information about its resources and services for citizens inquiring about human radiation experiments.

NASA's records concerning human radiation experiments are generally limited to summary reports from principal investigators and do not contain medical information on individuals, apart from the records of astronauts. Information about individual participation may be requested in writing under the Privacy Act using FOIA procedures and NASA's standard Human Radiation Exposure Log form. Inquiries should be directed to: NASA Johnson Space Center, Freedom of Information Coordinator, Public Affairs Office, Mail Code AP2, Houston, Texas 77058, Attn.: Direc-

tor, Space and Life Sciences Directorate. NASA's information retrieval systems in this area are limited, and success will largely depend on the quality and detail of the information provided to NASA. NASA will refer requests for information requiring access to non-NASA records to the appropriate individual or institution.

Central Intelligence Agency

General

The CIA does not maintain a public reading room but does issue several publications that may be of interest. For information, write: Central Intelligence Agency, Public Affairs Office, Washington, D.C. 20505, or telephone 703-351-2053.

Information on Human Radiation Experiments

In response to the Advisory Committee's request, the Central Intelligence Agency has provided the following information about its resources and services for citizens inquiring about human radiation experiments.

The CIA has no special facilities for handling requests concerning human radiation experiments nor any information resources specifically concerned with them. Privacy Act and Freedom of Information Act requests should be filed in the usual ways. The CIA is not prepared to facilitate the identification or the retrieval of nongovernment records that may be associated with government activities. Requests should be addressed in writing to: Information and Privacy Coordinator, CIA, Washington, D.C. 20505.

Nuclear Regulatory Commission

General

The Nuclear Regulatory Commission (NRC) Headquarters Public Document Room maintains an extensive collection of documents related to NRC licensing proceedings and other significant decisions and actions, and documents from the regulatory activities of the former Atomic Energy Commission. The reading room is located at 2120 L Street, N.W., Washington, D.C.; telephone 202-634-3273, toll-free 800-397-4209 or fax 202-634-3343. The Public Document Room is open Monday through Friday from 7:45 a.m. to 4:15 p.m., except on federal holidays. Reference librarians are available to assist users with information requests. A bibliographic database is available for on-line searching twenty-four hours a day. For additional information call the above telephone number or write: Nuclear Regulatory Commission, Public Document Room, Washington, D.C. 20555.

The commission also maintains eighty-eight local public document rooms in libraries in cities and towns near commercially operated nuclear power reactors and certain nonpower reactor facilities. A list of local public document rooms is available from the Director, Division of Freedom of Information and Publications Services, Nuclear Regulatory Commission, Washington, D.C. 20555–0001. To obtain specific information about the availability of documents at the local public document rooms, NRC's Local Public Document Room Program staff may be contacted directly by calling, toll-free, 800–638–8081. Citizens may also request the publication *Users' Guide for the NRC Public Document Room* (NUREG/BR-0004, Rev. 2).

Freedom of Information Act inquiries should be directed in writing to the Director, Division of Freedom of Information and Publications Services, Nuclear Regulatory Commission, Washington, D.C. 20555–0001. For further information, call 301-415-7175.

For general information, contact the Office of Public Affairs, Nuclear Regulatory Commission, Washington, D.C. 20555–0001; telephone 301-415-8200. Citizens may request the publication *Citizen's Guide to U.S. Nuclear Regulatory Commission Information* (NUREG/BR-0100, Rev. 2).

Information on Human Radiation Experiments

In response to the Advisory Committee's request, the Nuclear Regulatory Commission has provided the following information about its resources and services for citizens inquiring about human radiation experiments.

Although the NRC and its predecessor, the regulatory division of the Atomic Energy Com-

mission (AEC), have not conducted or sponsored human radiation experiments, their license files do contain some relevant information about the radioactive materials that were distributed and the purposes to which they were put, human radiation experiments among them. AEC and NRC records do not contain names or other identifying information about the subjects of such experiments and only rarely contain detailed information about the experiments themselves. The NRC also collects information about occupational exposures, medical misadministrations, and other cases of overexposure. This information is available to the public, subject to the restrictions of the Privacy Act and FOIA. Citizens may request agency documents under the Freedom of Information Act and/or the Privacy Act by writing to: Director, Division of Freedom of Information and Publication Services, Office of Administration, Nuclear Regulatory Commission, Washington, D.C. 20555–0001.

The agency will search all agency records, if requested to do so, and can search license files by institution.

PART III: PERSONAL MEDICAL RECORDS

Citizens who participated in experiments have medical records of the same type as those created by their personal physicians, whether the experiments were conducted in doctors' offices, research laboratories, or medical facilities such as hospitals and sanatoria. As discussed in part I, these records are distinct from the scientific records of the experiments and must be sought in different places. Medical records, unlike scientific records, will contain most of the information necessary to finding out what medical actions were taken and why specific procedures were followed.

Citizens share ownership of their medical records with their physicians and the medical facilities where they were treated and have the right to copies of these records. The records should be available to the individual or an authorized relative for the asking (though there may be a copying charge). In this part, we discuss how to find personal medical records and where those records may be located.

A Basic Distinction

Many individuals who contacted the Advisory Committee were understandably confused by the difference between the broad array of medical interventions involving radiation and the "human radiation experiments" that the Committee was chartered to review. The difference is this: While medical interventions are not expressly intended to accomplish anything more than therapy, "experiments" are designed to yield generalizable scientific knowledge. This is not to imply that experiments offer no therapeutic benefit (many do), only that they are organized in a different way, taking place in a controlled setting and potentially involving thousands of subjects.

It is not always easy for a patient to tell from circumstances whether he or she is involved in a larger study. One reason for the difficulty is semantic. Some ad hoc medical interventions are loosely called "experimental," meaning that they have not been proven effective or generally accepted as safe by the medical community. Experiments, meanwhile, are commonly known by another name: "human subject research."[28] Matters are complicated by the dual role many doctors play, rapidly switching hats between physician and investigator. Given all this potential for misunderstanding, those who conduct human research are under an acute ethical responsibility to clearly explain the purposes of a procedure in obtaining the subject's consent.

A citizen who believes that he or she or a relative may have been a subject in government-sponsored human research should begin the search for facts in the medical records, which provide the details of the patient's condition and the treatment administered for it. A medical professional should be asked to review these records and check for signs of a research purpose. In many cases, having the records reviewed by a professional will answer most questions and concerns. The next two sections give advice on finding one's records.

Personal Medical Records Created by Physicians and Hospitals

Physicians and medical facilities should be approached directly by the individual or by an authorized relative. As with any request for private information, a request for medical records should be formal, direct, and clear, and it should include significant personal details to assure the identity of the correspondent and, thus, the legitimacy of the request. These details are similar to those needed to request a birth certificate—date and place of birth, parent's names, and so forth. The letter should also include as many details as possible about the circumstances of interest, such as the dates of treatment, the names of the physicians, and any other information that will help locate the records. Institutions may have standard forms that need to be used; if the request occurs at some distance in time or geography, the identity of the correspondent may have to be certified in some way. These are common procedures, designed to protect an individual's privacy by preventing the unauthorized release of information.

If the name of the physician or medical facility that conducted the procedure in question is known but the address is unknown, one of the indexes of physicians and facilities available at a public library should be useful.[29] If the names are unknown, one place to start is with the individual's current physician and local hospital. They may have copies of older medical records because they were authorized to obtain medical histories. They are also likely to have (or to be able to get) information about how to contact physicians or medical facilities in other locations.

If the names of the physician and facility are not readily found, more extensive research in family papers and a broader correspondence with individuals who may have information will be necessary. Former friends, neighbors and co-workers, extended family members, clergy, and any other associates are all potential sources of information, as is the patient's health insurance company. Without the names of the medical personnel and facilities involved it will be very difficult to find records at nongovernmental facilities. If the treatment received occurred in a government facility, see part II of this appendix, which describes how to find those records. Outside the military services and large government research and social benefit programs, however, there are no large lists of individual citizens matched to their medical experiences that would provide the needed information.

Where Else Could
the Information Be?

In general, unless there are regulations or legal obligations that require other arrangements, records stay where they are created. For example, if a patient was treated at Hospital X, Hospital X is likely to be where those records are kept. It is possible that Hospital X destroys all records that are, say, thirty years old; it is also possible that Hospital X stores those records with a firm that specializes in document storage. In either case, the disposition and location of the records will be known to Hospital X and possibly to no one else.

Physicians and institutions, however, create records other than patient medical records that may also contain important medical information. When asked for medical records, Hospital X may not think of all the records that an individual might find valuable in reconstructing his or her medical history, other records that reflect activities under its sponsorship.

These records may not be coordinated or housed with any of the others. Departmental records at a hospital may be retired with those of the hospital generally or they may not. Departmental records at a university are typically retired to the university archives, usually housed in the university library; a hospital department's records at a university with a medical school may be retired to the medical school library. The academic records of faculty members are treated similarly. Records of private research and personal papers, however, are often given to the faculty member's alma mater rather than to the university where the research was done, so that both locations need to be searched. If the faculty member was a physician, it is also possible that such records were given to the institution where he or she attended medical school rather than to an undergraduate school (and then to the medical school, rather than to the university itself). In either case, it is unlikely that actual patient records would be included in an institution's archives. Many retiring physicians offer former patients (or their successor physicians) their files or may destroy these records if the patients cannot be reached.

As reported to the Advisory Committee, in some cases (see part II of this appendix) federal agencies will help citizens locate or retrieve records that were created or are held by nongovernment organizations or individuals.

PART IV: USING THE ACHRE COLLECTION AS A PLACE TO START

What Is in the Collection and What Is Not

The ACHRE research collection, which will be deposited in its entirety at the National Archives as part of Record Group 220, Presidential Committees, Commissions, and Boards,[30] is composed primarily of documents identified through agency search processes or selected by the Advisory Committee through requests or site visits to forty-five nonfederal as well as federal institutions. These efforts have not exhausted all research possibilities, but the volume of materials now identified and available to the public is very large. The Advisory Committee has not attempted to collect everything that might be pertinent, but has emphasized primary materials of wide importance. The resulting collection is rich in its breadth and variety, but frequently limited in the depth to which individual events or people are documented. Most records in the collection do not contain information about the individual subjects of human radiation experiments.

ACHRE records can make two significant contributions to the efforts of the individual researcher. First, there is no other collection in which pertinent materials from so many different sources are available in a scholarly arrangement with a substantial finding aid. Second, the collection deposited with the National Archives also includes the Advisory Committee's own research documents, including substantial unpublished notes, histories, analyses, and findings. The comprehensiveness of the collection and the added value of the Advisory Committee's scholarship make the ACHRE records a good starting point for citizens researching the public and private histories of human radiation experiments.

Experiments

The Advisory Committee's general charge was to provide advice on the character of historical and present-day policies and practices in human radiation research. The scope of such activities and the difficulty in identifying and retrieving relevant records were initially underestimated, but agency and Advisory Committee staff sought out and documented as many experiments as resources permitted. Two points should be emphasized here. First, the agencies and the Advisory Committee collected and recorded information about every experiment that could be documented. The inclusion of an experiment should not be taken as an indication that the experiment was ethically improper or likely to have caused harm to those involved.

Second, there was never an expectation that this effort would succeed in assembling a complete list of experiments or that full documentation for any large proportion of those identified could be discovered and retrieved within the time permitted. The Advisory Committee's research interest was focused on understanding the scope of activity (for example, the number and types of subjects typical for experiments of a certain character) and the policy context (for example, institutional procedures for the review of informed consent practices), than on accumulating the details of particular cases. As a result, although the Advisory Committee's log of such experiments is the most comprehensive and detailed assembled to date, the records of particular experiments are incomplete. Many experiments are documented entirely through a publication of results and many others are documented by references to even briefer descriptions of experiments in records reviewed by the Interagency Working Group or ACHRE.

The chief value of the Advisory Committee's experiment record series is in providing identifying information such as location, dates, and researchers' names—a good place to start. The experiment records are indexed by location, financial sponsorship, principal investigator (and his or her home institution), and other key pieces of information that could support extended research. Such information may be used to find additional information either in the ACHRE collection or elsewhere.

Finding Aids

There are two sets of finding aids for the Advisory Committee's records. The first is entitled *Sources and Documentation*, a supplement to this final report. The two-volume supplement features accounts of the agency and ACHRE research processes, descriptions of the record collections assembled by the Advisory Committee and of individual documents identified as significant, a complete bibliography of the published sources used in the Advisory Committee's research, brief descriptions of individual experiments, lists of testifiers and interviewees, indexes, collections of documents, and other research aids.

The second aid is the electronic record upon which much of the supplemental volume is based. Unfortunately, the National Archives is unable to make this information available in its original format, although it will be available there in simplified electronic formats with explanatory documentation. Copies of the original databases, documentation, and operating instructions will be available at the National Security Archive, an independent research institute whose offices are in the Gelman Library at George Washington University.[31] In addition to these facilities, both the National Archives and the National Security Archive provide access to the electronic records contained in the Advisory Committee's original gopher.[32] The gopher materials include electronic copies of the Advisory Committee meeting documents (briefing books, minutes, and transcripts), condensed descriptions of record collections and experiments, and other information.

How to Go From the ACHRE Collection to Agency Records

There are two sources of information that connect records contained in the ACHRE collection with those of the agencies and the National Archives. The first are document identifiers provided by the agencies; the second are transmittal records that identify the origins of the records.

Agency Document Identifiers

Most Department of Energy and a large proportion of Department of Defense documents are marked with unique identifiers that will allow location of those documents in DOE and DOD retrieval systems. Those retrieval systems include provenance information,[33] that is, information that identifies the record's office of origin and other information about its creation and current location.

DOE documents are stamped with a CIC number,[34] a six-digit accession number that uniquely identifies a document or document set (that is, documents described as a group rather than individually) that can be used to retrieve CIC records with their attendant provenance and other information management information. DOE's Internet facility can be used to identify these documents. Information is also available directly from the CIC, which also provides its index on CD-ROM using Folio Views text retrieval software.

Beginning in the fall of 1994, DOD documents supplied to the Advisory Committee were assigned accession numbers by the Radiation Experiments Command Center (RECC). These numbers denote a document's origin and the date it was sent to ACHRE. For example, records bearing numbers beginning "ARM" originated with the U.S. Army. Later in 1994 the RECC began to assign accession numbers retroactively to documents transmitted earlier. These accession numbers are available in the RECC library catalog, which was converted by ACHRE staff and is available among the Advisory Committee's records in both hard copy and electronic formats.[35]

Records of Agency Transmittal

Most records accessioned into the ACHRE collection were transmitted or deposited with documents indicating their origins. For example, materials obtained from the National Archives usually have notations indicating record group, series, and box numbers; agency records have accompanying documents indicating where materials were obtained; and donations from individuals include such information as the address of the donor. This information is collected in a ACHRE Records Management Series, Records Accession and Disposition File. Summaries of this information are included in the electronic records kept in the Document Collection database. Additional information concerning specially requested information is contained in the Agency Data Requests records file, which includes the Agency Data Requests Tracking database.

NOTES

1. There are intersections, naturally, as in contracts and grants, applications and responses, and so forth, but program history is not constituted of *cumulative* accounts of individual program *experiences* but, rather, *summary* accounts of overall program *performance*.

2. The scientific records of an experiment contain various medical facts about an individual subject, but generally only information pertinent to the conduct of the experiment and not the subject's medical history. The complete records of an experiment may include the medical records, but they will be handled separately from the scientific records. This may or may not mean that the medical records and the scientific records are the responsibility of different individuals and are stored in different places; it will certainly mean that they are created, controlled, and preserved under different guidelines.

3. For example, *Sources and Documentation* describes the contents and classifications of the record groups and entries examined at the various National Archives facilities, the record groups and accessions reviewed at the various federal records centers, and the record collections examined at various agency record storage facilities, history offices, and other locations.

4. Department of Energy, Office of Human Radiation Experiments, *Human Radiation Experiments: The Department of Energy Roadmap to the Story and the Records* (Washington, D.C.: Department of Energy, February 1995). For ordering information, write: U.S. Department of Commerce, Technology Administration, National Technical Information Service, Springfield, Virginia 22161; or telephone: 703–487–4650.

5. Although the National Archives and Records Administration (NARA) system includes records of the judicial and legislative branches of the federal government, most citizen researchers are looking for records created by agencies of the executive branch, and so the following information is generally limited to those records. A brief discussion of judicial and

legislative records is included in *Sources and Documentation.*

6. Because the National Archives was not established until 1934 and the records centers only came into existence in 1950, there are some instances where the records of federal officials and agencies are outside the "physical control" of the government. Also, unfortunately, no general rule can be applied to contractor records. The handling of the records of contract work is controlled by the terms of the contract, which may require anything from deposit of complete records with the contracting agency to complete retention of all records by the contractor. The citizen will need to research such situations on a case-by-case basis. Agency records should include copies of the contract or grant instruments, however, and research should begin with those.

7. *Provenance* refers to the origin, creation, and ownership (or chain of custody) of records or other items.

8. A folder listing is a list of the titles of the file folders (that is, what is on their labels) that are contained in the shipment. Because it reproduces the file labels more or less exactly, such a listing, while invaluable, is only as informative as the labels. SF-135s are unlikely to contain information on individual documents.

9. The most practical resource is *A Citizen's Guide on Using the Freedom of Information Act and the Privacy Act of 1974 to Request Government Records, House Report 104–156* (Washington, D.C.: GPO, 1995), prepared by the Committee on Government Management, Information, and Technology of the House of Representatives. Another important resource (for those interested in the administrative and legal details) is the annual Department of Justice publication *Freedom of Information Act Guide & Privacy Act Overview.* Both volumes are available from the U.S. Government Printing Office, Superintendent of Documents, Mail Stop: SSOP, Washington, D.C. 20401–9328. The American Civil Liberties Union (ACLU) also publishes an annual guide to FOIA and the Privacy Act; for information call (202) 544–1681.

10. These sorts of records are subject to Privacy Act controls whether they are in the keeping of the originating agency or the National Archives.

11. This is an area in which there is not agreement among the agencies. For example, the Advisory Committee was assured by one agency that records retrievable by the names of principal investigators were not subject to the Privacy Act—after all, officials said, these individuals were government contractors and grantees who had a practical relationship with the federal government that had to be substantiated by reports under the law. Under similar circumstances, however, another agency provided the Advisory Committee with information that it said could not be made public because it had been retrieved by the name of a principal investigator.

12. The act uses the phrase "any person," so that inquiry is not restricted to U.S. citizens.

13. "There are three basic elements to a FOIA request letter. First, the letter should state that the request is being made under the Freedom of Information Act. Second, the request should identify the records that are being sought as specifically as possible. Third, the name and address of the requester must be included." *A Citizen's Guide on Using the Freedom of Information Act,* 8.

14. The Department of Justice's *Overview,* 32 fn. 103, cites a Federal District Court decision: "FOIA creates only a right of access to records, not a right to personal services." *Hudgins v. IRS,* 620 F.Supp. 19, 21 (D.D.C. 1985), *aff'd,* 808 F.2d 137 (D.C. Cir.), *cert. denied,* 484 U.S. 803 (1987).

15. The agencies concerned with human radiation experiments have provided information on their procedures for filing FOIA requests, and these are included in part II of this appendix.

16. *A Citizen's Guide on Using the Freedom of Information Act,* 10, should be consulted on how fees may be waived.

17. The effect of this provision is potentially highly elastic because, under the act, the agency may lengthen the time it takes to provide records in order to look for the records, search through the records, or consult with another agency or office.

18. The requirements for MDRs under Executive Order 12958 are very similar to those of the FOIA described in the previous section, and accordingly, there is no separate discussion of this alternative procedure. Among the few differences are that only U.S. citizens may file MDRs, and that if there is a denial of an MDR in whole or in part there is a right to an administrative appeal, but no right of judicial redress.

19. The *U.S. Government Manual,* published annually as a special edition of the *Federal Register,* is available by writing: Superintendent of Public Documents, P.O. Box 317954, Pittsburgh, Pennsylvania 15250–7954; telephone (202) 783–3238.

20. A *FOIA reading room* is a publicly accessible facility that houses information that has been released to the public by the agency, either voluntarily or as a result of a citizen's FOIA request. Almost without exception, however, these repositories contain only a small fraction of the records that have been released over the years. FOIA reading rooms are generally managed and staffed by the agency library. But access to agency libraries varies, and many agencies do not have FOIA reading rooms. Most agencies, however, have an office of public affairs that may be contacted for general information about the agency and its programs. An agency's *FOIA office* handles all FOIA requests and is the primary source of information about the agency's FOIA procedures.

21. CIC is a records center operated by the Reynolds Electrical & Engineering Co., under contract with DOE. Reynolds's address is P.O. Box 98521, Las Vegas, Nevada 89193–8521. The CIC is the major source of the documents made available

by DOE through the Internet and provides reference services and copies of documents to help the public.

22. Some records transferred to the St. Louis facility were destroyed in a fire in 1973.

23. The World Wide Web is a network of Internet sites using graphical and hypertext formats permitting access to images and linking distant and diverse information sources.

24. Department of Energy, Assistant Secretary for Environment, Safety and Health, *Human Radiation Experiments Associated with the U.S. Department of Energy and Its Predecessors* (Washington, D.C.: U.S. Department of Energy, July 1995).

25. Some records transferred to the St. Louis facility were destroyed in a fire in 1973.

26. The operations of the Department of Health and Human Services (DHHS) are diverse and decentralized and include several large components, such as the Food and Drug Administration (FDA), the Public Health Service (PHS), and the National Institutes of Health (NIH), which are so well known that they sometimes may appear to be independent. PHS is one of the major subdivisions of the department; FDA and NIH are components of PHS.

27. The "institutes" that make up the National Institutes of Health are organized around medical specialties such as cancer and mental health, and physiological topics such as the heart and the kidneys. They are based in Bethesda, Maryland.

28. The Common Rule governing human experimentation in most federal government agencies uses this phrase. See 56 Fed. Reg. 28,012 (1991) (§ 101[a]).

29. Some reference books that might be useful: (1) *Directory of Physicians in the United States*, issued by the American Medical Association; (2) *Official ABCS Directory of Board Certified Medical Specialists*, issued by the American Board of Medical Specialties; (3) *The World of Learning*, which contains entries for major universities that include medical center faculty lists and addresses; and (4) *Directory of U.S. Hospitals*, published by Health Care Investment Analysts, Inc.

30. These records will be available at the National Archives in late 1995.

31. The National Security Archive, Gelman Library, Suite 701, 2130 H Street, N.W., Washington, D.C. 20037; telephone, 202-994-7000; fax, 202-994-7005; e-mail, archive@cap.gwu.edu.

32. A gopher is a software application that provides menu-driven access to electronic files, frequently over the Internet. The Advisory Committee maintained both a gopher and a World Wide Web home page.

33. For additional information on provenance, see the section on the National Archives in part I.

34. CIC numbers are assigned by the Coordination and Information Center. The CIC document number identifies the records series in which a document is indexed. The records of concern to the Advisory Committee are primarily from the human radiation series, which uses numbers 700,000–799,999. Other series cover such related topics as Enewetak Atoll, fallout, and Glenn T. Seaborg.

35. Hard copy format is available in the transmittal documents (see next section).

Index

Page numbers followed by n *indicate the entry will be found in a note; page numbers followed by* fn *indicate the entry will be found in a footnote.*

Printed in the USA/Agawam, MA
February 13, 2015